Advances in Structural Biology

Advances in Structural Biology

Applications in Protein Structure, Function, and Disease

Edited by

ANAS SHAMSI
Centre of Medical and Bio-allied Health Sciences Research, Ajman University, Ajman, United Arab Emirates

IMTAIYAZ HASSAN
Center for Interdisciplinary Research in Basic Sciences, Jamia Millia Islamia, New Delhi, India

MOYAD SHAHWAN
College of Pharmacy and Health Sciences, Ajman University, Ajman, United Arab Emirates

BELGIN SEVER
Department of Pharmaceutical Chemistry, Anadolu University, Eskişehir, Türkiye

ELSEVIER

ACADEMIC PRESS
An imprint of Elsevier

Academic Press is an imprint of Elsevier
125 London Wall, London EC2Y 5AS, United Kingdom
50 Hampshire Street, 5th Floor, Cambridge, MA 02139, United States

ISBN: 978-0-443-44988-8

For Information on all Academic Press publications
visit our website at https://www.elsevier.com/books-and-journals

Publisher: Megan Ball
Acquisitions Editor: Michelle Fisher
Editorial Project Manager: Harshita Bathla
Production Project Manager: Vijayaraj Purushothaman
Cover Designer: Mark Rogers

Typeset by MPS Limited, Chennai, India

Contents

Section IV Translational biology

List of contributors

Minhal Abidi
Department of Biochemistry, Jawaharlal Nehru Medical College, Aligarh Muslim University, Aligarh, Uttar Pradesh, India

Malika Adam Abeeda
Centre for Virology, Jamia Hamdard University, New Delhi, Delhi, India

Rizwan Ahmad
Department of Biochemistry, Jawaharlal Nehru Medical College, Aligarh Muslim University, Aligarh, Uttar Pradesh, India

Luqman Ahmad Khan
Department of Biosciences, Jamia Millia Islamia, Jamia Nagar, New Delhi, India

Neesar Ahmed
School of Life Sciences, BSA Crescent Institute of Science & Technology, Chennai, Tamil Nadu, India

Tofayel Ahmed
Saha Institute of Nuclear Physics, Biophysical Sciences Division, Kolkata, West Bengal, India; Homi Bhaba National Institute, Mumbai, Maharashtra, India

Ayesha Aiman
Centre for Interdisciplinary Research in Basic Sciences, Jamia Millia Islamia, New Delhi, India; Department of Biosciences, Jamia Millia Islamia, Jamia Nagar, New Delhi, India

Umme Aiman
Department of Bioengineering, Integral University, Lucknow, Uttar Pradesh, India

Salman Akhtar
Department of Bioengineering, Integral University, Lucknow, Uttar Pradesh, India

Ashish Anjankar
Department of Biochemistry, Jawaharlal Nehru Medical College, Datta Meghe Institute of Higher Education and Research, Wardha, Maharashtra, India

Vaibhav Anjankar
Department of Anatomy, Jawaharlal Nehru Medical College, Datta Meghe Institute of Higher Education and Research, Wardha, Maharashtra, India

Jaoud Ansari
Centre for Interdisciplinary Research in Basic Sciences, Jamia Millia Islamia, New Delhi, India

Neha Kausar Ansari
Department of Biochemistry, Faculty of Life Sciences, Aligarh Muslim University, Aligarh, Uttar Pradesh, India

Saleha Anwar
Department of Toxicology, School of Chemical and Life Sciences, Jamia Hamdard, New Delhi, Delhi, India

Sarika Bano
Centre for Interdisciplinary Research in Basic Sciences, Jamia Millia Islamia, New Delhi, India

Mohd Basheeruddin
Department of Biochemistry, Jawaharlal Nehru Medical College, Datta Meghe Institute of Higher Education and Research, Wardha, Maharashtra, India

Ishaan Bhardwaj
Department of Computer Science, Jamia Millia Islamia, South Delhi, Delhi, India

Gourav Choudhir
Centre for Interdisciplinary Research in Basic Sciences, Jamia Millia Islamia, New Delhi, India

Archana Dhok
Department of Biochemistry, Jawaharlal Nehru Medical College, Datta Meghe Institute of Higher Education and Research, Wardha, Maharashtra, India

Khuzin Dinislam
Department of General Chemistry, Bashkir State Medical University, Ufa, Republic of Bashkortostan, Russia

Shivani Dixit
Department of Biophysics, All India Institute of Medical Sciences, New Delhi, India

Jamal-E-Fatima
BMI-EP Lab, School of Electronic Science and Engineering, University of Electronic Science and Technology of Chine, Chengdu, Sichuan, P.R. China

Naseem Gaur
International Centre for Genetic Engineering and Biotechnology, New Delhi, India

Deepali Gupta
Department of Biophysics, All India Institute of Medical Sciences, New Delhi, India

Safia Habib
Department of Biochemistry, Jawaharlal Nehru Medical College, Aligarh Muslim University, Aligarh, Uttar Pradesh, India

Saira Hamid
Center for Interdisciplinary Research in Basic Sciences, Jamia Millia Islamia, New Delhi, India

Imtaiyaz Hassan
Center for Interdisciplinary Research in Basic Sciences, Jamia Millia Islamia, New Delhi, India

Arbab Husain
Department of Biotechnology and Life Sciences, Faculty of Sciences, Mangalayatan University, Aligarh, Uttar Pradesh, India

Faiza Iram
Centre for Interdisciplinary Research in Basic Sciences, Jamia Millia Islamia, New Delhi, India

Sana Iram
Department of Medical Biotechnology, Yeungnam University, Gyeongsan, Gyeongsangbuk-do, South Korea

Asimul Islam
Centre for Interdisciplinary Research in Basic Sciences, Jamia Millia Islamia, New Delhi, India

Shazia Jamal
School of Life Sciences, BSA Crescent Institute of Science & Technology, Chennai, Tamil Nadu, India

Aaliya Javed
Centre for Interdisciplinary Research in Basic Sciences, Jamia Millia Islamia, New Delhi, India

Sushma Jeyaraj
Department of Biotechnology, School of Bioengineering, College of Engineering and Technology, SRM Institute of Science and Technology, Chengalpattu, Tamil Nadu, India

MD Kaif
Centre for Interdisciplinary Research in Basic Sciences, Jamia Millia Islamia, New Delhi, India; Delhi Pharmaceutical Sciences and Research University, Delhi, India

Mohammad Affan Kareem
School of Allied Health Sciences, Datta Meghe Institute of Higher Education and Research, Wardha, Maharashtra, India

Punit Kaur
Department of Biophysics, All India Institute of Medical Sciences, New Delhi, India

Faez Iqbal Khan
Department of Biological Sciences, School of Science, Xi'an Jiaotong-Liverpool University, Suzhou, Jiangsu, P.R. China

Hamda Khan
Department of Biochemistry, Jawaharlal Nehru Medical College, Aligarh Muslim University, Aligarh, Uttar Pradesh, India

Maryam Khan
Department of Biological Sciences, Kuwait University, Shadadiya, Al Asimah, Kuwait

Md. Khurshid A. Khan
School of Life Sciences, BSA Crescent Institute of Science & Technology, Chennai, Tamil Nadu, India

Mohammad Kalim Ahmad Khan
Department of Bioengineering, Integral University, Lucknow, Uttar Pradesh, India

Sahar Afzal Khan
Department of Biotechnology, Jamia Hamdard, New Delhi, Delhi, India

Sumaiya Khan
Centre for Interdisciplinary Research in Basic Sciences, Jamia Millia Islamia, New Delhi, India

Afreen Khanam
Department of Biotechnology and Life Sciences, Faculty of Sciences, Mangalayatan University, Aligarh, Uttar Pradesh, India

Shabana Khatoon
Centre for Interdisciplinary Research in Basic Sciences, Jamia Millia Islamia, New Delhi, India; Indira Gandhi National Open University, New Delhi, India

Mukesh Kumar
Department of Optometry, School of Allied Health Sciences, Galgotias University, Noida, Uttar Pradesh, India

Sanjay Kumar
Structural and Molecular Biology, City College of New York, New York City, NY, United States

Dominic Skaria Mathew
Department of Computer Science, Jamia Millia Islamia, South Delhi, Delhi, India

Moinuddin
Department of Biochemistry, Jawaharlal Nehru Medical College, Aligarh Muslim University, Aligarh, Uttar Pradesh, India

Debalaya Mukherjee
Saha Institute of Nuclear Physics, Biophysical Sciences Division, Kolkata, West Bengal, India; Homi Bhaba National Institute, Mumbai, Maharashtra, India

Mohd Mustafa
Department of Biochemistry, Jawaharlal Nehru Medical College, Aligarh Muslim University, Aligarh, Uttar Pradesh, India

Aabgeena Naeem
Department of Biochemistry, Faculty of Life Sciences, Aligarh Muslim University, Aligarh, Uttar Pradesh, India

Neha
Department of Forensic Science, School of Basic and Applied Sciences, SGT University, Gurugram, Haryana, India

Sara Othman
Department of Molecular Biology and Genetics, Faculty of Engineering and Natural Sciences, Istinye University, Istanbul, Turkey

Zahoor Ahmad Parray
Department of Bio-Science and Technology, M.M College of Engineering, Maharishi Markandeshwar University (Deemed to be), Ambala, Haryana, India

Suhel Parvez
Department of Toxicology, School of Chemical and Life Sciences, Jamia Hamdard, New Delhi, Delhi, India

Shaheda Perween
Department of Biochemistry, Jawaharlal Nehru Medical College, Aligarh Muslim University, Aligarh, Uttar Pradesh, India

Sana Qausain
Department of Biochemistry, Jawaharlal Nehru Medical College, Datta Meghe Institute of Higher Education and Research, Wardha, Maharashtra, India

Aysha Afreen Rafeek
Department of Biotechnology, School of Bioengineering, College of Engineering and Technology, SRM Institute of Science and Technology, Chengalpattu, Tamil Nadu, India

Alpana Srishti Rai
Amity Institute of Biotechnology, Amity University Uttar Pradesh (AUUP), Noida, Uttar Pradesh, India

Shazia Rashid
Amity Institute of Biotechnology, Amity University Uttar Pradesh (AUUP), Noida, Uttar Pradesh, India

Rudrani Rath
Department of Biotechnology, School of Bioengineering, College of Engineering and Technology, SRM Institute of Science and Technology, Chengalpattu, Tamil Nadu, India

Khushi Raza
Department of Toxicology, Jamia Hamdard, New Delhi, New Delhi, India

Minal Richi
Department of Biotechnology, American International University, Jahra, Kuwait

Antara Saha
Saha Institute of Nuclear Physics, Biophysical Sciences Division, Kolkata, West Bengal, India; Homi Bhaba National Institute, Mumbai, Maharashtra, India

Seneha Santoshi
Amity Institute of Biotechnology, Amity University Uttar Pradesh (AUUP), Noida, Uttar Pradesh, India

Anas Shamsi
Centre of Medical and Bio-allied Health Sciences Research, Ajman University, Ajman, United Arab Emirates

Hanzla Shaz
Center for Interdisciplinary Research in Basic Sciences, Jamia Millia Islamia, New Delhi, India

Samarth Shukla
Department of Pathology, Jawaharlal Nehru Medical College, Datta Meghe Institute of Higher Education and Research, Wardha, Maharashtra, India

Sarah Siddiquah
Center for Interdisciplinary Research in Basic Sciences, Jamia Millia Islamia, New Delhi, India

Shubhi Singh
Department of Biotechnology, School of Bioengineering, College of Engineering and Technology, SRM Institute of Science and Technology, Chengalpattu, Tamil Nadu, India

Priya Swaminathan
Department of Biotechnology, School of Bioengineering, College of Engineering and Technology, SRM Institute of Science and Technology, Chengalpattu, Tamil Nadu, India

Aaliya Taiyab
Center for Interdisciplinary Research in Basic Sciences, Jamia Millia Islamia, New Delhi, India

Sana Tanweer
Department of Molecular Medicine, Jamia Hamdard University, New Delhi, Delhi, India

Netri Upadhyay
Department of Biochemistry, Jawaharlal Nehru Medical College, Datta Meghe Institute of Higher Education and Research, Wardha, Maharashtra, India

Sanchit Varma
Department of Molecular Medicine, Jamia Hamdard University, New Delhi, Delhi, India

About the editors

Dr. Anas Shamsi is an accomplished structural biochemist and drug-discovery expert with extensive academic and research experience in biomedical sciences. He currently serves as a Research Assistant Professor at the Center of Medical and Bio-Allied Health Sciences Research, Ajman University, United Arab Emirates. Dr. Shamsi's research portfolio reflects a profound commitment to the identification and validation of novel drug targets, particularly in the context of cancer and neurodegenerative diseases. His work integrates structural biology, protein folding, protein -ligand interactions, molecular modelling, biophysical techniques, and artificial intelligence-assisted drug discovery to identify promising therapeutic candidates. A prolific researcher, Dr. Shamsi has authored more than 260 peer-reviewed articles in leading international journals. His publications have received over 30,000 citations, and his h-index exceeds 50, reflecting the international impact of his research. He has led or contributed to more than 10 funded research projects and continues to make meaningful contributions to cancer biology, structural biochemistry, kinase-targeted drug discovery, and therapeutic innovation. He also serves in editorial roles for international scientific journals and has been recognized among the world's top 2% of scientists for four consecutive years. He has led more than 10 funded projects and made vital contributions to neurodegenerative diseases and cancer therapeutics.

Dr. Imtaiyaz Hassan is a renowned structural biologist and drug discovery expert. With over 17 years of distinguished academic and research experience, Dr. Hassan has established himself as a prominent figure in biomedical sciences, with deep expertise in structural biology, protein folding, drug design, and kinase-targeted therapeutics. Prof. Hassan's research portfolio reflects a profound commitment to the identification and validation of novel drug targets, particularly in the context of cancer and neurodegenerative diseases. His groundbreaking work in structure-based drug design and protein-ligand interactions has significantly advanced the development of selective and potent protein kinase inhibitors. A prolific researcher, Dr. Hassan has authored over 600 peer-reviewed research articles in leading international journals. His work has garnered more than 22,000 citations, and his current h-index stands at an impressive 72, reflecting his global impact in the scientific community. He has led more than 22 funded research projects and continues to make meaningful contributions to cancer biology, structural biochemistry, and therapeutic innovation.

Dr. Moyad Shahwan is a professor of clinical biochemistry at Ajman University, United Arab Emirates, with extensive expertise in biochemistry, drug design, molecular modeling, and the biochemical analysis of phytochemicals. He earned his PhD and MSc in biochemistry from Aligarh University, India. Dr. Shahwan has published over 300 research articles in peer-reviewed journals and supervised more than 37 master's theses. His work has been recognized with multiple awards, including Ajman University's Best Researcher Award in 2021. As the head of the Centre of Medical and Bio-Allied Health Sciences Research, Dr. Shahwan has led research initiatives and established new research groups and facilities. His deep expertise in clinical and biochemical research, particularly in diabetes mellitus and public health, makes him an invaluable editor for this book.

Dr. Belgin Sever is currently an associate professor at the Department of Pharmaceutical Chemistry, Faculty of Pharmacy, Anadolu University, Turkey. Her research is mainly focused on the development of anticancer, antimicrobial, antidiabetic, anti-HIV, and anti-ALS drug candidates, especially based on medicinal chemistry and computational chemistry disciplines. She received her MSc and PhD in pharmaceutical chemistry from Anadolu University (Faculty of Pharmacy), Turkey, in 2015 and 2019, respectively. After her PhD, she was accepted as a postdoctoral researcher of the Takeda Science Foundation International Fellowship Programme. During this time, she worked on anti-HIV drug development at Kumamoto University, Faculty of Life Sciences, Medicinal and Biological Chemistry Science Farm Joint Research Laboratory, Kumamoto, Japan. Dr. Sever has received various awards and has published many high-quality articles in prestigious international peer-reviewed journals. She has served on various journal editorial boards, including as topic editor for the journals *Pharmaceuticals*, *Molecules*, and *Life*.

Acknowledgments

I would like to express my sincere gratitude to all contributing authors for their valuable scholarly contributions and for their cooperation throughout the development of this volume. Their expertise and commitment have been instrumental in shaping the quality and scope of this book.

I am deeply thankful to my coeditors for their continuous support, insightful discussions, and collaborative efforts in bringing this work to completion.

I also extend my appreciation to the editorial and production teams at Elsevier for their guidance, professionalism, and assistance at every stage of the publication process.

Furthermore, I acknowledge the support of my colleagues, collaborators, and institution, whose academic environment and encouragement have played an important role in this endeavor.

On a personal note, I would like to express my heartfelt gratitude to my wife, my parents, and my family members for their unwavering support, patience, and encouragement throughout this journey.

Dr. Anas Shamsi

SECTION I

Foundations: computational and conceptual

CHAPTER 1

Unraveling protein function with structural biology approaches

Sana Qausain[1], Mohd Basheeruddin[1], Ashish Anjankar[1], Md. Khurshid A. Khan[2], Faez Iqbal Khan[3], Archana Dhok[1], Mohammad Affan Kareem[4], and Netri Upadhyay[1]

[1]Department of Biochemistry, Jawaharlal Nehru Medical College, Datta Meghe Institute of Higher Education and Research, Wardha, Maharashtra, India,
[2]School of Life Sciences, BSA Crescent Institute of Science & Technology, Chennai, Tamil Nadu, India,
[3]Department of Biological Sciences, School of Science, Xi'an Jiaotong-Liverpool University, Suzhou, Jiangsu, P.R. China,
[4]School of Allied Health Sciences, Datta Meghe Institute of Higher Education and Research, Wardha, Maharashtra, India

1.1 Introduction

1.1.1 Importance of proteins in cellular processes

The most universal group of biomolecules in the living system is proteins. They are involved in almost all the processes that the cell goes through, including structural scaffolding as well as complex signaling cascades. This functional diversity is a result of their capacity to change into a large number of conformations, making them essential in keeping homeostasis intact. Examples are enzymes, which catalyze biochemical reactions by reducing the activation energy, which allows life-sustaining processes to proceed within physiological time frames, for example, glycolysis and oxidative phosphorylation. The cell architecture, motility, and mechanical stability are achieved through structural proteins like actin, tubulin, and keratin, whereas regulatory proteins regulate gene expression, signal transduction, and the cell cycle (LaPelusa and Kaushik, 2022). Besides, cellular communication is mediated by proteins that are involved in cellular communication via the process of ligand-receptor interactions, which trigger immune recognition and hormonal regulation processes. Protein structural or functional dysregulation is often linked to disease (such as the amyloid-β aggregation in Alzheimer's disease, mutant p53 in cancer, and sickle-cell anemia, caused by misfolded hemoglobin). Thus a deep knowledge of protein structure and dynamics is the key to explaining the molecular processes underlying physiology and pathology (Su et al., 2024).

1.1.2 Evolution of structural biology as a discipline

The desire to learn the details of protein structure in relation to functionality led to the emergence of structural biology as a science. The first milestones were the discovery of

Advances in Structural Biology
DOI: https://doi.org/10.1016/B978-0-443-44988-8.00022-8

the α-helix and β-sheet secondary structures by Linus Pauling in the 1950s, which formed the conceptual basis of protein folding. A short time later, X-ray crystallography was developed, and the first high-resolution structures of myoglobin and hemoglobin were obtained to demonstrate the atomic basis of oxygen transport. These successes indicated that the work of proteins could be fully described only in terms of their three-dimensional structures. This was followed by the development of nuclear magnetic resonance (NMR) spectroscopy in the following decades, enabling researchers to study the structure of proteins in solution and examine dynamic processes that could not have been investigated through crystallography. Recently, the invention of cryo-electron microscopy (cryo-EM) has brought a revolution to structural biology since large macromolecular complexes, which were previously inaccessible, can now be imaged at near-atomic resolution. At the same time, methods based on computational science, including molecular dynamics (MD) simulations and, most recently, deep learning-based prediction models, have given unprecedented insight into protein folding and interactions. All of these experimental and computational developments have not just increased the structural repertoire of proteins that can be found in the Protein Data Bank (PDB), but have also made available integrative methods that can capture transient states, conformational flexibility, and multiprotein assemblies. The history of structural biology is marked by the transition from the static snapshots of biomolecules to the dynamic holistic perspectives on them, and from the basic findings to biomedical uses (Carugo and Djinović-Carugo, 2023).

1.1.3 Scope and objectives of this chapter

This chapter aims to provide a brief review of the existing data on the role of structural biology in our understanding of protein functions and simultaneously identify the methodological breakthroughs and translating practice. It begins with the general information on the protein structural hierarchy and its direct association with functionality, followed by a detailed discussion of each of the experimental techniques, such as X-ray crystallography, NMR spectroscopy, cryo-EM, and small-angle scattering. The methods of computational and integrative strategies will also be provided by demonstrating how hybrid strategies are expanding the horizons of structural analysis. Besides this approach, this chapter also brings up functional knowledge gained in structural biology, including enzymatic catalysis, ligand recognition, signal transduction, and protein–protein interactions (PPIs). Particular attention will be paid to translation, for example, the role of structural biology in the rational design of drugs, the discovery of biomarkers, and synthetic biology. Finally, the recent developments and the future will be discussed with a focus on how the discipline is still evolving with the new techs of time-resolved structural research, single-molecule techniques, and artificial intelligence (AI)-driven predictions.

This chapter is an attempt to provide a general outline of the knowledge of how structural biology can open up the molecular reasoning of protein activity by uniting the most fundamental ideas with the most modern developments. Such a type of knowledge is crucial in the area of creating fundamental science and diagnostics, therapeutics, and biotechnology inventions.

1.2 Protein structure as a dynamic hierarchy

Proteins are also vital biological macromolecules that regulate the structure, regulation, and signaling of cells. They are three-dimensionally structured, and this defines their catalytic, structural, and interaction properties, thus connecting genetic data with cellular activity. Knowledge of protein structure is thus key to the fields of biochemistry and molecular biology, which include hierarchical organization, structure-function relationships, and conformational dynamics (Alberts et al., 2002).

1.2.1 Structural organization and its functional limits

Proteins are traditionally characterized as hierarchically organized between primary and quaternary structures. It is the primary structure, which is determined by the sequence of amino acids, that determines folding and functionality, and even the replacement of a single amino acid can lead to drastic functional effects, as in the case of sickle-cell anemia due to a mutation in the hemoglobin β-chain (Sanvictores and Farci, 2025). Secondary structures such as α-helices and β-sheets, stabilized mainly by hydrogen bonds, provide structural frameworks essential for stability and activity (Rehman et al., 2025). Tertiary structure refers to the complete three-dimensional folding of the polypeptide chain via hydrophobic interactions, disulfide bonds, and electrostatic interactions that result into functional regions that include active and binding sites (Rehman and Botelho, 2025). Quaternary structure arises from the assembly of multiple subunits into functional complexes, exemplified by hemoglobin and ATP synthase, which are critical for oxygen transport and energy metabolism, respectively (Szczesny-Malysiak et al., 2020). The scaffolding is done by secondary and tertiary structures, and functional specificity is frequently due to the flexible loops, the transient contacts, and marginally stable regions, the static structures of which are not well seen. The further interpretation issue is that the states of oligomers are dynamic and may vary depending on the cellular signals in the case of quaternary organization (Van Der Lee et al., 2014), as illustrated in Fig. 1.1.

1.2.2 Relationship between protein structure and function

The structural conformation has a direct impact on protein functioning, as it allows selective interaction with substrates, cofactors, and other biomolecules. The active sites

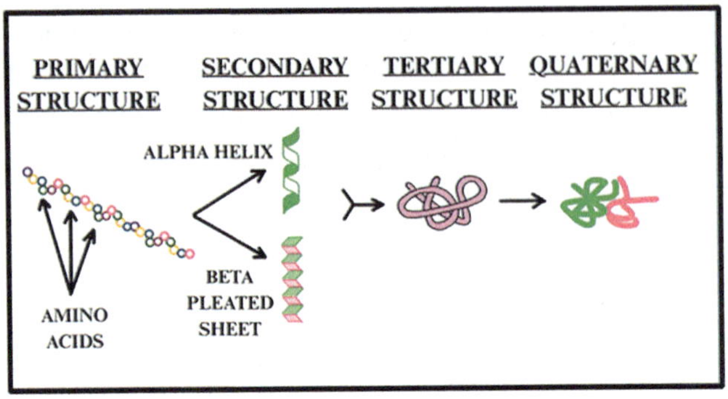

Figure 1.1 Schematic representation of protein structural hierarchy from primary to quaternary organization.

of enzymes are complementary to substrate transition states. Functional diversity is produced by domain organization and structural motifs, for example, the immunoglobulins and membrane transport proteins, including aquaporins. Such structural changes or misfolding may lead to disease, such as cystic fibrosis due to CFTR folding defects. Allosteric control also underscores the structure-function interactions where hemoglobin has a cooperative oxygen-binding behavior due to T and R state changes. These principles form the basis of rational drug design and protein engineering (Borkakoti et al., 2025; Marinko et al., 2019).

1.2.3 Conformational flexibility and dynamics

Proteins are dynamic in nature, and they exist in populations of conformational states and not as rigid structures. Investigations into experimental and computational biology have revealed that transient conformations have been found to mediate biological activity by induction-fit or conformational selection levels. IDPs are intrinsically disordered proteins that are important in signaling and regulation because of their structural plasticity. However, dysregulated conformational dynamics can promote pathological aggregation, as observed in amyloid-β in Alzheimer's disease and α-synuclein in Parkinson's disease. These findings have informed therapeutic strategies aimed at stabilizing native conformations or preventing aberrant protein interactions (Agback et al., 2023; Vadukul et al., 2023).

1.3 Experimental approaches in structural biology

Structural biology has become a multidisciplinary area that combines biophysics, chemistry, and biology to offer mechanistic information on proteins, nucleic acids, and macromolecular assemblies. Three-dimensional structural determination of biomedical structures has been a historical transformation in biomedical research with the ability to make discoveries around enzyme catalysis and drug-target interactions. This part explores the four main methods of experimentation, such as X-ray crystallography, NMR spectroscopy, Cryo-EM, and small-angle scattering methods, their principles, methodologies, uses, and limitations (Maiti et al., 2024) as summarized in Fig. 1.2 and Table 1.1.

1.3.1 X-ray crystallography

1.3.1.1 Principle and methodology

X-ray crystallography has continued to be the foundation of structural biology and has helped in most protein structures available in the PDB. The principle of the method is the diffraction of X-rays by the periodic distribution of atoms in a crystalline lattice. Patterns formed by constructive interference of diffracted beams are mathematically (using Fourier transforms) transformed into electron density maps that can be interpreted to form atomic models. Crystallization of proteins is a significant process and usually the limiting one. New techniques of high-throughput screening of crystallization, sources of synchrotron radiation, and computational refinement have gained prominence in accelerating the process of structural determination (Weber et al., 2019).

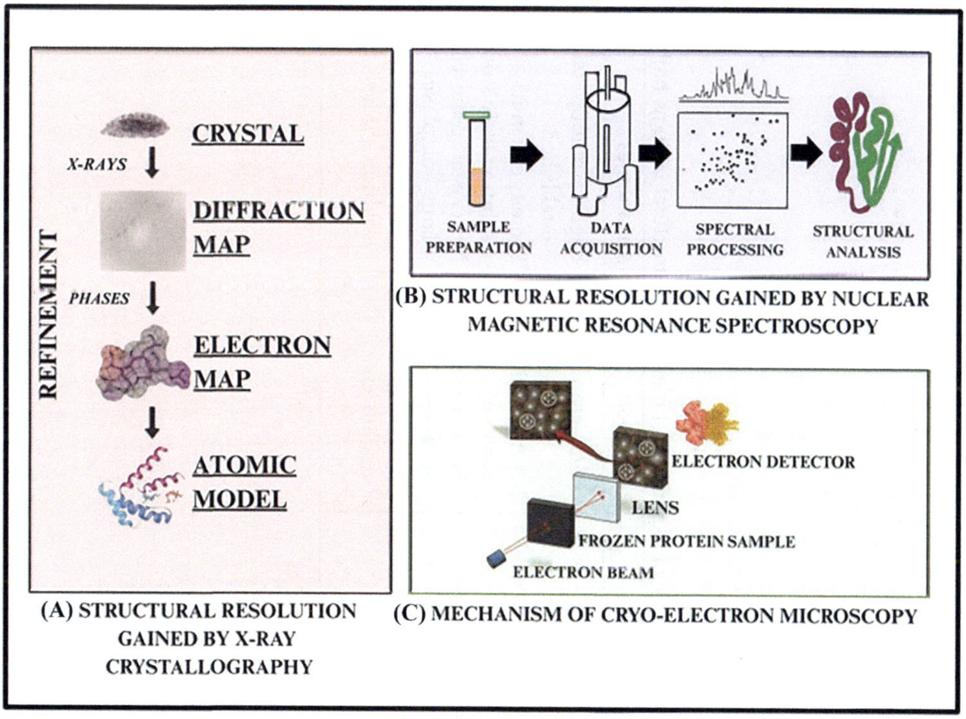

Figure 1.2 Comparative schematic of structural resolution obtained by X-ray crystallography, NMR spectroscopy, and cryo-electron microscopy.

Table 1.1 Overview of experimental methods in structural biology, highlighting resolution, advantages, and challenges.

Method	Resolution	Advantages	Challenges	Example applications
X-ray crystallography	1–3 Å (atomic)	High-resolution, well-established	Requires crystals, static snapshots	Enzyme active sites, ribosome structures
NMR spectroscopy	2–3 Å (small proteins)	Protein dynamics, solution-state studies	Limited to ~40–100 kDa	Folding pathways, conformational exchange
Cryo-EM	<3 Å (recent advances)	No crystals required, large complexes	Requires high particle homogeneity	Viral capsids, membrane proteins
SAXS/SANS	10–30 Å (low resolution)	Flexible proteins, in-solution studies	Limited structural detail	Intrinsically disordered proteins, conformational changes

1.3.1.2 Achievements and limitations

Structural biology has been a game changer by elucidating the DNA double helix, establishing the first enzyme structure, lysozyme, and elucidating the high-resolution structures of ribosomal complexes. There are also some limitations with crystallography, though: (1) (LaPelusa and Kaushik, 2022) it is important to have high-quality crystals, which with membrane proteins and large complexes is often difficult to achieve, (2) (Su et al., 2024) crystallization artifacts can occur, and (3) (Carugo and Djinović-Carugo, 2023) it cannot measure the dynamics of conformational changes as structures provide only snapshots (Birch et al., 2018; Wolff et al., 2023).

1.3.2 Nuclear magnetic resonance spectroscopy

1.3.2.1 Applications in studying protein dynamics

NMR takes advantage of the magnetic characteristics of atomic nuclei (usually, the indications of ^{1}H, ^{13}C, and ^{15}N) to offer detailed structural and dynamic data in solution. In contrast to crystallography, NMR does not need crystals, which enables one to evaluate proteins under conditions that are as similar to physiology as possible. By performing multidimensional NMR experiments, backbone and side-chain assignments can be performed to yield distance and angle information, resulting in high-resolution structures of proteins of up to 40 kDa. In addition to stationary structures, NMR is the only technique with unique strengths in the characterization of protein dynamics over timescales ranging from picoseconds to seconds. Studies of conformational change, folding intermediates, and ligand-induced transitions caused by a ligand can be done using relaxation dispersion and residual dipolar coupling methods (Narayanan et al., 2017).

1.3.2.2 Constraints in high-molecular-weight proteins

There are, nevertheless, inherent problems in using the technique with large proteins because of spectral overlap and signal broadening. These problems have been partially solved recently with techniques like the transverse relaxation-optimized spectroscopy and isotope labeling techniques, which allow expansion of the molecular weight limit to around 100 kDa in some systems. NMR is, however, not suitable for very large assemblies of macromolecules compared to cryo-EM (Zhu and Yao, 2008).

1.3.3 Cryo-electron microscopy

1.3.3.1 Single-particle analysis and advances in resolution

In the past 10 years, cryo-EM has transformed the field of structural biology and even received the Nobel Prize in Chemistry in 2017. In cryo-EM single-particle analysis, proteins are immediately frozen in vitreous ice, and this maintains their native conformations. Millions of two-dimensional images are computed to produce high-resolution three-dimensional structures by combining thousands of two-dimensional images. With technological advances, including direct electron detectors, energy filters, and image-processing algorithms, near-atomic resolutions (<3 Å) have been realized, competing with crystallography. This so-called resolution revolution has made it possible to study the structure of ribosomes, ion channels, and viral capsids at high resolution (Xu and Dang, 2022).

1.3.3.2 Applications in membrane protein structures

Cryo-EM can specifically be effective with membrane proteins, which are notoriously hard to crystallize. G-protein coupled receptors (GPCRs), ion channels, and transporter structures have given information on drug-binding mechanisms and conformational cycles. Cryo-EM also has the advantage of being able to examine heterogeneous or transient states, which is another distinguishing factor compared with crystallography (de Oliveira et al., 2021).

1.3.4 Small-angle scattering techniques

1.3.4.1 SAXS and SANS for flexible proteins and complexes

Small-angle scattering, such as small-angle X-ray scattering (SAXS) and small-angle neutron scattering (SANS), is a low-resolution (~10–30 Å) technique that gives information about the macromolecular shape and size in solution. The methods are especially useful with flexible or intrinsically disordered proteins, which are recalcitrant to crystallization or NMR characterization. SAXS involves the use of X-rays produced by a synchrotron, whereas SANS involves the use of neutron beams using contrast variation techniques to help differentiate between protein and solvent contributions. Parameters obtained from both of the approaches include the radius of gyration, molecular envelope, and conformational changes during the process of ligand binding (Chen et al., 2023). SAXS and SANS are strong complementary techniques, even though they cannot resolve as much as crystallography or cryo-EM. They are able to authenticate computational models, track conformational changes, and examine large macromolecular complexes under near-native conditions (Ogorzalek et al., 2018) (Table 1.1).

1.4 Computational and integrative approaches

Structural biology has evolved into a hybrid field, with computational tools and integrative methods becoming essential components of structural biology and no longer an experimental discipline. The enormous complexity of biological macromolecules, combined with experimental constraints in the investigation of dynamic or transient conditions, has necessitated the integration of molecular modeling, simulations, and AI. These methods complement traditional experimental data, but they are also capable of producing a mechanistic understanding of biomolecular functioning. In this part, the main computational strategies, and the development of integrative structural biology and the increasing role of AI and machine learning in protein science are discussed (Mondal et al., 2023).

1.4.1 Molecular modeling and simulations

1.4.1.1 Homology modeling

Homology modeling, also known as comparative modeling, is one of the oldest and most common computational approaches to structural biology. This principle is based on the fact that proteins that are highly similar in terms of sequence identity (>30%) tend to have similar three-dimensional structures. By matching a target sequence to one or more experimentally determined template structures, scientists can produce structural models that can be used as working hypotheses for functional annotation, mutagenesis studies, or docking studies. A number of automated platforms like SWISS-MODEL, MODELLER, and PHYRE2 have formalized the homology modeling workflow to allow its wide accessibility even to a nonspecialist. The models are widely used in rational drug design, especially in cases where experimental structures are not known. As an example, GPCR modeling without crystallographic data has been used to offer essential information about ligand-binding pockets and receptor activation (Meng et al., 2025). However, there are certain limitations to homology modeling. The quality of the alignment and the level of sequence identity with the template are important determinants of the accuracy of a model. Loops, flexible regions, and side-chain orientations tend to have poor reliability. Some of the efforts to mitigate these challenges involve the incorporation of fragment libraries, refinement of MD and hybrid methods that involve both modeling and experimental restraints (Seffernick and Lindert, 2020).

1.4.1.2 Molecular dynamics simulations

Although homology modeling gives static representations, the MD simulations can be used to investigate the movement of proteins over time. MD is based on Newtonian mechanics, with atomic interactions being represented by force fields, for example, AMBER, CHARMM, or OPLS. MD simulations using numerical integration of the equations of motion produce simulations of conformational fluctuations, ligand binding, and folding pathways (Sinha et al., 2022).

There are many areas of the use of MD: (1) Drug discovery: Simulations are applied to optimize the results of docking, approximate binding free energies, such as molecular mechanics (MM)/PBSA and FEP, and investigate induced-fit mechanisms.

(2) Protein folding: MD has enlightened us about intermediate and misfolding states in relation to disease (Alzheimer's and Parkinson's).

(3) Membrane proteins: Atomistic simulations are used to obtain an understanding of lipid-protein interactions, gating of ion channels, and dynamics of transport (Sinha et al., 2022).

Hardware and algorithm improvements, such as using a GPU and specialized computers, such as Anton, have brought MD down to the millisecond scale, allowing the recording of biologically relevant conformational events. Despite this advancement, there have been difficulties in sampling rare events, accuracy in force fields, and the cost of computing for large systems. A number of improved sampling methods (e.g., metadynamics, replica-exchange MD) and hybrid quantum mechanics/MM methods are now being applied to overcome these limitations (Jones et al., 2022).

1.4.2 Integrative structural biology

Recognition of the inability of any single experimental method to give full structural information of all biomolecular systems has triggered the development of integrative structural biology. This paradigm entails the integration of complementary datasets from numerous experimental sources alongside computational modeling to produce consistent, validated structural models. As an illustration, atomic data with high resolution, for example, X-ray crystallography or Cryo-EM can be used in combination with low-resolution envelopes measured by SAXS, cross-linking mass spectrometry, or fluorescence resonance energy transfer. Computational docking and MDs simulations can be used to coordinate these various datasets to give structural models that are both detailed at the atomic level and macromolecular level (Ziegler et al., 2021). The explanation of the nuclear pore complex (NPC), a mega-data-sized assembly that could not be analyzed by any single method, is a landmark case study. Integrative methods that used cryo-EM, subunit crystallography, cross-linking information, and computational models produced pseudo-atomic models that described transport pathways across the NPC. Integrative modeling has also been used to describe flexible assemblies, including the Mediator complex and dynamic viral capsids (Akey et al., 2023). The Integrative Modeling Platform has been a key computational platform that has made such studies possible. It offers statistical methods to combine the heterogeneous data into probabilistic models, to quantify uncertainties, and to verify predictions. Integrative methods are therefore a viable solution to the exploration of refractory systems that cannot be studied by conventional single-technique approaches (Webb et al., 2018).

1.4.3 Role of artificial intelligence and machine learning

1.4.3.1 DeepMind's AlphaFold and its impact

With the emergence of AI, structural biology has changed, and the breakthrough of AlphaFold2 by DeepMind was a milestone. Using large sequence-structure relationships trained through deep neural networks, AlphaFold2 demonstrated unprecedented performance in the Critical Assessment of Structure Prediction competition, competing with experimental techniques on a wide range of targets. In contrast to homology modeling, AlphaFold does not make use of known templates alone but employs attention-based architectures and evolutionary couplings to learn long-range interactions (Jumper et al., 2021). Millions of proteins in a wide variety of organisms have been provided with structural predictions in the AlphaFold Protein Structure Database, created in partnership with EMBL-EBI. The resource has made structural information accessible to all and has allowed the generation of hypotheses for proteins that were considered intractable. As an example, AlphaFold predictions have been used to aid experimental design in studies of SARS-CoV-2 to enable drug repurposing and vaccine development. AlphaFold has weaknesses, even though it has been a transformative project. It is more successful at predicting monomeric, and not complex, intrinsically disordered proteins and conformational ensembles. This has to be combined with experimental validation in order to prevent misinterpretations (Bertoline et al., 2023).

1.4.3.2 Integration into experimental workflows

Predictions with the use of AI are becoming part of experimental pipelines. AlphaFold models can be used to give templates for molecular replacement in crystallography, thus speeding up phase determination. In cryo-EM, the use of predicted models aids in the fitting of low-resolution maps as well as resolving flexible domains. AlphaFold-derived restraints are advantageous for NMR spectroscopy because they make resonance assignments more accurate. In addition to AlphaFold, machine learning models are being trained to perform a wide range of tasks, including cryo-EM image classification, optimization of crystallization conditions, and prediction of protein-ligand interactions. These advances imply that in the future, AI will complement, and not supplant, experimental structural biology and allow more effective and precise characterization of biomolecular systems (Laurents, 2022) (Fig. 1.3).

1.5 Structural insights into protein function

Proteins are not merely the most adaptable biomolecules in living organisms but also the molecular entities that ensure the presence of cellular homeostasis, the organization of biochemical pathways, and the reaction to environmental stimuli. Their tight association with their three-dimensional structures makes their functions closely related, and the recent developments in the field of structural biology have given huge knowledge about protein functionality at the atomic scale. In this section, it is described how structural studies contribute to the study of enzymatic catalysis, ligand binding, signal transduction, and macromolecular interactions.

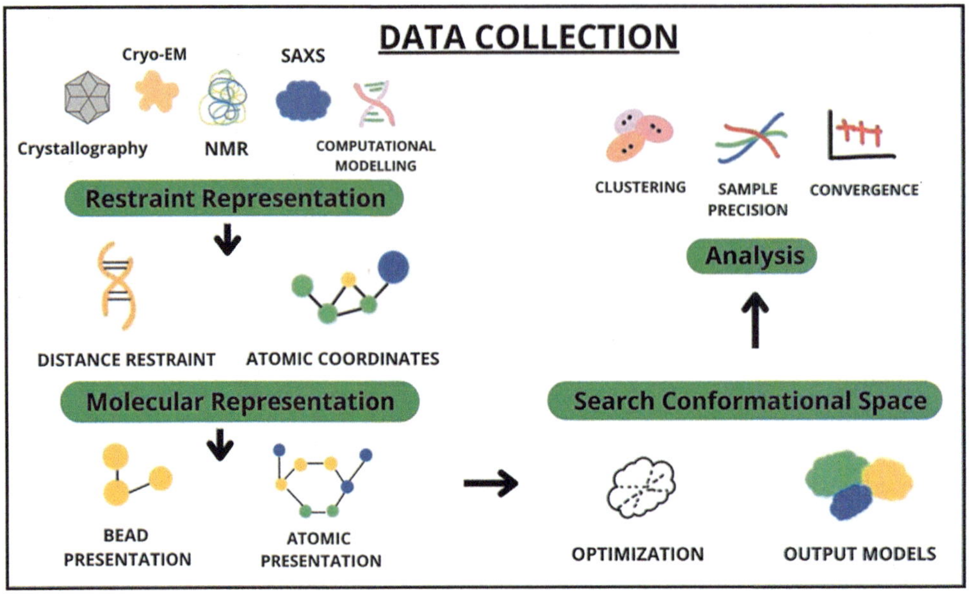

Figure 1.3 Workflow of integrative structural biology combining experimental and computational approaches. The schematic should illustrate integration of data from crystallography, cryo-electron microscopy (*cryo-EM*), nuclear magnetic resonance (*NMR*), small-angle X-ray scattering (*SAXS*), and computational modeling, culminating in validated structural models.

1.5.1 Enzymatic catalysis and active site architecture

Enzymes are the catalysts of nature, and they increase the speed of biochemical reactions several-fold. It is the three-dimensional structure of amino acid residues within the active sites that is closely linked to the catalytic power of enzymes. X-ray crystallographic and NMR structural studies have found that residues are often present in geometries that stabilize the transition state and therefore reduce the activation energy at the active sites. As an example, the catalytic triad of serine proteases (serine, histidine, and aspartate) has been used as a textbook example of how the positioning of specific residues facilitates the nucleophilic attack on peptide bonds. Crystal structures of trypsin and chymotrypsin depict how changes in conformation can be used to precisely position catalytic residues. On the same note, kinase studies present the mechanisms through which active site residues facilitate the phosphorylation of substrates by orienting substrates and positioning divalent cations (Arter et al., 2022).

A recent combination of time-resolved crystallography and MD simulations has shown that enzyme flexibility is a fundamental catalytic factor that allows enzymes to undergo dynamic rearrangements during substrate turnover. These observations bring out the importance of structure in the explanation of enzyme catalytic efficiency and specificity (Mao, 2025).

1.5.2 Ligand recognition and binding specificity

Ligand recognition forms the basis of metabolism, signal perception, and drug action. The mechanisms of protein discrimination among ligands have been elucidated using structural biology. The induced-fit and conformational-selection concepts were developed due to the structural evidence that proteins are not fixed templates, but dynamic systems that can alter themselves with the binding of a ligand (Du et al., 2016).

The hemoglobin-oxygen system is one of the most widely researched models. Cooperative oxygen binding is due to the transition between the T (tense) and R (relaxed) states of hemoglobin, and crystal structures give atomic detail about the conformational changes (Lavrinenko et al., 2023). Likewise, structural analyzes of GPCRs have defined how small ligands, peptides, or hormones stabilize specific receptor conformations that define downstream signaling pathways. These structural insights are very important in drug discovery. Atomic-level visualization of the binding pocket facilitated the rational design of HIV protease inhibitors, which ensured that medicinal chemists customized high-affinity and specific molecules. Therefore ligand recognition research is a foundation of translational research (Wingler and Lefkowitz, 2020).

1.5.3 Signal transduction and conformational changes

Signal transduction refers to the mechanism of transforming the signals from the external environment into internal cellular responses and is typically mediated by changes in protein structure. The activity switches are conformational switches that are achieved by phosphorylation or attachment of ligands and environmental factors. A good example is the receptor tyrosine kinases (RTKs), where the binding of ligands causes dimerization and autophosphorylation and subsequently activates the downstream pathways. The crystal structure of the epidermal growth factor receptor (EGFR) extracellular domain in the ligand-bound and ligand-free conditions has clarified the role of rearrangements of the structure in activating the kinase (Huang et al., 2021). Similarly, structural snapshots of heterotrimeric G-proteins depict that binding and hydrolysis of guanosine triphosphate cause conformational alterations required during the process of transitioning between the active and inactive states. Cellular signaling networks are made up of these conformational networks (Aranda-García et al., 2025).

This has been extended by advanced techniques like Cryo-EM enables one to visualize proteins in different functional states. One example is the Cryo-EM analysis of ion channels; it is possible to demonstrate how channels can be gated by a conformational change of transmembrane helices in response to voltage or ligand binding. The findings confirm the importance of dynamic structural biology in explaining signal transduction (Lau et al., 2018).

1.5.4 Protein–protein and protein-nucleic acid interactions

Proteins usually need to interact with other proteins or nucleic acids in order to carry out complex biological functions. These interactions can be explained structurally to give the molecular grammar of cellular networks. Surface patches mediate PPIs through hydrogen bonds, salt bridges, and hydrophobic contacts. As an example, the p53-MDM2 complex has been structurally characterized to elaborate on how MDM2 suppresses the tumor suppressor p53. Small-molecule inhibitors have been developed that disrupt the interaction and recover p53 activity in cancers by structural mapping of this interface (Zhu et al., 2022).

Protein-nucleic acid interactions are also basic. RNA polymerase and ribosome structures have shown how proteins identify the nucleotide sequences and catalyze transcription and translation with an extraordinary degree of fidelity. Helix-turn-helix motifs and zinc-finger domains frequently mediate the recognition of transcription factors by DNA, and structural observations have shown the mechanism by which sequence specificity can be attained by direct base interactions and DNA shape recognition (Hennig, 2025).

Such interactions not only form the basis of basic processes but also offer therapeutic avenues. The structural mapping of CRISPR-Cas9 assemblies, has radically changed the genome editing process, enabling the use of DNA-targeting proteins to be engineered precisely (Past, present, and future of CRISPR genome editing technologies: Cell, 2025) (Table 1.2).

1.6 Emerging paradigms and limitations of structure-function relationships

The classical principle in the history of molecular biology was that a well-defined three-dimensional structure is a direct determinant of biological function. Although this paradigm has been invaluable in developing our knowledge of biomolecular processes, accumulating experimental and theoretical data suggest that the structure-function connection is much more complicated than was initially studied. Modern structural biology is recognizing the fact that not only does protein behavior rely on the fixed structures that are obtained under artificially forced experimental conditions, but also on dynamic conformational ensembles, cellular contexts, and constantly changing energy landscapes that rearrange molecular behavior.

1.6.1 In-cell and in situ structural biology: moving beyond the test tube

The classical structural methods of high resolution, including X-ray crystallography and solution NMR spectroscopy, usually measure dilute and nonphysiological purified biomolecules. But the intracellular environment is highly macromolecular

Table 1.2 Structural features of proteins underlying their functional roles.

Functional category	Representative protein(s)	Key structural features	Biological significance
Enzymatic catalysis	Trypsin, chymotrypsin, kinases	Catalytic triad (Ser-His-Asp), active site geometry, divalent cation coordination, flexible loops	Facilitate substrate binding, stabilize the transition state, accelerate reaction rates
Ligand recognition and binding specificity	Hemoglobin, GPCRs, HIV protease	Allosteric conformational states (T/R), ligand-binding pockets, induced-fit conformational changes	Enable selective ligand binding, cooperative effects, and drug-target specificity
Signal transduction and conformational changes	EGFR (receptor tyrosine kinase), G-proteins, ion channels	Dimerization interfaces, switch regions, conformational flexibility, phosphorylation sites	Convert extracellular signals into intracellular responses, regulate enzyme activity, and cellular signaling pathways
Protein–protein interactions	p53-MDM2 complex, immunoglobulins	Surface patches forming hydrogen bonds, hydrophobic interactions, salt bridges	Mediate regulation of tumor suppressors, immune response, and assembly of protein complexes
Protein-nucleic acid interactions	RNA polymerase, ribosome, transcription factors, CRISPR-Cas9	DNA/RNA recognition motifs (helix-turn-helix, zinc finger), catalytic sites, sequence-specific binding regions	Facilitate transcription, translation, DNA editing, and regulation of gene expression

congested, heterotrophic molecular interactions, and spatially compartmentalized, with all having a significant effect on protein folding, stability, and function. In cell and in situ structural biology methods have arisen to fill this gap between in vitro determination of structure and in vivo biological reality.

Conventional high-resolution structural methods, including X-ray crystallography and solution NMR spectroscopy, are generally used to study purified biomolecules in dilute, nonphysiological conditions. Conversely, the intracellular environment is highly characterized by macromolecular crowding, heterogeneous molecular interactions, and spatial compartmentalization, all of which significantly impact protein folding, stability, and functioning. In cell and in situ structural biology methods have arisen in order to fill the gap between in vitro determination of structure and in vivo biological reality. Cryo-electron tomography (cryo-ET) has proven to be a game changer in this regard and can be used to visualize macromolecular complexes in intact cells under near-native conditions. In contrast to single-particle cryo-EM, cryo-ET maintains cell structure and allows studying proteins in functional biological relationships. Studies that have used cryo-ET have shown that a great number of protein complexes in cells can take up completely different forms than those observed in isolation systems, and hence point to the importance of the cellular context on structure and function. Besides that, in-cell NMR spectroscopy has provided residue-level data on protein dynamics, folding states, and PPIs in living cells, revealing changes in stability and conformational preference under crowding.

Another problem with the classical structure-function model is macromolecular crowding effects. The effects of crowding include changes in free-energy landscapes, such as favoring compact states, stabilizing transient interactions, and functional assemblies that would not be discovered in dilute solutions. Therefore these determinations of structures in vitro can only embody a subset of the functionally relevant conformations.

1.6.2 Energy landscape theory and conformational ensembles

The energy-landscape approach has revolutionized our perception of protein structure and functionality. Instead of a single fixed structure, proteins occupy a rugged free-energy surface that consists of a number of interchangeable conformational states. Often, functional activity can be generated during transitions between these states and not by an isolated conformation. Induced-fit and conformational selection models are now found together as complementary processes defining the binding of ligands and molecular recognition. Considering the energy-landscape perspective, ligands stabilize preexisting conformations in an ensemble, and they do not cause conformational changes, but they change the population distribution. The paradigm is especially relevant to intrinsically disordered proteins and regions, which are not present in a stable tertiary form but instead play key functions in signaling, regulation, and molecular recognition.

Complex computational algorithms, such as MD simulations and enhanced sampling techniques, have played a key role in mapping such free-energy surfaces as well as finding transient and low-population states, which are often not visible in experimental structure determination. These hidden folds are of importance in allosteric control, catalysis of enzymes, as well as pathological misfolding, thus supporting the idea that function is encoded in ensembles and not isolated folds.

1.6.3 When structure misleads function: failures and controversies

Though successful achievements have been seen, structural biology has had significant failures in finding biological function through association with nondynamic structures. Crystallographic snapshots can be nonphysiological in nature, and conformations of crystals stabilized by crystal packing forces can provide falsely interpretative mechanistic information. Likewise, functionally neutral mutations that were postulated by minimal perturbation of the structure have sometimes had catastrophic functional effects because of issues with dynamics or disturbed allosteric networks. Enzymes with very similar overall folds can have very different catalytic activities, and even proteins with almost identical structures can have very different biological functions. On the other hand, massive structural changes do not necessarily imply loss of functionality, especially when there are compensatory processes that are maintained. These findings indicate the shortcomings of the use of structural similarity as a predictor of functionality.

The drug discovery process continues to be controversial as well, with the idea of high-affinity ligand binding based on immobile structures not necessarily being followed by cellular efficacy. Such a disconnect frequently occurs due to the absence of flexibility in the protein, alternative binding modes, and cellular mechanisms like competing interactions and posttranslational modifications.

1.6.4 Integrating structure, dynamics, and cellular context

The integrative nature of modern structural biology is becoming more and more accepted, as experimental techniques, computational modeling, and cellular studies are integrated to give a broad picture of biomolecular behavior. Hybrid techniques combining cryo-EM, NMR, small-angle scattering, and molecular simulations allow assemblies of different sizes to be characterized at a variety of spatial and temporal levels. Finally, the structure-function relationship must be considered as a multidimensional concept that involves conformational dynamics, energy landscapes, and environmental effects. The understanding of these complexities is not only critical for streamlining our basic knowledge of biology, but also for improving translational biology, such as in rational drug design, and in understanding disease-related mutations.

1.7 Translational applications of structural biology

Structural biology has gone beyond its initial purpose of explaining macromolecular structure to form the foundation of biotechnology. Structural biology has triggered progress in drug discovery, disease biomarker discovery, and synthetic biology by offering atomic-level details about protein structures. The following section discusses these translational applications and how structural information can be used to drive innovation in medicine and biotechnology (Table 1.3).

1.7.1 Drug discovery and rational design

Structural biology has brought a revolution in the field of drug discovery and optimization of therapeutic agents. The concept of structure-based drug design (SBDD) uses high-resolution protein structures to guide the design of small molecules that interact with the target proteins in a specific manner. This method is the opposite of the conventional one because it pays attention to the three-dimensional structure of the target to enable the development of compounds with high specificity and potency. One of the achievements in SBDD is the creation of inhibitors of the EGFR in nonsmall cell

Table 1.3 Structural biology-driven insights into therapeutic protein targeting.

Therapeutic area	Target protein	Structural insight leading to discovery	Outcome
Cancer	EGFR (mutant form)	Structural analysis revealed a mutant-specific binding pocket	Development of EGFR inhibitors (e.g., gefitinib, erlotinib)
HIV/AIDS	HIV protease	Identification of the active site and substrate-binding pocket	Design of protease inhibitors (e.g., ritonavir, atazanavir)
Antimicrobial resistance	Bacterial ribosome	Structural elucidation of antibiotic binding sites	Development of new antibiotics targeting resistant strains
Autoimmune diseases	TNF-α	Structural mapping of TNF-α receptor interaction sites	Design of TNF-α inhibitors (e.g., infliximab, adalimumab)

lung cancer. The structural studies identified the existence of a mutant form of EGFR, which is constitutively active, and as a result, tyrosine kinase inhibitors that specifically bind to this mutant form were designed, which inhibited its activity, thus benefiting patients with this mutation (Wang et al., 2024). On the same note, the development of protease inhibitors in the management of HIV/AIDS is an example of the strength of SBDD. Through the study of the structure of the HIV protease enzyme, scientists were able to determine the critical binding sites and create inhibitors that are able to effectively inhibit the active form of the enzyme, preventing viral maturation and replication (Ghosh et al., 2006).

The given examples emphasize the effectiveness of SBDD in the creation of targeted therapeutic agents that are effective and minimally toxic, which is why structural understanding is essential in current drug discovery.

1.7.2 Disease biomarker identification

The structural variations (SVs) such as deletions, insertions, and duplications in the genome have become important contributors to a number of diseases. The detection and characterization of these SVs are essential in the study of the mechanisms of disease and the creation of diagnostic instruments. Progress in genomic technologies has enabled the detection of SVs related to neurological disorders. For example, some research has found that some SVs have the ability to interfere with genes that are related to the development of neurons, resulting in diseases like intellectual disability and autism spectrum disorders (Salnikov et al., 2024). Structural analysis of these variants gives information about the ability of genomic changes to influence protein function and lead to the development of disease pathology. Moreover, structural biology combined with genomic data has been used to discover new biomarkers that can be used to detect disease at an early stage. Through the structural implications of SVs, researchers will be in a position to design diagnostic assays that are sensitive and specific enough to determine such variants at an earlier stage, enabling early intervention and tailored treatment plans. The current study on the structural characterization of SVs is promising in terms of adding more knowledge about complex diseases and improving both diagnostic and therapeutic strategies (Lin et al., 2022).

1.7.3 Synthetic biology and protein engineering

Synthetic biology uses the concepts of engineering and biology to design and build new biological parts, devices, and systems. Structural biology is a central component in this field because it offers detailed data on the structure of proteins, which is critical for the rational design of new proteins with the intended functionality. The recent developments in protein engineering permit the production of enzymes that are highly stable and active, and specific to industrial applications. As an example, the modification of the cellulase engineering process to produce biofuels is achieved by modifying the structure of the enzyme in order to enhance its resistance to the inhibitors found in the biomass and make the conversion process more efficient (Contreras et al., 2020). Structural understanding in the field of therapeutic protein design has enabled the development of monoclonal antibodies, which have enhanced specificity and reduced immunogenicity. Through the three-dimensional structure of target antigens, researchers can designantibody binding sites with the highest affinity and the lowest adverse immune reactions (Chiu et al., 2019).

Besides, the introduction of computational instruments and AI-based platforms has expedited the development of new proteins. Programs such as AlphaFold have greatly increased the capacity to predict protein structures, whereby complex architectures and functions of proteins are now achievable. These advances in synthetic biology and protein engineering show the paradigm shift that structural biology has made in producing new solutions in medicine, industry, and environmental sustainability (Yang et al., 2023).

1.8 Recent advances and future perspectives in structural biology

Structural biology has experienced a paradigm shift in its development, which has been driven by technological advancements and interdisciplinary partnerships. The combination of different methodologies has helped researchers understand the dynamic character of biomolecules and no longer take snapshots of them but rather take a more holistic view of their functions. This part explores the recent developments and future of the field.

1.8.1 Hybrid approaches combining cryo-EM, crystallography, and simulations

The combination of the cryo-EM, X-ray crystallography, and computational simulations has led to the emergence of hybrid methods that provide a more detailed perspective on the macromolecular structures. Cryo-EM gives highly resolved pictures of large complexes in close to natural conditions, whereas X-ray crystallography gives atomic-level information on smaller and well-organized molecules. Computational simulations, especially MD, can be used to study how conformational change with time (Cassidy et al., 2018). A combination of these methods allows scientists to construct better models of biomolecules. As an example, the cryo-EM maps can be optimized with the help of MD simulations to resolve flexible areas in the still images, which are usually disordered. Such synergy increases the understanding of structural data and the design of specific therapeutics (Son et al., 2024).

1.8.2 Time-resolved structural biology for capturing transient states

The transient states of biological processes are not always easily captured through traditional approaches. Time-resolved structural biology has become an imperative method that is used to study these transient conformations. Time-resolved cryo-EM and serial femtosecond crystallography are techniques that enable the observation of structural changes on timescales of between microseconds and milliseconds. Such techniques have played a significant role in the explanation of enzyme catalysis, signal transduction, and protein folding mechanisms (Mäeots and Enchev, 2022). As an example, time-resolved cryo-EM has helped researchers understand the conformational dynamics of the anaphase-promoting complex/cyclosome, which is an important cell cycle regulator. Through the detection of intermediate states, researchers are able to recognize possible drug targets and develop molecules that regulate certain events in such processes (Banari et al., 2025).

1.8.3 Single-molecule approaches for dynamic measurements

Single-molecule methods have transformed the biomolecular dynamics field by allowing the real-time observation of single molecules. The techniques of single-molecule fluorescence resonance energy transfer (smFRET) and graphene-based energy transfer (GET) give an understanding of conformational changes, binding processes, and molecular interactions with a resolution never tested before (Agam et al., 2023).

For example, smFRET has been used to study the folding dynamics of RNA molecules and found intermediates that cannot be observed in ensemble measurements. Likewise, DNA-protein interactions have been studied with GET, with a spatial resolution at the angstrom scale and subsecond time resolution. Such single-molecule methods are crucial to the study of the heterogeneity of biological systems and to the creation of personalized therapeutic approaches (Szalai et al., 2025).

1.8.4 Future challenges and opportunities

In spite of the advances, there are still various challenges in structural biology. The amphipathic nature of membrane proteins is one of the primary obstacles, as the latter are not always easy to isolate and stabilize. Innovations in cryo-EM and the creation of new detergents and nano disks have increased the study of these proteins, though additional inventions are required to increase the level of resolution and throughput (Lee et al., 2022). The other issue is how to combine the structural data with functional information. Whereas structural methods can give finer details of biomolecules, kinetic and thermodynamic measurements are necessary to comprehend their roles in complex biological systems. To gain a comprehensive picture of the biomolecular functionality, the emergence of hybrid models incorporating structural, biochemical, and biophysical information will be crucial (Maestro et al., 2025). These challenges are the future of structural biology, as they can be overcome with the help of technological progress and collaboration between disciplines. The combination of machine learning and AI with structural methods has the potential to automate data analysis, predict structures, and create new biomolecules. Moreover, it will be possible to share data and resources because the creation of international databases and cooperative platforms will speed up the discovery process and promote innovation (Varadi et al., 2022).

1.9 Conclusion

Structural biology has come to be an essential field that connects the molecular architecture with biological activity. During the last decades, the developments in the methodologies, including X-ray crystallography, NMR, cryo-EM, and computational simulations, have enabled the unparalleled resolution of biomolecular structures. These structural studies have revolutionized our knowledge of enzyme catalysis, ligand recognition, signal transduction, protein–protein/protein-nucleic acid interactions, and have served as a basis for translational use in medicine, biotechnology, and synthetic biology.

1.9.1 Summary of key insights

The key discovery of structural biology is the complex correlation between the three-dimensional conformation and biological activity. An example is enzymes, which show that the active site residues are specific and the enzymes are capable of changing their conformation to catalyze reactions and be specific to a substrate. Dynamic changes in the protein structure direct ligand recognition, enabling high specificity, and regulation of cooperative effects as seen in hemoglobin and GPRCs. Likewise, the conformational changes that are involved in intracellular signal transduction have been explained through structural studies of signaling proteins, including RTKs and heterotrimeric G-proteins.

Furthermore, protein–protein and protein-nucleic acid interactions have been mapped on an atomic scale, which defines how complex formation is controlled by surface complementarity, electrostatic interactions, and conformational flexibility. These have led to new therapeutic advances, including small-molecule inhibitors of p53-MDM2 interactions and genome editing by CRISPR-Cas9. Translational research has also been developed using structural biology, which has been the basis of structure-based drug discovery, disease biomarker discovery, and protein engineering to be used in synthetic biology. New technological convergences, such as hybrid methods that merge cryo-EM, X-ray crystallography, and molecular dynamics simulations, have increased our ability to examine large, flexible, and transient biomolecular assemblies.

Now, time-resolved structural biology and single-molecule methods make it possible to capture intermediate states and dynamic processes, and shed light on mechanistic pathways that had been previously inaccessible. All of these developments indicate that biological functionality is an emergent characteristic of structure, dynamics, and interaction networks.

1.9.2 Contribution of structural biology to modern life sciences

Structural biology has a wide range of applications in the life sciences, such as molecular medicine, pharmacology, biotechnology, and synthetic biology. Structural understanding in drug discovery has been used to rationally design therapeutics that are highly target specific and have minimal off-target effects. Examples of landmarks include EGFR inhibitors against nonsmall cell lung cancer, anti-HIV protease inhibitors, and antibacterial ribosomal antibiotics. In addition to therapeutics, structural biology has helped in the discovery of biomarkers. Early identification of pathological conditions and the development of individualized medicine are possible due to the structural characterization of disease-associated proteins and variants. As an example, the knowledge of structural deviations in cancer- or neurodegeneration-linked proteins can be used to inform diagnostics and targeted treatment. Structural knowledge in synthetic biology and protein engineering can be used to rationally design new enzymes, antibodies, and protein complexes with improved stability, catalytic properties, or desired functions. The translational potential of structural knowledge has resulted in the creation of biomolecules that have been optimized to serve industrial, therapeutic, or environmental purposes through computational predictions with structural validation.

In addition, structural biology promotes essential knowledge of cellular processes. Through the mechanisms of macromolecule folding, interactions, and responses to environmental factors, structural studies help in understanding the molecular mechanisms of life, providing templates to systems biology and integrative models.

1.9.3 Outlook for future directions

The future of structural biology lies in its ability to elucidate the complexity of biomolecular activity in cellular environments. New methods, including in situ cryo-EM, correlative light and electron microscopy, and time-resolved spectroscopy with advanced resolution, are likely to provide high-resolution information in natural cellular contexts. Combining with AI and machine learning will speed up the prediction of structures, dynamic modeling, and functional annotation of new biomolecules. There are still problems, such as the investigation of highly dynamic, intrinsically disordered proteins and complex membrane assemblies. The solution to such challenges will involve novel sample preparation, imaging, and computational approaches. The integration of structural biology with genomics, metabolomics, and systems biology will further make possible interdisciplinary studies that provide a comprehensive view of cellular networks and pathophysiology.

In general, structural biology is still in the process of transitioning towards a discipline that is no longer concerned with static snapshots, but with dynamic, multiscale representations of biomolecules. Its impact on drug discovery, diagnostics, synthetic biology, and basic biology is extensive, and continued innovations are expected to further expand our molecular insight into life, as well as allow applications to be made that will radically change science and medicine.

References

Agam, G., Gebhardt, C., Popara, M., Mächtel, R., Folz, J., Ambrose, B., et al., 2023. Reliability and accuracy of single-molecule FRET studies for characterization of structural dynamics and distances in proteins. Nature Methods 20 (4), 523–535. Available at https://doi.org/10.1038/s41592-023-01807-0, https://www.nature.com/nmeth/.

Agback, T., Lesovoy, D., Han, X., Lomzov, A., Sun, R., Sandalova, T., et al., 2023. Combined NMR and molecular dynamics conformational filter identifies unambiguously dynamic ensembles of Dengue protease NS2B/NS3pro. Communications Biology 6 (1). Available at https://doi.org/10.1038/s42003-023-05584-6.

Akey, C.W., Echeverria, I., Ouch, C., Nudelman, I., Shi, Y., Wang, J., et al., 2023. Implications of a multiscale structure of the yeast nuclear pore complex. Molecular Cell 83 (18), 3283. Available at https://doi.org/10.1016/j.molcel.2023.08.025, www.molecule.org.

Alberts, B., Johnson, A., Lewis, J., Raff, M., Roberts, K., Walter, P., 2002. Analyzing protein structure and function. Molecular Biology of the Cell, 4th edition. Garland Science [cited 2025 Oct 10]. Available at https://www.ncbi.nlm.nih.gov/books/NBK26820/.

Aranda-García, D., Stepniewski, T.M., Torrens-Fontanals, M., García-Recio, A., Lopez-Balastegui, M., Medel-Lacruz, B., et al., 2025. Large scale investigation of GPCR molecular dynamics data uncovers allosteric sites and lateral gateways. Nature Communications 16 (1). Available at https://doi.org/10.1038/s41467-025-57034-y, https://www.nature.com/ncomms/.

Arter, C., Trask, L., Ward, S., Yeoh, S., Bayliss, R., 2022. Structural features of the protein kinase domain and targeted binding by small-molecule inhibitors. Journal of Biological Chemistry 298 (8), 102247. Available at https://doi.org/10.1016/j.jbc.2022.102247.

Banari, A., Samanta, A.K., Munke, A., Laugks, T., Bajt, S., Grünewald, K., et al., 2025. Advancing time-resolved structural biology: latest strategies in cryo-EM and X-ray crystallography. Nature Methods 22 (7), 1420–1435. Available at https://doi.org/10.1038/s41592-025-02659-6, https://www.nature.com/nmeth/.

Bertoline, L.M.F., Lima, A.N., Krieger, J.E., Teixeira, S.K., 2023. Before and after AlphaFold2: an overview of protein structure prediction. Frontiers in Bioinformatics 3. Available at https://doi.org/10.3389/fbinf.2023.1120370, www.frontiersin.org/journals/bioinformatics.

Birch, J., Axford, D., Foadi, J., Meyer, A., Eckhardt, A., Thielmann, Y., et al., 2018. The fine art of integral membrane protein crystallisation. Methods (San Diego, Calif.) 147, 150–162. Available at https://doi.org/10.1016/j.ymeth.2018.05.014, http://www.elsevier.com/inca/publications/store/6/2/2/9/1/4/index.htt.

Borkakoti, N., Ribeiro, A.J.M., Thornton, J.M., 2025. A structural perspective on enzymes and their catalytic mechanisms. Current Opinion in Structural Biology 92. Available at https://doi.org/10.1016/j.sbi.2025.103040, https://www.sciencedirect.com/science/journal/0959440X.

Carugo, O., Djinović-Carugo, K., 2023. Structural biology: a golden era. PLoS Biology 21 (6), e3002187. Available at https://doi.org/10.1371/journal.pbio.3002187.

Cassidy, C.K., Himes, B.A., Luthey-Schulten, Z., Zhang, P., 2018. CryoEM-based hybrid modeling approaches for structure determination. Current Opinion in Microbiology 43, 14–23. Available at https://doi.org/10.1016/j.mib.2017.10.002, http://www.elsevier.com/locate/mib.

Chen, S.H., Weiss, K.L., Stanley, C., Bhowmik, D., 2023. Structural characterization of an intrinsically disordered protein complex using integrated small-angle neutron scattering and computing. Protein Science 32 (10). Available at https://doi.org/10.1002/pro.4772, http://onlinelibrary.wiley.com/journal/10.1002/(ISSN)1469-896X.

Chiu, M.L., Goulet, D.R., Teplyakov, A., Gilliland, G.L., 2019. Antibody structure and function: the basis for engineering therapeutics. Antibodies 8 (4). Available at https://doi.org/10.3390/antib8040055, https://www.mdpi.com/2073-4468/8/4/55/pdf.

Contreras, F., Pramanik, S., Rozhkova, A.M., Zorov, I.N., Korotkova, O., Sinitsyn, A.P., et al., 2020. Engineering robust cellulases for tailored lignocellulosic degradation cocktails. International Journal of Molecular Sciences 21 (5), 1589. Available at https://doi.org/10.3390/ijms21051589.

de Oliveira, T.M., van Beek, L., Shilliday, F., Debreczeni, J., Phillips, C., 2021. Cryo-EM: the resolution revolution and drug discovery. SLAS Discovery 26 (1), 17–31. Available at https://doi.org/10.1177/2472555220960401, http://journals.sagepub.com/loi/jbx.

Du, X., Li, Y., Xia, Y.L., Ai, S.M., Liang, J., Sang, P., et al., 2016. Insights into protein–ligand interactions: mechanisms, models, and methods. International Journal of Molecular Sciences 17 (2). Available at https://doi.org/10.3390/ijms17020144, http://www.mdpi.com/1422-0067/17/2/144/pdf.

Ghosh, A.K., Sridhar, P.R., Leshchenko, S., Hussain, A.K., Li, J., Kovalevsky, A.Y., et al., 2006. Structure-based design of novel HIV-1 protease inhibitors to combat drug resistance. Journal of Medicinal Chemistry 49 (17), 5252–5261. Available at https://doi.org/10.1021/jm060561m.

Hennig, J., 2025. Structural biology of RNA and protein-RNA complexes after AlphaFold3. Chembiochem: a European Journal of Chemical Biology 26 (7). Available at https://doi.org/10.1002/cbic.202401047.

Huang, X.L., Khan, M.I., Wang, J., Ali, R., Ali, S.W., Zahra, Qu.A., et al., 2021. Role of receptor tyrosine kinases mediated signal transduction pathways in tumor growth and angiogenesis—New insight and futuristic vision. International Journal of Biological Macromolecules 180, 739–752. Available at https://doi.org/10.1016/j.ijbiomac.2021.03.075, www.elsevier.com/locate/ijbiomac.

Jones, D., Allen, J.E., Yang, Y., Drew Bennett, W.F., Gokhale, M., Moshiri, N., et al., 2022. Accelerators for classical molecular dynamics simulations of biomolecules. Journal of Chemical Theory and Computation 18 (7), 4047–4069. Available at https://doi.org/10.1021/acs.jctc.1c01214, http://pubs.acs.org/journal/jctcce.

Jumper, J., Evans, R., Pritzel, A., Green, T., Figurnov, M., Ronneberger, O., et al., 2021. Highly accurate protein structure prediction with AlphaFold. Nature 596 (7873), 583–589. Available at https://doi.org/10.1038/s41586-021-03819-2, http://www.nature.com/nature/index.html.

LaPelusa, A., Kaushik, R., 2022. Physiology, proteins. StatPearls. StatPearls Publishing [cited 2025 Oct 10]. Available at https://www.ncbi.nlm.nih.gov/books/NBK555990/.

Lau, C., Hunter, M.J., Stewart, A., Perozo, E., Vandenberg, J.I., 2018. Never at rest: insights into the conformational dynamics of ion channels from cryo-electron microscopy. Journal of Physiology 596 (7), 1107–1119. Available at https://doi.org/10.1113/JP274888, http://onlinelibrary.wiley.com/journal/10.1111/(ISSN)1469-7793.

Laurents, D.V., 2022. AlphaFold 2 and NMR spectroscopy: partners to understand protein structure, Dynamics and Function. Front Mol Biosci 9 May 17 [cited 2025 Oct 12]. Available at https://www.frontiersin.org/journals/molecular-biosciences/articles/10.3389/fmolb.2022.906437/full.

Lavrinenko, I.A., Vashanov, G.A., Hernández Cáceres, J.L., Nechipurenko, Y.D., 2023. Mathematical models describing oxygen binding by hemoglobin. Biophysical Reviews 15 (5), 1269–1278. Available at https://doi.org/10.1007/s12551-023-01110-4, https://www.springer.com/journal/12551.

Lee, H.J., Lee, H.S., Youn, T., Byrne, B., Chae, P.S., 2022. Impact of novel detergents on membrane protein studies. Chemistry 8 (4), 980–1013. Available at https://doi.org/10.1016/j.chempr.2022.02.007, http://www.cell.com/chem/home.

Lin, X., Yang, Y., Melton, P.E., Singh, V., Simpson-Yap, S., Burdon, K.P., et al., 2022. Integrating genetic structural variations and whole-genome sequencing into clinical neurology. Neurology: Genetics 8 (4). Available at https://doi.org/10.1212/NXG.0000000000200005, http://ng.neurology.org/.

Mäeots, M.E., Enchev, R.I., 2022. Structural dynamics: review of time-resolved cryo-EM. Acta Crystallogr D Struct Biol 78 (Pt 8), 927–935.

Maestro, A., Zaccai, N.R., Gonzalez-Martinez, J.F., Sanchez-Puga, P., Tajuelo, J., Rubio, M.A., et al., 2025. Combined thermodynamic and time-resolved structural analysis of interactions between AP2 and biomimetic plasma membranes provides insights into clathrin-mediated endocytosis. Communications Biology 8 (1). Available at https://doi.org/10.1038/s42003-025-08597-5, https://www.springer.com/journal/42003.

Maiti, S., Singh, A., Maji, T., Saibo, N.V., De, S., 2024. Experimental methods to study the structure and dynamics of intrinsically disordered regions in proteins. Current Research in Structural Biology 7. Available at https://doi.org/10.1016/j.crstbi.2024.100138, https://www.sciencedirect.com/science/journal/2665928X.

Mao, Y., 2025. Dynamics-based drug discovery by time-resolved cryo-EM. Current Opinion in Structural Biology 91, 103001. Available at https://doi.org/10.1016/j.sbi.2025.103001.

Marinko, J.T., Huang, H., Penn, W.D., Capra, J.A., Schlebach, J.P., Sanders, C.R., 2019. Folding and misfolding of human membrane proteins in health and disease: from single molecules to cellular proteostasis. Chemical Reviews 119 (9), 5537–5606. Available at https://doi.org/10.1021/acs.chemrev.8b00532, http://pubs.acs.org/journal/chreay.

Meng, Y., Zhang, Z., Zhou, C., Tang, X., Hu, X., Tian, G., et al., 2025. Protein structure prediction via deep learning: an in-depth review. Frontiers Pharmacology 16 Apr 3 [cited 2025 Oct 11]. Available at https://www.frontiersin.org/journals/pharmacology/articles/10.3389/fphar.2025.1498662/full.

Mondal, A., Lenz, S., MacCallum, J.L., Perez, A., 2023. Hybrid computational methods combining experimental information with molecular dynamics. Current Opinion in Structural Biology 81. Available at https://doi.org/10.1016/j.sbi.2023.102609, http://www.elsevier.com/locate/sbi.

Narayanan, C., Bafna, K., Roux, L.D., Agarwal, P.K., Doucet, N., 2017. Applications of NMR and computational methodologies to study protein dynamics. Archives of Biochemistry and Biophysics 628, 71–80. Available at https://doi.org/10.1016/j.abb.2017.05.002, http://www.elsevier.com/inca/publications/store/6/2/2/7/8/7/index.htt.

Ogorzalek, T.L., Hura, G.L., Belsom, A., Burnett, K.H., Kryshtafovych, A., Tainer, J.A., et al., 2018. Small angle X-ray scattering and cross-linking for data assisted protein structure prediction in CASP 12 with prospects for improved accuracy. Proteins: Structure, Function and Bioinformatics 86, 202–214. Available at https://doi.org/10.1002/prot.25452, http://onlinelibrary.wiley.com/journal/10.1002/(ISSN)1097-0134.

Past, present, and future of CRISPR genome editing technologies: Cell [Internet]. [cited 2025 Oct 12]. Available from: https://www.cell.com/cell/fulltext/S0092-8674(24)00111-9.

Rehman, I., Botelho, S., 2025. Biochemistry, tertiary protein structure. StatPearls. StatPearls Publishing, Treasure Island (FL) [cited 2025 Oct 10]. Available at http://www.ncbi.nlm.nih.gov/books/NBK470269/.

Rehman, I., Farooq, M., Botelho, S., 2025. Biochemistry, secondary protein structure. StatPearls. StatPearls Publishing, Treasure Island (FL) [cited 2025 Oct 10]. Available at http://www.ncbi.nlm.nih.gov/books/NBK470235/.

Salnikov, P., Korablev, A., Serova, I., Belokopytova, P., Yan, A., Stepanchuk, Y., et al., 2024. Structural variants in the Epb41l4a locus: TAD disruption and Nrep gene misregulation as hypothetical drivers of neurodevelopmental outcomes. Scientific Reports 14 (1). Available at https://doi.org/10.1038/s41598-024-52545-y.

Sanvictores, T., Farci, F., 2025. Biochemistry, primary protein structure. StatPearls. StatPearls Publishing, Treasure Island (FL) [cited 2025 Oct 10]. Available at http://www.ncbi.nlm.nih.gov/books/NBK564343/.

Seffernick, J.T., Lindert, S., 2020. Hybrid methods for combined experimental and computational determination of protein structure. Journal of Chemical Physics 153 (24). Available at https://doi.org/10.1063/5.0026025, http://scitation.aip.org/content/aip/journal/jcp.

Sinha, S., Tam, B., Wang, S.M., 2022. Applications of molecular dynamics simulation in protein study. Membranes 12 (9). Available at https://doi.org/10.3390/membranes12090844, http://www.mdpi.com/journal/membranes.

Son, A., Kim, W., Park, J., Lee, W., Lee, Y., Choi, S., et al., 2024. Utilizing molecular dynamics simulations, machine learning, Cryo-EM, and NMR spectroscopy to predict and validate protein dynamics. International Journal of Molecular Sciences 25 (17), 9725. Available at https://doi.org/10.3390/ijms25179725.

Su, J., Song, Y., Zhu, Z., Huang, X., Fan, J., Qiao, J., et al., 2024. Cell–cell communication: new insights and clinical implications. Signal Transduction and Targeted Therapy 9 (1). Available at https://doi.org/10.1038/s41392-024-01888-z.

Szalai, A.M., Ferrari, G., Richter, L., Hartmann, J., Kesici, M.Z., Ji, B., et al., 2025. Single-molecule dynamic structural biology with vertically arranged DNA on a fluorescence microscope. Nature Methods 22 (1), 135–144. Available at https://doi.org/10.1038/s41592-024-02498-x, https://www.nature.com/nmeth/.

Szczesny-Malysiak, E., Dybas, J., Blat, A., Bulat, K., Kus, K., Kaczmarska, M., et al., 2020. Irreversible alterations in the hemoglobin structure affect oxygen binding in human packed red blood cells. Biochimica et Biophysica Acta (BBA) - Molecular Cell Research 1867 (11), 118803. Available at https://doi.org/10.1016/j.bbamcr.2020.118803.

Vadukul, D.M., Papp, M., Thrush, R.J., Wang, J., Jin, Y., Arosio, P., et al., 2023. α-Synuclein aggregation is triggered by oligomeric amyloid-β 42 via heterogeneous primary nucleation. Journal of the American Chemical Society 145 (33), 18276–18285. Available at https://doi.org/10.1021/jacs.3c03212, http://pubs.acs.org/journal/jacsat.

Van Der Lee, R., Buljan, M., Lang, B., Weatheritt, R.J., Daughdrill, G.W., Dunker, A.K., et al., 2014. Classification of intrinsically disordered regions and proteins. Chemical Reviews 114 (13), 6589–6631. Available at https://doi.org/10.1021/cr400525m, http://pubs.acs.org/journal/chreay.

Varadi, M., Anyango, S., Deshpande, M., Nair, S., Natassia, C., Yordanova, G., et al., 2022. AlphaFold protein structure database: massively expanding the structural coverage of protein-sequence space with high-accuracy models. Nucleic Acids Research 50 (1), D439. Available at https://doi.org/10.1093/nar/gkab1061, https://academic.oup.com/nar/issue.

Wang, C., Wang, X., Wang, X., Tian, B., Zhang, S., Wang, T., et al., 2024. Design, synthesis and biological evaluation of potent epidermal growth factor receptor tyrosine kinase (EGFR-TK) inhibitors against resistance mutation for lung cancer treatment. Bioorganic Chemistry 143. Available at https://doi.org/10.1016/j.bioorg.2023.107004, https://www.sciencedirect.com/science/journal/00452068.

Webb, B., Viswanath, S., Bonomi, M., Pellarin, R., Greenberg, C.H., Saltzberg, D., et al., 2018. Integrative structure modeling with the integrative modeling platform. Protein Science 27 (1), 245–258. Available at https://doi.org/10.1002/pro.3311, http://onlinelibrary.wiley.com/journal/10.1002/(ISSN)1469-896X.

Weber, P., Pissis, C., Navaza, R., Mechaly, A.E., Saul, F., Alzari, P.M., et al., 2019. High-throughput crystallization pipeline at the crystallography core facility of the institut pasteur. Molecules (Basel, Switzerland) 24 (24). Available at https://doi.org/10.3390/molecules24244451, https://www.mdpi.com/1420-3049/24/24/4451/pdf.

Wingler, L.M., Lefkowitz, R.J., 2020. Conformational basis of g protein-coupled receptor signaling versatility. Trends in Cell Biology 30 (9), 736–747. Available at https://doi.org/10.1016/j.tcb.2020.06.002, www.elsevier.com/locate/tcb.

Wolff, A.M., Nango, E., Young, I.D., Brewster, A.S., Kubo, M., Nomura, T., et al., 2023. Mapping protein dynamics at high spatial resolution with temperature-jump X-ray crystallography. Nature Chemistry 15 (11), 1549–1558. Available at https://doi.org/10.1038/s41557-023-01329-4, https://www.nature.com/nchem/.

Xu, Y., Dang, S., 2022. Recent technical advances in sample preparation for single-particle Cryo-EM. Front Mol Biosci 9 Jun 24 [cited 2025 Oct 11]. Available at https://www.frontiersin.org/journals/molecular-biosciences/articles/10.3389/fmolb.2022.892459/full.

Yang, Z., Zeng, X., Zhao, Y., Chen, R., 2023. AlphaFold2 and its applications in the fields of biology and medicine. Signal Transduction and Targeted Therapy 8 (1). Available at https://doi.org/10.1038/s41392-023-01381-z.

Zhu, G., Yao, X., 2008. TROSY-based NMR experiments for NMR studies of large biomolecules. Progress in Nuclear Magnetic Resonance Spectroscopy 52 (1), 49–68. Available at https://doi.org/10.1016/j.pnmrs.2007.10.001.

Zhu, H., Gao, H., Ji, Y., Zhou, Q., Du, Z., Tian, L., et al., 2022. Targeting p53–MDM2 interaction by small-molecule inhibitors: learning from MDM2 inhibitors in clinical trials. Journal of Hematology & Oncology 15 (1). Available at https://doi.org/10.1186/s13045-022-01314-3.

Ziegler, S.J., Mallinson, S.J.B., St. John, P.C., Bomble, Y.J., 2021. Advances in integrative structural biology: towards understanding protein complexes in their cellular context. Computational and Structural Biotechnology Journal 19, 214–225. Available at https://doi.org/10.1016/j.csbj.2020.11.052, www.csbj.org.

CHAPTER 2

Bioinformatics tools for protein modeling in neurodegenerative diseases

Shubhi Singh, Rudrani Rath, Sushma Jeyaraj, Aysha Afreen Rafeek, and Priya Swaminathan
Department of Biotechnology, School of Bioengineering, College of Engineering and Technology, SRM Institute of Science and Technology, Chengalpattu, Tamil Nadu, India

2.1 Introduction

Neurodegenerative disorders are diseases caused by a gradual loss of selectively vulnerable populations of neurons, unlike static neuronal loss, which occurs due to toxic or metabolic disorders (Dugger and Dickson, 2017). Toxic disorders involve rapid neuronal damage caused by harmful substances rather than slow, progressive neurodegeneration. These may arise from external toxins such as heavy metals, pesticides, and industrial chemicals, or from internal metabolic toxins, such as ammonia or uremic compounds. Such toxic exposures cause acute, nonselective neuronal injury through mechanisms, such as oxidative stress and excitotoxicity, distinguishing them from the gradual, protein-misfolding–driven neuronal loss seen in neurodegenerative diseases (NDD). Based on protein pathology, they can be classified into amyloidoses (amyloid-β aggregates), tauopathies (abnormal tau protein deposits), α-synucleinopathies (α-synuclein accumulation), and TDP-43 proteinopathies (TAR DNA-binding protein 43 inclusions) (Rinauro et al., 2024). The most common NDD include Parkinson's disease (PD), Huntington's disease (HD), Alzheimer's disease (AD), prion disease, spinal muscular atrophy, amyotrophic lateral sclerosis (ALS), motor neuron disease, and spinocerebellar ataxia. Collectively affecting up to 15% of the global population, they represent a leading cause of physical and cognitive disability worldwide. AD and other dementias are the seventh leading cause of death worldwide, affecting more than 55 million people. In addition to ageing, genetics, and family history, the Lancet Commission estimates that around 45% of dementia cases are linked to 14 modifiable risk factors (Zhang et al., 2024). Similarly, the global prevalence of PD has been rising steadily since the 1980s, with a particularly marked increase over the past two decades. The proteins involved in these disorders have atypical conformational properties. The dysfunction of protein homeostasis creates the pathogenic basis of most neurodegenerative disorders, ultimately resulting in protein misfolding (Chiti and Dobson, 2017).

Advances in Structural Biology
DOI: https://doi.org/10.1016/B978-0-443-44988-8.00007-1

Misfolded proteins activate cellular stress pathways, leading to aggregation and accumulation that cause neurotoxicity and ultimately contribute to NDD. Fig. 2.1 shows how protein homeostasis failure leads to toxic oligomers, aggregation, and neurodegeneration. It also highlights the use of computational models to predict disease spread and understand progression. Fibrillar forms of amyloidogenic proteins were formerly considered the primary toxic aggregates in several protein misfolding diseases. They can either directly harm the cells and tissues of their origin or sequester functioning proteins. Through surface-catalyzed secondary nucleation, a positive feedback loop is established that further promotes oligomer multiplication. Growing evidence establishes that smaller, intermediate, and metastable soluble aggregates, known as misfolded protein oligomers, are often more toxic than their mature fibrillar forms (Rinauro et al., 2024). Protein structure prediction and modeling are powerful computational methodologies for interpreting protein function by the reconstruction of three-dimensional structures from amino acid sequences and their subsequent analysis. The onset and progression of NDD are highly dynamic processes. So, aligning real and simulated time is crucial for accurately representing protein spread and brain changes. The implementation of network diffusion model, evolutionary scale modeling, and susceptible-infected-recovered protein models has contributed to a greater understanding of disease progression (Kuhlman and Bradley, 2019).

2.2 Protein modelling approaches: foundations and methodologies

2.2.1 Homology modelling

Homology modeling, also known as comparative modeling, is a computational approach used to build a three-dimensional structure of a protein when its structure has not been solved experimentally. This approach is necessary because the number of available protein sequences exceeds the number of experimentally resolved structures, and accurate structural information is essential for understanding protein function, molecular interactions, and the basis of disease mechanisms. While protein sequences may vary a lot during evolution, their overall structures tend to remain more conserved (Hameduh et al., 2020). This means that if two proteins are related, chances are high that they will fold in a similar way. Fig. 2.2 outlines the key steps in a standard homology modeling workflow. It shows how a target sequence is aligned with a template structure, followed by model building and refinement. The process ends with validation steps to assess the quality and accuracy of the predicted protein model. As experimental methods such as X-ray crystallography, NMR spectroscopy, or cryo-EM have helped us discover the structures of a small fraction of all proteins, homology modeling has become an important tool to bridge this gap and provide structural insights from sequence data alone.

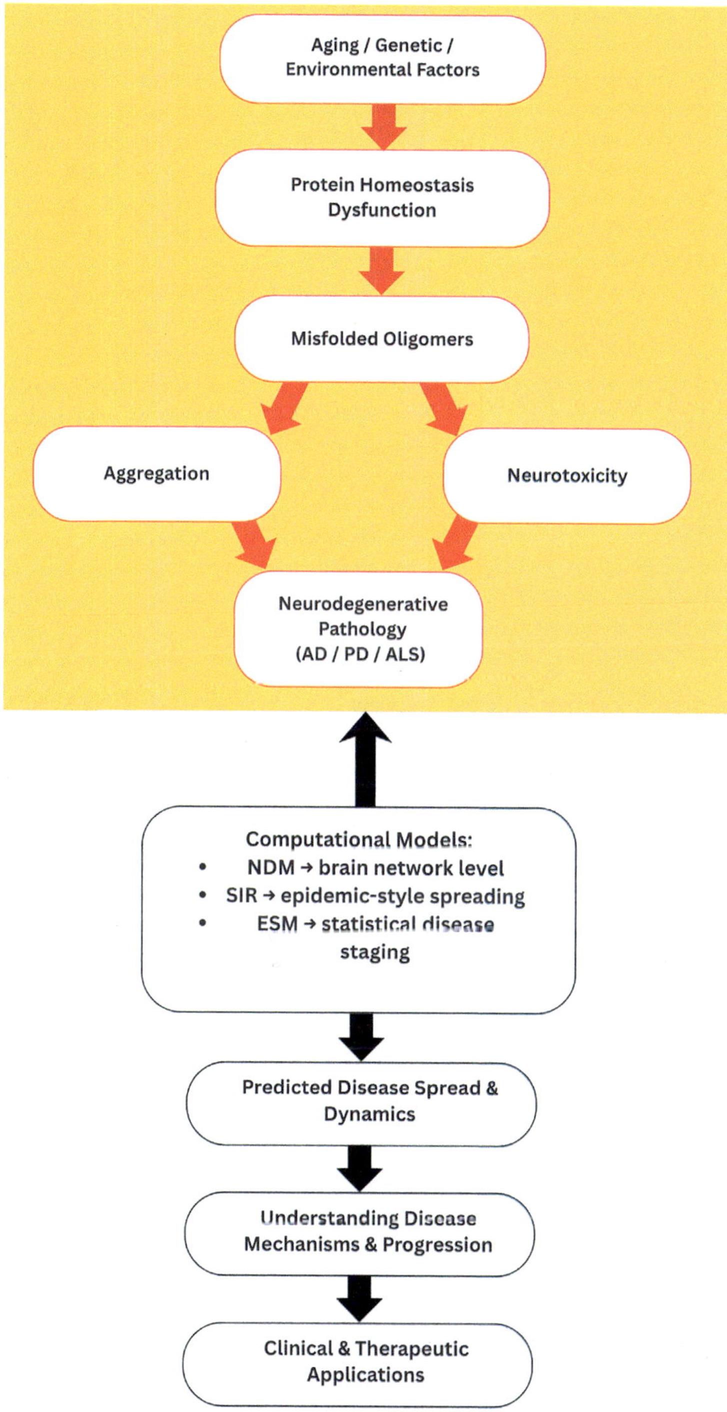

Figure 2.1 Mechanistic overview of protein homeostasis dysfunction leading to neurodegeneration.

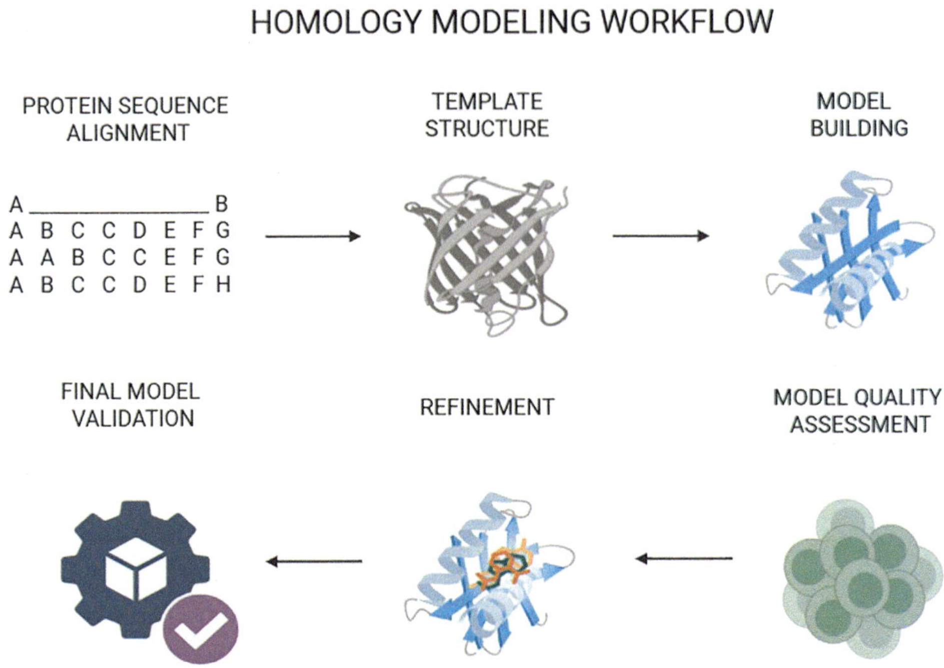

Figure 2.2 Homology modeling workflow overview.

Proteins play a central role in NDD such as AD, PD, HD, and ALS, where their tendency to misfold or clump together drives much of the pathology. To understand how these proteins behave, researchers need to look at their three-dimensional structures. Many of the proteins, such as tau, α-synuclein, and huntingtin, resist traditional structural techniques because they are unstable or contain long disordered regions. This is where homology modeling becomes essential; it provides a practical way to study structural features when traditional experimental methods fail. To overcome this problem, researchers often choose homology modeling. By using the structure of a related protein as a guide, it becomes possible to piece together a reasonable model of the protein of interest.

Such models have provided insights into the ways mutations can weaken protein stability, highlighted possible binding pockets for therapeutic compounds, and offered explanations for how protein–protein interactions (PPI) might contribute to the formation of toxic aggregates in NDD (Orry and Abagyan, 2016).

2.2.2 Threading and fold recognition

Fold recognition and threading are useful computational tools that help predict the three-dimensional structures of proteins, even when sequence alignment does not reveal

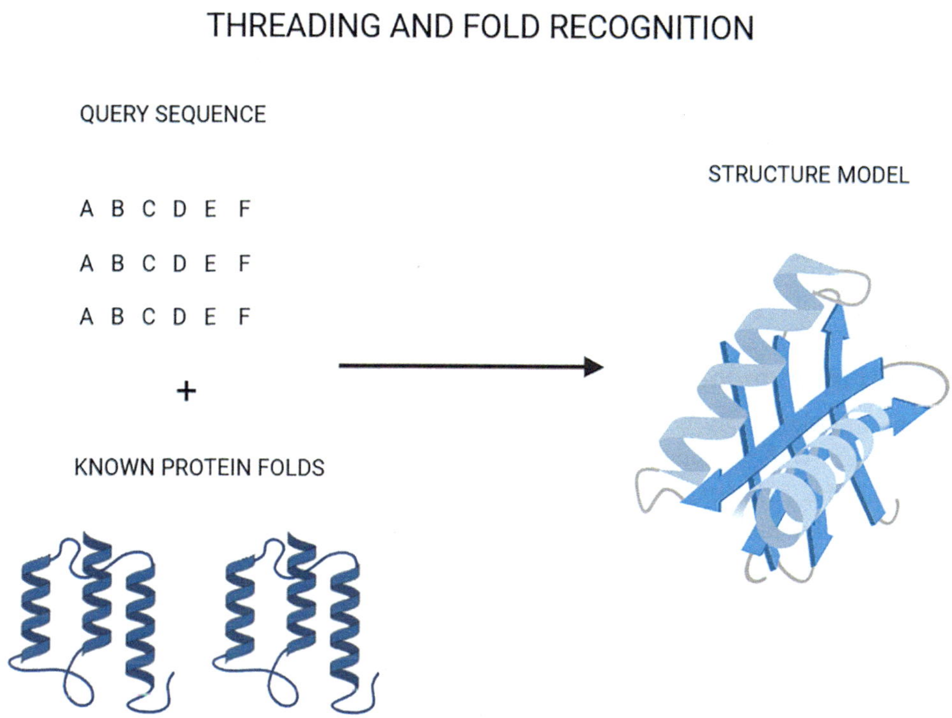

Figure 2.3 Threading and fold recognition workflow.

obvious structural relatives (McGuffin, 2008). These approaches are needed because a significant proportion of proteins share structural similarity despite lacking detectable sequence homology, and conventional alignment methods cannot reliably identify their templates. Fig. 2.3 illustrates the threading and fold recognition workflow, where a query sequence is matched against known protein folds. By identifying the closest structural template, the method predicts a suitable 3D model for the sequence. This approach is especially useful when homologous structures are limited or absent (McGuffin, 2002). They might not be as detailed as X-ray crystallography or cryo-EM, but they are much faster and cheaper. Their utility becomes particularly important when experimental structure determination is impractical or when no clear homologous template exists, enabling researchers to still identify compatible structural frameworks and build preliminary models. This saves considerable time and resources in situations where experimental methods are not feasible.

NDD exemplify the critical importance of fold recognition and threading, as protein misfolding constitutes a fundamental characteristic of these conditions. Proteins such as amyloid-beta, tau, α-synuclein, and huntingtin can undergo shape changes in diseases such as Alzheimer's, Parkinson's, and Huntington's. When this happens, they form incorrect shapes and start to stick together, which hurts nerve cells in the end (Krokidis et al., 2023). Researchers can get a better idea of how diseases progress by using computers to look at these misfolding events. This also suggests where new treatments might be needed. Scientists can find parts of these proteins that could be drug targets by predicting their structures (Roche et al., 2011). This helps keep the proteins in the right shape or stops them from forming dangerous aggregates. Thus fold recognition and threading are not only instrumental in characterizing misfolding mechanisms but also play a strategic role in guiding drug discovery and therapeutic design for NDD.

2.2.3 Ab initio modelling

Ab initio modeling is a way to identify the 3D shape of a protein based only on its amino acid sequence, without using any known templates. This approach is particularly necessary for proteins that lack detectable homologs or templates, ensuring that structural prediction remains possible even when comparative methods cannot be applied. It tries to explain how and why proteins fold into certain shapes based on the laws of physics and chemistry, which is different from homology modeling (Lee et al., 2009). There are three main parts to ab initio methods: energy functions to check for stability, conformational search strategies to look for possible folds, and model selection to find the best structure. Ongoing improvements in these areas are slowly making this method more reliable.

The core of NDD research is figuring out why proteins misfold and clump together in harmful ways. Experimental techniques such as solid-state NMR and X-ray diffraction can elucidate the structure of mature fibrils; however, the initial phases of aggregation, characterized by the formation of small, unstable oligomers, are significantly more challenging to observe in laboratory settings (Chiti and Dobson, 2017). This makes ab initio modeling a critical tool, as it enables exploration of folding and misfolding events that are experimentally inaccessible, particularly at the earliest stages of aggregation. At this point, ab initio modeling becomes important. Methods such as discrete molecular dynamics (DMD) begin from first principles, recreating protein behavior step by step through simplified but carefully designed models of structure and interactions. This is different from traditional methods that rely on prior structural knowledge (Urbanc et al., 2006). These simulations help find the main forces that cause proteins to fold and misfold by slowly making them more complicated and comparing the predictions to what happens in real life. For instance, DMD has been used to study amyloid-beta in AD, showing how small peptides fold, misfold, and group together to form oligomers (Sun and Ding, 2020). Even with simpler models, the predictions match up well with experimental data and give us new information about structures and mechanisms that experiments can't figure out yet. Thus ab initio modeling offers valuable insight into the earliest, fleeting stages of protein misfolding and aggregation, helping us understand how these processes occur and guiding the design of more precise therapies.

2.3 Computational tools for protein modelling

Computational tools for protein modeling are software and algorithm-based methods that start with the amino acid sequence and then predict or analyze the three-dimensional structure, dynamics, interactions, or other properties of proteins. We use them because experimental methods such as X-ray crystallography, NMR, or cryo-EM can be slow, expensive, and sometimes impossible to do in practice (Meng et al., 2025).

We use these tools when we want to infer a protein's likely shape, stability, binding sites, or how it might interact with other molecules (such as in drug design, enzyme engineering, or predicting function), without doing lab experiments. Advances in deep learning-based tools have recently made structure predictions much more accurate and efficient.

2.3.1 SWISS-MODEL

SWISS-MODEL was the first fully automated protein homology modeling server and has been continuously improved during the last 25 years. Homology modeling (comparative modeling) predicts the 3D structure of a protein based on how similar its sequence is to known structural templates.

It can now predict the 3D structures of both homomeric (multiple copies of the same protein unit) and heteromeric (different protein units) protein complexes using the amino acid sequences of the interacting units as input. Other recent upgrades include a new modeling engine (ProMod3) and a new quality-checking system (QMEANDisCo) (Waterhouse et al., 2018).

ProMod3 makes the predicted protein structures more accurate compared to older versions. QMEANDisCo stands for quality model energy analysis with distance constraints. It helps estimate this accuracy at a local level (Benkert et al., 2008). It is built on an upgraded version of the older QMEAN scoring system. This version of SWISS-MODEL can now model protein assemblies using information on quaternary structures available in the PDB (Biasini et al., 2014). However, this method works only if homologous complexes already exist in the PDB to use as templates. They added methods such as PPI fingerprint analysis and improved model quality estimation (Waterhouse et al., 2018).

2.3.2 MODELLER

MODELLER is a computer program designed to perform protein homology modeling. Unlike SWISS-MODEL, it requires the user to provide a detailed input consisting of the target–template alignment, the atomic coordinates of the chosen template structures, and a script file, thereby demanding greater manual effort and expertise, whereas SWISS-MODEL streamlines the process by only requiring the target sequence and performing template identification and alignment automatically (Eswar et al., 2006). It calculates a model with all nonhydrogen atoms without operator involvement in a few minutes.

In addition to building models, MODELLER can perform other tasks such as fold assignment, aligning two protein sequences or their profiles, multiple protein sequences and/or structures, clustering sequences and/or structures, and ab initio modeling of loops in protein structures (Šali and Blundell, 1993). MODELLER builds protein structures by applying spatial restraints derived from multiple sources, including homology with template structures, stereochemical rules, statistical patterns from known proteins, and optional experimental or manual inputs (Webb and Sali, 2017). These constraints are expressed as probability functions and refined through energy minimization and molecular dynamics (MD) using simulated annealing, a methodology comparable to NMR-based structural determination.

2.3.3 Phyre2

Phyre2 is one of the most widely used protein structure prediction servers. Phyre2 is the upgraded version of the original Phyre server. It was completely redesigned, with no shared components from the old system, making it faster and more accurate. The most common use is to predict the 3D structure of a protein from its amino acid sequence. The other facilities include uploading a known structure and searching against a range of genomes to find similar proteins (Backphyre), modeling many protein sequences at once, fold recognition, weekly scanning of complex proteins that are difficult to model (PhyreAlarm), and model quality analysis (Phyre investigator) (Kelley et al., 2015).

Phyre2 excels in detecting discrete homologous models because it uses hidden Markov models and advanced algorithms like HHsearch. In general, Phyre2 searches for templates and provides a list of possible matches (Söding, 2005). Phyre2 also offers a special feature called one-to-one threading, which enables users to upload both a protein sequence to be modeled and a specific structural template of their choice. Unlike the standard Phyre2 workflow, which provides a list of potential template hits, one-to-one threading directly produces an alignment between the input sequence and the chosen template, along with a 3D model of the protein and an associated confidence score (Kelley and Sternberg, 2009).

2.3.4 Iterative Threading ASSEmbly Refinement

Iterative Threading ASSEmbly Refinement (I-TASSER) can predict protein structures as well as their functions. Even when two proteins do not share the same overall structure, they can still perform similar biochemical functions. This is because local structural motifs are preserved across different proteins, and these motifs carry out the functional activity (Yang et al., 2015). The method was extended to predict biological function by comparing both local and global structural similarities between a predicted structure and proteins of known function.

This allowed I-TASSER to identify functional properties such as ligand-binding sites, enzyme commission numbers (which classify enzymes by their reactions), and gene ontology terms (standardized descriptions of molecular function, biological process, and cellular component). This works even when proteins are not homologous, which makes it much more powerful than sequence-based methods such as BLAST. Many tools can predict structures, but I-TASSER stands out because it predicts full-length structures accurately across easy and hard targets, and also integrates function prediction directly from the structures (Roy et al., 2010).

2.3.5 AlphaFold (AI-based tool)

AlphaFold uses a deep neural network that integrates evolutionary, physical, and geometric constraints of proteins, enabling it to predict accurate 3D structures directly from their amino acid sequences. It uses the primary sequence of the protein along with multiple sequence alignments (MSAs) of homologous proteins and, when available, a structural template. The key innovations of AlphaFold include the joint embedding of MSAs and pairwise features, which enable the model to simultaneously capture evolutionary relationships and residue–residue interactions. AlphaFold introduces a new output representation and loss function that enable accurate, end-to-end prediction of the 3D coordinates of all heavy atoms in a protein.

It also employs equivariant attention mechanisms, ensuring that predictions remain consistent with the geometric properties of three-dimensional space (Jumper et al., 2021) The network functions in two main stages. The first stage is called the Evoformer, which processes the inputs. It produces two essential representations: an MSA representation (Nseq × Nres), which encodes evolutionary information, and a pair representation (Nres × Nres), which captures residue–residue interactions. Here, Nseq refers to the number of sequences in the MSA, and Nres denotes the number of residues in the protein sequence (Thangamalai et al., 2025). Within the Evoformer, a combination of attention-based and nonattention mechanisms enables information exchange between these two representations, allowing the network to reason about both spatial relationships and evolutionary constraints (Jumper et al., 2021). Importantly, during training, structural hypotheses emerge early within the Evoformer blocks and are progressively refined as the network layers deepen, leading to increasingly accurate predictions.

Table 2.1 summarizes the comparison of protein-modeling tools.

Table 2.1 Comparison summary of protein-modeling tools.

Tool	Type	Key features	Strengths	Limitations	References
SWISS-MODEL	Homology modeling	Automated template search; ProMod3 engine; QMEANDisCo quality scoring	Easy to use; reliable quality checks; model complexes	Template-dependent; limited manual control	(Waterhouse et al., 2018)
MODELLER	Homology modeling	Builds models using spatial restraints and energy minimization	Flexible; supports multitemplate and loop modeling	Requires scripting; less user-friendly	(Webb and Sali, 2017)
Phyre2	Threading/ homology	Hidden markov model-based fold recognition; one-to-one threading; model confidence scoring	Detects distant homologs; fast and web-based	Limited refinement; lower accuracy for novel folds	(Kelley et al., 2015)
I-TASSER	Hybrid (threading + ab initio)	Iterative fragment assembly; structure and function prediction	High accuracy; integrates function and ligand site prediction	Slow; computationally demanding	(Roy et al., 2010)
AlphaFold	Deep learning (AI-based)	Evoformer network; end-to-end prediction; pLDDT confidence scoring	State-of-the-art accuracy; works without templates	Requires high computing power; complex architecture	(Thangamalai et al., 2025)

2.4 Tool selection criteria for neurodegenerative disease research

NDD research consists of various data types, including genomics, transcriptomics, proteomics, and imaging, necessitating meticulous tool selection. To capture disease-relevant heterogeneity, methods like single-cell sequencing need special pipelines for quality control, normalization, integration, and clustering. In addition to compatibility, reproducibility, and clear workflows are essential, as shown by systematic methods for variant annotation and prioritization that improve interpretation (Rossi et al., 2023). Integrative tools that adhere to FAIR principles and amalgamate various data sources are especially beneficial for biomarker identification and sustaining clinical relevance (Waury et al., 2022). As AI and ML are used more and more in NDD research, it is important to think about benchmarking, accuracy, and interpretability to make sure that the results are strong and easy to understand. It is also suggested to use hybrid (Galal et al., 2025) methods that combine hypothesis-driven models with data-driven exploration to cut down on bias and make results more reliable. In NDD bioinformatics, putting data-type fit, reproducibility, interpretability, and community support first helps make sure that the work is both biologically relevant and has the potential to be used in the real world.

2.5 Case studies: protein modelling in specific neurodegenerative diseases

2.5.1 Alzheimer's disease

In AD, amyloid beta-protein (Aβ) (Yun et al., 2007) can misfold and form small clusters known as oligomers, which are harmful to brain cells. It is hard to study these oligomers in the lab because they are not stable and come in different sizes.

Scientists looked at the two main forms, Aβ40 and Aβ42 (Urbanc et al., 2004), using computer simulations called DMD. Scientists found that charged interactions make the oligomers bigger. Aβ42 makes oligomers of different sizes and has a special fold at the C-terminal end that makes them stick together strongly. Aβ40, in contrast, tends to form more stable and uniform clusters, as presented in Fig. 2.4. This figure compares healthy and Alzheimer's-affected neurons. In a healthy brain, stable tau proteins support intact microtubules and normal neuronal structure. In AD, tau tangles cause microtubule disintegration, leading to neurofibrillary tangles and neuronal degeneration. The C-terminal end of Aβ42, however (Illodo et al., 2024), plays a big role in making toxic oligomers. If scientists can target this region, they might be able to stop harmful clusters from forming or even change existing ones to make them less dangerous, opening up a possible way to treat AD.

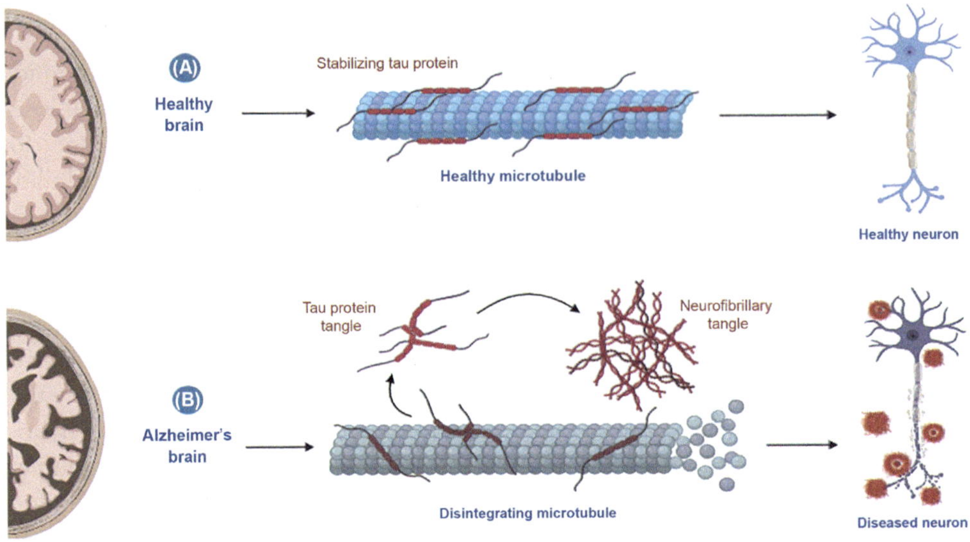

Figure 2.4 Brain degeneration in Alzheimer's disease.

2.5.2 Parkinson's disease

PD is a progressive neurodegenerative disorder that affects movement over time and causes symptoms such as tremors, stiffness, and slowed motion. The condition primarily results from the degeneration of dopamine-producing neurons in a brain region known as the substantia nigra. A major sign of PD is the buildup of α-synuclein (Bartels et al., 2011) protein aggregates in neurons. These aggregates form structures called Lewy bodies and Lewy neurites, which are characteristic of the disease. α-synuclein is a naturally disordered protein. It can be a folded tetramer or a monomer. Imbalances in these forms promote misfolding into toxic fibrils and oligomers abundant in β-sheets, as shown in (Fig. 2.5). This figure contrasts normal basal ganglia pathways with those altered in PD. Loss of dopaminergic input weakens the direct pathway and overactivates the indirect pathway. These changes reduce thalamic stimulation, resulting in hypokinetic motor symptoms.

Computational models are widely used to study the mechanisms of α-synuclein aggregation. Conventional MD shows structural flexibility, but replica-exchange MD (Semenyuk, 2022) finds rare, aggregation-prone conformations. Coarse-grained MD enables the simulation of larger systems and extended timescales, while ensemble docking or ligand-bound MD identifies small molecules that influence aggregation. Together, these techniques help us better understand PD and aid in the creation of novel treatments.

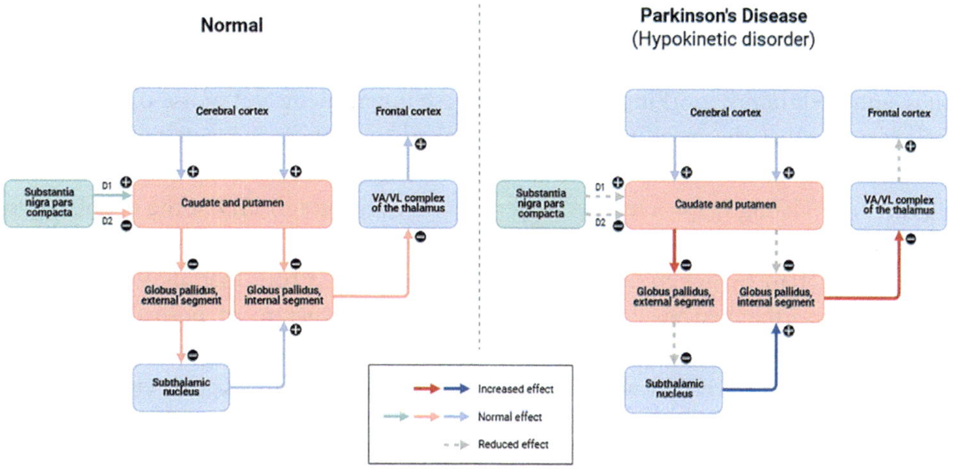

Figure 2.5 Normal versus Parkinson's disease basal ganglia pathways.

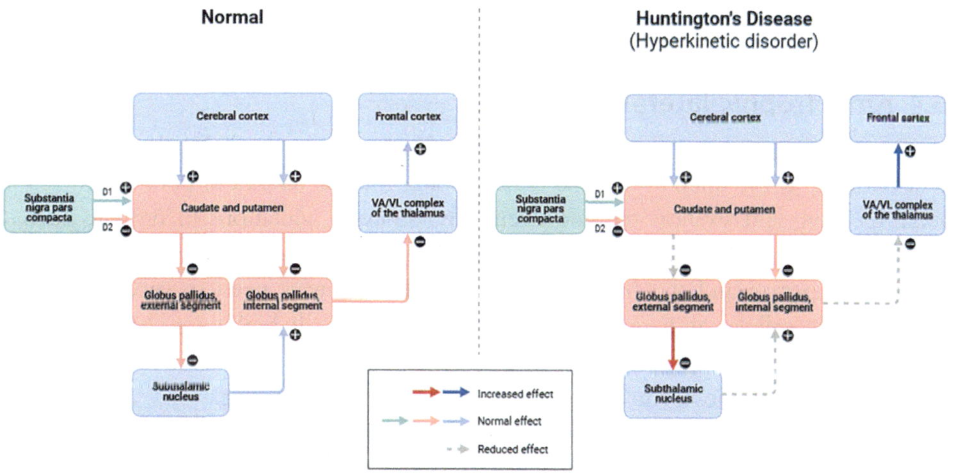

Figure 2.6 Basal ganglia circuitry in normal conditions and Huntington's disease.

2.5.3 Huntington's disease

HD is a neurological disorder characterized by an aberrant expansion of glutamine repeats, referred to as polyQ (Hatters, 2008), inside the huntingtin gene. This mutation alters the huntingtin protein, causing it to misfold and aggregate within neurons. These aggregates destroy neurons over time, especially in the basal ganglia, which is the portion of the brain that governs movement, as shown in Fig. 2.6 shows the basal ganglia circuitry in normal versus HD conditions. Degeneration of striatal neurons reduces inhibitory control in the indirect pathway. This imbalance increases thalamic output, giving rise to hyperkinetic movements. They also stop cells from working normally. Because of this, individuals with HD have issues with moving, remembering things, and their brains working in general. These impairments get worse over time. We know for sure that the polyQ expansion is linked to protein aggregation, but we still don't know how these clumps make the disease worse.

Protein modeling helps scientists understand the changes in HD at the cellular level. They can observe the effect of extra polyQ repeats on the shape and stability of the huntingtin protein by generating 3D maps. AlphaFold is a tool that allows researchers (Hatters, 2008; Ethirajulu et al., 2024) to compare the structure of normal and mutant huntingtin and predict how the protein will fold. These predictions are validated in the laboratory by utilizing cells or animals such as mice and zebrafish (Kumar et al., 2021) to observe the extent to which the defective protein induces changes in behavior and damages brain cells. By combining experiments with computer models, scientists can establish an association between protein modifications and disease symptoms. This information is also beneficial in the development of new treatments, such as gene therapies or medications, that can prevent the protein from clumping and protect brain cells.

2.5.4 Amyotrophic lateral sclerosis

ALS is a progressive disease (Kumar et al., 2021; Grassmann et al., 2021) that damages the motor neurons in the brain and spinal cord, causing muscle weakness, poor coordination, trouble speaking and swallowing, and eventually paralysis. Mutations in proteins such as SOD1, TDP-43, and FUS cause the misfolding and make it stick together, which disrupts normal cell function and the balance of proteins that neurons need to survive. Fig. 2.7 summarizes the cellular and molecular defects driving ALS. It highlights abnormalities in RNA processing, protein homeostasis, vesicle transport, and mitochondrial function. Together, these disruptions contribute to motor neuron degeneration.

Protein modeling has become a crucial tool to understand how these ALS-related proteins misfold and clump together. Scientists study how mutations affect the structure, stability, and interactions of proteins like SOD1 and TDP-43 (Das and Plotkin, 2013) using computational tools such as structural modeling and MD simulations. These observations help to explain how the disease develops and guide the design of treatments to prevent protein clumps, protect neurons, and slow ALS progression. By combining computer models with lab experiments, protein modeling gives a clear picture of the molecular changes that drive the disease and points to possible ways to intervene.

2.6 Molecular docking and dynamics

Molecular docking and MD are ways to use computers to look at how a ligand interacts with a protein. Docking finds the best binding pose and an approximate affinity, while

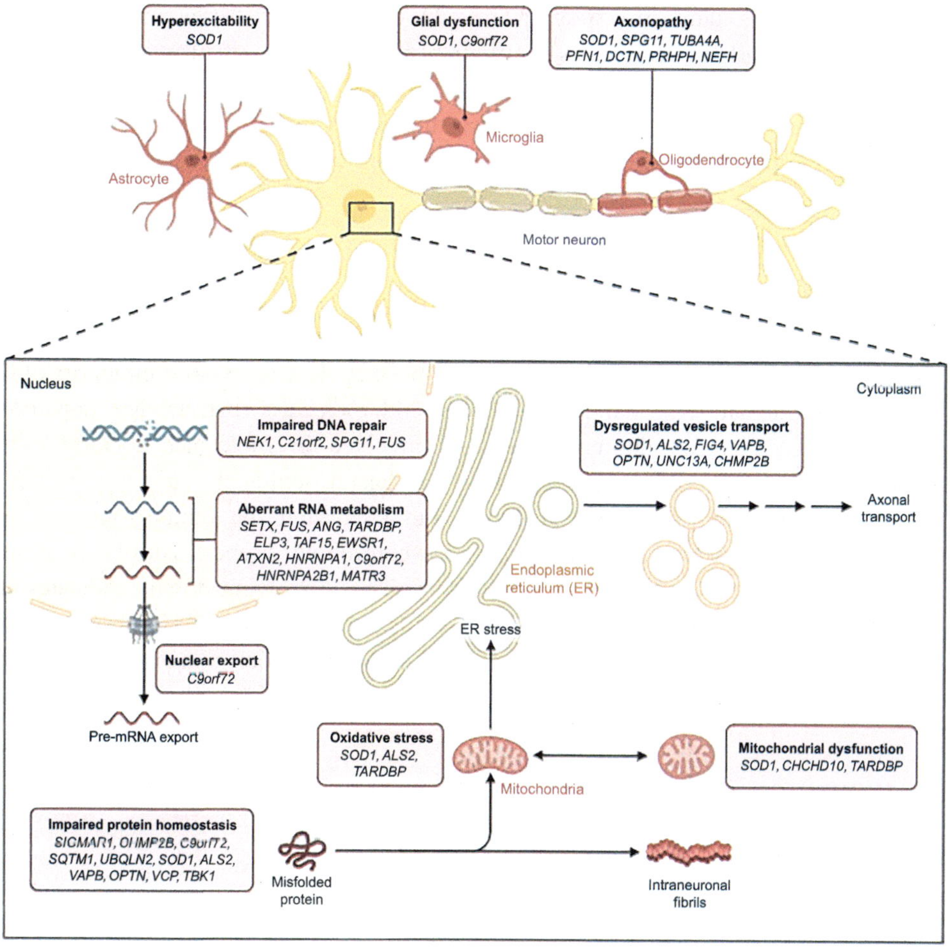

Figure 2.7 Cellular and molecular mechanisms underlying amyotrophic lateral sclerosis (*ALS*).

dynamics shows how atoms move over time to see how stable and realistic the protein–ligand complex is. We use these tools because they are faster and cheaper than doing experiments. They help us find drug candidates, improve binding predictions, and learn about molecular stability before testing in the lab. They are usually used for virtual screening (docking) at the beginning of drug discovery and for validating and analyzing stability (dynamics) later on.

2.6.1 Docking methodologies

Molecular docking is very useful in ND research because it helps scientists find drugs that can target key proteins involved in illnesses, such as Alzheimer's and Parkinson's, such as amyloid-β, tau, and α-synuclein. Docking allows screening of natural compounds and synthetic molecules for their potential to inhibit protein aggregation or modulate receptor activity. For example, docking has been widely applied to identify small molecules that can cross the blood–brain barrier and interact with AD targets, thereby providing a cost-effective first step before experimental validation (Meng et al., 2011).

2.6.1.1 AutoDock

AutoDock and AutoDock Vina are frequently used in NDD studies to virtually screen ligands against disease-relevant targets. They have been applied to explore inhibitors of β-secretase (BACE1), tau aggregation modulators, and molecules that stabilize dopamine receptors in PD. Fig. 2.8 illustrates the steps involved in molecular docking using AutoDock, starting from preparing the protein and ligand, setting the grid box, running the docking simulation, and finally analyzing and visualizing the binding interactions. Because of its efficient search algorithms and broad community support, AutoDock allows researchers to evaluate thousands of compounds in silico, narrowing down candidates for further biochemical testing (Trott and Olson, 2010; Ferreira et al., 2015). For instance, flavonoids and phytochemicals have been screened using AutoDock to assess their binding with acetylcholinesterase, a therapeutic target in AD.

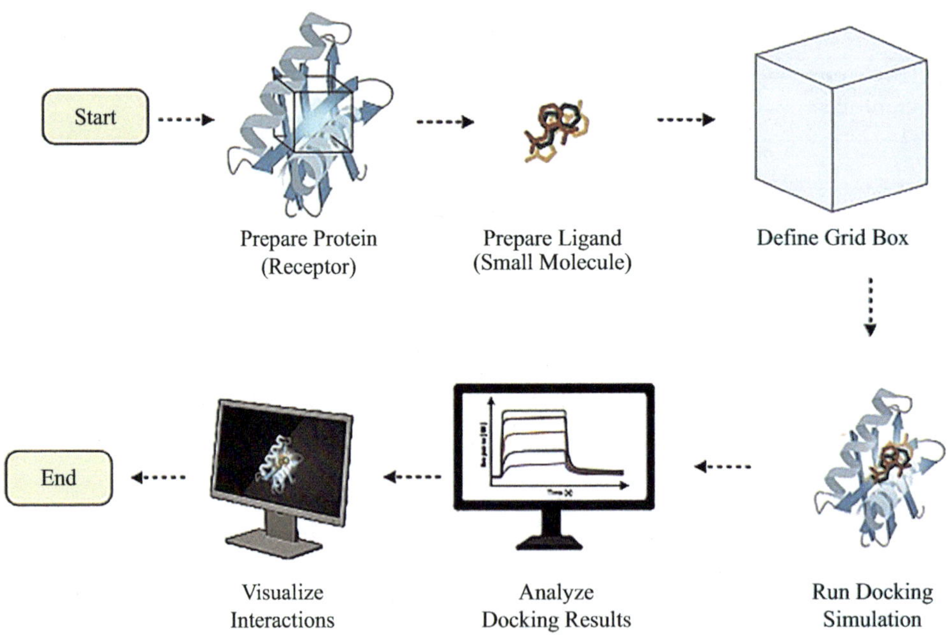

Figure 2.8 Overview of the AutoDock molecular docking workflow.

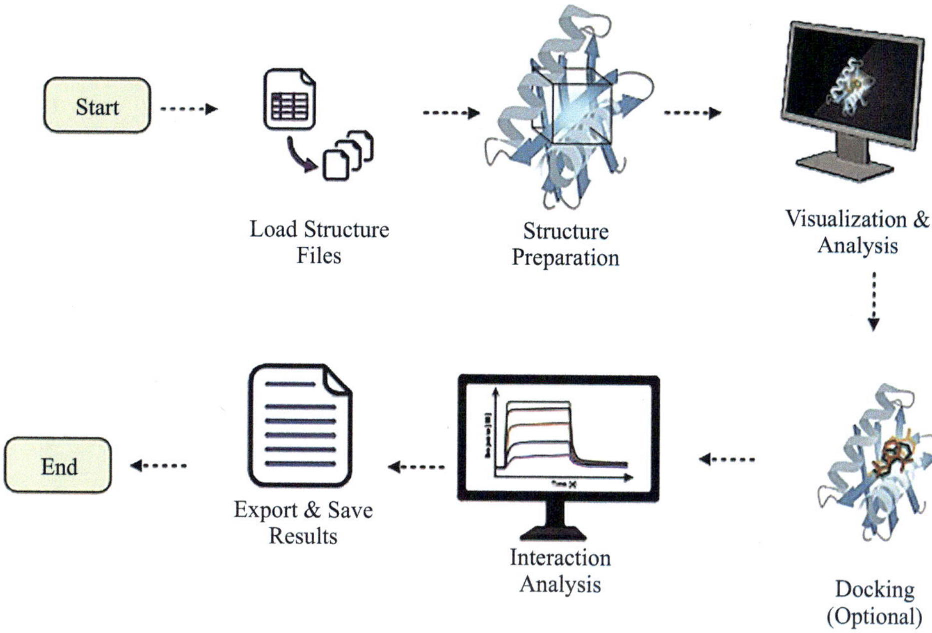

Figure 2.9 Workflow of structural modeling and analysis in UCSF Chimera.

2.6.1.2 UCSF Chimera

UCSF Chimera is commonly used in docking studies for NDD because it helps prepare both ligands and protein targets. Its DockPrep tool cleans and optimizes structures, such as acetylcholinesterase, prion proteins, or tau fibrils, before they are used for docking. Fig. 2.9 shows a workflow of UCSF Chimera showing steps from structure loading and preparation to visualization, interaction analysis, and result export. Chimera is also widely used to visualize how ligands bind to key ND proteins, making it easier for researchers to see important binding hotspots and hydrophobic interactions (Pettersen et al., 2004). In Alzheimer's research, UCSF Chimera is often used together with AutoDock Vina for virtual screening of natural compounds, creating a simple and efficient workflow that supports drug discovery in academic labs.

2.6.2 Molecular dynamics

2.6.2.1 GROningen MAchine for Chemical Simulations

GROningen MAchine for Chemical Simulations (GROMACS) MD simulations are crucial for validating docking results in NDD research, as protein misfolding and aggregation are highly dynamic processes. Docking provides just a fixed picture of how a ligand might bind, but GROMACS shows what really happens over time, revealing

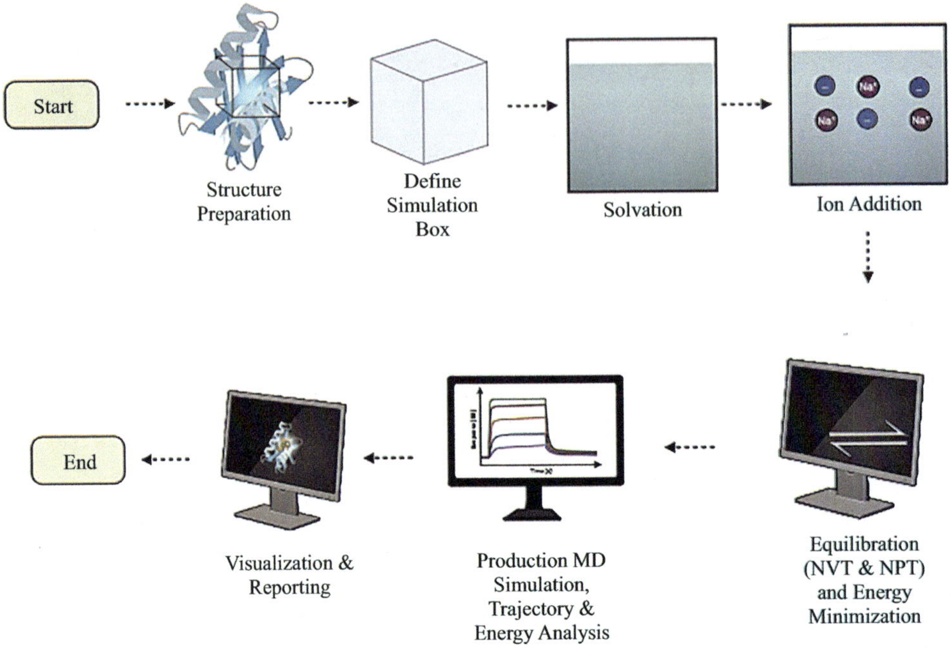

Figure 2.10 Overview of the GROMACS molecular dynamics (*MD*) workflow.

whether a ligand can truly stabilize or destabilize dynamic targets, such as amyloid-β oligomers or α-synuclein fibrils.

Fig. 2.10 illustrates the steps involved in running MD simulations with GROMACS. Several studies have combined docking with GROMACS to confirm the stability of lead compounds and to explore drug–protein interactions at the atomic level under physiological conditions (Johnsson et al., 2024). For example, GROMACS simulations have helped refine docking-based predictions for inhibitors that block amyloid fibril growth, strengthening the real-world relevance of computational results in ND research.

2.6.3 Protein structure prediction and modeling

2.6.3.1 Rosetta and Robetta

The Rosetta and Robetta computer programs are useful for figuring out how proteins fold, misfold, and group together. These are important steps in diseases that cause the brain to break down, such as Alzheimer's, Parkinson's, and Huntington's. These diseases often break down the structure of proteins such as amyloid-β (Aβ), tau, and α-synuclein, making them toxic and killing neurons. David Baker's lab at the University of Washington made the Rosetta suite, which is a great tool for predicting, improving, and designing the shapes of proteins. Robetta is the web interface for automated modeling that is easy to use (https://robetta.bakerlab.org). The first step in the process is to find the amino acid sequence of a protein. Robetta then uses Rosetta's fragment assembly method to process the sequence. It does this by either template-based modeling (when a similar structure already exists) or ab initio prediction (for new proteins) (Kim et al., 2004). The software creates models by linking together short peptide pieces. It uses an energy function to find the shapes that are the most biologically relevant and have the least amount of energy by balancing physical forces and statistical potentials. After the first model is made, Rosetta refinement protocols do side-chain repacking, hydrogen-bond optimization, and energy minimization in Cartesian space. This results in atomic-level accuracy for the predicted structures (DiMaio et al., 2009). This improved modeling is especially useful for studying NDD, where even small changes in shape can affect protein aggregation, like when β-sheets in tau fibrils or α-synuclein oligomers rearrange. We use metrics such as root mean square deviation (RMSD), global distance test–high accuracy (GDT-HA), and energy scoring to see how well models work. After that, we check them again with outside tools such as PROCHECK and MolProbity. Rosetta has special parts, such as RosettaDock and RosettaDesign, that let scientists look at how proteins interact with each other and make smart variants or peptide inhibitors that don't stick together (Gainza et al., 2020). The Rosetta–Robetta pipeline shows scientists how diseases change the shape of proteins, how mutations (such as tau P301L or α-synuclein A53T) change the structure, and how to treat the disease. When used together, these tools make a powerful computer platform that connects molecular structure to how diseases work. They offer structural insights that improve our understanding of neurodegeneration and aid in the formulation of targeted interventions.

2.7 Protein structure refinement

Protein structure refinement is a computational method used for making predicted protein models better by fixing structural mistakes, optimizing side-chain packing, and improving the overall stereochemical quality. Refinement helps make models that are more realistic and trustworthy because the first predicted structures may not be correct. This step is important before performing downstream analyzes such as docking, MD, or functional studies, where structural accuracy is very important.

2.7.1 Protein structure refinement using the GalaxyRefine web server in neurodegenerative disorder research

ADs, PDs, and HDs are neurodegenerative disorders that cause proteins to misfold and stick together, which kills neurons and stops them from working. Proteins such as amyloid-β (Aβ), tau, and α-synuclein form strange clumps because they don't stay in the same shape. We need to know how their 3D structures work to understand how diseases work and how to make therapeutic inhibitors (Selkoe and Hardy, 2016).

It is often necessary to improve predicted protein models to fix mistakes in the packing of side chains, the shape of the backbone, and the orientation of loops. The GalaxyRefine web server is a powerful and simple tool that can improve protein structures by repacking side chains and using MD to relax. This process makes models that are better because they have more accurate stereochemistry, better rotamer distributions, and fewer steric clashes. All of these things are important for neurodegenerative protein research that looks at docking, aggregation, and drug binding. Researchers can upload PDB models of proteins that are likely to misfold, such as the tau repeat domain or α-synuclein monomer, to the GalaxyRefine interface (http://galaxy.seoklab.org). The server makes a number of refined structures, and each one is compared to GDT-HA, MolProbity, and RMSD (Lee et al., 2019). These better models help us better test compounds that stop aggregation or simulate aggregation pathways.

2.7.2 Workflow of protein structure refinement using the GalaxyRefine web server

Protein structure refinement is essential for improving the accuracy and stability of predicted or modeled structures before downstream analysis. The GalaxyRefine web server provides an automated approach to optimize side-chain conformations and overall structural geometry. Fig. 2.11 shows the workflow of protein structure refinement using the GalaxyRefine web server, beginning with input preparation, followed by side-chain rebuilding, evaluation, and final model verification.

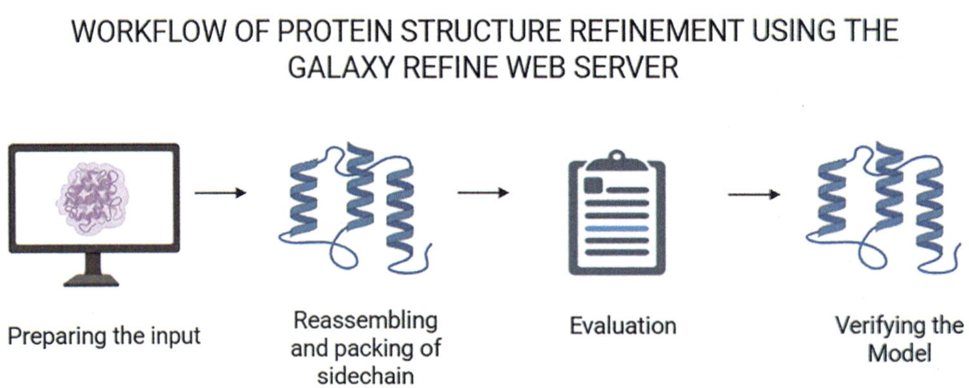

Figure 2.11 Workflow of protein structure refinement using the GalaxyRefine server.

2.7.2.1 Preparing the input

The researcher initially needs a protein structure file in PDB format. Prediction tools such as AlphaFold or I-TASSER usually provide this starting model. Before uploading it to GalaxyRefine, the structure must be checked to confirm that no atoms are missing and no chains are broken. In neurodegeneration studies, this is especially important for flexible proteins, such as tau or α-synuclein, which are notoriously difficult to capture experimentally.

2.7.2.2 Reassembling and packing of sidechains

When amino acids are uploaded, GalaxyRefine begins by reconstructing their side chains. These side chains are the flexible portions of each residue that can rotate into different positions, and they play a major role in determining how the protein folds and how it interacts with surrounding molecules. The software uses a Monte Carlo simulation to test out different orientations and choose the one that causes the fewest problems and is the most stable (Heo et al., 2013). This step is very important for neurodegenerative proteins because even small misalignments in the side chains can change how the protein sticks together or binds to other molecules.

2.7.2.3 Letting the structure relax

GalaxyRefine runs a short MD simulation that slowly lets the protein settle into a more stable shape. This is like letting a crumpled piece of paper become smoother. The algorithm can use either mild or strong relaxation, depending on how much movement the structure needs (Lee et al., 2019). The result is a structure that better shows how the protein might exist in real biological settings.

2.7.2.4 Making and testing models

After relaxing the protein model, GalaxyRefine makes five better versions of it. The researcher can choose the version that has the best balance of accuracy and stability because each one is slightly different. The server automatically gives the models scores based on how much the structure has improved using the GDT-HA, RMSD, and MolProbity score. Then, you can use the model that got the best scores for more tests, such as molecular docking or aggregation studies.

2.7.2.5 Verifying the model

The refined structure should then be evaluated using additional validation tools such as PROCHECK or MolProbity to ensure its quality before it is used in further analyzes. These tools check things like bond angles, side-chain positions, and other details about the structure.

2.8 Challenges and limitations

2.8.1 Modeling intrinsically disordered proteins

In NDD, several key proteins, such as tau in Alzheimer's and α-synuclein in Parkinson's, have unstable three-dimensional shapes and behave as intrinsically disordered proteins (IDPs). Their dynamic conformational ensembles make them hard to study in the lab, which is why computational models, such as MD and coarse-grained simulations, are so popular. These techniques can uncover transient structures, forecast aggregation-prone motifs, and facilitate drug discovery for disordered targets (Robustelli et al., 2018).

2.8.2 Protein aggregation complexity

Protein aggregation is a defining characteristic of NDDs, but it is a very complicated and varied process. Proteins like amyloid-β or α-synuclein can make many toxic intermediates, such as oligomers and protofibrils, before coming together to form fibrils. Because it is hard to study these temporary states in the lab, simulations give us useful information about aggregation pathways, energetics, and how changes in the environment or mutations affect toxicity (Robustelli et al., 2018; Knowles et al., 2014).

2.8.3 Computational resource requirements

In silico research on IDPs and protein aggregation needs a lot of computer power. All-atom MD simulations require lengthy trajectories to observe gradual conformational alterations, whereas larger systems, such as oligomers, necessitate substantial memory and processing capabilities. Coarse-grained models and enhanced sampling methods are two ways to lower the cost of computing, but they often come at the cost of resolution (Notova et al., 2020).

2.8.4 Limitations of current methodologies

Even with all the progress so far, computational methods for NDD research still struggle with some challenges. Most force fields are built for stable, folded proteins, so they don't always capture the behavior of disordered proteins very well. And rare events like the early steps of nucleation are still hard for simulations to sample properly. Simplified computer models, like those using implicit solvents or smaller systems, don't fully capture what happens in real life. To truly understand how proteins clump together in NDD, researchers need to compare their simulation results with real laboratory experiments.

2.9 Future perspectives and emerging technologies

2.9.1 Multiscale modelling approaches

Protein aggregation plays a very crucial role in many NDD, such as AD, PD, and HD. Researchers use a variety of systems (complex organism models to test-tube experiments) to study its effects, but it is difficult to associate the findings of the experiments directly with the various diseases. A major reason is the mismatch in timescales. The aggregation in vitro can occur within a few hours to a few days, whereas disease progression in humans takes decades. Due to recent advancements in imaging and biochemical techniques, protein aggregation can be monitored at high resolutions and manipulated in specific targeted ways. To understand the complex outcomes produced by these techniques, researchers need quantitative models. Chemical kinetics provides the framework that links experimental observations to the microscopic mechanisms of aggregation (Meisl et al., 2022).

The rate-determining step is central to kinetics, since only a few key processes dictate the overall behavior of complex reaction networks. These "control switch" steps govern the system, while nonrate-limiting stages have little effect. Identifying them reveals the key drivers of aggregation and the most promising drug targets. Therapies targeting these critical steps are far more likely to be effective than those aimed at minor processes.

2.9.2 Quantum computing applications

Supercomputers have helped us learn a lot more about NDD by letting us run detailed molecular simulations, like modeling the release of neurotransmitters, supporting AI-driven drug screenings to see how drugs affect neurons, and using machine learning to look at how the brains of Parkinson's patients age. Quantum computing is still in its early stages, but it promises even bigger breakthroughs. It could speed up drug discovery by simulating complicated molecular interactions, help us learn more about Alzheimer's and Parkinson's by looking at how proteins fold and misfold, and find early biomarkers by processing huge genomic and proteomic datasets. Also, it could help create personalized treatment plans based on a person's medical and genetic history. Early research is looking into quantum methods for beta-secretase inhibitors in AD. Nonetheless, obstacles such as error correction, scalability, and accessibility indicate that the practical, routine application of quantum computing in neurodegeneration research remains distant (Stefano, 2024).

2.9.3 Personalized medicine and precision therapeutics

NDD are very hard to understand, and people may react differently to the same drugs. A person's genes can make a medicine work differently, so each person needs their own treatment. Doctors can make medicines work better and cause fewer side effects by looking at genes and small chemical changes that affect them (Griñán-Ferré et al., 2024).

Changes in DNA or in the proteins that help organize it can also change how a person responds to a drug. When doctors know about these changes, they can better guess how people will react and change their treatments for each person. This is very important for diseases such as Alzheimer's and Parkinson's because it makes medicines safer and more effective (Griñán-Ferré et al., 2024; Leonard et al., 2024).

Genetic tests and computer models of patients are two simple tools that personalized medicine uses to plan treatments. These tools help doctors choose the right medicine, dose, and mix for each patient. This makes treatments safer, more accurate, and more likely to work (Rossi et al., 2023; Griñán-Ferré et al., 2024; Leonard et al., 2024).

2.9.4 Collaborative platforms and data sharing

The European Platform for Neurodegenerative Diseases (EPND) lets researchers share data and biological samples from studies on AD, PD, and other diseases that damage the brain. It lets hospitals and research centers across Europe safely upload, access, and analyze datasets. By combining information from many studies, scientists can find patterns and insights that smaller studies might miss (Hermans et al., 2025).

EPND also makes sure all data is handled securely and follows privacy and ethical rules. This encourages open science and prevents researchers from repeating the same work, allowing them to focus on discoveries. Overall, the platform speeds up research on NDD by making data easier to reach and use (Bose et al., 2022).

2.10 Conclusion

Protein misfolding and aggregation are central to the pathology of NDD, yet their structural characterization remains experimentally challenging. Bioinformatics and computational protein modeling offer robust methodologies for investigating these processes through structure prediction, aggregation analysis, and therapeutic target discovery. Advances in homology modeling, threading, ab initio methods, and AI-driven tools, such as AlphaFold, have significantly enhanced the precision and breadth of structural investigations. When integrated with molecular docking and dynamics simulations, these methodologies provide significant insights into disease mechanisms and facilitate rational drug discovery. Despite current limitations in modeling IDPs and aggregation dynamics, the ongoing convergence of computational and experimental methodologies offers significant potential for enhancing precision therapies in NDD.

References

Bartels, T., Choi, J.G., Selkoe, D.J., 2011. α-Synuclein occurs physiologically as a helically folded tetramer that resists aggregation. Nature 477 (7362), 107–111. Available at https://doi.org/10.1038/nature10324.

Benkert, P., Tosatto, S.C.E., Schomburg, D., 2008. QMEAN: a comprehensive scoring function for model quality assessment. Proteins: Structure, Function and Genetics 71 (1), 261–277. Available at https://doi.org/10.1002/prot.21715.

Biasini, M., Bienert, S., Waterhouse, A., Arnold, K., Studer, G., Schmidt, T., et al., 2014. SWISS-MODEL: modelling protein tertiary and quaternary structure using evolutionary information. Nucleic Acids Research 42 (W1), W252. Available at https://doi.org/10.1093/nar/gku340.

Bose, N., Brookes, A.J., Scordis, P., Visser, P.J., 2022. Data and sample sharing as an enabler for large-scale biomarker research and development: The EPND perspective. Frontiers in Neurology 13. Available at https://doi.org/10.3389/fneur.2022.1031091, http://www.frontiersin.org/Neurology.

Chiti, F., Dobson, C.M., 2017. Protein misfolding, amyloid formation, and human disease: A summary of progress over the last decade. Annual Review of Biochemistry 86, 27–68. Available at https://doi.org/10.1146/annurev-biochem-061516-045115, http://arjournals.annualreviews.org/loi/biochem.

Das, A., Plotkin, S.S., 2013. Mechanical probes of SOD1 predict systematic trends in metal and dimer affinity of ALS-associated mutants. Journal of Molecular Biology 425 (5), 850–874. Available at https://doi.org/10.1016/j.jmb.2012.12.022, https://www.journals.elsevier.com/journal-of-molecular-biology.

DiMaio, F., Tyka, M.D., Baker, M.L., Chiu, W., Baker, D., 2009. Refinement of Protein Structures into Low-Resolution Density Maps Using Rosetta. Journal of Molecular Biology 392 (1), 181–190. Available at https://doi.org/10.1016/j.jmb.2009.07.008, https://www.journals.elsevier.com/journal-of-molecular-biology.

Dugger, B.N., Dickson, D.W., 2017. Pathology of neurodegenerative diseases. Cold Spring Harbor Perspectives in Biology 9 (7). Available at https://doi.org/10.1101/cshperspect.a028035, http://cshperspectives.cshlp.org/content/9/7/a028035.full.pdf.

Eswar, N., Webb, B., Marti-Renom, M.A., Madhusudhan, M.S., Eramian, D., Shen, M.Y., et al., 2006. Comparative protein structure modeling using Modeller. Current Protocols in Bioinformatics. 5, 5.6. https://doi.org/10.1002/0471250953.bi0506s15.

Ethirajulu, A.K., Sriramoju, V., Bhat, A.G., Ramanathan, M., 2024. Evaluating alphafold for clinical pharmacology and pharmacogenetics: a case-study of huntingtin variants linked to Huntington's disease. AAPS Journal 26 (6). Available at https://doi.org/10.1208/s12248-024-00969-9, https://www.springer.com/journal/12248.

Ferreira, L.G., Dos Santos, R.N., Oliva, G., Andricopulo, A.D., 2015. Molecular docking and structure-based drug design strategies. Molecules (Basel, Switzerland) 20 (7), 13384–13421. Available at https://doi.org/10.3390/molecules200713384, http://www.mdpi.com/1420-3049/20/7/13384/pdf.

Gainza, P., Sverrisson, F., Monti, F., Rodolà, E., Boscaini, D., Bronstein, M.M., et al., 2020. Deciphering interaction fingerprints from protein molecular surfaces using geometric deep learning. Nature Methods 17 (2), 184–192. Available at https://doi.org/10.1038/s41592-019-0666-6, http://www.nature.com/nmeth/.

Galal, A., Moustafa, A., Salama, M., 2025. Transforming neurodegenerative disorder care with machine learning: strategies and applications. Neuroscience 573, 272–285. Available at https://doi.org/10.1016/j.neuroscience.2025.03.036.

Grassmann, G., Miotto, M., Di Rienzo, L., Salaris, F., Silvestri, B., Zacco, E., et al., 2021. A computational approach to investigate TDP-43 C-terminal fragments aggregation in amyotrophic lateral sclerosis. Biomolecules 11 (12), 1905.

Griñán-Ferré, C., Bellver-Sanchis, A., Guerrero, A., Pallàs, M., 2024. Advancing personalized medicine in neurodegenerative diseases: the role of epigenetics and pharmacoepigenomics in pharmacotherapy. Pharmacological Research 205, 107247. Available at https://doi.org/10.1016/j.phrs.2024.107247.

Hameduh, T., Haddad, Y., Adam, V., Heger, Z., 2020. Homology modeling in the time of collective and artificial intelligence. Computational and Structural Biotechnology Journal 18, 3494–3506. Available at https://doi.org/10.1016/j.csbj.2020.11.007.

Hatters, D.M., 2008. Protein misfolding inside cells: the case of Huntingtin and Huntington's disease. IUBMB Life 60 (11), 724–728. Available at https://doi.org/10.1002/iub.111.

Heo, L., Park, H., Seok, C., 2013. GalaxyRefine: protein structure refinement driven by side-chain repacking. Nucleic Acids Research 41 (W1), W384. Available at https://doi.org/10.1093/nar/gkt458.

Hermans, A.M.M., Bakker, E., Starokozhko, V., den Otter, L., Elferink, A.J.A., Bradshaw, A., et al., 2025. Biomarkers for neurodegenerative diseases in regulatory decision-making by the European Medicines Agency. Alzheimer's and Dementia: Translational Research and Clinical Interventions 11 (1). Available at https://doi.org/10.1002/trc2.70072, https://alz-journals.onlinelibrary.wiley.com/doi/full/10.1016/j.trci.2015.06.002.

Illodo, S., Al-Soufi, W., Novo, M., 2024. Critical aggregation concentration and reversibility of amyloid-β (1–40) oligomers. Archives of Biochemistry and Biophysics 761, 110179. Available at https://doi.org/10.1016/j.abb.2024.110179.

Johnsson, F., Karagöl, T., Karagöl, A., Zhang, S., 2024. Structural bioinformatic study of six human olfactory receptors and their AlphaFold3 predicted water-soluble QTY variants and OR1A2 with an odorant octanoate and TAAR9 with spermidine. QRB Discovery 6. Available at https://doi.org/10.1017/qrd.2024.18, http://www.cambridge.org/core/journals/qrb-discovery.

Jumper, J., Evans, R., Pritzel, A., Green, T., Figurnov, M., Ronneberger, O., et al., 2021. Highly accurate protein structure prediction with AlphaFold. Nature 596 (7873), 583–589. Available at https://doi.org/10.1038/s41586-021-03819-2.

Kelley, L.A., Sternberg, M.J.E., 2009. Protein structure prediction on the Web: a case study using the Phyre server. Nature Protocols 4 (3), 363–371. Available at https://doi.org/10.1038/nprot.2009.2.

Kelley, L.A., Mezulis, S., Yates, C.M., Wass, M.N., Sternberg, M.J.E., 2015. The Phyre2 web portal for protein modeling, prediction and analysis. Nature Protocols 10 (6), 845–858. Available at https://doi.org/10.1038/nprot.2015.053, http://www.natureprotocols.com/.

Kim, D.E., Chivian, D., Baker, D., 2004. Protein structure prediction and analysis using the Robetta server. Nucleic Acids Research 32 (Web Server), W526. Available at https://doi.org/10.1093/nar/gkh468.

Knowles, T.P.J., Vendruscolo, M., Dobson, C.M., 2014. The amyloid state and its association with protein misfolding diseases. Nature Reviews. Molecular Cell Biology 15 (6), 384–396. Available at https://doi.org/10.1038/nrm3810, http://www.nature.com/molcellbio.

Krokidis, M.G., Exarchos, T.P., Avramouli, A., Vrahatis, A.G., Vlamos, P., 2023. Computational and functional insights of protein misfolding in neurodegeneration. Advances in Experimental Medicine and Biology 1423, 201–206. Available at https://doi.org/10.1007/978-3-031-31978-5_18, https://www.springer.com/series/5584.

Kuhlman, B., Bradley, P., 2019. Advances in protein structure prediction and design. Nature Reviews. Molecular Cell Biology 20 (11), 681–697. Available at https://doi.org/10.1038/s41580-019-0163-x.

Kumar, V., Singh, C., Singh, A., 2021. Zebrafish an experimental model of Huntington's disease: molecular aspects, therapeutic targets and current challenges. Molecular Biology Reports 48 (12), 8181–8194. Available at https://doi.org/10.1007/s11033-021-06787-y.

Lee, J., Wu, S., Zhang, Y., 2009. Ab initio protein structure prediction. From Protein Structure to Function with Bioinformatics. Springer Netherlands, United States, pp. 3–25. Available at http://doi.org/10.1007/978-1-4020-9058-5_1, http://springerlink.com/openurl.asp?genre=book&isbn=978-1-4020-9057-8.

Lee, G.R., Won, J., Heo, L., Seok, C., 2019. GalaxyRefine2: simultaneous refinement of inaccurate local regions and overall protein structure. Nucleic Acids Research 47 (W1), W451. Available at https://doi.org/10.1093/nar/gkz288.

Leonard, H.L., Nalls, M.A., Day-Williams, A., Esmaeeli, S., Jarreau, P., Bandres-Ciga, S., et al., 2024. Open science in precision medicine for neurodegenerative diseases. Nature Reviews. Drug Discovery 23 (4), 233–234. Available at https://doi.org/10.1038/d41573-024-00017-3, https://www.nature.com/nrd/.

McGuffin L.J., 2002. Fully automated methods for protein fold recognition using predicted secondary structure.

McGuffin, L.J., 2008. Protein fold recognition and threading. Computational Structural Biology. WORLD SCIENTIFIC, pp. 37–60.

Meisl, G., Knowles, T.P.J., Klenerman, D., 2022. Mechanistic models of protein aggregation across length-scales and time-scales: from the test tube to neurodegenerative disease. Frontiers in Neuroscience 16. Available at https://doi.org/10.3389/fnins.2022.909861, https://www.frontiersin.org/journals/neuroscience#.

Meng, X.Y., Zhang, H.X., Mezei, M., Cui, M., 2011. Molecular docking: a powerful approach for structure-based drug discovery. Current Computer-Aided Drug Design 7 (2), 146–157. Available at https://doi.org/10.2174/157340911795677602, http://www.eurekaselect.com/582/journal/current-computer-aided-drug-design.

Meng, Y., Zhang, Z., Zhou, C., Tang, X., Hu, X., Tian, G., et al., 2025. Protein structure prediction via deep learning: an in-depth review. Frontiers in Pharmacology 16. Available at https://doi.org/10.3389/fphar.2025.1498662.

Notova, S., Bonnardel, F., Lisacek, F., Varrot, A., Imberty, A., 2020. Structure and engineering of tandem repeat lectins. Current Opinion in Structural Biology 62, 39–47. Available at https://doi.org/10.1016/j.sbi.2019.11.006.

Orry, A.J.W., Abagyan, R., 2016. Homology Modeling: Methods and Protocols. Humana Press.

Pettersen, E.F., Goddard, T.D., Huang, C.C., Couch, G.S., Greenblatt, D.M., Meng, E.C., et al., 2004. UCSF Chimera - a visualization system for exploratory research and analysis. Journal of Computational Chemistry 25 (13), 1605–1612. Available at https://doi.org/10.1002/jcc.20084.

Rinauro, D.J., Chiti, F., Vendruscolo, M., Limbocker, R., 2024. Misfolded protein oligomers: mechanisms of formation, cytotoxic effects, and pharmacological approaches against protein misfolding diseases. Molecular Neurodegeneration 19 (1). Available at https://doi.org/10.1186/s13024-023-00651-2, https://molecularneurodegeneration.biomedcentral.com.

Robustelli, P., Piana, S., Shaw, D.E., 2018. Developing a molecular dynamics force field for both folded and disordered protein states. Proceedings of the National Academy of Sciences of the United States of America 115 (21) E4758, Available from. Available at https://doi.org/10.1073/pnas.1800690115, http://www.pnas.org/content/pnas/115/21/E4758.full.pdf.

Roche, D.B., Buenavista, M.T., Tetchner, S.J., McGuffin, L.J., 2011. The IntFOLD server: an integrated web resource for protein fold recognition, 3D model quality assessment, intrinsic disorder prediction, domain prediction and ligand binding site prediction. Nucleic Acids Research 39 (2), W171. Available at https://doi.org/10.1093/nar/gkr184.

Rossi, S.L., Subramanian, P., Bovenkamp, D.E., 2023. The future is precision medicine-guided diagnoses, preventions and treatments for neurodegenerative diseases. Frontiers in Aging Neuroscience 15. Available at https://doi.org/10.3389/fnagi.2023.1128619, https://www.frontiersin.org/journals/aging-neuroscience#.

Roy, A., Kucukural, A., Zhang, Y., 2010. I-TASSER: a unified platform for automated protein structure and function prediction. Nature Protocols 5 (4), 725–738. Available at https://doi.org/10.1038/nprot.2010.5.

Šali, A., Blundell, T.L., 1993. Comparative protein modelling by satisfaction of spatial restraints. Journal of Molecular Biology 234 (3), 779–815. Available at https://doi.org/10.1006/jmbi.1993.1626.

Selkoe, D.J., Hardy, J., 2016. The amyloid hypothesis of Alzheimer's disease at 25 years. EMBO Molecular Medicine 8 (6), 595–608. Available at https://doi.org/10.15252/emmm.201606210.

Semenyuk, P.I., 2022. REMD simulations of full-length alpha-synuclein together with ligands reveal binding region and effect on amyloid conversion. International Journal of Molecular Sciences 23 (19), 11545. Available at https://doi.org/10.3390/ijms231911545.

Stefano, G.B., 2024. Quantum computing and the future of neurodegeneration and mental health research. Republic Brain Sciences 14 (1). Available at https://doi.org/10.3390/brainsci14010093, http://www.mdpi.com/journal/brainsci.

Sun, Y., Ding, F., 2020. αB-Crystallin chaperone inhibits Aβ aggregation by capping the β-sheet-rich oligomers and fibrils. The Journal of Physical Chemistry. B 124 (45), 10138–10146. Available at https://doi.org/10.1021/acs.jpcb.0c07256.

Söding, J., 2005. Protein homology detection by HMM–HMM comparison. Bioinformatics (Oxford, England) 21 (7), 951–960. Available at https://doi.org/10.1093/bioinformatics/bti125.

Thangamalai, M.S., Desai, D., Selvaraj, C., 2025. Revolutionizing structural biology: AI-driven protein structure prediction from AlphaFold to next-generation innovations. Advances in Protein Chemistry and Structural Biology 147, 1–19. Available at https://doi.org/10.1016/bs.apcsb.2025.04.002, https://www.sciencedirect.com/science/bookseries/18761623.

Trott, O., Olson, A.J., 2010. AutoDock Vina: improving the speed and accuracy of docking with a new scoring function, efficient optimization, and multithreading. Journal of Computational Chemistry 31 (2), 455–461. Available at https://doi.org/10.1002/jcc.21334.

Urbanc, B., Cruz, L., Yun, S., Buldyrev, S.V., Bitan, G., Teplow, D.B., et al., 2004. In silico study of amyloid β-protein folding and oligomerization. Proceedings of the National Academy of Sciences 101 (50), 17345–17350. Available at https://doi.org/10.1073/pnas.0408153101.

Urbanc, B., Borreguero, J.M., Cruz, L., Stanley, H.E., 2006. Ab initio discrete molecular dynamics approach to protein folding and aggregation. Methods in Enzymology 412, 314–338. Available at https://doi.org/10.1016/S0076-6879(06)12019-4.

Waterhouse, A., Bertoni, M., Bienert, S., Studer, G., Tauriello, G., Gumienny, R., et al., 2018. SWISS-MODEL: homology modelling of protein structures and complexes. Nucleic Acids Research 46 (W1), W296. Available at https://doi.org/10.1093/nar/gky427.

Waury, K., Willemse, E.A.J., Vanmechelen, E., Zetterberg, H., Teunissen, C.E., Abeln, S., 2022. Bioinformatics tools and data resources for assay development of fluid protein biomarkers. Biomarker Research 10 (1). Available at https://doi.org/10.1186/s40364-022-00425-w, https://biomarkerres.biomedcentral.com.

Webb, B., Sali, A., 2017. Protein Structure Modeling with MODELLER1654. Springer Science and Business Media LLC, pp. 39–54. Available at http://doi.org/10.1007/978-1-4939-7231-9_4.

Yang, J., Yan, R., Roy, A., Xu, D., Poisson, J., Zhang, Y., 2015. The I-TASSER Suite: protein structure and function prediction. Nature Methods 12 (1), 7–8. Available at https://doi.org/10.1038/nmeth.3213.

Yun, S., Urbanc, B., Cruz, L., Bitan, G., Teplow, D.B., Stanley, H.E., 2007. Role of electrostatic interactions in amyloid β-protein (Aβ) oligomer formation: a discrete molecular dynamics study. Biophysical Journal 92 (11), 4064–4077. Available at https://doi.org/10.1529/biophysj.106.097766.

Zhang, X., Guo, T., Zhang, Y., Jiao, M., Ji, L., Dong, Z., et al., 2024. Global burden of Alzheimer's disease and other dementias attributed to metabolic risks from 1990 to 2021: results from the global burden of disease study 2021. BMC Psychiatry 24 (1). Available at https://doi.org/10.1186/s12888-024-06375-x.

CHAPTER 3

Bioinformatics tools for protein modeling in cancer

Alpana Srishti Rai, Seneha Santoshi, and Shazia Rashid
Amity Institute of Biotechnology, Amity University Uttar Pradesh (AUUP), Noida, Uttar Pradesh, India

3.1 Introduction

Cancer, a group of diseases defined by uncontrolled cell proliferation, remains one of the most significant global health challenges of the 21st century. At its core, cancer is a disease that is initiated and propelled by somatic and germline mutations that disrupt normal cellular function (Hasan et al., 2020). These genetic alterations, however, exert their pathological effects almost entirely through the proteins they encode. The intricate machinery of the cell—governing processes from DNA replication and repair to signal transduction, metabolic control, and apoptosis—is executed by a vast and dynamic network of proteins. When the genes encoding these proteins are altered, the resulting aberrant protein products can disrupt cellular homeostasis, leading to malignant transformation. In practice, the biological consequences of these protein-level changes vary widely depending on cellular context and tissue type.

The central tenet of molecular biology, the structure-function paradigm, posits that the specific three-dimensional (3D) conformation of a protein is inextricably linked to its biological role (Zheng et al., 2020). A protein's function, whether it is enzymatic catalysis, signal mediation, or structural support, is dictated by the precise spatial arrangement of its constituent amino acids. Consequently, structural perturbations, even those originating from a single amino acid substitution (a missense mutation), can have profound functional consequences (Roy et al., 2024). Such a change might destabilize the protein's core, obliterate a critical phosphorylation site, alter the geometry of an enzyme's active site, or disrupt a key protein–protein interaction (PPI) interface. Not all structural changes result in measurable functional disruption, which makes structural interpretation particularly important in cancer-associated variants. A detailed, atomic-level understanding of the architecture of oncoproteins (e.g., RAS, MYC), tumor suppressors (e.g., p53, BRCA1), and other cancer-implicated proteins is therefore not merely an academic pursuit but a critical prerequisite for deciphering disease mechanisms and for the rational design of targeted, next-generation therapeutics (Qiu et al., 2024).

For decades, the determination of high-resolution protein structures has been the exclusive domain of experimental biophysical techniques. X-ray crystallography,

Advances in Structural Biology
DOI: https://doi.org/10.1016/B978-0-443-44988-8.00010-1

nuclear magnetic resonance (NMR) spectroscopy, and, more recently, cryogenic electron microscopy (cryo-EM) have collectively populated the Protein Data Bank (PDB) with over 200,000 macromolecular structures, providing the foundational knowledge for much of modern medicine (Berman et al., 2000). While these methods are the undisputed gold standard, they are often laborious, time-consuming, and resource-intensive. Furthermore, they are not universally applicable. Many proteins, particularly large dynamic complexes or membrane-bound proteins (which constitute a majority of drug targets), remain recalcitrant to crystallization or are too large for traditional NMR (Kühlbrandt, 2014). These limitations are encountered frequently in cancer research, where many clinically relevant targets fall into these challenging categories. This "experimental gap" between the vastness of known protein sequences—expanding exponentially through genomic sequencing—and the comparatively small number of solved structures has created a need for robust computational approaches to predict protein structures.

Bioinformatics, situated at the intersection of computer science, statistics, and molecular biology, offers a powerful arsenal of tools and databases to bridge this gap (Huang et al., 2024). These computational methods translate primary amino acid sequences into structural and functional insights. Historically, the field was dominated by comparative (homology) modeling, which leverages evolutionary conservation to build models based on existing structural templates. While effective, this approach failed when a suitable template was unavailable. The recent emergence of deep learning and artificial intelligence (AI) has dramatically reshaped this landscape, culminating in tools like AlphaFold, which demonstrate predictive accuracy that can rival experimental methods for many monomeric proteins (Jumper et al., 2021). This revolution has opened the door to in silico structural biology on a proteome-wide scale. At the same time, the predictive power of these approaches depends strongly on the availability of sequence data and does not eliminate the need for careful biological interpretation.

This chapter presents an overview of widely used bioinformatics tools for protein modeling, with particular emphasis on their relevance to cancer research. Classical homology-based approaches, contemporary AI-driven predictors, and supporting tools for template identification, validation, and interaction analysis are discussed in the context of cancer-associated proteins. Selected examples are used to illustrate how these computational methods contribute to molecular-level understanding and support therapeutic research.

3.2 The evolving landscape of protein structure prediction

The journey to accurately predict a protein's 3D structure from its one-dimensional sequence has been a central challenge in biology for over half a century. The methods developed to tackle this "protein folding problem" have evolved significantly, moving

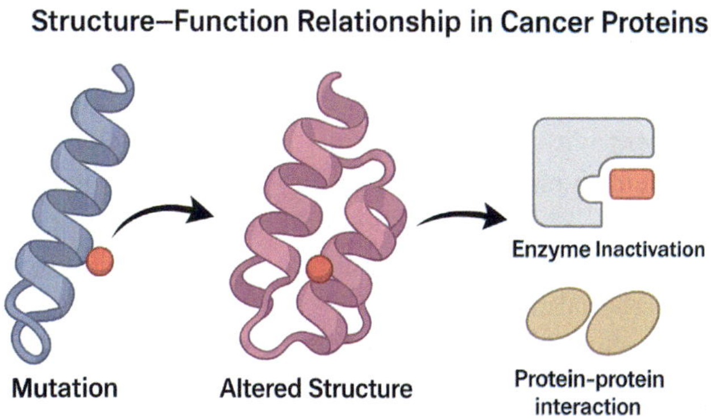

Figure 3.1 Changes in protein structure and function by missense mutation. Schematic representation of how missense mutations can influence protein structure and function. Depending on their position and physicochemical properties, single amino acid substitutions may affect folding stability, active-site geometry, or protein–protein interaction interfaces, thereby contributing to oncogenic processes.

from first-principles physics simulations to data-driven comparative methods, and most recently, to sophisticated deep learning architectures (Meng et al., 2025).

Fig. 3.1 depicts the impact of missense mutations on protein structure and function. A single amino acid change can destabilize folding, disrupt active sites, or interfere with PPIs, initiating oncogenic pathways. The persistence of this challenge reflects both the inherent complexity of protein folding and the limitations of any single methodological approach.

3.2.1 The experimental gold standard

An understanding of experimental structure-determination techniques provides essential context for evaluating the strengths and limitations of computational protein modeling.

- X-ray crystallography: This has historically been the most prolific method. Pioneered for proteins by John Kendrew and Max Perutz (who solved the structures of myoglobin and hemoglobin in the late 1950s), this technique requires a protein to be purified and coaxed into forming a highly ordered crystal (Kendrew et al., 1958). This crystal is then exposed to X-rays, and the resulting diffraction pattern is mathematically reconstructed into an electron density map, from which an atomic model is built. Its primary limitations are the significant bottleneck of crystallization and the fact that the resulting structure is a static, often "averaged" conformation within a nonphysiological crystal lattice. As a result, crystallographic structures often provide an invaluable reference point, but they may not fully capture the dynamic behavior of proteins in living cells.
- NMR spectroscopy: Unlike X-ray crystallography, NMR is performed on proteins in solution, which is closer to their native state. Established as a method

for protein structure determination by Kurt Wüthrich in the 1980s, NMR measures the magnetic properties of atomic nuclei to derive a set of distance constraints between nearby atoms (Wüthrich, 1989). A family or ensemble of structures consistent with these constraints is then calculated. NMR is powerful for studying protein dynamics and flexibility, but is typically limited to smaller, soluble proteins (e.g., < 30 kDa). These size and sensitivity constraints limit its routine application to many cancer-relevant proteins, despite its strengths in probing conformational flexibility.
- Cryo-EM: This technique has undergone a "resolution revolution" in the last decade. Developed by Jacques Dubochet, JoachimFrank, and Richard Henderson (who were awarded the 2017 Nobel Prize in Chemistry), this method involves flash-freezing purified proteins in a thin layer of vitreous ice and imaging thousands of individual particles with an electron microscope (Cressey and Callaway, 2017). These two-dimensional (2D) projections are then computationally averaged and reconstructed into a 3D density map. Cryo-EM is exceptionally powerful for elucidating the structures of large, complex, or dynamic assemblies that are intractable by other methods, such as ribosomes, spliceosomes, and membrane-bound ion channels (Kühlbrandt, 2014). Its growing relevance to cancer biology lies in the ability to resolve macromolecular assemblies that are difficult to study using other experimental techniques.

The data generated by these methods form the PDB, which serves as the foundational "training data" for nearly all computational prediction tools. Fig. 3.2 shows the schematic representation of the three major experimental techniques for protein structure determination.

3.2.2 The classical era: comparative and template-based modeling

Before the AI revolution, computational prediction was broadly divided into two categories: template-based modeling (TBM) and ab initio (or de novo) modeling.

3.2.2.1 Template-based modeling

Which includes homology modeling and protein threading, has historically been the most reliable and widely used approach. It is built on the empirical observation that protein structure is more conserved during evolution than protein sequence (Kihara et al., 2006). Two proteins descending from a common ancestor will often maintain the same 3D fold even after their sequences have diverged significantly. If a protein (the "target") has a detectable sequence similarity to a protein of known structure (the "template"), one can build a model of the target based on the template's backbone.

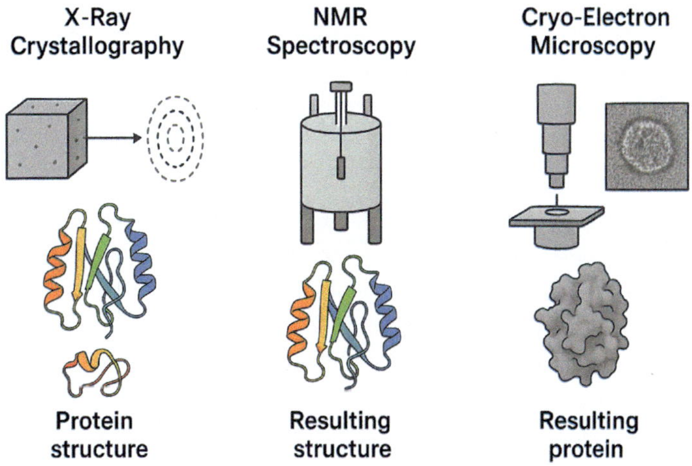

Figure 3.2 Experimental techniques for protein structure determination. Schematic representation of the three major experimental techniques for protein structure determination—X-ray crystallography, nuclear magnetic resonance (NMR) spectroscopy, and cryo-electron microscopy.

The typical homology modeling workflow consists of several discrete steps (Webb and Sali, 2016) (Fig. 3.3).

1. Template identification: The target sequence is searched against the PDB using sequence alignment tools like BLAST or PSI-BLAST. If a good template (e.g., >30%–40% sequence identity) is found, the process proceeds.
2. Sequence alignment: This is the most critical step. The target and template sequences are meticulously aligned. Errors in this alignment (e.g., misplacing a gap) will be directly propagated into the final model.
3. Backbone generation: The backbone coordinates (N, Cα, C) of the aligned regions in the target are copied directly from the template.
4. Loop modeling: Regions where the target and template sequences do not align, such as insertions or deletions (indels), must be modeled from scratch. This is a major challenge, as these loops are often flexible and lack a single defined conformation.
5. Side-chain packing: The amino acid side chains of the target are "packed" onto the template backbone, typically using a library of known rotamer (side-chain) conformations and an energy function to select the most favorable combination.
6. Model refinement and validation: The final model is subjected to energy minimization to relieve atomic clashes and is then assessed using a variety of validation tools to check its stereochemical quality and overall plausibility.

Protein threading (or fold recognition) is a more advanced form of TBM used when no obvious homolog can be found via sequence alignment (e.g., in the <30% "twilight

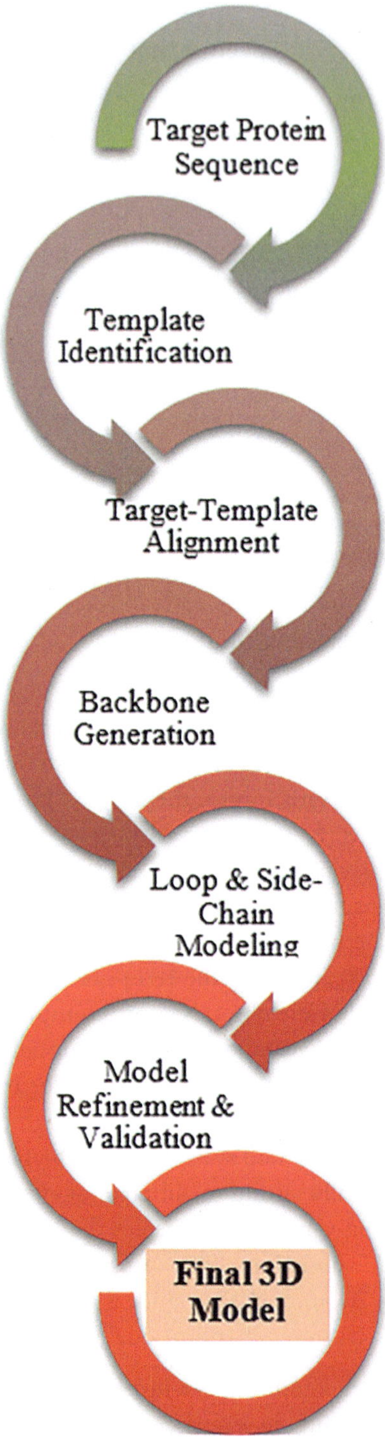

Figure 3.3 Process of template-based (homology) modeling. Workflow for template-based (homology) modeling: The process begins by identifying structurally similar proteins (templates)

(continued)

zone"). Instead of comparing sequence-to-sequence, threading methods compare the target sequence against a library of known protein folds, using scoring functions that evaluate how well the target sequence "fits" onto a given 3D backbone (Bowie et al., 1991).

3.2.2.2 *Ab initio modeling*

In its classical sense, attempts to predict structure from sequence alone, without any template, by simulating the physical process of folding. This typically involves using a physics-based potential energy function and conformational sampling algorithms (like molecular dynamics) to find the lowest-energy state (the native structure). For decades, this approach was computationally prohibitive and largely unsuccessful for all but the smallest peptides, due to the astronomical size of the conformational search space and inaccuracies in the energy functions.

3.2.3 The AI revolution: deep learning and coevolution

The field was fundamentally and irrevocably altered between 2018 and 2020, primarily through the Critical Assessment of Structure Prediction (CASP) competition. At CASP13 (2018) and CASP14 (2020), DeepMind's AlphaFold program demonstrated a significant leap in predictive accuracy, solving structures with an accuracy comparable to that of experimental methods for many targets (Jumper et al., 2021).

Fig. 3.4 depicts the workflow for AI-driven protein structure prediction (Cressey and Callaway, 2017) using coevolutionary data (e.g., AlphaFold). A target sequence is used to build a deep multiple sequence alignment (MSA). Deep learning networks analyze the MSA for patterns of correlated mutations (coevolution) to predict interresidue distances and orientations. A separate structure module then folds the polypeptide chain based on these predicted constraints and learned stereochemical principles, yielding a high-accuracy 3D model typically accompanied by a per-residue confidence score (Jumper et al., 2021).

in the Protein Data Bank (*PDB*) for a target sequence using tools like BLAST or hidden Markov model–hidden Markov model prediction. This is followed by carefully aligning the target and template sequences, a critical step for model accuracy. A 3D model is built using the template backbone coordinates, incorporating challenging steps like modeling nonaligned loops and packing side chains. The resulting model undergoes energy minimization (refinement) and rigorous validation (e.g., stereochemistry checks using WHAT IF) to ensure structural plausibility (Roy et al., 2024; Waterhouse et al., 2018).

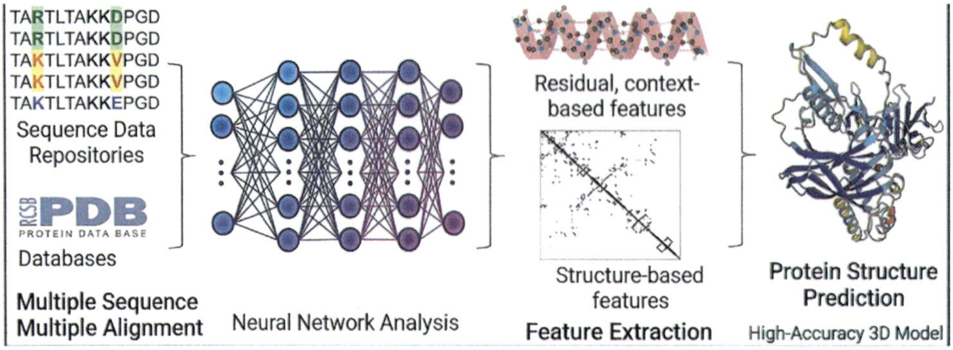

Figure 3.4 Protein structure prediction using tools like AlphaFold. Workflow for AI-driven protein structure prediction using coevolutionary data (e.g., AlphaFold).

This breakthrough was not based on simulating physics. Instead, it was a triumph of deep learning, building on a key evolutionary insight: coevolution.

The logic is as follows: if a protein is to maintain its specific 3D fold over evolutionary time, amino acids that are in direct contact in 3D space must be "compatible." If one of these residues mutates (e.g., changing a large hydrophobic residue to a small polar one), the protein's fold will be destabilized unless the residue it contacts also mutates to compensate (e.g., changing from a small polar to a large hydrophobic). Therefore, by analyzing an MSA of thousands of homologous sequences for a given protein family, one can find pairs of residues (columns in the alignment) that mutate in a correlated fashion. These coevolving pairs are highly likely to be in contact in the 3D structure.

AlphaFold, RoseTTAFold, and similar AI tools use deep neural networks (specifically, "transformers" or "attentional networks") to analyze massive MSAs and identify these subtle coevolutionary patterns (Baek et al., 2021). The network then predicts a highly accurate "distogram"—a 2D map of the predicted distances and orientations between all pairs of residues. This 2D map is then passed to another neural network that, guided by learned stereochemical principles, translates this set of spatial restraints into a full 3D atomic model.

This approach effectively bypasses the classical physics simulation, replacing it with a data-driven process learned from the entire PDB and all known sequence data. Crucially, it allows for high-accuracy ab initio prediction even when no close (or even remote) template exists, as long as a "deep" MSA with many diverse sequences can be built. This has opened the door to modeling vast, previously "dark" regions of the proteome, which is of particular importance in cancer research, where many key proteins are novel or poorly characterized (Guo et al., 2024).

3.3 Bioinformatics tools for protein modeling

Modern bioinformaticians have access to a wide array of tools, ranging from automated web servers for nonexperts to sophisticated standalone software for in-depth research. These tools can be broadly categorized by their primary function.

Fig. 3.5 depicts the schematic overview of the bioinformatics "toolbox" for protein modeling in cancer research. Tools are broadly categorized based on their primary function: template detection (e.g., hidden Markov model [HMM]–HMM prediction [HHpred]), template-based model building (e.g., MODELLER, SWISS-MODEL), AI-driven/ab initio model building (e.g., AlphaFold, RoseTTAFold), model validation and quality assessment (e.g., WHAT IF), structural analysis and comparison (e.g., DALI, LigPlot+), and supporting databases or integrated workbenches (e.g., ModBase).

3.3.1 Template-based modeling platforms

These tools remain highly relevant, particularly when good templates exist, as they are computationally fast and built on decades of refinement.

3.3.1.1 MODELLER

- Description: MODELLER is not a web server but a powerful, scriptable, standalone software package primarily used in academic and research settings. Its core methodology is "comparative modeling by satisfaction of spatial restraints." The user provides a sequence alignment between the target and one or more templates, and MODELLER calculates a model by blending the template structures and satisfying a set of spatial restraints (e.g., Cα-Cα distances, dihedral angles) derived from the templates. Its flexibility is its greatest strength. Because MODELLER is implemented in Python, it can be adapted to a wide range of custom modeling tasks, including multichain modeling, the incorporation of experimental restraints (for example, from NMR or cross-linking data), and the generation of model ensembles

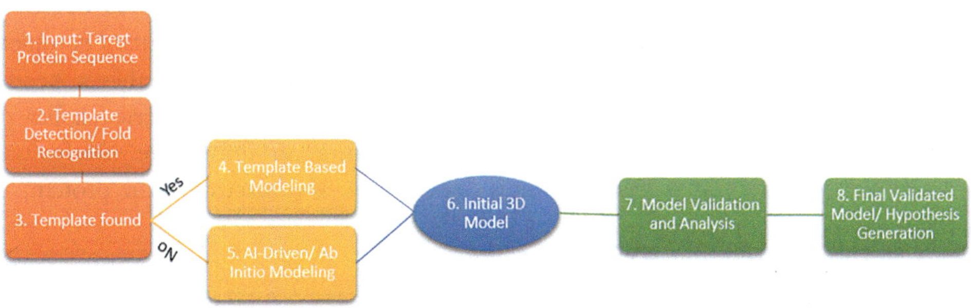

Figure 3.5 Steps involved in protein modeling in cancer research using bioinformatics toolbox. A schematic overview of the bioinformatics 'toolbox' for protein modeling in cancer research.

for flexible regions. In practice, this flexibility has made MODELLER particularly useful for detailed structural analysis of cancer-related proteins with established structural templates.

- Case application: A critical application of MODELLER in oncology is the structural characterization of drug-resistant variants where experimental data are missing. A prominent example is the T315I "gatekeeper" mutation in the BCR-ABL oncoprotein, which renders the drug imatinib (an oral targeted therapy medication used to treat cancer) ineffective. Researchers utilize MODELLER to construct a high-precision 3D model of this mutant based on the known wild-type crystal structure. Such models provide a structural reference that allows the predicted steric effects of the mutation on drug binding to be examined in the absence of direct experimental data. Furthermore, it can support in silico screening campaigns to identify novel compounds capable of accommodating the mutant side chain, directly facilitating the design of second- and third-generation kinase inhibitors.

3.3.1.2 SWISS-MODEL

- Description: SWISS-MODEL is one of the most established and user-friendly fully automated web servers for homology modeling (Waterhouse et al., 2018). It guides the user through the entire workflow: template search, model building, and quality estimation. A key feature is the SWISS-MODEL Template Library, a curated database of potential templates derived from the PDB, which are preprocessed and annotated to improve selection. SWISS-MODEL is particularly strong in modeling oligomeric structures (protein complexes) if the template itself is an oligomer, and it can automatically model the position of ligands and cofactors if present in the template. SWISS-MODEL is often the first tool used by cancer biologists who need a quick, reliable structural model.
- Case application: Following the identification of a novel missense mutation in a dimer interface of a tumor suppressor protein, the SWISS-MODEL server serves as an ideal tool for structural assessment. Unlike many other platforms, SWISS-MODEL automatically leverages dimeric templates to construct a model of the mutant complex. This modeling capability allows researchers to visually demonstrate whether a specific mutation (e.g., changing an arginine to a tryptophan) introduces steric clashes or disrupts critical salt bridges at the interface, offering an immediate mechanistic hypothesis for how the mutation disrupts dimerization and inactivates the protein.

3.3.1.3 Iterative threading ASSEmbly refinement

- Description: Iterative threading ASSEmbly refinement (I-TASSER) is a highly ranked, automated server that functions as a hybrid method, combining threading, fragment assembly, and ab initio refinement (Yang and Zhang, 2015). It first attempts to find templates by "threading" the target sequence against a library of known folds. It then excises structurally conserved fragments from these templates and assembles them. For regions without a template (e.g., loops), it uses ab initio modeling (based on the Rosetta framework) to build the structure. The resulting models are then refined. A key output of I-TASSER is that it also provides functional predictions, such as Gene Ontology terms and potential ligand-binding sites, by comparing the predicted structure to a library of functional proteins. I-TASSER is commonly applied to targets of moderate difficulty, particularly in cases where a complete structural template is unavailable but conserved structural fragments can be identified.

- Case application: In cancer-related studies, I-TASSER has been used to analyze multidomain proteins in situations where full-length templates are not available. For example, in studies of tumor suppressors such as RASSF2, I-TASSER can generate a homology model for conserved domains alongside ab initio models for novel regions, which are subsequently assembled into a full-length structure. Such models allow interdomain interactions to be examined and provide a structural basis for assessing the potential effects of cancer-associated mutations within regions that lack experimental characterization.

3.3.1.4 Phyre2 (protein homology/analogY recognition engine V 2.0)

- Description: The Phyre2 server is widely used for remote homology detection and fold recognition in protein structure analysis (Kelley et al., 2015). It applies a profile–profile matching strategy, derived from HHpred methodology, to compare a target sequence against databases of known protein structures. It is highly effective at identifying a likely fold for a protein, even when sequence identity is in the "twilight zone" (<30%). It presents results visually, showing which parts of the sequence were modeled and with what level of confidence, and can supplement its homology modeling with a now-outdated ab initio method for regions without a template. For many studies, Phyre2 is used as an initial step in the structural and functional assessment of proteins that lack close homologs.
- Case application: An application of Phyre2 is reported in the structural analysis of the CYLD tumor suppressor gene (Roy et al., 2024). In the absence of an experimentally determined structure, Phyre2 is used to generate a structural scaffold that supports the mapping of 18 high-risk missense mutations. This approach allows physicochemical features, including predicted stability changes and residue contacts, to be examined to assess how specific single-nucleotide polymorphisms may disrupt stabilizing interactions or catalytic regions.

3.3.1.5 Robetta/Rosetta

- Description: Robetta is a web server that provides access to the Rosetta software suite, an academic platform developed for macromolecular modeling (Baek et al., 2021). Rosetta comprises a collection of protocols for tasks such as ab initio structure prediction, loop modeling, protein–protein docking, and structural refinement, largely based on fragment assembly and physics-based energy functions. In recent implementations, the Robetta server incorporates the deep learning predictor RoseTTAFold, allowing hybrid modeling approaches that combine data-driven prediction with energy-based refinement. In cancer-related research, Rosetta has been applied not only to structure prediction but also to problems involving molecular interaction analysis and protein design.
- Case application: A major goal in cancer therapy is to inhibit PPIs that drive oncogenesis, such as the p53-MDM2 interaction (where MDM2 tags p53 for destruction). These large, flat interfaces are notoriously "undruggable" by small molecules. Rosetta software is used to design protocols, from scratch, novel peptides or small proteins (e.g., "stapled peptides") that are precisely shaped to bind to the MDM2 pocket with high affinity, thereby "rescuing" p53 from degradation. This is a frontier of rational drug design (Arkin and Wells, 2004).

3.3.2 AI-driven, ab initio, and fold recognition tools

This class of tools is defined by its ability to generate high-quality models even in the absence of templates, relying on deep learning or highly sensitive profile-comparison methods.

3.3.2.1 AlphaFold

- Description: Developed by DeepMind, AlphaFold (and its successor, AlphaFold 2) represents the current state-of-the-art in monomeric protein structure prediction (Jumper et al., 2021). As described in section 2.3, "transformer-based" deep neural network is used to integrate information from MSAs, coevolutionary signals, and a "recycling" process to iteratively refine a 3D model. Its output includes a predicted local distance difference test (pLDDT) score for each residue, a highly reliable internal confidence metric (pLDDT > 90 is considered high accuracy, < 70 is low, and < 50 is likely disordered). Perhaps most impactfully, DeepMind and EMBL-EBI have used it to create the AlphaFold Protein Structure Database (AFDB), which provides free, high-accuracy predictions for over 200 million proteins, including the entire human proteome (Varadi et al., 2022).
- Case application 1 (modeling the "Unmodellable"): High-quality structural models for "orphan" human proteins—those with no homologs in the PDB—that have

been implicated in cancer through genomic screens can be easily obtained. This provides an immediate structural context for understanding their function.

- Case application 2 (fusion proteins): As highlighted in the literature (Kumar and Kim, 2024), a major challenge in oncology is modeling cancer-specific fusion proteins (e.g., BCR-ABL, EML4-ALK). These chimeras are unique to the disease and have no single template. AlphaFold is being successfully applied to predict the structures of these fusions, revealing how the two fused domains are oriented and how this fusion might create a new, constitutively active enzyme or a novel druggable interface.
- Case application 3 (drug discovery): AlphaFold models are being used to identify novel "cryptic" or allosteric binding pockets on oncoproteins that were not visible in static experimental structures, opening new avenues for drug design (Ren et al., 2023).

3.3.2.2 *trRosetta/RoseTTAFold*

- Description: Developed concurrently with AlphaFold, trRosetta (and its evolution, RoseTTAFold from the Baker lab) is another deep learning-based predictor (Baek et al., 2021). It operates on similar principles, using neural networks to analyze MSAs. However, instead of just predicting distances (a distogram), it predicts a full 6D transformation (distance and orientation) between all residue pairs. This additional geometric information is incorporated into the folding stage to constrain model assembly. In comparative assessments, RoseTTAFold has shown strong predictive performance and reduced computational demands for certain classes of targets. It has also been applied with particular emphasis on the modeling of protein–protein complexes.
- Case application: In cancer-related studies, RoseTTAFold has been used to examine regulatory interactions between oncogenic proteins and their inhibitors. By predicting the 3D structure of such complexes, including X–Y interaction pairs, the resulting models can be analyzed to identify interface regions that may be relevant for mutagenesis experiments or inhibitor design (Baek et al., 2021).

3.3.2.3 *Hidden Markov model–Hidden Markov model prediction*

- Description: HHpred is not a 3D structure prediction tool, but a server designed for template identification through remote homology detection (Zimmermann et al., 2018). It compares the HMM of a query sequence against databases of HMMs derived from experimentally determined protein structures. By relying on profile–profile comparisons, which capture family-level conservation patterns, HHpred can identify statistically significant evolutionary relationships even when sequence identity is low. As a result, it is frequently used during the initial stages of comparative modeling workflows, particularly when suitable templates are not readily apparent from sequence-based searches.

- Case application: In cancer-related genomic studies, HHpred has been applied to assist in the functional annotation of genes that lack clear sequence similarity to known proteins. For example, when large-scale sequencing identifies recurrently mutated genes that produce BLAST-negative results, HHpred can be used to search for remote evolutionary relationships. In some cases, profile–profile comparisons suggest similarity to established enzyme families, such as DNA glycosylases, despite low overall sequence identity. Such findings provide a starting point for functional interpretation and support the selection of templates for subsequent 3D modeling.

3.3.2.4 3D-position-specific scoring matrix/mGenThreader

- Description: These are classic and well-regarded threading (fold recognition) servers. 3D-position-specific scoring matrix (3D-PSSM) was a pioneering method that created a library of structural "superfamilies" and used PSSMs (sequence profiles) to score the compatibility of a query sequence with each fold. mGenThreader is another advanced threading method that combines sequence profile alignment with additional structural information, such as predicted secondary structure, to improve the accuracy of fold recognition. Like HHpred, these tools are used for functional annotation of novel proteins.
- Case application: Before the advent of AlphaFold, an oncogene with no known template would use a metaserver (a server that runs multiple tools at once). This server would run mGenThreader, Phyre2, and other threading algorithms. If all three tools independently returned a "hit" for the same fold (e.g., a "TIM barrel"), even with low confidence, this "consensus" prediction would give enough confidence to pursue that functional hypothesis experimentally.

3.3.2.5 Fold and function assignment server

- Description: Fold and function assignment server (FFAS) is another sensitive fold recognition server that operates by comparing sequence profiles (Jaroszewski et al., 2005). A key feature of FFAS is its calibration and the reliability of its Z-score, which provides a robust statistical measure of the likelihood that a match is a true positive (a true structural homolog) rather than a chance alignment. FFAS is used to find distant relationships and infer function.
- Case application: The FFAS server is instrumental in inferring the function of novel proteins secreted by cancer cells based on structural compatibility. In scenarios where a protein is suspected to aid in metastasis (e.g., a protease) but lacks sequence homology to known enzymes, FFAS can be used to scan for compatible protein folds. A statistically significant Z-score linking the query protein to a known family,

such as chymotrypsin-like serine proteases, indicates a likely structural relationship. This fold assignment can support functional interpretation and provide a basis for modeling the putative active site in subsequent inhibitor-design studies (Jaroszewski et al., 2005).

3.3.3 Model validation, analysis, and visualization

The generation of a structural model represents only one stage of the workflow, as careful validation and analysis are required before meaningful biological conclusions can be drawn.

3.3.3.1 WHAT IF

- Description: WHAT IF is a long-established software package used for macromolecular structure analysis and validation (Vriend, 1990). It comprises a collection of tools for assessing basic stereochemical parameters, such as bond lengths, bond angles, and backbone torsion angles, as well as more complex features including atomic packing, hydrogen-bond networks, and surface accessibility. These analyses are commonly applied to both experimental structures and computational models in order to identify stereochemical inconsistencies or implausible features prior to downstream applications such as molecular dynamics simulations or virtual screening.
- Case application: In high-throughput modeling workflows, WHAT IF is frequently used as an initial validation step. For example, when large ensembles of models are generated, such as multiple conformations of a p53 mutant, it is impractical to subject all structures to computationally intensive molecular dynamics simulations. WHAT IF can be applied to screen models for major stereochemical violations, including severe atomic clashes or disallowed Ramachandran angles, allowing only physically plausible structures to be retained for further analysis (Vriend, 1990).

3.3.3.2 DALI

- Description: The DALI server is widely used for structural alignment and comparison of protein structures. It compares the 3D coordinates of a query protein (an experimental structure or a model) with structures deposited in the PDB by analyzing patterns of internal Cα-Cα distances. Because this approach does not rely on sequence similarity, DALI can identify proteins with related folds even in the absence of detectable sequence homology, supporting functional interpretation when sequence-based methods are uninformative.
- Case application: In practice, DALI can be applied to support functional interpretation based on structural similarity rather than sequence identity. This is

particularly valuable when analyzing high-confidence AI models of "unknown" proteins. If an AlphaFold prediction for a novel cancer-associated protein yields a unique 3D fold, submitting this coordinate file to DALI allows for a global search against the PDB. DALI may reveal that the protein shares a fold with a known superfamily, such as kinases, despite having no sequence similarity. This "convergent evolution" insight immediately classifies the protein as a putative kinase, establishing it as a potential target for existing kinase inhibitors (Holm, 2020).

3.3.3.3 Pcons

- Description: Pcons is a consensus-based metaserver for evaluating the quality of protein models. The underlying principle is that if multiple different prediction servers (e.g., I-TASSER, Phyre2, and Robetta) all agree on the structure of a particular region, that region is likely to be modeled correctly. Pcons takes a set of models for a single target and calculates a "quality score" for each residue, based on how similar it is to the consensus of all other models. Pcons is used to increase confidence in models of difficult targets.
- Case application: Pcons is utilized to assess model reliability in difficult modeling scenarios, such as large multidomain tumor suppressors. When different prediction algorithms (e.g., I-TASSER vs. Phyre2) yield conflicting structures for specific regions, Pcons evaluates the consensus across multiple methods. By calculating a residue-specific quality score, it allows researchers to distinguish between high-confidence regions (where all models agree) and unreliable, potentially disordered regions. This stratification prevents the overinterpretation of low-confidence structural features during drug screening or functional analysis (Larsson et al., 2009).

3.3.4 Databases and integrated workbenches

These resources either store precomputed models or provide integrated environments to streamline the modeling process.

3.3.4.1 ModBase

- Description: ModBase is one of the largest and oldest databases of precomputed comparative protein models (Pieper et al., 2014). It is maintained by the Sali lab (creators of MODELLER) and automatically generates models for all sequences in the UniProt database that have a detectable template in the PDB. It is important to distinguish this from the AFDB; ModBase models are exclusively homology models, and their quality is entirely dependent on the best available template. Before AlphaFold, ModBase was the primary resource for obtaining a structural model quickly.

- Case application: ModBase serves as a primary resource for large-scale systems biology studies where generating individual models is impractical. For conducting proteome-wide analyses—such as mapping all known mutations across the entire human kinase family—ModBase provides immediate access to thousands of precomputed comparative models. This repository is particularly useful for retrieving template-based models that may include ligand or cofactor coordinates inherited from the template, information that is often absent in ab initio AI predictions (Pieper et al., 2014).

3.3.4.2 Molecular interactive development environment

- Description: Molecular interactive development environment (MolIDE) is a graphical user interface application developed to support and integrate different steps of the homology modeling workflow (Ghorbani et al., 2016). It provides a single interface that links external tools for sequence alignment, secondary structure prediction, loop construction, and model refinement. By bringing these components together within one environment, MolIDE reduces the need for command-line interaction and facilitates the use of comparative modeling approaches by researchers with limited computational training.
- Case application: In cancer-related research, MolIDE is used to support structural modeling tasks by experimental biologists who may not routinely use scripting-based software. Its integrated interface allows users to perform template selection, alignment editing, and model construction within a single workflow. This setup enables relatively rapid generation and visualization of 3D models of mutant proteins, which can assist in preliminary structural analysis and hypothesis development (Ghorbani et al., 2016).

3.3.4.3 Biskit (bioinformatics toolkits in Python)

- Description: Biskit is a Python-based library designed for structural bioinformatics analysis, with functionality that overlaps partially with Biopython while focusing more strongly on structure-related workflows (Grünberg et al., 2007). It includes modules for tasks such as parsing PDB files, coordinating homology modeling procedures, analyzing molecular dynamics trajectories, and performing protein–-protein docking. Biskit is often used to link independent software tools, such as MODELLER or docking programs, into reproducible computational workflows.
- Case application: In systems-level cancer studies, Biskit is applied to support automated structural bioinformatics pipelines. For example, in analyses involving large sets of Table 3.1 variants, such as multiple TP53 missense mutations, Biskit can be used to manage scripted workflows that retrieve structural data, execute comparative modeling steps, and analyze simulation outputs. This approach

Table 3.1 A summary of key bioinformatics tools in cancer protein modeling.

Tool category	Tool name	Primary function	Specific application in cancer research
AI/deep learning	AlphaFold (Jumper et al., 2021)	High-accuracy ab initio 3D structure prediction from sequence via deep learning.	Modeling novel oncogenes, tumor suppressors, and cancer-specific fusion proteins (e.g., BCR–ABL) lacking experimental structures (Meng et al., 2025; Kumar and Kim, 2024).
	trRosetta/RoseTTAFold (Baek et al., 2021)	Deep learning-based prediction of interresidue geometry (6D) to build 3D models.	Alternative to AlphaFold; strong in predicting the structure of protein–protein complexes, such as oncoprotein-inhibitor pairs.
Template-based modeling	MODELLER (Webb and Sali, 2016)	Standalone software for homology modeling by "satisfaction of spatial restraints."	Building high-resolution, user-controlled models of mutants (e.g., drug-resistant kinases) for detailed simulation or SBDD.
	SWISS-MODEL (Waterhouse et al., 2018)	Automated web server for homology modeling; strong in modeling complexes and ligands.	Rapidly modeling mutations in known protein families (e.g., BRAF V600E) to visualize impact on active sites or dimer interfaces.
	I-TASSER (Yang and Zhang, 2015)	Hybrid threading and ab initio fragment assembly server for structure and function prediction.	Modeling multidomain proteins where templates are incomplete, used to study mutations in tumor suppressors like RASSF2.
	Phyre2 (Kelley et al., 2015)	Homology/analogy recognition server using sensitive profile-profile matching.	Fold recognition and model building; used to generate the 3D model of the CYLD tumor suppressor for mutation analysis (Roy et al., 2024).
	Robetta/Rosetta (Baek et al., 2021)	Server/software suite for ab initio, TBM, and (critically) protein design.	De novo design of peptide-based inhibitors to block oncogenic protein–protein interactions, such as p53-MDM2.

Template detection	HHpred (Zimmermann et al., 2018)	Highly sensitive remote homology detection using HMM–HMM comparison.	The essential first step for finding templates for novel cancer proteins in the "twilight zone" (<30% identity), enabling functional annotation.
	3D-PSSM/FFAS (Jaroszewski et al., 2005)	Classic threading/fold recognition servers for functional annotation.	Used in consensus-based approaches to confidently assign a fold (e.g., "kinase-like") to a new protein of interest from a cancer screen.
Model validation	WHAT IF (Vriend, 1990)	Comprehensive software suite for stereochemical and geometrical validation of models.	Critical quality control step to filter out physically implausible models before computationally expensive docking or simulation.
Analysis & visualization	DALI (Holm, 2020)	Server for comparing 3D protein structures (sequence-independent).	Functional annotation by structural homology; e.g., identifying that an AlphaFold model of an "unknown" protein has a kinase fold.
Database/ workbench	ModBase (Pieper et al., 2014)	Database of precomputed comparative (homology) models.	A repository for rapidly obtaining homology models, often used in large-scale bioinformatics pipelines.
	MolIDE/Biskit (Grünberg et al., 2007)	GUI/Python library for integrating and automating modeling workflows.	Accelerating modeling tasks for biologists (MolIDE) or for building complex, large-scale analysis pipelines (Biskit).

reduces manual intervention and supports reproducible analysis across large datasets (Grünberg et al., 2007).

3.4 Key applications in cancer biology and therapeutics

The structural models generated by the tools discussed above serve as atomic-level representations that can be used to examine disease mechanisms and to guide the development of potential interventions.

3.4.1 Elucidating the structural impact of cancer-associated mutations

The most direct and widespread application of protein modeling in oncology is to understand the consequences of the thousands of somatic and germline mutations identified in patient tumors. The Cancer Genome Atlas and other large-scale sequencing projects have revealed a vast landscape of "variants of uncertain significance"—missense mutations whose role in tumorigenesis is unknown. Modeling provides a direct, mechanistic path to stratify them.

A clear case study is the computational analysis of the CYLD gene, which encodes a tumor suppressor protein (Roy et al., 2024). This study serves as a model workflow:

1. Variant collection: Researchers identified 446 nsSNPs (missense mutations) in the CYLD gene from public databases.
2. Deleteriousness prediction: They first used a panel of sequence-based tools (e.g., SIFT, PolyPhen-2, PROVEAN) to predict which variants were likely to be "deleterious" versus "benign." This shortlisted 18 high-risk variants.
3. Structural modeling: Lacking an experimental structure, they used the Phyre2 server to generate a high-quality 3D homology model of the CYLD protein.
4. Mechanistic analysis: This 3D model was the key. They mapped the 18 high-risk mutations onto the structure and used further computational tools (e.g., I-Mutant, SDM) to analyze the physicochemical reasons for the damage. They could demonstrate, for example, that the G572A mutation was predicted to destabilize the protein by breaking essential hydrogen bonds, while the R395G mutation disrupted a critical salt bridge.

This in silico workflow allowed variants to be prioritized based on predicted structural and physicochemical effects, providing a rationale for subsequent functional investigation of selected mutations. Similar multistep strategies have been applied to other cancer-associated proteins, including the p53 tumor suppressor, where mutations within the DNA-binding domain frequently alter structural stability and functional integrity (Kastenhuber and Lowe, 2017).

3.4.2 Structure-based drug design and virtual screening

Structural models are commonly used in structure-based drug design workflows. Once a 3D model of an oncoprotein has been generated, for example using AlphaFold or MODELLER, it can serve as a target for virtual screening studies. These workflows typically employ molecular docking programs, such as AutoDock Vina (Trott and Olson, 2010), to computationally evaluate libraries of chemical compounds for potential binding. Interaction analysis tools, including LigPlot+, are then used to examine predicted ligand–protein contacts, supporting the prioritization of compounds for further experimental evaluation (Laskowski and Swindells, 2011).

3.4.3 Modeling protein–protein interactions as cancer targets

Many oncogenic proteins, such as MYC, exert their effects through PPIs rather than through well-defined ligand-binding pockets, which complicates conventional drug discovery approaches (Kozakov et al., 2017). Computational modeling methods, including AlphaFold-Multimer, RoseTTAFold, and docking platforms such as ClusPro (Kozakov et al., 2017), have been used to investigate the structural features of these interfaces.

One well-studied example is the interaction between p53 and MDM2, in which MDM2 negatively regulates p53 activity. Structural modeling of this interaction revealed a defined binding region on MDM2 that could be targeted by inhibitory molecules. These insights informed the development of peptide-based inhibitors, including stapled peptides generated using Rosetta, as well as small-molecule compounds such as nutlins (Arkin and Wells, 2004), which disrupt the interaction and restore p53 activity.

3.4.4 Understanding aberrant fusion proteins in oncology

Fusion proteins represent a distinct class of oncogenic drivers formed through chromosomal rearrangements that join segments of two separate genes. Examples include BCR–ABL in chronic myeloid leukemia and EML4–ALK in non-small-cell lung cancer. Although these fusion proteins are clinically relevant drug targets, their structural characterization is challenging due to their size, conformational flexibility, and limited experimental data.

- Case application: As discussed by Kumar et al. (2024), AI-based structure prediction tools such as AlphaFold and RoseTTAFold have been applied to model full-length fusion proteins. In the case of BCR–ABL, predicted structures suggest that the BCR-derived region promotes dimerization and constitutive activation of the ABL kinase domain. Similarly, models of EML4–ALK indicate that the EML4 component contributes to kinase activation through oligomerization. These structural interpretations support functional hypotheses and can inform the design of inhibitors that account for fusion-specific conformations.

3.5 Current challenges and methodological limitations

Despite recent advances, computational protein models require careful interpretation and validation before being used to draw biological conclusions. Computational protein structures provide useful insights, but their limitations should be carefully considered when interpreting biological function.

Static structures versus dynamic ensembles: Most predicted structures, including those generated by AlphaFold, represent a single static conformation rather than the full range of dynamic states adopted by a protein. Real proteins are dynamic, flexible machines that breathe, bend, and adopt multiple conformations to function. AI predictors often fail to capture this inherent flexibility and cannot, by default, model how a protein's structure changes upon binding to a ligand (induced fit) or a partner protein (Pons et al., 2024).

1. Insensitivity to mutations: While AI models are trained on wild-type sequences, they are surprisingly insensitive to the effects of single-point mutations (Qiu et al., 2024). Because the prediction is so heavily weighted by the coevolutionary signal from thousands of other sequences in the MSA, the model will often "ignore" the single mutation and produce a structure nearly identical to the wild-type, even when that mutation is known to be destabilizing. This is a major bottleneck for personalized medicine.
2. Absence of nonprotein components: Models are trained on polypeptide chains. They do not intrinsically predict the presence or position of essential nonprotein components. This includes:

- Ligands and cofactors: AlphaFold does not predict where a drug, a metabolite, or a metal ion (e.g., zinc in a zinc-finger domain) will bind.
- Post-translational modifications (PTMs): Cancer-associated signaling pathways are influenced by PTMs such as phosphorylation, ubiquitination, and glycosylation (Qiu et al., 2024). Most current structure-prediction models do not explicitly account for these modifications, despite their potential to affect protein structure and function.

3. Modeling intrinsically disordered proteins: A substantial proportion of the human proteome, particularly proteins involved in signaling and transcription such as p53 and MYC, contains intrinsically disordered regions (Qiu et al., 2024). Although AlphaFold's pLDDT scores are useful for identifying regions of low structural confidence, these predictions do not capture the ensemble of conformations adopted by flexible or disordered regions.

4. Accuracy of PPIs: Although tools such as AlphaFold-Multimer have expanded the scope of complex prediction, modeling large, transient, or multicomponent protein assemblies remains challenging. In such cases, predictive accuracy is variable and generally lower than that achieved for individual protein chains.

3.6 Future research outlook

Recent studies suggest that protein modeling research is increasingly extending beyond static representations of isolated protein structures. One area of active development involves the integration of structure prediction with molecular dynamics simulations, with the aim of capturing conformational flexibility and generating ensembles of protein states rather than single static models (Szczepski and Jaremko, 2025). Such approaches are particularly relevant for understanding allosteric regulation and transient or cryptic binding sites.

Parallel efforts focus on improving sensitivity to sequence variation by training predictive models to account for point mutations, ligand binding, and PTMs. In addition, there is growing interest in extending structure-based approaches from individual proteins to pathway-level analyses, linking structural biology with systems-level perspectives (Szczepski and Jaremko, 2025). These developments include the application of modeling tools to cancer-specific fusion proteins, which present unique structural and functional challenges (Kumar and Kim, 2024). Together, these directions support the gradual development of integrated computational pipelines for drug discovery that combine target identification, binding-site analysis, and compound design (Qiu et al., 2024; Ren et al., 2023).

Beyond these established research directions, there is increasing discussion around the potential for more personalized applications of protein modeling. One proposed concept is the use of patient-specific molecular data to inform structural analyses, in which tumor sequencing results are combined with computational modeling to examine the potential effects of individual mutations on protein structure and drug binding (Szczepski and Jaremko, 2025). In principle, such approaches could allow in silico evaluation of therapeutic options by comparing predicted interactions between mutant proteins and available drugs (Gendoo, 2020). While these ideas remain largely exploratory, they illustrate how structural modeling may contribute to precision oncology research in the future.

To achieve this, the field must pivot from static prediction to dynamic perturbation. The key question will no longer be, "What does this protein look like?" but rather, "What happens to this protein's dynamic behavior when this specific mutation occurs, or when this specific drug binds?" This requires a new generation of tools that are sensitive enough to capture the subtle energetic and conformational shifts caused by a single atomic change.

A further challenge concerns the interpretability of deep learning-based prediction methods. Many current models function as complex systems with limited transparency, which complicates the biological interpretation of their outputs. Ongoing work in explainable artificial intelligence seeks to address this limitation by developing models that can provide interpretable rationales for predicted structures, such as identifying key residue interactions that influence conformational outcomes. Improved interpretability is expected to support error identification, increase confidence in predictions, and facilitate closer integration between data-driven modeling and hypothesis-driven molecular biology.

3.7 Conclusion

Bioinformatics approaches to protein structure modeling have evolved substantially over recent decades and now represent a central component of cancer research. Classical template-based methods, such as MODELLER and SWISS-MODEL, continue to be relevant for the detailed analysis and refinement of proteins with well-characterized structural homologs. At the same time, advances in deep learning-based prediction, particularly through tools such as AlphaFold and RoseTTAFold, have expanded the range of proteins that can be examined structurally, including fusion proteins and domains for which experimental data remain limited.

The availability of predicted 3D structures has facilitated the interpretation of cancer-associated mutations by providing a structural context for changes in stability, intermolecular interactions, and binding-site geometry. These models also support structure-based drug discovery workflows by enabling computational analysis of potential ligand interactions before experimental validation (Qiu et al., 2024; Ren et al., 2023). When combined with model validation tools (e.g., WHAT IF), structural comparison platforms (e.g., DALI), and interaction analysis software (e.g., LigPlot+), protein modeling contributes to an integrated computational framework that links sequence variation to functional and therapeutic hypotheses.

Collectively, these developments underscore the growing role of computational structure prediction in addressing the molecular complexity of cancer, while also highlighting the continued need for careful validation and biological interpretation as these methods are applied to increasingly complex systems.

References

Arkin, M.M.R., Wells, J.A., 2004. Small-molecule inhibitors of protein-protein interactions: progressing towards the dream. Nature Reviews. Drug Discovery 3 (4), 301–317. Available at https://doi.org/10.1038/nrd1343, http://www.nature.com/nrd/index.html.

Baek, M., DiMaio, F., Anishchenko, I., Dauparas, J., Ovchinnikov, S., Lee, G.R., et al., 2021. Accurate prediction of protein structures and interactions using a three-track neural network. Science (New York, N.Y.) 373 (6557), 871–876. Available at https://doi.org/10.1126/science.abj8754, https://science.sciencemag.org/content/373/6557/871.full.

Berman, H.M., Westbrook, J., Feng, Z., Gilliland, G., Bhat, T.N., Weissig, H., et al., 2000. P.E. Bourne, the protein data bank. Nucleic Acids Research 28 (1), 235–242. Available at https://doi.org/10.1093/nar/28.1.235, https://academic.oup.com/nar/issue.

Bowie, J.U., Lüthy, R., Eisenberg, D., 1991. A method to identify protein sequences that fold into a known three-dimensional structure. Science (New York, N.Y.) 253 (5016), 164–170. Available at www.sciencemag.org.

Cressey, D., Callaway, E., 2017. Cryo-electron microscopy wins chemistry Nobel. Nature 550 (7675). Available at https://doi.org/10.1038/nature.2017.22738.

Gendoo, D.M.A., 2020. Bioinformatics and computational approaches for analyzing patient-derived disease models in cancer research. Computational and Structural Biotechnology Journal 18, 375–380. Available at https://doi.org/10.1016/j.csbj.2020.01.010, www.csbj.org.

Ghorbani, M., Karimi, H., Ghorbani, F., 2016. MolIDE: a server for protein structure modeling. International Journal of Scientific Research in Science and Technology 2 (4), 75–79.

Grünberg, R., Nilges, M., Leckner, J., 2007. Biskit - A software platform for structural bioinformatics. Bioinformatics (Oxford, England) 23 (6), 769–770. Available at https://doi.org/10.1093/bioinformatics/btl655, http://bioinformatics.oxfordjournals.org/.

Guo, S.B., Meng, Y., Lin, L., Zhou, Z.Z., Li, H.L., Tian, X.P., et al., 2024. Artificial intelligence alphafold model for molecular biology and drug discovery: a machine-learning-driven informatics investigation. Molecular Cancer 23 (1). Available at https://doi.org/10.1186/s12943-024-02140-6, https://molecular-cancer.biomedcentral.com.

Hasan, M.R., Paul, B.K., Ahmed, K., Bhuyian, T., 2020. Design protein-protein interaction network and protein-drug interaction network for common cancer diseases: a bioinformatics approach. Informatics in Medicine Unlocked 18, 100311. Available at https://doi.org/10.1016/j.imu.2020.100311.

Holm, L., 2020. DALI and the persistence of protein shape. Protein Science: a Publication of the Protein Society 29 (1), 128–140. Available at https://doi.org/10.1002/pro.3749.

Huang, J., Mao, L., Lei, Q., Guo, A.-Y., 2024. Bioinformatics tools and resources for cancer and application. Chinese Medical Journal 137 (17), 2052–2064. Available at https://doi.org/10.1097/CM9.0000000000003254.

Jaroszewski, L., Rychlewski, L., Li, Z., Li, W., Godzik, A., 2005. FFAS03: a server for profile–profile sequence alignments. Nucleic Acids Research 33 (Web Server issue).

Jumper, J., Evans, R., Pritzel, A., Green, T., Figurnov, M., Ronneberger, O., et al., 2021. Highly accurate protein structure prediction with AlphaFold. Nature 596 (7873), 583–589. Available at https://doi.org/10.1038/s41586-021-03819-2.

Kastenhuber, E.R., Lowe, S.W., 2017. Putting p53 in context. Cell 170 (6), 1062–1078.

Kelley, L.A., Mezulis, S., Yates, C.M., Wass, M.N., Sternberg, M.J.E., 2015. The Phyre2 web portal for protein modeling, prediction and analysis. Nature Protocols 10 (6), 845–858. Available at https://doi.org/10.1038/nprot.2015.053.

Kendrew, J.C., Bodo, G., Dintzis, H.M., Parrish, R.G., Wyckoff, H., Phillips, D.C., 1958. A three-dimensional model of the myoglobin molecule obtained by x-ray analysis. Nature 181 (4610), 662–666. Available at https://doi.org/10.1038/181662a0.

Kihara, D., Yang, Y.D., Hawkins, T., 2006. Bioinformatics resources for cancer research with an emphasis on gene function and structure prediction tools. Cancer Informatics 2, 25–35. Available at https://doi.org/10.1177/117693510600200020, http://www.la-press.com/CI-2-Kihara(Pr).pdf.

Kozakov, D., Hall, D.R., Xia, B., Porter, K.A., Padhorny, D., Yueh, C., et al., 2017. The ClusPro web server for protein-protein docking. Nature Protocols 12 (2), 255–278. Available at https://doi.org/10.1038/nprot.2016.169, http://www.natureprotocols.com/.

Kühlbrandt, W., 2014. The resolution revolution. Science (New York, N.Y.) 343 (6178), 1443–1444. Available at https://doi.org/10.1126/science.1251652.

Kumar, H., Kim, P., 2024. Artificial intelligence in fusion protein three-dimensional structure prediction: review and perspective. Clinical and Translational Medicine 14 (8). Available at https://doi.org/10.1002/ctm2.1789.

Larsson, P., Skwark, M.J., Wallner, B., Elofsson, A., 2009. Assessment of global and local model quality in CASP8 using Pcons and ProQ. Proteins 77 (Suppl 9), 167–172. Available at https://doi.org/10.1002/prot.22476.

Laskowski, R.A., Swindells, M.B., 2011. LigPlot+: Multiple ligand-protein interaction diagrams for drug discovery. Journal of Chemical Information and Modeling 51 (10), 2778–2786. Available at https://doi.org/10.1021/ci200227u, http://pubs.acs.org/journal/jcisd8.

Meng, Y., Zhang, Z., Zhou, C., Tang, X., Hu, X., Tian, G., et al., 2025. Protein structure prediction via deep learning: an in-depth review. Frontiers in Pharmacology 16. Available at https://doi.org/10.3389/fphar.2025.1498662.

Pieper, U., Webb, B.M., Dong, G.Q., Schneidman-Duhovny, D., Fan, H., Kim, S.J., et al., 2014. ModBase, a database of annotated comparative protein structure models and associated resources. Nucleic Acids Research 42 (1), D336. Available at https://doi.org/10.1093/nar/gkt1144.

Pons, J.L., Reys, V., Grand, F., Moreau, V., Gracy, J., Exner, T.E., et al., 2024. @TOME 3.0: interfacing protein structure modeling and ligand docking. Journal of Molecular Biology 436 (17). Available at https://doi.org/10.1016/j.jmb.2024.168704, https://www.sciencedirect.com/science/journal/00222836.

Qiu, X., Li, H., Ver Steeg, G., Godzik, A., 2024. Advances in AI for protein structure prediction: implications for cancer drug discovery and development. Biomolecules 14 (3), 339. Available at https://doi.org/10.3390/biom14030339.

Ren, F., Ding, X., Zheng, M., Korzinkin, M., Cai, X., Zhu, W., et al., 2023. AlphaFold accelerates artificial intelligence powered drug discovery: efficient discovery of a novel CDK20 small molecule inhibitor. Chemical Science 14 (6), 1443–1452. Available at https://doi.org/10.1039/d2sc05709c, http://pubs.rsc.org/en/journals/journal/sc.

Roy, A.S., Feroz, T., Islam, M.K., Munim, M.A., Supti, D.A., Antora, N.J., et al., 2024. A computational approach for structural and functional analyses of disease-associated mutations in the human CYLD gene. Genomics & Informatics 22 (1). Available at https://doi.org/10.1186/s44342-024-00007-2.

Szczepski, K., Jaremko, Ł., 2025. AlphaFold and what is next: bridging functional, systems and structural biology. Expert Review of Proteomics 22 (2), 45–58. Available at https://doi.org/10.1080/14789450.2025.2456046.

Trott, O., Olson, A.J., 2010. AutoDock Vina: improving the speed and accuracy of docking with a new scoring function, efficient optimization, and multithreading. Journal of Computational Chemistry 31 (2), 455–461. Available at https://doi.org/10.1002/jcc.21334.

Varadi, M., Anyango, S., Deshpande, M., Nair, S., Natassia, C., Yordanova, G., et al., 2022. AlphaFold protein structure database: massively expanding the structural coverage of protein-sequence space with high-accuracy models. Nucleic Acids Research 50 (D1), D439. Available at https://doi.org/10.1093/nar/gkab1061.

Vriend, G., 1990. WHAT IF: a molecular modeling and drug design program. Journal of Molecular Graphics 8 (1), 52–56. Available at https://doi.org/10.1016/0263-7855(90)80070-v.

Waterhouse, A., Bertoni, M., Bienert, S., Studer, G., Tauriello, G., Gumienny, R., et al., 2018. SWISS-MODEL: homology modelling of protein structures and complexes. Nucleic Acids Research 46 (W1), W296. Available at https://doi.org/10.1093/nar/gky427.

Webb, B., Sali, A., 2016. Comparative protein structure modeling using MODELLER. Current protocols in bioinformatics 54. Available at https://doi.org/10.1002/cpbi.3.

Wüthrich, K., 1989. Protein structure determination in solution by nuclear magnetic resonance spectroscopy. Science (New York, N.Y.) 243 (4887), 45–50. Available at https://doi.org/10.1126/science.2911719.

Yang, J., Zhang, Y., 2015. I-TASSER server: new development for protein structure and function predictions. Nucleic Acids Research 43 (W1), W174. Available at https://doi.org/10.1093/nar/gkv342.

Zheng, H., Zhang, G., Zhang, L., Wang, Q., Li, H., Han, Y., et al., 2020. Comprehensive review of web servers and bioinformatics tools for cancer prognosis analysis. Frontiers in Oncology 10. Available at https://doi.org/10.3389/fonc.2020.00068.

Zimmermann, L., Stephens, A., Nam, S.Z., Rau, D., Kübler, J., Lozajic, M., et al., 2018. A completely reimplemented mpi bioinformatics toolkit with a new hhpred server at its core. Journal of Molecular Biology 430 (15), 2237–2243. Available at https://doi.org/10.1016/j.jmb.2017.12.007, https://www.journals.elsevier.com/journal-of-molecular-biology.

CHAPTER 4

Unraveling protein folding and structural dynamics

Faiza Iram[1], Ayesha Aiman[1,2], Luqman Ahmad Khan[2], and Asimul Islam[1]
[1]Centre for Interdisciplinary Research in Basic Sciences, Jamia Millia Islamia, New Delhi, India,
[2]Department of Biosciences, Jamia Millia Islamia, Jamia Nagar, New Delhi, India

4.1 Introduction

The integration of protein folding and dynamics provides a modern perspective in structural biology, enabling a deeper understanding of proteins' versatile nature. Proteins are involved in various biological processes, including catalysis (Kokkinidis et al., 2012), structural support (Morris et al., 2022), cell signaling (Westermarck et al., 2013), and molecular recognition (Persch et al., 2015). The functional adaptability of a protein relies on its compactly folded native conformation (Haliloglu and Bahar, 2015). Protein folding is a unique phenomenon through which a polypeptide chain folds into a compact and functional conformation, known as the native state of the protein (Shaitan, 2023). While the dynamic nature of proteins explains how these molecular entities fluctuate within and around their native conformation to perform biological functions (Wei et al., 2016). Both of these phenomena are governed by the same physical principles and together provide the foundation for understanding protein structure and its function (Gianni et al., 2021). Earlier, proteins were believed to follow "lock-and-key" or "induced-fit" models, based on allosteric regulation that emphasizes the static nature of proteins (Zhuravlev and Papoian, 2010). It has now been established that, even in the folded state, proteins are not rigid entities but exist as ensembles of conformations that shift across a wide range of timescales. This timescale can range from femtoseconds to seconds and beyond (Wei et al., 2016; Livesay, 2010). The characteristic dynamic motions allow proteins to explore a continuum of conformations (Micheletti, 2013). Furthermore, these motions are directly linked to basic biological processes such as enzyme catalysis, allostery, ligand recognition, and molecular signaling, as discussed earlier (Yang et al., 2014). Understanding these dynamic motions is crucial to elucidating the underlying molecular mechanisms of the biological processes. Additionally, it can be rationalized in disease-causing mutations, in the design of allosteric drugs, and in the engineering of stable or functionally enhanced proteins. Presently, the 3D structures of almost 2,300,000 proteins can be freely accessed in the protein data bank (Berman and Burley, 2025). This has become possible due to advances in experimental methods and computational approaches. Conventional techniques,

Advances in Structural Biology
DOI: https://doi.org/10.1016/B978-0-443-44988-8.00019-8

namely X-ray crystallography, can provide a high-resolution picture of a folded protein, while more advanced nuclear magnetic resonance (NMR) can characterize conformational ensembles and transition intermediates of a protein in solution. More recently, cryo-electron microscopy (cryo-EM) has emerged as a powerful tool for probing conformational changes and large-scale structural transitions in proteins (Vant et al., 2022). Among advanced computational approaches, molecular dynamics simulations (MDS) can provide insights into conformational transitions that are difficult to detect under in vitro conditions. MDS links structure, energetics, and function by quantifying changes in parameters such as the root mean square deviation, root mean square fluctuation, radius of gyration, and solvent-accessible surface area. These parameters can describe how proteins maintain structural integrity or undergo transitions essential for catalysis and regulation (Childers and Daggett, 2017). To complement experimental techniques such as NMR, X-ray crystallography, and cryo-EM, MD simulations offer a powerful approach for atomistic insights into the observed phenomena. All of the above-mentioned approaches comprehensively aid in deciphering the protein dynamics. In this regard, the present chapter aims to unravel the principles governing protein motions and to highlight the crucial experimental and computational tools used to explore protein dynamics. It further emphasizes the transition from a static, structure-centric paradigm to a dynamic ensemble perspective, in which protein function arises from a continuum of conformations that change across energy

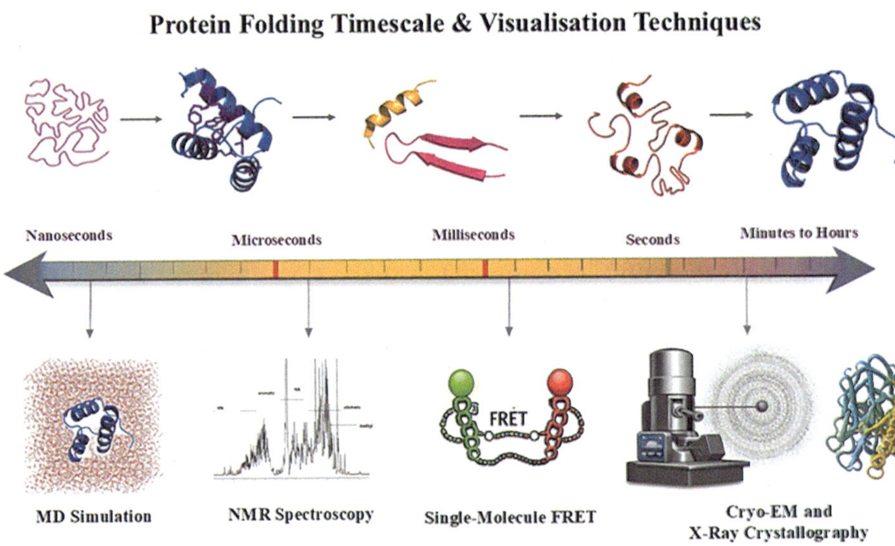

Figure 4.1 Diagrammatic representation of different biophysical and computational tools used to probe conformational changes in the protein along the protein folding timescales.

landscapes. The relationship between protein folding timescales and analytical tools is summarized in Fig. 4.1.

4.2 Protein folding problem

The dynamic process of protein folding talks about three fundamental questions. The first aspect is how a single chain of amino acids encodes the formation of the protein's compactly packed native state. In 1962 Christian B. Anfinsen showed that certain proteins can spontaneously refold in vitro to recover their native structure and biological activity even after chemical unfolding (Childers and Daggett, 2017). Then he proposed the thermodynamic hypothesis, which states that all the information required for protein folding is encoded in the protein's sequence (Levinthal, 1968). The second aspect of the folding problem focuses on the folding pathway. A typical polypeptide chain has an astronomically large number of possible conformations, making a random search for the native structure impossible on biological timescales. Yet, proteins fold within seconds to minutes, depending on their chain length, structure, and environmental conditions (Levinthal, 1968). This apparent paradox led to the development of modern free-energy landscape (FEL) theory. According to this theory, protein folding is a biased, stochastic process in which the protein descends towards the lowest-energy native state (Levinthal, 1968; Dinner et al., 2000). The protein achieves this through innumerable number of possible pathways, which are directed by intramolecular interactions. Instead of folding along one fixed pathway, a protein folds through ensembles of routes that funnel toward the native basin (Wang et al., 2012). This unique tendency of proteins has begun as a result of evolutionary pressures that favor energetically favorable configurations. It necessarily preserves the degree of flexibility required for proper protein function (Dutta et al., 2022). The third component of the protein folding problem is the prediction challenge: whether a protein's 3D structure and dynamic behavior can be determined directly from its amino-acid sequence, without relying on experimental structural data (Dill et al., 2007). In addition, if a sequence truly encodes structure, then, in principle, it should be possible to computationally decode that information. Early efforts used physics-based models, in which atomic interactions were described by empirical force fields, and folding was simulated by molecular dynamics, but the huge conformational space and the long timescales involved made accurate prediction difficult for all but the smallest proteins. Over time, hybrid strategies emerged that combined physical principles with statistical knowledge extracted from known protein structures (Seffernick and Lindert, 2020), enabling researchers to recognize recurring structural motifs and fold topologies. In recent times, advancements in machine- learning-based approaches have significantly enhanced predictive accuracy by uncovering complex relationships between amino-acid sequence and 3D structure from large structural databases (Straiton, 2023). However, the overall process of the prediction problem is not only restricted to identifying a single static fold, but rather encompasses variables, viz., conformational flexibility, alternative conformations, and folding pathways that ultimately direct the protein function. This progress in prediction directly enables practical advances in structural biology, drug discovery, and protein design by providing access to structural and dynamic information.

4.3 Thermodynamics and kinetics of protein folding

4.3.1 Thermodynamics

Protein folding is a thermodynamic process driven by the tendency of a system to minimize its free energy (Yang et al., 2013). As discussed in the previous section, a newly synthesized polypeptide chain explores its many possible conformations. It does not do so randomly; it moves on to a complex FEL shaped by the equilibrium of stabilizing interactions (Rose et al., 2006). The native form of a protein corresponds to the ensemble of conformations with the lowest Gibbs free-energy under physiological conditions (Cohen and Pielak, 2017). It is considered thermodynamically favored over the unfolded ensemble (Sorokina et al., 2022). Several physical forces contribute to stabilizing the folded state of a protein. Among all of these forces, the hydrophobic forces play a dominant role. Nonpolar side chains of amino acids are energetically unfavorable when exposed to water. Consequently, these amino acids move into the interior of the protein during folding. As a result, reducing the exposure of surrounding water molecules to these amino acids provides a major thermodynamic driving force for core formation of the protein. This hydrophobic collapse helps to define the overall topology of the protein (van Dijk et al., 2015). Furthermore, hydrogen bonding between the peptide backbone and polar side chains stabilizes the folded state of proteins (Powers et al., 2005). The formation of H-bonds between side chains adds another layer of precision, structural stability (Powers et al., 2005), and ensures that the final folded structure is well organized. Moreover, Van der Waals interactions arise from the close packing of atoms in the protein interior (Roth et al., 1996). Although these forces are individually weak, they become collectively significant when residues are precisely arranged. Additionally, electrostatic interactions, including salt bridges between oppositely charged residues, can further stabilize folded structures (Pace et al., 2009). The strength of these interactions depends strongly on the local environment, such as the dielectric constant and solvent exposure. Additionally, in some proteins, disulfide bonds provide further stabilization. These covalent interactions between cysteine residues reduce the entropy of the unfolded state and help stabilize the native folded state (Matsumura et al., 1989; Pace et al., 2014). Among different covalent and noncovalent interactions, disulfide bonds tend to have an important role, especially in extracellular proteins that function in oxidizing environments. Although these stabilizing forces are present, the folded state of most proteins is only marginally more stable than the unfolded state. The free energy difference between these two states is about 5–15 kcal mol^{-1} (Magliery, 2015). Due to the dynamic nature, proteins can undergo conformational changes required for function. In this way, the thermodynamics of protein folding provide both stability and adaptability. This enables proteins to function reliably while remaining responsive to the needs of the cellular environment.

4.3.2 Kinetics

Protein folding is a time-dependent process in which a polypeptide chain moves across its FEL. During this process, the crosses barriers that separate unfolded, intermediate, and native states (Englander and Mayne, 2014). While thermodynamics determines whether a protein can fold into a stable native state, kinetics determines how fast and by what routes folding occurs. The time span of protein folding follows an extraordinary range, from microseconds for small, fast-folding proteins to seconds or longer for larger and multidomain systems. Earlier, it was believed that protein folding follows a single well-defined path, but it is now clear that most proteins fold through ensembles of pathways. Primarily, proteins follow two-state kinetics, where only unfolded and native states are significantly populated, and folding appears as a single cooperative transition (Malhotra and Udgaonkar, 2016). In more complex proteins, intermediate states become transiently or stably populated. These intermediates may represent partially folded domains, collapsed but disordered structures, or near-native conformations that lack final compact conformation or native topology (Veitshans et al., 1997). The presence of intermediates can slow down the process of protein folding, introduce kinetic traps, or increase the likelihood of misfolding and aggregation (Lu et al., 2022) A central concept in folding kinetics is the transition state ensemble, which represents the collection of high-free-energy conformations that must be crossed for folding to progress (Ding et al., 2002). Rather than being a single structure, the transition state is an ensemble of conformations that share similar energetic and structural features. Experimental approaches such as φ-value analysis, combined with mutational studies, provide insight into which regions of a protein are structured in the transition state (Fersht and Sato, 2004). Therefore they play key roles in determining folding rates. Several conceptual models (Fig. 4.2) have also been proposed to describe how folding proceeds along these landscapes. In nucleation-condensation models (Table 4.1), folding begins with the formation of a small, cooperative nucleus of native-like structure. This is followed by the rapid folding and joining of the rest of the protein. In addition, framework models emphasize the early formation of secondary-structure elements (Table 4.1). Later, these elements assemble into a compactly packed native form. Hydrophobic collapse models propose an initial nonspecific compaction driven by

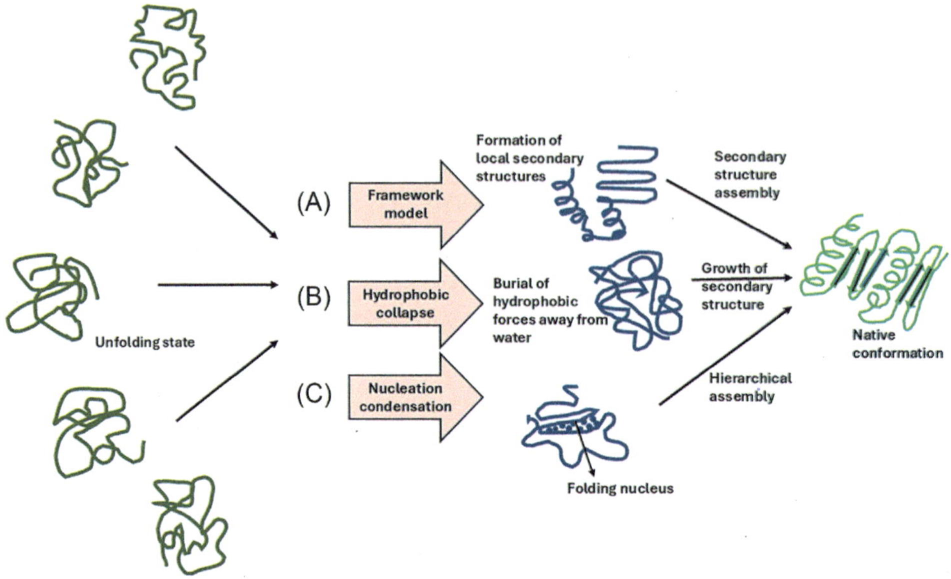

Figure 4.2 Various types of protein folding models: (A) framework model, (B) hydrophobic collapse model, and (C) nucleation model.

hydrophobic interactions (Table 4.1). Subsequently, rearrangements refine the native structure. These models are not mutually exclusive. Rather, they highlight different folding features that may dominate under different conditions or for different proteins. During protein folding, intermediates often arise. Some of these intermediates are on-pathway, indicating that they are part of the correct folding pathway that facilitate efficient folding. Others are off-pathway, representing misfolded conformations that must partially or fully unfold before the correct folding. Off-pathway intermediates are particularly important because they can slow protein folding and increase the likelihood of aggregation (Bhatia and Udgaonkar, 2022).

4.4 The energy landscape of protein folding

The conventional kinetic models used to study protein folding help us understand how different proteins fold under optimal and suboptimal conditions. Each of these models demonstrates a unique facet of this dynamic process. Some proteins show the formation of key secondary-structure elements that eventually aid in the formation of tertiary structures (Rehman et al., 2024). On the other hand, some follow the generation of a cooperative nucleus first and later the formation of native contacts (Shao and Zhu, 2019). To add, hydrophobic collapse can also lead to the formation of precise secondary and tertiary structures during protein folding (Jamadagni et al., 2011). Initially, it was

Table 4.1 Major theoretical models of protein folding, their mechanistic interpretations, and inherent limitations.

Protein folding model	Proposed explanation	Limitations	References
Framework (sequential) model	Secondary structures (helices/sheets) form first; these elements then pack together to build the tertiary native fold	The folding order is not well explained; some proteins don't show early stable secondary structure before collapse; it doesn't handle concerted tertiary formation	(Arai, 2018; Baldwin, 1989)
Hydrophobic collapse model	Folding is driven by rapid nonspecific collapse of the chain due to hydrophobic interactions; secondary structure forms after collapse	Not able to explain cases where local structures form early; doesn't specify how correct native packing emerges from nonspecific collapse	(Arai, 2018; Sadqi, Lapidus and Muñoz, 2003; Cheung, García and Onuchic, 2002)
Nucleation-condensation model	A small folding nucleus with some native contacts forms early, and the rest of the structure "condenses" around it in a cooperative way	Determining the nucleus-stage is experimentally not possible; not applicable to large multidomain or intrinsically disordered proteins	(Nölting and Agard, 2008)
Folding funnel/ energy-landscape theory	Folding is guided by a multidimensional energy landscape shaped like a funnel; many microscopic paths lead downhill to the native state	Very abstract; difficult to convert into simple predictive kinetics; relies on a statistical viewpoint rather than concrete mechanistic steps	(Onuchic, Luthey-Schulten and Wolynes, 1997; Bryngelson et al., 1995)

believed that a protein chooses one mechanism over the other. However, it has now become clear that these theories do not provide mutually exclusive explanations and work on the same physical phenomenon. Each model demonstrates the state that may dominate in different stages of the folding process or for different protein sequences. The broader understanding of the process eventually led to the FEL framework Fig. 4.3. This makes the kinetics and thermodynamics of protein folding together into one

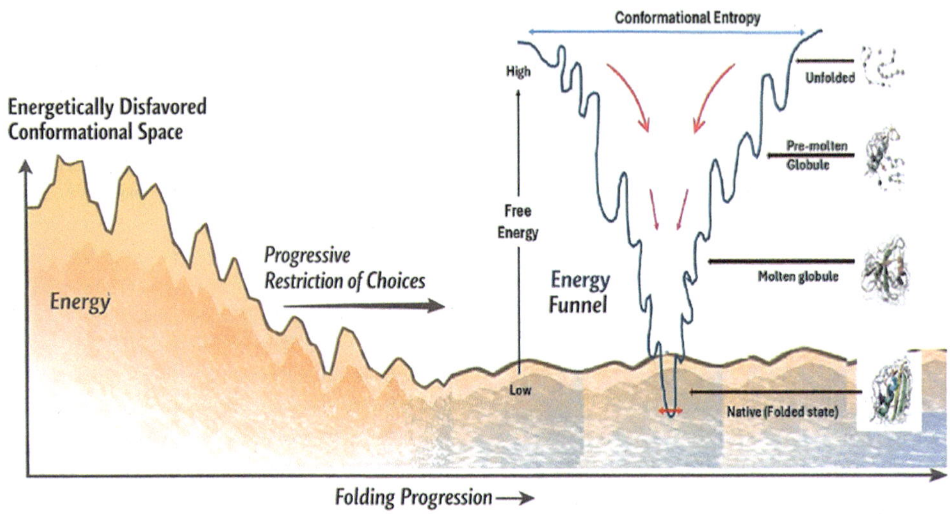

Figure 4.3 Illustration of the free-energy landscape model showing protein folding through continuing reduction in the conformational entropy towards the thermodynamically stable native state.

cohesive model. Henceforth, it can be said that a protein does not follow the predetermined path. Instead, a protein explores many possible conformations across a multidimensional FEL that eventually leads it toward the native state (Mallamace et al., 2016). In addition, the FEL model also explains how proteins resolve Levinthal's paradox (Martínez, 2014). It further sheds light on the heterogeneous folding pathways, intermediates, and kinetic traps that arise from landscape ruggedness (Martínez, 2014). Levinthal pointed out that a protein cannot fold by randomly sampling all possible conformations, because doing so would take an astronomically long time. In the energy-landscape model, this paradox disappears because it states that folding is not an unbiased search. The free-energy surface is shaped so that interactions consistent with the native structure are energetically favored from the very beginning of folding (Liu et al., 2012). These forces bias the conformational search toward compact, partially ordered states and progressively restrict the number of accessible configurations. As a result, large regions of conformational space are energetically disfavored, and the protein is guided downhill toward its native state through progressively restricted choices, allowing folding to occur rapidly and reliably on biological timescales. Hence, it can be said that protein folding, stability, and function are interrelated phenomena on a shared free-energy surface. Even in the properly folded state, a protein continues to explore its energy landscape across a wide range of timescales. It can happen from rapid atomic vibrations to slow and large-scale conformational transitions (Wei et al., 2016). These motions underpin enzyme catalysis, molecular recognition, allostery, and regulation.

4.5 Protein dynamics: motion and function

The marginal stability of proteins imparts a dynamic nature and allows them to adopt multiple conformations (Teilum et al., 2011). It further facilitates interactions with other biomolecules that are responsible for their diverse biological roles. As stated earlier, protein motions occur over a wide range of timescales, spanning many orders of magnitude and driving different structural features of the protein. The fastest movement occurs at the scale of femtoseconds to picoseconds, which includes bond vibrations and rapid side-chain rotations. These ultrafast motions reflect the intrinsic flexibility of chemical bonds and contribute to the local adaptability of a protein structure (Ojeda-May et al., 2021). On nanosecond to microsecond timescales, proteins can undergo larger-amplitude motions such as loop rearrangements and fluctuations of secondary-structure elements. This phenomenon is known as the domain "breathing" (Mariño Pérez et al., 2022). Several biological processes, namely ligand access to buried binding sites, communication between distant regions of a protein, and early steps in allosteric regulation, are influenced by these motions (Kumar and Jernigan, 2021). Moreover, at the timescale of milliseconds to seconds, proteins switch between a range of functional states and complete folding or unfolding processes. Such slow motions from milliseconds to seconds are often directly coupled with biological function. For instance, enzymes cycle between open and closed conformations or in signaling proteins that act as molecular switches (Bryan and Orban, 2010). Together, these motions form a hierarchy in which fast, local fluctuations enable and support slower, larger-scale rearrangements. This creates a dynamic continuum rather than discrete categories of motion. In enzyme catalysis, conformational changes facilitate substrate binding by transiently opening access pathways, aligning catalytic residues, and stabilizing reaction intermediates (Callender and Dyer, 2015; Schwartz, 2023). The phenomenon of allostery further underscores the functional importance of protein dynamics. The binding of a ligand at one site can modulate the conformational ensemble at another side of the protein (Campitelli et al., 2025). In addition, molecular recognition also relies greatly on dynamics (Boehr et al., 2009). Proteins often exist in multiple conformations even before binding a ligand. This allows recognition to proceed through conformational selection and is frequently followed by an induced-fit mechanism (Mamidi et al., 2022). Dynamic flexibility thus enables both specificity and adaptability in molecular interactions.

4.6 Experimental and computational approaches to study protein folding and dynamics

To understand the complex process of protein folding, it is imperative to have a deeper insight. For this, sophisticated instruments are required in a bid to decipher the intricate

3D structures of proteins and their dynamic motions across a diverse array of spatial and temporal scales. As of now, no single experimental or computational approach can serve this purpose. Instead, the current picture of protein folding and dynamics has emerged from the integration of complementary approaches, each of which reveals a different aspect of the process. In this section, we will provide a comprehensive overview of the experimental and computational tools used to study protein folding and dynamics, highlighting how each method provides a different aspect of structure and motion, and how their integration has shaped the modern understanding of proteins as dynamic systems.

4.6.1 Ensemble spectroscopic techniques

The experimental ensemble techniques examine protein folding and dynamics by analyzing the average signal from a large population of molecules (ensemble) in controlled conditions. These techniques help to understand the underlying physical principles of protein folding. Moreover, it offers quantitative insights into the kinetics, stability, and thermodynamics of the process. The most widely used ensemble techniques are discussed in the present section. (1) Circular dichroism (CD) spectroscopy (Glick, 2006). CD spectroscopy can be particularly employed to monitor changes in the secondary structure of proteins under optimal and suboptimal conditions (Zavrtanik et al., 2024). The changes in CD signals indicate the folding and unfolding transitions of a protein (Linhares and Ramos, 2023; Ramos et al., 2004). In addition, by studying CD signals as a function of temperature, pH, or chemical denaturant concentration, we can determine melting temperatures (T_m), folding free energies, and the cooperativity of folding transitions (Greenfield, 2006; Benjwal et al., 2006; Woody and Koslowski, 2002; Aiman et al., 2025). CD spectroscopy can also be used to rapidly assess the stability of mutants or engineered variants of a protein (Oyama et al., 2026). (2) Fluorescence spectroscopy provides an additional depth to the process of protein folding (Kumar et al., 2023). Proteins contain aromatic residues that absorb light at specific wavelengths and denote the configurational integrity of a protein. Specifically, tryptophan, as its fluorescence changes markedly with the change in its local environment (Fossum et al., 2023). During unfolding, when hydrophobic residues earlier buried in the protein core become exposed, their fluorescence intensity and emission wavelength change. This serves as a marker of conformational changes of a protein. Numerous extrinsic fluorescent probes, such as thioflavin T dye and 8-anilino-1-naphthalenesulfonic acid dye, are also attached to the protein for monitoring tructural changes under different conditions (Hawe et al., 2008). Another approach, coupled with rapid-mixing methods such as stopped-flow, fluorescence experiments have been used for resolving folding and unfolding kinetics on millisecond timescales (Chan et al., 1997). (3) UV-visible absorption spectroscopy and (4) Fourier-transform infrared (FTIR) spectroscopy. UV-visible spectroscopy can track the changes in absorbance of aromatic residues or cofactors of a protein (Singhal et al., 2024), while FTIR spectroscopy measures the vibrational modes of functional groups within the protein (Siddique, 2024), where the peptide backbone is the primary source of the infrared signal. The protein infrared spectrum can be characterized by several amide bands arising from the peptide bond (CONH) of the backbone. The most sensitive band to protein secondary structure is Amide I (1700–1600 cm^{-1}) because the C=O bonds are involved in hydrogen bonding patterns that define protein folding. The Amide II band (1560–1520 cm^{-1}) is primarily derived from N–H bending and C–N stretching, but it is less commonly used for quantification than Amide I. Amide III and Others (1300–600 cm^{-1}) involve complex combinations of C–N stretching and N–H bending that are often used for detailed studies of local conformational changes (Chatterley et al., 2022; Stani et al., 2020). Furthermore, (5) calorimetric techniques like differential scanning calorimetry (DSC) can be employed to directly measure the energetics of protein folding. DSC measures the heat absorbed or released during folding and unfolding transitions and provides information on changes inchange in enthalpy (ΔH), entropy (ΔS), and heat capacity (C_p) associated with the process (Demarest and Frasca, 2019). (6) Dynamic light scattering (DLS); it is used to determine the particle size (hydrodynamic radius, Rh) of proteins in solution. DLS can detect the state of compaction, oligomerization, or aggregation of the proteins (Stetefeld et al., 2016). (7) Small-angle X-ray scattering and (8) small-angle neutron scattering can be used for analyzing the structure, size, shape, and ordering of materials at the nanoscale (1–100 nm) (Cheng et al., 2022). These techniques are particularly useful for studying partially folded states and proteins that are challenging to crystallize. (9) Another important ensemble approach includes the hydrogen-deuterium exchange (HDX), which is usually monitored by mass spectrometry (HDX-MS) or NMR spectrometry. HDX-MS reveals the local stability and native-state dynamics by measuring the rate at which backbone amide hydrogens exchange with deuterium in a solvent (Tsutsui and Wintrode, 2012). When a protein is exposed to deuterated buffer (D_2O), the regions that are well-folded or buried (α-helices and β-sheets) exchange slowly, while fast exchange suggests flexibility, exposed regions, or transient unfolding (Weis, 2016). Apart from the benefits of using these techniques, a limitation of ensemble techniques is that the results represent averages of all molecules in the sample, and this may mask the heterogeneity in folding pathways (Sugiki et al., 2017; Hodge et al., 2020). Yet, ensemble measurements yield highly reproducible, quantitative parameters that aid in understanding protein dynamics (Day and Daggett, 2005).

4.6.2 Nuclear magnetic resonance spectroscopy

NMR spectroscopy is based on the absorption of electromagnetic radiation in the radio frequency region 4–900 MHz by the nuclei of an atom. It can be used to study changes in protein structure, stability, and motion at atomic resolution in solution (Hu et al., 2021). A protein is dissolved in a suitable buffer (often an H_2O/D_2O mixture) and placed in a strong magnetic field. The atoms in proteins, such as hydrogen, carbon, and nitrogen, possess nuclear spins that can behave like tiny magnets (Kowalsky, 1962). When a protein sample is placed in a strong magnetic field within an NMR instrument, its tiny magnetic moments align in different ways, establishing distinct energy states. One of the important benefits of NMR spectroscopy is its ability to characterize proteins in their native states as an ensemble of interconverting conformations, instead of as a single static structure (Alderson and Kay, 2020). To study protein folding, NMR can be employed to detect the signals from individual residues. As a result, NMR can tell us whether a protein is tightly packed, flexible, or disordered, and at which stage of folding it is. This ability of NMR to resolve folding at the residue level has enabled the mapping of folding pathways and the characterization of transient intermediates that are elusive to ensemble spectroscopic methods.

4.6.3 Single-molecule techniques

The ensemble experiments provide valuable average thermodynamic data, but these techniques obscure the diverse and individual behaviors of individual molecules. During measurements, the inherent averaging of signals makes it difficult to detect multiple folding pathways, intermediate states, and rare phenomena (Sun et al., 2021). Single-molecule techniques address this limitation by detecting short-lived reaction intermediates. They allows researchers to directly observe pathway heterogeneity, rare events, and stochastic transitions (Sun et al., 2021). These techniques have revolutionized our ability to view folding as an ensemble of stochastic trajectories, rather than as a single predetermined path. Single-molecule methods have shown that even after folding is complete, proteins often exhibit spontaneous transitions between different conformations (Sonar and Mishra, 2025). Single-molecule Förster resonance energy transfer (smFRET) is a sensitive technique to study individual biomolecules, such as proteins, without ensemble averaging (Sasmal et al., 2016). It can measure precise, nanometer-scale distances and conformational dynamics during protein folding (Kuzmenkina et al., 2005). The protein molecule is site-specifically labeled with a donor and an acceptor fluorophore at defined positions. Then, individual molecules are detected either by immobilizing them on the surface or by freely diffusing molecules through confocal microscopy. FRET efficiency is calculated by measuring donor and acceptor fluorescence intensities over time, which are directly proportional to the distance between the labeled sites (Wu et al., 2020). Numerous smFRET experiments have shown pathway heterogeneity, coexistence of multiple folding routes, and the occurrence of partially folded states even under native conditions (Durham et al., 2020; Islam and Mishra, 2024; Schuler et al., 2016). Another approach is single-molecule force spectroscopy, which detects the forces and motions of biomolecules (Neuman and Nagy, 2008). During the process, a piconewton force is applied to a single protein molecule to induce unfolding, which is then relaxed to allow refolding (Petrosyan et al., 2021). Researchers can determine folding and unfolding rates by measuring changes in extension and force responses (Neupane et al., 2018; Hoffer et al., 2019). It further helps to detect the occurrence of intermediate states (Cecconi et al., 2005; Stigler et al., 2011) and to determine free-energy barriers (Sen Mojumdar et al., 2017). Atomic force spectroscopy is especially valuable for studying multidomain, mechanically active proteins such as ion channels and receptors (Müller et al., 2021).

4.6.4 Structure-based techniques

Structure-based research approaches are fundamental to studying protein folding and dynamics. These methods provide detailed information on the arrangement of atoms and the changes that occur over time throughout protein folding. In earlier times, these techniques were considered simple tools for determining the static, or single-state, structure of a protein. These techniques are now regarded as much more useful for examining conformational divergence, intermediate formation, and the dynamics of transitions. Since its discovery, X-ray crystallography has always been considered the backbone of structural biology because of its ability to provide atomic-level resolution of protein structure (Shi, 2014). With the evaluation of how X-rays interact with a crystal containing a protein sample using diffraction methods, researcher can more accurately determine the location of most of the atoms in a protein when it is in its native state (Ringe and Petsko, 1985). Protein structures obtained through X-ray crystallography are important for identifying active sites in proteins (Handing et al., 2018; Cotelesage et al., 2012) and establishing structure/function relationships between the structure and activity of proteins (Maveyraud and Mourey, 2020). X-ray crystallographic data can also be used as part of a greater understanding of the complex processes of protein folding and dynamics. This is primarily because X-ray crystal structures reveal multiple forms of the same protein (e.g., multiple conformations, flexible loops, and alternate side-chain conformations) (Cotelesage et al., 2012). The use of B-factor (temperature factor) in the analysis of X-ray crystallographic data also provides information on the dynamics of individual atoms, allowing the identification of regions within a protein that may exhibit significant flexibility or disorder (de Sá Ribeiro and Lima, 2025). With advances in time-resolved crystallographic techniques, conformational transitions can be visualized by capturing transient snapshots of intermediate molecular states along functional routes (Khusainov et al., 2024; Trincao et al., 2013). Rows of individual proteins or macromolecular structures can now be observed using in situ techniques, such as cryo-EM, which is gaining significant attention in this era of research and allows the study of molecules and cells in near-native conditions. This reduces the hectic work of going through the much more laborious process of obtaining high-concentration protein crystals in their purest forms for the study of solution-phase conformational dynamics (Chang et al., 2021). To achieve this, molecules are trapped in a very thin layer of amorphous glass and cooled quickly with liquid nitrogen, allowing them to be viewed in various states of native conformational equilibrium. Following that, thousands or millions of particle images are computationally superimposed to create a 3D structural reconstruction of the particles under study (Zhong et al., 2020; Cheng, 2015). When viewed, these images appear to be among the highest-resolution images of native structures obtained from a cell. A significant advantage of cryo-EM is that it can be used to simultaneously observe multiple conformational states of proteins or macromolecules in complex, as they are being studied (Tang et al., 2023). Due to its many benefits, cryo-EM has become a cutting-edge microscopic tool for exploring large-scale domain shifts, assembly, and folding processes within highly complex biological systems (Zhou and Li, 2022; Videbæk et al., 2025; Jeyaraj et al., 2025). In addition, time-resolved structural methods, including pump-probe crystallography and rapid-mixing cryo-EM, are now enabling scientists to characterize structural changes that occur after the initiation of folding, binding, or catalysis over a transition time interval of interest. Thus these techniques provide direct structural support for the intermediate states that energy-landscape models have previously predicted.

4.6.5 Molecular dynamics simulations

MD simulation is a computer-based in silico research approach that predicts the behavior of proteins and ligands in a medium (such as water) using predefined algorithms and software. The most commonly used software for MD simulation is GROMACS, which uses Newton's equations of motion to make predictions and produce output (Hollingsworth and Dror, 2018). The forces that act on each protein and ligand are predicted by employing force fields that contain a set of bonded as well as nonbonded interactions. These force fields can be either AMBER, OPLS, or CHARMM, etc. Examples of bonded interactions include atomic bonds, angles between two atoms, and torsions. Nonbonded interactions include Van der Waals forces, electrostatic interactions, and others (Delemotte et al., 2012). Using MD simulation, one can understand folding, unfolding, structural reordering, and other alterations without the use of state-of-the-art microscopic techniques (Childers and Daggett, 2018; Ferina and Daggett, 2019). Due to limitations in computational resources, in earlier times it was very difficult to run MD simulations beyond a few nanoseconds, but as time has evolved, so have the researchers, and now we have more accurate force fields, with much more effective algorithms, and highly powerful computing resources. Due to these advancements, MD simulations have expanded their scope to make additional complex predictions. Nowadays, MD simulations are performed over microseconds to milliseconds, allowing the study of

Table 4.2 Molecular and pathological features of protein misfolding and aggregation-related diseases (Dobson, 1999; Chiti and Dobson, 2006; Bashir et al., 2024).

Disease	Target protein	Aggregated form	Key pathological features
Alzheimer's disease	Amyloid-β (Aβ), tau	Amyloid fibrils, soluble oligomers	Extracellular plaques, intracellular neurofibrillary tangles, synaptic dysfunction
Parkinson's disease	α-synuclein	Lewy bodies, oligomers	Dopaminergic neuron loss, motor impairment
Huntington's disease	Huntingtin (polyQ-expanded)	Intracellular aggregates	Toxic gain of function, transcriptional dysregulation
Amyotrophic lateral sclerosis	TDP-43, SCD1, FUS	Cytoplasmic inclusions	Progressive muscle weakness and paralysis
Prion diseases (e.g., Creutzfeldt–Jakob disease)	Prion protein	Infectious amyloid aggregates	Rapid neurodegeneration, transmissible aggregation
Type II diabetes mellitus	Islet amyloid polypeptide	Amyloid fibrils	β-cell dysfunction and loss
Systemic amyloidosis	Immunoglobulin light chains (AL), transthyretin	Amyloid fibrils	Organ failure due to amyloid deposition
Cataract	Crystallin	Amorphous aggregates	Loss of lens transparency
Cystic fibrosis	CFTR (misfolded)	ER-retained misfolded protein	Loss of function due to degradation (*aggregation, not fibrillar*)

proteins in time frames nearly identical to the original ones (Shaw et al., 2010; Karplus and McCammon, 2002). More modern and efficient MD simulation methods are available, such as replica-exchange MD (Parrill and Lipkowitz, 2015) and metadynamics. These are improved versions of the conventional ones and allow simulations to escape local energy minima and better explore conformational space (Ríos-Rozas et al., 2025). Such advancements have enabled the characterization of folding intermediate states of proteins and their FELs, which could not have been studied within a feasible time frame (Promoting transparency and reproducibility in enhanced molecular simulations, 2019). Coarse-grained MD simulations have contributed significantly to developing general folding principles (e.g., exploring the relationship between topology and folding rates (Clementi, 2008). In addition, Markov state models (MSMs), which combine multiple short simulations into a single long simulation, provide an alternative to running a long simulation for analyzing kinetic networks of conformational states and their associated transition probabilities (Clementi, 2008). MSMs enable the derivation of folding kinetics, pathway fluxes, and long-time behavior from a statistically sampled dataset (Bose et al., 2025).

4.7 Protein misfolding, aggregation, and disease

Protein misfolding and aggregation are common phenomena that arise from the dynamic nature of proteins and are at the core of various disorders. There are molecular chaperones inside the cell that guide proper folding, thereby preventing misfolding and reducing the risk of protein aggregation. Additionally, the proteostasis system within the cell helps prevent misfolding. Under certain conditions, it may become compromised, leading to the accumulation of misfolded and aggregated proteins. Once aggregation occurs, it further impairs proteostasis by sequestering chaperones and degradation machinery. Aggregation can progress through multiple pathways; for instance, small oligomers form and later grow into toxic fibrils. Some proteins form amorphous (Borgia et al., 2015), disordered aggregates, whilst others assemble into highly ordered amyloid fibrils (Chiti and Dobson, 2017). In amyloid fibril arrangement, β-sheet regions from different protein molecules align to form extended β-sheets that stack along the fibril axis (Chiti and Dobson, 2017). In the amyloid formation, usually follows nucleation-dependent kinetics, in which a slow and unfavorable nucleation step is followed by rapid fibril growth (Chang et al., 2024). Table 4.2 summarizes various diseases that arise as a result of protein misfolding and aggregation.

4.8 Conclusion and future perspectives

Protein folding and dynamics are among the most fundamental concepts in science, bridging the gap between a simple amino-acid sequence and its biological function. As discussed throughout this chapter, proteins are not rigid molecules but dynamic entities. Proteins continuously explore a range of conformations on complex FELs. The transition from an unfolded polypeptide chain to a functional native state is governed by a well-regulated interplay of thermodynamic forces, kinetic pathways, and environmental changes. These processes ensure that proteins achieve both structural stability and the flexibility required for catalysis, signaling, and molecular recognition. Advances in experimental techniques, including spectroscopic tools, NMR, cryo-EM, and single-molecule methods, have significantly strengthened our understanding of protein structural heterogeneity and folding pathways. At the same time, computational approaches such as MDS and machine-learning-based structure prediction provide atomistic insights into conformational landscapes that are often difficult to obtain under in vitro conditions. Under certain circumstances, improper folding or misfolding of proteins can lead to aggregation that underlies numerous neurodegenerative and metabolic disorders. However, the phenomenon of protein folding within a complex cellular environment and how it fails remains poorly understood. The continuous integration of high-resolution experimental techniques, advanced computational modeling, and artificial intelligence will help understand the mechanisms governing protein folding and dynamics. In due course, understanding the principles of protein folding and motion will not only deepen our understanding of fundamental biological processes but also open new avenues for improving drug discovery and therapeutic interventions.

References

Parrill, A.L., Lipkowitz, K.B., 2015. Free-Energy Calculations with Metadynamics: Theory and Practice Reviews in Computational Chemistry Volume 28. Wiley, pp. 1–49. Available at http://doi.org/10.1002/9781118889886.ch1.

Aiman, A., Iram, F., Alam, D., Basir, S.F., Khan, L.A., Shahid, M., et al., 2025. Biophysical insights into pH-induced structural dynamics of recombinant interferon β-1b: aggregation under acidic conditions. International Journal of Biological Macromolecules 318 (Pt 3), 145128. Available at https://doi.org/10.1016/j.ijbiomac.2025.145128.

Alderson, T.R., Kay, L.E., 2020. Unveiling invisible protein states with NMR spectroscopy. Current Opinion in Structural Biology 60, 39–49. Available at https://doi.org/10.1016/j.sbi.2019.10.008, http://www.elsevier.com/locate/sbi.

Arai, M., 2018. Unified understanding of folding and binding mechanisms of globular and intrinsically disordered proteins. Biophysical Reviews 10 (2), 163–181. Available at https://doi.org/10.1007/s12551-017-0346-7.

Baldwin, R.L., 1989. How does protein folding get started? Trends in Biochemical Sciences 14 (7), 291–294. Available at https://doi.org/10.1016/0968-0004(89)90067-4.

Bashir, S., Aiman, A., Shahid, M., Chaudhary, A.A., Sami, N., Basir, S.F., et al., 2024. Amyloid-induced neurodegeneration: a comprehensive review through aggregomics perception of proteins in health and pathology. Ageing Research Reviews 96, 102276. Available at https://doi.org/10.1016/j.arr.2024.102276.

Benjwal, S., Verma, S., Röhm, K.H., Gursky, O., 2006. Monitoring protein aggregation during thermal unfolding in circular dichroism experiments. Protein Science 15 (3), 635–639. Available at https://doi.org/10.1110/ps.051917406, http://www.proteinscience.org/cgi/reprint/15/3/635.

Berman, H.M., Burley, S.K., 2025. Protein data bank (PDB): fifty-three years young and having a transformative impact on science and society. Quarterly Reviews of Biophysics 58. Available at https://doi.org/10.1017/S0033583525000034, http://journals.cambridge.org/action/displayJournal?jid=QRB.

Bhatia, S., Udgaonkar, J.B., 2022. Heterogeneity in protein folding and unfolding reactions. Chemical Reviews 122 (9), 8911–8935. Available at https://doi.org/10.1021/acs.chemrev.1c00704, http://pubs.acs.org/journal/chreay.

Boehr, D.D., Nussinov, R., Wright, P.E., 2009. The role of dynamic conformational ensembles in biomolecular recognition. Nature Chemical Biology 5 (11), 789–796. Available at https://doi.org/10.1038/nchembio.232, http://www.nature.com/nchembio.

Borgia, A., Kemplen, K.R., Borgia, M.B., Soranno, A., Shammas, S., Wunderlich, B., et al., 2015. Transient misfolding dominates multidomain protein folding. Nature Communications 6. Available at https://doi.org/10.1038/ncomms9861, https://www.nature.com/ncomms/.

Bose, S., Kilinc, C., Dickson, A., 2025. Markov state models with weighted ensemble simulation: how to eliminate the trajectory merging bias. Journal of Chemical Theory and Computation 21 (4), 1805–1816. Available at https://doi.org/10.1021/acs.jctc.4c01141.

Bryan, P.N., Orban, J., 2010. Proteins that switch folds. Current Opinion in Structural Biology 20 (4), 482–488. Available at https://doi.org/10.1016/j.sbi.2010.06.002, https://www.sciencedirect.com/science/journal/0959440X.

Bryngelson, J.D., Onuchic, J.N., Socci, N.D., Wolynes, P.G., 1995. Funnels, pathways, and the energy landscape of protein folding: a synthesis. Proteins: Structure, Function, and Bioinformatics 21 (3), 167–195. Available at https://doi.org/10.1002/prot.340210302.

Callender, R., Dyer, R.B., 2015. The dynamical nature of enzymatic catalysis. Accounts of Chemical Research 48 (2), 407–413. Available at https://doi.org/10.1021/ar5002928, http://pubs.acs.org/journal/achre4.

Campitelli, P., Kazan, I.C., Hamilton, S., Ozkan, S.B., 2025. Dynamic allostery: evolution's double-edged sword in protein function and disease. Journal of Molecular Biology 437 (20), 169175. Available at https://doi.org/10.1016/j.jmb.2025.169175.

Cecconi, C., Shank, E.A., Bustamante, C., Marqusee, S., 2005. Direct observation of the three-state folding of a single protein molecule. Science (New York, N.Y.) 309 (5743), 2057–2060. Available at https://doi.org/10.1126/science.1116702.

Chan, C.-K., Hu, Y., Takahashi, S., Rousseau, D.L., Eaton, W.A., Hofrichter, J., 1997. Submillisecond protein folding kinetics studied by ultrarapid mixing. Proceedings of the National Academy of Sciences 94 (5), 1779–1784. Available at https://doi.org/10.1073/pnas.94.5.1779.

Chang, R., Yuan, C., Zhou, P., Xing, R., Yan, X., 2024. Peptide self-assembly: from ordered to disordered. Accounts of Chemical Research 57 (3), 289–301. Available at https://doi.org/10.1021/acs.accounts.3c00592.

Chang, W.H., Huang, S.H., Lin, H.H., Chung, S.C., Tu, I.P., 2021. Cryo-EM analyses permit visualization of structural polymorphism of biological macromolecules. Frontiers in Bioinformatics 1. Available at https://doi.org/10.3389/fbinf.2021.788308, www.frontiersin.org/journals/bioinformatics.

Chatterley, A.S., Laity, P., Holland, C., Weidner, T., Woutersen, S., Giubertoni, G., 2022. Broadband multidimensional spectroscopy identifies the amide II vibrations in silkworm films. Molecules (Basel, Switzerland) 27 (19), 6275. Available at https://doi.org/10.3390/molecules27196275.

Cheng, L., Ye, A., Yang, Z., Gilbert, E.P., Knott, R., de Campo, L., et al., 2022. Small-angle X-ray scattering (SAXS) and small-angle neutron scattering (SANS) study on the structure of sodium caseinate in dispersions and at the oil-water interface: effect of calcium ions. Food Structure 32, 100276. Available at https://doi.org/10.1016/j.foostr.2022.100276.

Cheng, Y., 2015. Single-particle cryo-EM at crystallographic resolution. Cell 161 (3), 450–457. Available at https://doi.org/10.1016/j.cell.2015.03.049.

Cheung, M.S., García, A.E., Onuchic, J.N., 2002. Protein folding mediated by solvation: water expulsion and formation of the hydrophobic core occur after the structural collapse. Proceedings of the National Academy of Sciences 99 (2), 685–690. Available at https://doi.org/10.1073/pnas.022387699.

Childers, M.C., Daggett, V., 2018. Validating molecular dynamics simulations against experimental observables in light of underlying conformational ensembles. The Journal of Physical Chemistry. B 122 (26), 6673–6689. Available at https://doi.org/10.1021/acs.jpcb.8b02144.

Childers, M.C., Daggett, V., 2017. Insights from molecular dynamics simulations for computational protein design. Molecular Systems Design and Engineering 2 (1), 9–33. Available at https://doi.org/10.1039/c6me00083e, www.rsc.org/journals-books-databases/about-journals/molecular-systems-design-engineering/.

Chiti, F., Dobson, C.M., 2006. Protein misfolding, functional amyloid, and human disease. Annual Review of Biochemistry 75, 333–366. Available at https://doi.org/10.1146/annurev.biochem.75.101304.123901.

Chiti, F., Dobson, C.M., 2017. Protein misfolding, amyloid formation, and human disease: a summary of progress over the last decade. Annual Review of Biochemistry 86, 27–68. Available at https://doi.org/10.1146/annurev-biochem-061516-045115, http://arjournals.annualreviews.org/loi/biochem.

Clementi, C., 2008. Coarse-grained models of protein folding: toy models or predictive tools? Current Opinion in Structural Biology 18 (1), 10–15. Available at https://doi.org/10.1016/j.sbi.2007.10.005.

Cohen, R.D., Pielak, G.J., 2017. Quinary interactions with an unfolded state ensemble. Protein Science 26 (9), 1698–1703. Available at https://doi.org/10.1002/pro.3206, http://onlinelibrary.wiley.com/journal/10.1002/(ISSN)1469-896X.

Cotelesage, J.J.H., Pushie, M.J., Grochulski, P., Pickering, I.J., George, G.N., 2012. Metalloprotein active site structure determination: synergy between X-ray absorption spectroscopy and X-ray crystallography. Journal of Inorganic Biochemistry 115, 127–137. Available at https://doi.org/10.1016/j.jinorgbio.2012.06.019.

Day, R., Daggett, V., 2005. Ensemble versus single-molecule protein unfolding. Proceedings of the National Academy of Sciences 102 (38), 13445–13450. Available at https://doi.org/10.1073/pnas.0501773102.

Delemotte, L., Klein, M.L., Tarek, M., 2012. Molecular dynamics simulations of voltage-gated cation channels: insights on voltage-sensor domain function and modulation. Frontiers in Pharmacology 3. Available at https://doi.org/10.3389/fphar.2012.00097.

Demarest, S.J., Frasca, V., 2019. Differential Scanning Calorimetry in the Biopharmaceutical Sciences Biophysical Characterization of Proteins in Developing Biopharmaceuticals. Elsevier, United States, pp. 311–332. Available at http://www.sciencedirect.com/science/book/9780444641731.

de Sá Ribeiro, F., Lima, L.M.T.R., 2025. Temperature-resolved crystallography reveals rigid-body dominance over local flexibility in B-factors. ACS Omega 10 (34), 38871–38881. Available at https://doi.org/10.1021/acsomega.5c04454, pubs.acs.org/journal/acsodf.

Dill, K.A., Ozkan, S.B., Weikl, T.R., Chodera, J.D., Voelz, V.A., 2007. The protein folding problem: when will it be solved? Current Opinion in Structural Biology 17 (3), 342–346. Available at https://doi.org/10.1016/j.sbi.2007.06.001.

Ding, F., Dokholyan, N.V., Buldyrev, S.V., Stanley, H.E., Shakhnovich, E.I., 2002. Direct molecular dynamics observation of protein folding transition state ensemble. Biophysical Journal 83 (6), 3525–3532. Available at https://doi.org/10.1016/S0006–3495(02)75352-6, http://www.sciencedirect.com/science/journal/00063495.

Dinner, A.R., Šali, A., Smith, L.J., Dobson, C.M., Karplus, M., 2000. Understanding protein folding via free-energy surfaces from theory and experiment. Trends in Biochemical Sciences 25 (7), 331–339. Available at https://doi.org/10.1016/s0968-0004(00)01610-8.

Dobson, C.M., 1999. Protein misfolding, evolution and disease. Trends in Biochemical Sciences 24 (9), 329–332. Available at https://doi.org/10.1016/S0968-0004(99)01445-0.

Durham, R.J., Latham, D.R., Sanabria, H., Jayaraman, V., 2020. Structural dynamics of glutamate signaling systems by smFRET. Biophysical Journal 119 (10), 1929–1936. Available at https://doi.org/10.1016/j.bpj.2020.10.009, http://www.sciencedirect.com/science/journal/00063495.

Dutta, P., Roy, P., Sengupta, N., 2022. Effects of external perturbations on protein systems: a microscopic view. ACS Omega 7 (49), 44556–44572. Available at https://doi.org/10.1021/acsomega.2c06199.

Englander, S.W., Mayne, L., 2014. The nature of protein folding pathways. Proceedings of the National Academy of Sciences 111 (45), 15873–15880. Available at https://doi.org/10.1073/pnas.1411798111.

Ferina, J., Daggett, V., 2019. Visualizing protein folding and unfolding. Journal of Molecular Biology 431 (8), 1540–1564. Available at https://doi.org/10.1016/j.jmb.2019.02.026.

Fersht, A.R., Sato, S., 2004. Φ-Value analysis and the nature of protein-folding transition states. Proceedings of the National Academy of Sciences 101 (21), 7976–7981. Available at https://doi.org/10.1073/pnas.0402684101.

Fossum, C.J., Johnson, B.O.V., Golde, S.T., Kielman, A.J., Finke, B., Smith, M.A., et al., 2023. Insights into the mechanism of tryptophan fluorescence quenching due to synthetic crowding agents: a combined experimental and computational study. ACS Omega 8 (47), 44820–44830. Available at https://doi.org/10.1021/acsomega.3c06006, pubs.acs.org/journal/acsodf.

Gianni, S., Freiberger, M.I., Jemth, P., Ferreiro, D.U., Wolynes, P.G., Fuxreiter, M., 2021. Fuzziness and frustration in the energy landscape of protein folding, function, and assembly. Accounts of Chemical Research 54 (5), 1251–1259. Available at https://doi.org/10.1021/acs.accounts.0c00813, http://pubs.acs.org/journal/achre4.

Glick, D., 2006. Circular dichroism and its empirical application to biopolymers. Methods of Biochemical Analysis 31. Wiley, pp. 61–163. Available at http://doi.org/10.1002/9780470110522.ch2.

Greenfield, N.J., 2006. Using circular dichroism collected as a function of temperature to determine the thermodynamics of protein unfolding and binding interactions. Nature Protocols 1 (6), 2527–2535. Available at https://doi.org/10.1038/nprot.2006.204.

Haliloglu, T., Bahar, I., 2015. Adaptability of protein structures to enable functional interactions and evolutionary implications. Current Opinion in Structural Biology 35, 17–23. Available at https://doi.org/10.1016/j.sbi.2015.07.007.

Handing, K.B., Niedzialkowska, E., Shabalin, I.G., Kuhn, M.L., Zheng, H., Minor, W., 2018. Characterizing metal-binding sites in proteins with X-ray crystallography. Nature Protocols 13 (5), 1062–1090. Available at https://doi.org/10.1038/nprot.2018.018, http://www.natureprotocols.com/.

Hawe, A., Sutter, M., Jiskoot, W., 2008. Extrinsic fluorescent dyes as tools for protein characterization. Pharmaceutical Research 25 (7), 1487–1499. Available at https://doi.org/10.1007/s11095-007-9516-9.

Hodge, E.A., Benhaim, M.A., Lee, K.K., 2020. Bridging protein structure, dynamics, and function using hydrogen/deuterium-exchange mass spectrometry. Protein Science 29 (4), 843–855. Available at https://doi.org/10.1002/pro.3790, http://onlinelibrary.wiley.com/journal/10.1002/(ISSN)1469-896X.

Hoffer, N.Q., Neupane, K., Woodside, M.T., 2019. Measuring the average shape of transition paths during the folding of a single biological molecule. Biophysical Journal 116 (3), 325a. Available at https://doi.org/10.1016/j.bpj.2018.11.1766.

Hollingsworth, S.A., Dror, R.O., 2018. Molecular dynamics simulation for All. Neuron 99 (6), 1129–1143. Available at https://doi.org/10.1016/j.neuron.2018.08.011, http://www.cell.com/neuron/home.

Hu, Y., Cheng, K., He, L., Zhang, X., Jiang, B., Jiang, L., et al., 2021. NMR-based methods for protein analysis. Analytical Chemistry 93 (4), 1866–1879. Available at https://doi.org/10.1021/acs.analchem.0c03830.

Islam, F., Mishra, P.P., 2024. Unraveling Neurological Enigmas with SmFRET Exploring the Intricacies of Neuroscience with Single-Molecule Förster Resonance Energy Transfer (SmFRET) Nanobiotechnology: Applications of Nanomaterials in Biotechnology, Medicine and Healthcare. CRC Press, India, pp. 318–337. Available at http://www.tandfebooks.com/doi/book/9781040015735.

Jamadagni, S.N., Godawat, R., Garde, S., 2011. Hydrophobicity of proteins and interfaces: insights from density fluctuations. Annual Review of Chemical and Biomolecular Engineering 2 (1), 147–171. Available at https://doi.org/10.1146/annurev-chembioeng-061010–114156.

Jeyaraj, G., Rajendran, A.K., Sathishkumar, K., Almutairi, B.O., Vadivelu, A., Chokkakula, S., et al., 2025. High-resolution protein modeling through Cryo-EM and AI: current trends and future perspectives – a review. Frontiers in Molecular Biosciences 12. Available at https://doi.org/10.3389/fmolb.2025.1688455, journal.frontiersin.org/journal/molecular-biosciences.

Karplus, M., McCammon, J.A., 2002. Molecular dynamics simulations of biomolecules. Nature Structural Biology 9 (9), 646–652. Available at https://doi.org/10.1038/nsb0902-646.

Khusainov, G., Standfuss, J., Weinert, T., 2024. The time revolution in macromolecular crystallography. Structural Dynamics 11 (2). Available at https://doi.org/10.1063/4.0000247.

Kokkinidis, M., Glykos, N.M., Fadouloglou, V.E., 2012. Protein flexibility and enzymatic catalysis. Advances in Protein Chemistry and Structural Biology 87, 181–218. Available at https://doi.org/10.1016/B978-0-12-398312-1.00007-X, https://www.elsevier.com/books/advances-in-protein-chemistry-and-structural-biology/eisenberg/978-0-12-374827-0.

Kowalsky, A., 1962. Nuclear magnetic resonance studies of proteins. Journal of Biological Chemistry 237 (6), 1807–1819. Available at https://doi.org/10.1016/s0021–9258(19)73941-6.

Kumar, A., Jernigan, R.L., 2021. Ligand binding introduces significant allosteric shifts in the locations of protein fluctuations. Frontiers in Molecular Biosciences 8. Available at https://doi.org/10.3389/fmolb.2021.733148, journal.frontiersin.org/journal/molecular-biosciences.

Kumar, R., Tripathi, T., Saudagar, P., 2023. Fluorescence Spectroscopy-Based Methods to Study Protein Folding Dynamics Protein Folding Dynamics and Stability: Experimental and Computational Methods. Springer Nature, United States, pp. 25–36. Available at https://link.springer.com/book/10.1007/978–981-99–2079-2.

Kuzmenkina, E.V., Heyes, C.D., Nienhaus, G.U., 2005. Single-molecule Förster resonance energy transfer study of protein dynamics under denaturing conditions. Proceedings of the National Academy of Sciences 102 (43), 15471–15476. Available at https://doi.org/10.1073/pnas.0507728102.

Levinthal, C., 1968. Are there pathways for protein folding? Journal de Chimie Physique 65, 44–45. Available at https://doi.org/10.1051/jcp/1968650044.

Linhares, L.A., Ramos, C.H.I., 2023. Unlocking insights into folding, structure, and function of proteins through circular dichroism spectroscopy—a short review. Applied Biosciences 2 (4), 639–655. Available at https://doi.org/10.3390/applbiosci2040040, https://www.mdpi.com/journal/applbiosci.

Liu, S.-Q., Tan, D.-Y., Zhang, K.-Q., Ji, X.-L., Tao, Y., Fu, Y.-X., 2012. Protein Folding, Binding and ENERGY Landscape: a synthesis. INTECH Open Access Publisher.

Livesay, D.R., 2010. Protein dynamics: dancing on an ever-changing free energy stage. Current Opinion in Pharmacology 10 (6), 706–708. Available at https://doi.org/10.1016/j.coph.2010.09.015.

Lu, Y., Zhu, B., Li, Q., Du, J., Chen, T., 2022. The folding and misfolding mechanisms of multidomain proteins. Medicine in Drug Discovery 14, 100126. Available at https://doi.org/10.1016/j.medidd.2022.100126.

Magliery, T.J., 2015. Protein stability: computation, sequence statistics, and new experimental methods. Current Opinion in Structural Biology 33, 161–168. Available at https://doi.org/10.1016/j.sbi.2015.09.002, http://www.elsevier.com/locate/sbi.

Malhotra, P., Udgaonkar, J.B., 2016. How cooperative are protein folding and unfolding transitions? Protein Science 25 (11), 1924–1941. Available at https://doi.org/10.1002/pro.3015, http://onlinelibrary.wiley.com/journal/10.1002/(ISSN)1469-896X.

Mallamace, F., Corsaro, C., Mallamace, D., Vasi, S., Vasi, C., Baglioni, P., et al., 2016. Energy landscape in protein folding and unfolding. Proceedings of the National Academy of Sciences 113 (12), 3159–3163. Available at https://doi.org/10.1073/pnas.1524864113.

Mamidi, A.S., Surolia, A., MacKerell, A., 2022. Mixed mechanism of conformational selection and induced fit as a molecular recognition process in the calreticulin family of proteins. PLoS Computational Biology 18 (12), e1010661. Available at https://doi.org/10.1371/journal.pcbi.1010661.

Mariño Pérez, L., Ielasi, F.S., Bessa, L.M., Maurin, D., Kragelj, J., Blackledge, M., et al., 2022. Visualizing protein breathing motions associated with aromatic ring flipping. Nature 602 (7898), 695–700. Available at https://doi.org/10.1038/s41586-022-04417-6, http://www.nature.com/nature/index.html.

Martínez, L., 2014. Introducing the levinthal's protein folding paradox and its solution. Journal of Chemical Education 91 (11), 1918–1923. Available at https://doi.org/10.1021/ed300302h.

Matsumura, M., Signor, G., Matthews, B.W., 1989. Substantial increase of protein stability by multiple disulphide bonds. Nature 342 (6247), 291–293. Available at https://doi.org/10.1038/342291a0.

Maveyraud, L., Mourey, L., 2020. Protein X-ray crystallography and drug discovery. Molecules (Basel, Switzerland) 25 (5), 1030. Available at https://doi.org/10.3390/molecules25051030.

Micheletti, C., 2013. Comparing proteins by their internal dynamics: exploring structure–function relationships beyond static structural alignments. Physics of Life Reviews 10 (1), 1–26. Available at https://doi.org/10.1016/j.plrev.2012.10.009.

Sen Mojumdar, S., N. Scholl, Z., Dee, D.R., Rouleau, L., Anand, U., Garen, C., et al., 2017. Partially native intermediates mediate misfolding of SOD1 in single-molecule folding trajectories. Nature Communications 8 (1). Available at https://doi.org/10.1038/s41467-017-01996-1.

Morris, R., Black, K.A., Stollar, E.J., 2022. Uncovering protein function: from classification to complexes. Essays in Biochemistry 66 (2), 255–285. Available at https://doi.org/10.1042/EBC20200108, https://portlandpress.com/essaysbiochem/article/66/3/255/231650/Uncovering-protein-function-from-classification-to.

Müller, D.J., Dumitru, A.C., Lo Giudice, C., Gaub, H.E., Hinterdorfer, P., Hummer, G., et al., 2021. Atomic force microscopy-based force spectroscopy and multiparametric imaging of biomolecular and cellular systems. Chemical Reviews 121 (19), 11701–11725. Available at https://doi.org/10.1021/acs.chemrev.0c00617, http://pubs.acs.org/journal/chreay.

Neuman, K.C., Nagy, A., 2008. Single-molecule force spectroscopy: optical tweezers, magnetic tweezers and atomic force microscopy. Nature Methods 5 (6), 491–505. Available at https://doi.org/10.1038/nmeth.1218, https://www.nature.com/nmeth/.

Neupane, K., Hoffer, N.Q., Woodside, M.T., 2018. Measuring the local velocity along transition paths during the folding of single biological molecules. Physical Review Letters 121 (1). Available at https://doi.org/10.1103/PhysRevLett.121.018102, http://harvest.aps.org/v2/bagit/articles/10.1103/PhysRevLett.121.018102/apsxml.

Nölting, B., Agard, D.A., 2008. How general is the nucleation-condensation mechanism? Proteins: Structure, Function, and Bioinformatics 73 (3), 754–764. Available at https://doi.org/10.1002/prot.22099.

Ojeda-May, P., Mushtaq, A.U.I., Rogne, P., Verma, A., Ovchinnikov, V., Grundström, C., et al., 2021. Dynamic connection between enzymatic catalysis and collective protein motions. Biochemistry 60 (28), 2246–2258. Available at https://doi.org/10.1021/acs.biochem.1c00221, http://pubs.acs.org/journal/bichaw.

Onuchic, J.N., Luthey-Schulten, Z., Wolynes, P.G., 1997. Theory of protein folding: the energy landscape perspective. Annual Review of Physical Chemistry 48 (1), 545–600. Available at https://doi.org/10.1146/annurev.physchem.48.1.545, http://arjournals.annualreviews.org/loi/physchem.

Oyama, T., Suzuki, S., Akao, K.-ichi, 2026. Circular dichroism spectroscopy in protein engineering and pharmaceutical development: applications in structural characterization and quality assessment. Protein Expression and Purification 237, 106826. Available at https://doi.org/10.1016/j.pep.2025.106826.

Pace, C.N., Grimsley, G.R., Scholtz, J.M., 2009. Protein ionizable groups: pK values and their contribution to protein stability and solubility. Journal of Biological Chemistry 284 (20), 13285–13289. Available at https://doi.org/10.1074/jbc.R800080200, http://www.jbc.org/cgi/reprint/284/20/13285.

Pace, C.N., Scholtz, J.M., Grimsley, G.R., 2014. Forces stabilizing proteins. FEBS Letters 588 (14), 2177–2184. Available at https://doi.org/10.1016/j.febslet.2014.05.006.

Persch, E., Dumele, O., Diederich, F., 2015. Molecular recognition in chemical and biological systems. Angewandte Chemie International Edition 54 (11), 3290–3327. Available at https://doi.org/10.1002/anie.201408487.

Petrosyan, R., Narayan, A., Woodside, M.T., 2021. Single-molecule force spectroscopy of protein folding. Journal of Molecular Biology 433 (20), 167207. Available at https://doi.org/10.1016/j.jmb.2021.167207.

Powers, E.T., Deechongkit, S., Kelly, J.W., 2005. Backbone-Backbone H-bonds make context-dependent contributions to protein folding kinetics and thermodynamics: lessons from amide-to-ester mutations. Advances in Protein Chemistry 72, 39–78. Available at https://doi.org/10.1016/S0065–3233(05)72002-7.

Promoting transparency and reproducibility in enhanced molecular simulations, 2019. Nature Methods 16 (8), 670–673.

Ramos, C.H.I., Lima, M.V., Silva, S.L.F., Borin, P.F.L., Régis, W.C.B., Santoro, M.M., 2004. Stability and folding studies of the N-domain of troponin C. Evidence for the formation of an intermediate. Archives of Biochemistry and Biophysics 427 (2), 135–142. Available at https://doi.org/10.1016/j.abb.2004.05.002.

Rehman, I., Kerndt, C.C., Botelho, S., 2024. Biochemistry, Tertiary Protein Structure. StatPearls Publishing [Internet].

Ringe, D., Petsko, G.A., 1985. Mapping protein dynamics by X-ray diffraction. Progress in Biophysics and Molecular Biology 45 (3), 197–235. Available at https://doi.org/10.1016/0079–6107(85)90002-1.

Rose, G.D., Fleming, P.J., Banavar, J.R., Maritan, A., 2006. A backbone-based theory of protein folding. Proceedings of the National Academy of Sciences 103 (45), 16623–16633. Available at https://doi.org/10.1073/pnas.0606843103.

Roth, C.M., Neal, B.L., Lenhoff, A.M., 1996. Van der Waals interactions involving proteins. Biophysical Journal 70 (2), 977–987. Available at https://doi.org/10.1016/s0006–3495(96)79641-8.

Ríos-Rozas, S.E., Morales, N., Valdés-Muñoz, E., Urra, G., Flores-Morales, C.A., Farías-Abarca, J., et al., 2025. Integrative computational approaches for the discovery of triazole-based urease inhibitors: a machine learning, virtual screening, and meta-dynamics framework. International Journal of Molecular Sciences 26 (23). Available at https://doi.org/10.3390/ijms262311576, http://www.mdpi.com/journal/ijms.

Sadqi, M., Lapidus, L.J., Muñoz, V., 2003. How fast is protein hydrophobic collapse? Proceedings of the National Academy of Sciences 100 (21), 12117–12122. Available at https://doi.org/10.1073/pnas.2033863100.

Sasmal, D.K., Pulido, L.E., Kasal, S., Huang, J., 2016. Single-molecule fluorescence resonance energy transfer in molecular biology. Nanoscale 8 (48), 19928–19944. Available at https://doi.org/10.1039/c6nr06794h, http://www.rsc.org/publishing/journals/NR/Index.asp.

Schuler, B., Soranno, A., Hofmann, H., Nettels, D., 2016. Single-molecule FRET spectroscopy and the polymer physics of unfolded and intrinsically disordered proteins. Annual Review of Biophysics 45 (1), 207–231. Available at https://doi.org/10.1146/annurev-biophys-062215-010915.

Schwartz, S.D., 2023. Protein dynamics and enzymatic catalysis. The Journal of Physical Chemistry. B 127 (12), 2649–2660. Available at https://doi.org/10.1021/acs.jpcb.3c00477.

Seffernick, J.T., Lindert, S., 2020. Hybrid methods for combined experimental and computational determination of protein structure. The Journal of Chemical Physics 153 (24). Available at https://doi.org/10.1063/5.0026025.

Shaitan, K.V., 2023. Why do proteins fold into unique 3D structures? And other questions. Federation Russian Journal of Physical Chemistry B 17 (3), 550–570. Available at https://doi.org/10.1134/S1990793123030259, https://www.springer.com/journal/11826.

Shao, Q., Zhu, W., 2019. Nonnative contact effects in protein folding. Physical Chemistry Chemical Physics 21 (22), 11924–11936. Available at https://doi.org/10.1039/c8cp07524g.

Shaw, D.E., Maragakis, P., Lindorff-Larsen, K., Piana, S., Dror, R.O., Eastwood, M.P., et al., 2010. Atomic-level characterization of the structural dynamics of proteins. Science (New York, N.Y.) 330 (6002), 341–346. Available at https://doi.org/10.1126/science.1187409, http://www.sciencemag.org/cgi/reprint/330/6002/341.pdf.

Shi, Y., 2014. A glimpse of structural biology through X-ray crystallography. Cell 159 (5), 995–1014. Available at https://doi.org/10.1016/j.cell.2014.10.051.

Siddique, I., 2024. Exploring functional groups and molecular structures: a comprehensive analysis using FTIR spectroscopy. SSRN Electronic Journal. Available at https://doi.org/10.2139/ssrn.4886526.

Singhal, A., Saini, U., Chopra, B., Dhingra, A.K., Jain, A., Chaudhary, J., 2024. UV-visible spectroscopy: a review on its pharmaceutical and bio-allied sciences applications. Current Pharmaceutical Analysis 20 (3), 161–177. Available at https://doi.org/10.2174/0115734129300562240408042614, https://www.eurekaselect.com/download_file.

Sonar, R., Mishra, P.P., 2025. Exploring Dynamics at the Molecular Scale: Advances in Single-Molecule Biophysics Biochemical and Biophysical Methods in Molecular and Cellular Biology. Springer Nature, India, pp. 163–184. Available at https://link.springer.com/book/9789819620883.

Sorokina, I., Mushegian, A.R., Koonin, E.V., 2022. Is protein folding a thermodynamically unfavorable, active, energy-dependent process? International Journal of Molecular Sciences 23 (1). Available at https://doi.org/10.3390/ijms23010521, https://www.mdpi.com/1422-0067/23/1/521/pdf.

Stani, C., Vaccari, L., Mitri, E., Birarda, G., 2020. FTIR investigation of the secondary structure of type I collagen: new insight into the amide III band. Spectrochimica Acta Part A: Molecular and Biomolecular Spectroscopy 229, 118006. Available at https://doi.org/10.1016/j.saa.2019.118006.

Stetefeld, J., McKenna, S.A., Patel, T.R., 2016. Dynamic light scattering: a practical guide and applications in biomedical sciences. Biophysical Reviews 8 (4), 409–427. Available at https://doi.org/10.1007/s12551-016-0218-6, http://www.springer.com/life+sciences/biochemistry+%26+biophysics/journal/12551.

Stigler, J., Ziegler, F., Gieseke, A., Gebhardt, J.C.M., Rief, M., 2011. The complex folding network of single calmodulin molecules. Science (New York, N.Y.) 334 (6055), 512–516. Available at https://doi.org/10.1126/science.1207598, http://www.sciencemag.org/content/334/6055/512.full.pdf.

Straiton, J., 2023. The path to solving the protein folding problem. Biotechniques 74 (5), 199–201. Available at https://doi.org/10.2144/btn-2023-0031.

Sugiki, T., Kobayashi, N., Fujiwara, T., 2017. Modern technologies of solution nuclear magnetic resonance spectroscopy for three-dimensional structure determination of proteins open avenues for life scientists. Computational and Structural Biotechnology Journal 15, 328–339. Available at https://doi.org/10.1016/j.csbj.2017.04.001.

Sun, H., Guo, Z., Hong, H., Yu, P., Xue, Z., Chen, H., 2021. Protein folding mechanism revealed by single-molecule force spectroscopy experiments. Biophysics Reports 7 (5), 399–412. Available at https://doi.org/10.52601/bpr.2021.210024, http://www.biophysics-reports.org/article/exportPdf?id=2d5e9761-d96b-49cb-91ff-5bdf8ccb8b83&language=en.

Tang, W.S., Zhong, E.D., Hanson, S.M., Thiede, E.H., Cossio, P., 2023. Conformational heterogeneity and probability distributions from single-particle cryo-electron microscopy. Current Opinion in Structural Biology 81. Available at https://doi.org/10.1016/j.sbi.2023.102626, http://www.elsevier.com/locate/sbi.

Teilum, K., Olsen, J.G., Kragelund, B.B., 2011. Protein stability, flexibility and function. Biochimica et Biophysica Acta - Proteins and Proteomics 1814 (8), 969–976. Available at https://doi.org/10.1016/j.bbapap.2010.11.005.

Trincao, J., Hamilton, M.L., Christensen, J., Pearson, A.R., 2013. Dynamic structural science: recent developments in time-resolved spectroscopy and X-ray crystallography. Biochemical Society Transactions 41 (5), 1260–1264. Available at https://doi.org/10.1042/BST20130125, http://www.biochemsoctrans.org/bst/041/1260/0411260.pdf.

Tsutsui, Y., Wintrode, P.L., 2012. Hydrogen/deuterium exchange-mass spectrometry: a powerful tool for probing protein structure, dynamics and interactions. Frontiers in Medicinal Chemistry 6, 291–323. Available at https://doi.org/10.2174/978160805464011306014, http://www.eurekaselect.com/node/53553/frontiers-in-medicinal-chemistry.

van Dijk, E., Hoogeveen, A., Abeln, S., Grubmüller, H., 2015. The hydrophobic temperature dependence of amino acids directly calculated from protein structures. PLoS Computational Biology 11 (5), e1004277. Available at https://doi.org/10.1371/journal.pcbi.1004277.

Vant, J.W., Sarkar, D., Nguyen, J., Baker, A.T., Vermaas, J.V., Singharoy, A., 2022. Exploring cryo-electron microscopy with molecular dynamics. Biochemical Society Transactions 50 (1), 569–581. Available at https://doi.org/10.1042/BST20210485, https://portlandpress.com/biochemsoctrans/article/50/1/569/230848/Exploring-cryo-electron-microscopy-with-molecular.

Veitshans, T., Klimov, D., Thirumalai, D., 1997. Protein folding kinetics: timescales, pathways and energy landscapes in terms of sequence-dependent properties. Folding and Design 2 (1), 1–22. Available at https://doi.org/10.1016/s1359-0278(97)00002-3.

Videbæk, T.E., Hayakawa, D., Hagan, M.F., Grason, G.M., Fraden, S., Rogers, W.B., 2025. Measuring multisubunit mechanics of geometrically programmed colloidal assemblies via cryo-EM multi-body refinement. Proceedings of the National Academy of Sciences 122 (37). Available at https://doi.org/10.1073/pnas.2500716122.

Wang, J., Oliveira, R.J., Chu, X., Whitford, P.C., Chahine, J., Han, W., et al., 2012. Topography of funneled landscapes determines the thermodynamics and kinetics of protein folding. Proceedings of the National Academy of Sciences 109 (39), 15763–15768. Available at https://doi.org/10.1073/pnas.1212842109.

Wei, G., Xi, W., Nussinov, R., Ma, B., 2016. Protein ensembles: how does nature harness thermodynamic fluctuations for life? The diverse functional roles of conformational ensembles in the cell. Chemical Reviews 116 (11), 6516–6551. Available at https://doi.org/10.1021/acs.chemrev.5b00562.

Weis, D.D., 2016. Hydrogen Exchange Mass Spectrometry of Proteins: Fundamentals, Methods, and Applications. Wiley. Available at http://doi.org/10.1002/9781118703748.

Westermarck, J., Ivaska, J., Corthals, G.L., 2013. Identification of protein interactions involved in cellular signaling. Molecular and Cellular Proteomics 12 (7), 1752–1763. Available at https://doi.org/10.1074/mcp.R113.027771, http://www.mcponline.org/content/12/7/1752.full.pdf+html.

Woody, R.W., Koslowski, A., 2002. Recent developments in the electronic spectroscopy of amides and α-helical polypeptides. Biophysical Chemistry 101–102, 535–551. Available at https://doi.org/10.1016/S0301-4622(02)00187-4.

Wu, L., Huang, C., Emery, B.P., Sedgwick, A.C., Bull, S.D., He, X.P., et al., 2020. Förster resonance energy transfer (FRET)-based small-molecule sensors and imaging agents. Chemical Society Reviews

49 (15), 5110–5139. Available at https://doi.org/10.1039/c9cs00318e, http://pubs.rsc.org/en/journals/journal/cs.

Yang, L.Q., Ji, X.L., Liu, S.Q., 2013. The free energy landscape of protein folding and dynamics: a global view. Journal of Biomolecular Structure and Dynamics 31 (9), 982–992. Available at https://doi.org/10.1080/07391102.2012.748536, http://www.tandfonline.com/loi/tbsd20.

Yang, L.Q., Sang, P., Tao, Y., Fu, Y.X., Zhang, K.Q., Xie, Y.H., et al., 2014. Protein dynamics and motions in relation to their functions: several case studies and the underlying mechanisms. Journal of Biomolecular Structure and Dynamics 32 (3), 372–393. Available at https://doi.org/10.1080/07391102.2013.770372.

Zavrtanik, U., Lah, J., Hadži, S., 2024. Estimation of peptide helicity from circular dichroism using the ensemble model. Journal of Physical Chemistry B 128 (11), 2652–2663. Available at https://doi.org/10.1021/acs.jpcb.3c07511, http://pubs.acs.org/journal/jpcbfk.

Zhong, E.D., Bepler, T., Davis, J.H., Berger, B., 2020. Reconstructing continuous distributions of 3D protein structure from cryo-EM images. In: 8th International Conference on Learning Representations (ICLR 2020). International Conference on Learning Representation.

Zhou, X., Li, Y., Zhang, C., Zheng, W., Zhang, G., Zhang, Y., 2022. Progressive assembly of multi-domain protein structures from cryo-EM density maps. Nature Computational Science 2 (4), 265–275. Available at 10.1038/s43588-022-00232-1.

Zhuravlev, P.I., Papoian, G.A., 2010. Protein functional landscapes, dynamics, allostery: a tortuous path towards a universal theoretical framework. Quarterly Reviews of Biophysics 43 (3), 295–332. Available at https://doi.org/10.1017/S0033583510000119.

SECTION II

Methodologies: experimental and integrative techniques

CHAPTER 5

Cryo-electron microscopy in membrane protein structural biology

Tofayel Ahmed[1,2], Antara Saha[1,2], and Debalaya Mukherjee[1,2]
[1]Saha Institute of Nuclear Physics, Biophysical Sciences Division, Kolkata, West Bengal, India,
[2]Homi Bhaba National Institute, Mumbai, Maharashtra, India

5.1 Introduction

Membrane proteins perform essential cellular functions including signal transduction, energy conversion, molecular transport, and cell–cell communication. They constitute approximately 30% of the human proteome and represent over 60% of current drug targets, yet their structural characterization has historically been slow due to many technical challenges (Loll, 2003; Overington, Al-Lazikani and Hopkins, 2006; Cheng, 2018a; Levental and Lyman, 2023). The most fundamental challenge for structural work on membrane proteins is that they need to be extracted into an enclosure of detergent micelles that is considered much harsher than their native lipid bilayer environment. This harsh treatment often causes protein unfolding and aggregation, creating a barrier to obtaining a folded and homogeneous protein population. If the folding is retained after successful extraction, the yield is often very low, rendering structural work impossible. This is especially the predominant event when working with native membranes. Therefore scientists resort to expressing the proteins in heterologous systems (Birch et al., 2020). However, these systems are not ideal because they do not accurately mimic the lipid composition of the native source, lack necessary machinery for proper posttranslational modifications, and, often, the overexpression may cause cellular stress and toxicity in the host organism. Additionally, these systems may lack essential chaperones or cofactors required for correct protein folding and function.

Membrane proteins are inherently dynamic, and they exist in multiple conformational states that are essential to their functional role. This conformational heterogeneity, while biologically important, severely hampered traditional crystallographic approaches requiring homogeneous samples for structure determination. It is already well accepted that hydrophobic surfaces and detergent micelles prevent the formation of stable crystal contacts. Therefore X-ray crystallography, the dominant structural biology technique for most of the 20th century, was not widely adopted for structural studies on membrane proteins. In contrast, cryo-electron microscopy (cryo-EM) offers several distinct and well-defined advantages for membrane protein structure determination

Advances in Structural Biology
DOI: https://doi.org/10.1016/B978-0-443-44988-8.00018-6

(Benjin and Ling, 2020). First, cryo-EM typically requires only microliter-scale sample volumes, often at concentrations of ~0.5–5 mg mL^{-1}, which is substantially lower than the amounts historically required for crystallization-based approaches, particularly for large or heterogeneous membrane protein complexes. Second, specimens are preserved by rapid vitrification, capturing proteins in near-native states within detergent micelles, lipid nanodiscs, amphipols, or native membrane fragments (Dubochet et al., 1988). This preservation minimizes dehydration and lattice-induced artifacts, thereby maintaining physiologically relevant conformations and lipid–protein interactions. Third, cryo-EM bypasses the need for crystal growth, eliminating a major time-consuming bottleneck in membrane protein structural biology, and it allows structure determination on reasonable timescales, often within weeks to a few months once a stable biochemical preparation is available. Finally, cryo-EM is uniquely well-suited to conformationally dynamic and flexible membrane proteins, as modern image classification strategies enable the separation and reconstruction of multiple coexisting structural states from a single experimental dataset. On a more technical note, at comparable nominal resolution, cryo-EM maps can frequently exhibit clearer density for certain fine features in the macromolecular complex, including ordered solvent and small ligands, than is typically observed in X-ray electron-density maps. This is due to the direct retrieval of both amplitudes and phases in electron micrographs, unlike X-ray data, where phase retrieval is tricky, which often leads to phase errors that strongly disrupt the electron density of finer macromolecular features (Cheng, 2015; Kühlbrandt, 2019). Together, these capabilities have transformed cryo-EM from a static imaging method into a powerful tool for elucidating mechanistic landscapes, conformational equilibria, and functional transitions in membrane protein systems. The so-called "resolution revolution" in cryo-EM, driven primarily by the introduction of direct electron detectors (DEDs) together with advances in image-processing algorithms, microscope stability, and automation, fundamentally reshaped structural biology, especially the field of membrane proteins (Kühlbrandt, 2014). DEDs provided a step change in detector quantum efficiency and enabled dose-fractionated movie acquisition, allowing correction of beam-induced specimen motion and radiation damage at the single-particle level. These developments substantially improved signal-to-noise ratios and made near-atomic (~2–3 Å) 3D reconstructions achievable for many membrane proteins that had previously resisted high-resolution analysis. In parallel, major advances in image-processing methodologies, particularly Bayesian maximum-likelihood–based alignment and three-dimensional classification algorithms, have enabled the computational separation of structurally heterogeneous particle populations (Frank, 2009; Scheres, 2012a, 2012b; Carroni and Saibil, 2016; Cheng, 2018b). This capacity to resolve multiple coexisting conformational states from a single dataset directly addressed the long-standing problem of conformational and compositional heterogeneity that often precluded membrane protein crystallization.

The first structural insights into a membrane protein were obtained in 1975 at ~7 Å resolution using electron crystallography, a method based on EM imaging of 2D crystals and a direct conceptual precursor of modern-day cryo-EM. In this landmark study, Richard Henderson and Nigel Unwin analyzed naturally occurring 2D crystalline arrays of bacteriorhodopsin embedded within the native cell membrane of *Halobacterium*, revealing the arrangement of seven transmembrane (7TM) α-helices and establishing membrane protein architecture at the level of secondary structure (Henderson and Unwin, 1975). This work provided the first experimental demonstration that high-resolution structural information could be extracted from membrane proteins without removing them from a lipid environment. Over the subsequent decade and a half, continued methodological advances in electron optics, specimen preparation, and image reconstruction progressively improved the attainable resolution. By 1990, the electron crystallography structure of bacteriorhodopsin had beeen resolved at ~3.5 Å resolution, enabling unambiguous side-chain placement and providing a detailed description of the proton translocation pathway and retinal binding pocket. This study represented one of the earliest atomic-level mechanism-oriented structural descriptions of a membrane protein (Henderson et al., 1990). This achievement represented a decisive proof that electron-based approaches could rival X-ray crystallography in fetching structural detail, while still being able to retaining a membrane-embedded context. The broader applicability of electron crystallography to complex membrane assemblies was firmly established in 1994 with the determination of the light-harvesting chlorophyll a/b–protein complex at ~3.4 Å resolution by Kühlbrandt et al. (1994). This structure constituted the first multisubunit membrane protein complex solved at near-atomic resolution by electron crystallography, demonstrating that this structural approach was not limited to small or single-component biological systems. Consequently, until 2013, all the deposited PDB coordinates of cryo EM structures were, in fact, derived through the application of electron crystallography (Cheng, 2018a). In a parallel trajectory, the development of single-particle cryo-EM ushered in a new era in 2014, removing the requirement for crystalline order altogether (Cheng, 2018b; Kühlbrandt, 2022). Single-particle cryo-EM enabled high-resolution structure determination across a broad spectrum of membrane protein families, including transient receptor potential (TRP) channels, potassium (K^+) channels, G-protein-coupled receptors (GPCRs), gamma-aminobutyric acid type A receptor ($GABA_AR$), bacterial secretion systems, NADH:ubiquinone oxidoreductase (complex I), and ATP synthase. These examples were selected to span diverse biological functions, structural complexities, and evolutionary origins, and together they highlight how cryo-EM has provided unprecedented insights into membrane protein architecture, conformational dynamics, and molecular mechanisms. The following sections examine each system separately, emphasizing the understanding of the structure–function relationship in these membrane proteins and membrane protein complexes, as enabled by the cryo-EM revolution.

5.2 Transient receptor potential channels

In 2013, near-atomic resolution structures of the capsaicin receptor TRPV1, a member of the TRP family, were solved at ~3.4–4.2 Å resolution using cryo-EM (Cao et al., 2013; Liao et al., 2013). These landmark studies represented the first detailed visualization of a eukaryotic ion channel by cryo-EM, revealing its tetrameric domain-swapped architecture, the ion-conducting pore, and the ligand-binding pocket that accommodates capsaicin and resiniferatoxin (RTX). Following this TRPV1 breakthrough, the structural exploration of ion channels via cryo-EM rapidly expanded. Within the next five years, near-atomic resolution structures of other mammalian TRP channels, for example, TRPA1, TRPV2, TRPM4, and TRPC3 were reported, providing molecular insights into the structure, gating, ion selectivity, lipid modulation, and pharmacology of this channel family (Huynh et al., 2016; Fan et al., 2018; Sierra-Valdez et al., 2018; Guo et al., 2017; Winkler et al., 2017; Zubcevic et al., 2016; Paulsen et al., 2015; Zhao, McVeigh and Moiseenkova-Bell, 2021; Pumroy et al., 2022; Rocereta et al., 2025).

Cryo-EM has transformed our understanding of TRP channel activation by providing high-resolution views of how diverse ligands engage distinct structural domains to modulate gating. A total of forty sites have been reported so far across different mammalian TRP channel subfamilies, which aid in ligand binding (Fig. 5.1). Across TRP channel subfamilies, four principal ligand-binding regions have emerged: (1) the membrane-facing vanilloid pocket, (2) the voltage sensor–like (VSL) domain comprising S1–S4, (3) the pore domain formed by S5–S6 and the selectivity filter, and (4) cytoplasmic domains including the ankyrin repeat domain (ARD), linker helices, N-linker, and calmodulin-binding motifs. Each region exhibits unique pharmacological and structural signatures, reflecting the polymodal sensory capabilities of TRP channels.

The vanilloid pocket constitutes a conserved ligand-binding cavity in multiple TRP channel clades (Zhao, McVeigh, and Moiseenkova-Bell, 2021). Structurally, it is formed at the interface of the S3 helix and S4–S5 linker of one subunit and the S6 helix of the adjacent subunit, with direct continuity to the cytosolic leaflet of the plasma membrane. Initial cryo-EM visualizations in TRPV1 channels established its presence and functional relevance (Cao et al., 2013; Liao et al., 2013). Endogenously, the site is commonly occupied by phosphatidylinositol-class lipids, which bias the channel toward a closed conformation. Exogenous ligands tune gating; for example, capsaicin induces partial opening, whereas combined binding of RTX and DkTx captures the pore in a maximally open state (Cao et al., 2013). In contrast, competitive antagonists such as AMG-9810 and SB366791 bind the vanilloid pocket to promote a closed state of the channel, demonstrating how different chemotypes bias the gating energy landscape in opposing directions (Neuberger et al., 2023; Gao et al., 2026).

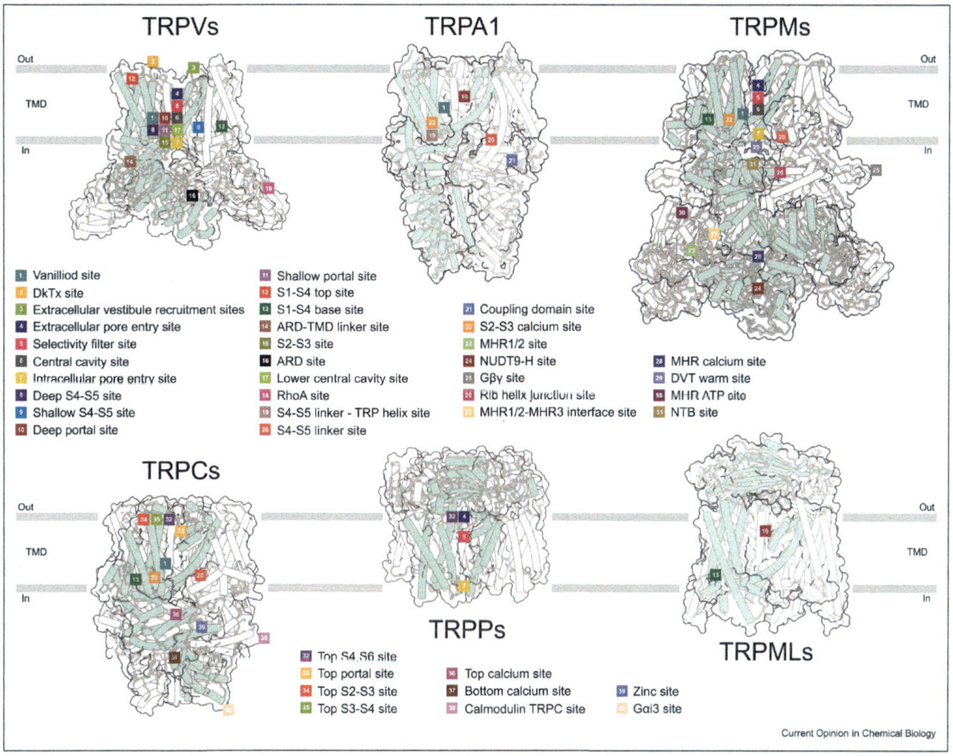

Figure 5.1 Ligand binding sites in transient receptor potential (*TRP*) channels. The 40 ligand-binding sites in TRP channels are shown and marked. The structures bear PDB IDs: 8GF8 (TRPV1), 6PQQ (TRPA1), 6PUO (TRPM2), 7WDB (TRPC5), 8HK7 (TRPP2) and 7SQ8 (TRPML1) (Talyzina et al., 2025). *Reproduced from Talyzina, I.A., Nadezhdin, K.D., Sobolevsky, A.I., 2025. Forty sites of TRP channel regulation. Current Opinion in Chemical Biology. 84, 102550. 10.1016/j.cbpa.2024.102550., licensed under a Creative Commons Attribution 4.0 International (CC BY 4.0) license.*

The VSL domain, comprising helices S1–S4, forms a second major regulatory site. Although TRP channels mostly lack the canonical S4 arginine "gating charge" characteristic of voltage-gated cation channels, cryo-EM revealed that the VSL bundle acts as a docking platform for multiple classes of modulators. For example, cooling agents such as menthol and icilin, and phosphoinositides bind within interfacial cavities formed by S1–S4, the preS1 helix, the S4–S5 linker, and the TRP helix in TRPM8 (Yin et al., 2022). Similarly, in TRPV3, structural and mutagenesis studies have identified multiple putative 2-aminoethoxydiphenyl borate (2-APB) binding sites, including one at the base of the VSLD and another in the preS1/ARD-TMD interface (Singh, McGoldrick, and Sobolevsky, 2018). These interactions do not directly open the pore; instead, they reshape the coupling between the VSL and the pore domain, shifting thermal activation thresholds and ligand sensitivity.

Compared to the allosteric mechanism, a direct pore-blocking mechanism in TRP channels is less studied by cryo-EM. Cryo-EM structures of TRPV2 and TRPV5 with the inorganic molecule ruthenium red (RR) reveal how this molecule can act as a bona fide pore blocker in the TRPV subfamily (Pumroy et al., 2024). It was shown that RR is capable of blocking both channels in the apo as well as activator-bound states by binding at the selectivity filter region (Pumroy et al., 2024). A few molecules have been reported to modulate TRP channel functionality by binding to the cytoplasmic domains. One such molecule is ATP, which acts as an inhibitor for the TRPM4 channel. Interestingly, a cryo-EM study on the TRPM4 channel reveals two distinct ATP-binding sites in the cytoplasmic domain of the channel, and the occupancy of these sites was dependent on whether the channel was studied at cold or physiological temperatures (Hu et al., 2024).

Thus across all four ligand-binding regions, cryo-EM has revealed that TRP channels populate multiple conformational states—apo-closed, ligand-bound closed, sensitized, open, and desensitized/inactivated—reflecting the distributed allostery required for polymodal integration. This emergent structural framework explains how diverse chemical and physical stimuli converge on a conserved S6 activation gate through domain-specific ligand engagement and interdomain coupling.

5.3 Potassium (K^+) channels

For decades, the functional mechanism of potassium (K^+) channels puzzled researchers, as they are capable of conducting ions at near-diffusion rates (10^8 ions sec^{-1}) while maintaining varying degrees of exquisite selectivity against smaller sodium (Na^+) ions (Furini and Domene, 2009). The answer to this paradox required determining the atomic architecture of the channel pore, a challenge that first yielded to crystallography before being extensively described by cryo-EM.

The structural era began in 1998 with the elucidation of the KcsA channel from *Streptomyces lividans* by Roderick MacKinnon and colleagues using X-ray crystallography (Doyle et al., 1998). This work was pivotal because it provided the first atomic view of the selectivity filter, the specialized and conserved sequence motif (TVGYG) responsible for ion discrimination. The electron-density maps revealed that the backbone carbonyl oxygen atoms of the filter residues project into the pore to form four potential K^+ binding sites. These oxygen atoms are arranged to closely resemble the hydration shell of a K^+ ion in aqueous solution. When a hydrated K^+ ion enters the pore, it sheds its water molecules and is coordinated by eight carbonyl oxygen atoms in a "snug-fit" geometry (Doyle et al., 1998; Bezanilla and Armstrong, 1972). This coordination provides an energetic compensation that balances the dehydration penalty. In contrast, the smaller Na^+ ion cannot contact all the oxygen ligands simultaneously, making its dehydration energetically unfavorable, which effectively excludes it from the pore (Doyle et al., 1998).

Early interpretations of KcsA structures favored the "soft knock-on" model, in which K^+ ions alternate with water molecules inside the selectivity filter, reducing electrostatic repulsion and allowing concomitant water copermeation (Doyle et al., 1998; Zhou et al., 2001; Morais-Cabral, Zhou, and MacKinnon, 2001). This was the canonical view through the 1990s–early 2010s based on X-ray crystallography and electrophysiology. The subsequent "hard/direct knock-on" model emerged later (post 2010), from long molecular dynamics (MD) simulations and newer structural data, showing fully dehydrated K^+ ions in direct contact of each other ("KKKK"), which move through the selectivity filter solely by Coulombic repulsion without water copermeation (Köpfer et al., 2014; Kopec et al., 2018; Li, Rietmeijer, and Brohawn, 2020; Lee et al., 2025; Lee et al., 2025). Today, the "direct knock-on" model is often considered the dominant mechanism in wild-type K^+ channels, consistent with multiple studies based on MD, solid-state NMR, and time-resolved works, while the "soft knock-on" appears in mutated or inactivated states where water molecules enter the selectivity filter (Köpfer et al., 2014; Lee et al., 2025). X-ray crystallography was also instrumental in defining the basic architecture of these voltage-gated channels. The structure of the mammalian Kv1.2 channel confirmed that the voltage-sensing domains (VSDs) are independent units that are peripherally attached to the pore and are mechanically coupled to the channel gate via a helical linker (Long et al., 2005a, 2005b). These structures provided the framework for the "sliding helix" or "helical screw" models of voltage gating, where positive charges on the S4 helix move through a focused electric field to open the pore (Long et al., 2005a, 2005b; Catterall, 1986, 2010).

However, the X-ray studies faced intrinsic biophysical limitations. To achieve the ordered lattices required for diffraction, researchers frequently had to truncate flexible intracellular domains, remove auxiliary subunits, and extract the proteins into detergents that stripped away essential boundary lipids (Doyle et al., 1998; Long et al., 2005a; Bajaj, 2025). This created a "reductionist" view of the channel. Also, the crystal structures of Kv channels were often captured in nonnative, detergent-stabilized environments, which lacked a membrane lateral pressure profile, thus raising controversy about whether the VSDs corresponded to resting or activated states (Long et al., 2005a, 2005b; Cantor, 1997; Jiang et al., 2003a; Jiang et al., 2003b; Bezanilla, 2008) of the channels. Furthermore, many K^+ channels, such as the high-conductance Ca^{2+}-activated K^+ (BK) channel or the ATP-sensitive K^+ (K_{ATP}) channel, had proved resistant to crystallization, and they finally yielded to cryo-EM structure determination. Although the precise reasons are difficult to pinpoint, it is well recognized that large molecular size, multidomain architecture, and pronounced conformational heterogeneity significantly hinder crystallization of ion channels (Lee and Cui, 2010; Li et al., 2017).

The advent of the "resolution revolution" in cryo-EM around 2013 dismantled these barriers, fundamentally altering our understanding of channel structure, gating, and regulation (Kühlbrandt, 2014). Cryo-EM allows for the imaging of proteins in a frozen-hydrated state without the need for crystallization, preserving them in near-native conformations. Crucially, this method facilitated the use of lipid nanodiscs, patches of lipid bilayer scaffolded by proteins, enabling researchers to visualize channels within a membrane environment (Fig. 5.2). The impact of cryo-EM on elucidating

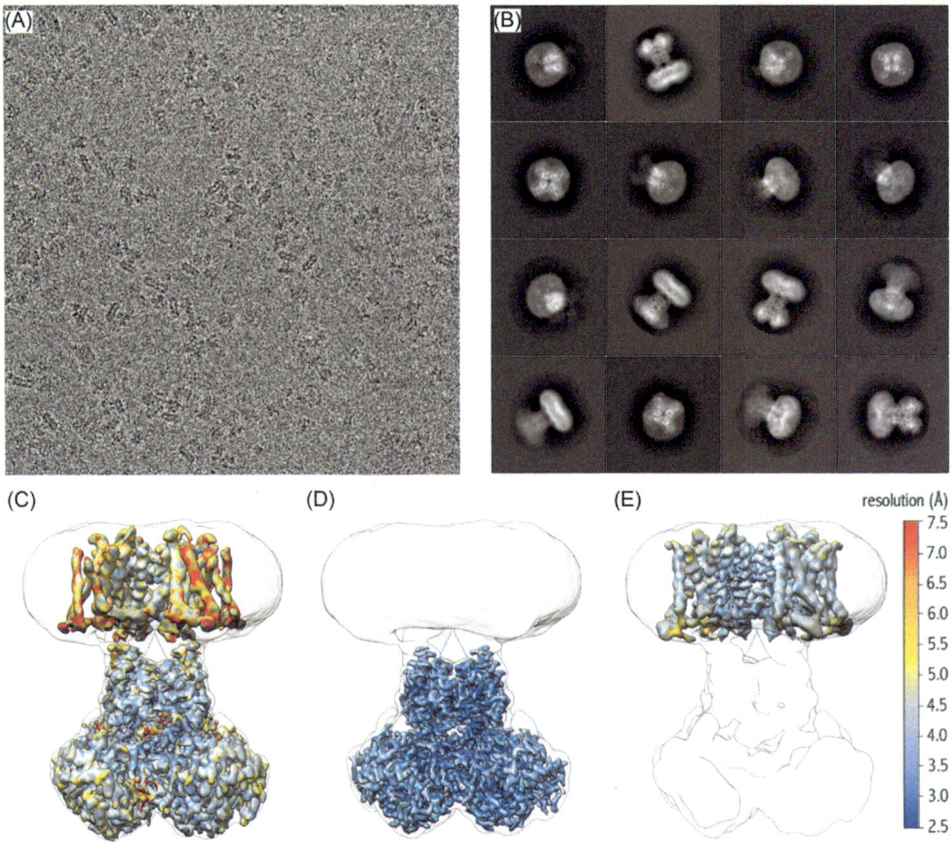

Figure 5.2 Cryo-electron microscopy data and 3D reconstruction of K_V chimera in lipid nanodisc. (A) Cryo-EM image of K_V chimera collected at −1.7 μm defocus. (B) Good 2D class averages showing top, tilted, and side views of the channel. (C) K_V chimera cryo-EM map colored according to local resolution. (D) The cytosolic domain of the K_V chimera was refined separately and shown according to local resolution coloring. (E) The transmembrane domain of the K_V chimera was refined separately and displayed by local resolution coloring. In (C), (D), and (E), low-resolution 3D class envelopes in transparent gray densities are overlaid onto high-resolution maps. *Reproduced from Matthies, D., Bae, C., Toombes, G.E.S., Fox, T., Bartesaghi, A., Subramaniam, S., et al., 2018. Single-particle cryo-EM structure of a voltage-activated potassium channel in lipid nanodiscs. eLife 7, public domain (CC0 1.0).*

functional mechanisms is perhaps best exemplified by the BK channel (Tao, Hite, and MacKinnon, 2017; Tao and MacKinnon, 2019). For years, the mechanism by which intracellular Ca^{2+} binding opened the TM pore was speculative because the massive intracellular "gating ring" (RCK domains) was too flexible to be captured by crystallography. Cryo-EM structures of the full-length *Aplysia californica* BK channel revealed a literal "pulley" mechanism. High-resolution maps showed that Ca^{2+} binding causes a rigid-body rotation of the gating ring, which physically pulls on the S6 TM helices via peptide linkers to expand the inner pore (Tao, Hite, and MacKinnon, 2017; Tao and MacKinnon, 2019). This direct visualization of electromechanical coupling in a massive macromolecular machine was a feat impossible with X-ray methods.

Similarly, cryo-EM resolved the complex assembly of the K_{ATP} channel, an octamer of more than 800 kDa in size consisting of four pore-forming Kir6.2 subunits and four regulatory SUR1 subunits. X-ray crystallography had not successfully captured the interface between these distinct modules. Cryo-EM structures not only visualized this interface but also clarified the mechanism of nucleotide gating (Li et al., 2017; Lee, Chen, and MacKinnon, 2017; Martin et al., 2017). They showed how ATP binding to the Kir6.2 subunit induces a tightening of the cytoplasmic domain that constricts the pore, while Mg^{2+}-nucleotide binding to the SUR1 subunit causes a conformational change that mechanically pulls the pore open.

Furthermore, cryo-EM has reshaped our understanding of C-type inactivation, a time-dependent process in which the channel stops conducting even while the activation gate remains open. Crystallographic studies of KcsA suggested that inactivation involved a "collapse" of the selectivity filter, physically blocking ion flux (Zhou et al., 2001; Cuello et al., 2010; Ostmeyer et al., 2013). However, cryo-EM structures of eukaryotic K_v channels in lipid nanodiscs have revealed a more multifaceted reality. In native-like lipids, the inactivated state often involves subtle side-chain reorientations behind the selectivity filter rather than a total collapse, suggesting that the "collapsed" structures seen in crystals may have been artifacts of the detergent environment or crystal packing forces (Matthies et al., 2018; Tan et al., 2022; Ong et al., 2022; Tyagi et al., 2022; Stix et al., 2023; Fernández-Mariño et al., 2023).

Cryo-EM structures of several eukaryotic K^+ channels have directly visualized lipids wedged into defined pockets between the VSD and pore domain, supporting the view that lipids act as structural cofactors rather than inert solvents. In human Kv3.1a, high-resolution cryo-EM maps reveal a well-defined native lipid density in the groove at the pore domain/VSD interface, located between the pore helix and the S1–S2 linker. The lipid headgroup forms specific polar interactions with His212 in the VSD and residues in S6, while its acyl chains pack against hydrophobic side chains in the pore helix, allowing the authors to explicitly relate this pocket to known lipid-modulatory sites in KcsA and the Kv1.2/2.1 chimera (Chi et al., 2022). Cryo-EM structures of the cardiac KCNQ1–CaM complex, captured with PIP_2 and the activator ML277, reveal PIP_2

bound in the cleft between the VSD and the pore domain. In this position, PIP_2 bridges the S4–S5 linker and S6 to stabilize the open activation gate. These findings provide a structural basis for the "uncoupled" state observed in functional studies, where the absence of PIP_2 prevents the pore from opening, even when the VSD is activated (Zaydman et al., 2013; Sun and MacKinnon, 2017; Ma et al., 2022). Together, these structures provide direct evidence that specific lipids occupy nonannular binding sites at the VSD–pore interface and participate in electromechanical coupling in voltage-gated K^+ channels.

In conclusion, by resolving multiple functional states of K^+ channels in native-like lipid environments, cryo-EM has expanded the structural view of these proteins from largely static, filter-centric architectures toward dynamic machines whose gating is shaped by electromechanical coupling, lipid cofactors, ions, and small-molecule modulators. The structural perspectives obtained by cryo-EM unify decades of electrophysiology and biophysics, indicating how K^+ channels integrate chemical and electrical cues to tune membrane excitability.

5.4 Gamma-aminobutyric acid type A receptors, 5-HT3A, and glycine receptors

$GABA_AR$ are a Cys-loop family of ligand-gated chloride channels in our central nervous system, and they are central to brain activity due to their involvement in GABAergic inhibitory neurotransmission (Ghit et al., 2021). These receptors are targeted in various neuropsychiatric ailments such as anxiety, insomnia, seizures, muscle spasms, and epilepsy, thus making them attractive candidates for structural studies involving cryo-EM. However, in the initial years, they could only be studied by homology modeling and not by experimental structural techniques. Obtaining crystals of $GABA_ARs$ is formidably challenging due to the inherent heterogeneity arising from heteromeric subunit architecture, multiple N-linked glycans, and long extracellular loops. In 2014, the first crystal structure of the human homo-pentameric $GABA_AR$, constituted of β3 subunits, was solved at 3 Å resolution (Miller and Aricescu, 2014). This provided the first glimpse into the pore architecture and subunit interfaces of any $GABA_AR$ assembly. The pentameric assembly of $GABA_AR$ is constructed from a pool of nineteen monomeric subunits, encoded by nineteen different genes. These subunits are structurally very similar, but they possess different pharmacological and physiological properties.

Structural exploration of $GABA_AR$ gained momentum after the resolution revolution in cryo-EM. We now have near-atomic maps of multiple full-length heteromeric $GABA_ARs$ in a lipid environment, revealing unexpected stoichiometries, subunit arrangements, and multiple ligand-bound states (Fig. 5.3). For example, we now have pentameric cryo-EM structures assembled from α1, β2, γ2 and α1, β3, γ2 subunits that

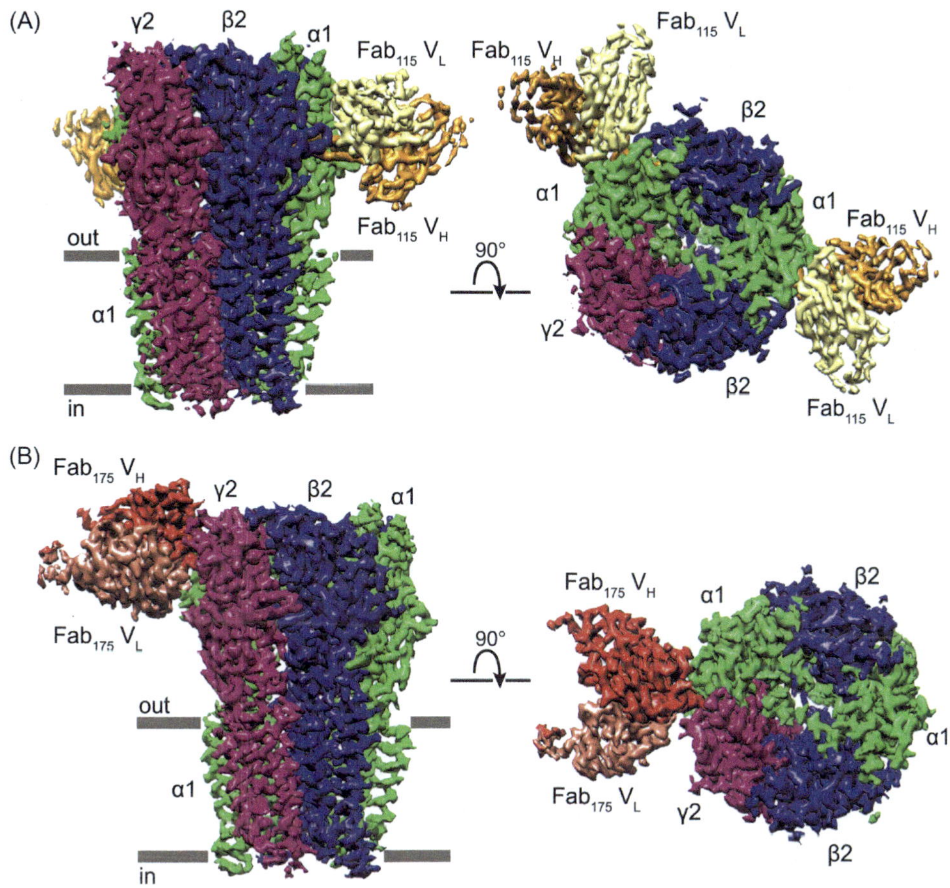

Figure 5.3 Cryo-electron microscopy (cyro-EM) reconstructions of $GABA_AR$ bound to different antibodies. (A) Cryo-EM reconstructions of $GABA_AR$ bound to Fab_{115} antibodies, shown from the side and top. (B) Same as (A), but Fab_{175} is used instead of Fab_{115}. Reproduced from Noviello et al. (2022). *Reprinted from Noviello, C.M., Kreye, J., Teng, J., Prüss, H., Hibbs, R.E., 2022. Structural mechanisms of GABAA receptor autoimmune encephalitis. Cell. 185, 2469–2477.e13. 10.1016/j.cell.2022.06.025, Copyright (2022), with permission from Elsevier.*

are reconstituted in lipid nanodiscs, capturing the receptor bound to a physiological agonist (GABA), classical benzodiazepines (e.g., diazepam, alprazolam, and flumazenil), and channel blockers (picrotoxin and bicuculline). These structures have enabled direct visualization of chemically and functionally distinct ligands in their receptor-bound forms (Zhu et al., 2018; Masiulis et al., 2019; Kim and Hibbs, 2021).

Among many such success stories involving membrane proteins, especially ion channels, two more families have captured attention: the 5-HT_{3A} receptor and glycine receptor (GlyR). The 5-HT_{3A} receptors are clinically important due to their implication in nausea and vomiting caused by radiation and chemotherapy. The serotonin-bound activated conformations were the first cryo-EM-enabled structures for this family, which were followed by structures with serotonin antagonists (palonosetron, ondansetron, and alosetron) in lipidic environments (Basak et al., 2018, 2020). Together, these structures have revealed the global conformational changes that underpin gating in the 5-HT_{3A} receptor and the presence of lateral portals that are implicated in ion permeation. Similarly, cryo-EM has transformed the field of GlyRs by enabling high-resolution structures of both homomeric and heteromeric GlyRs in multiple conformational states (Yu, Zhu, et al., 2021; Zhu, 2022; Kumar et al., 2020). These structural snapshots have revealed precise ligand-binding sites, subunit interfaces, and gating pathways, providing unprecedented insight into how the agonizts glycine and taurine, the partial agonist GABA, the antagonist strychnine, the noncompetitive pore-blocker picrotoxin, and the positive allosteric modulator ivermectin interact with GlyRs (Yu, Zhu, et al., 2021; Kumar et al., 2020, 2022; Gibbs et al., 2023; Du et al., 2015). With this atomic-level understanding, researchers now have a structural framework for rational drug design aimed at GlyRs, paving the way for more targeted screening of small-molecule modulators against disorders linked to GlyR dysfunction, including hyperekplexia (startle disease), chronic pain, spasticity, and certain forms of epilepsy.

5.5 Ion channel complexes from a native source

A significant conceptual shift has also gained momentum in the field to study channels in their native molecular environments using cryo-EM and single-particle analysis. This approach is different from those discussed, so far overexpressed and detergent-purified constructs that often lack native lipids, auxiliary proteins, or all the necessary posttranslational modifications. Very recently, native ion channels were isolated from mammalian brain tissue using gentle immuno-affinity purification, enabling high-resolution cryo-EM analysis of their near-physiological architectures (Yu, Rao, et al., 2021; Zhao et al., 2019; Sun et al., 2023). The AMPA-receptor structures using such an approach have revealed heterogeneous subunit stoichiometries and the precise arrangement of auxiliary proteins, demonstrating how these partners modulate pore shape, ligand-binding domain coupling, and gating mechanisms inside a native environment. Subsequent work on hippocampal AMPA receptors has defined the structural basis of allosteric inhibition by mapping modulator-binding pockets and showing how these molecules constrain conformational transitions required for channel opening (Yu, Rao, et al., 2021). Additionally, native $GABA_AR$ structures were captured in physiologically relevant subunit assemblies, which were bound to the neurotransmitter agonist GABA and positive allosteric modulators zolpidem, flurazepam, and allopregnanolone, thereby elucidating subtype-selective binding sites of these ligands and the conformational states of the receptor that underlie the pharmacological specificity of these ligands (Sun et al., 2023). Collectively, these structures demonstrate that cryo-EM of endogenous ion channel complexes can resolve physiologically relevant stoichiometries, auxiliary subunit arrangements, and ligand-stabilized conformations, informing subtype-selective pharmacology and structure-based therapeutic design. However, this approach to structure determination using cryo-EM is still rare, mainly due to sample scarcity and antibody dependence.

5.6 G-protein coupled receptors

In humans, GPCRs are generally categorized into several main families: class A (rhodopsin-like), class B (secretin-like), class C (glutamate-like), class F (frizzled/smoothened), and class T (adhesion GPCRs), among which class A is the largest and accounts for ~80% of the ~800 unique human GPCRs (Liu et al., 2024; Shihoya et al., 2024). GPCRs transduce extracellular signals into intracellular responses by engaging heterotrimeric G proteins belonging to four major families: Gi/o, Gq/11, Gs, and G12/13. These G proteins are composed of three subunits—Gα, Gβ, and Gγ—which assemble as a heterotrimer at the cytoplasmic face of the membrane. The Gα subunit, which defines the signaling family, binds guanine nucleotides and possesses intrinsic GTPase activity, whereas Gβ and Gγ form a stable obligate dimer (Gβγ) that functions as a signaling unit in its own right.

For much of the twentieth century, mechanistic understanding of signaling via GPCRs was derived primarily from pharmacology, mutagenesis, spectroscopy, and biochemical reconstitution, while high-resolution structural information on these receptors was lacking. The earliest structural explorations of GPCR structures that described the 7TM helical architecture for these receptors were done on 2D crystals made from frog rhodopsins (Unger et al., 1997). A major milestone was achieved with the elucidation of the X-ray crystallography structure of inactive-state bovine rhodopsin, which for the first time provided high-resolution insights into any GPCR structure (Palczewski et al., 2000). In the following decade, X-ray crystallography structures of additional class A GPCRs, including the β-adrenergic and adenosine receptors, confirmed this canonical fold and provided insights into ligand recognition within this TM core (Cherezov et al., 2007; Jaakola et al., 2008). However, these achievements typically relied on substantial protein engineering, such as thermostabilizing point mutations (e.g., StaRs for the A2A adenosine receptor and alanine/valine substitutions to rigidify TM helices), truncation of flexible intracellular loops (ICL) and terminal tails, and insertion of fusion partners such as T4 lysozyme or BRIL into ICL3—to obtain well-ordered crystals (Cherezov et al., 2007; Jaakola et al., 2008; Rasmussen et al., 2011; Doré et al., 2011; Chun et al., 2012; Heydenreich et al., 2015). While they are effective for structure determination, such modifications often stabilize receptors in inactive or preactive conformations and reduce their conformational flexibility. Consequently, X-ray crystallography seemed to be poorly suited for routinely capturing the native, fully assembled GPCR signaling complexes. In 2011, Rasmussen et al. published a landmark structural study on a GPCR–G-protein complex through exceptional stabilization strategies (Rasmussen et al., 2011). The authors provided the first crystal structure of an active GPCR–G-protein complex, the β_2-adrenergic receptor bound to G proteins (β_2AR–Gs), revealing how a high-affinity agonist BI-167107 binding drives intracellular conformational changes that enable G-protein engagement. This engagement, as pointed out by the authors, is purely mediated by the Ras-like-GTPase domain of Gα, and the Gβ and Gγ subunits refrain from this participation. The application of cryo-EM eventually entered the scene to rapidly broaden access to signaling-competent GPCR assemblies across multiple G-protein subtypes, including Gi/o- and Gq/11-coupled receptors (Shihoya et al., 2024; Draper-Joyce et al., 2018; García-Nafría and Tate, 2019; Maeda et al., 2019; García-Nafría and Tate, 2021; Gusach, García-Nafría, and Tate, 2023; Duan et al., 2024). These structures together revealed a conserved activation geometry characterized by outward displacement of TM helix 6, formation of a cytoplasmic cavity, and the insertion of the α5-helix located at the Gα C-terminal end into the GPCR core. Additionally, they uncovered receptor- and subtype-specific variations in ICL2 and ICL3 engagement with the G-protein and in the architecture of the α5-helix binding interface that determine coupling selectivity and influence the efficiency of nucleotide exchange and G-protein activation. Importantly, cryo-EM enabled determination of these complexes with a reduced reliance on fusion partners for receptors or on stabilizing nanobodies and permitted analysis in detergent micelles and lipid nanodiscs that better recapitulate the bilayer milieu and lipid-mediated allosteric interactions (Fig. 5.4).

Beyond visualizing active GPCR–G protein complexes, cryo-EM has fundamentally advanced our understanding of GPCR signaling by resolving arrestin-bound states and alternative effector engagements. While initial structural insights were obtained via X-ray crystallography (e.g., β1AR–β-arrestin1), subsequent cryo-EM studies on rhodopsin and β2AR complexes have revealed that arrestins engage both the phosphorylated C-terminal tail and the receptor core through multiple, topologically distinct configurations (Shukla et al., 2013; Kang et al., 2015; Zhou et al., 2017; Tsai et al., 2019; Staus et al., 2020). These studies demonstrate that arrestin-bound receptors adopt conformations distinct from those stabilized by G proteins. Furthermore, biophysical analyzes reveal significant structural heterogeneity, driven by both variable C-tail phosphorylation patterns and the depth of arrestin core engagement. Collectively, these structures provide a structural framework for understanding GPCR desensitization, internalization, and arrestin-mediated signaling and offer mechanistic support for the long-standing concept of ligand bias by

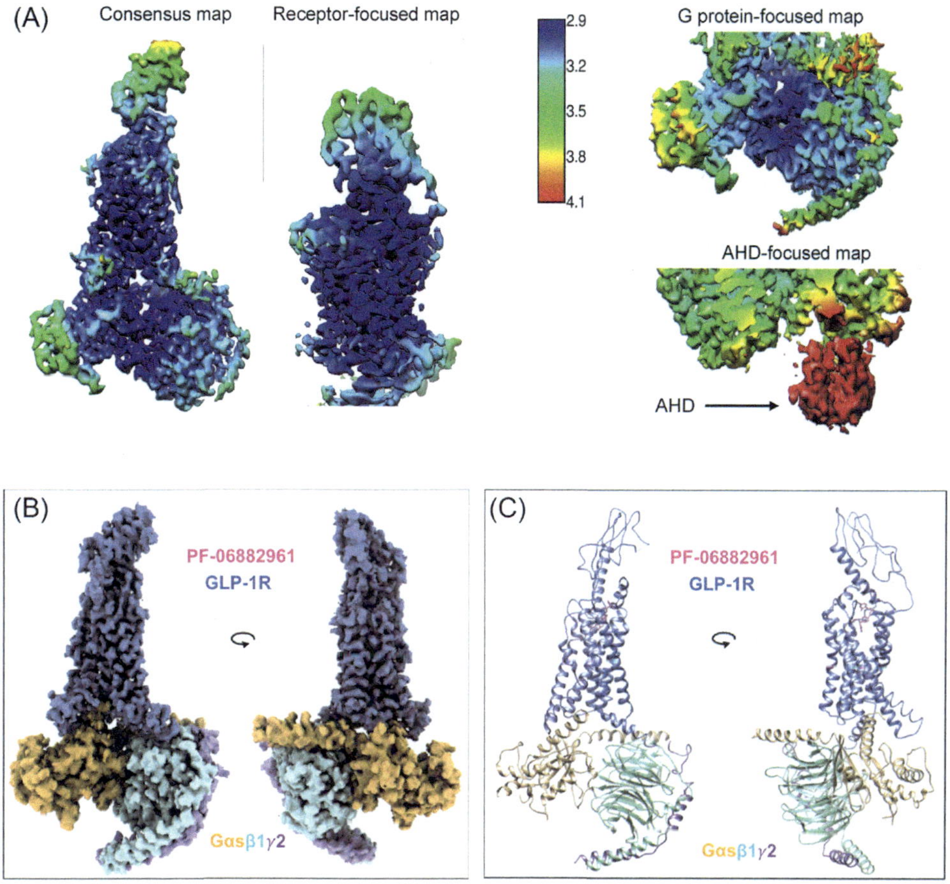

Figure 5.4 Cryo-electron microscopy (Cryo-EM) density of GLP-1R, a GPCR, with dominant-negative G protein and Pfizer-developed small molecule 06882961. (A) Cryo-EM reconstruction of the complex between GLP-1R, G-protein, and molecule 06882961, colored according to local resolution. (B) Orthogonal views of the cryo-EM density obtained by combining consensus and locally refined maps, using a 2 Å zone and individual component masks for model fitting in UCSF ChimeraX. (C) Ribbon representations of the PF-06882961–GLP-1R–Gs complex backbone, with coloring indicating individual structural elements as annotated in the figure. Reproduced from Zhang et al. (2021). *Reprinted from Zhang, X., Johnson, R.M., Drulyte, I., Yu, L., Kotecha, A., Danev, R., Wootten, D., Sexton, P.M., Belousoff, M.J., 2021. Evolving cryo-EM structural approaches for GPCR drug discovery. Structure. 29, 963–974.e6. 10.1016/j.str.2021.04.008, Copyright (2021), with permission from Elsevier.*

demonstrating that differential effector coupling can be encoded through receptor conformational states and C-tail phosphorylation patterns.

In parallel, advances in three-dimensional classification, focused refinement, and heterogeneity analysis enabled cryo-EM datasets to be partitioned into structurally distinct subpopulations, revealing that GPCRs occupy ensembles of conformations rather than a single, fully active state. Analyzes of agonist-bound receptors and receptor–effector complexes identified coexisting intermediate and active-like conformations under identical biochemical conditions, consistent with models in which receptor activation proceeds through multiple conformational steps. Taken together, these findings established cryo-EM as an important tool to decipher fundamental mechanisms involving GPCRs and also guide drug discovery targeting these receptors.

5.7 Bacterial secretion systems

Cryo-EM has fundamentally reshaped our understanding of bacterial secretion systems by overcoming long-standing barriers posed by their size, membrane embedding, and dynamic nature. Historically, bacterial secretion complexes were studied through genetics, biochemistry, and low-resolution EM, which together established component lists and rough architectural cartoons but failed to explain how these machines were assembled, energized, or operated. X-ray crystallography provided atomic detail for isolated soluble domains, yet could not capture intact secretion systems that span multiple membranes while simultaneously containing flexible, transient subunits. Cryo-EM emerged as the ideal method to study these systems precisely because the secretion systems are large, oligomeric, and often symmetric, making them amenable to particle averaging for 3D reconstruction.

Broadly, the secretion systems are classified into two types based on their functional mechanism: one-step and two-step secretion systems. In the former, substrates are transported directly from the cytoplasm to the exterior of the cell, or straight into a target host cell—in a single, continuous process. These systems form a trans-envelope channel that bypasses the periplasm entirely. The substrate does not need to fold in the periplasm and is typically secreted in an unfolded or partially unfolded state. Classic examples include the Type I secretion system (T1SS), Type III secretion system (T3SS), the canonical variant of the Type IV secretion system (T4SS), and the Type VI secretion system (T6SS). Despite their very different architectures and biological roles, they share a common conceptual principle of tight coupling between energy consumption and translocation, and secretion occurs through a single, continuous conduit spanning both membranes. Cryo-EM has been particularly powerful for structure determination of these systems because it helps visualize the entire trans-envelope assembly and, in some cases, capture distinct functional states such as contraction (T6SS) or needle engagement (T3SS).

Hu and colleagues reported a high-resolution in situ structure of the complete Type III secretion system (T3SS) injectisome in intact *Salmonella* cells using cryo-ET and subtomogram averaging (Hu et al., 2017). The injectisome was visualized spanning the inner membrane, periplasm, and outer membrane within the native bacterial envelope. A continuous axial density connected the cytoplasmic components to the extracellular needle, demonstrating an uninterrupted secretion conduit. The basal body was resolved as a multiring assembly embedded in both membranes. The inner-membrane export apparatus was localized beneath the basal body and shown to be closely associated with the secretion channel. A large cytosolic sorting platform was clearly visualized below the basal body. This sorting platform appeared as a cage-like structure aligned coaxially with the secretion channel. Genetic tagging and deletion analyzes confirmed that it contains the ATPase and associated scaffold proteins (Hu et al., 2017; Soto, Galán, and Lara-Tejero, 2022). Comparison of injectisomes with and without the sorting platform revealed significant structural differences. Assembly of the sorting platform induced conformational remodeling of the basal body, including compression of the inner-membrane rings. These observations suggested a functional coupling between cytosolic and membrane-embedded components. The study revealed how symmetry mismatches between the sorting platform, basal body, and needle are accommodated by flexible adapter regions. Importantly, no periplasmic chamber or release site was detected. The absence of any structural discontinuity is consistent with a one-step secretion model of the T3SS.

In contrast to one-step secretion systems, two-step secretion systems such as the Type II secretion system (T2SS) operate through a sequential mechanism that is now firmly grounded in cryo-EM single-particle analysis and cryo-ET observations. In the first step, secretion substrates are translocated across the inner membrane by the Sec or Tat pathways and released into the periplasm, where they fold into their native conformations. Importantly, cryo-EM studies show that the T2SS machinery is spatially and structurally disengaged from this inner-membrane translocation event, confirming that the system does not participate in substrate export at this stage. Thus these studies establish a clear mechanistic separation between inner-membrane transport and secretion (Korotkov and Sandkvist, 2019). The second step of secretion is carried out by the T2SS apparatus itself, whose modular architecture has been directly visualized by cryo-EM single-particle analysis and cryo-ET (Korotkov and Sandkvist, 2019; Kooger et al., 2018; Chernyatina and Low, 2019; Ghosal et al., 2019; Yu et al., 2023). High-resolution reconstructions reveal that, at the outer membrane, a large oligomeric secretin forms a gated pore that opens to the extracellular space. Beneath this pore, cryo-EM maps uncovered a spacious periplasmic vestibule, providing a structural explanation for how fully folded substrates can be accommodated prior to secretion. This vestibule represents a bona fide periplasmic intermediate that had long been hypothesized but could not be visualized before the application of cryo-EM.

Cryo-EM further resolved the inner-membrane assembly platform, including the cytoplasmic ATPase that powers the system. Structural work based on cryo-ET analyzes demonstrated that ATP hydrolysis drives the polymerization of a short pseudopilus that extends into the periplasm but does not traverse the outer membrane (Korotkov and Sandkvist, 2019). These observations provided decisive support for the "pseudopilus piston model," in which cycles of pseudopilus extension push folded substrates from the periplasmic vestibule into and through the secretin pore. Critically, cryo-EM reconstructions show no continuous channel spanning both membranes, ruling out a one-step conduit mechanism and explaining why secretion necessarily proceeds in two discrete stages (Chernyatina and Low, 2019; Ghosal et al., 2019; Dazzoni et al., 2023). Together, these cryo-EM–derived structural insights reveal how the spatial separation of Sec/Tat-mediated translocation, periplasmic folding, and outer-membrane export enables the T2SS to specialize in the secretion of fully folded proteins, fundamentally distinguishing it from one-step secretion system.

5.8 NADH:ubiquinone oxidoreductase (complex I)

A major area of membrane protein biology that has been transformed by advances in cryo-EM is the study of the mitochondrial respiratory chain, with particular emphasis on NADH:ubiquinone oxidoreductase (complex I) (Parey et al., 2018; Caruana and Stroud, 2020). Complex I is the largest and most intricate enzyme of the oxidative phosphorylation system, functioning as the primary entry point for electrons from NADH into the respiratory chain while simultaneously driving proton translocation across the inner mitochondrial membrane. Early structural insights into this enzyme were derived from X-ray crystallography studies of bacterial homologs from *Thermus thermophilus* and *Escherichia coli*, which revealed the conserved L-shaped architecture, and they also defined the fourteen core subunits responsible for redox chemistry and proton pumping (Efremov and Sazanov, 2011; Baradaran et al., 2013). These studies established the basic framework of a hydrophilic peripheral arm, housing the flavin mononucleotide and iron–sulfur clusters, which is coupled to a membrane arm that contains the proton-translocating machinery.

However, the mammalian enzyme is substantially more complex, containing an additional thirty or thirty one supernumerary subunits that nearly double its molecular mass and introduce extensive regulatory, stabilizing, and assembly interfaces (Fig. 5.5). The size, flexibility, and compositional heterogeneity of mammalian complex I rendered it essentially intractable to X-ray crystallography. This barrier was overcome by single-particle cryo-EM, which enabled the first structures of the bovine enzyme at subnanometer to near-atomic resolution (Fig. 5.5) (Vinothkumar, Zhu, and Hirst, 2014; Zhu, Vinothkumar, and Hirst, 2016). These early cryo-EM reconstructions not only revealed the complete set of core and supernumerary subunits within an intact assembly but also demonstrated the power of image classification to resolve distinct conformational states, including active and inactive forms, thereby linking structure directly to catalytic regulation. Notably, the complete structure of *Escherichia coli* complex I was first reported much later, in 2021, by cryo-EM (Kolata and Efremov, 2021).

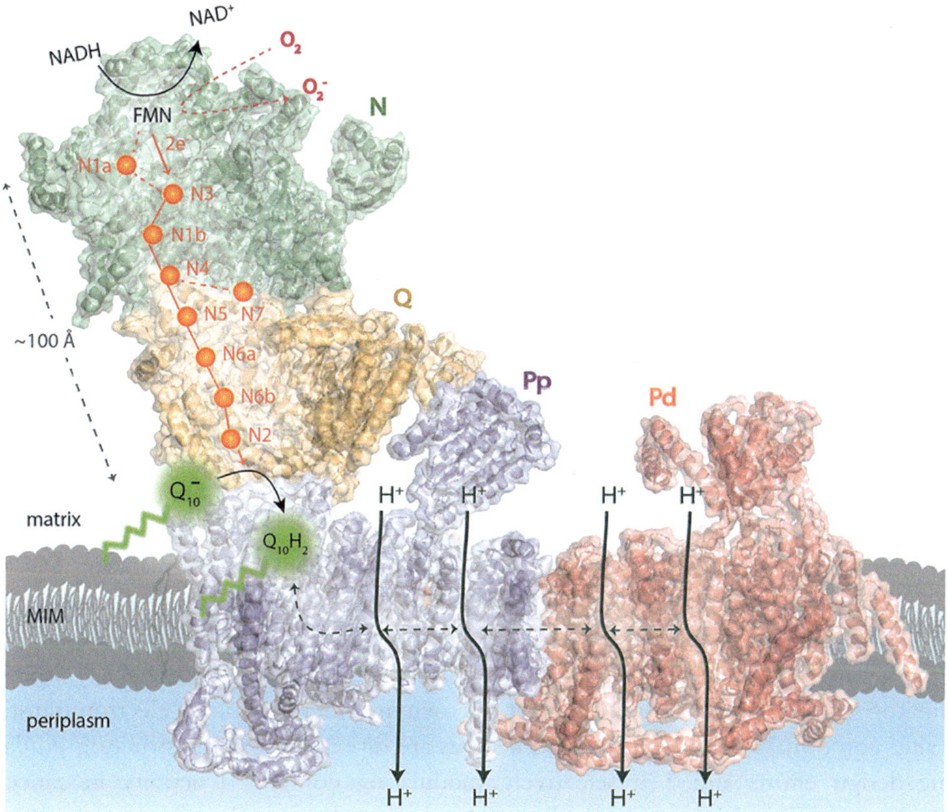

Figure 5.5 Schematic representation of the functional modules in mitochondrial complex I. In the N-module, NADH oxidation at FMN donates electrons through Fe–S clusters to reduce ubiquinone in the Q-module. This reduction drives conformational changes in the P-modules that couple to four proton pump sites. Reduced FMN can also generate reactive oxygen species. Model based on complex I of bovine heart by Vinothkumar et al. (PDB ID: 4UQ8). Reproduced from Giachin et al. (2016) under a Creative Commons Attribution 4.0 International (CC BY 4.0) license. *Reproduced from Giachin, G., Bouverot, R., Acajjaoui, S., Pantalone, S., Soler-López, M., 2016. Dynamics of Human Mitochondrial Complex I Assembly: Implications for Neurodegenerative Diseases. Frontiers in Molecular Biosciences 3, 10.3389/fmolb.2016.00043, Under a Creative Commons Attribution 4.0 International (CC BY 4.0) license.*

From around 2020 onward, rapid methodological advances in detectors, image-processing algorithms, and sample preparation led to a surge in higher-resolution cryo-EM structures of complex I from multiple species (Parey et al., 2019; Agip et al., 2023; Kühlbrandt, Carreira, and Yildiz, 2025). These datasets allowed reliable side-chain assignment, identification of bound lipids and quinone species, and increasingly confident modeling of ordered water molecules within the protein interior. Such details have been crucial for refining mechanistic models that couple long-range conformational changes in the peripheral arm to proton translocation along the membrane arm, particularly through conserved antiporter-like subunits and hydrated channels that may support proton transfer via Grotthuss-type mechanisms (Kühlbrandt, Carreira, and Yildiz, 2025; Cukierman, 2006; de Grotthuss, 1806).

Cryo-EM has also been instrumental in visualizing respiratory chain supercomplexes, including assemblies such as $I_1III_2IV_1$, $I_1III_2IV_2$, and I_1III_2 (Caruana and Stroud, 2020; Kühlbrandt, Carreira, and Yildiz, 2025). These structures suggest how complex I physically associates with downstream complexes to optimize electron transfer efficiency, stabilize individual components, and potentially reduce reactive oxygen species formation. While the physiological significance of supercomplex formation remains debated, cryo-EM has provided a structural platform for addressing these questions in a native-like context.

From a pharmacological perspective, complex I is an especially compelling drug target, with more than sixty known small-molecule modulators (Zhou et al., 2023). Cryo-EM studies have mapped the binding modes of classical inhibitors such as piericidin A and rotenone, as well as clinically relevant compounds including IACS-010759, biguanides, and pyridaben (Parey et al., 2019; Bridges et al., 2020; Chung et al., 2021, 2022; Bridges et al., 2023). These structures consistently highlight the central role of the quinone-binding cavity and adjacent channels, revealing how chemically diverse inhibitors exploit overlapping yet distinct microenvironments to block electron transfer or alter conformational coupling. Such insights are directly informing structure-guided drug design efforts aimed at selectively modulating complex I activity in cancer, mitochondrial disorders, metabolic diseases, and neurodegeneration (Yap et al., 2023; Pujalte-Martin et al., 2024; Rai et al., 2025).

Looking ahead, continued improvements in cryo-EM resolution and time-resolved approaches hold promise for visualizing extended networks of ordered water molecules within the membrane arm and along putative proton pathways. Achieving this level of detail would provide a definitive, experimentally grounded mechanism of proton pumping in complex I. Ultimately, this deeper mechanistic understanding is expected to translate into more precise therapeutic strategies that target mitochondrial energy metabolism with greater efficacy and safety.

5.9 ATP synthase

ATP synthases are rotary chemo-mechanical enzymes that harness electrochemical proton gradients to drive ATP synthesis, constituting the terminal energy-conversion step in oxidative and photophosphorylation (Fig. 5.6).

Their conserved architecture consists of a catalytic F_1 sector and a membrane-integrated F_o proton-translocating motor, which are connected by central and peripheral stalks and anchored by a stator that prevents corotation of the $\alpha_3\beta_3$ catalytic

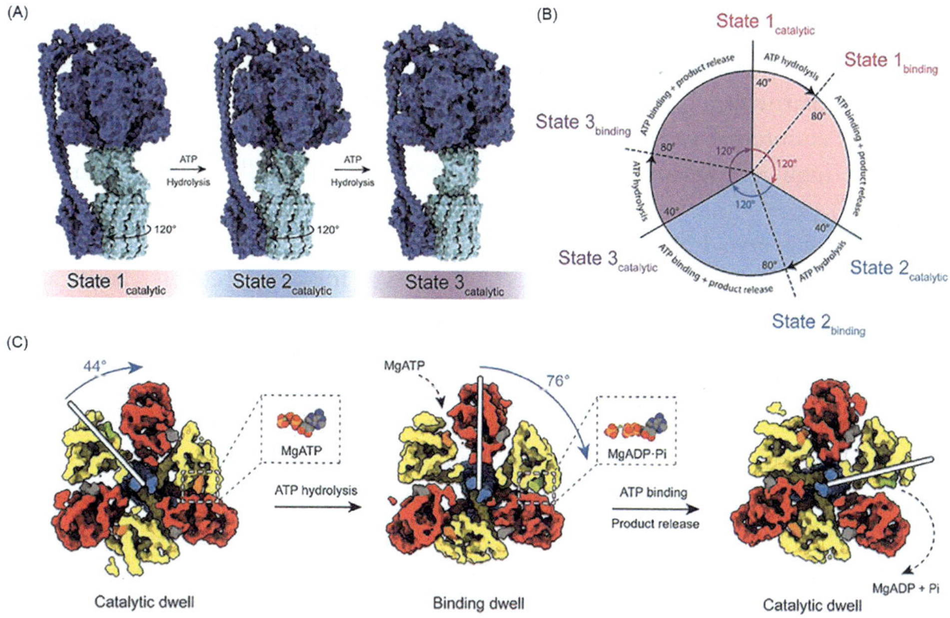

Figure 5.6 Functional mechanism of rotary ATP synthase. (A) Cryo-electron microscopy (cryo-EM) atomic models of *Bacillus PS3* ATP synthase reveal three principal rotational states of the complex (PDB: 6N2Y, 6N30, 6N2Z; Guo et al. (2019)). The light-blue components rotate relative to the dark-blue components, and the direction of rotation during ATP hydrolysis is indicated. (B) Schematic of the rotary catalytic cycle of the F1 region during ATP hydrolysis. (C) Cryo-EM structures of *Bacillus PS3* F1 resolve catalytic substeps and corresponding γ-subunit angular positions (PDB: 7L1R, 7L1Q; Sobti et al. (2021)). ATP, ADP, and noncatalytic nucleotides are shown in orange, green, and dark gray, respectively. Based on Sobti et al. (2021). Reproduced from Courbon and Rubinstein (2022) under a Creative Commons Attribution 4.0 International (CC BY 4.0) license. *Reproduced from Courbon, G.M., Rubinstein, J.L., 2022. CryoEM Reveals the Complexity and Diversity of ATP Synthases. Frontiers in Microbiology. 13, 10.3389/fmicb.2022.864006, under a Creative Commons Attribution 4.0 International (CC BY 4.0) license.*

head with the rotor. Foundational biochemical works, rotation assays, and early EM projections in the late 20th century established the rotary binding-change mechanism, yet structural resolution of intact membrane assemblies remained incomplete for decades (Boyer, 1993, 1997; Noji et al., 1997; Rubinstein, Walker, and Henderson, 2003). X-ray crystallography provided high-resolution snapshots of isolated F_1 complexes, notably from bacterial and mitochondrial sources, but the entire structure of the F_1F_o ATP synthase remained recalcitrant to crystallization due to conformational heterogeneity, detergent-induced destabilization, and species-specific modes of oligomerization. To the best of our knowledge, the 4 Å structure of *Paracoccus denitrificans* ATP synthase is the only known report of a successful attempt to capture the entire complex by X-ray crystallography (Morales-Rios et al., 2015; Morales-Rios et al., 2015), and none have been reported from mitochondrial or chloroplastic sources by the same technique. In particular, bacterial ATP synthases function as monomers, just like their chloroplastic counterparts. The mitochondrial ATP synthases, on the other hand, form dimers and higher-order oligomers (Kühlbrandt, 2019; Kühlbrandt, Carreira, and Yildiz, 2025).

Single-particle cryo-EM accelerated progress by enabling structural analysis of intact membrane assemblies in near-native states. Advances in image processing, classification, and membrane mimetics stabilized F_1F_o complexes and preserved rotational and regulatory intermediates. Cryo-EM visualization of dimeric and super-assembled mitochondrial ATP synthase complexes demonstrated that cristae bending is driven by ATP synthase-induced membrane curvature, thereby promoting local proton retention along the inner mitochondrial membrane and supporting oxidative phosphorylation flux (Hahn et al., 2016; Kühlbrandt, 2019; Mühleip et al., 2021; Kühlbrandt, Carreira, and Yildiz, 2025). Cryo-EM structures of bacterial ATP synthase, exemplified by *Bacillus PS3*, revealed several rotational states of the intact enzyme at near-atomic resolution (Guo, Suzuki, and Rubinstein, 2019). These structures allowed direct counting of the c-ring subunits and clarified how the rotor interacts with the stator components. They also showed that the peripheral stalk can flex slightly, providing mechanical cushioning as the rotary motion of the c-ring is transmitted to the catalytic head. On the other hand, cryo-EM structures of chloroplast ATP synthase uncovered distinctive regulatory features, most notably the γ-subunit redox loop, which forms a disulfide bond under oxidizing conditions that blocks rotation and inhibits ATP hydrolysis, providing a direct structural basis for redox regulation in the dark when the proton motive force collapses (Hahn et al., 2018).

5.10 Conclusion

The "resolution revolution" in cryo-EM has fundamentally transformed membrane protein structural biology from a field constrained by crystallization challenges to one empowered by methodological versatility and native-state preservation. By enabling structure determination in detergents and lipid-embedded environments without crystallization, cryo-EM has revealed the intricate mechanistic landscapes of ion channels, receptors, transporters, and energy-transducing complexes. The technique's unique capacity to resolve conformational heterogeneity from single datasets has unveiled the dynamic nature of membrane proteins, establishing them as allosteric machines rather than static structures. From TRP channels' polymodal sensing to ATP synthase's rotary catalysis, cryo-EM has integrated structural snapshots with functional mechanisms, providing atomic-level frameworks for understanding physiology and disease. Looking forward, continued advances in detector technology, image-processing algorithms, and time-resolved approaches promise to capture increasingly transient intermediates and subtle conformational transitions in membrane protein complexes, ultimately enabling structure-based therapeutic design with unprecedented precision and mechanistic insight.

Acknowledgements

This work is supported by funding for Basic and Applied Research in Biophysics and Material Science (RSI 4002) from the Department of Atomic Energy (DAE), Govt. of India. AS and DM are thankful to the University Grants Commission (UGC) and DAE, Govt. of India, respectively, for fellowship support.

Conflict of interests

The authors have no conflicts of interest to declare. All coauthors have seen and agreed with the content of the manuscript, and there are no financial interests to report.

References

Agip, A.N.A., Chung, I., Sanchez-Martinez, A., Whitworth, A.J., Hirst, J., 2023. Cryo-EM structures of mitochondrial respiratory complex I from *Drosophila melanogaster*. eLife 12. Available at https://doi.org/10.7554/eLife.84424.

Bajaj, R., 2025. Hurdles and advancements in experimental membrane protein structural biology. Journal of Structural Biology 217 (4), 108251. Available at https://doi.org/10.1016/j.jsb.2025.108251.

Baradaran, R., Berrisford, J.M., Minhas, G.S., Sazanov, L.A., 2013. Crystal structure of the entire respiratory complex i. Nature 494 (7438), 443–448. Available at https://doi.org/10.1038/nature11871.

Basak, S., Gicheru, Y., Rao, S., Sansom, M.S.P., Chakrapani, S., 2018. Cryo-EM reveals two distinct serotonin-bound conformations of full-length 5-HT3A receptor. Nature 563 (7730), 270–274. Available at https://doi.org/10.1038/s41586-018-0660-7, http://www.nature.com/nature/index.html.

Basak, S., Kumar, A., Ramsey, S., Gibbs, E., Kapoor, A., Filizola, M., et al., 2020. High-resolution structures of multiple 5-HT3aR-setron complexes reveal a novel mechanism of competitive inhibition. eLife. 9, 1–56. Available at https://doi.org/10.7554/eLife.57870, https://elifesciences.org/articles/57870.

Benjin, X., Ling, L., 2020. Developments, applications, and prospects of cryo-electron microscopy. Protein Science 29 (4), 872–882. Available at https://doi.org/10.1002/pro.3805, http://onlinelibrary.wiley.com/journal/10.1002/(ISSN)1469-896X.

Bezanilla, F., 2008. Ion channels: from conductance to structure. Neuron 60 (3), 456–468. Available at https://doi.org/10.1016/j.neuron.2008.10.035.

Bezanilla, F., Armstrong, C.M., 1972. Negative conductance caused by entry of sodium and cesium ions into the potassium channels of squid axons. Journal of General Physiology 60 (5), 588–608. Available at https://doi.org/10.1085/jgp.60.5.588.

Birch, J., Cheruvara, H., Gamage, N., Harrison, P.J., Lithgo, R., Quigley, A., 2020. Changes in membrane protein structural biology. Biology 9 (11), 1–31. Available at https://doi.org/10.3390/biology9110401, https://www.mdpi.com/2079–7737/9/11/401/pdf.

Boyer, P.D., 1993. The binding change mechanism for ATP synthase - some probabilities and possibilities. BBA - Bioenergetics 1140 (3), 215–250. Available at https://doi.org/10.1016/0005–2728(93)90063-L.

Boyer, P.D., 1997. The ATP synthase - a splendid molecular machine. Annual Review of Biochemistry 66, 717–749. Available at https://doi.org/10.1146/annurev.biochem.66.1.717.

Bridges, H.R., Blaza, J.N., Yin, Z., Chung, I., Pollak, M.N., Hirst, J., 2023. Structural basis of mammalian respiratory complex I inhibition by medicinal biguanides. Science (New York, N.Y.) 379 (6630), 351–357. Available at https://doi.org/10.1126/science.ade3332.

Bridges, H.R., Fedor, J.G., Blaza, J.N., Di Luca, A., Jussupow, A., Jarman, O.D., et al., 2020. Structure of inhibitor-bound mammalian complex I. Nature Communications 11 (1). Available at https://doi.org/10.1038/s41467-020-18950-3, http://www.nature.com/ncomms/index.html.

Cantor, R.S., 1997. The lateral pressure profile in membranes: a physical mechanism of general anesthesia. Biochemistry 36 (9), 2339–2344. Available at https://doi.org/10.1021/bi9627323.

Cao, E., Liao, M., Cheng, Y., Julius, D., 2013. TRPV1 structures in distinct conformations reveal activation mechanisms. Nature 504 (7478), 113–118. Available at https://doi.org/10.1038/nature12823.

Carroni, M., Saibil, H.R., 2016. Cryo electron microscopy to determine the structure of macromolecular complexes. Methods (San Diego, Calif.) 95, 78–85. Available at https://doi.org/10.1016/j.ymeth.2015.11.023, http://www.elsevier.com/inca/publications/store/6/2/2/9/1/4/index.htt.

Caruana, N.J., Stroud, D.A., 2020. The road to the structure of the mitochondrial respiratory chain supercomplex. Biochemical Society Transactions 48 (2), 621–629. Available at https://doi.org/10.1042/BST20190930, https://doi.org/10.1042/BST20190930.

Catterall, W.A., 1986. Molecular properties of voltage-sensitive sodium channels. Annual Review of Biochemistry 55, 953–985.

Catterall, W.A., 2010. Ion channel voltage sensors: structure, function, and pathophysiology. Neuron 67 (6), 915–928. Available at https://doi.org/10.1016/j.neuron.2010.08.021.

Cheng, Y., 2015. Single-particle cryo-EM at crystallographic resolution. Cell 161 (3), 450–457. Available at https://doi.org/10.1016/j.cell.2015.03.049, https://www.sciencedirect.com/journal/cell.

Cheng, Y., 2018a. Membrane protein structural biology in the era of single particle cryo-EM. Current Opinion in Structural Biology 52, 58–63. Available at https://doi.org/10.1016/j.sbi.2018.08.008, http://www.elsevier.com/locate/sbi.

Cheng, Y., 2018b. Single-particle cryo-EM-How did it get here and where will it go. Science (New York, N.Y.) 361 (6405), 876–880. Available at https://doi.org/10.1126/science.aat4346, http://science.sciencemag.org/content/361/6405/876/tab-pdf.

Cherezov, V., Rosenbaum, D.M., Hanson, M.A., Rasmussen, S.G.F., Foon, S.T., Kobilka, T.S., et al., 2007. High-resolution crystal structure of an engineered human β2-adrenergic G protein-coupled receptor. Science (New York, N.Y.) 318 (5854), 1258–1265. Available at https://doi.org/10.1126/science.1150577.

Chernyatina, A.A., Low, H.H., 2019. Core architecture of a bacterial type II secretion system. Nature Communications 10 (1). Available at https://doi.org/10.1038/s41467-019-13301-3, http://www.nature.com/ncomms/index.html.

Chi, G., Liang, Q., Sridhar, A., Cowgill, J.B., Sader, K., Radjainia, M., et al., 2022. Cryo-EM structure of the human Kv3.1 channel reveals gating control by the cytoplasmic T1 domain. Nature Communications 13 (1). Available at https://doi.org/10.1038/s41467-022-29594-w.

Chun, E., Thompson, A.A., Liu, W., Roth, C.B., Griffith, M.T., Katritch, V., et al., 2012. Fusion partner toolchest for the stabilization and crystallization of G protein-coupled receptors. Structure (London, England: 1993) 20 (6), 967–976. Available at https://doi.org/10.1016/j.str.2012.04.010.

Chung, I., Serreli, R., Cross, J.B., Di Francesco, M.E., Marszalek, J.R., Hirst, J., 2021. Cork-in-bottle mechanism of inhibitor binding to mammalian complex I. Science Advances 7 (20). Available at https://doi.org/10.1126/sciadv.abg4000, https://advances.sciencemag.org/content/7/20/eabg4000.

Chung, I., Wright, J.J., Bridges, H.R., Ivanov, B.S., Biner, O., Pereira, C.S., et al., 2022. Cryo-EM structures define ubiquinone-10 binding to mitochondrial complex I and conformational transitions accompanying Q-site occupancy. Nature Communications 13 (1). Available at https://doi.org/10.1038/s41467-022-30506-1, https://www.nature.com/ncomms/.

Courbon, G.M., Rubinstein, J.L., 2022. CryoEM Reveals the Complexity and Diversity of ATP Synthases. Frontiers in Microbiology 13. Available at 10.3389/fmicb.2022.864006.

Cuello, L.G., Jogini, V., Cortes, D.M., Perozo, E., 2010. Structural mechanism of C-type inactivation in K + channels. Nature 466 (7303), 203–208. Available at https://doi.org/10.1038/nature09153.

Cukierman, S., 2006. Et tu, Grotthuss! and other unfinished stories. Biochimica et Biophysica Acta (BBA) - Bioenergetics 1757 (8), 876–885. Available at https://doi.org/10.1016/j.bbabio.2005.12.001.

Dazzoni, R., Li, Y., López-Castilla, A., Brier, S., Mechaly, A., Cordier, F., et al., 2023. Structure and dynamic association of an assembly platform subcomplex of the bacterial type II secretion system. Structure (London, England: 1993) 31 (2), 152. Available at https://doi.org/10.1016/j.str.2022.12.003, http://www.journals.elsevier.com/structure/.

Doré, A.S., Robertson, N., Errey, J.C., Ng, I., Hollenstein, K., Tehan, B., et al., 2011. Structure of the adenosine A 2 A receptor in complex with ZM241385 and the xanthines XAC and caffeine. Structure (London, England: 1993) 19 (9), 1283–1293. Available at https://doi.org/10.1016/j.str.2011.06.014.

Doyle, D.A., Cabral, J.M., Pfuetzner, R.A., Kuo, A., Gulbis, J.M., Cohen, S.L., et al., 1998. The structure of the potassium channel: molecular basis of K^+ conduction and selectivity. Science (New York, N.Y.) 280 (5360), 69–77. Available at https://doi.org/10.1126/science.280.5360.69.

Draper-Joyce, C.J., Khoshouei, M., Thal, D.M., Liang, Y.L., Nguyen, A.T.N., Furness, S.G.B., et al., 2018. Structure of the adenosine-bound human adenosine A1 receptor-Gi complex. Nature 558 (7711), 559–565. Available at https://doi.org/10.1038/s41586-018-0236-6, http://www.nature.com/nature/index.html.

Du, J., Lü, W., Wu, S., Cheng, Y., Gouaux, E., 2015. Glycine receptor mechanism elucidated by electron cryo-microscopy. Nature 526 (7572), 224–229. Available at https://doi.org/10.1038/nature14853, http://www.nature.com/nature/index.html.

Duan, J., He, X.H., Li, S.J., Xu, H.E., 2024. Cryo-electron microscopy for GPCR research and drug discovery in endocrinology and metabolism. Nature Reviews Endocrinology 20 (6), 349–365. Available at https://doi.org/10.1038/s41574-024-00957-1, https://www.nature.com/nrendo/.

Dubochet, J., Adrian, M., Chang, J.-J., Homo, J.-C., Lepault, J., McDowall, A.W., et al., 1988. Cryo-electron microscopy of vitrified specimens. Quarterly Reviews of Biophysics 21 (2), 129–228. Available at https://doi.org/10.1017/s0033583500004297.

Efremov, R.G., Sazanov, L.A., 2011. Structure of the membrane domain of respiratory complex i. Nature 476 (7361), 414–421. Available at https://doi.org/10.1038/nature10330.

Fan, C., Choi, W., Sun, W., Du, J., 2018. W. Lu, Structure of the human lipid-gated cation channel TRPC3. eLife. 7. Available at https://doi.org/10.7554/eLife.36852, https://elifesciences.org/articles/36852.

Fernández-Mariño, A.I., Tan, X.F., Bae, C., Huffer, K., Jiang, J., Swartz, K.J., 2023. Inactivation of the Kv2.1 channel through electromechanical coupling. Nature 622 (7982), 410–417. Available at https://doi.org/10.1038/s41586-023-06582-8, https://www.nature.com/nature/.

Frank, J., 2009. Single-particle reconstruction of biological macromolecules in electron microscopy-30 years. Quarterly Reviews of Biophysics 42 (3), 139–158. Available at https://doi.org/10.1017/S0033583509990059.

Furini, S., Domene, C., 2009. Atypical mechanism of conduction in potassium channels. Proceedings of the National Academy of Sciences of the United States of America 106 (38), 16074–16077. Available at https://doi.org/10.1073/pnas.0903226106, http://www.pnas.org/content/106/38/16074.full.pdf.

Gao, Y.-H., Huang, Y.-Z., Li, Z.-X., Chen, X.-Y., Shao, C.-Y., Li, H.-W., et al., 2026. Structures of TRPV1 bound by hyperthermia-inducing analgesics. Cell Reports 45 (1), 116765. Available at https://doi.org/10.1016/j.celrep.2025.116765.

García-Nafría, J., Tate, C.G., 2019. Cryo-EM structures of GPCRs coupled to G s, G i and G o. Molecular and Cellular Endocrinology 488, 1–13. Available at https://doi.org/10.1016/j.mce.2019.02.006, www.elsevier.com/locate/mce.

García-Nafría, J., Tate, C.G., 2021. Structure determination of GPCRs: Cryo-EM compared with X-ray crystallography. Biochemical Society Transactions 49 (5), 2345–2355. Available at https://doi.org/10.1042/BST20210431, https://portlandpress.com/biochemsoctrans/article/49/5/2345/229854/Structure-determination-of-GPCRs-cryo-EM-compared.

Ghit, A., Assal, D., Al-Shami, A.S., Hussein, D.E.E., 2021. GABAA receptors: structure, function, pharmacology, and related disorders. Journal of Genetic Engineering and Biotechnology 19 (1). Available at https://doi.org/10.1186/s43141-021-00224-0, https://jgeb.springeropen.com/.

Ghosal, D., Kim, K.W., Zheng, H., Kaplan, M., Truchan, H.K., Lopez, A.E., et al., 2019. In vivo structure of the Legionella type II secretion system by electron cryotomography. Nature Microbiology 4 (12), 2101–2108. Available at https://doi.org/10.1038/s41564-019-0603-6, www.nature.com/nmicrobiol/.

Giachin, G., Bouverot, R., Acajjaoui, S., Pantalone, S., Soler-López, M., 2016. Dynamics of Human Mitochondrial Complex I Assembly: Implications for Neurodegenerative Diseases. Frontiers in Molecular Biosciences 3. Available at 10.3389/fmolb.2016.00043.

Gibbs, E., Klemm, E., Seiferth, D., Kumar, A., Ilca, S.L., Biggin, P.C., et al., 2023. Conformational transitions and allosteric modulation in a heteromeric glycine receptor. Nature Communications 14 (1). Available at https://doi.org/10.1038/s41467-023-37106-7, https://www.nature.com/ncomms/.

de Grotthuss, C., 1806. Sur la décomposition de l'eau et des corps qu'elle tient en dissolution à l'aide de l'électricité galvanique. Annali di Chimica 58, 54–74.

Guo, H., Suzuki, T., Rubinstein, J.L., 2019. Structure of a bacterial ATP synthase. eLife 8, e43128. Available at https://doi.org/10.7554/eLife.43128.

Guo, J., She, J., Zeng, W., Chen, Q., Bai, X.C., Jiang, Y., 2017. Structures of the calcium-activated, non-selective cation channel TRPM4. Nature 552 (7684), 205–209. Available at https://doi.org/10.1038/nature24997, http://www.nature.com/nature/index.html.

Gusach, A., García-Nafría, J., Tate, C.G., 2023. New insights into GPCR coupling and dimerisation from cryo-EM structures. Current Opinion in Structural Biology 80. Available at https://doi.org/10.1016/j.sbi.2023.102574, http://www.elsevier.com/locate/sbi.

Hahn, A., Parey, K., Bublitz, M., Mills, D.J., Zickermann, V., Vonck, J., et al., 2016. Structure of a complete ATP synthase dimer reveals the molecular basis of inner mitochondrial membrane morphology. Molecular Cell 63 (3), 445–456. Available at https://doi.org/10.1016/j.molcel.2016.05.037, www.molecule.org.

Hahn, A., Vonck, J., Mills, D.J., Meier, T., Kühlbrandt, W., 2018. Structure, mechanism, and regulation of the chloroplast ATP synthase. Science (New York, N.Y.) 360 (6389). Available at https://doi.org/10.1126/science.aat4318, http://science.sciencemag.org/content/360/6389/eaat4318/tab-pdf.

Henderson, R., Baldwin, J.M., Ceska, T.A., Zemlin, F., Beckmann, E., Downing, K.H., 1990. Model for the structure of bacteriorhodopsin based on high-resolution electron cryo-microscopy. Journal of Molecular Biology 213 (4), 899–929. Available at https://doi.org/10.1016/S0022–2836(05)80271-2.

Henderson, R., Unwin, P.N.T., 1975. Three-dimensional model of purple membrane obtained by electron microscopy. Nature 257 (5521), 28–32. Available at https://doi.org/10.1038/257028a0.

Heydenreich, F.M., Vuckovic, Z., Matkovic, M., Veprintsev, D.B., 2015. Stabilization of G protein-coupled receptors by point mutations. Frontiers in Pharmacology 6. Available at https://doi.org/10.3389/fphar.2015.00082, http://journal.frontiersin.org/article/10.3389/fphar.2015.00082.pdf.

Hu, B., Lara-Tejero, M., Kong, Q., Galán, J.E., Liu, J., 2017. In situ molecular architecture of the *Salmonella* type III secretion machine. Cell 168 (6), 1065. Available at https://doi.org/10.1016/j.cell.2017.02.022, https://www.sciencedirect.com/journal/cell.

Hu, J., Park, S.J., Walter, T., Orozco, I.J., O'Dea, G., Ye, X., et al., 2024. W. Lü, Physiological temperature drives TRPM4 ligand recognition and gating. Nature 630 (8016), 509–515. Available at https://doi.org/10.1038/s41586-024-07436-7, https://www.nature.com/nature/.

Huynh, K.W., Cohen, M.R., Jiang, J., Samanta, A., Lodowski, D.T., Zhou, Z.H., et al., 2016. Structure of the full-length TRPV2 channel by cryo-EM. Nature Communications 7. Available at https://doi.org/10.1038/ncomms11130, http://www.nature.com/ncomms/index.html.

Jaakola, V.P., Griffith, M.T., Hanson, M.A., Cherezov, V., Chien, E.Y.T., Lane, J.R., et al., 2008. The 2.6 angstrom crystal structure of a human A2A adenosine receptor bound to an antagonist. Science (New York, N.Y.) 322 (5905), 1211–1217. Available at https://doi.org/10.1126/science.1164772.

Jiang, Y., Ruta, V., Chen, J., Lee, A., MacKinnon, R., 2003a. The principle of gating charge movement in a voltage-dependent K+ channel. Nature 423 (6935), 42–48. Available at https://doi.org/10.1038/nature01581.

Jiang, Y., Lee, A., Chen, J., Ruta, V., Cadene, M., Chait, B.T., et al., 2003b. X-ray structure of a voltage-dependent K^+ channel. Nature 423 (6935), 33–41. Available at https://doi.org/10.1038/nature01580.

Kang, Y., Zhou, X.E., Gao, X., He, Y., Liu, W., Ishchenko, A., et al., 2015. Crystal structure of rhodopsin bound to arrestin by femtosecond X-ray laser. Nature 523 (7562), 561–567. Available at https://doi.org/10.1038/nature14656, http://www.nature.com/nature/index.html.

Kim, J.J., Hibbs, R.E., 2021. Direct structural insights into GABAA receptor pharmacology. Trends in Biochemical Sciences 46 (6), 502–517. Available at https://doi.org/10.1016/j.tibs.2021.01.011, www.elsevier.com/locate/tibs.

Kolata, P., Efremov, R.G., 2021. Structure of *Escherichia coli* respiratory complex I reconstituted into lipid nanodiscs reveals an uncoupled conformation. eLife 10. Available at https://doi.org/10.7554/elife.68710.

Kooger, R., Szwedziak, P., Böck, D., Pilhofer, M., 2018. CryoEM of bacterial secretion systems. Current Opinion in Structural Biology 52, 64–70. Available at https://doi.org/10.1016/j.sbi.2018.08.007, http://www.elsevier.com/locate/sbi.

Kopec, W., Köpfer, D.A., Vickery, O.N., Bondarenko, A.S., Jansen, T.L.C., de Groot, B.L., et al., 2018. Direct knock-on of desolvated ions governs strict ion selectivity in K+ channels. Nature Chemistry 10 (8), 813–820. Available at https://doi.org/10.1038/s41557-018-0105-9, http://www.nature.com/nchem/index.html.

Korotkov, K.V., Sandkvist, M., 2019. Architecture, function, and substrates of the type II secretion system. EcoSal Plus 8 (2), 1–14. Available at https://doi.org/10.1128/ecosalplus.ESP-0034–2018, http://www.asmscience.org/docserver/fulltext/ecosalplus/8/2/ESP-0034–2018.pdf?expires=1553920700&id=id&accname=45052&checksum=DEFEE45542ECB201DF885A5B2B179A87.

Kumar, A., Basak, S., Rao, S., Gicheru, Y., Mayer, M.L., Sansom, M.S.P., et al., 2020. Mechanisms of activation and desensitization of full length glycine receptor in lipid nanodiscs. Nature Communications 11 (1). Available at https://doi.org/10.1038/s41467-020-17364-5, http://www.nature.com/ncomms/index.html.

Kumar, A., Kindig, K., Rao, S., Zaki, A.M., Basak, S., Sansom, M.S.P., et al., 2022. Structural basis for cannabinoid-induced potentiation of alpha1-glycine receptors in lipid nanodiscs. Nature Communications 13 (1). Available at https://doi.org/10.1038/s41467-022-32594-5, http://www.nature.com/ncomms/index.html.

Köpfer, D.A., Song, C., Gruene, T., Sheldrick, G.M., Zachariae, U., De Groot, B.L., 2014. Ion permeation in K^+ channels occurs by direct Coulomb knock-on. Science (New York, N.Y.) 346 (6207), 352–355. Available at https://doi.org/10.1126/science.1254840, http://www.sciencemag.org/content/346/6207/352.full.pdf.

Kühlbrandt, W., 2019. Structure and mechanisms of F-Type ATP synthases. Annual Review of Biochemistry 88, 515–549. Available at https://doi.org/10.1146/annurev-biochem-013118–110903, http://www.annualreviews.org/journal/biochem.

Kühlbrandt, W., Carreira, L.A.M., Yildiz, Ö., 2025. Cryo-EM of mitochondrial complex I and ATP synthase. Annual Review of Biophysics 54 (1), 209–226. Available at https://doi.org/10.1146/annurev-biophys-060724–110838, http://www.annualreviews.org/journal/biophys.

Kühlbrandt, W., Wang, D.N., Fujiyoshi, Y., 1994. Atomic model of plant light-harvesting complex by electron crystallography. Nature 367 (6464), 614–621. Available at https://doi.org/10.1038/367614a0.

Kühlbrandt, W., 2014. The resolution revolution. Science (New York, N.Y.) 343 (6178), 1443–1444. Available at https://doi.org/10.1126/science.1251652.

Kühlbrandt, W., 2022. Forty years in cryoEM of membrane proteins. Microscopy: The Journal of the Quekett Microscopical Club 71 (Supplement_1), i30. Available at https://doi.org/10.1093/jmicro/dfab041.

Lee, B.R., White, K.I., Socolich, M., Klureza, M.A., Henning, R., Srajer, V., et al., 2025. Direct visualization of electric-field-stimulated ion conduction in a potassium channel. Cell 188 (1), 77. Available at https://doi.org/10.1016/j.cell.2024.12.006, https://www.sciencedirect.com/science/journal/00928674.

Lee, K.P.K., Chen, J., Mackinnon, R., 2017. Molecular structure of human katp in complex with ATP and ADP. eLife. 6. Available at https://doi.org/10.7554/eLife.32481, https://elifesciences.org/content/6/e32481-download.pdf.

Lee, S.J., Schillings, J., De Vries, R., Matamoros, M., De Groot, B.L., Nichols, C.G., 2025. From selective permeation to physiology in potassium channels. Function 6 (6). Available at https://doi.org/10.1093/function/zqaf050, https://academic.oup.com/function.

Lee, U.S., Cui, J., 2010. BK channel activation: structural and functional insights. Trends in Neurosciences 33 (9), 415–423. Available at https://doi.org/10.1016/j.tins.2010.06.004.

Levental, I., Lyman, E., 2023. Regulation of membrane protein structure and function by their lipid nano-environment. Nature Reviews. Molecular Cell Biology 24 (2), 107–122. Available at https://doi.org/10.1038/s41580-022-00524-4, https://www.nature.com/molcellbio.

Li, B., Rietmeijer, R.A., Brohawn, S.G., 2020. Structural basis for pH gating of the two-pore domain K^+ channel TASK2. Nature 586 (7829), 457–462. Available at https://doi.org/10.1038/s41586-020-2770-2, http://www.nature.com/nature/index.html.

Li, N., Wu, J.X., Ding, D., Cheng, J., Gao, N., Chen, L., 2017. Structure of a pancreatic ATP-sensitive potassium channel. Cell 168 (1–2), 101. Available at https://doi.org/10.1016/j.cell.2016.12.028, https://www.sciencedirect.com/journal/cell.

Liao, M., Cao, E., Julius, D., Cheng, Y., 2013. Structure of the TRPV1 ion channel determined by electron cryo-microscopy. Nature 504 (7478), 107–112. Available at https://doi.org/10.1038/nature12822.

Liu, S., Anderson, P.J., Rajagopal, S., Lefkowitz, R.J., Rockman, H.A., 2024. G protein-coupled receptors: a century of research and discovery. Circulation Research 135 (1), 174–197. Available at https://doi.org/10.1161/CIRCRESAHA.124.323067, https://www.ovid.com/product-details.454.html.

Loll, P.J., 2003. Membrane protein structural biology: The high throughput challenge. Journal of Structural Biology 142 (1), 144–153. Available at https://doi.org/10.1016/S1047–8477(03)00045-5, http://www.elsevier.com/inca/publications/store/6/2/2/9/0/0/index.htt.

Long, S.B., Campbell, E.B., MacKinnon, R., 2005a. Crystal structure of a mammalian voltage-dependent Shaker family K^+ channel. Science (New York, N.Y.) 309 (5736), 897–903. Available at https://doi.org/10.1126/science.1116269.

Long, S.B., Campbell, E.B., MacKinnon, R., 2005b. Voltage sensor of Kv1.2: structural basis of electromechanical coupling. Science (New York, N.Y.) 309 (5736), 903–908. Available at https://doi.org/10.1126/science.1116270.

Ma, D., Zhong, L., Yan, Z., Yao, J., Zhang, Y., Ye, F., et al., 2022. Structural mechanisms for the activation of human cardiac KCNQ1 channel by electro-mechanical coupling enhancers. Proceedings of the National Academy of Sciences of the United States of America. 119 (45). Available at https://doi.org/10.1073/pnas.2207067119, https://www.pnas.org/doi/10.1073/pnas.2207067119.

Maeda, S., Qu, Q., Robertson, M.J., Skiniotis, G., Kobilka, B.K., 2019. Structures of the M1 and M2 muscarinic acetylcholine receptor/G-protein complexes. Science (New York, N.Y.) 364 (6440), 552–557. Available at https://doi.org/10.1126/science.aaw5188, https://science.sciencemag.org/content/364/6440/552/tab-pdf.

Martin, G.M., Yoshioka, C., Rex, E.A., Fay, J.F., Xie, Q., Whorton, M.R., et al., 2017. Cryo-EM structure of the ATP-sensitive potassium channel illuminates mechanisms of assembly and gating. eLife. 6. Available at https://doi.org/10.7554/eLife.24149, https://elifesciences.org/content/6/e24149-download.pdf.

Masiulis, S., Desai, R., Uchański, T., Serna Martin, I., Laverty, D., Karia, D., et al., 2019. GABAA receptor signalling mechanisms revealed by structural pharmacology. Nature 565 (7740), 454–459. Available at https://doi.org/10.1038/s41586-018-0832-5, http://www.nature.com/nature/index.html.

Matthies, D., Bae, C., Toombes, G.E.S., Fox, T., Bartesaghi, A., Subramaniam, S., et al., 2018. Single-particle cryo-EM structure of a voltage-activated potassium channel in lipid nanodiscs. eLife 7. Available at https://doi.org/10.7554/eLife.37558, https://elifesciences.org/articles/37558.

Miller, P.S., Aricescu, A.R., 2014. Crystal structure of a human GABAA receptor. Nature 512 (7514), 270–275. Available at https://doi.org/10.1038/nature13293, http://www.nature.com/nature/index.html.

Morais-Cabral, J.H., Zhou, Y., MacKinnon, R., 2001. Energetic optimization of ion conduction rate by the K^+selectivity filter. Nature 414 (6859), 37–40. Available at https://doi.org/10.1038/35102000.

Morales-Rios, E., Watt, I.N., Zhang, Q., Ding, S., Fearnley, I.M., Montgomery, M.G., et al., 2015. Purification, characterization and crystallization of the F-ATPase from *Paracoccus denitrificans*. Open Biology 5 (9), 150119. Available at https://doi.org/10.1098/rsob.150119.

Morales-Rios, Montgomery, E., Leslie, M.G., Andrew, G.W., Walker, J.E., 2015. Structure of ATP synthase from *Paracoccus denitrificans* determined by X-ray crystallography at 4.0 Å resolution. PNAS [Internet] [cited 2026 Jan 21]. Available at https://www.pnas.org/doi/10.1073/pnas.1517542112?url_ver=Z39.88–2003&rfr_id=ori%3Arid%3Acrossref.org&rfr_dat=cr_pub++0pubmed.

Mühleip, A., Kock Flygaard, R., Ovciarikova, J., Lacombe, A., Fernandes, P., Sheiner, L., et al., 2021. ATP synthase hexamer assemblies shape cristae of *Toxoplasma mitochondria*. Nature Communications 12 (1). Available at https://doi.org/10.1038/s41467-020-20381 z.

Neuberger, A., Oda, M., Nikolaev, Y.A., Nadezhdin, K.D., Gracheva, E.O., Bagriantsev, S.N., et al., 2023. Human TRPV1 structure and inhibition by the analgesic SB-366791. Nature Communications 14 (1). Available at https://doi.org/10.1038/s41467-023-38162-9, https://www.nature.com/ncomms/.

Noji, H., Yasuda, R., Yoshida, M., Kinosita, K., 1997. Direct observation of the rotation of F1-ATPase. Nature 386 (6622), 299–302. Available at https://doi.org/10.1038/386299a0.

Noviello, C.M., Kreye, J., Teng, J., Prüss, H., Hibbs, R.E., 2022. Structural mechanisms of GABAA receptor autoimmune encephalitis. Cell 185, 2469–2477.e13. Available at 10.1016/j.cell.2022.06.025.

Ong, S.T., Tyagi, A., Chandy, K.G., Bhushan, S., 2022. Mechanisms underlying C-type Inactivation in Kv channels: lessons from structures of human Kv1.3 and fly Shaker-IR channels. Frontiers in Pharmacology 13. Available at https://doi.org/10.3389/fphar.2022.924289, http://www.frontiersin.org/Pharmacology.

Ostmeyer, J., Chakrapani, S., Pan, A.C., Perozo, E., Roux, B., 2013. Recovery from slow inactivation in K^+ channels is controlled by water molecules. Nature 501 (7465), 121–124. Available at https://doi.org/10.1038/nature12395.

Overington, J.P., Al-Lazikani, B., Hopkins, A.L., 2006. How many drug targets are there? Nature Reviews. Drug Discovery 5 (12), 993–996. Available at https://doi.org/10.1038/nrd2199.

Palczewski, K., Kumasaka, T., Hori, T., Behnke, C.A., Motoshima, H., Fox, B.A., et al., 2000. Crystal structure of rhodopsin: a G protein-coupled receptor. Science (New York, N.Y.) 289 (5480), 739–745. Available at https://doi.org/10.1126/science.289.5480.739.

Parey, K., Brandt, U., Xie, H., Mills, D.J., Siegmund, K., Vonck, J., et al., 2018. Cryo-EM structure of respiratory complex I at work. eLife. 7. Available at https://doi.org/10.7554/eLife.39213, https://elifesciences.org/magazine.

Parey, K., Haapanen, O., Sharma, V., Köfeler, H., Züllig, T., Prinz, S., et al., 2019. High-resolution cryo-EM structures of respiratory complex I: mechanism, assembly, and disease. Science Advances 5 (12). Available at https://doi.org/10.1126/sciadv.aax9484, https://advances.sciencemag.org/content/5/12/eaax9484/tab-pdf.

Paulsen, C.E., Armache, J.P., Gao, Y., Cheng, Y., Julius, D., 2015. Structure of the TRPA1 ion channel suggests regulatory mechanisms. Nature 520 (7548), 511–517. Available at https://doi.org/10.1038/nature14367, http://www.nature.com/nature/index.html.

Pujalte-Martin, M., Belaïd, A., Bost, S., Kahi, M., Peraldi, P., Rouleau, M., et al., 2024. Targeting cancer and immune cell metabolism with the complex I inhibitors metformin and IACS-010759. Molecular Oncology 18 (7), 1719–1738. Available at https://doi.org/10.1002/1878-0261.13583, http://onlinelibrary.wiley.com/journal/10.1002/(ISSN)1878-0261.

Pumroy, R.A., De Jesús-Pérez, J.J., Protopopova, A.D., Rocereta, J.A., Fluck, E.C., Fricke, T., et al., 2024. Molecular details of ruthenium red pore block in TRPV channels. EMBO Reports 25 (2), 506–523. Available at https://doi.org/10.1038/s44319-023-00050-0, https://www.embopress.org/page/journal/14693178/about.

Pumroy, R.A., Protopopova, A.D., Fricke, T.C., Lange, I.U., Haug, F.M., Nguyen, P.T., et al., 2022. Structural insights into TRPV2 activation by small molecules. Nature Communications 13 (1). Available at https://doi.org/10.1038/s41467-022-30083-3, http://www.nature.com/ncomms/index.html.

Rai, N.K., Venugopal, H., Rajesh, R., Ancha, P., Venkatesh, S., 2025. Mitochondrial complex-1 as a therapeutic target for cardiac diseases. Molecular and Cellular Biochemistry 480 (2), 869–890. Available at https://doi.org/10.1007/s11010-024-05074-1, https://www.springer.com/journal/11010.

Rasmussen, S.G.F., DeVree, B.T., Zou, Y., Kruse, A.C., Chung, K.Y., Kobilka, T.S., et al., 2011. Crystal structure of the β2 adrenergic receptor–Gs protein complex. Nature 477 (7366), 549–555. Available at https://doi.org/10.1038/nature10361.

Rocereta, J.A., Sturhahn, T., Pumroy, R.A., Fricke, T.C., Herzog, C., Leffler, A., et al., 2025. Structural insights into TRPV2 modulation by probenecid. Nature Structural and Molecular Biology 32 (6), 1019–1029. Available at https://doi.org/10.1038/s41594-025-01494-9, https://www.nature.com/nsmb/.

Rubinstein, J.L., Walker, J.E., Henderson, R., 2003. Structure of the mitochondrial ATP synthase by electron cryomicroscopy. EMBO Journal 22 (23), 6182–6192. Available at https://doi.org/10.1093/emboj/cdg608.

Scheres, S.H.W., 2012a. A bayesian view on cryo-EM structure determination. Journal of Molecular Biology 415 (2), 406–418. Available at https://doi.org/10.1016/j.jmb.2011.11.010, https://www.journals.elsevier.com/journal-of-molecular-biology.

Scheres, S.H.W., 2012b. RELION: implementation of a bayesian approach to cryo-EM structure determination. Journal of Structural Biology 180 (3), 519–530. Available at https://doi.org/10.1016/j.jsb.2012.09.006.

Shihoya, W., Iwama, A., Sano, F.K., Nureki, O., 2024. Cryo-EM advances in GPCR structure determination. Journal of Biochemistry 176 (1), 1–10. Available at https://doi.org/10.1093/jb/mvae029, http://jb.oxfordjournals.org/.

Shukla, A.K., Manglik, A., Kruse, A.C., Xiao, K., Reis, R.I., Tseng, W.C., et al., 2013. Structure of active β-arrestin-1 bound to a G-protein-coupled receptor phosphopeptide. Nature 497 (7447), 137–141. Available at https://doi.org/10.1038/nature12120.

Sierra-Valdez, F., Azumaya, C.M., Romero, L.O., Nakagawa, T., Cordero-Morales, J.F., 2018. Structure–function analyses of the ion channel TRPC3 reveal that its cytoplasmic domain allosterically modulates channel gating. Journal of Biological Chemistry 293 (41), 16102–16114. Available at https://doi.org/10.1074/jbc.RA118.005066, http://www.jbc.org/content/293/41/16102.full.pdf.

Singh, A.K., McGoldrick, L.L., Sobolevsky, A.I., 2018. Structure and gating mechanism of the transient receptor potential channel TRPV3. Nature Structural and Molecular Biology 25 (9), 805–813. Available at https://doi.org/10.1038/s41594-018-0108-7, http://www.nature.com/nsmb/index.html.

Sobti, M., Ueno, H., Noji, H., Stewart, A.G., 2021. The six steps of the complete F1-ATPase rotary catalytic cycle. Nature Communications 12. Available at 10.1038/s41467-021-25029-0.

Soto, J.E., Galán, J.E., Lara-Tejero, M., 2022. Assembly and architecture of the type III secretion sorting platform. Proceedings of the National Academy of Sciences 119 (51). Available at https://doi.org/10.1073/pnas.2218010119.

Staus, D.P., Hu, H., Robertson, M.J., Kleinhenz, A.L.W., Wingler, L.M., Capel, W.D., et al., 2020. Structure of the M2 muscarinic receptor–β-arrestin complex in a lipid nanodisc. Nature 579 (7798), 297–302. Available at https://doi.org/10.1038/s41586-020-1954-0, http://www.nature.com/nature/index.html.

Stix, R., Tan, X.F., Bae, C., Fernández-Mariño, A.I., Swartz, K.J., Faraldo-Gómez, J.D., 2023. Eukaryotic Kv channel Shaker inactivates through selectivity filter dilation rather than collapse. Science Advances 9 (49). Available at https://doi.org/10.1126/SCIADV.ADJ5539, https://www.science.org/doi/10.1126/sciadv.adj5539.

Sun, C., Zhu, H., Clark, S., Gouaux, E., 2023. Cryo-EM structures reveal native GABAA receptor assemblies and pharmacology. Nature 622 (7981), 195–201. Available at https://doi.org/10.1038/s41586-023-06556-w, https://www.nature.com/nature/.

Sun, J., MacKinnon, R., 2017. Cryo-EM structure of a KCNQ1/CaM complex reveals insights into congenital long QT syndrome. Cell 169 (6), 1042. Available at https://doi.org/10.1016/j.cell.2017.05.019, https://www.sciencedirect.com/journal/cell.

Talyzina, I.A., Nadezhdin, K.D., Sobolevsky, A.I., 2025. Forty sites of TRP channel regulation. Current Opinion in Chemical Biology 84, 102550. Available at 10.1016/j.cbpa.2024.102550.

Tan, X.F., Bae, C., Stix, R., Fernández-Mariño, A.I., Huffer, K., Chang, T.H., et al., 2022. Structure of the Shaker Kv channel and mechanism of slow C-type inactivation. Science Advances 8 (11). Available at https://doi.org/10.1126/sciadv.abm7814, https://www.science.org/doi/10.1126/sciadv.abm7814.

Tao, X., Hite, R.K., MacKinnon, R., 2017. Cryo-EM structure of the open high-conductance Ca^{2+} -activated K+channel. Nature 541 (7635), 46–51. Available at https://doi.org/10.1038/nature20608, http://www.nature.com/nature/index.html.

Tao, X., Mackinnon, R., 2019. Molecular structures of the human slo1 k+channel in complex with b4. eLife. 8. Available at https://doi.org/10.7554/eLife.51409, https://elifesciences.org/articles/51409.

Tsai, C.-J., Marino, J., Adaixo, R., Pamula, F., Muehle, J., Maeda, S., et al., 2019. Cryo-EM structure of the rhodopsin-Gαi-βγ complex reveals binding of the rhodopsin C-terminal tail to the gβ subunit. eLife 8. Available at https://doi.org/10.7554/elife.46041.

Tyagi, A., Ahmed, T., Jian, S., Bajaj, S., Ong, S.T., Goay, S.S.M., et al., 2022. Rearrangement of a unique Kv1.3 selectivity filter conformation upon binding of a drug. Proceedings of the National Academy of Sciences of the United States of America. 119 (5). Available at https://doi.org/10.1073/pnas.2113536119, https://www.pnas.org/content/119/5/e2113536119.

Unger, V.M., Hargrave, P.A., Baldwin, J.M., Schertler, G.F.X., 1997. Arrangement of rhodopsin transmembrane α-helices. Nature 389 (6647), 203–206. Available at https://doi.org/10.1038/38316.

Vinothkumar, K.R., Zhu, J., Hirst, J., 2014. Architecture of mammalian respiratory complex I. Nature 515 (7525), 80–84. Available at https://doi.org/10.1038/nature13686, http://www.nature.com/nature/index.html.

Winkler, P.A., Huang, Y., Sun, W., Du, J., 2017. Wei Lü, Electron cryo-microscopy structure of a human TRPM4 channel. Nature 552 (7684), 200–204. Available at https://doi.org/10.1038/nature24674.

Yap, T.A., Daver, N., Mahendra, M., Zhang, J., Kamiya-Matsuoka, C., Meric-Bernstam, F., et al., 2023. Complex I inhibitor of oxidative phosphorylation in advanced solid tumors and acute myeloid leukemia: phase I trials. Nature Medicine 29 (1), 115–126. Available at https://doi.org/10.1038/s41591-022-02103-8, https://www.nature.com/nm/.

Yin, Y., Zhang, F., Feng, S., Butay, K.J., Borgnia, M.J., Im, W., et al., 2022. Activation mechanism of the mouse cold-sensing TRPM8 channel by cooling agonist and PIP2. Science (New York, N.Y.) 378 (6616). Available at https://doi.org/10.1126/science.add1268, https://www.science.org/doi/10.1126/science.add1268.

Yu, J., Rao, P., Clark, S., Mitra, J., Ha, T., Gouaux, E., 2021. Hippocampal AMPA receptor assemblies and mechanism of allosteric inhibition. Nature 594 (7863), 448–453. Available at https://doi.org/10.1038/s41586-021-03540-0, http://www.nature.com/nature/index.html.

Yu, J., Zhu, H., Lape, R., Greiner, T., Du, J., Lü, W., et al., 2021. Mechanism of gating and partial agonist action in the glycine receptor. Cell 184 (4), 957. Available at https://doi.org/10.1016/j.cell.2021.01.026, https://www.sciencedirect.com/journal/cell.

Yu, Z., Wu, Y., Chen, M., Huo, T., Zheng, W., Ludtke, S.J., et al., 2023. Membrane translocation process revealed by in situ structures of type II secretion system secretins. Nature Communications 14 (1). Available at https://doi.org/10.1038/s41467-023-39583-2, https://www.nature.com/ncomms/.

Zaydman, M.A., Silva, J.R., Delaloye, K., Li, Y., Liang, H., Larsson, H.P., et al., 2013. Kv7.1 ion channels require a lipid to couple voltage sensing to pore opening. Proceedings of the National Academy of Sciences of the United States of America 110 (32), 13180–13185. Available at https://doi.org/10.1073/pnas.1305167110, http://www.pnas.org/content/110/32/13180.full.pdf+html.

Zhang, X., Johnson, R.M., Drulyte, I., Yu, L., Kotecha, A., Danev, R., Wootten, D., Sexton, P.M., Belousoff, M.J., 2021. Evolving cryo-EM structural approaches for GPCR drug discovery. Structure 29, 963–974.e6. Available at 10.1016/j.str.2021.04.008.

Zhao, Y., Chen, S., Swensen, A.C., Qian, W.J., Gouaux, E., 2019. Architecture and subunit arrangement of native AMPA receptors elucidated by cryo-EM. Science (New York, N.Y.) 364 (6438), 355–362. Available at https://doi.org/10.1126/science.aaw8250, https://science.sciencemag.org/content/364/6438/355/tab-pdf.

Zhao, Y., McVeigh, B.M., Moiseenkova-Bell, V.Y., 2021. Structural pharmacology of TRP channels. Journal of Molecular Biology 433 (17). Available at https://doi.org/10.1016/j.jmb.2021.166914, https://www.journals.elsevier.com/journal-of-molecular-biology.

Zhou, X.E., He, Y., de Waal, P.W., Gao, X., Kang, Y., Van Eps, N., et al., 2017. Identification of phosphorylation codes for arrestin recruitment by G protein-coupled receptors. Cell 170 (3), 457. Available at https://doi.org/10.1016/j.cell.2017.07.002, https://www.sciencedirect.com/journal/cell.

Zhou, Y., Morais-Cabral, J.H., Kaufman, A., Mackinnon, R., 2001. Chemistry of ion coordination and hydration revealed by a K+channel-Fab complex at 2.0 Å resolution. Nature 414 (6859), 43–48. Available at https://doi.org/10.1038/35102009.

Zhou, Y., Zou, J., Xu, J., Zhou, Y., Cen, X., Zhao, Y., 2023. Recent advances of mitochondrial complex I inhibitors for cancer therapy: current status and future perspectives. European Journal of Medicinal Chemistry 251, 115219. Available at https://doi.org/10.1016/j.ejmech.2023.115219.

Zhu, H., 2022. Structure and mechanism of glycine receptor elucidated by cryo-electron microscopy. Frontiers in Pharmacology 13. Available at https://doi.org/10.3389/fphar.2022.925116, http://www.frontiersin.org/Pharmacology.

Zhu, J., Vinothkumar, K.R., Hirst, J., 2016. Structure of mammalian respiratory complex i. Nature 536 (7616), 354–358. Available at https://doi.org/10.1038/nature19095, http://www.nature.com/nature/index.html.

Zhu, S., Noviello, C.M., Teng, J., Walsh, R.M., Kim, J.J., Hibbs, R.E., 2018. Structure of a human synaptic GABAA receptor. Nature 559 (7712), 67–88. Available at https://doi.org/10.1038/s41586-018-0255-3, http://www.nature.com/nature/index.html.

Zubcevic, L., Herzik, M.A., Chung, B.C., Liu, Z., Lander, G.C., Lee, S.Y., 2016. Cryo-electron microscopy structure of the TRPV2 ion channel. Nature Structural and Molecular Biology 23 (2), 180–186. Available at https://doi.org/10.1038/nsmb.3159, http://www.nature.com/nsmb/index.html.

CHAPTER 6

Computational and artificial intelligence methodologies for big data analytics in structural biology

Mohammad Kalim Ahmad Khan, Umme Aiman, and Salman Akhtar
Department of Bioengineering, Integral University, Lucknow, Uttar Pradesh, India

6.1 Introduction

The explosive growth of high-throughput biological technologies has deeply impacted the life sciences world, making it a data-intensive field. Modern platforms such as next-generation sequencing (NGS), cryo-electron microscopy (cryo-EM), high-resolution mass spectrometry (MS), and large-scale interactomes have been capable of generating enormous genomic, transcriptomic, proteomic, structural, and imaging datasets at extremely high speeds and scales (Park et al., 2021). As a result, there have been major analytical and computational challenges, primarily in the domain of structural biology and bioinformatics, where comprehending the functional, structural, and evolutionary impact of these data requires the use of advanced computational methods. The biological data have several core features, such as their massive scale, high throughput, natural heterogeneity, a high degree of uncertainty, and the need for biological significance. These features call for the creation of strong analytical frameworks that can connect different types of data, reduce noise, guarantee reproducibility, and yield biologically meaningful results (Auslander et al., 2021).

Biomedical major repositories are a good example of exponential data growth due to multimodality and strong dependence on biological context. Protein sequences, three-dimensional structures, and multiomics profiles ought to be processed using computational strategies that handle the noise and enable reproducible inference. The enormous growth of protein sequences illustrated in Figs. 6.1, 6.2, and 6.3 demonstrates that current sequencing technologies consistently outperform traditional computational pipelines.

Conventional computer-based methods or heuristic computational approaches have had a hard time trying to address these challenges at scale. To address this gap, machine learning (ML), deep learning (DL), and advanced statistical modeling frameworks have gradually become the main tools and emerged as central pillars of modern structural biology. DL has brought about a significant change in structural biology (Cui et al.,

Advances in Structural Biology
DOI: https://doi.org/10.1016/B978-0-443-44988-8.00012-5

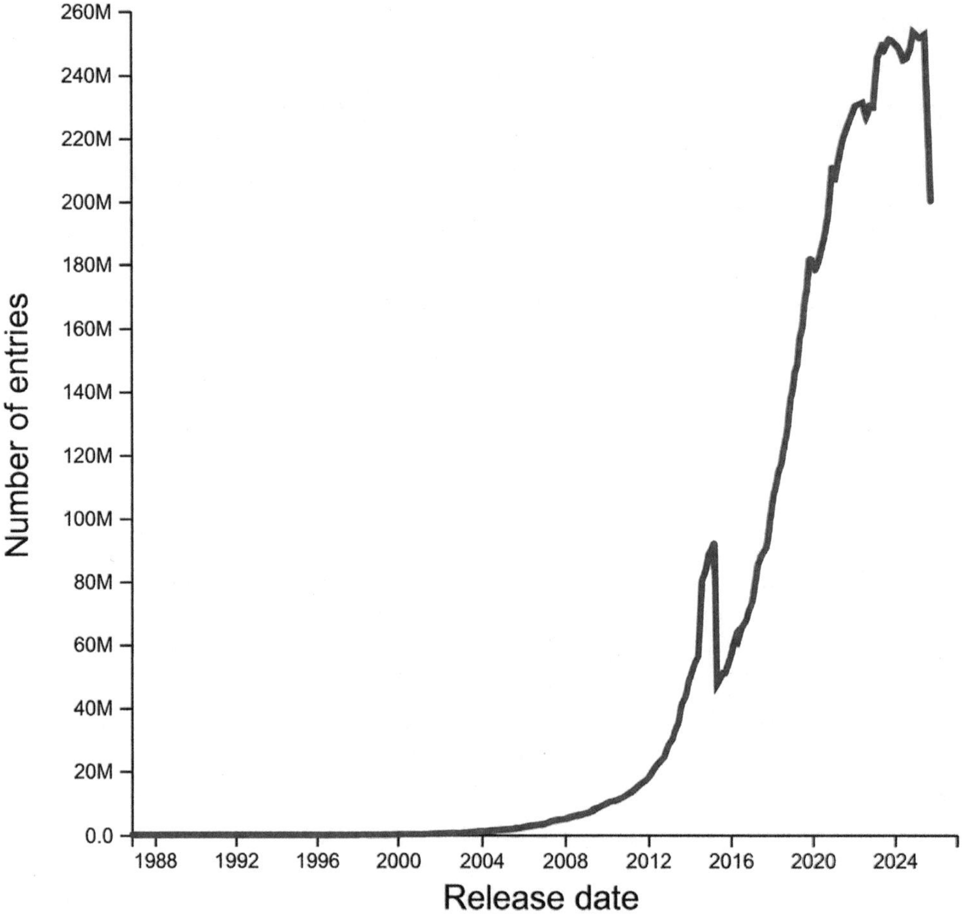

Figure 6.1 Growth of protein sequences over time: UniProtKB release 2025_04 data retrieved on November 30, 2025, depicting the growth trends in the UniProt database over time, which is considered as major source of biological sequence data.

2022), particularly through transformer models and graph neural networks (GNNs), which have enabled breakthroughs in protein structure prediction, cryo-EM image processing, molecular interaction, molecular docking, and drug discovery (Chen et al., 2024; Li et al., 2023). Such artificial intelligence (AI)-based approaches not only improve predictive models but also are used to complement sequence analysis, structural modeling, and experimental pipelines in structural determination, target selection, and interpretation of biological findings, allowing biological hypotheses to be tested at the proteome scale (Krokidis et al., 2025) and supporting drug design via structured AI platforms (Mukherjee et al., 2025).

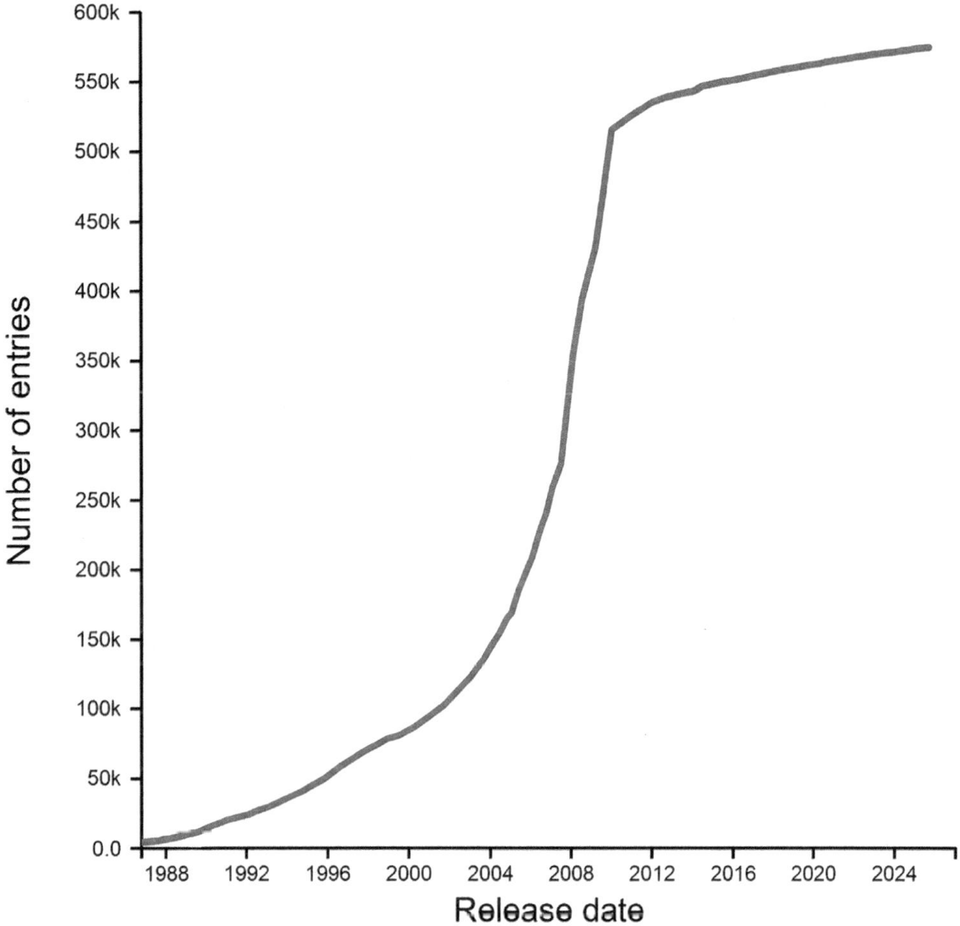

Figure 6.2 Growth of protein sequences over time in Swiss-Prot: notable growth trends in the Swiss-Prot database over time, reflecting the accumulation of manually curated protein sequences retrieved on November 30, 2025, from UniProtKB release 2025_04.

Nevertheless, there are both significant theoretical and practical challenges facing bioinformatics and structural biology (Valous et al., 2024). Additionally, advances in cloud computing, GPU acceleration, and explainable AI (XAI) have made these frameworks increasingly scalable and accessible. Such pipelines, by integrating high-throughput data with state-of-the-art computational techniques, accelerate biological discovery, mechanistic interpretation, and translational research, which are difficult to resolve using a single technique (Karim et al., 2023).

The chapter focuses on computational methodologies that enable big data analytics in bioinformatics, computational biology, and structural biology, rather

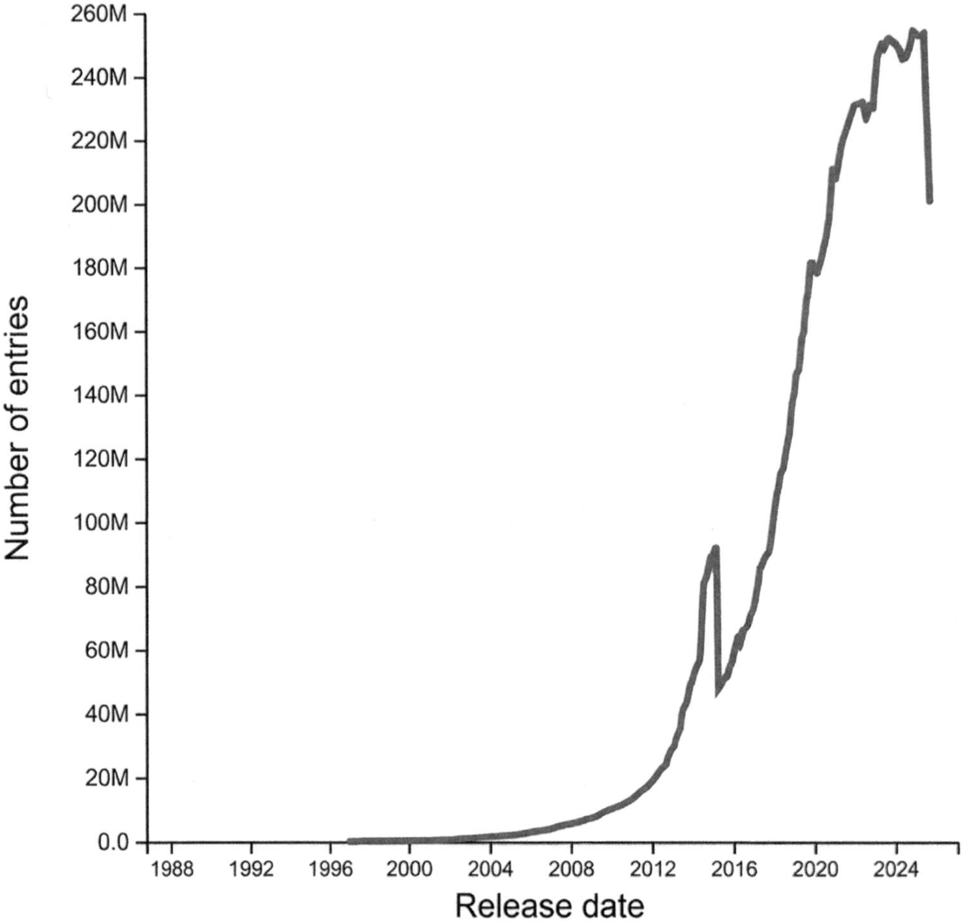

Figure 6.3 Growth of protein sequences over time in TrEMBL: growth trends in the TrEMBL database over time, reflecting the rapid increase in computationally annotated protein sequences. Data accessed on November 30, 2025, from UniProtKB release 2025_04.

than only on experimental data itself. The following sections highlight algorithmic strategies and AI-driven framework that support the sources of biological big data, the defining characteristics of these datasets, the computational methodologies used to analyze them, and the diverse applications that shape current and future research in this field. By emphasizing computational methodologies, analytical pipelines, and an integrated framework rather than descriptive background, the chapter aims to provide a conceptual view of how big data analytics is shaping structural biology into a predictive science.

6.2 Sources and types of biological big data

The advent and rapid progress of high-throughput technologies have led to a substantial increase in the diversity, scale, and complexity of biological data. For instance, in structural biology and bioinformatics, huge amounts of data are generated by several platforms on molecular, structural, cellular, and systems-level phenomena. The categories that describe biological data and their importance in structural biology are outlined below (Shen et al., 2023; Suster et al., 2024).

6.2.1 Genomic and transcriptomic data

Modern NGS platforms have made it possible to perform large-scale production and handling of high-resolution genomic and transcriptomic data. A huge amount of multi-dimensional data is generated by whole-genome sequencing, exome sequencing, long-read sequencing, single-cell sequencing, and spatial transcriptomics for the analysis of structure-function relationships at a molecular level. The genomic findings serve as a vital resource for determining evolutionary links, finding structural variants, and establishing genotype-phenotype associations. Transcriptomic data, including bulk RNA-seq, single-cell RNA-seq, and spatially defined expression estimates, shed light on context-dependent gene expression and protein abundance (Liu et al., 2025). Incorporating genome and transcriptome data helps structural modeling algorithms identify isoform-specific sequences, physiologically relevant mutations, and tissue-specific protein expression patterns. The growing demand for long-read sequencing (PacBio HiFi and Oxford Nanopore) adds levels of structural complexity, allowing for precise measurement of complete transcripts and uncovering novel splicing events that influence protein configuration and stability (Mair et al., 2020; Dong et al., 2021). Collectively, genomic and transcriptomic data constitute the foundation for structural and functional predictions.

6.2.1.1 Next-generation sequencing technologies and big-data considerations

NGS platforms enable high-throughput generation of genomic and transcriptomic data with varying read lengths, accuracy, and throughput. Short-read systems such as Illumina offer high accuracy for large cohort studies, while long-read technologies like Oxford Nanopore and PacBio HiFi resolve structural variants, repetitive regions, and full-length transcript isoforms. Targeted platforms such as Ion Torrent provide rapid, cost-effective sequencing for focused applications. A comparative summary of major platforms is provided in Table 6.1.

NGS datasets are large and heterogeneous, requiring efficient computational pipelines for quality control, alignment or assembly, quantification, variant detection, and annotation. High-performance computing and cloud-based infrastructures are essential for managing these data-intensive workflows. Together, NGS technologies and

Table 6.1 Comparison of major next-generation sequencing (*NGS*) platforms.

NGS platform	Typical read length	Accuracy	Cost level	Key advantages	Limitations	Common applications
Illumina	50–300 bp (short reads)	>99.9%	Moderate	High accuracy, high throughput, robust for large cohort studies	Limited ability to resolve long repeats/structural variants	WGS, WES, RNA-seq, metagenomics
Oxford Nanopore	>10 kb to ultra-long (>100 kb)	~95% raw (higher with polishing)	Moderate	Real-time sequencing, portability, excellent for structural variant analysis	Lower raw accuracy; requires postprocessing	De novo assembly, long-read transcriptomics, structural variant analysis
PacBio HiFi	10–25 kb	>99.9%	High	Long, highly accurate reads ideal for complex genomes	Higher cost; lower throughput	High-accuracy assemblies, isoform sequencing, clinical genomics
Ion Torrent	100–400 bp	~98%–99%	Low	Fast and cost-effective for targeted panels	Homopolymer errors; limited scalability	Targeted sequencing, microbial panels
NGS platform	Typical read length	Accuracy	Cost level	Key advantages	Limitations	Common applications

big-data analytics convert raw sequencing reads into actionable biological information that supports downstream structural, functional, and clinical studies.

6.2.2 Proteomics and structural big data

Proteomics datasets generated using MS, quantitative proteomics, crosslinking MS, and cryo-EM are directly relevant to structural biology. These datasets capture information on protein abundance, posttranslational modifications, conformational dynamics, and protein–protein or protein–ligand interactions. Protein structural data represent one of the fastest-growing categories of big biological data (Stahl et al., 2024). The advent of the AlphaFold Protein Structure Database, RoseTTAFold-based repositories, and other AI-driven structural prediction platforms has resulted in millions of computed structure models (CSMs) encompassing vast regions of the protein universe (Gilep et al., 2024; Wang et al., 2024). The integration of crosslinking data, hydrogen–deuterium exchange experiments, nuclear magnetic resonance (NMR) restraints, and high-resolution cryo-EM maps into structural proteomics generates highly multidimensional information. These datasets support the identification of conformational states, allosteric hotspots, structural motifs, and ligand-binding pockets, thereby advancing structure-based drug discovery (Cerofolini et al., 2019).

One of the major issues that emphasizes the necessity for structural big data is the increased gap between the tremendous growth of biological sequence databases and the rather slow collection of experimentally determined three-dimensional protein structures. While UniProt now contains over 250 million protein sequences (Fig. 6.4), the Protein Data Bank (PDB) includes only approximately 245,000 experimentally resolved structures (www.rcsb.org) (Fig. 6.5).

This gap illustrates that biological sequence databases are rapidly developing, considerably exceeding the capabilities of classic structural biology approaches (Fig. 6.6). To overcome this mismatch, big data analytics and AI-powered structural prediction have become critical. AlphaFold DB and ModelArchive now have over 1 million high-quality computed structural models, with AlphaFold DB alone predicting more than 200 million UniProt-derived proteins (Fig. 6.7). These computational models frequently attain accuracies close to, and even exceeding, those of experimental approaches for stable, well-folded proteins. Collectively, these tools demonstrate how ML, DL, and high-performance computing are changing structural biology by enabling quick, scalable, and proteome-wide structure inference, therefore drastically narrowing the sequence–structure gap.

6.2.3 Molecular interaction networks

Biological interactions collectively form extensive and interconnected networks that encompass various data types, including protein–protein interaction (PPI) networks,

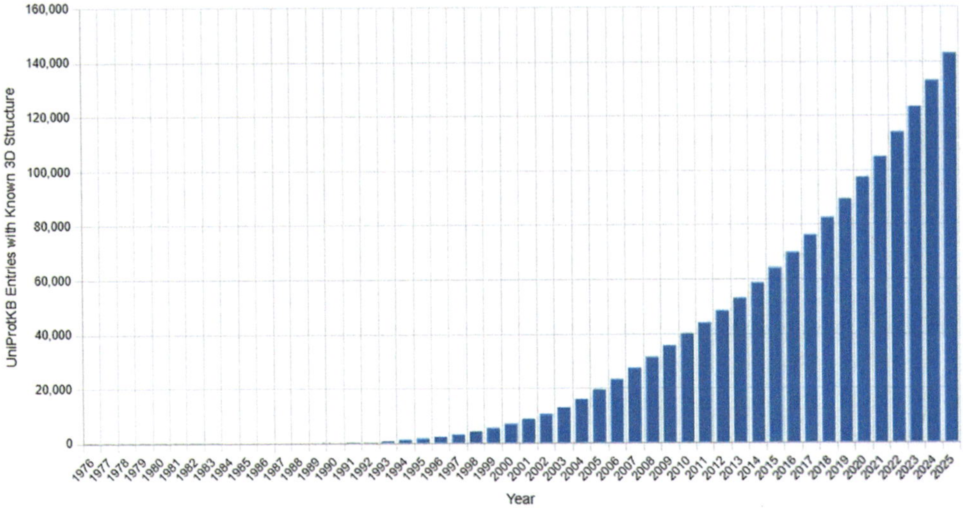

Figure 6.4 Three-dimensional structures of UniProt data. Annual growth of UniProtKB entries with experimentally determined three-dimensional structures in the Protein Data Bank (*PDB*) from 1976–2025, illustrating a steady rise driven by advances in structural biology. Data retrieved on November 30, 2025, from RCSB PDB. *Data sourced from the RCSB Protein Data Bank (PDB) and UniProtKB; accessed on November 30, 2025.*

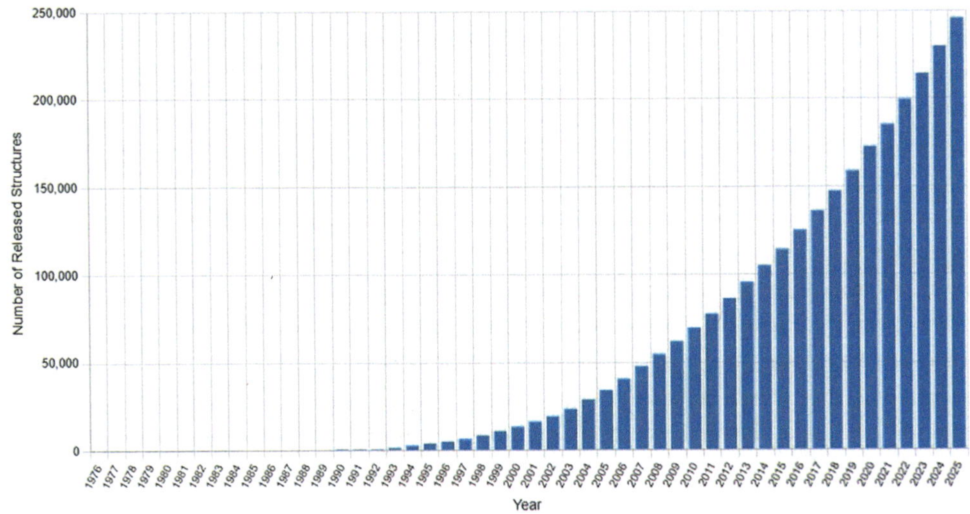

Figure 6.5 Growth trend of three-dimensional structures. From 1976 to 2025, the annual growth in the number of structures deposited in the Protein Data Bank (*PDB*) is shown. The line illustrates the increasing prevalence of experimentally determined macromolecular structures. Data sourced from the RCSB PDB (RCSB PDB; https://www.rcsb.org), accessed on November 30, 2025. *Data sourced from the RCSB Protein Data Bank (PDB; https://www.rcsb.org), accessed on November 30, 2025.*

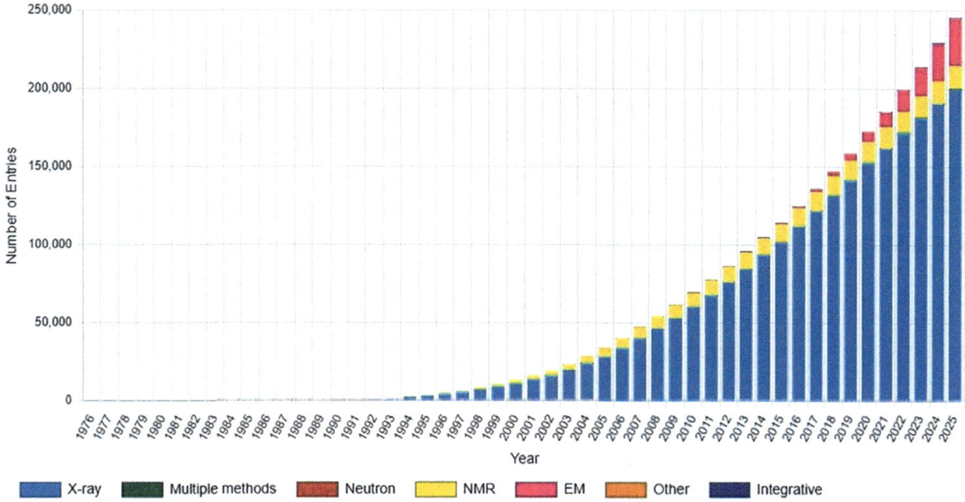

Figure 6.6 Growth of Protein Data Bank (*PDB*) structures over time: trends in three-dimensional structures determined using experimental methods, including X-ray crystallography, nuclear magnetic resonance spectroscopy, three-dimensional electron microscopy, and other techniques (e.g., solution scattering, EPR, IR, fluorescence transfer). Data sourced from the RCSB PDB (PDB; https://www.rcsb.org), accessed on November 30, 2025. *Data sourced from the RCSB Protein Data Bank (PDB; https://www.rcsb.org), accessed on November 30, 2025.*

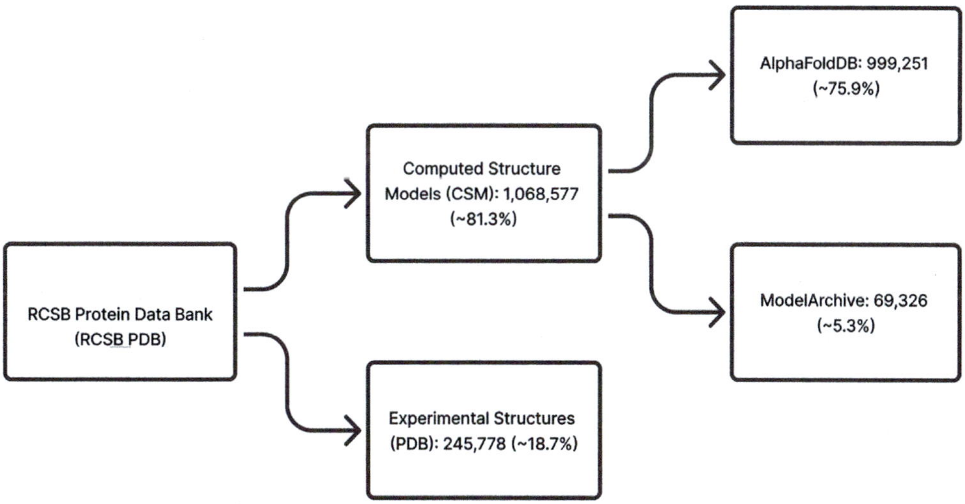

Figure 6.7 Experimental and computed structure model (*CSM*) structures: distribution of protein structures in the RCSB Protein Data Bank (*PDB*), highlighting the predominance of CSMs (~81.3%) over experimentally determined structures (~18.7%). Within CSMs, AlphaFold DB accounts for most entries (~75.9%), while ModelArchive contributes a smaller fraction (~5.4%). Illustration based on data retrieved from RCSB PDB on November 30, 2025. *Illustration based on data retrieved from RCSB PDB on November 30, 2025.*

gene regulatory networks (GRNs), signaling pathways, metabolic networks, epigenetic modification networks, and host–pathogen interaction networks (Kiouri et al., 2025). These datasets are typically curated from major biological resources, such as STRING, IntAct, BioGRID, KEGG, and Reactome, all of which have significantly expanded with the advent of high-throughput interactomics (Wang et al., 2022; Ouma et al., 2018). The integration of network analytics with structural information facilitates the identification of critical nodes, structural motifs that govern protein interactions, and system-level determinants of disease. Recently, network-based structural modeling approaches, such as NetFold and GNN frameworks, have gained prominence for predicting protein complexes and deriving structural constraints directly from the underlying network topologies (Milano et al., 2022).

6.2.4 Imaging and spatial biological data

Advancements in imaging technologies have transformed structural biology into a highly data-intensive discipline, with major sources of high-volume imaging data including cryo-EM, cryo-electron tomography (cryo-ET), superresolution microscopy, next-generation X-ray crystallography detectors, four-dimensional live-cell imaging platforms, and spatial proteomics and metabolomics techniques. Cryo-EM and cryo-ET experiments are capable of producing data in the range of terabytes, in the form of thousands to millions of particle images and tomograms, which necessitate various preprocessing operations such as denoising, particle classification, and three-dimensional reconstruction (Coleman et al., 2025; Groen et al., 2019). Similarly, superresolution microscopy approaches like STORM, PALM, and SIM generate highly detailed data that map the molecular arrangement at nanometer-scale resolution. When combined, these imaging datasets provide vital information on structural organization, dynamic behavior, conformational heterogeneity, and subcellular localization. Thus they are indispensable for integrative and multiscale structural modeling (Sanches et al., 2024).

6.2.5 Multiomics and integrative datasets

Contemporary biological research integrates multiple omics layers, including genomics, transcriptomics, proteomics, epigenomics, metabolomics, and phenotypic profiling. These modalities produce signals that require harmonization through integrative frameworks. Multiomics datasets enable system-level models linking molecular structures to cellular phenotypes. Structural bioinformatics investigations focus on incorporating multiomics features into structure predictions. Combining mutation datasets with protein structural information enables variant modeling, particularly in cancer, where missense mutations affect protein structure (Peng et al., 2020). Integration follows data acquisition, preprocessing, normalization, and feature refinement. Integration may occur through early, joint, late, or hybrid approaches. Datasets undergo computational modeling for classification, regression, clustering, or biomarker discovery, with performance evaluation. Interpretability methods connect the molecular characteristics that are identified to the regulatory process. Public repositories contain datasets that have been curated across various omics domains. AI copes with the vast amount of omics data, as such data are characterized by their volume, variety, and velocity. AI models can be harnessed to study large-scale biological data, allowing for personalized medicine and drug discovery. Such technologies facilitate the integration of multiomics data produced by NGS, MS, and single-cell platforms. Whereas ML algorithms identify relationships among genes, proteins, and metabolites. Table 6.2 displays the principal AI methodologies.

6.2.6 High-throughput experimental and computational platforms

These days, various experimental and computational platforms produce large datasets, which are indispensable for modern structural biology. These platforms include high-throughput crystallography and fragment screening, deep mutational scanning,

Table 6.2 Representative artificial intelligence (*AI*) tools used in multiomics data analysis.

Tool	Description	Resource
PandaOmics	AI-driven platform for multimodal omics and biomedical text mining, enabling therapeutic target and biomarker discovery; developed by Insilico Medicine.	https://insilico.com/pandaomics
Statistical Omics Pattern Hunter Integrative Engine	Separates common and specific transcriptional responses using generative neural networks.	https://github.com/greenelab/sophie
Single-cell epigenomic and transcriptomic MIXed-data annotation via labeling	Provides universal, source-free single-cell RNA-seq annotations using advanced cell-type perception models.	https://github.com/aster-wx/scEMAIL
Transcriptome and image spatial transcriptomics	Integrates transcriptomic data with histopathological images for spatial transcriptomics analysis.	http://lifeome.net/software/tist/
Triplet-based Gene Ontology prediction	Predicts gene functions by integrating expression profiles with protein homology information.	https://zhanggroup.org/TripletGO/
Drug similarity learning	Uses similarity learning to support transcriptional phenotypic drug discovery.	https://github.com/bm2-lab/DrSim/

molecular dynamics (MD) simulations, metadynamics, and replica exchange as enhanced sampling techniques, large-scale docking and virtual screening (VS) campaigns, and AI-based structure prediction spanning the proteome scale (Agajanian et al., 2023). Crucially, gigabytes of trajectory data that capture atomic-level motions over microsecond periods can be produced using MD simulations. Consequently, this necessitates the utilization of specific computational resources and promotes the development of GPU-accelerated and cloud-based architectures (Seffernick and Lindert, 2020). Each of these systems works together to create the multimodal data environment that underpins modern computational and structural biology.

6.3 Characteristics of biological big data

The distinctive features of biological data differentiate them from traditional large-scale datasets in the physical sciences or engineering. These characteristics present various data processing problems, but they also direct the creation of storage systems, ML techniques, and computational workflows. A deep knowledge of the inherent characteristics of big biological data is necessary to get the most out of analytical methods in structural biology and bioinformatics.

6.3.1 Scale

One of the main features of biological data is its enormity and the rapid growth of its scale. The amount of genomic data by itself has outpaced the growth rate of astronomy and social-media data, reaching exabyte-scale production yearly. The constant production of sequences, high-throughput instrument-generated data, spectra, images, and interaction profiles has led to the exponential expansion of archives like NCBI, UniProt, PRIDE, PDB, and the AlphaFold Protein Structure database (Segura et al., 2023). Cryo-EM investigations in structural biology frequently generate many terabytes per session, including intermediate reconstructions, raw movies, micrographs, and classification results. Likewise, AI-driven protein structure prediction platforms produce millions of computed structural models, expanding the availability of structural data at a never-before-seen rate. Scalable storage systems, distributed computing architectures, and effective data compression methods are required due to this enormous scale (Wankowicz and Fraser, 2024; Song et al., 2024).

6.3.2 Throughput

High-throughput technologies are capable of producing data at exceedingly fast rates. For instance, NGS machines can sequence human genomes entirely in under a day by yielding billions of structurally and functionally informative reads. Moreover, single-cell technologies enable transcriptome analysis of millions of individual cells from a single experiment, thereby greatly boosting throughput. In the era of structural biology, automated crystallography pipelines as well as high-flux X-ray beamlines make it possible to gather data at a rate of several hundred crystals per day. Additionally, cryo-EM devices coupled with automated data collection software have the capacity to capture and record thousands of micrographs per hour (Chung et al., 2022). Such levels of throughput require algorithms capable of performing computational processing in real time or near real time, along with efficient scheduling and automated preprocessing pipelines. The inflow of data is so rapid that it frequently exceeds the capacity for analysis, thus necessitating the advent of AI-based preprocessing and cloud computing solutions (David et al., 2019).

6.3.3 Heterogeneity

Biological records are basically very different from each other because of changes in origin, format, scale, and biological meaning. This heterogeneity can be seen in different aspects of biology, such as the presence of various data types like sequences, structures, images, and interaction maps; variability at the species and tissue levels; experimental noise; as well as temporal and spatial variations. One of the major challenges in structural bioinformatics is unifying NMR constraints, cryo-EM, crosslinking data, predicted structures, and molecular simulation results, which require very adaptable systems that can effectively combine different types of information. If the analysis is based on multiomics, things get even more complicated because it is necessary to reconcile genetic variations, gene expression profiles, protein abundance, metabolite levels, and structural annotations. Therefore ML algorithms have to be developed in such a way that they are able to work with multimodal input, missing or unbalanced data, as well as unevenly distributed data (Athaya et al., 2023).

6.3.4 Uncertainty

Biological data are often noisy mainly due to variability in sample preparation, reagents, and instrument limitations. There can be discrepancies and technical errors in genomic sequencing depending on the different read lengths, GC content, and sequencing chemistry that are utilized. Electron beam scattering, induced motion, and radiation damage also trigger noise in imaging data, which in turn affects structural interpretation. Both experimentally resolved and AI-predicted structural models can be considered to have uncertainties. For instance, AlphaFold predictions come with per-residue confidence metrics and predicted aligned errors), which indicate the uncertainty of the model at the atomic and domain levels (Song et al., 2020). Uncertainties in MD simulations arise from force-field limitations, sampling bias, and sensitivity to initial conditions. These uncertainties must be addressed using statistical modeling, probabilistic inference, and robust validation approaches.

6.3.5 Biological relevance

To properly understand biological datasets, they must be interpreted within a suitable biological context, thus making biological relevance the key feature that differentiates big biological data from other types. Structural features, for example, conformational flexibility, domain interactions, mutation effects, and ligand-binding properties, can only be considered important if they are combined with functional evidence from proteomics, transcriptomics, or phenotypic assays. Hence, biological relevance chiefly revolves around interpretation in a specific context, that is, the elucidation of the relationship between molecular structure and function, the discovery of disease-associated variants, the prediction of physiological outcomes, and the support of translational research. Ultimatelyy, what really counts regarding the role of big data in structural biology is not how much data there are but how well the data can be used to generate mechanistic insights that connect molecular behavior and system-level phenotypes (Fleetwood et al., 2020).

6.3.6 Complexity and high dimensionality

Biological systems are inherently complex and are marked by nonlinear interactions, overlapping pathways, and dependencies that extend across multiple spatial and temporal scales. As a result, data coming from such systems are typically of extremely high dimensionality, consisting of thousands or even millions of features (Kopp et al., 2022). To illustrate, single-cell transcriptomic data can reveal the expression of more than 20,000 genes across millions of cells, cryo-EM images consist of millions of pixels showing several hidden conformational states, and MD trajectories record millions of atomic coordinates over time. All these characteristics warrant the use of dimensionality reduction methods, require supercomputing facilities, and involve the utilization of sophisticated learning architectures like GNNs, variational autoencoders (VAEs), and transformer-based models to sift out significant patterns from huge and complicated datasets (Fleetwood et al., 2020).

6.3.7 Dynamic nature

Biological processes are dynamic and change continuously with age, environmental conditions, and changes in cellular state. Thus the temporal dimension has greatly complicated biological data. Structural biology has undergone a revolution, changing its focus from solely static structures to dynamic ensembles, because cryo-EM can show multiple conformational states, and MD simulations can capture the transitions between them. This kind of dynamic data is capable of generating energy landscapes, predicting allosteric regulation, describing ligand-induced conformational changes, and modeling time-resolved biological mechanisms. Hence, biological big data are not only huge in volume but also temporally rich and naturally dynamic (Ivanisenko et al., 2024).

6.3.8 Interconnectedness

Living organisms are not self-contained, sufficient systems. Their physical features, gene expression, protein interactions, epigenetic modifications, and physiological characteristics are all very much intertwined. Representations in the form of networks reflect this connectedness and help to understand the system at a level beyond the individual components. Hence, the analysis of huge datasets should allow for the presence of graph structures, network dependencies, and integrative models. This high degree of interconnectedness makes large biological datasets excellent candidate for graph theory, network science, and relational ML methods (Sherif et al., 2025).

6.4 Computational methodologies for big data analytics

The rapid growth of biological big data has created a demand for highly efficient computational methods that can not only deal with but also analyze and understand such complex datasets. In modern structural bioinformatics workflows, big data typically must be processed through multiple sequential steps of preprocessing, feature extraction, predictive modeling, and integrative interpretation. ML and DL serve as core analytical engines within this pipeline, enabling computational analysis of proteomic-scale and multiomics datasets. Together, these methods provide a unified framework for transforming raw biological data into predictions of secondary structure elements, solvent accessibility, intrinsically disordered regions, and other biologically meaningful insights. Here, we discuss the major computational models and learning strategies that have been instrumental in shaping modern big data analytics (Kumar and Srivastava, 2024).

6.4.1 Machine learning approaches

Classical methods like support vector machines, random forests, gradient boosting, logistic regression, and k-means clustering have been widely used for sequence classification, structural feature prediction, protein function annotation, and drug–target interaction prediction (Avanija, 2025). Moreover, ensemble learning techniques have often led to improved accuracy across different datasets. Recently, ML has been applied to various tasks such as predicting the effects of mutations on protein stability, detecting binding pockets and hotspot residues, revealing coevolutionary residue relationships, sorting cryo-EM particle images, and forecasting the best drug candidates from huge VS libraries. While classical ML methods depend heavily on handcrafted features, they still have merits such as efficiency, interpretability, and comparatively low computational cost.

6.4.2 Deep learning frameworks

DL has revolutionized structural biology and bioinformatics. Various cutting-edge DL architectures, including convolutional neural networks (CNNs), recurrent neural networks, transformers, VAEs, and GNNs, can automatically learn hierarchical and latent representations from raw sequences, structures, images, and molecular graphs (Zeng et al., 2024). While the basic principles of DL architectures are explained generally, this section focuses on their application to structure prediction, graph neural networks, and integrative biological modeling.

6.4.2.1 Transformer models for protein structure and function

Transformer architectures, which were originally designed for processing natural language, now provide the foundation for current protein sequence and structure prediction. Models like ESMFold, AlphaFold2, RoseTTAFold2, Evoformer, and ProtT5 use attention mechanisms to resolve long-range relationships and explicitly infer structural constraints from primary sequences (Tan et al., 2024). These transformer-based architectures are gradually becoming popular in a variety of fields such as predicting the structure of proteins, understanding protein sequence representations, evaluating the effects of mutations, performing docking and complex modeling, and rational protein design. The capacity of transformers to encapsulate millions of parameters enables them to accurately generalize over a wide range of protein families and fold types, outperforming earlier approaches (Jiao et al., 2023).

6.4.2.2 Graph neural networks

GNNs represent proteins as graphs; in such graphs, nodes correspond to atoms or residues, while edges represent the spatial or functional relationships between nodes. This is an especially effective method for representing structural data and interaction networks. GNNs have been applied to various protein-related tasks such as ligand-binding prediction, interface residue classification, structural refinement, and molecular property prediction. A few platforms using GNNs, namely, EquiBind, DiffDock, and ProteinMPNN, have shown excellent results in docking and protein design tasks, leveraging equivariant architectures that preserve the symmetry of molecules (Chou et al., 2024).

6.4.2.3 Deep learning for cryo-electron microscopy and structural imaging

Denoising, particle picking, classification, and three-dimensional reconstruction are all examples of cryo-EM tasks that have been extensively performed using DL. CNN architectures have outperformed standard algorithms for detecting particles in noisy micrographs, whereas autoencoder frameworks make conformational landscape mapping easier. In addition, DL has accelerated superresolution microscopy image reconstruction, automated tomogram segmentation, real-time motion correction, and structural state grouping. These advances enable the generation of high-quality structural maps, even under challenging experimental settings (Dhakal et al., 2025; Heebner et al., 2022).

6.4.3 Network-based computational methods

Network science offers a solid foundation for studying interrelated biological systems. Networks in structural bioinformatics can represent a variety of interactions such as residue interaction network, PPIs, GRNs, metabolic networks, and drug–target interaction networks. Key residues, crucial proteins, functional modules, and regulatory hubs are routinely identified using graph-theoretical criteria including centrality, modularity, and shortest paths (Zhang et al., 2024). Network-based prediction techniques include functional annotation, network propagation for disease-associated gene prioritization, graph embedding for representation learning, topology-based mutation effect prediction, and protein complex interface mapping. Modern models combine network topology with structural constraints, allowing them to accurately predict binding interfaces, allosteric pathways, and mutational tolerance (Yang et al., 2021).

6.4.4 Scalable and high-performance computing platforms

Because of the size and complexity of today's biological datasets, high-performance, cloud-enabled, distributed computational systems are required (Tirupati et al., 2024). The key technologies include the following:

6.4.4.1 High-performance computing

Large-scale simulations, structural modeling, and image processing are all made possible by HPC infrastructure. MD trajectories, DL model training, and cryo-EM reconstruction frequently require GPU-enabled clusters to meet computational demands (Lai et al., 2020).

6.4.4.2 Cloud computing

AWS, Google Cloud, and Microsoft Azure are examples of platforms that offer flexible computational environments that can be scaled in response to demand. Cloud-native solutions make it easier to train DL models in a distributed manner, process sequencing datasets on demand, and store and share massive structural datasets remotely. Cloud computing has made high-performance computational infrastructure more accessible by providing scalable resources and facilitating seamless global collaboration (Moriya et al., 2024).

6.4.4.3 Parallel and distributed computing

Frameworks like Apache Spark, Dask, and Ray enable parallel processing of huge datasets, making them suitable for a variety of applications such as large-scale omics integration, rapid image preprocessing, and large-scale VS. These technologies, powered by parallel architectures, can perform rapid data analytics while saving time and making the greatest use of computational resources (Fatima et al., 2023).

6.4.4.4 GPU and TPU acceleration

The prediction and visualization of molecular structures have become deeply dependent on GPU acceleration nowadays. Large-scale protein language models can be trained more efficiently using TPUs, especially for transformer-based models (Chen et al., 2025).

6.4.5 Molecular simulation and artificial intelligence-driven modeling

Molecular simulations consist of a wide spectrum of methods, such as classical MD, coarse-grained simulations, Monte Carlo methods, and enhanced sampling techniques, which are all powerful computational tools for the study of structural biology. These simulations produce high-dimensional temporal data through which the dynamics of proteins, ligand binding, and conformational transitions can be revealed. The coupling of AI with molecular simulations has facilitated force-field optimization, enhanced sampling, trajectory clustering, prediction of free-energy landscapes, and automatic recognition of metastable states. Such AI-boosted simulations have made it possible to gain a deeper understanding of protein folding, allosteric regulation, and drug-binding mechanisms (Sidky et al., 2020).

6.4.6 Integrative and hybrid modeling

Contemporary structural biology places great emphasis on implementing hybrid approaches that merge various experimental constraints and computational predictions. Integrative modeling platforms gather data from cryo-EM maps, NMR restraints, crosslinking MS, SAXS, predicted structures, and interaction networks. Multimodal integration like this enables the building of large membrane protein complexes and flexible assemblies that are not easily characterized by experimental methods alone (Son et al., 2024). Merging AI-powered pipelines, hybrid modeling, and experimental constraints within unified pipelines improves precision, diminishes noise, and facilitates the interpretation of partial or ambiguous data. Such an integrated workflow exemplifies the shift from single-method analyses to holistic data-driven biology.

6.5 Applications in structural biology and bioinformatics

The fusion of big data analytics and structural biology has opened up new dimensions for biological discoveries. Advanced AI-based methods, high-throughput experiments, and computational hybrid approaches have fundamentally changed the landscape of structure prediction, mechanism discovery, drug development, and biological system modeling. The following sections highlight the major areas where big data techniques are being used.

6.5.1 Artificial intelligence-driven protein structure prediction

A significant breakthrough in the field of big data analytics is the development of AI-based protein structure prediction. The introduction of AlphaFold2 and RoseTTAFold has changed the game completely. This change refers to the use of DL and attention-based architectures that can infer structural constraints from evolutionary sequence patterns without relying on templates, in contrast to the old template-based methods for obtaining protein structures. Thanks to the large sequence databases such as UniRef, MGnify, and metagenomic datasets, these models can be trained on hundreds of millions of sequences. Due to this, they are capable of capturing coevolutionary signals, which are important features for accurately predicting the three-dimensional structures of proteins. Subsequently, AI-based protein structure prediction has advanced markedly to proteome-scale modeling. Such modeling enables the generation of structural information for a wide range of organisms, thereby supporting complex structure predictions such as AlphaFold-Multimer and RoseTTAFold2 Complex, prediction of ligand-bound conformations, assessment of foldability and designability, and structural refinement through AI-enhanced MD simulations (Díaz-Rovira et al., 2023; Liu et al., 2023, Nguyen et al., 2024). Combining cryo-EM density maps with DL-based models has made de novo modeling of large assemblies such as ribosomes, membrane channels, and viral capsids even more possible. AI-generated CSMs are the basis of a plethora of downstream drug discovery, functional annotation, and systems biology applications.

6.5.1.1 Key protein structure and target databases

Protein structure prediction and analysis depend heavily on the availability of well-curated structural and target databases. The PDB is the main source of experimentally derived structures, while platforms like AlphaFold DB provide substantial coverage with their high-confidence computational models. Homology modeling databases such as the SWISS-MODEL Repository and ModBase, together with fold classification systems like SCOP and CATH, help in functional annotation and evolutionary studies. In addition, other resources such as the Electron Microscopy Data Bank for cryo-EM maps and Orientations of Proteins in Membranes database for membrane protein topology provide added structural understanding. All these databases, taken together, form the bedrock of today's structural bioinformatics and AI-driven modeling pipelines (Table 6.3).

6.5.2 Structural big data in drug discovery

Big data analytics has significantly transformed drug discovery by enabling large-scale VS, multitarget profiling, and structure-based optimization. The combination of structural databases, chemical libraries, and AI-powered predictive models is at the heart of every step in drug discovery. Various public AI platforms have been created to facilitate structural prediction, VS, drug–target interaction modeling, conformational sampling, and ADMET profiling, among other tasks, in support of structural big data analytics (Wang et al., 2024). The most commonly used AI resources related to structural bioinformatics and drug discovery are summarized in Table 6.4.

6.5.2.1 Chemical and ligand databases for drug discovery

Chemical and ligand databases provide the crucial molecular framework for VS, QSAR modeling, and AI-driven drug development. Public repositories, such as PubChem, ChEMBL, ZINC, and DrugBank, provide access to millions of bioactive compounds, annotated activities, and drug-like libraries. In addition, specialized databases facilitate fragment screening, natural product exploration, and pharmacokinetic profiling. Collectively, these resources enable efficient hit identification, target mapping, and computational optimization in drug discovery workflows (see Table 6.5).

6.5.2.2 High-throughput virtual screening

Contemporary VS can evaluate millions of compounds using GPU-accelerated docking algorithms, deep learning-based scoring functions, and parallel computing frameworks (Glaser et al., 2021). Structural ensembles derived from MD simulations and cryo-EM intermediate states facilitate screening against multiple conformations of a target, thereby increasing the hit-discovery rates.

6.5.2.3 Artificial intelligence-based lead optimization

DL models, including GNNs, protein–ligand convolutional models, and generative models such as GANs and diffusion models, have been employed to identify chemical modifications that optimize binding affinity, selectivity, pharmacokinetic properties, and toxicity profiles. Generative AI platforms have enabled the de novo design of small molecules, peptides, and macrocycles that adhere to specific structural constraints (de Souza Neto et al., 2020).

Table 6.3 Target structure databases.

Database	Description	Size (structure)	Key features	URL
Protein Data Bank (PDB)	Primary repository for experimentally determined protein and nucleic acid three-dimensional structures	~247,000 structures	X-ray diffraction, electron microscopy, solution nuclear magnetic resonance (NMR), electron crystallography, neutron diffraction, solid-state NMR, solution scattering, fiber diffraction, powder diffraction, electron paramagnetic resonance, infrared spectroscopy, fluorescence transfer	https://www.rcsb.org/
AlphaFold DB	Artificial intelligence-predicted protein structures by DeepMind	>200 million predicted models	Includes per-residue confidence scores	https://alphafold.ebi.ac.uk/
SWISS-MODEL Repository	Collection of automated homology-modeled structures	~1.0 million + models	Template-based modeling	https://swissmodel.expasy.org/repository
ModBase	Comparative protein structure models	Millions of models (varies by release)	Built on sequence homology	https://modbase.compbio.ucsf.edu/
PDB-REDO	Automatically re-refined PDB structures	>180,000 refined models	Improved accuracy and resolution	https://pdb-redo.eu/
Orientations of Proteins in Membranesdatabase	Orientation of proteins in membranes	~8000+ membrane protein structures	Provides membrane orientation	https://opm.phar.umich.edu/
SCOPe	Structural classification database	~250,000 domains	Fold classification	https://scop.berkeley.edu/
CATH	Structural classification database	~150,000 domains	Fold classification	https://www.cathdb.info/
Electron Microscopy Data Bank	Repository of cryo-electron microscopy density maps	~52,000 entries	Three-dimensional volumetric maps	https://www.ebi.ac.uk/emdb/

Table 6.4 Overview of major public artificial intelligence (*AI*) resources for drug discovery.

AI platform	Purpose	Applications	URL
TxGemma	Open-source multimodal foundation model for molecular property prediction, target identification, and drug–target interaction modeling	Hit discovery, target prediction, generative molecule design	https://deepmind.google/models/gemma/txgemma/
Boltz-1	Open-source biomolecular three-dimensional structure prediction model for proteins, RNA, and DNA; AlphaFold 3-level accuracy	Structure prediction, protein engineering, target identification	https://jclinic.mit.edu/boltz-1/
Artificial Intelligence-Ready CHEmiCal Knowledge Base	Cloud-based curated chemical knowledge base for machine learning	Virtual screening (VS), predictive modeling, dataset preparation	https://aircheck.ai/
Protein–Ligand Interaction Dataset Resource	Large-scale dataset of ~449,000 protein–ligand complexes with >500 annotations	DTI modeling, VS benchmarking, docking evaluation	https://www.plinder.sh/
PolarisHub	Open-source benchmarking hub with standardized datasets and evaluation tools	Model comparison, reproducibility studies	https://polarishub.io/
BioEmu-1	Diffusion-based AI model for rapid protein conformational ensemble sampling (100,000 × faster than molecular dynamics)	Protein flexibility modeling, conformational sampling, drug design	https://github.com/microsoft/bioemu
Dyna-1	Deep learning model inferring millisecond-	Protein dynamics analysis, binding/	https://github.com/WaymentSteeleLab/Dyna-1

(*Continued*)

Table 6.4 (Continued)

AI platform	Purpose	Applications	URL
	timescale protein dynamics from incomplete nuclear magnetic resonance data	mechanism prediction	
TamGen	Target-aware generative model (GPT-like) for drug-like molecule generation	De novo drug design, target-specific inhibitor generation	https://github.com/microsoft/TamGen
LigandMPNN (University of Washington + Baker Lab) - Ligand-conditioned Message Passing Neural Network	Ligand-conditioned version of ProteinMPNN for protein sequence design at protein–ligand interfaces	Protein–ligand interface design, therapeutic protein engineering	https://www.ipd.uw.edu/2025/03/introducing-ligandmpnn/
BoltzDesign1	AI-driven generative model for protein engineering and optimization	Enzyme design, synthetic biology	https://github.com/yehlincho/BoltzDesign1
MolPhenix	Foundation model predicting molecular effects on cell morphology using contrastive learning	Phenotypic screening, mechanism-of-action studies, lead optimization	https://www.recursion.com/

6.5.2.4 *Fragment-based and structure-guided design*

High-throughput crystallography and cryo-EM support fragment identification and refinement. The integration of structural data with ML-based free-energy predictions accelerates lead optimization (Isigkeit et al., 2024). The ability to model protein flexibility through big-data-enabled simulations enhances the identification of inducible pockets and allosteric sites.

6.5.2.5 *Multitarget and polypharmacology profiling*

Big data approaches facilitate polypharmacology by analyzing global interaction networks, identifying off-target interactions, and predicting toxicological liabilities (Pinzi and Rastelli, 2020). These models are essential for combating complex diseases like cancer, neurodegeneration, and infectious disorders.

Table 6.5 Major ligand databases used in drug discovery.

Database	Description	Size (compounds)	Key features	URL
ZINC	Free database of commercially available compounds	~1 billion	Drug-like, lead-like, and fragment-like molecules	https://zinc.docking.org/
ChEMBL	Bioactive molecules with drug-like properties	~2.9 million	Curated bioactivity data; drug discovery relevance	https://www.ebi.ac.uk/chembl/
PubChem	Open database of small molecules	~123 million	Experimental and computed chemical properties	https://pubchem.ncbi.nlm.nih.gov/
DrugBank	Approved and investigational drugs	~15,000	Detailed drug–target interactions and clinical annotations	https://go.drugbank.com/
Enamine REAL	Largest commercially available compound library	~30 billion	Virtual and physically accessible screening compounds	https://enamine.net/compound
MolPort	Commercial chemical compound marketplace	~7 million	Aggregated supplier catalogs for easy sourcing	https://www.molport.com/
Mcule	High-quality purchasable compounds	~500 million	Vendor-curated library designed for screening	https://mcule.com/
ChemSpace	Expansive molecular space for drug discovery	~15 billion	Ultra-diverse virtual and purchasable screening libraries	https://chem-space.com/

6.5.3 Functional annotation of proteins

Despite advancements in sequencing technology, a substantial proportion of newly identified proteins remains uncharacterized. Big data analytics facilitates functional annotation using integrative and predictive methodologies.

6.5.3.1 Sequence- and structure-based annotation

AI models trained on extensive protein family datasets can predict Enzyme Commission numbers, Gene Ontology terms, ligand-binding residues, protein–protein interfaces, and protein subcellular localization. Structural models generated by AlphaFold have further expedited functional annotation by elucidating conserved motifs, fold families, and catalytic geometries (Nauman et al., 2019).

6.5.3.2 Evolutionary and coevolutionary analysis

Extensive sequence alignments from metagenomic datasets provide coevolutionary signals that identify functional residues, allosteric pathways, and interaction interfaces (Xie et al., 2023). These signals are particularly beneficial for proteins that lack known homologs.

6.5.3.3 Variant interpretation

Big data enhances the prediction of the effects of mutations on protein stability, folding efficiency, interaction specificity, and disease association. AI-based variant effect predictors utilize millions of natural and engineered sequences to improve prediction accuracy, whereas structural insights offer critical support for clinical interpretation in areas such as cancer genomics, rare diseases, and pharmacogenomics (Taherdoost and Ghofrani, 2024).

6.5.4 Evolutionary and comparative analysis

The easy accessibility of large genomic and proteomic datasets allows for comparative evolutionary analysis across species and protein families.

6.5.4.1 Phylogenomics and structural evolution

The integrative models use sequence, structural, and functional annotations to predict evolutionary pathways. Structural alignments of AI-generated models uncover remote fold linkages and ancestral structural traits (Chen et al., 2024).

6.5.4.2 *Evolutionary couplings and constraints*

Coadapted residue pairs were discovered by statistical coupling analysis. These couplings inform protein folding pathways, stability determinants, and epistatic interactions (Colavin et al., 2022).

6.5.4.3 *Tracking protein family expansion*

Massive data collection facilitates the tracking of protein diversification across phylogenetic trees, which improves our understanding of adaptive evolution, horizontal gene transfer, and domain shuffling (Caetano-Anollés et al., 2021).

6.5.5 Systems biology and network modeling

Systems biology leverages the convergence of multiomics landscapes and structural data to connect individual molecular interactions with system behavior. By mapping these high-dimensional inputs onto network models, scientists can recreate the regulatory circuits and pathways that govern cellular phenotypes. This combination of methods is not just a set of gene or protein lists; it aims to depict the cell's spatiotemporal dynamics. Whereas multiomics gives the inventory and the concentration of the components, structural data supplies the mechanistic constraints for understanding how these elements interact physically (Fan et al., 2025).

6.5.5.1 *Multiomics integration*

In order to develop mechanistic models that explain the link between genotype and phenotype, integrative frameworks connect genomic, transcriptomics, proteomics, epigenomics, metabolomics, and structural data (Fan et al., 2025).

6.5.5.2 *Network-based modeling*

Big data analytics facilitates the extraction of GRNs, PPI networks, metabolic pathways, and signaling cascades. Network analysis makes use of these networks to identify critical nodes, druggable targets, and context-specific regulatory modules (Dahal et al., 2020).

6.5.5.3 *Structural systems biology*

Network structure stability, protein complex architecture, and allosteric modulation can all be simulated by integrating structural information into system models. These techniques help forecast the pathway-level consequences of mutations and pharmacological treatments (O'Connor et al., 2023).

6.5.6 Precision medicine and clinical applications

Big data serves as the foundational engine of precision medicine, providing the computational power necessary to translate massive biological datasets into actionable clinical insights. This is particularly transformative in fields such as oncology, infectious diseases, and rare genetic disorders, where the sheer volume of genomic, transcriptomic, and clinical variables exceeds the capacity of traditional diagnostic work (Dlamini et al., 2022).

6.5.6.1 Structural interpretation of clinical variants

Investigating disease mutations in the context of protein structure offers a powerful means of predicting how these mutations might affect the protein's function, stability, and how well it will respond to treatment (Dlamini et al., 2022).

6.5.6.2 Personalized drug targeting

By combining structural and omics data, it becomes possible to tailor the choice of drug targets to an individual, which is particularly important in cancer, where structural alterations caused by mutations determine the response to a drug (Hassan et al., 2022).

6.5.6.3 Artificial intelligence for biomarker discovery

Deep learning-based analysis of imaging, proteomics, and liquid biopsy data works to discover disease biomarkers that can be used for early detection and prognostic classification (Hartman et al., 2023).

6.5.6.4 Predicting drug resistance

The integration of structural dynamics and ML makes it possible to forecast resistance mutations, viral escape variants, and antibiotic resistance mechanisms. Such knowledge is tremendously helpful in therapeutic design and clinical decision-making (Afzal et al., 2022).

6.5.6.5 Artificial intelligence-designed drugs in clinical development

The incorporation of big data analytics, bioinformatics, and computational biology has laid the groundwork for AI-based drug discovery approaches. By exploiting biological datasets such as genomics, proteomics, structural data, chemical libraries, and clinical records, AI systems are capable of discovering new targets, refining molecular structures, and identifying repurposing opportunities, thus opening the door for drug discovery beyond the capabilities of conventional drug discovery methods. ML algorithms, knowledge graphs, generative chemistry, and AI-driven simulations transform these multifaceted biological datasets into modern therapeutic solutions. Thus facilitating the rapid discovery of new medications in oncology, infectious diseases, neurology, and rare

Table 6.6 Successful and clinically advanced artificial intelligence (*AI*)-designed drug candidates.

Drug	Company	Modality	Indication	AI role	Clinical status
Halicin	MIT and Broad Institute	Small-molecule antibiotic	Multidrug-resistant bacterial infections	Deep learning (MPNN), virtual screening of >100 M compounds	Preclinical success; landmark AI discovery
Baricitinib (repurposed)	BenevolentAI/Eli Lilly	Small molecule (JAK1/2 inhibitor)	COVID-19 antiviral + antiinflammatory	Knowledge-graph mining, graph neural networks, NLP-based literature mining	Successfully repurposed; globally used
DSP-1181	Exscientia + Sumitomo Dainippon Pharma	Small molecule (5-HT1A agonist)	Obsessive-compulsive disorder	Generative AI design, ADMET modeling, rapid design cycles	Entered phase 1 (2020)
Rentosertib (ISM001–055)	Insilico Medicine	Small molecule	Idiopathic pulmonary fibrosis	AI-based target discovery + de novo drug design	Ongoing phase I
BXCL-501 (Dexmedetomidine film)	BioXcel Therapeutics	Small molecule (re-innovation)	Agitation in schizophrenia, bipolar disorder, Alzheimer's disease	AI-driven repurposing and indication expansion	FDA-approved as IGALMI™
Talabostat (BXCL-701)	BioXcel Therapeutics	Small molecule	Oncology: prostate cancer	AI repurposing	Phase IIb
BMF-219	Biomea Fusion/ A2A Pharma	Small molecule	Diabetes, oncology	AI-driven ligand optimization	Phase I/II
BDTX-1535	Black Diamond Therapeutics	Small molecule	EGFR-mutant cancers	AI target discovery	Phase I/II
NDI-034858	Nimbus Therapeutics	Small molecule	Autoimmune diseases	AI-based design and optimization	Successful (phase II)

REC-994	Recursion Pharma	Small molecule	Cerebral cavernous malformations	Phenotypic screening + AI target discovery	Phase II
RLY-4008	Relay Therapeutics	Small molecule	FGFR2-driven cancers	Physics-machine learning hybrid molecular design	Phase I/II
Bevantolol (SOM-3355)	SOM Biotech	Small molecule (repurposed)	Stimulant-use disorder	AI repurposing	Phase IIa completed
Tolcapone (CRX-1008)	SOM Biotech	Small molecule	Transthyretin amyloidosis	AI repurposing	Phase II
GRANITE-001	Gritstone Bio	Personalized neoantigen vaccine	Solid tumors	AI neoantigen prediction (EDGE™)	Phase II/III
SLATE-001	Gritstone Bio	Neoantigen vaccine	Mutant KRAS cancers	AI-based epitope selection	Phase I/II
EVX-01	Evaxion Biotech	Cancer vaccine	Melanoma	AI immunogenic epitope prediction	Phase I/II
AU-007	Biolojic Design	Antibody	Solid tumors	AI antibody design and optimization	Phase I/II
Absci + Owkin Antibodies	Absci + Owkin	De novo antibodies	Immuno-oncology/ inflammation	Generative AI + multimodal AI	Preclinical
COVID Moonshot – DNDI-6510	Open Science Consortium	Small- molecule antiviral	SARS-CoV-2 Mpro inhibitor	AI-guided fragment merging, prioritization	Advanced preclinical

disorders. Consequently, several AI-derived small molecules, biologics, and personalized vaccines have progressed from In silico designs to experimental animal models and clinical trials, thus demonstrating the influence of big-data-driven AI pipelines. A detailed compilation of these AI-enabled therapeutics agents and their development stage is shown in Table 6.6 (Jayatunga et al., 2024).

6.6 Case studies

The following case studies illustrate how computational methodologies and integrative pipelines are applied in real-world biology. Each example demonstrates recent groundbreaking applications that depict the combination of large datasets, AI-driven modeling, and structural interpretation to address complex structural challenges.

6.6.1 Case study 1: artificial intelligence-driven protein structure prediction transforming global structural coverage

A classic illustration of how big data has impacted structural biology is the inception and worldwide application of AlphaFold2, AlphaFold-Multimer, and RoseTTAFold2. These models, which were trained on colossal datasets consisting of millions of sequences, structural templates, and coevolutionary matrices, have thoroughly changed protein structure determination.

6.6.1.1 Methodological integration

This case is an example of how large-scale biological data, metagenomic sequences, UniProtKB reference clusters, and PDB structures are being used to train transformer-based architectures that can predict three-dimensional protein structures with an accuracy comparable to that of experimental methods (Bertoline et al., 2023). By using attention mechanisms and pair representation modeling, AlphaFold2 is able to grasp highly complex long-range interactions that traditional methods fail to recognize (Segura et al., 2023).

6.6.1.2 Impact on structural biology

These big-data-driven advancements have had a significant impact, allowing for structural coverage of whole proteomes in humans, plants, bacteria, and viruses. They have produced hundreds of millions of high-confidence CSMs accessible in public databases, thereby accelerating functional annotation and evolutionary analysis across protein families and providing structural insights into proteins otherwise resistant to characterization by crystallography, NMR, or cryo-EM (Uzoeto et al., 2024).

6.6.1.3 Transformations in drug discovery and systems biology

AI-driven structures have facilitated the discovery of new ligand-binding pockets, identification of allosteric sites, and supported the modeling of protein–protein and protein–ligand complexes. Structural models of challenging targets, including GPCRs, ion channels, and membrane-bound complexes, underpin hypothesis-driven investigations in drug design and systems biology (Conflitti et al., 2025).

6.6.1.4 Limitations and ongoing developments

Despite their efficacy for globular proteins, challenges remain in modeling intrinsically disordered regions, dynamic conformational ensembles, and ligand-bound protein states. Next-generation models focus on integrating cryo-EM maps, MD trajectories, and multiomics data to address these limitations (Song et al., 2024).

6.6.2 Case study 2: cryo-electron microscopy and big data pipelines revealing SARS-CoV-2 spike dynamics

The COVID-19 pandemic catalyzed an unprecedented global scientific effort, generating one of the largest structural biology datasets to date. Cryo-EM of the SARS-CoV-2 spike glycoprotein demonstrated the effectiveness of big data pipelines in structural bioinformatics, revealing spike conformational dynamics, receptor-binding transitions, and antibody interactions. Variant monitoring, antibody engineering, and vaccine development have all been influenced by these structural databases. AI-driven drug repurposing initiatives have accelerated the development of COVID-19 therapeutics. BenevolentAI used a biomedical knowledge graph with ML to analyze molecular pathways and host–virus interactions. Their AI system identified baricitinib, a JAK1/JAK2 inhibitor, as a promising therapeutic candidate. The platform predicted that baricitinib could mitigate hyperinflammation through JAK–STAT pathway inhibition and potentially block viral entry by inhibiting AAK1, a regulator of clathrin-mediated endocytosis. The ACTT-2 clinical trial showed that the combination of baricitinib with remdesivir improved recovery time compared to remdesivir alone. Baricitinib received Emergency Use Authorization from the FDA in November 2020, followed by full approval for COVID-19 patients. These examples demonstrate how computational biology, ML, and big data platforms accelerated understanding and therapeutic discovery during the pandemic (Monteleone et al., 2022).

6.6.2.1 High-throughput cryo-electron microscopy data acquisition

Cryo-EM facilities worldwide have generated terabytes of raw micrographs capturing various conformational states of spike proteins. Automated particle picking, motion correction, and classification pipelines enhanced by ML models have expedited the analysis of heterogeneous datasets (Dhakal et al., 2025). Tools such as CryoSPARC, Warp, and Topaz make use of AI-based denoising and particle identification. The aim is to minimize the duration from data acquisition to structural reconstruction (DiIorio and Kulczyk, 2023).

6.6.2.2 Discovery of conformational states

Big data analytics has opened the door to the discovery of several structural states, such as the on and off conformations (RBD up and RBD down), fusion intermediate states, and glycan shielding patterns. Mapping thousands of spike particles has elucidated the conformational pathways associated with viral entry, immune evasion, and host receptor recognition (Calvaresi et al., 2023).

6.6.2.3 Structural basis for vaccine and therapeutic design

The discovery of the structural basis for vaccine and drug design has led to the stabilization of the spike protein in its prefusion form, which has greatly facilitated the development of mRNA vaccines and the design of monoclonal antibodies. The study also provided the basis for epitope mapping, the design of neutralizing antibodies, and the prediction of variant escape mutations. This example demonstrates the potential of integrating high-throughput structural imaging with AI-driven analytics to accelerate global therapeutic development (Cui et al., 2024).

6.6.2.4 Continued relevance in variant surveillance

Delta, Omicron, XBB, and other variants have necessitated rapid structural reassessment to evaluate their immune escape. By combining scanning mutagenesis, DL, variant prediction, and structural modeling, researchers have been able to monitor antigenic drift almost in real time (Yan et al., 2023).

6.6.3 Case study 3: artificial intelligence-guided drug discovery for kinase inhibitors using multiomics and structural data

Kinases are major drug targets in cancer and inflammation therapy. Recently, big-data-enabled pipelines have shown that combining structural models, molecular simulations, and multiomics datasets can accelerate the discovery of selective kinase inhibitors.

6.6.3.1 Data integration framework

A striking example is the integration of proteome-wide kinase structure prediction (AlphaFold and PDB data) with cancer genomic mutation datasets (TCGA and ICGC), massive chemical libraries containing more than 100 million compounds, and multiomics data, including transcriptomics and phosphoproteomics. These multimodal data were integrated using DL-based multitask models to predict kinase inhibitor binding profiles and mutation-specific sensitivities (Mantini et al., 2021).

6.6.3.2 Artificial intelligence-enhanced virtual screening and binding prediction

GNNs, in combination with transformer-based ligand models, serve as the primary drivers for high-throughput virtual screening of potential kinase inhibitors, including those targeting CDK2, JAK2, BRAF, EGFR, and BTK. Structural ensembles obtained from MD simulations improve the effectiveness of VS by providing dynamic binding-site conformations that capture protein flexibility, thereby improving binding predictions. The compounds listed in public databases like PubChem, ZINC (noncommercial), and National Cancer Institute Synthetic Accessible Virtual Inventory, which contains approximately 10^{12} molecules, represent only a small fraction of the accessible chemical space. Commercially enumerated libraries and proprietary pharmaceutical chemical spaces expand this number up to 10^{20} compounds. Fig. 6.8 illustrates how these accessible chemical libraries constitute only a minor fraction of the theoretical chemical universe. The difference basically points to the big-data-driven character of the modern drug discoveries. As a result of experimental screening only being able to test a limited subset of chemical space, bioinformatics and computational biology, backed by HPC and advanced AI, have to be used to explore it. Among other ML frameworks, GNNs, transformer-based molecular language models, and RL-enabled generative chemistry contribute to swift navigation prediction and design in this vast molecular space. It's crucial to combine big data analytics and AI-powered computer-based workflows in order to find biologically relevant chemical regions, improve ligand structures, and speed up VS pipelines (Carpenter et al., 2018; Li et al., 2021).

6.6.3.3 Predicting drug resistance and mutation effects

For the prediction of resistance-conferring mutations and their possible side effects on inhibitor binding, AI models are trained on global clinical genomic data. Structural annotations provide insights into steric disruptions, allosteric perturbations, and conformational shifts, thereby guiding the refinement of inhibitors with improved resistance profiles (Agajanian et al., 2023).

6.6.3.4 Experimental validation and translational applications

Candidate molecules identified through big-data-enabled computational pipelines have advanced to biochemical assays, cell-based evaluations, and animal models. These pipelines have significantly reduced early discovery timelines from years to months and now support precision oncology initiatives that focus on patient-specific kinase mutations (Zhai et al., 2023).

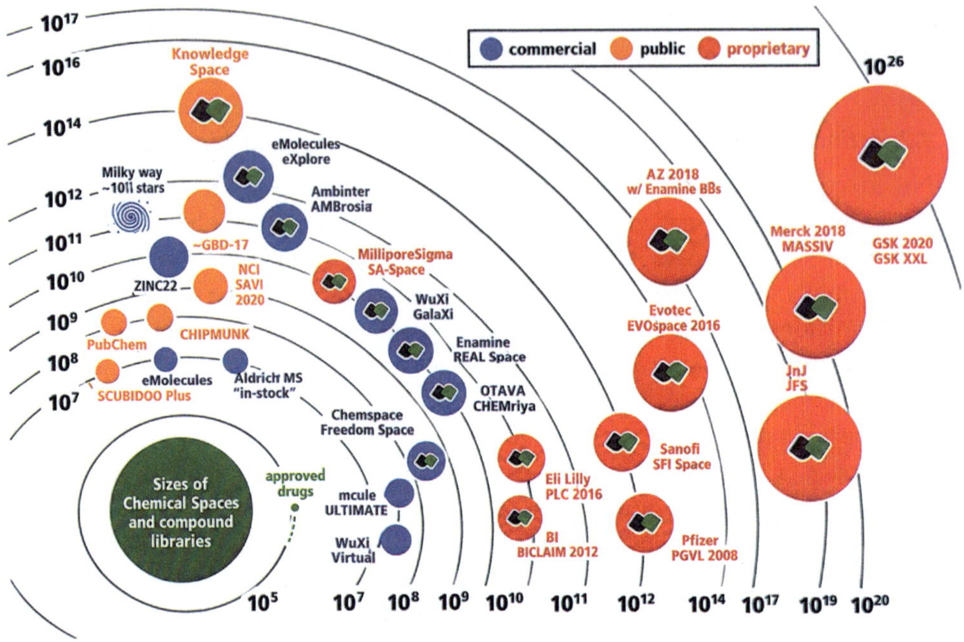

Figure 6.8 Possible chemical universe: Relative sizes of public, commercial, and proprietary chemical spaces, ranging from approved drugs and public compound repositories to large commercial and pharmaceutical libraries, illustrating the vast molecular diversity utilized in artificial intelligence-driven drug discovery. Reproduced with permission from BioSolveIT (https://www.biosolveit.de/). *Reproduced with permission from BioSolveIT GmbH (https://www.biosolveit.de/).*

6.6.4 Case study 4: warfarin dosing with artificial intelligence and big data integration

Warfarin exemplifies the complexities in dosing medications with narrow therapeutic indices and variable patient responses. While clinicians traditionally used trial-and-error adjustments guided by the INR, big data analytics and AI have transformed warfarin dosing into precision medicine. Warfarin response depends on clinical, environmental, and genetic factors, including age, diet, liver function, comorbidities, drug interactions, and polymorphisms in CYP2C9, VKORC1, and CYP4F2. Big data analytics processes datasets from warfarin studies, EHRs, pharmacogenomic consortia, and patient monitoring. ML models analyze variables for dose predictions, with AI identifying nonlinear relationships. In one case, a 62-year-old man with atrial fibrillation initiated warfarin therapy. Genetic testing showed CYP2C93/3 and VKORC1 −1639 AA genotypes, indicating reduced metabolism and heightened sensitivity. An AI model integrated clinical, biochemical, and genomic features to predict a 1 mg per day maintenance dose, confirmed by INR monitoring (Gan et al., 2025). AI systems synthesize EHR and genomic data into dynamic dosing frameworks, using federated learning (FL) and GNNs to capture drug–gene–environment interactions. While direct oral anticoagulants have reduced warfarin use, it remains widely prescribed where genomic testing and monitoring are feasible. The AI workflow for warfarin dosing is shown in Fig. 6.9.

These case studies demonstrate the impact of big data analytics in structural biology and bioinformatics. They show how AI has revolutionized structure prediction, accelerated structural virology and vaccine development, and integrated structural models with multiomics approaches in drug discovery. These advances increasingly depend on interpreting large-scale heterogeneous datasets.

6.7 Future directions and emerging trends in integrative drug discovery

The swift advancement of computational technologies, multiomics platforms, and AI/ML frameworks is transforming the conceptualization and execution of therapeutic discovery. As biological systems are increasingly characterized at the cellular, molecular, and structural levels, future drug discovery pipelines are anticipated to depend more heavily on data-rich, multiscale, and mechanistically interpretable computational methodologies. Several promising trajectories have emerged.

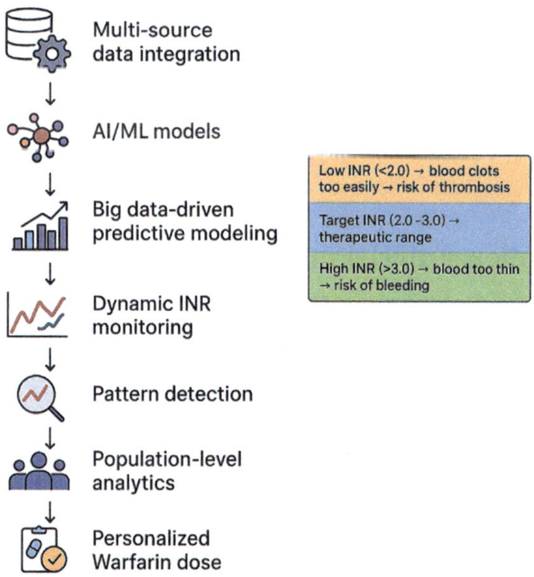

Figure 6.9 Artificial intelligence (*AI*) workflow for warfarin dosing: the flow chart illustrates how multisource data and INR monitoring are integrated to predict safe, personalized warfarin doses, with INR levels indicating risks of thrombosis (<2.0), therapeutic control (2.0–3.0), or bleeding (>3.0).

6.7.1 Expansion of foundation models and large biological models

The success of large language models, including GPT-based architectures and protein-specific transformers, has catalyzed the development of large biological models (LBMs) capable of simultaneously learning rich representations from sequences, structures, chemical graphs, reactions, and literature (Afrose et al., 2024). These models are expected to elucidate generalizable rules of protein folding, dynamics, and ligand interactions, expedite de novo protein and ligand design, enhance the prediction of mutational impacts and disease mechanisms, and integrate multilayer omics data, such as transcriptomics, proteomics, and metabolomics, for precision target identification. Recent LBMs, including ESM-2, ProtGPT-2, and AlphaFold 3-derived models, have demonstrated unprecedented performance in predicting higher-order biomolecular structures and their interactions (Ocana et al., 2025).

6.7.2 Multiscale digital twins of cell and tissue systems

Digital twins of biological systems—computational replicas that mirror cellular- or organism-level processes—are emerging as powerful tools for simulating drug responses prior to synthesis or clinical testing. Upcoming future models should be able to combine GRNs, metabolic flux simulations, PPI, dynamics, spatiotemporal drug distribution, and immune cell response. With the help of these multiscale computational twins, it would make it possible to conduct in silico clinical trials, make personalized dosing predictions, and, most importantly, accurately predict possible toxicity with the highest degree of fidelity (Mikołajewska et al., 2025; The Era of Preemptive Medicine: Developing Medical Digital Twins through Omics, 2025).

6.7.3 Next-generation artificial intelligence-native structure prediction

The evolution in the field of drug discovery has made a notable change. Nowadays, the area of structure prediction has changed from relying on template-based methods to using complete deep learning solutions in an end-to-end manner. In view of that, upcoming developments will concentrate on the following terms:

6.7.3.1 Conformational ensembles

Being able to predict conformational ensembles rather than just static structures significantly expands the applicability of the method to such dynamic phenomena as allosteric regulation, induced fit, formation of cryptic pockets, and conformational states modulated by temperature or pH (Hu et al., 2024).

6.7.3.2 Real-time folding/unfolding pathways

Diffusion models and neural differential equation solvers are two technologies whose advancements are expected to allow the reconstruction of folding pathways and transition states, which can be further utilized for the mechanistic design of stabilizers and inhibitors.

6.7.4 Autonomous laboratories and closed-loop artificial intelligence systems

Innovations in the combination of robotics with AI-driven design systems have resulted in "self-driving" laboratories or autonomous laboratories that can design, synthesize, test, and optimize compounds iteratively. These labs couple several processes, such as model retraining on the fly, automated synthesis of chemicals, high-throughput biological screening, and feedback loops at real-time pace. Systems like these have already shown a more than 10-fold acceleration in hit-to-lead timeline (Nussinov et al., 2022).

6.7.5 Quantum computing in drug discovery

Quantum computers have started tackling some biomolecular problems that are beyond the power of classical computers. It is anticipated that in the future, these applications will simulate the electronic structures of highly charged proteins or those with metal cofactors, predict reaction mechanisms and energy barriers, optimize multivariable landscapes for drug design, and perform docking with great accuracy through quantum Monte Carlo-inspired models. Despite the fact that present quantum hardware is still quite restricted, the combination of quantum and classical methods seems highly suitable for the study of catalytic mechanisms and the optimization of ligand fields (Li et al., 2024).

6.7.6 Integration of human variability and population data

Typically, the drug development industry tends to rely on "one-size-fits-all" approaches. However, newly integrative models want to combine population genomics datasets, ethnic and regional allele frequency variations, patient-specific pharmacogenomics, and age- and sex-dependent metabolic patterns. AI-powered methods can project diverse drug reactions, side effects, and the emergence of resistance for different segments of the population, thereby helping to devise therapeutic strategies that are more effective (Kant et al., 2025; Shamim et al., 2024).

6.7.7 Enhanced interpretability and causality-driven artificial intelligence

Amidst the mounting calls for openness from regulatory bodies in the use of AI for drug development, more and more weight is being put on neural networks that can be understood by humans, models that are aware of causality, counterfactual analysis of drug efficacy, and mechanistic rationalization of predicted interactions. It is anticipated that next-generation frameworks will not only be able to make precise predictions but also explain the reasons for them in biological terms, with these explanations being backed by structural and experimental evidence (Sarvesh et al., 2024).

6.7.8 Sustainable and cost-effective artificial intelligence pipelines

Considering the enormous computational resources that large models require, the next generation of solutions will be adapted to use as little energy as possible for training, model distillation and compression, using FL for data privacy, and applying green chemistry-guided computational optimization. These techniques are not only in line with international sustainability standards but also make AI-based drug discovery more affordable for developing countries (Yang et al., 2025; Haripriya et al., 2025).

6.7.9 Convergence of artificial intelligence, biomolecular simulation, and target-specific precision therapeutics

The highlighted trend in computational drug discovery is convergence, where AI models gradually become the main drivers of MD simulation, QM/MM calculations, and high-fidelity docking workflows. These unified platforms can predict binding, allostery, and conformational changes, score ligands with sub-kcal/mol accuracy, design multitarget or modality-switching therapeutics, and select optimal combinations of small molecules, antibodies, peptides, or PROTACs. Such convergent pipelines are expected to significantly reduce attrition rates in preclinical drug discovery (Chou et al., 2024; Rutherford and McManus, 2024).

6.8 Future directions

Contemporary drug discovery is driven by biological data from genomics, proteomics, structural biology, molecular simulations, and biomedical literature. Future advances will depend on integrating this heterogeneous big data with emerging computational technologies, including XAI (e.g., SHapley Additive exPlanations, Local Interpretable Model-agnostic Explanations), foundation models, quantum-enhanced computation, hybrid AI–quantum workflows, and digital-twin systems. Foundation models, including language models (e.g., GPT), structural models (e.g., AlphaFold, ESMFold), cheminformatics transformers (e.g., MolBERT), protein language models (e.g., ProtBERT, ProtTrans), and multimodal scientific models (e.g., BioGPT for literature mining, Galactica for scientific reasoning, ProtGPT-2 for protein design), offer transferable molecular, structural, and textual representations. XAI enhances mechanistic transparency, quantum computing expedites biophysical simulations, and digital twins enable individualized therapeutic predictions through virtual patient simulations. These converging methodologies represent a future in which big data analytics, AI-driven foundation models, quantum simulations, and virtual biological modeling foster interpretable, scalable, and clinically significant advances in computational biology and drug discovery.

6.8.1 Explainable and interpretable artificial intelligence

Understanding how these models create their predictions is becoming increasingly important as DL gets more integrated into structure prediction, ligand design, and biological inference. Traditional black-box models, while strong, have poor interpretability, which limits their application in mechanistic biology and clinical decision-making. XAI approaches seek to fill this void by offering clear and biologically meaningful interpretations. In structural biology, XAI can uncover critical residues that influence model predictions, structural motifs that control binding affinity, and interaction networks that drive functional inference. Frameworks for protein–ligand scoring, de novo binder design, mutational effect prediction, and variant pathogenicity modeling are increasingly incorporating methodologies like feature attribution, attention-based visualization, gradient-based saliency mapping, and surrogate modeling (Karim et al., 2023; Brima and Atemkeng, 2024). In drug discovery, interpretability is one of the ways in which trust and regulatory acceptance may be enhanced, as it, in addition to providing a mechanistic rationale, plays a part in AI-guided decisions. Transparent models are capable of uncovering the reasons why certain ligands are prioritized, why predicted toxicity arises from the structural features, as well as which molecular descriptors contribute the most to selectivity. As the integration of AI into experimental pipelines becomes standard, XAI will be central to ensuring reproducibility, reducing false positives, and strengthening the cross-validation between computational predictions and biochemical data (Karako and Tang, 2024).

6.8.2 Quantum computing in structural biology

Quantum computing has emerged as a promising frontier for solving complex biological problems that are intractable using classical algorithms. Although still in its early developmental stage, quantum applications are advancing rapidly and show significant potential in structural biology. Quantum algorithms may eventually accelerate protein folding and conformational sampling by evaluating energy landscapes with exponential speed up, enhance molecular docking and optimization through quantum annealing approaches that more efficiently solve NP-hard combinatorial search problems, and improve QM/MM and electronic structure calculations to enable high-accuracy simulations of enzymatic catalysis, charge transfer, and transition-state modeling (Pal et al., 2024). They may also provide more precise free-energy estimations than classical methods. Hybrid quantum–classical workflows have already been tested for small-molecule docking, fragment assembly, and force-field optimization. As qubit stability, error correction, and hardware scalability continue to improve, quantum-enhanced simulations are expected to become integral components of computational pipelines, particularly for targets with substantial conformational complexity or systems that require high-resolution electronic details (Lappala, 2024).

6.8.3 Federated learning and privacy-preserving analytics

A significant challenge in biomedical AI is the distributed, sensitive, and often confidential nature of biological and clinical datasets. FL addresses this issue by enabling AI models to be trained across multiple decentralized datasets, including hospital systems, research institutes, and sequencing centers, without requiring direct data sharing (Yang et al., 2024). In structural bioinformatics, FL can facilitate the collaborative refinement of protein–ligand scoring functions, support the distributed learning of mutational effect predictors across diverse clinical cohorts, enable the consensus development of generative molecular design frameworks, and enhance models for variant interpretation and disease stratification. Privacy-preserving techniques, such as differential privacy, secure multiparty computation, and homomorphic encryption, ensure that highly sensitive patient and genomic data remain confidential even in these joint ventures. This model is in line with international regulatory standards and greatly improves the possibility of large-scale cross-institutional biomolecular modeling projects (Abbas et al., 2024).

6.8.4 Integrative multiomics and structural systems biology

The future of structural biology is expected to overcome the limitations of single-data-type analysis and move toward the use of integrative multiscale models that bridge molecular structure with cellular physiology and organismal behavior. The integration of multiomics, such as genomics, transcriptomics, epigenomics, proteomics, metabolomics, and interactomes, offers a complete picture of the biological systems. Combined with structural knowledge, it can reveal more profound mechanisms of molecular function, regulatory dynamics, and disease-associated perturbations. Novel approaches, such as structural multiomics modeling (Kant et al., 2025), contextualize protein structures within pathway regulation and network rewiring, whereas AI-driven fusion architectures connect diverse datasets for improved biomarker identification and therapeutic targeting. In addition, omics-derived constraints have started to be extensively used to guide simulations, thereby enabling MD or docking workflows to be tailored to disease states/patients-specific mutations. Hybrid pipelines combining structure prediction, MD simulations, and experimental data allow iterative model refinement, and the integration of simulations, AI, and experiments is rapidly becoming a vital paradigm. Closed-loop systems, where AI-generated hypotheses are tested experimentally and the results are fed back into the training datasets, are expected to greatly facilitate discovery, significantly increase prediction accuracy, and, at the same time, drastically reduce the experimental workload. Overall, integrative structural systems biology is on the verge of transforming descriptive biology into predictive biology and thereby enabling personalized therapeutic strategies based on molecular-level details (Facchiano et al., 2024).

6.9 Conclusion

Nowadays, in structural biology and bioinformatics, big data analytics plays an essential role in the understanding of biological systems and their conversion into therapeutic insights. The creation of highly efficient sequencing, proteomics, cryo-EM, imaging, and multiomics technologies has generated large amounts of data that require very powerful computational solutions. AI-based structure prediction, network biology, and integrative modeling of biological processes rapidly provide scientists with new ideas to explore. The coupling of ML with molecular simulations and computer-aided platforms has quickened the pace of drug discovery and systems-level understanding. Data heterogeneity, uncertainty, and privacy, among other issues, require the development of clear analytical frameworks. XAI, FL, quantum computing, and multiomics systems biology are revolutionizing traditional research methods and transforming them into a process of continuous discovery, thus offering deeper knowledge and therapies tailored for cancer patients. Big data analytics is redefining structural biology and bioinformatics, positioning these fields for breakthroughs in biological understanding and precision medicine.

Acknowledgments

The authors gratefully acknowledge Integral University for providing the necessary support, resources, and facilities that contributed to the successful preparation of this book chapter.

References

Avanija, V.A., 2025. AI-Driven Heart Disease Prediction Using Machine Learning And Deep Learning Techniques. International Journal of Computational and Experimental Science and Engineering 11 (2). Available at https://doi.org/10.22399/ijcesen.1669.

Abbas, S.R., Abbas, Z., Zahir, A., Lee, S.W., 2024. Federated learning in smart healthcare: a comprehensive review on privacy, security, and predictive analytics with IoT integration. Healthcare 12 (24), 2587. Available at https://doi.org/10.3390/healthcare12242587.

Afrose, N., Chakraborty, R., Hazra, A., Bhowmick, P., Bhowmick, M., 2024. AI-Driven Drug Discovery and Development. IGI Global, pp. 259–277. Available at http://doi.org/10.4018/979-8-3693-3629-8.ch013.

Afzal, S., Hassan, M., Ullah, S., Abbas, H., Tawakkal, F., Khan, M.A., 2022. Breast cancer; discovery of novel diagnostic biomarkers, drug resistance, and therapeutic implications. Frontiers in Molecular Biosciences 9. Available at https://doi.org/10.3389/fmolb.2022.783450.

Agajanian, S., Alshahrani, M., Bai, F., Tao, P., Verkhivker, G.M., 2023. Exploring and learning the universe of protein allostery using artificial intelligence augmented biophysical and computational approaches. Journal of Chemical Information and Modeling 63 (5), 1413–1428. Available at https://doi.org/10.1021/acs.jcim.2c01634, http://pubs.acs.org/journal/jcisd8.

Athaya, T., Ripan, R.C., Li, X., Hu, H., 2023. Multimodal deep learning approaches for single-cell multi-omics data integration. Briefings in Bioinformatics 24 (5). Available at https://doi.org/10.1093/bib/bbad313, http://bib.oxfordjournals.org.

Auslander, N., Gussow, A.B., Koonin, E.V., 2021. Incorporating machine learning into established bioinformatics frameworks. International Journal of Molecular Sciences 22 (6), 2903. Available at https://doi.org/10.3390/ijms22062903.

Bertoline, L.M.F., Lima, A.N., Krieger, J.E., Teixeira, S.K., 2023. Before and after AlphaFold2: an overview of protein structure prediction. Frontiers in Bioinformatics 3. Available at https://doi.org/10.3389/fbinf.2023.1120370, www.frontiersin.org/journals/bioinformatics.

Brima, Y., Atemkeng, M., 2024. Saliency-driven explainable deep learning in medical imaging: bridging visual explainability and statistical quantitative analysis. BioData Mining 17 (1). Available at https://doi.org/10.1186/s13040-024-00370-4.

Caetano-Anollés, G., Aziz, M.F., Mughal, F., Caetano-Anollés, D., 2021. Tracing protein and proteome history with chronologies and networks: folding recapitulates evolution. Expert Review of Proteomics 18 (10), 863–880. Available at https://doi.org/10.1080/14789450.2021.1992277, https://www.tandfonline.com/loi/ieru20?open=9&year=2012&repitition=0#vol_9_2012.

Calvaresi, V., Wrobel, A.G., Toporowska, J., Hammerschmid, D., Doores, K.J., Bradshaw, R.T., et al., 2023. Structural dynamics in the evolution of SARS-CoV-2 spike glycoprotein. Nature Communications 14 (1). Available at https://doi.org/10.1038/s41467-023-36745-0, https://www.nature.com/ncomms/.

Carpenter, K.A., Cohen, D.S., Jarrell, J.T., Huang, X., 2018. Deep learning and virtual drug screening. Future Medicinal Chemistry 10 (21), 2557–2567. Available at https://doi.org/10.4155/fmc-2018-0314, http://www.future-science.com/loi/fmc.

Cerofolini, L., Fragai, M., Ravera, E., Diebolder, C.A., Renault, L., Calderone, V., 2019. Integrative approaches in structural biology: a more complete picture from the combination of individual techniques. Biomolecules 9 (8), 370. Available at https://doi.org/10.3390/biom9080370.

Chen, J.Z., Bisardi, M., Lee, D., Cotogno, S., Zamponi, F., Weigt, M., et al., 2024. Understanding epistatic networks in the B1 β-lactamases through coevolutionary statistical modeling and deep mutational scanning. Nature Communications 15 (1). Available at https://doi.org/10.1038/s41467-024-52614-w, https://www.nature.com/ncomms/.

Chen, K.C., Li, T.Y., Wang, Y.Y., See, S., Wang, C.C., Wille, R., et al., 2025. Validating large-scale quantum machine learning: efficient simulation of quantum support vector machines using tensor networks. Machine Learning: Science and Technology 6 (1). Available at https://doi.org/10.1088/2632-2153/adb4ba, https://iopscience.iop.org/journal/2632-2153.

Chen, L., Li, Q., Nasif, K.F.A., Xie, Y., Deng, B., Niu, S., et al., 2024. AI-driven deep learning techniques in protein structure prediction. International Journal of Molecular Sciences 25 (15), 8426. Available at https://doi.org/10.3390/ijms25158426.

Chou, C.L., Lin, C.T., Kao, C.T., Lin, C.C., 2024. A novel rational PROTACs design and validation via AI-driven drug design approach. ACS Omega 9 (37), 38371–38384. Available at https://doi.org/10.1021/acsomega.3c10183, pubs.acs.org/journal/acsodf.

Chou, Y.T., Chang, W.T., Jean, J.G., Chang, K.H., Huang, Y.N., Chen, C.S., 2024. StructGNN: An efficient graph neural network framework for static structural analysis. Computers and Structures 299. Available at https://doi.org/10.1016/j.compstruc.2024.107385, https://www.sciencedirect.com/science/journal/00457949.

Chung, J., Durie, C., Lee, J., 2022. Artificial intelligence in cryo-electron microscopy. Life (Chicago, Ill: 1978) 12 (8), 1267. Available at https://doi.org/10.3390/life12081267.

Colavin, A., Atolia, E., Bitbol, A.F., Huang, K.C., 2022. Extracting phylogenetic dimensions of coevolution reveals hidden functional signals. Scientific Reports 12 (1). Available at https://doi.org/10.1038/s41598-021-04260-1, www.nature.com/srep/index.html.

Coleman, K., Schroeder, A., Loth, M., Zhang, D., Park, J.H., Sung, J.Y., et al., 2025. Resolving tissue complexity by multimodal spatial omics modeling with MISO. Nature Methods 22 (3), 530–538. Available at https://doi.org/10.1038/s41592-024-02574-2, https://www.nature.com/nmeth/.

Conflitti, P., Lyman, E., Sansom, M.S.P., Hildebrand, P.W., Gutiérrez-de-Terán, H., Carloni, P., et al., 2025. Functional dynamics of G protein-coupled receptors reveal new routes for drug discovery. Nature Reviews. Drug Discovery 24 (4), 251–275. Available at https://doi.org/10.1038/s41573-024-01083-3, https://www.nature.com/nrd/.

Cui, F., Zhang, Z., Cao, C., Zou, Q., Chen, D., Su, X., 2022. Protein–DNA/RNA interactions: Machine intelligence tools and approaches in the era of artificial intelligence and big data. Proteomics 22 (8). Available at https://doi.org/10.1002/pmic.202100197.

Cui, L., Liu, H., Wang, H., Li, S., Xue, W., Xia, N., et al., 2024. Comprehensive Overview of Broadly Neutralizing Antibodies against SARS-COV-2 Variants. Available at https://doi.org/10.20944/preprints202404.0708.v1.

Dahal, S., Yurkovich, J.T., Xu, H., Palsson, B.O., Yang, L., 2020. Synthesizing systems biology knowledge from omics using genome-scale models. Proteomics 20 (17-18). Available at https://doi.org/10.1002/pmic.201900282, http://onlinelibrary.wiley.com/journal/10.1002/(ISSN)1615-9861.

de Souza Neto, L.R., Moreira-Filho, J.T., Neves, B.J., Maidana, R.L.B.R., Guimarães, A.C.R., Furnham, N., et al., 2020. In silico strategies to support fragment-to-lead optimization in drug discovery. Frontiers in Chemistry 8. Available at https://doi.org/10.3389/fchem.2020.00093, http://journal.frontiersin.org/journal/chemistry.

David, L., Arús-Pous, J., Karlsson, J., Engkvist, O., Bjerrum, E.J., Kogej, T., et al., 2019. Applications of deep-learning in exploiting large-scale and heterogeneous compound data in industrial pharmaceutical research. Frontiers in Pharmacology 10. Available at https://doi.org/10.3389/fphar.2019.01303, http://www.frontiersin.org/Pharmacology.

Dhakal, A., Gyawali, R., Wang, L., Cheng, J., 2025. Artificial intelligence in cryo-EM protein particle picking: recent advances and remaining challenges. Briefings in Bioinformatics 26 (1). Available at https://doi.org/10.1093/bib/bbaf011.

Diaz-Rovira, A.M., Martín, H., Beuming, T., Díaz, L., Guallar, V., Ray, S.S., 2023. Are deep learning structural models sufficiently accurate for virtual screening? Application of docking algorithms to AlphaFold2 predicted structures. Journal of Chemical Information and Modeling 63 (6), 1668–1674. Available at https://doi.org/10.1021/acs.jcim.2c01270, http://pubs.acs.org/journal/jcisd8.

DiIorio, M.C., Kulczyk, A.W., 2023. Novel artificial intelligence-based approaches for Ab initio structure determination and atomic model building for cryo-electron microscopy. Micromachines 14 (9), 1674. Available at https://doi.org/10.3390/mi14091674.

Dlamini, Z., Skepu, A., Kim, N., Mkhabele, M., Khanyile, R., Molefi, T., et al., 2022. AI and precision oncology in clinical cancer genomics: from prevention to targeted cancer therapies-an outcomes based patient care. Informatics in Medicine Unlocked 31, 100965. Available at https://doi.org/10.1016/j.imu.2022.100965.

Dong, X., Tian, L., Gouil, Q., Kariyawasam, H., Su, S., De Paoli-Iseppi, R., et al., 2021. The long and the short of it: Unlocking nanopore long-read RNA sequencing data with short-read differential expression analysis tools. NAR Genomics and Bioinformatics 3 (2). Available at https://doi.org/10.1093/nargab/lqab028, https://academic.oup.com/nargab.

Facchiano, A., Heider, D., Mutarelli, M., 2024. Editorial: artificial intelligence and bioinformatics applications for omics and multi-omics studies. Frontiers in Genetics 15. Available at https://doi.org/10.3389/fgene.2024.1371473.

Fan, B.-L., Chen, L.-H., Chen, L.-L., Guo, H., 2025. Integrative multi-omics approaches for identifying and characterizing biological elements in crop traits: current progress and future prospects. International Journal of Molecular Sciences 26 (4), 1466. Available at https://doi.org/10.3390/ijms26041466.

Fatima, S.A., Faiza Nasim, S., Ahmed, S., 2023. Enhancing agricultural operations: big data analytics using distributed and parallel computing. International Journal of Emerging Engineering and Technology 2 (2), 1–7. Available at https://doi.org/10.57041/v3jj9f69.

Fleetwood, O., Kasimova, M.A., Westerlund, A.M., Delemotte, L., 2020. Molecular insights from conformational ensembles via machine learning. Biophysical Journal 118 (3), 765–780. Available at https://doi.org/10.1016/j.bpj.2019.12.016, http://www.sciencedirect.com/science/journal/00063495.

Gan, T.R.X., Tan, L.W.J., Egermark, M., Truong, A.T.L., Kumar, K., Tan, S.B., et al., 2025. AI-assisted warfarin dose optimisation with CURATE.AI for clinical impact: Retrospective data analysis. Bioengineering and Translational Medicine 10 (3). Available at https://doi.org/10.1002/btm2.10757, http://onlinelibrary.wiley.com/journal/10.1002/(ISSN)2380-6761.

Gilep, K., Obarska-Kosinska, A., Kosinski, J., 2024. Improving AlphaFold 3 structural modeling by incorporating explicit crosslinks. bioRxiv, Germany bioRxiv. Available from: https://www.biorxiv.org. doi: 10.1101/2024.12.03.626671.

Glaser, J., Vermaas, J.V., Rogers, D.M., Larkin, J., LeGrand, S., Boehm, S., et al., 2021. High-throughput virtual laboratory for drug discovery using massive datasets. The International Journal of High Performance Computing Applications 35 (5), 452–468. Available at https://doi.org/10.1177/10943420211001565.

Groen, J., Conesa, J.J., Valcárcel, R., Pereiro, E., 2019. The cellular landscape by cryo soft X-ray tomography. Biophysical Reviews 11 (4), 611–619. Available at https://doi.org/10.1007/s12551-019-00567-6, http://www.springer.com/life+sciences/biochemistry+%26+biophysics/journal/12551.

Haripriya, R., Khare, N., Pandey, M., 2025. Privacy-preserving federated learning for collaborative medical data mining in multi-institutional settings. Scientific Reports 15 (1). Available at https://doi.org/10.1038/s41598-025-97565-4.

Hartman, E., Scott, A.M., Karlsson, C., Mohanty, T., Vaara, S.T., Linder, A., et al., 2023. Interpreting biologically informed neural networks for enhanced proteomic biomarker discovery and pathway analysis. Nature Communications 14 (1). Available at https://doi.org/10.1038/s41467-023-41146-4, https://www.nature.com/ncomms/.

Hassan, M., Awan, F.M., Naz, A., deAndrés-Galiana, E.J., Alvarez, O., Cernea, A., et al., 2022. Innovations in genomics and big data analytics for personalized medicine and health care: a review. International Journal of Molecular Sciences 23 (9), 4645. Available at https://doi.org/10.3390/ijms23094645.

Heebner, J.E., Purnell, C., Hylton, R.K., Marsh, M., Grillo, M.A., Swulius, M.T., 2022. Deep learning-based segmentation of cryo-electron tomograms. Journal of Visualized Experiments 2022 (189). Available at www.jove.com, doi: 10.3791/64435.

Hu, Y., Yang, H., Li, M., Zhong, Z., Zhou, Y., Bai, F., et al., 2024. Exploring protein conformational changes using a large-scale biophysical sampling augmented deep learning strategy. Advanced Science 11 (44). Available at https://doi.org/10.1002/advs.202400884.

Isigkeit, L., Hörmann, T., Schallmayer, E., Scholz, K., Lillich, F.F., Ehrler, J.H.M., et al., 2024. Automated design of multi-target ligands by generative deep learning. Nature Communications 15 (1). Available at https://doi.org/10.1038/s41467-024-52060-8.

Ivanisenko, T.V., Demenkov, P.S., Ivanisenko, V.A., 2024. An accurate and efficient approach to knowledge extraction from scientific publications using structured ontology models, graph neural networks, and large language models. International Journal of Molecular Sciences 25 (21), 11811. Available at https://doi.org/10.3390/ijms252111811.

Jayatunga, M.K.P., Ayers, M., Bruens, L., Jayanth, D., Meier, C., 2024. How successful are AI-discovered drugs in clinical trials? A first analysis and emerging lessons. Drug Discovery Today 29 (6), 104009. Available at https://doi.org/10.1016/j.drudis.2024.104009.

Jiao, P., Wang, B., Wang, X., Liu, B., Wang, Y., Li, J., et al., 2023. Struct2GO: protein function prediction based on graph pooling algorithm and AlphaFold2 structure information. Bioinformatics (Oxford, England) 39 (10). Available at https://doi.org/10.1093/bioinformatics/btad637.

Kant, S., Deepika, Roy, S., 2025. Artificial intelligence in drug discovery and development: transforming challenges into opportunities. Discover Pharmaceutical Sciences 1 (1). Available at https://doi.org/10.1007/s44395-025-00007-3.

Kant, S., Deepika, Roy, S., 2025. Integrative multi-omics and artificial intelligence: a new paradigm for systems biology. OMICS: A Journal of Integrative Biology 29 (12), 576–587. Available at https://doi.org/10.1177/15578100251392371.

Karako, K., Tang, W., 2024. Applications of and issues with machine learning in medicine: bridging the gap with explainable AI. BioScience Trends 18 (6), 497–504. Available at https://doi.org/10.5582/bst.2024.01342.

Karim, M.R., Islam, T., Shajalal, M., Beyan, O., Lange, C., Cochez, M., et al., 2023. Explainable AI for bioinformatics: methods, tools and applications. Briefings in Bioinformatics 24 (5). Available at https://doi.org/10.1093/bib/bbad236, http://bib.oxfordjournals.org.

Kiouri, D.P., Batsis, G.C., Chasapis, C.T., 2025. Structure-based approaches for protein–protein interaction prediction using machine learning and deep learning. Biomolecules 15 (1). Available at https://doi.org/10.3390/biom15010141, http://www.mdpi.com/journal/biomolecules.

Kopp, W., Akalin, A., Ohler, U., 2022. Simultaneous dimensionality reduction and integration for single-cell ATAC-seq data using deep learning. Nature Machine Intelligence 4 (2), 162–168. Available at https://doi.org/10.1038/s42256-022-00443-1.

Krokidis, M.G., Koumadorakis, D.E., Lazaros, K., Ivantsik, O., Exarchos, T.P., Vrahatis, A.G., et al., 2025. AlphaFold3: an overview of applications and performance insights. International Journal of Molecular Sciences 26 (8), 3671. Available at https://doi.org/10.3390/ijms26083671.

Kumar, N., Srivastava, R., 2024. Deep learning in structural bioinformatics: current applications and future perspectives. Briefings in Bioinformatics 25 (3). Available at https://doi.org/10.1093/bib/bbae042.

Lai, J., Yu, H., Tian, Z., Li, H., 2020. Hybrid MPI and CUDA parallelization for CFD applications on multi-GPU HPC clusters. Scientific Programming 2020, 1–15. Available at https://doi.org/10.1155/2020/8862123.

Lappala, A., 2024. The next revolution in computational simulations: harnessing AI and quantum computing in molecular dynamics. Current Opinion in Structural Biology 89, 102919. Available at https://doi.org/10.1016/j.sbi.2024.102919.

Li, E.H., Spaman, L.E., Tejero, R., Janet Huang, Y., Ramelot, T.A., Fraga, K.J., et al., 2023. Blind assessment of monomeric AlphaFold2 protein structure models with experimental NMR data. Journal of Magnetic Resonance 352. Available at https://doi.org/10.1016/j.jmr.2023.107481, http://www.elsevier.com/inca/publications/store/6/2/2/8/8/4/index.htt.

Li, S., Zhou, J., Xu, T., Huang, L., Wang, F., Xiong, H., et al., 2021. Proceedings of the ACM SIGKDD International Conference on Knowledge Discovery and Data Mining 9781450383325 975 985 Association for Computing Machinery China Structure-aware Interactive Graph Neural Networks for the Prediction of Protein-Ligand Binding Affinity, Available at https://doi.org/10.1145/3447548.3467311

Li, W., Yin, Z., Li, X., Ma, D., Yi, S., Zhang, Z., et al., 2024. A hybrid quantum computing pipeline for real world drug discovery. Scientific Reports 14 (1). Available at https://doi.org/10.1038/s41598-024-67897-8.

Liu, H., Li, Y., Karsidag, M., Tu, T., Wang, P., 2025. Technical and biological biases in bulk transcriptomic data mining for cancer research. Journal of Cancer 16 (1), 34–43. Available at https://doi.org/10.7150/jca.100922.

Liu, J., Zhao, K., Zhang, G., 2023. Improved model quality assessment using sequence and structural information by enhanced deep neural networks. Briefings in Bioinformatics 24 (1). Available at https://doi.org/10.1093/bib/bbac507.

Mair, F., Erickson, J.R., Voillet, V., Simoni, Y., Bi, T., Tyznik, A.J., et al., 2020. A targeted multi-omic analysis approach measures protein expression and low-abundance transcripts on the single-cell level. Cell Reports 31 (1), 107499. Available at https://doi.org/10.1016/j.celrep.2020.03.063.

Mantini, G., Pham, T.V., Piersma, S.R., Jimenez, C.R., 2021. Computational analysis of phosphoproteomics data in multi-omics cancer studies. Proteomics 21 (3-4). Available at https://doi.org/10.1002/pmic.201900312, http://onlinelibrary.wiley.com/journal/10.1002/(ISSN)1615-9861.

Mikołajewska, E., Mikołajewski, D., Mikołajczyk, T., Paczkowski, T., 2025. Generative AI in AI-based digital twins for fault diagnosis for predictive maintenance in industry 4.0/5.0. Applied Sciences 15 (6), 3166. Available at https://doi.org/10.3390/app15063166.

Milano, M., Agapito, G., Cannataro, M., 2022. Challenges and limitations of biological network analysis. BioTech 11 (3), 24. Available at https://doi.org/10.3390/biotech11030024.

Monteleone, S., Kellici, T.F., Southey, M., Bodkin, M.J., Heifetz, A., 2022. Fighting COVID-19 with artificial intelligence. Methods in Molecular Biology 2390, 103–112. Available at https://doi.org/10.1007/978-1-0716-1787-8_3, http://www.springer.com/series/7651.

Moriya, T., Yamada, Y., Yamamoto, M., Senda, T., 2024. GoToCloud optimization of cloud computing environment for accelerating cryo-EM structure-based drug design. Communications Biology 7 (1). Available at https://doi.org/10.1038/s42003-024-07031-6.

Mukherjee, A., Abraham, S., Singh, A., Balaji, S., Mukunthan, K.S., 2025. From data to cure: a comprehensive exploration of multi-omics data analysis for targeted therapies. Molecular Biotechnology 67 (4), 1269–1289. Available at https://doi.org/10.1007/s12033-024-01133-6.

Nauman, M., Rehman, H.U.R., Politano, G., Benso, A., 2019. Beyond homology transfer: deep learning for automated annotation of proteins. Journal of Grid Computing 17 (2), 225–237. Available at https://doi.org/10.1007/s10723-018-9450-6.

Nguyen, P.T., Harris, B.J., Mateos, D.L., González, A.H., Murray, A.M., Yarov-Yarovoy, V., 2024. Structural modeling of ion channels using AlphaFold2, RoseTTAFold2, and ESMFold. Channels 18 (1). Available at https://doi.org/10.1080/19336950.2024.2325032, http://www.tandfonline.com/loi/kchl20.

Nussinov, R., Zhang, M., Liu, Y., Jang, H., 2022. AlphaFold, artificial intelligence (AI), and allostery. The Journal of Physical Chemistry. B 126 (34), 6372–6383. Available at https://doi.org/10.1021/acs.jpcb.2c04346.

Ocana, A., Pandiella, A., Privat, C., Bravo, I., Luengo-Oroz, M., Amir, E., et al., 2025. Integrating artificial intelligence in drug discovery and early drug development: a transformative approach. Biomarker Research 13 (1). Available at https://doi.org/10.1186/s40364-025-00758-2.

O'Connor, L.M., O'Connor, B.A., Lim, S.B., Zeng, J., Lo, C.H., 2023. Integrative multi-omics and systems bioinformatics in translational neuroscience: a data mining perspective. Journal of Pharmaceutical Analysis 13 (8), 836–850. Available at https://doi.org/10.1016/j.jpha.2023.06.011, http://www.journals.elsevier.com/journal-of-pharmaceutical-analysis.

Ouma, W.Z., Pogacar, K., Grotewold, E., Ioshikhes, I., 2018. Topological and statistical analyses of gene regulatory networks reveal unifying yet quantitatively different emergent properties. PLoS Computational Biology 14 (4), e1006098. Available at https://doi.org/10.1371/journal.pcbi.1006098.

Pal, S., Bhattacharya, M., Lee, S.S., Chakraborty, C., 2024. Quantum computing in the next-generation computational biology landscape: from protein folding to molecular dynamics. Molecular Biotechnology 66 (2), 163–178. Available at https://doi.org/10.1007/s12033-023-00765-4, https://www.springer.com/journal/12033.

Park, Y., Heider, D., Hauschild, A.-C., 2021. Integrative analysis of next-generation sequencing for next-generation cancer research toward artificial intelligence. Cancers 13 (13), 3148. Available at https://doi.org/10.3390/cancers13133148.

Peng, A., Mao, X., Zhong, J., Fan, S., Hu, Y., 2020. Single-cell multi-omics and its prospective application in cancer biology. Proteomics 20 (13). Available at https://doi.org/10.1002/pmic.201900271.

Pinzi, L., Rastelli, G., 2020. Identification of target associations for polypharmacology from analysis of crystallographic ligands of the protein data Bank. Journal of Chemical Information and Modeling 60 (1), 372–390. Available at https://doi.org/10.1021/acs.jcim.9b00821.

Rutherford, K.A., McManus, K.J., 2024. PROTACs: current and future potential as a precision medicine strategy to combat cancer. Molecular Cancer Therapeutics 23 (4), 454–463. Available at https://doi.org/10.1158/1535-7163.MCT-23-0747, https://aacrjournals.org/mct/article/23/4/454/741867/PROTACs-Current-and-Future-Potential-as-a.

Sanches, P.H.G., de Melo, N.C., Porcari, A.M., de Carvalho, L.M., 2024. Integrating molecular perspectives: strategies for comprehensive multi-omics integrative data analysis and machine learning applications in transcriptomics, proteomics, and metabolomics. Biology 13 (11). Available at https://doi.org/10.3390/biology13110848, www.mdpi.com/journal/biology.

Sarvesh Koli Komal Bhat Prajwal Korade, D.M.A.M.O.K., 2024. Unlocking machine learning model decisions: a comparative analysis of LIME and SHAP for enhanced interpretability. Journal of Electrical Systems 20 (2s), 598–613. Available at https://doi.org/10.52783/jes.1480.

Seffernick, J.T., Lindert, S., 2020. Hybrid methods for combined experimental and computational determination of protein structure. The Journal of Chemical Physics 153 (24). Available at https://doi.org/10.1063/5.0026025.

Segura, J., Rose, Y., Bi, C., Duarte, J., Burley, S.K., Bittrich, S., 2023. RCSB protein data bank: visualizing groups of experimentally determined PDB structures alongside computed structure models of proteins. Frontiers in Bioinformatics 3. Available at https://doi.org/10.3389/fbinf.2023.1311287.

Shamim, M.A., Shamim, M.A., Arora, P., Dwivedi, P., 2024. Artificial intelligence and big data for pharmacovigilance and patient safety. Journal of Medicine, Surgery, and Public Health 3, 100139. Available at https://doi.org/10.1016/j.glmedi.2024.100139.

Shen, Z., Wei, K., Zang, H., Li, L., Wang, G., 2023. The application of artificial intelligence to the bayesian model algorithm for combining genome data. Academic Journal of Science and Technology 8 (3), 132–135. Available at https://doi.org/10.54097/ykhccb53.

Sherif, Z.A., Ogunwobi, O.O., Ressom, H.W., 2025. Mechanisms and technologies in cancer epigenetics. Frontiers in Oncology 14. Available at https://doi.org/10.3389/fonc.2024.1513654.

Sidky, H., Chen, W., Ferguson, A.L., 2020. Machine learning for collective variable discovery and enhanced sampling in biomolecular simulation. Molecular Physics 118 (5). Available at https://doi.org/10.1080/00268976.2020.1737742, www.tandf.co.uk/journals/titles/00268976.asp.

Son, A., Kim, W., Park, J., Lee, W., Lee, Y., Choi, S., et al., 2024. Utilizing molecular dynamics simulations, machine learning, cryo-EM, and NMR spectroscopy to predict and validate protein dynamics. International Journal of Molecular Sciences 25 (17), 9725. Available at https://doi.org/10.3390/ijms25179725.

Song, M., Behmanesh, I., Moaveni, B., Papadimitriou, C., 2020. Accounting for modeling errors and inherent structural variability through a hierarchical bayesian model updating approach: an overview. Sensors 20 (14), 3874. Available at https://doi.org/10.3390/s20143874.

Song, X., Bao, L., Feng, C., Huang, Q., Zhang, F., Gao, X., et al., 2024. Accurate prediction of protein structural flexibility by deep learning integrating intricate atomic structures and cryo-EM density information. Nature Communications 15 (1). Available at https://doi.org/10.1038/s41467-024-49858-x.

Stahl, K., Warneke, R., Demann, L., Bremenkamp, R., Hormes, B., Brock, O., et al., 2024. Modelling protein complexes with crosslinking mass spectrometry and deep learning. Nature Communications 15 (1). Available at https://doi.org/10.1038/s41467-024-51771-2.

Suster, C.J.E., Pham, D., Kok, J., Sintchenko, V., 2024. Emerging applications of artificial intelligence in pathogen genomics. Frontiers in Bacteriology 3. Available at https://doi.org/10.3389/fbrio.2024.1326958.

Taherdoost, H., Ghofrani, A., 2024. AI's role in revolutionizing personalized medicine by reshaping pharmacogenomics and drug therapy. Intelligent Pharmacy 2 (5), 643–650. Available at https://doi.org/10.1016/j.ipha.2024.08.005.

Tan, Y., Li, M., Zhou, B., Zhong, B., Zheng, L., Tan, P., et al., 2024. Simple, efficient, and scalable structure-aware adapter boosts protein language models. Journal of Chemical Information and Modeling 64 (16), 6338–6349. Available at https://doi.org/10.1021/acs.jcim.4c00689.

The Era of Preemptive Medicine: Developing Medical Digital Twins through Omics, 2025. IoT, and AI Integration. JMA Journal 8 (1), 1–10. Available at https://doi.org/10.31662/jmaj.2024-0213.

Tirupati, K.K., Mahadik, S., Abul Khair, M.D., Goel, O., Jain, A., 2024. Optimizing machine learning models for predictive analytics in cloud environments. International Journal for Research Publication and Seminar 13 (5), 611–642. Available at https://doi.org/10.36676/jrps.v13.i5.1530.

Uzoeto, H.O., Cosmas, S., Bakare, T.T., Durojaye, O.A., 2024. AlphaFold-latest: revolutionizing protein structure prediction for comprehensive biomolecular insights and therapeutic advancements. Nigeria

Beni-Suef University Journal of Basic and Applied Sciences 13 (1). Available at https://doi.org/10.1186/s43088-024-00503-y, https://www.springer.com/journal/43088.

Valous, N.A., Popp, F., Zörnig, I., Jäger, D., Charoentong, P., 2024. Graph machine learning for integrated multi-omics analysis. British Journal of Cancer 131 (2), 205–211. Available at https://doi.org/10.1038/s41416-024-02706-7, https://www.nature.com/bjc/.

Wang, J., Watson, J.L., Lisanza, S.L., 2024. Protein design using structure-prediction networks: AlphaFold and RoseTTAFold as protein structure foundation models. Cold Spring Harbor Perspectives in Biology 16 (7), a041472. Available at https://doi.org/10.1101/cshperspect.a041472.

Wang, K., Huang, Y., Wang, Y., You, Q., Wang, L., 2024. Recent advances from computer-aided drug design to artificial intelligence drug design. RSC Medicinal Chemistry 15 (12), 3978–4000. Available at https://doi.org/10.1039/d4md00522h.

Wang, S., Wu, R., Lu, J., Jiang, Y., Huang, T., Cai, Y.-D., 2022. Protein-protein interaction networks as miners of biological discovery. Proteomics 22 (15-16). Available at https://doi.org/10.1002/pmic.202100190.

Wankowicz, S.A., Fraser, J.S., 2024. Comprehensive encoding of conformational and compositional protein structural ensembles through the mmCIF data structure. IUCrJ 11 (4), 494–501. Available at https://doi.org/10.1107/S2052252524005098, http://www.iucrj.org/m/.

Xie, J., Zhang, W., Zhu, X., Deng, M., Lai, L., 2023. Coevolution-based prediction of key allosteric residues for protein function regulation. eLife 12. Available at https://doi.org/10.7554/elife.81850.

Yan, T., Zheng, R., Li, Y., Sun, S., Zeng, X., Yue, Z., et al., 2023. Epidemiological insights into the omicron outbreak via MeltArray-assisted real-time tracking of SARS-CoV-2 variants. Viruses 15 (12), 2397. Available at https://doi.org/10.3390/v15122397.

Yang, X., Yang, S., Lian, X., Wuchty, S., Zhang, Z., Przytycka, T., 2021. Transfer learning via multi-scale convolutional neural layers for human–virus protein–protein interaction prediction. Bioinformatics (Oxford, England) 37 (24), 4771–4778. Available at https://doi.org/10.1093/bioinformatics/btab533.

Yang, Z.-X., Li, Y., Wang, R.-F., Hu, P., Su, W.-H., 2025. Deep learning in multimodal fusion for sustainable plant care: a comprehensive review. Sustainability 17 (12), 5255. Available at https://doi.org/10.3390/su17125255.

Yang, M., Huang, D., Zhan, X., 2024. Federated learning for privacy-preserving medical data sharing in drug development. https://doi.org/10.20944/preprints202410.1641.v1.

Zeng, X., Li, S.J., Lv, S.Q., Wen, M.L., Li, Y., 2024. A comprehensive review of the recent advances on predicting drug-target affinity based on deep learning. Frontiers in Pharmacology 15. Available at https://doi.org/10.3389/fphar.2024.1375522, http://www.frontiersin.org/Pharmacology.

Zhai, K., Yousef, M.S., Mohammed, S., Al-Dewik, N.I., Qoronfleh, M.W., 2023. Optimizing clinical workflow using precision medicine and advanced data analytics. Processes 11 (3). Available at https://doi.org/10.3390/pr11030939, http://www.mdpi.com/journal/processes.

Zhang, Y., Dong, M., Deng, J., Wu, J., Zhao, Q., Gao, X., et al., 2024. Graph masked self-distillation learning for prediction of mutation impact on protein–protein interactions. Communications Biology 7 (1). Available at https://doi.org/10.1038/s42003-024-07066-9.

CHAPTER 7

Big data enabled discovery in bioinformatics and structural biology

Jaoud Ansari[1], Shabana Khatoon[1,2], MD Kaif[1,3], Gourav Choudhir[1], Naseem Gaur[4], and Asimul Islam[1]
[1]Centre for Interdisciplinary Research in Basic Sciences, Jamia Millia Islamia, New Delhi, India,
[2]Indira Gandhi National Open University, New Delhi, India,
[3]Delhi Pharmaceutical Sciences and Research University, Delhi, India,
[4]International Centre for Genetic Engineering and Biotechnology, New Delhi, India

List of abbreviations

AI	Artificial intelligence
BAM	Binary Alignment Map
BCF	Binary Call Format
BED	Browser Extensible Data Format
BigWig	Binary Wiggle Format for Genomic Data
BinaryCIF	Binary Crystallographic Information File
CCP4	Collaborative Computational Project No. 4 format
CCPN	Collaborative Computing Project for NMR
CG	Coarse grained
DCD	CHARMM/NAMD Trajectory Format
EM	Electron microscopy
EMDB	Electron Microscopy Data Bank
FAIR	Findable, accessible, interoperable, reusable
FASTA	Fast All Sequence Format
FASTQ	Fast Quality Sequence Format
GFF	General Feature Format
GNN	Graph Neural Networks
GO	Gene Ontology
GPU	Graphics Processing Unit
GTF	Gene Transfer Format
HDF5	Hierarchical Data Format Version 5
IMP	Integrative modeling platforms
JCAMP-DX	Joint Committee on Atomic and Molecular Physical Data Exchange Format
MI	Mutual information
MicroED	Microcrystal Electron Diffraction
ML	Machine learning
mmCIF	Macromolecular Crystallographic Information File
MMTF	Macromolecular Transmission Format
MRC	Medical Research Council format

Advances in Structural Biology
DOI: https://doi.org/10.1016/B978-0-443-44988-8.00006-X

NetCDF	Network Common Data Form
NMR	Nuclear magnetic resonance
NMR-STAR	NMR Self Defining Text Archival and Retrieval format
nmrML	NMR Markup Language
PDB	Protein Data Bank
PDC	Protein Data Collection
PPI	Protein–Protein Interaction
RMSD	Root mean square deviation
SAM	Sequence Alignment/Map format
SASA	Solvent accessible surface area
VCF	Variant Call Format

7.1 Introduction

In the earlier era of medical research, no one might have imagined that the data would increase so much that a new dedicated field of bioinformatics would emerge to help understand data relevance. With the advancements in science and technology, various experimental structural biology and bioinformatics techniques have made it possible to study even those structures that were once considered impossible, but this has come with multiple costs. One of which is the vast amount of data, increasing at a logarithmic rate, often called big data, and is generated because of the results of experimental work (Gharajeh, 2018). The analytics of data is not easy, and the importance was felt with the completion of the Human Genome Project. This was a 13-year-long multiinstitutional research project, which made possible the mapping of sequence of proteome (Nurk et al., 2022). The results were stunning with the massive amount of information, which was hard to evaluate in that era. These findings necessitated the development of advanced computers, tools, and algorithms for the storage, analysis, and understandable interpretation of the outcomes (Greene et al., 2014). Now, more and more research areas are making huge contributions to increase this ocean of data currently available (Fig. 7.1) (Pongor and Landsman, 1999). Most significantly contributing techniques within structural biology are proteomics data gathered using techniques such as X-ray crystallography, nuclear magnetic resonance (NMR) spectroscopy, cryo-electron microscopy (cryo-EM), and molecular dynamics simulation (MD simulation), etc. (Shome et al., 2024). Obtained information from these techniques is then used to predict and understand the dynamic behavior of proteins and other macromolecules. The Protein Data Bank (PDB) is one such repository, which was specifically created and used for storage of protein structure-related information, containing hundreds of thousands of structures (more precisely, 247,287 as of December 31, 2025, source: https://www.rcsb.org/stats/summary), each with its own extensive set of associated data (Berman and Burley, 2025). However, things are more challenging than they

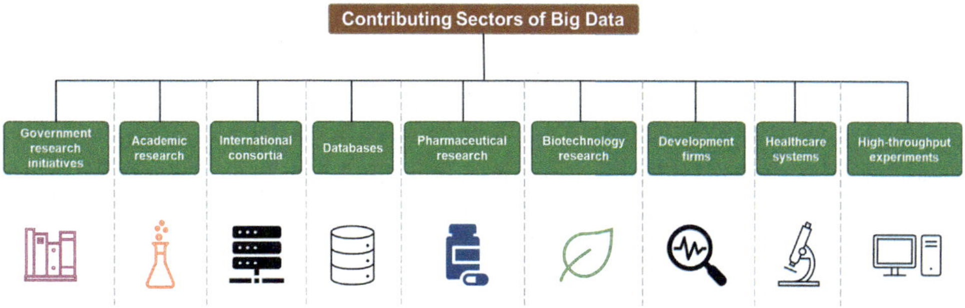

Figure 7.1 Major big data contributing sectors in structural biology.

appear, as this enormous amount of data with key problems including maintaining quality data, storage, linking it with other technologies, processing, quality issues, noise, finding translational relevance etc. (Fillinger et al., 2019).

The three defining features of big data, significant volume, speed of accumulation (velocity), and diverse types of data (variety), create major potential, but also major obstacles, for the current generation of researchers (Abdalla, 2022). When disparate molecular and clinical datasets are combined by structural biologists and bioinformaticians, they have a new way of approaching the study of human biological systems and can describe molecular aspects of various diseases/pathologic states. The scale and speed of this type of analysis are unprecedented (O'Connor et al., 2023; Zitnik et al., 2019); therefore, researchers can identify molecules with the greatest potential to be successful candidates for development into safe and effective therapy via specific biological targets. For example, the use of big data from structural biology and bioinformatics presents significant methodological and analytical challenges. Researchers will need to be trained in specialized big-data analysis software, be able to analyze and interpret very large amounts of high-throughput data, and have harmonized or standardized heterogeneous biomedical datasets to enable comprehensive and innovative objective data-driven discoveries (Fillinger et al., 2019; Gupta, Kumar and Kumar, 2024). Researchers who lack adequate analytical frameworks and expertise may feel overwhelmed by the large amounts of quantitative data and therefore may hinder rather than help clarify biological insights that are relevant to the study (Galetsi, Katsaliaki and Kumar, 2019).

7.2 Types of big data in structural biology and bioinformatics

Structural biology and bioinformatics experiments generate an enormous amount of data, ranging from atomic coordinates to 3D density volumes, raw experimental image stacks, multimicrosecond conformational trajectories, and linked omics and network layers that help in linking structures and functions (Li and Chen, 2014).

7.2.1 Protein structural data

Protein structural data form a hierarchical set of representations that begin with primary sequence, which contains linear amino acids tightly bonded to each other and extend through secondary structural elements, tertiary fold geometries, and quaternary assembly topologies. Primary sequence information anchors all higher-order descriptions and is stored as plain sequence records (Fig. 7.2) (Burley et al., 2022; Varadi et al., 2022). These sequences provide the residue identities and order from which secondary structure elements, such as α-helices and β-strands, are assigned by geometric or hydrogen-bond criteria. Tertiary structures are the 3D arrangement of residues into definable domains and active sites, and are distributed as atomic coordinate files, along with annotation and validation metadata, in community archives such as the PDB and its members worldwide. Quaternary protein structure refers to biologically relevant oligomeric states and larger macromolecular assemblies formed by the linking of two or more tertiary structures (Burley et al., 2022; Varadi et al., 2022).

7.2.2 Nucleic acid structural data

Nucleic acid structural data includes duplex DNA conformations, regulatory and structured noncoding RNA 2D motifs, and compact RNA tertiary structures such as internal loops and pseudoknots. Experimental atomic coordinates for DNA and RNA are deposited alongside protein structures in the PDB. Specialized sequence and 2D structure databases provide high-coverage details of noncoding RNA sequences and

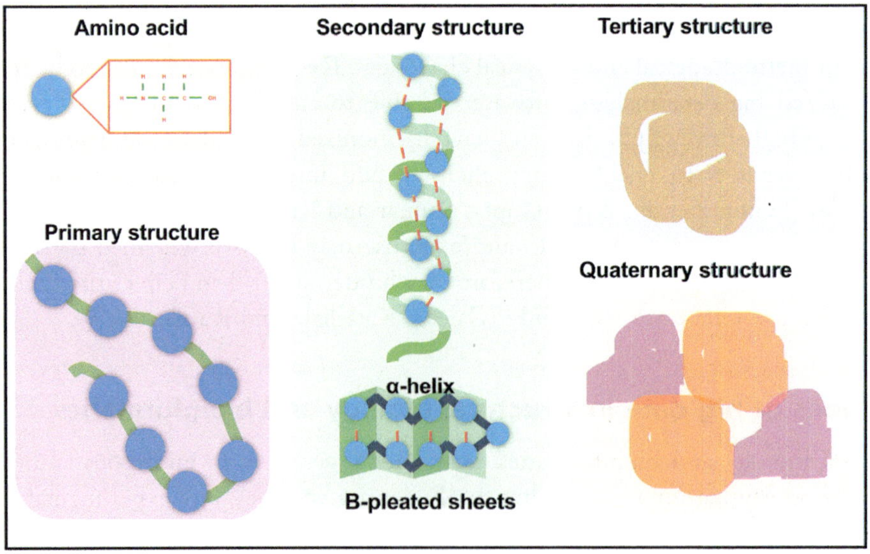

Figure 7.2 Various levels of protein structure.

predicted secondary structures, which are used as scaffolds for 3D modeling (Burley et al., 2022; Sweeney et al., 2021; Hong et al., 2021; Shen et al., 2024). When these nucleic acids are sequenced using genomic sequencing techniques, they generate data in astronomical figures, with a single human genome producing roughly 100–200 GB of raw digital information (Vidal et al., 2019; He, Ge and He, 2017; Greenfield, Wittorff and Hultner, 2019). According to earlier predictions, it was projected that by 2025, annual genomic data output would reach 40 exabytes, potentially competing with the data footprints of YouTube and Twitter (now, known as X) (Greenfield, Wittorff and Hultner, 2019; Muir et al., 2016). However, accurate current figures are not yet available to confirm this. This creates a massive burden on big data analytics, as the cost of storing these files now often exceeds the cost of sequencing them (Muir et al., 2016). This enormous quantity of data is a computational bottleneck, and aligning billions of short DNA sequences requires immense processing power (graphics processing units [GPUs]) and specialized compression. The substantial file sizes also pose a significant logistical hurdle for high-speed data transfer in global research (O'Connell et al., 2023).

7.2.3 Interaction data

Interactions are a central structural layer because many biological functions arise from specific macromolecular contacts. Protein–protein association and direct interaction evidence are curated at scale by community resources and serve as substrates for docking and structural modeling of complexes (Kundrotas et al., 2018; Ooi et al., 2010). Public interaction resources provide millions of binary evidences and probabilistic association scores that are frequently integrated with structural coordinates to propose interfaces and to guide mutational studies. Protein–small-molecule binding records and experimentally measured affinities are archived in dedicated binding databases and in curated subsets of the structural archive that pair coordinate models with biochemical potency data. This forms the empirical backbone for structure-based drug discovery and for training predictive scoring algorithms. Structural information for nucleic acid and protein interactions is similarly accumulated in the structural archive and in specialized annotation resources that catalog contact motifs and binding preferences (Oughtred et al., 2021; del Toro et al., 2022; Liu et al., 2025).

7.2.4 Experimental imaging data

Experimental systems produce fundamentally different data objects. X-ray diffraction experiments yield reflection intensities from which atomic coordinates are solved and archived (Poger, Yen and Braet, 2023; Mu et al., 2021). NMR experiments generate spectral data, chemical shifts, and restraint sets that describe ensembles rather than single static models, and these data are collected in a central spectroscopic repository

(Hoch et al., 2023). 3D density maps from single-particle cryo-EM and tomography are stored as voxel grids in dedicated map archives (Valentini et al., 2015), while the underlying raw multiframe detector image stacks are preserved in raw image archives to enable reprocessing and method development. Typical cryo-EM experiments now produce TBs of raw movies per project, motivating community infrastructure for long-term storage and transfer (Poger, Yen and Braet, 2023). Microcrystal electron diffraction extends electron-based methods to microcrystals, generating diffraction data that can enable high-resolution structures from crystals that are too small for conventional X-ray approaches (Mu et al., 2021). Small-angle X-ray and small-angle neutron scattering generate low-resolution solution scattering curves and reconstructed bead or envelope models that complement high-resolution coordinates by reporting global shapes and conformational ensembles in solution. Each of these experimental streams has distinct metadata and provenance requirements and therefore places different demands on archiving, reusability, and standardized reporting (Burley et al., 2022; Mu et al., 2021; Iudin et al., 2023; Valentini et al., 2015).

7.2.5 Computational data

Computational studies produce large, structured outputs that both mirror and extend experimental data. MD simulations generate dynamic 3D coordinate trajectories that can accumulate terabytes (TBs) for long, replicated, or large-system simulations and that demand streaming, compression, and automated feature extraction workflows for analysis (Hildebrand, Rose and Tiemann, 2019). Comparative modeling and proteome-scale structural prediction generate large sets of coordinates, accompanied by per-model confidence metrics; these predicted collections enable proteome-level structural surveys but require clustering and benchmarking. Docking and virtual screening generate large enumerations of candidate binding poses and scored complexes; when combined with experimental binding affinity data, these computational outputs inform ML pipelines for pose ranking and affinity prediction. Community efforts for standardized deposition, searchable metadata, and shared simulation hubs are maturing but remain an active area of methodological development due to the scale and heterogeneity of the data (Varadi et al., 2022; Valentini et al., 2015; Tiemann et al., 2024).

7.2.6 Omics and network data

Structural interpretation is frequently embedded in broader omics and network contexts, expression atlases and sequencing repositories provide the tissue, developmental and perturbation contexts that moderate whether a structure is biologically relevant; curated sequence knowledgebases furnish canonical sequences and functional annotations used to generate and map models; evolutionary covariance and comparative genomics resources supply the covariation signals exploited by modern structure prediction methods and interface inference; and pathway and disease network databases aggregate interaction and phenotype data that link structural hypotheses to cellular programs and clinical phenotypes. Integration across these layers converts isolated 3D records into mechanistic models that can be tested experimentally or applied to translational problems, such as target prioritization and mechanism-based therapeutic design (Bateman et al., 2023; Gillespie et al., 2022; Szklarczyk et al., 2023).

7.3 Challenges in big data analytics

The structural biology and bioinformatics data are vast and complex, yet they can be grouped into a few broad categories. These include sequence data (nucleotide and protein), structural data, genomic data, functional and derived data, as well as high-throughput experimental data (such as transcriptomics, proteomics, and metabolomics). All of these data are accompanied by essential provenance and experimental metadata that determine reuse value and reproducibility. The scope and examples of these categories are reflected across primary community resources and the annual Nucleic Acids Research database compilations (Sayers et al., 2023). Table 7.1 summarizes major challenges in big data analytics.

7.3.1 Data volume, velocity, and heterogeneity

Structural biology research generates vast quantities of data in a relatively short period. High-throughput imaging, for instance, can produce TBs of data per experiment, and

Table 7.1 Major challenges in big data analytics.

Category	Key issues	References
Volume and velocity	TB-scale data floods	Berger et al. (2016); Holzinger, Dehmer and Jurisica (2014)
Poor data quality	Noise, bias, resolution limits	Guo et al. (2022); Schwalbe et al. (2024)
Interoperability	Fragmented file formats	Niu et al. (2022); Katayama et al. (2010)
Storage and retrieval	Inefficient indexing	Philippe et al. (2011); Masseroli et al. (2014)
Integration	Cross-domain linking difficulties	Chen et al. (2010)
Reproducibility	Lack of standardized pipelines	Kämpf et al. (2019)
Ethics and privacy	Sensitive biomedical data	Wang et al. (2020)
ML challenges	Overfitting, interpretability, and resource demands	Rickert and Lieleg (2022); Yousef and Allmer (2023); Han and Liu (2022)

modern cryo-EM sessions often yield on the order of 1 TB of raw images in a single run (Poger, Yen and Braet, 2023). Similarly, MD simulations (e.g., long MD trajectories) produce huge trajectory files reaching many gigabytes. Databases of predicted models are also expanding rapidly. The AlphaFold protein structure database grew from only tens of gigabytes in 2021 to approximately 23 TB in 2022 and continues to grow as more structures are predicted (Zhang and Pyle, 2023). This flood of data illustrates the classic "three V's" of big data, namely volume, velocity, and variety (Hildebrand, Rose and Tiemann, 2019), as applied to structural biology. Unless systematic analysis methods and storage infrastructure keep upgrading with increasing demands, the steep volume of data can become devastating, hindering scientific insight (Poger, Yen and Braet, 2023).

7.3.2 Data quality

Various factors, including experimental noise, modeling bias, and resolution constraints, have severely affected the ability to accurately assess and characterize 3D structural data (Wankowicz, 2025; Sapoval et al., 2022). Having access to a considerable amount of data does not mean there is a corresponding increase in quality, as experimental measurements exhibit a high degree of noise and are subject to systematic errors (for example, low-contrast cryo-EM sample particles or syndromal disorder in crystals), which leads to degraded data quality (Li et al., 2024). Resolution is an inherent limitation; very high-resolution (e.g., ≤2.5 Å) structures are required to place atoms with precision, while low-resolution models will provide only coarse locations of individual atoms (Rosenthal and Henderson, 2003). Although a cryo-EM entry may specify a single nominal resolution, it usually does not indicate the level of local variability associated with that cryo-EM entry. In addition to these issues with the source of structural data, structural models have inherent biases. For example, in terms of the structural data available in the PDB, there is considerable underrepresentation of membrane proteins and disordered complexes, while soluble enzymes and drug targets are heavily overrepresented in PDB (Wankowicz, 2025). Even if all data in the same dataset is obtained through either of two modeling or refinement processes, subtly different models will be expected to emerge from that same raw data (Wankowicz, 2024). For this reason, validation and benchmarking are an important component of model evaluation; almost all models built from data at finite resolution will have local errors, and the quality of these structural models varies substantially within the archive (Wankowicz, 2025).

7.3.3 File format, fragmentation, and lack of interoperability

Structural biology relies on many specialized formats, and this fragmentation hinders data exchange. Macromolecular coordinates moved from the old fixed-format PDB files to the more extensible PDBx/Macromolecular Crystallographic Information File (mmCIF) format (to allow larger complexes, more metadata fields, etc.) (The UniProt Consortium, 2015; Westbrook and Fitzgerald, 2003). However, many tools still assume the legacy PDB format or use ad hoc file types, so format conversion is often needed. Experimental data have their own standards; for example, cryo-EM and tomography maps are typically stored as Collaborative Computational Project No. 4/Medical Research Council (MRC) or Hierarchical Data Format Version 5 (HDF5) volumes, often accompanied by metadata in XML or mmCIF files (Turner et al., 2024). MD simulation processed visual media, NMR ensembles, and integrative models, each of which uses distinct conventions. Thus methodologies that integrate data must juggle multiple schemas and risk loss of metadata during conversion. Even basic storage is inefficient; PDB/mmCIF files are text-heavy and redundant, so general compression only helps to a certain extent. This has motivated the development of new binary formats (e.g., Macromolecular Transmission Format [MMTF], BinaryCIF, the Protein Data Collection [PDC] format) that can reduce file size by ~70% to 80% compared to gzipped text (Zhang and Pyle, 2023). Until the community adopts common standards and rich metadata schemas across modalities, the proliferation of formats will continue to impede seamless data interoperability. Additional file formats and their uses are provided in the below Table 7.2.

7.3.4 Storage, indexing, and data retrieval

High-dimensional data places heavy demands on storage and databases. An MD simulation trajectory is effectively a 4D array (atoms × coordinates × time) that can reach hundreds of gigabytes for a modest protein and microsecond simulation. Traditional relational databases or flat files cannot be indexed this efficiently; retrieving a single frame or atomic time course may require scanning TBs of data. To address this, researchers are exploring specialized solutions, such as advanced coordinate compression (for example, the PDC format can compress proteins to ~22% of the original file size [Zhang and Pyle, 2023]) and spatio-temporal partitioning schemes (utilizing HDF5, key-value stores, or optimized indexing). Some groups propose purpose-built MD repositories and search engines. However, it is worth noting that these systems are still emerging, and efficient querying, indexing, and analytics on PB-scale trajectory and volumetric datasets remain active areas of development.

7.3.5 Integration problems linking structural data with sequences, networks, or phenotypes

Linking 3D structural data to other biological data (e.g., sequences, interaction networks, phenotypes) presents additional challenges. At the sequence level, progress has been made; PDB entries are mapped to UniProt sequences, and resources like

Table 7.2 Common file formats and their features.

Format	Data type	Use case	Pros	Cons	References
PDB	Atomic coordinates	Protein structures, legacy use	Human-readable, widely accepted	Limited metadata, fragile, file size constraints	Adams et al. (2019)
mmCIF/PDBx	Extended coordinates/ metadata	Modern structural data deposition	Rich metadata, handles large/ complex structures	Large file sizes, more complex parsing	Adams et al. (2019); Westbrook et al. (2022)
MMTF	Binary coordinates	High-performance visualization	Highly compressed, fast parsing	Limited adoption, tool support is still growing	Bradley et al. (2017)
BinaryCIF	Compressed mmCIF	Efficient streaming and visualization	Compression supports web-based viewers	Needs broader standardization	Sehnal et al. (2020)
PDC	Compact coordinates	Storage-heavy MD or structure modeling	~22% size of original text	Niche format, limited tool support	Zhang and Pyle (2023)
MRC/CCP4	3D density maps	Cryo-EM volumes, X-ray maps	EM industry standard, metadata support	Large file sizes, format variants	Cheng et al. (2015)
HDF5	Volumetric or trajectory data	MD simulations, cryo-EM stacks	Efficient indexing handles high-dimensional data	Complex structure, learning curve	Folk et al. (2011)
DCD, XTC, TRR	MD trajectories	Recording atom positions over time	Compressed storage, widely supported	Not human-readable, tool dependency	McGibbon et al. (2015)
NetCDF	MD trajectories	Scientific data exchange	Self-describing, portable	Requires specialized libraries	Davis, and Rew (1990)
NMR-STAR	NMR spectral and structural	NMR experimental data deposition	Standardized format, rich metadata	Verbose requires conversion for downstream use	Ulrich et al. (2019)
JCAMP-DX, nmrML	NMR spectral data	Raw and processed NMR spectra	Interoperable, FAIR-aligned	Partial adoption, heterogeneous content	Schober et al. (2018)
FASTA	Sequence data	DNA/protein sequence storage	Simple, universal	No quality info, limited metadata	Pearson and Lipman (1988)
FASTQ	Sequencing reads and quality	Raw NGS output	Contains base quality scores	Large size, human–machine readability trade-off	Cock et al. (2009)
SAM/BAM	Aligned reads	Sequence alignment, storage, and processing	BAM is compressed/indexed; random access enabled	SAM is bulky; BAM is tied to the reference genome	Li et al. (2009)
VCF/BCF	Variant calls	SNPs and structural variant data	Rich annotations, standardized	Complex specs, variable metadata completeness	Danecek et al. (2011)
BED/GTF/GFF	Genomic regions/ annotations	Gene models, features, and genome tracks	Lightweight, text-based, widely used	Limited structure, relies on external metadata	Kent et al. (2002); Quinlan and Hall (2010)
BigWig	Coverage tracks	Genome-wide data visualization	Efficient storage, random access	Binary format, tool-dependent	Kent et al. (2010)

SIFTS provide detailed residue-level correspondence between structures and UniProtKB (Wankowicz, 2025). This enables direct integration of structure with genomics and proteomics data. However, connecting structures to networks or phenotypes is significantly more challenging. Protein–Protein Interaction (PPI) networks and pathway databases typically label nodes by gene or protein, rather than by structure. Mapping a structure onto such a network requires custom cross-references or a loss of spatial detail (Chen et al., 2010; Mathivanan et al., 2006; Milano, Agapito and Cannataro, 2022). Phenotype or disease databases rarely include structural context, so one must infer links (e.g., mapping a mutation in a gene to its protein's structure). Because each domain (structural, genomic, and clinical) uses its own identifiers and standards, automated cross-domain researchers like us find integrating them difficult, making the process complex. In general practice, the structural, network, and phenotype datasets have often remained seen as siloed, which limits the ability to perform truly integrative, multiscale analyses (Chen et al., 2010; Mathivanan et al., 2006; Milano, Agapito and Cannataro, 2022).

7.3.6 Barriers to reproducibility, benchmarking, and model validation

The problems related to data reproducibility and reliability have been exacerbated by the lack of uniformity in how data, metadata, or analysis methods are recorded across different experiments. As a result, performing the same experiment multiple times using differing sources of data can lead to inconclusive results. Researchers have recognized this issue, and a recent commentary noted there are a "large amount of resources wasted on producing unrepeatable" compute models if there is no prevalent use of standard data creation, which could promote collaboration; there is a strong desire to build standards or shared databases and ranchos for evaluation of created structures while utilizing the principles of findable, accessible, interoperable, and reusable (Koehler Leman et al., 2021). In this regard, though compliance is not universal, efforts are being made to establish standard procedures. Regardless, it is important to validate a new model by ensuring its predictions during structural analysis are accurate when measured against previously utilized experimental data or proof of concept controls. If these validations (and established) checks do not occur, and no use of additional reliable and trusted data is included for establishing provenance, then researchers have limited ability to establish confidence when making conclusions using accumulated data sets of structural biology (Wankowicz, 2025; Koehler Leman et al., 2021).

7.4 Analytical techniques and tools for structural biology big data

Data on large-scale structures in structural biology may originate from many different types of structures (i.e., files indicating atomic positions within space, density maps, molecular trajectories, etc.) and will typically require a significant amount of preprocessing before they can be analyzed. Raw coordinate files need to be converted to a common format (which includes an extensive amount of associated metadata). An illustrative example of this is that the legacy version of the PDB (Research Collaboratory for Structural Bioinformatics PDB; 2005) has now been superseded by the newer, extensible PDBx/mmCIF (van Ginkel et al., 2021). Open-source tools such as PDBeCIF are available to facilitate the parsing and conversion of structural data (PDB, mmCIF, CIF) and to provide access to the metadata within those formats (van Ginkel et al., 2021). NMR structures and spectra can also be represented in an extensible fashion using STAR/NMR-Self Defining Text Archival and Retrieval (NMR-STAR), with available tools (e.g., STARch or Collaborative Computing Project for NMR FormatConverter) to convert from older formats to the STAR/NMR-STAR formats. In instances where predicted models are being developed or combined with experimentally determined structures, there is a specific extension of mmCIF, called ModelCIF, that provides a schema for consistently representing computational models (Vallat et al., 2023). In practice, archives such as the wwPDB's Next Generation System provide a unified approach to mapping and annotating structural data across different sources of data. They provide templates that can enable the mapping of data from different sources by using the same set of reference databases (e.g., PDB to UniProt, PDB to Pfam, PDB to SCOP/CATH domains, etc.) and the creation of central repositories of metadata across multiple databases (Choudhary et al., 2024). These converters and schemas are all complementary and work together so that large-scale structure data resulting from the use of X-ray crystallography, NMR spectroscopy, cryo-EM, and prediction techniques are expressed in a format that is consistent and can be read by computers for analysis downstream.

7.4.1 Data preprocessing and standardization

Raw structural data need various forms of processing and optimization because they need to be cleaned, normalized, and expressed as standardized representations. Atoms, chains, residues, and experimentally derived parameters (e.g., occupancies and B-factors) are all extracted from three coordinate file types (i.e., PDB files, mmCIF files, and MMTF files) using parser scripts. Existing format conversion libraries (e.g., BioPython PDB, PDBeCIF and others) allow for automating the conversion of the older PDB text file format to a more modern mmCIF/CIF or JSON formats while still preserving the annotations (van Ginkel et al., 2021). There are established conventions within the structural biology community for each data type. For example, for NMR data, chemical shifts and restraints are stored in the STAR format, and these need to be converted using converter tools (BMRB NMR-STAR converter) that have been developed for the BMRB database. Cryo-EM maps conform to the MRC/HDF5 standard of data formats and use Electron Microscopy Data Bank data to provide metadata. Typically, automated protocols are used to perform various tasks such as renumbering atoms, validating chemistry (i.e., checking for atoms that have not been added, checking that there are no steric clashes), consistently assigning residue/atom labels, and aligning chain IDs. After all of the structures that have been cleaned have been assigned standardized identifiers (i.e., UniProt accession numbers, gene ontologies, enzyme codes), they are cross-referenced with other structural identifiers (e.g., WWPDB NextGen Archive). The mention of cross-references provides additional support that coordinate data from any source can be combined into a single dataset. Annotations also provide links between each structure and the sequence databases (i.e., UniProt, Pfam, SCOP2/CATH domains), as well as between structures and associated functional data (Choudhary et al., 2024). As a result, a clean set of validated and interoperable structural data occurs that can be used for structural analysis.

7.4.2 Structural analysis methods

To analyze 3D structures, one must rely on derived geometric or other types of descriptors that summarize both the geometric and interactional properties of the structure. Geometric descriptors can describe the quantitative aspects of a 3D structure through the metrics of its geometric properties, such as global metrics, for example, molecular weight, solvent-accessible surface area, volume, and radius of gyration (Li et al., 2024), and also through the distributions of bond lengths or angles. Shape fingerprints (e.g.,: 3D Zernike descriptors) represent the shape of the surface of a protein as a vector that also allows for quick shape comparisons (Han et al., 2019; Daberdaku and Ferrari, 2018; Sael et al., 2008). Other types of descriptors can include the counts of secondary structure elements, counts of hydrophobic patches on the surface of the molecule, or counts of the number of electrostatic potentials on the surface of a protein, as well as counts of the various sequence-derived features (i.e., amino acid composition, predicted disorder). By converting the 3D structure into an array of features, which includes geometric and other types of descriptors, these descriptors can then be used as input to statistical modeling or for performing similarity searches.

Contact maps are a depiction of the spatial relationship between residues or atoms of proteins. A contact map is a binary (or distance-weighted) matrix that indicates which residues or atoms are in close proximity to one another (less than a certain distance). In a sense, contact maps represent the 3D geometry of the protein in a 2D form using a dot or ellipse at position (i,j) to represent that residues (i) and (j) are spatially close (Hrabe and Godzik, 2014). With the use of a contact map, it is possible to easily identify important aspects of protein structure, such as secondary structures (diagonal bands of dots for helix) and long-range contacts between the residue pairs (off-diagonal band). Contact maps can also be used to compare structural similarities and analyze intra- and intermolecular interactions. For example, the NAPS tool uses individual nodes to represent each residue and draws an edge or line between two nodes if they are located within a specified distance from each other. The collective set of all atomic nodes can be analyzed as a global network of atomic contacts using various graph theory methodologies. Distance networks extend this idea; atoms or residues are viewed as vertices of a graph connected to other vertices by edges weighted according to distance or strength of interaction, and network analysis techniques can be applied to identify the modules or communication pathways in a protein (Hrabe and Godzik, 2014). Overall, geometric descriptors, contact maps, and distance-based graphs provide complementary ways to capture structural features (shape, packing, and topology) in a quantitative form, enabling statistical or graph algorithms to probe structure-function relationships.

7.4.3 Visualization and scalable rendering approaches

Interactive visualization tools are essential for exploring complex structures and assemblies at any scale. Desktop programs like VMD, PyMOL, and ChimeraX have long provided rich graphical rendering capabilities, but recent web-based viewers bring these features to browsers. For example, the web-native Mol* Viewer (used by PDBe and RCSB PDB) streams coordinate and experimental data in the browser and can display thousands of structures with annotations (Sehnal et al., 2021). It supports multiscale rendering: for extensive models, detailed geometry is shown only in focused regions, while distant parts are rendered coarsely (levels of detail generated on the fly) (Rose et al., 2018). Modern viewers utilize compressed binary formats (e.g., MMTF or BinaryCIF) and GPU-accelerated WebGL to handle large datasets. In the NGL Viewer, a highly compressed binary representation (MMTF) and WebGL rendering enable the interactive display of complexes with millions of atoms on typical hardware (Rose et al., 2018). Over the past years, many new web tools have emerged, like JSmol, 3Dmol.js, LiteMol, PV, iCn3D, Molmil, and NGL. Each of these enables high-performance graphics in standard browsers (Hanson, 2010; Rego and Koes, 2015; Sehnal et al., 2017; Rose and Hildebrand, 2015). Mol* combines advances from LiteMol and NGL into a unified library, achieving rapid rendering of huge assemblies (for example, it can visualize an entire HIV capsid or virus particle interactively) (Sehnal et al., 2021). Analyzing software will often also support fundamental analysis through the use of patterns when looking at items on-screen (like distance measurements, surface cutting, etc.), which means that they are great ways of providing a resourceful platform for large amounts of structural visualization.

7.4.4 Simulation and modeling techniques

Simulations or modeling of structures beyond that of simple analysis could produce structural data on a large scale through the use of many types of computational simulation and modeling as a whole. Example methods of enhanced sampling (metadynamics, umbrella sampling, accelerated MD, replica exchange) that create very quick explorations of conformational space by basically trying to either change or remove the energy landscape (Chen et al., 2026). Either way, these methods are a way that one can overcome the slow kinetics of biomolecules and provide a greater representation of rare states. Using coarse-grained (CG) models provides an additional method of generating models; these can take the very fine detail of molecular models (atoms) and group those atoms into fewer (for example, many groups of atoms into one bead) and therefore create a free-energy landscape that can be simulated over an even longer time (Chen et al., 2026). Popular CG force fields (e.g., MARTINI) and structure-based models allow efficient study of large complexes and assemblies that would be intractable atomistically.

A related strategy is hybrid or integrative modeling, which combines experimental data with computation. In these methods, information from X-ray, NMR, cryo-EM, SAXS, and crosslinking, among others, is integrated with modeling algorithms (docking, MD, Monte Carlo) to build a coherent model (Chen et al., 2026). For example, domains determined by crystallography can be flexibly fitted into lower-resolution electron microscopy density, or predicted domains can be docked according to sparse distance constraints; this hybrid approach extends structure determination to very large or complex systems. In fact, as the PDBeCIF guidelines note, multiscale/hybrid structures are explicitly accommodated by adopting unified data standards (van Ginkel et al., 2021). Enhanced sampling and coarse graining expand the accessible modeling scales, while hybrid modeling pipelines leverage all available data to construct and refine models of large biological machines (Chen et al., 2026; Jonic, Miyashita and Callebaut, 2018).

7.4.5 Network and pathway analysis methods

Structural data can also be analyzed through the lens of biological networks. At the molecular level, residue or atom interaction networks (discussed above) are interrogated with graph algorithms: centrality measures (degree, betweenness, closeness, eigenvector) identify key residues (Chakrabarty and Parekh, 2016), community detection finds modules of strongly interacting residues, and shortest-path or network-flow algorithms probe allosteric communication pathways within a protein. For example, residue-interaction networks in NAPS provide metrics (e.g., clustering coefficient) that correlate with functional sites. At a higher level, proteins themselves form interaction networks (protein–protein interactomes) or are nodes in metabolic/regulatory pathways. Graph-based analyses, such as graph clustering, motif search, or random walk models, can reveal functional modules or disease modules in these networks (Chakrabarty and Parekh, 2016).

In parallel, pathway enrichment analysis is used to interpret lists of proteins or genes in structural studies. Here, statistical tests identify known pathways or Gene Ontology terms that are overrepresented in a set of proteins more than expected by chance (Zeng et al., 2021). For example, one might take proteins with structurally determined mutations and ask which signaling or metabolic pathways they belong to, highlighting possible mechanisms. Tools like EnrichmentMap (Cytoscape), GSEA, and gProfiler implement this automation process by generating graphs of biological pathways, which allow the summarization of extensive protein lists into actionable pathway hypotheses. Specifically, network algorithms and enrichment analyses link structural findings (e.g., "functioning" structures) to systems-level biology through various means. Graph-based algorithms can be used to extract important structural motifs (e.g., parts of structures that have discrete bodies) and communication pathways (e.g., structural types and functions) represented by large graphs, while enrichment analyses connect these structures with the larger systems of biochemistry (Reimand et al., 2019; Aittokallio and Schwikowski, 2006; Cowen et al., 2017; Cho, Kim and Przytycka, 2012; Creixell et al., 2015; Wu et al., 2015).

7.4.6 Tools and platforms

An extensive ecosystem of software supports the execution of big data analytics within the structural biology discipline. Some examples of tools that support the formatting and visualization of structural biology data include BioPython, MDAnalysis, MDtraj, and public-domain software (e.g., LiteMol, PyMOL, VMD, NGL, ChimeraX, and Mol* etc.) (Sehnal et al., 2021; Rose et al., 2018; Crowther, Henderson and Smith, 1996). Examples of software that perform structural analysis include DSSP/STRIDE (secondary structure analysis), PyMOL/ProDy (fluctuations and normal modes), TM-align/STRUCTAL (structural alignment), CMView (contact map), and the NAPS software for protein-network analyses. Simulation packages range from classical MD engines (GROMACS, NAMD, AMBER, OpenMM) to Monte Carlo and coarse-graining toolkits (SIRAH, Martini, Rosetta). Integrative modeling platforms include themselves and HADDOCK (with Martini for CG docking), which combine data from multiple sources. Network analysis often utilizes Cytoscape (for PPI networks) and graph libraries (such as NetworkX and igraph), along with gene set enrichment tools (e.g., Profiler, Enrichr). Many of these tools are actively maintained with web or API interfaces. For instance, PDBe, RCSB PDB, and EMDataResource provide access to structural databases and utilize viewers (NGL/Mol*) for browser integration (Sehnal et al., 2021; Rose et al., 2018; Crowther, Henderson and Smith, 1996). By mapping data types to appropriate software format converters for raw data, feature extraction libraries for analysis, visualization engines for rendering, and ML frameworks for modeling, researchers can build end-to-end pipelines that scale from TB-scale structure collections to predictive models. More tools and platforms are included in the table below (Table 7.3).

Table 7.3 Tools and platforms for big data analytics in structural biology and bioinformatics.

Category	Tool/platform	Description	Strengths	Sources
Visualization	Mol*	Web-based viewer for large biomolecular structures and MD trajectories	High performance, scalable via WebGL	Sehnal et al. (2021)
	NGL	WebGL-based molecular graphics viewer for large complexes	Scalable rendering supports MMTF	Rose et al. (2018)
	JSmol/LiteMol/3Dmol.js	Browser-based structure viewers	Plugin-free, interactive	Crowther, Henderson and Smith (1996)
MD Engines and analysis	GROMACS/NAMD/AMBER/OpenMM	MD simulation suites with GPU acceleration	High-performance, broad force-field support	Lee et al. (2016)
	MDAnalysis/MDtraj	Python libraries for MD trajectory handling and analysis	Multiformat support, efficient data processing	Scherer et al. (2015)
	VMD	Visualizes MD simulations with powerful scripting support	Highly customizable, widely used	Humphrey, Dalke and Schulten (1996)
	CPPTRAJ (AmberTools)	Trajectory processing and analysis for Amber simulations	Comprehensive scripting for advanced trajectory analysis	Roe and Cheatham (2013)
	PyEMMA	Statistical analysis (Markov state modeling) for MD data	Kinetics modeling and clustering	Scherer et al. (2015)
	PLUMED		Integrates with multiple MD engines	Tribello et al. (2025)

(*Continued*)

Table 7.3 (Continued)

Category	Tool/platform	Description	Strengths	Sources
		Plugin for enhanced sampling and free energy calculations		
	mdciao	CLI/Python tool for automated MD contact analysis	Generates annotated figures/tables	Saltzberg et al. (2019)
Integrative modeling	IMP	Platform integrating multisource data for modeling large biomolecular assemblies	Flexible, Python/C++ modular API	Saltzberg et al. (2019)
Structural software platform	SBGrid	Curated structural biology software environments	Reproducibility via encapsulated environments	Herre et al. (2024)
Bioinformatics platforms	Galaxy	Web-based platform for reproducible bioinformatics workflows	GUI workflows, cloud integration	Abueg et al. (2024)

7.5 Machine learning and artificial intelligence

ML and AI methods offer powerful alternatives to traditional bioinformatics, especially for big data. Unlike rule-based or explicitly programmed approaches, ML can learn complex patterns directly from data. In structural biology, ML (and particularly deep learning) excels at extracting features from high-dimensional inputs (coordinates, graphs, images) without manual crafting. ML models can implicitly capture nonlinear relationships and high-order dependencies in the data, often achieving higher accuracy than simpler methods. Moreover, AI can integrate heterogeneous data (such as sequences, structures, and annotations) and improve as more data accumulates. ML/AI brings the strengths of representation learning, scalability to large datasets, and adaptive model improvement, which can complement (or surpass) traditional statistical or physics-based methods in analyzing structural big data (Glielmo et al., 2021; Kumar and Srivastava, 2024). However, working with techniques or training new ML-based models is not easy, as each one has to face numerous problems, as shown in Fig. 7.3.

7.5.1 Feature engineering and selection strategies for structural data

ML could typically increase training efficiency based on particularly engineered features. Depending on structure, structural datasets can be represented as hand-engineered descriptors (including distances, angles, and surface properties), graph-based features (including node degree/edge type), or as learned embeddings using similar/same structures. In one example, >18,000 features have been generated from protein tertiary structures (including counts of hydrogen bonds, surface tension, number of hydrophobic contacts, frequency of fragment types, etc.) (Silva et al., 2025). To reduce dimensionality and prevent overfitting, feature selection is very important for the development of effective models. Common strategies could include filter methods (including mutual Information [MI], ANOVA scores, etc.) and wrapper methods (including recursive feature elimination, etc.). InteracTor utilized MI to rank features by relevance, retaining those with the highest MI to the target variable, thereby drastically reducing the set while preserving predictive power (Silva et al., 2025). In practice, one might compute MI or correlation of each descriptor with the output, then select the top *k* features. Alternatively, regularization methods (L1/L2 penalties) or model-based importance (such as tree feature importance) automatically focus on informative features during training. The goal is to keep a compact set of nonredundant, high-signal features for structural data. These often include physical interactions (e.g., hydrogen bonds, salt bridges), global size/shape metrics (surface area, volume), and key sequence motifs (Silva et al., 2025). The selected features can then be scaled/normalized (e.g., z-scores) and fed into ML models. In summary, feature

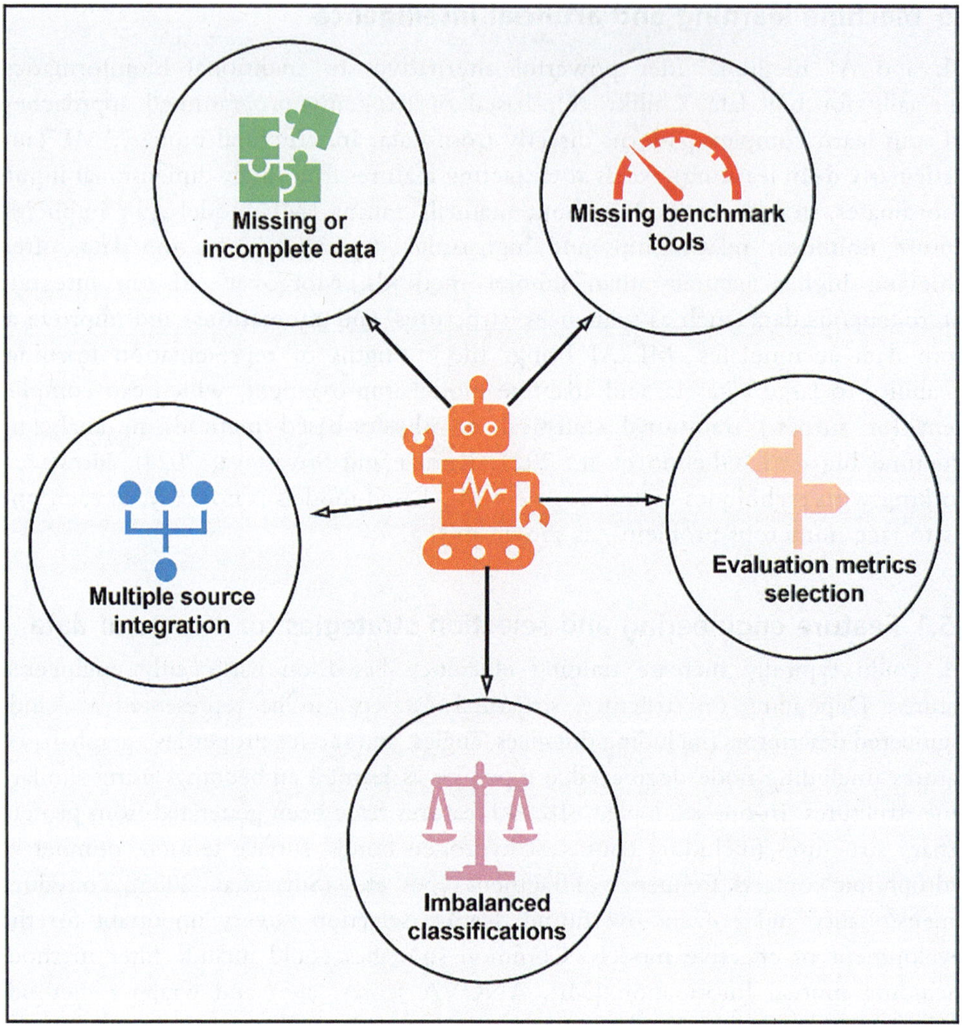

Figure 7.3 Problems working with machine learning.

engineering for structures combines domain-driven descriptors (geometry, physics, composition) with data-driven selection (MI scoring, PCA) to yield an efficient representation of the conformational data (Silva et al., 2025).

7.5.2 Supervised learning

In supervised learning, models are trained on labeled data to perform tasks like classification (predicting discrete classes) or regression (predicting continuous values). Structural biology applications include classifying protein folds or families, predicting enzyme function or ligand binding, and regressing binding affinities or stability.

Standard algorithms include random forests, support vector machines, gradient boosting, and feedforward neural networks (Kumar and Srivastava, 2024). For example, ensemble tree models (e.g., XGBoost or gradient boosting) have been applied to predict protein family membership from structural features, achieving high accuracy (F1-scores ~0.7–0.8 in a validation study) (Silva et al., 2025). Feedforward neural networks are also used; once a suitable feature vector is provided (sequence or structural features), a multilayer perceptron can learn complex, nonlinear decision boundaries (Tahmasebi and Hezarkhani, 2011; Balabin, Safieva and Lomakina, 2007). In regression tasks, supervised models map structure-derived features to numerical targets (binding energies, melting points), often using regularized linear models or kernel methods. Ensuring that model training occurs with cross-validation and hyperparameter tuning could help prevent overfitting because there are usually many features. Metrics such as accuracy, ROC-AUC or RMSE could be used to evaluate how well the model performed on held-out data. Supervised learning, thus, transforms structural descriptors into predictors of the system by using a large enough set of curated examples so that it can be used to automate the task of classifying a structure and/or creating a regression with quantified levels of accuracy, which is an alternative method to applying traditional computational methods (Kumar and Srivastava, 2024; Silva et al., 2025).

7.5.3 Unsupervised learning

Unsupervised learning analyses structural data that does not have any labels but contains many repetitions, making it ideal for use in the analysis of MD ensembles or of conformational libraries. Data analysis techniques would often make use of two key techniques, dimensionality reduction and clustering. The first decision required for dimensionality reduction and clustering is to choose numeric representations (features) of each conformation (e.g., principal torsion angles, pairwise distance, graph embeddings, etc.) (Conev et al., 2023). Linear techniques such as PCA and time-lagged ICA (TICA) would take the above features and project them into fewer dimensions where the largest amount of variance or slowest amount of dynamics occur in that space. Nonlinear methods (UMAP, diffusion maps, variational autoencoders) can capture more complex manifolds. For instance, the EnGens pipeline applies both PCA/TICA and nonlinear UMAP/SRV to embed static or trajectory data (Chen, Sidky and Ferguson, 2019; Schwantes and Pande, 2015; Trozzi, Wang and Tao, 2021). Once embedded, clustering algorithms (such as hierarchical agglomerative clustering, k-means, Gaussian mixture models, or density-based methods) partition the ensemble into distinct states. EnGens, for example, then extracts a representative structure from each cluster to form a reduced "conformational ensemble" that summarizes the data (Conev et al., 2023).

Cluster algorithms group similar conformations based on a distance metric (e.g., root mean square deviation or feature-space distance) (Glielmo et al., 2021). The choice of clustering scheme can be guided by expected biology: k-means finds compact, spherical clusters, hierarchical clustering builds a tree of states, and density-based methods find arbitrary-shaped basins (Glielmo et al., 2021). In all cases, the result is an unsupervised grouping of structures into states that can be analyzed (e.g., representative geometry, population). These methods effectively reduce high-dimensional structural data to a few canonical conformers. As a broad justification, unsupervised learning is essential for making sense of big simulation datasets: although the raw trajectory may contain millions of coordinates, its essential behavior often lies in a low-dimensional manifold. By combining dimensionality reduction with clustering, one can thus reveal central conformational states in an automatic, data-driven way.

7.5.4 Deep learning approaches

Deep neural networks have revolutionized structural analysis by automatically learning hierarchical features. 3D convolutional neural networks (3D CNNs) apply volumetric convolution kernels to grid representations of molecules (e.g., 3D voxelized electron density or atom density maps) (Wen, Huang and Guo, 2021). This captures local spatial patterns: analogous to image CNNs but in three dimensions. 3D CNNs have been used for tasks such as binding-site classification or ligand docking by learning from numerous examples of how the spatial patterns of atoms correlate with function. Graph Neural Networks (GNNs) operate on graph representations of structures (atoms or residues as nodes, bonds or contacts as edges) and learn node embeddings through message passing. GNNs naturally handle the irregular connectivity of molecular structures. They have achieved great success in chemistry and biology: by treating a protein as a graph of residues or a ligand as a graph of atoms, GNNs can predict properties (e.g., binding affinity, function) directly from raw graphs (Veličković, 2023). Indeed, GNNs have "powered significant recent advances" in molecular modeling, with architectures such as graph convolutions, graph attention networks, and geometric equivariant networks (Veličković, 2023).

Transformer architectures (self-attention models) have also been adapted for structural data. Initially developed for sequences, transformers can model 3D data by attending to all pairs of elements; for instance, graph transformers incorporate attention on edges to capture long-range dependencies. Recent work shows that GNNs themselves can be viewed as a generalization of transformers in graph form (Veličković, 2023). The key advantage is that attention-based models can globally contextualize each atom or residue based on the whole structure. Overall, deep learning frees us from manual feature design: convolutional filters and learned weights automatically extract multiscale structural motifs, while attention mechanisms emphasize the most relevant parts of a molecule. These models, however, require large training sets and benefit from GPU acceleration.

7.5.5 Semisupervised and transfer learning

In many structural problems, labeled data are scarce (e.g., experimentally measured binding affinities or functional annotations), but unlabeled structures are plentiful (Pallikkavaliyaveetil and Chandrasekaran, 2026). Semisupervised learning leverages this by training on a mixture of labeled and unlabeled data. Approaches include self-training (e.g., pseudo-labeling), cotraining on multiple views, and graph-based label propagation (Phan, Brouard and Mourad, 2024). These enable a model to learn general structural features from unlabeled examples and fine-tune on the limited labeled data. Transfer learning is also key: models pretrained on one domain (e.g., protein sequences or known structures) can be adapted to a related task (Heinzinger et al., 2024). For instance, a convolutional network trained on thousands of enzyme structures could be retrained (or used as a feature extractor) to predict a different property on a smaller dataset (Vinod, Chen and Das, 2025). In practice, one might use a large protein dataset to learn embeddings (as in protein language models or graph encoders) and then fine-tune them on a specific structural prediction task. Both strategies aim to overcome data paucity by exploiting related information.

7.5.6 Model evaluation, interpretability, and uncertainty quantification

For trustworthiness of ML models, rigorous evaluation and interpretation of the model are key, and the common evaluation method is cross-validation with held-out data and utilizing different metrics (accuracy for classification, RMSE/R^2 for regression) (Silva et al., 2025). Classification models will also report out various performance matrices (i.e., F1, precision, recall) across the held-out test set. For example, the F1 score is a common way to evaluate the performance of classification models (Silva et al., 2025). Analysis methods (SHAP, LIME, attention visualizations) are utilized to interpret the desired results from the model based on structural features associated with the data used to train the model. For example, SHAP analysis on InteracTor revealed that there were many intramolecular hydrogen bonds, and dispersion forces and surface hydrophobicity were considered to be among the main contributing features for driving the classifications of protein families (Silva et al., 2025). Additionally, feature attribution allows us to identify which method of structural elements (i.e., specific chemical bonds and/or motifs) principally affect the model's predicted outcome.

Quantifying the uncertainty in ML predictions is also critical. Uncertainty quantification can be used with ML predictions, as well as confidence intervals from Bayesian neural networks, dropout approximation methods, or ensemble models. In structural models, the uncertainty of data can come from insufficient data and/or model bias. The recent literature review says that uncertainty is one of the biggest issues when it comes to model reliability and interpretability. Some of the other main categories of methods used in the literature review for quantify and mitigate uncertainty are Bayesian inference and ensemble methods (Balsa-Canto et al., 2025). There are some specific ways that Monte Carlo dropout and bootstrap ensembles are used to calculate prediction intervals and propagate uncertainty throughout a structural ML model. Documenting uncertainty in a structural ML model and creating calibration plots or using the likelihood score to provide a measure of trustworthiness in the structural ML model will help to make sure that structural ML tools will provide both a prediction and a measure of confidence in the prediction. Using a combination of cross-validated performance metrics along with an interpretability analysis of ML models, as well as estimating the uncertainty of a model, is important for creating reliable and trustworthy ML structures in structural biology (Silva et al., 2025; Balsa-Canto et al., 2025).

7.6 Applications of big data analytics

7.6.1 Structural prediction and modeling

Predicting 3D structures from amino acid sequences and producing proteome-scale models have become practical through the utilization of large biological databases and novel data-driven algorithms. Modern structural predictors rely not only on neural-network architectures but also critically on the scale and diversity of training and auxiliary data, including manually curated experimental coordinates, massive sequence collections, metagenomic data, and predicted structures, which together supply the statistical signal that these models learn to exploit (Jumper et al., 2021; Tunyasuvunakool et al., 2021). AlphaFold exemplifies how large-scale data engineering converts raw biological records into dependable models. First, supervised learning utilized high-quality, experimentally solved structures from the PDB as ground truth, providing atomic coordinates and validated annotations that established the core mapping from sequence to structure. Second, evolutionary information derived from very large multiple-sequence alignments constructed from comprehensive sequence repositories provided coevolutionary constraints that indicate residue contacts and folding tendencies. Extracting those constraints at scale amplified the signal for the learning algorithms. Third, the team applied a semisupervised augmentation strategy after initial training, the model predicted structures for hundreds of thousands of protein sequences drawn from clustered sequence databases and then used a filtered, high-confidence subset of those predictions to further train the network (self-distillation), thereby expanding effective training data beyond experimentally solved cases and improving generalization to sequences with few homologs. Fourth, the systematic engineering of metadata, template retrieval pipelines, and per-model confidence scoring enabled the identification and filtering of lower-quality inputs and outputs, thereby increasing the practical reliability for end users. These combined data strategies, which included experimental labels, massive evolutionary alignments, self-distilled synthetic labels, and reliable quality metrics, explain much of AlphaFold's usefulness in diverse targets (Jumper et al., 2021; Tunyasuvunakool et al., 2021; Mirdita et al., 2017). Other recent systems demonstrate alternative ways that big data enables structure prediction. RoseTTAFold introduced three-track network architectures that integrate sequence, pairwise, and structural signals, benefiting from publicly available structural and sequence repositories to achieve competitive accuracy while remaining computationally accessible. Transformer-based protein language models and derived predictors (for example, ESMFold, built on large protein language models and metagenomic atlases) demonstrate that representations learned from billions of unlabeled sequences can directly provide structural priors, thereby reducing the reliance on explicit multiple-sequence alignment depth for specific proteins. By scaling to hundreds of millions to billions of sequences and predictions, these language-model approaches create massive atlases of putative structures that both broaden coverage of the protein universe and serve as new training or benchmarking corpora (Mirdita et al., 2017; Lin et al., 2023). The downstream value of these large datasets is already evident in proteome-wide databases of predicted models, which enable global clustering and structural annotation at scales previously impossible, revealing previously unannotated folds and guiding focused experimental work. Scalable clustering algorithms applied to hundreds of millions of predicted structures, identifying recurrent folds and novelty at the structural level. The public release of extensive predicted-structure collections closes a virtuous loop, where predicted structures become data that improve tools for annotation, homology detection, and even future predictors, thereby reinforcing the accumulation of data and model accuracy (Varadi et al., 2022; Jumper et al., 2021; Lin et al., 2023; Barrio-Hernandez et al., 2023; Varadi et al., 2024).

7.6.2 Disease network integration

Network medicine uses big data to connect genotype/structure with phenotype. Disease networks overlay genetic and interaction data, allowing "mutations in only a few genes may propagate to other components in the network," thereby forming a disease module (Sonawane et al., 2019). Such analyses reveal shared pathways and novel targets underlying complex diseases.

7.6.3 Emerging areas

Integrative structural biology combines diverse data sources (X-ray, NMR, cryo-EM, crosslinking MS, etc.) to compute the architectures of large assemblies (Sali, 2021). In cell structural biology aim remains to map macromolecules in native perspectives using cryo-electron tomography (cryo-ET). These efforts depend on big data amalgamation, such as merging cryo-ET, crosslinking/MS, and ML analyses, which requires huge efforts in data curation and then its standardization for reliability (Kyrilis et al., 2024).

7.7 Conclusions

The term "big data analytics" is commonly used to refer to a key element of bioinformatics and structural biology research (i.e., a huge amount of data), which helps researchers analyze their study results in ways that enable better use of their findings. As big data continues to increase exponentially, this topic continues to garner attention and will require continued investment in data infrastructure, standardization, and curation to realize its true potential. High-quality, standardized data and collaborations across such disciplines can influence big data analytics in significant ways. The rapid pace at which ML, AI, and other technologies for DL are improving makes these processes increasingly efficient. Another key component of the big data analytics process is resolving the issue of data heterogeneity, which is created by the existence of many different research networked sources in bioinformatics. Putting all of these disparate types of complex information together in an appropriate and uniform interface requires multiple techniques such as format and metadata standardization, schema mapping, ontology alignment, data fusions, ML-based harmonization, middleware and workflow platforms. By employing a combination of different practical techniques/strategies, researchers can integrate disparate/multiple formats for datasets, creating continuous sets of data that can be analyzed at scale (i.e., easily migrate from large-scale computing to smaller-scale computing) and allow researchers to make discoveries much faster than with traditional methods. As technology advances rapidly, researchers will generate more valuable data (i.e., data that is useful to society) than ever before and thus will provide the scientific community with the tools necessary to grasp and address the difficulty of biological macromolecule interactions (currently viewed as some of the hardest problems to solve).

References

Abdalla, H.B., 2022. A brief survey on big data: technologies, terminologies and data-intensive applications. Journal of Big Data 9 (1). Available at https://doi.org/10.1186/s40537-022-00659-3, https://journalofbigdata.springeropen.com.

Abueg, L.A.L., Afgan, E., Allart, O., Awan, A.H., Bacon, W.A., Baker, D., et al., 2024. The galaxy platform for accessible, reproducible, and collaborative data analyses: 2024 update. Nucleic Acids Research 52 (W1), W83. Available at https://doi.org/10.1093/nar/gkae410.

Adams, P.D., Afonine, P.V., Baskaran, K., Berman, H.M., Berrisford, J., Bricogne, G., et al., 2019. Announcing mandatory submission of PDBx/mmCIF format files for crystallographic depositions to the protein data bank (PDB). Acta Crystallographica Section D: Structural Biology 75, 451–454. Available at https://doi.org/10.1107/S2059798319004522, https://journals.iucr.org/d/.

Aittokallio, T., Schwikowski, B., 2006. Graph-based methods for analysing networks in cell biology. Briefings in Bioinformatics 7 (3), 243–255. Available at https://doi.org/10.1093/bib/bbl022.

Balabin, R.M., Safieva, R.Z., Lomakina, E.I., 2007. Comparison of linear and nonlinear calibration models based on near infrared (NIR) spectroscopy data for gasoline properties prediction. Chemometrics and Intelligent Laboratory Systems 88 (2), 183–188. Available at https://doi.org/10.1016/j.chemolab.2007.04.006.

Balsa-Canto, E., Campo-Manzanares, N., Moimenta, A.R., Roudaut, G., Troitiño-Jordedo, D., 2025. Quantifying and managing uncertainty in systems biology: mechanistic and data-driven models. Current Opinion in Systems Biology 42, 100557. Available at https://doi.org/10.1016/j.coisb.2025.100557.

Barrio-Hernandez, I., Yeo, J., Jänes, J., Mirdita, M., Gilchrist, C.L.M., Wein, T., et al., 2023. Clustering predicted structures at the scale of the known protein universe. Nature 622 (7983), 637–645. Available at https://doi.org/10.1038/s41586-023-06510-w, https://www.nature.com/nature/.

Bateman, A., Martin, M.-J., Orchard, S., Magrane, M., Ahmad, S., Alpi, E., et al., 2023. Jian Zhang, uniprot: the universal protein knowledgebase in 2023. Nucleic Acids Research 51 (D1), D523. Available at https://doi.org/10.1093/nar/gkac1052.

Berger, B., Daniels, N.M., Yu, Y.W., 2016. Computational biology in the 21st century: scaling with compressive algorithms. Communications of the ACM 59 (8), 72–80. Available at https://doi.org/10.1145/2957324, http://www.acm.org/pubs/cacm/.

Berman, H.M., Burley, S.K., 2025. Protein data bank (PDB): fifty-three years young and having a transformative impact on science and society. Quarterly Reviews of Biophysics 58. Available at https://doi.org/10.1017/S0033583525000034, http://journals.cambridge.org/action/displayJournal?jid=QRB.

Bradley, A.R., Rose, A.S., Pavelka, A., Valasatava, Y., Duarte, J.M., Prlić, A., et al., 2017. MMTF—An efficient file format for the transmission, visualization, and analysis of macromolecular structures. PLoS Computational Biology 13 (6). Available at https://doi.org/10.1371/journal.pcbi.1005575, http://www.ploscompbiol.org/article/browseVolume.action?field=volume.

Burley, S.K., Berman, H.M., Duarte, J.M., Feng, Z., Flatt, J.W., Hudson, B.P., et al., 2022. Protein data bank: a comprehensive review of 3D structure holdings and worldwide utilization by researchers, educators, and students. Biomolecules 12 (10). Available at https://doi.org/10.3390/biom12101425, http://www.mdpi.com/journal/biomolecules.

Chakrabarty, B., Parekh, N., 2016. NAPS: network analysis of protein structures. Nucleic Acids Research 44 (W1), W375. Available at https://doi.org/10.1093/nar/gkw383.

Chen, C., McGarvey, P.B., Huang, H., Wu, C.H., 2010. Protein bioinformatics infrastructure for the integration and analysis of multiple high-throughput omics data. Advances in Bioinformatics 2010. Available at https://doi.org/10.1155/2010/423589.

Chen, W., Sidky, H., Ferguson, A.L., 2019. Nonlinear discovery of slow molecular modes using state-free reversible VAMPnets. Journal of Chemical Physics 150 (21). Available at https://doi.org/10.1063/1.5092521, http://scitation.aip.org/content/aip/journal/jcp.

Chen, W., Görlich, F., Fuchs, P., Zavadlav, J., 2026. Enhanced sampling for efficient learning of coarse-grained machine learning potentials. Journal of Chemical Theory and Computation 22 (1), 219–230. Available at https://doi.org/10.1021/acs.jctc.5c01712.

Cheng, A., Henderson, R., Mastronarde, D., Ludtke, S.J., Schoenmakers, R.H.M., Short, J., et al., 2015. MRC2014: extensions to the MRC format header for electron cryo-microscopy and tomography. Journal of Structural Biology 192 (2), 146–150. Available at https://doi.org/10.1016/j.jsb.2015.04.002, http://www.elsevier.com/inca/publications/store/6/2/2/9/0/0/index.htt.

Cho, D.Y., Kim, Y.A., Przytycka, T.M., 2012. Chapter 5: network biology approach to complex diseases. PLoS Computational Biology 8 (12). Available at https://doi.org/10.1371/journal.pcbi.1002820, http://www.ploscompbiol.org/article/fetchObjectAttachment.action?uri=info%3Adoi%2F10.1371%2Fjournal.pcbi.1002820&representation=PDF.

Choudhary, P., Feng, Z., Berrisford, J., Chao, H., Ikegawa, Y., Peisach, E., et al., 2024. PDB NextGen Archive: centralizing access to integrated annotations and enriched structural information by the Worldwide Protein Data Bank. Database 2024. Available at https://doi.org/10.1093/database/baae041, http://database.oxfordjournals.org/.

Cock, P.J.A., Fields, C.J., Goto, N., Heuer, M.L., Rice, P.M., 2009. The Sanger FASTQ file format for sequences with quality scores, and the Solexa/Illumina FASTQ variants. Nucleic Acids Research 38 (6), 1767–1771. Available at https://doi.org/10.1093/nar/gkp1137.

Conev, A., Rigo, M.M., Devaurs, D., Fonseca, A.F., Kalavadwala, H., de Freitas, M.V., et al., 2023. EnGens: a computational framework for generation and analysis of representative protein conformational ensembles. Briefings in Bioinformatics 24 (4). Available at https://doi.org/10.1093/bib/bbad242, http://bib.oxfordjournals.org.

Cowen, L., Ideker, T., Raphael, B.J., Sharan, R., 2017. Network propagation: a universal amplifier of genetic associations. Nature Reviews. Genetics 18 (9), 551–562. Available at https://doi.org/10.1038/nrg.2017.38, http://www.nature.com/reviews/genetics.

Creixell, P., Reimand, J., Haider, S., Wu, G., Shibata, T., Vazquez, M., et al., 2015. Pathway and network analysis of cancer genomes. Nature Methods 12 (7), 615–621. Available at https://doi.org/10.1038/nmeth.3440, http://www.nature.com/nmeth/.

Crowther, R.A., Henderson, R., Smith, J.M., 1996. MRC image processing programs. Journal of Structural Biology 116 (1), 9–16. Available at https://doi.org/10.1006/jsbi.1996.0003.

Daberdaku, S., Ferrari, C., 2018. Exploring the potential of 3D Zernike descriptors and SVM for protein-protein interface prediction. BMC Bioinformatics 19 (1). Available at https://doi.org/10.1186/s12859-018-2043-3, http://www.biomedcentral.com/bmcbioinformatics/.

Danecek, P., Auton, A., Abecasis, G., Albers, C.A., Banks, E., DePristo, M.A., et al., 2011. The variant call format and VCFtools. Bioinformatics (Oxford, England) 27 (15), 2156–2158. Available at https://doi.org/10.1093/bioinformatics/btr330.

Davis, G., Rew, R., 1990. Data Management: NetCDF: An Interface for Scientific Data Access. IEEE Computer Society, pp. 76–82. Available at http://doi.org/10.1109/38.56302.

Fillinger, S., de la Garza, L., Peltzer, A., Kohlbacher, O., Nahnsen, S., 2019. Challenges of big data integration in the life sciences. Analytical and Bioanalytical Chemistry 411 (26), 6791–6800. Available at https://doi.org/10.1007/s00216-019-02074-9, link.springer.de/link/service/journals/00216/index.htm.

Folk, M., Heber, G., Koziol, Q., Pourmal, E., Robinson, D., 2011. An overview of the HDF5 technology suite and its applications. In: Proceedings of the EDBT/ICDT 2011 Workshop on Array Databases. Association for Computing Machinery, Uppsala, Sweden, pp. 36–47. https://doi.org/10.1145/1966895.1966900

Galetsi, P., Katsaliaki, K., Kumar, S., 2019. Values, challenges and future directions of big data analytics in healthcare: a systematic review. Social Science & Medicine 241, 112533. Available at https://doi.org/10.1016/j.socscimed.2019.112533.

Gharajeh, M.S., 2018. Chapter eight - biological big data analytics. In: Raj, P., Deka, G.C. (Eds.), Advances in Computers. Elsevier, pp. 321–355. Available at https://doi.org/10.1016/bs.adcom.2017.08.002.

Gillespie, M., Jassal, B., Stephan, R., Milacic, M., Rothfels, K., Senff-Ribeiro, A., et al., 2022. The reactome pathway knowledgebase 2022. Nucleic Acids Research 50 (D1), D687. Available at https://doi.org/10.1093/nar/gkab1028.

van Ginkel, G., Pravda, L., Dana, J.M., Varadi, M., Keller, P., Anyango, S., et al., 2021. PDBeCIF: an open-source mmCIF/CIF parsing and processing package. BMC Bioinformatics 22 (1). Available at https://doi.org/10.1186/s12859-021-04271-9, http://www.biomedcentral.com/bmcbioinformatics/.

Glielmo, A., Husic, B.E., Rodriguez, A., Clementi, C., Noé, F., Laio, A., 2021. Unsupervised learning methods for molecular simulation data. Chemical Reviews 121 (16), 9722–9758. Available at https://doi.org/10.1021/acs.chemrev.0c01195, http://pubs.acs.org/journal/chreay.

Greene, C.S., Tan, J., Ung, M., Moore, J.H., Cheng, C., 2014. Big data bioinformatics. Journal of Cellular Physiology 229 (12), 1896–1900. Available at https://doi.org/10.1002/jcp.24662, http://onlinelibrary.wiley.com/journal/10.1002/(ISSN)1097-4652.

Greenfield, D., Wittorff, V., Hultner, M., 2019. The importance of data compression in the field of genomics. IEEE Pulse 10 (2), 20–23. Available at https://doi.org/10.1109/MPULS.2019.2899747, http://pulse.embs.org/pastissues.html.

Guo, J., Yu, H., Xing, S., Huan, T., 2022. Addressing big data challenges in mass spectrometry-based metabolomics. Chemical Communications 58 (72), 9979–9990. Available at https://doi.org/10.1039/d2cc03598g, http://pubs.rsc.org/en/journals/journal/cc.

Gupta, A., Kumar, S., Kumar, A., 2024. Big data in bioinformatics and computational biology: basic insights. Methods in Molecular Biology 2719, 153–166. Available at https://doi.org/10.1007/978-1-0716-3461-5_9, https://www.springer.com/series/7651.

Han, H., Liu, X., 2022. The challenges of explainable AI in biomedical data science. BMC Bioinformatics 22. Available at https://doi.org/10.1186/s12859-021-04368-1, http://www.biomedcentral.com/bmcbioinformatics/.

Han, X., Sit, A., Christoffer, C., Chen, S., Kihara, D., Kolodny, R., 2019. A global map of the protein shape universe. PLoS Computational Biology 15 (4), e1006969. Available at https://doi.org/10.1371/journal.pcbi.1006969.

Hanson, R.M., 2010. Jmol – a paradigm shift in crystallographic visualization. Journal of Applied Crystallography 43 (5), 1250–1260. Available at https://doi.org/10.1107/s0021889810030256.

He, K.Y., Ge, D., He, M.M., 2017. Big data analytics for genomic medicine. International Journal of Molecular Sciences 18 (2). Available at https://doi.org/10.3390/ijms18020412, http://www.mdpi.com/1422-0067/18/2/412/pdf.

Heinzinger, M., Weisseno w, K., Sanchez, J.G., Henkel, A., Mirdita, M., Steinegger, M., et al., 2024. Bilingual language model for protein sequence and structure. NAR Genomics and Bioinformatics 6 (4). Available at https://doi.org/10.1093/nargab/lqae150, https://academic.oup.com/nargab.

Herre, C., Ho, A., Eisenbraun, B., Vincent, J., Nicholson, T., Boutsioukis, G., et al., 2024. Introduction of the Capsules environment to support further growth of the SBGrid structural biology software collection. Acta Crystallographica Section D: Structural Biology 80 (6), 439–450. Available at https://doi.org/10.1107/S2059798324004881, https://journals.iucr.org/d/.

Hildebrand, P.W., Rose, A.S., Tiemann, J.K.S., 2019. Bringing molecular dynamics simulation data into view. Trends in Biochemical Sciences 44 (11), 902–913. Available at https://doi.org/10.1016/j.tibs.2019.06.004, www.elsevier.com/locate/tibs.

Hoch, J.C., Baskaran, K., Burr, H., Chin, J., Eghbalnia, H.R., Fujiwara, T., et al., 2023. Biological magnetic resonance data bank. Nucleic Acids Research 51 (1), D368. Available at https://doi.org/10.1093/nar/gkac1050, https://academic.oup.com/nar/issue.

Holzinger, A., Dehmer, M., Jurisica, I., 2014. Knowledge discovery and interactive data mining in bioinformatics - state-of-the-art, future challenges and research directions. BMC Bioinformatics 15 (6). Available at https://doi.org/10.1186/1471-2105-15-S6-I1, http://www.biomedcentral.com/bmcbioinformatics/.

Hong, X., Zheng, J., Xie, J., Tong, X., Liu, X., Song, Q., Sen, et al., 2021. RR3DD: an RNA global structure-based RNA three-dimensional structural classification database. RNA Biology 18 (sup2), 738–746. Available at https://doi.org/10.1080/15476286.2021.1989200.

Hrabe, T., Godzik, A., 2014. ConSole: using modularity of contact maps to locate solenoid domains in protein structures. BMC Bioinformatics 15 (1). Available at https://doi.org/10.1186/1471-2105-15-119, http://www.biomedcentral.com/1471-2105/15/119.

Humphrey, W., Dalke, A., Schulten, K., 1996. VMD: visual molecular dynamics. Journal of Molecular Graphics 14 (1), 33–38. Available at https://doi.org/10.1016/0263-7855(96)00018-5.

Iudin, A., Korir, P.K., Somasundharam, S., Weyand, S., Cattavitello, C., Fonseca, N., et al., 2023. EMPIAR: the electron microscopy public image archive. Nucleic Acids Research 51 (1), D1503. Available at https://doi.org/10.1093/nar/gkac1062, https://academic.oup.com/nar/issue.

Jonic, S., Miyashita, O., Callebaut, I., 2018. Editorial: hybrid biomolecular modeling. Frontiers in Molecular Biosciences 5. Available at https://doi.org/10.3389/fmolb.2018.00098, https://www.frontiersin.org/articles/10.3389/fmolb.2018.00098/full.

Jumper, J., Evans, R., Pritzel, A., Green, T., Figurnov, M., Ronneberger, O., et al., 2021. Highly accurate protein structure prediction with AlphaFold. Nature 596 (7873), 583–589. Available at https://doi.org/10.1038/s41586-021-03819-2, http://www.nature.com/nature/index.html.

Kämpf, C., Specht, M., Scholz, A., Puppel, S.H., Doose, G., Reiche, K., et al., 2019. Uap: reproducible and robust HTS data analysis. BMC Bioinformatics 20 (1). Available at https://doi.org/10.1186/s12859-019-3219-1, http://www.biomedcentral.com/bmcbioinformatics/.

Katayama, T., Arakawa, K., Nakao, M., Ono, K., Aoki-Kinoshita, K.F., Yamamoto, Y., et al., 2010. The DBCLS biohackathon: standardization and interoperability for bioinformatics web services and workflows. Journal of Biomedical Semantics 1. Available at https://doi.org/10.1186/2041-1480-1-8, http://www.jbiomedsem.com/content/1/1/8.

Kent, W.J., Zweig, A.S., Barber, G., Hinrichs, A.S., Karolchik, D., 2010. BigWig and BigBed: enabling browsing of large distributed datasets. Bioinformatics (Oxford, England) 26 (17), 2204–2207. Available at https://doi.org/10.1093/bioinformatics/btq351.

Kent, W.J., Sugnet, C.W., Furey, T.S., Roskin, K.M., Pringle, T.H., Zahler, A.M., et al., 2002. The human genome browser at UCSC. Genome Research 12 (6), 996–1006. Available at https://doi.org/10.1101/gr.229102, http://www.genome.org/.

Koehler Leman, J., Lyskov, S., Lewis, S.M., Adolf-Bryfogle, J., Alford, R.F., Barlow, K., et al., 2021. Ensuring scientific reproducibility in bio-macromolecular modeling via extensive, automated benchmarks. Nature Communications 12 (1). Available at https://doi.org/10.1038/s41467-021-27222-7, https://www.nature.com/ncomms/.

Kumar, N., Srivastava, R., 2024. Deep learning in structural bioinformatics: current applications and future perspectives. Briefings in Bioinformatics 25 (3). Available at https://doi.org/10.1093/bib/bbae042, http://bib.oxfordjournals.org.

Kundrotas, P.J., Anishchenko, I., Dauzhenka, T., Kotthoff, I., Mnevets, D., Copeland, M.M., et al., 2018. Dockground: a comprehensive data resource for modeling of protein complexes. Protein Science 27 (1), 172–181. Available at https://doi.org/10.1002/pro.3295, http://onlinelibrary.wiley.com/journal/10.1002/(ISSN)1469-896X.

Kyrilis, F.L., Low, J.K.K., Mackay, J.P., Kastritis, P.L., 2024. Structural biology in cellulo: minding the gap between conceptualization and realization. Current Opinion in Structural Biology 87. Available at https://doi.org/10.1016/j.sbi.2024.102843, https://www.sciencedirect.com/science/journal/0959440X.

Lee, J., Cheng, X., Swails, J.M., Yeom, M.S., Eastman, P.K., Lemkul, J.A., et al., 2016. CHARMM-GUI input generator for NAMD, GROMACS, AMBER, OpenMM, and CHARMM/OpenMM simulations using the CHARMM36 additive force field. Journal of Chemical Theory and Computation 12 (1), 405–413. Available at https://doi.org/10.1021/acs.jctc.5b00935, http://pubs.acs.org/journal/jctcce.

Li, H., Handsaker, B., Wysoker, A., Fennell, T., Ruan, J., Homer, N., et al., 2009. The sequence alignment/map format and SAMtools. Bioinformatics (Oxford, England) 25 (16), 2078–2079. Available at https://doi.org/10.1093/bioinformatics/btp352.

Li, Q., Hu, Z., Wang, Y., Li, L., Fan, Y., King, I., et al., 2024. Progress and opportunities of foundation models in bioinformatics. Briefings in Bioinformatics 25 (6). Available at https://doi.org/10.1093/bib/bbae548, http://bib.oxfordjournals.org.

Li, Y., Chen, L., 2014. Big biological data: challenges and opportunities. Genomics, Proteomics and Bioinformatics 12 (5), 187–189. Available at https://doi.org/10.1016/j.gpb.2014.10.001, http://www.elsevier.com/wps/find/journaldescription.cws_home/707637/description#description.

Lin, Z., Akin, H., Rao, R., Hie, B., Zhu, Z., Lu, W., et al., 2023. Evolutionary-scale prediction of atomic-level protein structure with a language model. Science (New York, N.Y.) 379 (6637), 1123–1130. Available at https://doi.org/10.1126/science.ade2574, https://www.science.org/doi/10.1126/science.ade2574.

Liu, T., Hwang, L., Burley, S.K., Nitsche, C.I., Southan, C., Walters, W.P., et al., 2025. BindingDB in 2024: a FAIR knowledgebase of protein-small molecule binding data. Nucleic Acids Research 53 (1), D1633. Available at https://doi.org/10.1093/nar/gkae1075, https://academic.oup.com/nar/issue.

Masseroli, M., Mons, B., Bongcam-Rudloff, E., Ceri, S., Kel, A., Rechenmann, F., et al., 2014. Integrated bio-search: challenges and trends for the integration, search and comprehensive processing of biological information. BMC Bioinformatics 15. Available at https://doi.org/10.1186/1471-2105-15-S1-S2, http://www.biomedcentral.com/bmcbioinformatics/.

Mathivanan, S., Periaswamy, B., Gandhi, T.K.B., Kandasamy, K., Suresh, S., Mohmood, R., et al., 2006. An evaluation of human protein-protein interaction data in the public domain. BMC Bioinformatics 7 (5). Available at https://doi.org/10.1186/1471-2105-7-S5-S19.

McGibbon, R.T., Beauchamp, K.A., Harrigan, M.P., Klein, C., Swails, J.M., Hernández, C.X., et al., 2015. MDTraj: a modern open library for the analysis of molecular dynamics trajectories. Biophysical Journal 109 (8), 1528–1532. Available at https://doi.org/10.1016/j.bpj.2015.08.015, http://www.sciencedirect.com/science/journal/00063495.

Milano, M., Agapito, G., Cannataro, M., 2022. Challenges and limitations of biological network analysis. BioTech 11 (3). Available at https://doi.org/10.3390/biotech11030024, https://www.mdpi.com/2673-6284/11/3/24/pdf?version=1657268083.

Mirdita, M., von den Driesch, L., Galiez, C., Martin, M.J., Söding, J., Martin Steinegger, 2017. Uniclust databases of clustered and deeply annotated protein sequences and alignments. Nucleic Acids Research 45 (D1), D170. Available at https://doi.org/10.1093/nar/gkw1081.

Mu, X., Gillman, C., Nguyen, C., Gonen, T., 2021. An overview of microcrystal electron diffraction (MicroED). Annual Review of Biochemistry 90, 431–450. Available at https://doi.org/10.1146/annurev-biochem-081720-020121, http://www.annualreviews.org/journal/biochem.

Muir, P., Li, S., Lou, S., Wang, D., Spakowicz, D.J., Salichos, L., et al., 2016. The real cost of sequencing: scaling computation to keep pace with data generation. Genome Biology 17 (1). Available at https://doi.org/10.1186/s13059-016-0917-0, http://genomebiology.com/.

Niu, Y.N., Roberts, E.G., Denisko, D., Hoffman, M.M., 2022. Assessing and assuring interoperability of a genomics file format. Bioinformatics (Oxford, England) 38 (13), 3327–3336. Available at https://doi.org/10.1093/bioinformatics/btac327, http://bioinformatics.oxfordjournals.org/.

Nurk, S., Koren, S., Rhie, A., Rautiainen, M., Bzikadze, A.V., Mikheenko, A., et al., 2022. The complete sequence of a human genome. Science (New York, N.Y.) 376 (6588), 44–53. Available at https://doi.org/10.1126/science.abj6987, https://www.science.org/doi/10.1126/science.abj6987.

O'Connor, L.M., O'Connor, B.A., Lim, S.B., Zeng, J., Lo, C.H., 2023. Integrative multi-omics and systems bioinformatics in translational neuroscience: a data mining perspective. Journal of Pharmaceutical Analysis 13 (8), 836–850. Available at https://doi.org/10.1016/j.jpha.2023.06.011, http://www.journals.elsevier.com/journal-of-pharmaceutical-analysis.

Ooi, H.S., Schneider, G., Chan, Y.L., Lim, T.T., Eisenhaber, B., Eisenhaber, F., 2010. Databases of protein-protein interactions and complexes. Methods in molecular biology (Clifton, N.J.) 609, 145–159. Available at https://doi.org/10.1007/978-1-60327-241-4_9.

Oughtred, R., Rust, J., Chang, C., Breitkreutz, B.J., Stark, C., Willems, A., et al., 2021. The BioGRID database: a comprehensive biomedical resource of curated protein, genetic, and chemical interactions. Protein Science 30 (1), 187–200. Available at https://doi.org/10.1002/pro.3978, http://onlinelibrary.wiley.com/journal/10.1002/(ISSN)1469-896X.

O'Connell, K.A., Yosufzai, Z.B., Campbell, R.A., Lobb, C.J., Engelken, H.T., Gorrell, L.M., et al., 2023. Accelerating genomic workflows using NVIDIA Parabricks. BMC Bioinformatics 24 (1). Available at https://doi.org/10.1186/s12859-023-05292-2, https://bmcbioinformatics.biomedcentral.com/.

Pallikkavaliyaveetil, N., Chandrasekaran, S., 2026. Small data, big challenges: machine- and deep-learning strategies for data-limited drug discovery. Advanced Drug Delivery Reviews 229, 115762. Available at https://doi.org/10.1016/j.addr.2025.115762.

Pearson, W.R., Lipman, D.J., 1988. Improved tools for biological sequence comparison. Proceedings of the National Academy of Sciences of the United States of America 85 (8), 2444–2448. Available at https://doi.org/10.1073/pnas.85.8.2444.

Phan, H., Brouard, C., Mourad, R., 2024. Semi-supervised learning with pseudo-labeling compares favorably with large language models for regulatory sequence prediction. Briefings in Bioinformatics 25 (6). Available at https://doi.org/10.1093/bib/bbae560, http://bib.oxfordjournals.org.

Philippe, N., Salson, M., Lecroq, T., Léonard, M., Commes, T., Rivals, E., 2011. Querying large read collections in main memory: a versatile data structure. BMC Bioinformatics 12. Available at https://doi.org/10.1186/1471-2105-12-242France, http://www.biomedcentral.com/1471-2105/12/242.

Poger, D., Yen, L., Braet, F., 2023. Big data in contemporary electron microscopy: challenges and opportunities in data transfer, compute and management. Histochemistry and Cell Biology 160 (3),

169–192. Available at https://doi.org/10.1007/s00418-023-02191-8, https://www.springer.com/journal/418.
Pongor, S., Landsman, D., 1999. Bioinformatics and the developing world. Biotechnology and Development Monitor (40), 10–13.
Quinlan, A.R., Hall, I.M., 2010. BEDTools: a flexible suite of utilities for comparing genomic features. Bioinformatics (Oxford, England) 26 (6), 841–842. Available at https://doi.org/10.1093/bioinformatics/btq033.
Rego, N., Koes, D., 2015. 3Dmol.js: molecular visualization with WebGL. Bioinformatics (Oxford, England) 31 (8), 1322–1324. Available at https://doi.org/10.1093/bioinformatics/btu829, http://bioinformatics.oxfordjournals.org/.
Reimand, J., Isserlin, R., Voisin, V., Kucera, M., Tannus-Lopes, C., Rostamianfar, A., et al., 2019. Pathway enrichment analysis and visualization of omics data using g:profiler, GSEA, Cytoscape and EnrichmentMap. Nature Protocols 14 (2), 482–517. Available at https://doi.org/10.1038/s41596-018-0103-9, http://www.natureprotocols.com/.
Rickert, C.A., Lieleg, O., 2022. Machine learning approaches for biomolecular, biophysical, and biomaterials research. Biophysics Reviews 3 (2). Available at https://doi.org/10.1063/5.0082179, aip.scitation.org/journal/bpr.
Roe, D.R., Cheatham, T.E., 2013. PTRAJ and CPPTRAJ: software for processing and analysis of molecular dynamics trajectory data. Journal of Chemical Theory and Computation 9 (7), 3084–3095. Available at https://doi.org/10.1021/ct400341p.
Rose, A.S., Hildebrand, P.W., 2015. NGL viewer: a web application for molecular visualization. Nucleic Acids Research 43 (1), W576. Available at https://doi.org/10.1093/nar/gkv402, http://nar.oxfordjournals.org/.
Rose, A.S., Bradley, A.R., Valasatava, Y., Duarte, J.M., Prlic, A., Rose, P.W., 2018. NGL viewer: web-based molecular graphics for large complexes. Bioinformatics (Oxford, England) 34 (21), 3755–3758. Available at https://doi.org/10.1093/bioinformatics/bty419, http://bioinformatics.oxfordjournals.org/.
Rosenthal, P.B., Henderson, R., 2003. Optimal determination of particle orientation, absolute hand, and contrast loss in single-particle electron cryomicroscopy. Journal of Molecular Biology 333 (4), 721–745. Available at https://doi.org/10.1016/j.jmb.2003.07.013, https://www.journals.elsevier.com/journal-of-molecular-biology.
Sael, L., Li, B., La, D., Fang, Y., Ramani, K., Rustamov, R., et al., 2008. Fast protein tertiary structure retrieval based on global surface shape similarity. Proteins: Structure, Function and Genetics 72 (4), 1259–1273. Available at https://doi.org/10.1002/prot.22030.
Sali, A., 2021. From integrative structural biology to cell biology. Journal of Biological Chemistry 296, 100743. Available at https://doi.org/10.1016/j.jbc.2021.100743.
Saltzberg, D., Greenberg, C.H., Viswanath, S., Chemmama, I., Webb, B., Pellarin, R., et al., 2019. Modeling biological complexes using integrative modeling platform. Methods in Molecular Biology 2022, 353–377. Available at https://doi.org/10.1007/978-1-4939-9608-7_15, http://www.springer.com/series/7651.
Sapoval, N., Aghazadeh, A., Nute, M.G., Antunes, D.A., Balaji, A., Baraniuk, R., et al., 2022. Current progress and open challenges for applying deep learning across the biosciences. Nature Communications 13 (1). Available at https://doi.org/10.1038/s41467-022-29268-7, http://www.nature.com/ncomms/index.html.
Sayers, E.W., Bolton, E.E., Brister, J.R., Canese, K., Chan, J., Comeau, D.C., et al., 2023. Database resources of the national center for biotechnology information in 2023. Nucleic Acids Research 51 (1), D29. Available at https://doi.org/10.1093/nar/gkac1032, https://academic.oup.com/nar/issue.
Scherer, M.K., Trendelkamp-Schroer, B., Paul, F., Pérez-Hernández, G., Hoffmann, M., Plattner, N., et al., 2015. PyEMMA 2: a software package for estimation, validation, and analysis of markov models. Journal of Chemical Theory and Computation 11 (11), 5525–5542. Available at https://doi.org/10.1021/acs.jctc.5b00743, http://pubs.acs.org/journal/jctcce.
Schober, D., Jacob, D., Wilson, M., Cruz, J.A., Marcu, A., Grant, J.R., et al., 2018. NmrML: a community supported open data standard for the description, storage, and exchange of NMR data. Analytical

Chemistry 90 (1), 649–656. Available at https://doi.org/10.1021/acs.analchem.7b02795, http://pubs.acs.org/journal/ancham.

Schwalbe, H., Audergon, P., Haley, N., Amaro, C.A., Agirre, J., Baldus, M., et al., 2024. The future of integrated structural biology. Structure (London, England: 1993) 32 (10), 1563–1580. Available at https://doi.org/10.1016/j.str.2024.08.014, https://www.sciencedirect.com/science/journal/09692126.

Schwantes, C.R., Pande, V.S., 2015. Modeling molecular kinetics with tICA and the kernel trick. Journal of Chemical Theory and Computation 11 (2), 600–608. Available at https://doi.org/10.1021/ct5007357, http://pubs.acs.org/journal/jctcce.

Sehnal, D., Deshpande, M., Vařeková, R.S., Mir, S., Berka, K., Midlik, A., et al., 2017. LiteMol suite: interactive web-based visualization of large-scale macromolecular structure data. Nature Methods 14 (12), 1121–1122. Available at https://doi.org/10.1038/nmeth.4499, http://www.nature.com/nmeth/.

Sehnal, D., Bittrich, S., Velankar, S., Koča, J., Svobodová, R., Burley, S.K., et al., 2020. BinaryCIF and CIFTools—Lightweight, efficient and extensible macromolecular data management. PLoS Computational Biology 16 (10). Available at https://doi.org/10.1371/journal.pcbi.1008247, https://journals.plos.org/ploscompbiol/article/file?id=10.1371/journal.pcbi.1008247&type=printable.

Sehnal, D., Bittrich, S., Deshpande, M., Svobodová, R., Berka, K., Bazgier, V., et al., 2021. Mol*Viewer: modern web app for 3D visualization and analysis of large biomolecular structures. Nucleic Acids Research 49 (1), W431. Available at https://doi.org/10.1093/nar/gkab314, https://academic.oup.com/nar/issue.

Shen, T., Hu, Z., Sun, S., Liu, D., Wong, F., Wang, J., et al., 2024. Accurate RNA 3D structure prediction using a language model-based deep learning approach. Nature Methods 21 (12), 2287–2298. Available at https://doi.org/10.1038/s41592-024-02487-0, https://www.nature.com/nmeth/.

Shome, M., MacKenzie, T.M.G., Subbareddy, S.R., Snyder, M.P., 2024. The importance, challenges, and possible solutions for sharing proteomics data while safeguarding individuals' privacy. Molecular and Cellular Proteomics 23 (3). Available at https://doi.org/10.1016/j.mcpro.2024.100731, https://www.mcponline.org/current.

Silva, J.C.F., Schuster, L., Sexson, N., Erdem, M., Hulke, R., et al., 2025. Guo, InteracTor: feature engineering and explainable AI for profiling protein structure-interaction function relationships. PLoS Computational Biology 21 (10), e1013038. Available at https://doi.org/10.1371/journal.pcbi.1013038.

Sonawane, A.R., Weiss, S.T., Glass, K., Sharma, A., 2019. Network medicine in the age of biomedical big data. Frontiers in Genetics 10. Available at https://doi.org/10.3389/fgene.2019.00294, https://www.frontiersin.org/journals/genetics#.

Sweeney, B.A., Petrov, A.I., Ribas, C.E., Finn, R.D., Bateman, A., Szymanski, M., et al., 2021. RNAcentral 2021: secondary structure integration, improved sequence search and new member databases. Nucleic Acids Research 49 (D1), D212. Available at https://doi.org/10.1093/nar/gkaa921.

Szklarczyk, D., Kirsch, R., Koutrouli, M., Nastou, K., Mehryary, F., Hachilif, R., et al., 2023. The STRING database in 2023: protein-protein association networks and functional enrichment analyses for any sequenced genome of interest. Nucleic Acids Research 51 (1 D), D638. Available at https://doi.org/10.1093/nar/gkac1000, https://academic.oup.com/nar/issue.

Tahmasebi, P., Hezarkhani, A., 2011. Application of a modular feedforward neural network for grade estimation. Natural Resources Research 20 (1), 25–32. Available at https://doi.org/10.1007/s11053-011-9135-3.

The UniProt Consortium, 2015. UniProt: a hub for protein information. Nucleic Acids Research 43 (Database issue), D204–D212. Available at https://doi.org/10.1093/nar/gku989.

Tiemann, J.K.S., Szczuka, M., Bouarroudj, L., Oussaren, M., Garcia, S., Howard, R.J., et al., 2024. MDverse, shedding light on the dark matter of molecular dynamics simulations. eLife. Available at https://doi.org/10.7554/eLife.90061.3, https://elifesciences.org/.

del Toro, N., Shrivastava, A., Ragueneau, E., Meldal, B., Combe, C., Barrera, E., et al., 2022. The IntAct database: efficient access to fine-grained molecular interaction data. Nucleic Acids Research 50 (1), D648. Available at https://doi.org/10.1093/nar/gkab1006, https://academic.oup.com/nar/issue.

Tribello, G.A., Bonomi, M., Bussi, G., Camilloni, C., Armstrong, B.I., Arsiccio, A., et al., 2025. PLUMED Tutorials: a collaborative, community-driven learning ecosystem. Journal of Chemical Physics 162 (9). Available at https://doi.org/10.1063/5.0251501, http://scitation.aip.org/content/aip/journal/jcp.

Trozzi, F., Wang, X., Tao, P., 2021. UMAP as a dimensionality reduction tool for molecular dynamics simulations of biomacromolecules: a comparison study. Journal of Physical Chemistry B 125 (19), 5022–5034. Available at https://doi.org/10.1021/acs.jpcb.1c02081, http://pubs.acs.org/journal/jpcbfk.

Tunyasuvunakool, K., Adler, J., Wu, Z., Green, T., Zielinski, M., Žídek, A., et al., 2021. Highly accurate protein structure prediction for the human proteome. Nature 596 (7873), 590–596. Available at https://doi.org/10.1038/s41586-021-03828-1, http://www.nature.com/nature/index.html.

Turner, J., Abbott, S., Fonseca, N., Pye, R., Carrijo, L., Duraisamy, A.K., et al., 2024. Xiaodan Ma, EMDB—the electron microscopy data bank. Nucleic Acids Research 52 (D1), D456. Available at https://doi.org/10.1093/nar/gkad1019.

Ulrich, E.L., Baskaran, K., Dashti, H., Ioannidis, Y.E., Livny, M., Romero, P.R., et al., 2019. NMR-STAR: comprehensive ontology for representing, archiving and exchanging data from nuclear magnetic resonance spectroscopic experiments. Journal of Biomolecular NMR 73 (1-2), 5–9. Available at https://doi.org/10.1007/s10858-018-0220-3, www.wkap.nl/journalhome.htm/0925-2738.

Valentini, E., Kikhney, A.G., Previtali, G., Jeffries, C.M., Svergun, D.I., 2015. SASBDB, a repository for biological small-angle scattering data. Nucleic Acids Research 43 (1), D357. Available at https://doi.org/10.1093/nar/gku1047, http://nar.oxfordjournals.org/.

Vallat, B., Tauriello, G., Bienert, S., Haas, J., Webb, B.M., Žídek, A., et al., 2023. ModelCIF: an extension of PDBx/mmCIF data representation for computed structure models. Journal of Molecular Biology 435 (14). Available at https://doi.org/10.1016/j.jmb.2023.168021, https://www.journals.elsevier.com/journal-of-molecular-biology.

Varadi, M., Anyango, S., Deshpande, M., Nair, S., Natassia, C., Yordanova, G., et al., 2022. AlphaFold protein structure database: massively expanding the structural coverage of protein-sequence space with high-accuracy models. Nucleic Acids Research 50 (D1), D439. Available at https://doi.org/10.1093/nar/gkab1061.

Varadi, M., Bertoni, D., Magana, P., Paramval, U., Pidruchna, I., Radhakrishnan, M., et al., 2024. AlphaFold protein structure database in 2024: providing structure coverage for over 214 million protein sequences. Nucleic Acids Research 52 (D1), D368. Available at https://doi.org/10.1093/nar/gkad1011.

Veličković, P., 2023. Everything is connected: graph neural networks. Current Opinion in Structural Biology 79, 102538. Available at https://doi.org/10.1016/j.sbi.2023.102538.

Vidal, M.-E., Endris, K.M., Jozashoori, S., Karim, F., Palma, G., 2019. Semantic data integration of big biomedical data for supporting personalised medicine. In: Alor-Hernández, G., Sánchez-Cervantes, J.L., Rodríguez-González, A., Valencia-García, R. (Eds.), Current Trends in Semantic Web Technologies: Theory and Practice. Springer International Publishing, Cham, pp. 25–56. Available at http://doi.org/10.1007/978-3-030-06149-4_2.

Vinod, R., Chen, P.Y., Das, P., 2025. Reprogramming pretrained language models for protein sequence representation learning. Digital Discovery 4 (6), 1591–1601. Available at https://doi.org/10.1039/d4dd00195h, https://www.rsc.org/journals-books-databases/about-journals/digital-discovery/.

Wang, S., Bonomi, L., Dai, W., Chen, F., Cheung, C., Bloss, C.S., et al., 2020. Big data privacy in biomedical research. IEEE Transactions on Big Data 6 (2), 296–308. Available at https://doi.org/10.1109/TBDATA.2016.2608848, https://www.ieee.org/membership-catalog/productdetail/showProductDetailPage.html?product=PER472-ELE.

Wankowicz, S.A., 2024. Modeling bias toward binding sites in PDB structural models. bioRxiv. Available at https://doi.org/10.1101/2024.12.14.628518, https://www.biorxiv.org.

Wankowicz, S.A., 2025. Ten rules for a structural bioinformatic analysis. PLoS Computational Biology 21 (10), e1013094. Available at https://doi.org/10.1371/journal.pcbi.1013094.

Wen, H., Huang, C., Guo, S., 2021. The application of convolutional neural networks (CNNs) to recognize defects in 3D-printed parts. Materials 14 (10). Available at https://doi.org/10.3390/ma14102575, https://www.mdpi.com/1996-1944/14/10/2575/pdf.

Westbrook, J.D., Fitzgerald, P.M., 2003. The PDB format, mmCIF, and other data formats. Methods Biochem Anal 44, 161–179.

Westbrook, J.D., Young, J.Y., Shao, C., Feng, Z., Guranovic, V., Lawson, C.L., et al., 2022. PDBx/mmCIF ecosystem: foundational semantic tools for structural biology. Journal of Molecular Biology 434 (11), 167599. Available at https://doi.org/10.1016/j.jmb.2022.167599.

Wu, H., Manu, Jiao, R., Ma, J., 2015. Temporal and spatial dynamics of scaling-specific features of a gene regulatory network in Drosophila. Nature Communications 6. Available at https://doi.org/10.1038/ncomms10031, https://www.nature.com/ncomms/.

Yousef, M., Allmer, J., 2023. Deep learning in bioinformatics. Turkish Journal of Biology 47 (6), 366–382. Available at https://doi.org/10.55730/1300-0152.2671, https://journals.tubitak.gov.tr/cgi/viewcontent.cgi?article=2671&context=biology.

Zeng, H., Zhang, J., Preising, G.A., Rubel, T., Singh, P., Ritz, A., 2021. Graphery: interactive tutorials for biological network algorithms. Nucleic Acids Research 49 (1), W257. Available at https://doi.org/10.1093/nar/gkab420, https://academic.oup.com/nar/issue.

Zhang, C., Pyle, A.M., 2023. PDC: a highly compact file format to store protein 3D coordinates. Database 2023. Available at https://doi.org/10.1093/database/baad018, http://database.oxfordjournals.org/.

Zitnik, M., Nguyen, F., Wang, B., Leskovec, J., Goldenberg, A., Hoffman, M.M., 2019. Machine learning for integrating data in biology and medicine: principles, practice, and opportunities. Information Fusion 50, 71–91. Available at https://doi.org/10.1016/j.inffus.2018.09.012, http://www.elsevier.com/inca/publications/store/6/2/0/8/6/2/index.htt.

CHAPTER 8

AI and machine learning in structural biology

Sanjay Kumar
Structural and Molecular Biology, City College of New York, New York City, NY, United States

8.1 Introduction

A major discovery in 2020 surprised scientists worldwide. DeepMind's AlphaFold2 algorithm was able to predict protein structure with almost atomic accuracy using only amino acid sequences. Such predictions were something that biologists had not been able to do for more than 50 years (Jumper et al., 2021). Years of difficult research were now just hours and hours of work on a computer. The first thing that happened was that thousands of new protein structures that had never been seen before were put into open databases (Tunyasuvunakool et al., 2021). These changes facilitated the study of new enzymes, receptors, and complexes crucial to health and disease. Not only is one tool changing, but artificial intelligence (AI) and machine learning (ML) are also changing structural biology in ways that were once thought to be impossible (Aithani et al., 2023). Structural biology is a branch of the life sciences that focuses on determining the three-dimensional (3D) structure of proteins, DNA/RNA, and their complexes (Carugo and Djinović-Carugo, 2023). The shape and movement of an enzyme, a receptor, or a viral protein determine how they work. For example, the way an enzyme catalyzes a reaction, the way a receptor recognizes a drug, or the way a viral protein interacts with host machinery. To have structural knowledge is to have a full understanding of biology, as it involves trying to learn something that one does not already know. Some of the biggest scientific breakthroughs of the last 100 years have come from traditional structural biology, such as the double helix structure of DNA, the atomic structure of the ribosome, and membrane protein complexes (Ban et al., 2000; Hegde and Keenan, 2022; Watson and Crick, 1953). However, despite these successes, experimental procedures require significant time and energy and are not feasible for certain molecules. Here come AI/ML models that can predict, improve, and understand structures faster and better than traditional methods by finding patterns that are difficult to see in large biological datasets.

Before AI, structural biologists mostly used three main types of experiments:

- X-ray crystallography: this method has been the "gold standard" for a long time. Scientists can make atomic models by crystallizing proteins and shooting X-rays

Advances in Structural Biology
DOI: https://doi.org/10.1016/B978-0-443-44988-8.00020-4

through them. This method made it possible to make thousands of Protein Data Bank (PDB) entries and win nobel prizes. But there are problems with it: many proteins won't crystallize, membrane proteins are notoriously hard, and dynamic states often get "frozen out" of the picture (Drenth, 2007).

- Nuclear magnetic resonance (NMR) spectroscopy: NMR lets scientists see proteins in solutions and record their movements. But NMR can only be used on proteins that are less than 40 kDa, needs special equipment, and can take months to analyze (Wüthrich, 2001).
- Cryo-electron microscopy (cryo-EM): In the last ten years, cryo-EM has made a lot of progress, and it is often called the resolution revolution. It can see big complexes and membrane proteins without having to crystallize them. Cryo-electron tomography (cryo-ET) enhances this capability in situ by reconstructing 3D cellular landscapes to situate macromolecules within their native context. On the other hand, the system makes vast datasets that have noise and different shapes, so they need advanced computer methods to process and understand (Kühlbrandt, 2014).

Even with these improvements, there were still substantial issues. Many proteins were still not understood, and it could take years to figure out how genes turn into structures. Classical methods often didn't work whatsoever for systems that are dynamic and flexible, like intrinsically disordered proteins (IDPs) (Trivedi and Nagarajaram, 2022). Scientists understood that to decipher the complicated structure of the proteome, they required more than experimental methodologies: they needed smart computing. AI and ML work best when they can identify patterns in enormous, complicated datasets. This is precisely what structural biology is all about. Proteins are made up of 20 amino acids that can be arranged in billions of different ways. However, evolutionary rules dictate how they fold into 3D shapes. Researchers have started to predict folding pathways, residue contacts, and even folding landscapes with an accuracy that has never been seen before by training deep learning (DL) models on millions of sequences and known structures. AlphaFold2 (Jumper et al., 2021) and RoseTTAFold (Baek et al., 2021) are two tools that can now predict protein structures that are very close to the atomic level directly from sequence. This kind of technology speeds up discovery by a lot, and it works well with experimental methods instead of replacing them. AI models like cryoDRGN use noisy single-particle cryo-EM images to figure out continuous conformational heterogeneity, which shows groups of biomolecular conditions (Zhong et al., 2021). AI is connecting structural biology with genomics, transcriptomics, and proteomics in a more general way to make it possible to see biological processes from many different angles. Tasks that previously required years of repeated experimentation can frequently be completed in days, facilitating global access to high-quality structural data. AI enhances rather than replaces experiments. Predicted models help crystallographers in phase determination, help cryo-EM scientists fit atomic models into density maps, and let drug designers virtually test compounds against structures that didn't exist before.

8.2 Fundamentals of artificial intelligence in structural biology

AI and ML are no longer tools of the future in structural biology; they are now essential for how scientists study, predict, and understand complex structures. In essence, AI refers to algorithms that are capable of taking instructions that typically call for human intelligence, such as pattern recognition, prediction, and data-driven learning. ML features enable computers to discover patterns in data without explicit instruction. ML algorithms fail to conform to rigid guidelines. Rather, they adjust their parameters according to examples, which improves them as more data becomes available. In structural biology, this capacity to "learn from examples" is extremely advantageous. Although proteins worldwide adhere to the same biophysical principles, it can be challenging to translate sequence into structure. Human-made rules and simulations struggle to capture every interaction. However, ML models are able to make precise generalizations after learning patterns from thousands or even millions of examples.

8.2.1 Why structural biology fits artificial intelligence

Three main issues in structural biology make it an ideal candidate for AI solutions:

- **High-dimensional complexity:** Proteins are made up of thousands of atoms that interact with one another through physical forces. Conventional simulations can model these interactions, but they incur a significant computational cost. AI learns statistical shortcuts to simplify this complexity.
- **Big data:** Millions of sequences and structures are available in databases like the AlphaFold protein structure database (Varadi et al., 2022), RCSB PDB (Berman et al., 2000), and UniProt (Bateman et al., 2025) that can be used as training data. Such a wealth of data is ideal for ML.
- **Unseen patterns:** Evolution has encoded information in protein sequences. Coevolutionary signals hint at structural constraints, where mutations in one location offset those in another. AI models are far more adept at identifying and exploiting these subtle correlations than rule-based algorithms.

8.2.2 Artificial intelligence algorithms

In biological research, three ML techniques are frequently used:

1. **Supervised learning** uses labeled data, such as amino acid sequences matched with experimentally determined structures, to train models. This serves as the foundation

for programs such as AlphaFold2, which predicts new structures by training on the RCSB PDB, and Prop3D, a platform to provide access to ML-trained data (Draizen et al., 2024).

2. **Unsupervised learning** uses unlabeled data; models reveal hidden patterns. This helps structural biologists find repeating patterns in protein sequences or conformational states in cryo-EM data. The method has various clustering techniques and variational autoencoders (VAEs).
3. **Reinforcement learning** (RL) models learn by making mistakes and getting feedback on what they do. RL has potential in protein design, where models are rewarded for producing stable or functional folds, even though it is less prevalent in structural biology than drug design (Angermueller et al., 2016).

Now, it is possible to combine these methods. Because experimentally solved structures are relatively uncommon in the vast sequence space, semisupervised learning, for instance, makes use of both labeled and unlabeled protein sequences.

8.2.3 Neural network algorithms

The neural network, inspired by the structure of the human brain, has enabled modern AI breakthroughs. Neural networks are made up of layers of nodes (neurons) that are all connected to each other. Each node does a simple math operation. Networks can learn more abstract representations by stacking layers. In structural biology, the following algorithms are often used:

- **Convolutional neural networks** (CNNs): CNNs were first made for recognizing images, but they are great at finding patterns in space. People have used them to guess how residues will interact, read cryo-EM density maps, and sort protein folds.
- **Recurrent neural networks** (RNNs): These networks work with data that comes in a sequence, which makes them useful for looking at protein sequences. RNNs were the first to use sequence-based modeling, but transformers have mostly taken their place (Kumar and Srivastava, 2024). Transformers are the revolutionary architecture that powers AlphaFold2 and ESMFold (Lin et al., 2023). They use attention mechanisms to identify long-range dependencies in sequences. This methodology enables them to comprehend the interactions of distant amino acids in 3D space (Vaswani et al., 2017).
- **Graph neural networks** (GNNs): You can use graphs to show proteins, with residues as nodes and edges showing how they interact with each other. GNNs are great for modeling how proteins interact with each other, how ligands bind to proteins, and how proteins fold (Gainza et al., 2020).

These architectures let AI models use biological data in different ways, like sequences, images (density maps), and graphs of interaction.

8.2.4 Proteins as a "Language"

It was a big step forward in understanding to realize that protein sequences are like language. Just as grammar rules govern words, so do evolutionary rules govern amino acid sequences. Large language models changed the way we process natural language. They can be retrained with protein sequences to learn the "grammar of life." For example, ESMFold used language modeling on 65 million protein sequences to predict structures without needing direct structural training data. In the same way, models like ProtBERT and tasks assessing protein embeddings benchmark treat protein sequences like sentences, where the relationships between residues help them fold (Rao et al., 2019). This analogy has made AI tools easier to use. Proteins are no longer just thought of as chemical molecules; they are now considered "texts" that algorithms can read, translate, and understand.

8.2.5 Sources of training data

Large, high-quality datasets are essential for AI's success in structural biology. In order to train models for image interpretation, it needs density maps, a thorough database of protein sequences, and experimentally solved structures. The training challenge is defined by the interaction of sequence data, which is plentiful, and structure data, which is limited but accurate. In a vast sequence universe, AI must learn to generalize from little structural supervision. Here, we listed latest databases that can be used to train AI models (Table 8.1).

8.2.6 Early versus modern artificial intelligence Approaches

AI has been part of structural biology well before AlphaFold2. In the 1990s and early 2000s, hidden Markov models popularized through the HMMER webserver and databases like Pfam, which powered sensitive sequence alignment and domain detection beyond simple BLAST-style searching (Eddy, 1998; Potter et al., 2018; Sonnhammer et al., 1998). Around the same time, support vector machines predicted properties such as secondary structure and solvent accessibility, with widely cited examples including SVM-based secondary structure and exposure predictors (Kim and Park, 2003). PSIPRED used position-specific scoring matrices and small neural networks to predict secondary structure. Early contact-map predictors used small feed-forward networks that were trained on a small number of features (Fariselli and Casadio, 1999). The computer and data limitations of the time, such as CPU-bound training, small, curated datasets, and relatively shallow models, limited the usefulness of these approaches.

The convergence of three synergies likely explains the shift from 2010 to the 2020s:

- DL models that have the capacity to model long-range dependencies.
- CUDA-style hardware achieves cost-effective GPU acceleration.
- The system leverages extensive sequence and structural repositories, including UniProt, metagenomic data, and the constantly expanding PDB.

Table 8.1 List of data sources.

Database name	Purpose	Web address
RCSB Protein Data Bank (PDB)	Primary archive of 3D macromolecular structures	https://www.rcsb.org
PDBe (EBI)	European portal to the PDB with analysis tools	https://www.ebi.ac.uk/pdbe/
PDBj	Japan portal to the PDB	https://pdbj.org
EMDB	Database for density maps	https://www.ebi.ac.uk/emdb/
EMPIAR	Raw EM image datasets	https://www.ebi.ac.uk/empiar/
AlphaFold DB	Predicted protein structures database from AlphaFold	https://alphafold.ebi.ac.uk
SWISS-MODEL Repository	Curated homology models database	https://swissmodel.expasy.org/repository
ModelArchive	Depository for computed structure models	https://www.modelarchive.org
SASBDB	Database for small-angle scattering (SAXS) data/models	https://www.sasbdb.org
BMRB	Database for NMR restraints and chemical shifts	https://bmrb.io/
PDB-Dev	Database for integrative/hybrid structural models	https://pdb-dev.wwpdb.org
NDB	Database for DNA/RNA structures	https://ndbserver.rutgers.edu
SIFTS	Database for structure–sequence mapping	https://www.ebi.ac.uk/pdbe/docs/sifts/
UniProtKB	Gold-standard protein sequence and annotation	https://www.uniprot.org
UniRef	Nonredundant UniProt clusters	https://www.uniprot.org/uniref
UniParc	Comprehensive nonredundant sequence archive	https://www.uniprot.org/uniparc
NCBI Protein	NCBI protein sequences	https://www.ncbi.nlm.nih.gov/protein
RefSeq	Curated reference sequences	https://www.ncbi.nlm.nih.gov/refseq/
GenBank	Primary nucleotide archive	https://www.ncbi.nlm.nih.gov/genbank/
European nucleotide archive	Nucleotide sequences database	https://www.ebi.ac.uk/ena
DNA data bank of Japan	Protein sequences database	https://www.ddbj.nig.ac.jp
Ensembl	Vertebrate genomes and annotations	https://www.ensembl.org

(*Continued*)

Table 8.1 (Continued)

Database name	Purpose	Web address
Ensembl Genomes	Bacteria, fungi, plants, protists, metazoa	https://ensemblgenomes.org
InterPro	Integrated protein signatures and annotations	https://www.ebi.ac.uk/interpro
Pfam	Database of protein families	https://www.ebi.ac.uk/interpro/entry/pfam/
PROSITE	Motifs and profiles	https://prosite.expasy.org
SMART	Domain architectures	https://smart.embl.de
NCBI CDD	Conserved domains	https://www.ncbi.nlm.nih.gov/Structure/cdd/cdd.shtml
PANTHER	Families, trees, pathways, GO	http://www.pantherdb.org
eggNOG	Orthologous groups and annotations	http://eggnog5.embl.de
SCOPe	Database of domains	https://scop.berkeley.edu
CATH	Database of domains	https://www.cathdb.info
ECOD	Database of domains	http://prodata.swmed.edu/ecod/

No Permission Required.

Together, these elements made end-to-end structural prediction systems possible, including AlphaFold2 and other programs, including RoseTTAFold switch the field of view to data-driven representations and high-accuracy predictions on a near-atomic scale on many targets.

8.3 How artificial intelligence works

AI and ML have significantly altered structural biology by speeding up calculations and making predictions all along the molecular discovery process. They learn to recognize complicated, nonlinear patterns that classic physics-based algorithms can't easily find by training on huge datasets of known biological structures and sequences (Fig. 8.1).

Here, we elucidate each stage of AI and demonstrate its potential applications in various projects.

Step 1: What do you want the model to do? Define your problem, for example, predict a protein's 3D shape, model a complex, clean cryo-EM data, find a binding site, or design a new protein.

Step 2: Collect the raw inputs that your AI method needs, such as protein sequences from UniProt, NCBI RefSeq, 3D structures from PDB, cryo-EM/ET maps from EMDB or EMPIAR (The wwPDB Consortium, 2024). Select predicted models such as AlphaFold DB.

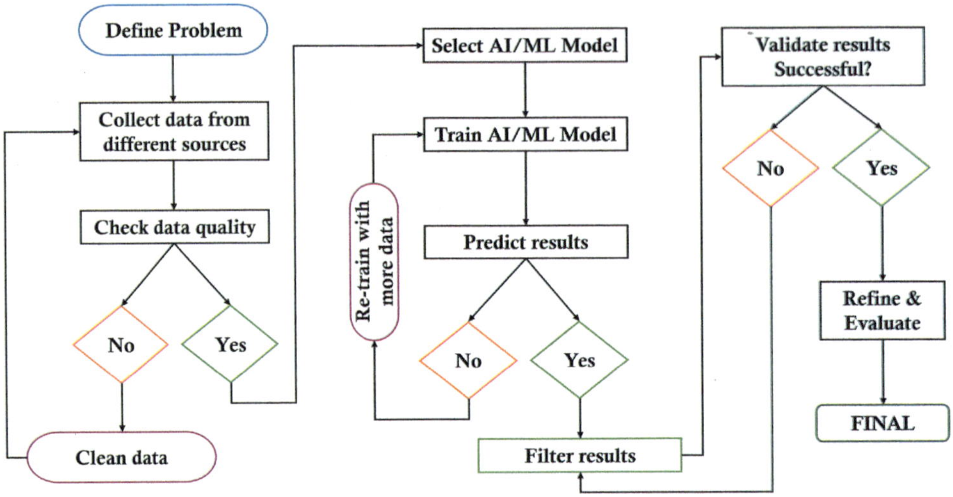

Figure 8.1 Artificial intelligence (*AI*)/Machine learning (*ML*) workflow for creating an AI trainee model to construct protein structure from sequences. Define the problem (predict three-dimensional structure from sequence) for the end-to-end pipeline from data to deployment. Putting together the data and preprocessing. The model architecture, which could be a transformer construct, and deploy and iteration (watch new targets, retrain with new structures, and package weights and version data). *No Permission Required.*

Step 3: To obtain better AI output and make the inputs usable and consistent, clean and prepare the data. Sequence preparation is done by removing duplicates and making multiple-sequence alignments (MSAs) with MMseqs2 (Mirdita et al., 2019), HHblits (Steinegger et al., 2019), and Clustal Omega (Sievers and Higgins, 2018). Images and maps are prepared through motion/CTF correction using MotionCor2 (Zheng et al., 2017), patch motion correction in CryoSPARC (Punjani et al., 2017), CTFFIND4 (Rohou and Grigorieff, 2015), and Warp (Tegunov and Cramer, 2019); and denoise/restoration processes with Topaz-Denoise (Bepler et al., 2020), IsoNet, cryoCARE (Buchholz et al., 2019), and DeepDeWedge (Wiedemann and Heckel, 2024).

Step 4: Select AI programs such as AlphaFold2 for protein structure prediction.

Step 5: To configure how the AI runs, set up the model. Such as picking the model and version, setting options like MSA depth and number of recycles, and choosing hardware (GPU if possible). If possible, use local installs or web servers (ColabFold for AF2/RF) and CryoSPARC for EM tasks.

Step 6: Run predictions that give you 3D models (PDB/mmCIF), confidence scores (pLDDT, PAE), complexes with interface scores/ranks, and enhanced EM maps.

Step 7: Evaluate quality quickly to keep the good and drop the bad. Compare model confidence pLDDT (per residue), PAE (domain/domain), model ranking, MolProbity (clashscore, Ramachandran), and ChimeraX for visual inspection (Meng et al., 2023).

Step 8: Design or optimize (if your goal is making things); for example, to design a protein loop, generate a backbone with RFdiffusion (Watson et al., 2023) and design a sequence with (Dauparas et al., 2022) and re-score with AlphaFold2.

Step 9: Treat AI as a hypothesis generator by planning experiments based on its output, such as cloning domains with high pLDDT scores, mutating interface residues predicted by AF-Multimer, choosing constructs that avoid low-confidence flexible linkers, and designing EM experiments guided by the predicted architecture.

Step 10: Validate experimentally. If the facility allows you, such as for protein models, use X-ray crystallography, NMR, or cryo-EM/cryo-ET (map comparison, local fit); for mutation analysis, use cross-link MS and HDX-MS; and for model building, use Coot (Emsley et al., 2010).

Step 11: To improve and make results useful to others, fix weak regions, try alternative models, add new data, and repeat steps 6–11. Deposit structures in PDB, maps in EMDB, and raw data in EMPIAR; publish code/pipelines on GitHub; and keep notes on parameters so others can reproduce your work.

8.4 How to create an artificial intelligence model?

To build an AI model for any project, you need to follow each step discussed in Section 8.4, Fig. 8.1. For example, to predict protein structures, you need to obtain clean training pairings of sequences known structures from public databases like UniProt and RCSB PDB. Transform each sequence into inputs that the model can learn from. Most recent systems use MSAs and transformer embeddings. Some also include equivariant geometric layers that take 3D rotations into account. Train the network to directly predict interresidue distances/angles or 3D coordinates, using losses that include incorrect geometry (clashes, bond lengths) and correct precise folds. To help with learning, add basic side tasks (secondary structure, contacts). Utilize GPUs, set aside test proteins that are significantly different from those in your existing dataset, and apply TM-score/RMSD along with confidence metrics (such as per-residue scores) for evaluation. When you use the model, run it on a fresh sequence (with or without an MSA), relax the output using a quick physics step, and check it against any experimental data you have. Keep the loop tight: improve how you organize your data, adjust the size and number of your models, and retrain them when you add more difficult instances, including membrane proteins and complexes.

8.4.1 Example: human β2-adrenergic receptor

You won't recreate AlphaFold2 or any big AI model on small systems; you need to configure a larger amount of CPU, GPU, and memory. Here, we build a small AI model to predict 3D structure from sequences. β2-Adrenergic Receptor is a GPCR, 413 aa, with many known structures for sanity checks. Protein sequences were downloaded from the UniProt database (Uniprot ID: P07550). Given a sequence + MSA, predict a contact map (L × L grid). Contacts help guide folding and validate predicted structures.

- **Step 1:** Pick ~5000 protein structure sequences in FASTA format for training from PDB with <30% mutual identity.
- **Step 2:** Perform MSA via MMseqs2 for each protein and determine the true contact map from the PDB (define contact if Cβ–Cβ distance ≤ 8 Å; use Cα for Gly).
- **Step 3:** Create split sets to train based on 80%/10%/10% homology with no close homology across splits.
- **Step 4:** Check features based on per-residue (L × F1). These features include one-hot amino acid encoding, position index, and predicted secondary structure (PSIPRED). Perform pairwise alignment (L × L × F2) and verify or compare coevolution scores from the MSA.
- Sequence separation |i−j| bucketed.
- Symmetry: enforce features (i, j)=features (j, i).
- **Step 5:** Creating an AI model using a 2D CNN or a small transformer over the L × L pairwise grid, such as a CNN block: [Conv(3 × 3) → BatchNorm → ReLU] × 6–10 layers + Residual skips. This process produces L × L logits, which are then passed through a sigmoid function to calculate the probability of contact for each pair (i, j). The loss is calculated by binary cross-entropy + class weighting (contacts are sparse). Two metrics can be used that are standard in contact prediction, such as precision@L, precision@L/5.
- **Step 6:** Run the training loop using PyTorch sketch for batch proteins with the Adam optimizer, lr = 1e - 3; early stopping on validation precision@L. 10–30 epochs are enough for a class train on CPU/GPU.
- **Step 7:** Use the trained model created at step 5 to predict contacts on ADRB2 using its MSA. Visualize the contact map; check if it shows the near-diagonal helical patterns typical of 7-TM bundles, and compare them to contacts extracted from your AlphaFold2 model as a sanity check.

8.5 Landmark breakthroughs

Decades of incremental progress have shaped the journey of structural biology. However, in recent years, significant advances in AI and ML have made previously impossible tasks achievable. These new developments have solved long-standing issues and given biologists new ways to work.

8.5.1 AlphaFold2: solving the protein folding problem

AlphaFold2 is the most renowned AI innovation for structural biology. The protein folding problem involves predicting a protein's 3D structure from its linear amino acid sequence. Accuracy was still far below experimental standards, even with improvements in statistical models and physics-based simulations. When AlphaFold2 participated in the critical assessment of protein structure prediction (CASP14) in 2020, that changed. On the global distance test, it obtained a median accuracy of 92.4, which indicates that its predictions were almost identical to those of experimental structures (Jumper et al., 2021). This achievement was revolutionary and warmly praised as the answer to one of the most significant problems facing biology. It models the links between amino acids in a sequence using a transformer-based neural network designed with an "attention mechanism". It combines data from evolutionary links, MSAs, and 3D structural templates. The approach improves predictions by using an iterative "structure module" to generate atomic-resolution models. The AlphaFold protein structure database now comprises millions of predicted structures, spanning most of the proteins that are known to exist in humans and many other species (Varadi et al., 2022). Its predictions are directing the development of drugs, enzyme modification, and the comprehension of disease mutations. AlphaFold2 models are used for molecular replacement (MR) in Phenix, which now solves a large majority of previously tough structures directly through prediction. These models are now used by structural biologists as starting points for crystallography and cryo-EM, which speeds up experiments.

8.5.2 RoseTTAFold: democratizing protein prediction

A group at the University of Washington's Baker Lab also came up with RoseTTAFold at around the same time (Baek et al., 2021). AlphaFold2's code and procedures were originally private, but RoseTTAFold provided an open-source, easy-to-use option (https://github.com/RosettaCommons/RoseTTAFold?tab=readme-ov-file). RoseTTAFold uses a three-track neural network that simultaneously analyzes 1. protein sequence data, 2. evolutionary links, and 3. 3D structural limits. By combining these tracks, RoseTTAFold might provide very accurate results without spending as much on computing power as AlphaFold2. It worked perfectly with the Rosetta software suite, which is famous for modeling proteins. It may now be used to predict not only protein structures but also protein–protein complexes and designer proteins, which makes it more useful.

8.5.3 ESMFold: proteins as a language

Meta AI created ESMFold, which was another significant step forward. Instead of depending significantly on evolutionary alignments (like AlphaFold2 and RoseTTAFold do), ESMFold was inspired by natural language processing. Meta AI created ESM-2, an LLM, by treating 65 million protein sequences as biological language phrases (Lin et al., 2023). The model acquired the protein sequence regularities that control folding and function without the need for structural supervision. ESMFold then translated the sequences directly into 3D models. It predicted millions of metagenomic proteins, broadening our understanding of the "dark proteome" (proteins that lack recognized structures). It works faster and is lighter than AlphaFold2, making it ideal for large-scale annotation tasks. It demonstrated the promise of language models in biology, laying the groundwork for the development of cross-disciplinary AI applications.

8.5.4 CryoDRGN: unveiling molecular flexibility

While AlphaFold2 and similar algorithms aim to predict structures from sequences, a major challenge in structural biology is understanding the data from experimental cryo-EM. Cryo-EM generates images of molecules frozen in small layers of ice, recording them in various orientations and conformations. However, standard approaches average these pictures into a single shape, resulting in a loss of molecular dynamics diversity. CryoDRGN is an AI algorithm that uses deep neural networks to map conformational heterogeneity in cryo-EM datasets (Zhong et al., 2021). CryoDRGN uses a VAE architecture to represent cryo-EM particles as points in a latent space. By sampling this region, scientists can recreate several conformations of the same protein or complex. It enabled structural biologists to look beyond static images and see structural ensembles. CryoDRGN symbolizes a move from static to dynamic structural biology, with AI assisting in capturing the "movie" of molecular life rather than just a single frame (https://github.com/ml-struct-bio/cryodrgn).

8.5.5 Topaz: particle picking

Topaz is a cryo-EM DL toolkit that is excellent at particle selecting using positive unlabeled learning. You only need a small collection of clean, hand-picked positives and no clear negatives. It learns from your raw micrographs and can pick from the complete dataset. It usually improves recall compared to LoG/template approaches, which you then clean up in 2D classification. It comes with Topaz-Denoise, a Noise2Noise-style model that helps increase SNR in micrographs and tomograms so they are easier to see. It is built into CryoSPARC and RELION (Scheres, 2012). The best way to do these tasks is to choose a wide, high-quality positive set that includes views, defocus, and ice. Then, train on raw photos (even if you glanced at denoised ones to help you pick the right ones) and adjust the score threshold to determine the right balance between recall and false positives.

8.5.6 DeepTracer: automating model building

Another bottleneck in cryo-EM is the building of atomic models, which includes fitting protein backbones and side chains into density maps. Traditionally, this required specialist intervention and may take several weeks. DeepTracer used AI to streamline this process (Pfab et al., 2021). DeepTracer uses a convolutional neural network to autonomously trace protein backbones in cryo-EM density maps and assign amino acid identities. It is capable of handling enormous assemblies and multichain complexes. It drastically reduced the time required to make a model. DeepTracer demonstrates how AI may alleviate structural biology's time-consuming bottlenecks, allowing pipelines to run more quickly and automatically.

8.5.7 IsoNet and DeepDeWedge: tomogram enhancement

By addressing the "missing-wedge" issue and lowering noise, the AI algorithms IsoNet and DeepDeWedge improve the clarity and evenness of cryo-ET tomograms (Wiedemann and Heckel, 2024; Liu et al., 2022). To increase signal-to-noise and make features like membranes and complexes easier to identify and interpret, IsoNet learns from the tomogram itself and iteratively fills in the lost directional information. Without any ground-truth training data, DeepDeWedge uses a self-supervised network to simultaneously denoise and reconstruct the missing wedge from the raw tilt series. Research indicates that DeepDeWedge matches IsoNet's performance in wedge recovery and advanced denoising, contributing to the creation of tomograms with greater contrast for subsequent tasks.

8.5.8 Find my sequence

It is a neural-network-based method that finds the protein sequence directly from electron-density maps from X-ray and cryo-EM reconstructions when the sequence is not known or is not clear. For example, when there are contaminants or truncated constructs, it starts with a model that only has a backbone and ignores side chains. Then it scores residue-type likelihoods against a database and delivers ranked sequence hits. This algorithm makes it possible to automatically assign and quickly rectify mistakes in automatically produced yet fragmented models. It was able to discover sequences for models built into crystallographic and cryo-EM maps in benchmarks. It was then turned into check my sequence for detecting and fixing register shifts. The results made it a useful tool in Phenix workflows when MR presents a model, but the sequence identity is unclear (Chojnowski et al., 2022).

8.5.9 DeepCentering

It uses a convolutional neural network trained to find crystals in loops and capillaries from multiview microscope images to automate synchrotron crystal centering (Ito et al., 2019). This replaces manual joystick centering or simple image-processing heuristics that have trouble with ice, glare, or small crystals. The technology is used on beamlines like SPring-8 to find loops, locate crystals, and align the beam axis so that the crystal is at the X-ray intersection with high accuracy and speed. This process makes it possible to gather data without supervision and speeds up high-volume screening. Reports reveal that the system works well with all types of samples and lighting circumstances, and that beamline control software includes end-to-end automation.

These breakthroughs have changed the appearance of structural biology. They collectively represent a paradigm shift from slow, labor-intensive structural determination to rapid, AI-assisted research. Recently, a growing ecosystem of platforms has arisen, encompassing practically every facet of structure-based research, from sequence-to-structure prediction to cryo-EM map interpretation to drug-target interaction analysis. Here we present a curated list of recent AI/ML tools in structural biology, along with a thorough table of platforms Table 8.2.

Table 8.2 Table of recent artificial intelligence (*AI*)/machine learning (*ML*) tools (2020–25).

Tool/platform	Purpose	Web address
AlphaFold2	The prediction of protein structures from sequences is achieved with near-atomic accuracy.	https://www.deepmind.com/research/highlighted-research/alphafold
CryoDRGN	Deep learning (DL) for cryo-EM heterogeneity and continuous conformations.	https://github.com/zhonge/cryodrgn
DeepTracer	Automated de novo atomic modeling from cryo-EM density maps.	https://deeptracer.uw.edu
Topaz	CNN-based cryo-EM particle picking.	https://github.com/tbepler/topaz
GraphSite	Graph neural network for ligand-binding site prediction.	https://github.com/huhlim/GraphSite
DeepRank	Framework for analyzing protein–protein interactions using DL.	https://deeprank.github.io
MaSIF	Geometric DL for biomolecular surface interactions.	https://github.com/LPDI-EPFL/masif
Atomwise	AI-driven small-molecule virtual screening.	https://www.atomwise.com
DeepDock	Geometric DL for ligand pose prediction.	https://github.com/DeepDock/deepdock
PocketMiner	Predicts hidden/cryptic binding pockets in protein structures.	https://github.com/deepdruglab/pocketminer

(*Continued*)

Table 8.2 (Continued)

Tool/platform	Purpose	Web address
ESM Metagenomic Atlas	Large-scale structural atlas of metagenomic proteins.	https://esmatlas.com
ModelArchive	Repository for AI and computational structural models.	https://www.modelarchive.org
crYOLO	CNN (YOLO) particle picking (incl. general models)	https://cryolo.readthedocs.io/
CryoTransformer	Transformer-based particle picking	https://github.com/jianlin-cheng/CryoTransformer
CryoSegNet	Segmentation-style particle picking in micrographs	https://github.com/jianlin-cheng/CryoSegNet
DRPnet	Deep net for automated particle center detection	https://github.com/emcoregit/DRPnet
Joint denoise + pick	Joint micrograph denoising and particle localization (DL)	https://github.com/nextpyp/spr_pick
Diffusion-model denoiser	Diffusion-based restoration/denoising of micrographs	https://arxiv.org/html/2401.01097v1 (arXiv)
DeepEMhancer	DL postprocessing (map sharpening/masking)	https://github.com/rsanchezgarc/deepEMhancer
IsoNet	Deep missing-wedge compensation/isotropic tomogram restoration	https://github.com/IsoNet-cryoET/IsoNet
CryoCARE	Content-aware DL denoising for cryo-TEM/ET	https://github.com/juglab/cryoCARE_T2T
DeepDeWedge	Self-supervised simultaneous denoising + missing-wedge correction	https://github.com/MLI-lab/DeepDeWedge
REST	DL restoration (denoise + wedge compensation) of cryo-ET volumes	https://github.com/Zhang-hn1125/REST
TomoTwin	Representation-learning particle picking in tomograms	https://github.com/MPI-Dortmund/tomotwin-cryoet
DeepFinder	CNN-based macromolecule localization/segmentation in tomograms	https://gitlab.inria.fr/serpico/deep-finder
MemBrain	DL pipeline for detecting membrane-bound complexes in tomograms	https://github.com/CellArchLab/MemBrain
DeePiCt	Supervised DL segmentation + particle localization (cellular context)	https://github.com/ZauggGroup/DeePiCt

(*Continued*)

Table 8.2 (Continued)

Tool/platform	Purpose	Web address
"AIS" platform	ML platform for segmentation and picking; shareable trained models	https://github.com/bionanopatterning/Ais
RoseTTAFold	Three-track network for folding	https://github.com/RosettaCommons/RoseTTAFold
ESMFold	MSA-free, LLM-based folding	https://esmatlas.com/resources?action=fold
OmegaFold	Single-sequence (language-model + transformer)	https://github.com/HeliXonProtein/OmegaFold
HelixFold-Single	MSA-free folding with protein LMs	https://paddlehelix.baidu.com/app/drug/protein-single/forecast
trRosettaX-Single	Single-sequence folding (s-ESM-1b embeddings)	https://yanglab.qd.sdu.edu.cn/trRosetta/benchmark_single/
RaptorX-Single	Single-sequence folding (PLM-integrated)	https://opig.stats.ox.ac.uk/webapps/sabdab-sabpred/sabdab/
OpenFold	Trainable open re-implementation of AF2	https://github.com/aqlaboratory/openfold
Uni-Fold	Open platform for AF2/multimer-like training and inference	https://github.com/dptech-corp/Uni-Fold
FastFold	Speed-optimized AF2 training/inference	https://github.com/hpcaitech/FastFold

No Permission Required.

8.6 Artificial intelligence applications in structural biology

The 3D structures of macromolecules help us understand how life works and are the basis for any structure-based discovery. Experimental methods that required a significant amount of time and effort in the past limited the study of 3D structures. The use of AI, especially DL, has entirely changed the scenario. AI models can now take big data of complex structural information, find subtle, nonlinear patterns, and make predictions that are almost as accurate as experiments. This accelerates the process of discovering new structures, molecules, and pathways. Here, we discuss major uses of AI in structural biology.

8.6.1 Sequence to structure

The most well-known use of AI in structural biology is figuring out a protein's 3D structure from its sequence. This accomplishment signifies a paradigm change, converting a decades-long grand challenge into computer work. AlphaFold2's structures are as accurate as experimental ones, especially for single-domain proteins. It uses an attention-based neural network architecture to estimate the distances and orientations between residues at the same time. It approaches the protein folding challenge as a geometric problem within a spatial graph. It can predict not only single chains but also complexes, including protein–protein interactions, protein–DNA interactions, and other multimers using methods such as AlphaFold-Multimer and RoseTTAFold. These models are essential for understanding how cells send signals and how complex biological pathways work, which are often controlled by enormous groups of molecules. The recent release of AlphaFold3 has also made it possible to predict the structure and interactions of proteins with a wider range of molecules, including DNA, RNA, and small-molecule ligands. This has opened up new areas for molecular modeling. Generative protein design is another important area of research. New models don't ask if a sequence folds into a structure; they ask how to make sequences and backbones that have the right shapes, symmetries, epitopes, or catalytic geometries. For example, RFdiffusion fine-tunes a structure prediction group for denoising operations. This method lets you conditionally generate backbones for monomers, symmetric oligomers, enzyme active-site scaffolds, and targeted binders, all of which have been tested in the lab (Watson et al., 2023). ProteinMPNN works with atomic backbones to create sequences that encode the target structures more robustly than earlier energy-based methods and works for a wide range of design problems (Dauparas et al., 2022). The new ecosystem now combines these features: first, use a similar diffusion/backbone generator to suggest a form and interface; then, use ProteinMPNN-style sequence optimization; and finally, utilize AlphaFold-class predictors and experimental tests to check. In reality, this loop has already produced de novo binders and symmetric nanomaterials, and it is being used increasingly for medicines and enzyme scaffolds.

8.6.2 Imaging data analysis and interpretation

Cryo-EM produces massive datasets, often reflecting proteins in multiple conformational states. Traditional reconstruction methods are used to make average particles into a single model, missing flexibility. In cryo-EM, AI and DL have become essential for automating and improving data processing:

- **Particle picking and 2D classification:** Deep CNNs are currently utilized to quickly and correctly find individual protein particle pictures in noisy raw micrographs. This process is called particle picking and classification. ML is used by tools like Warp, Topaz, and cryoSPARC to separate signals from noise, which speeds up the collection and curation of data by a lot.
- **Conformational heterogeneity:** Numerous proteins exist in several functional states (conformers). CryoDRGN's VAEs can sort these small structural changes inside a single dataset. This provides dynamic insights into protein function that are difficult to obtain through other methods.

- **Model building and interpretation**: AI-predicted structures, such as AlphaFold models, are being used more and more as high-quality starting points for fitting into experimental cryo-EM density maps. This speeds up and automates the final model construction and refinement process. DeepTracer accelerates atomic model building, fitting protein backbones into density maps.

AI also enters cryo-ET from start to finish with useful tools at every step. For example, denoisers like Topaz-Denoise, cryoCARE, and IsoNet boost SNR and reduce missing-edge artefacts in tilt series/tomograms; learnt pickers like TomoTwin and crYOLO-tomo find particles in crowded cellular volumes; and voxel-wise segmenters in EMAN2 (neural network segmenter), ilastik (Berg et al., 2019), SuRVoS2 (Pennington et al., 2022), and DeepFinder (Moebel et al., 2021) outline membranes, organelles, and complexes for targeted subtomogram averaging. IsoNet can also "restore" tomograms to get rid of missing-wedge blur, and map enhancers from single-particle procedures can assist in making features sharper for creating models. Warp does preprocessing on the fly as the data is collected. Then, downstream packages like RELION and Dynamo (Coray et al., 2024) employ these AI-prepared selects and masks to align and average subtomograms. When used together, these tools make cryo-ET a smoother, faster way to look at molecular machines inside cells.

8.6.3 Interactions analysis

Proteins normally function together, not by themselves. These interactions do a lot of work in cells. It's challenging to identify these contact sites in the lab, but AI can help by generating accurate estimates. DeepRank looks at where two proteins connect and how alterations (mutations) might impact that contact (Réau et al., 2023). MaSIF uses the structure of protein surfaces to determine appropriate docking partners (Gainza et al., 2020). AlphaFold-Multimer creates models of protein clusters to provide a comprehensive view (Homma et al., 2023). These tools are useful in many different fields. For example, they can help design vaccines by predicting how antibodies stick to antigens. They can also help explain how signals move by modeling kinases and adapter proteins. In synthetic biology, they can help make new materials and therapies by predicting contacts in designed protein circuits. AI also speeds up medication work by demonstrating how small chemicals could connect to a protein. GraphSite shows where molecules are likely to bond (Yuan et al., 2022). DeepDock (https://github.com/OptiMaL-PSE-Lab/DeepDock) and atomwise use DL to quickly locate promising hits in numerous molecules (Wallach et al., 2024). PocketMiner can even find "cryptic" pockets that are difficult to identify and that make it easier to drug tough targets (Meller et al., 2023). Researchers apply these approaches to find kinase inhibitors for cancer, to mimic enzyme targets for infectious diseases, and to repurpose medications by testing them on different targets. AI-made structures and AI-driven docking collaborate to build comprehensive pipelines that accelerate virtual screening and enhance hit improvement.

8.7 Challenges and limitations

AI has many advantages, such as quick structure prediction, annotation, flexible insights, and methods to combine various data types. However, it also has drawbacks, such as unknown decision-making behavior, biased training sets, a requirement for meticulous validation, high computational and energy requirements, and limited dynamic handling. AlphaFold2 and RoseTTAFold are all-powerful tools, but they function best on the small, stable, and soluble proteins that constitute the PDB. Membrane proteins, large assemblies, and disordered sections are still hard classes (Fig. 8.2). Even newer single-sequence models like ESMFold only help somewhat because the bias in the training data is still there. Another problem is that these systems are making decisions unknowingly, with no clarification from the AI models. Scores like pLDDT inform us whether we can trust a model, but they don't explain why a fold is correct or what physics caused it. That makes it tougher to learn how anything works just based on a guess. AI output should therefore be regarded as a compelling hypothesis rather than evidence. Checks using crystal, NMR, or cryo-EM are still required. A poor decision on a mutation in the

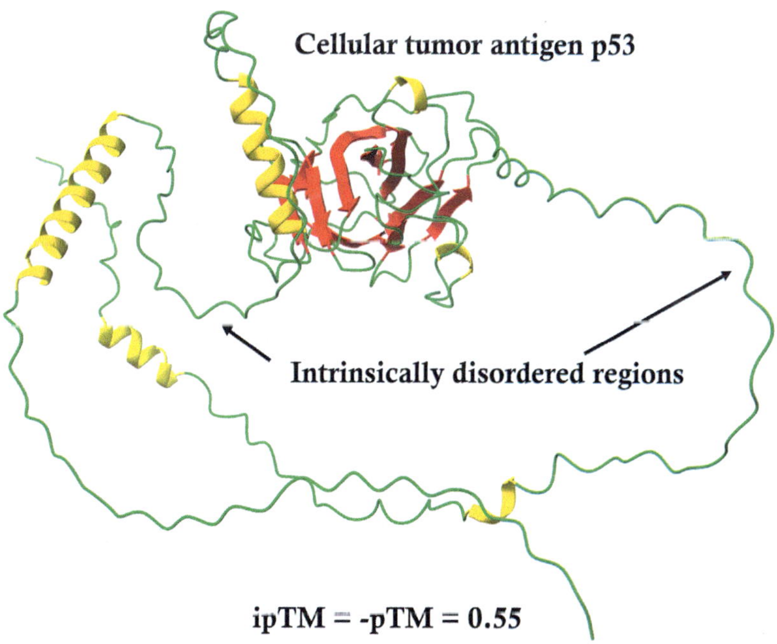

Figure 8.2 AlphaFold model of intrinsically disordered proteins from human p53. It shows a large region that cannot be modeled correctly. This is the major challenge in the prediction of protein structures by artificial intelligence algorithms. The reported ipTM (≈ pTM) = 0.55 shows moderate to poor global confidence, which is common when there are a lot of IDRs. This implies that the relative positioning of domains and lengthy flexible segments is ambiguous, but the folded core is relatively reliable. *No Permission Required.*

clinic could endanger patients, while a poor model in drug development can cost time and money. Computer needs are also important: training requires a lot of hardware and energy, and even normal runs might impose a load on labs lacking powerful GPUs or high-performance computers. Open resources like AlphaFold DB and ESM Atlas lessen the barrier, but not everyone has equal access.

8.8 Future directions

Although we are just beginning, AI has already influenced structural biology. Beyond isolated, static structures, the next wave will integrate various data types, real motion, the creation of novel proteins and medications, and even patient care. To show how a protein's structure relates to its function in the cell, models will first become multimodal, combining genomes, transcriptomics, proteomics, and metabolomics rather than just sequence or pictures. Second, generative AI will not just forecast what nature has created; it will also assist in the deliberate design of new proteins and small molecules, such as stronger antibodies, better enzymes, and intelligent components for synthetic biology. Third, AI will collaborate with molecular dynamics to record motion, including the energy landscape, the routes between states, and the way disorder and allostery evolve, providing us with "molecular movies" rather than still images. Last but not least, the discipline needs to develop responsibly by lowering the cost of running models, maintaining access to essential resources for labs worldwide, and enhancing interpretability to prevent outputs from being black boxes. By doing this, AI will transition from a useful helper to a central force in healthcare and discovery, creating novel biomolecules, elucidating illness, and directing therapy while remaining fair, transparent, and tested.

8.9 Conclusion

Structural biology used to be a slow, data-poor subject, but thanks to AI and ML, it can now predict and analyze structures quickly. Much of the folding difficulty was resolved by tools like AlphaFold2, and tools like CryoDRGN made it simpler to evaluate cryo-EM data, especially for flexible and moving proteins. This ability has made it possible for numerous labs to investigate goals that previously seemed unattainable. However, there are restrictions. Models are difficult to explain, can be expensive to operate, can inherit bias from their training data, and still require lab testing to produce reliable results. In the future, AI can be brought from the lab to the patient's bedside by combining various data types, employing generative design to create novel proteins and medications, and connecting with clinical data. AI will not replace structural biology as long as we develop these tools fairly, openly, and carefully while continuing to validate them. Instead, AI will continue to reshape structural biology, transforming long-standing issues into useful jobs and creating new avenues for research and medical advancement.

References

Aithani, L., Alcaide, E., Bartunov, S., Cooper, C.D.O., Doré, A.S., Lane, T.J., et al., 2023. Advancing structural biology through breakthroughs in AI. Current Opinion in Structural Biology 80. Available at https://doi.org/10.1016/j.sbi.2023.102601, http://www.elsevier.com/locate/sbi.

Angermueller, C., Pärnamaa, T., Parts, L., Stegle, O., 2016. Deep learning for computational biology. Molecular systems Biology 12 (7). Available at https://doi.org/10.15252/msb.20156651.

Baek, M., DiMaio, F., Anishchenko, I., Dauparas, J., Ovchinnikov, S., Lee, G.R., et al., 2021. Accurate prediction of protein structures and interactions using a three-track neural network. Science 373 (6557), 871–876. Available at https://doi.org/10.1126/science.abj8754, https://science.sciencemag.org/content/373/6557/871.full.

Ban, N., Nissen, P., Hansen, J., Moore, P.B., Steitz, T.A., 2000. The complete atomic structure of the large ribosomal subunit at 2.4 Å resolution. Science 289 (5481), 905–920. Available at https://doi.org/10.1126/science.289.5481.905.

Bateman, A., Martin, M.-J., Orchard, S., Magrane, M., Adesina, A., Ahmad, S., et al., 2025. Zhang, uniprot: the universal protein knowledgebase in 2025. Nucleic Acids Research 53 (D1). Available at https://doi.org/10.1093/nar/gkae1010.

Bepler, T., Kelley, K., Noble, A.J., Berger, B., 2020. Topaz-Denoise: general deep denoising models for cryoEM and cryoET. Nature Communications 11 (1). Available at https://doi.org/10.1038/s41467-020-18952-1.

Berg, S., Kutra, D., Kroeger, T., Straehle, C.N., Kausler, B.X., Haubold, C., et al., 2019. ilastik: interactive machine learning for (bio)image analysis. Nature Methods 16 (12), 1226–1232. Available at https://doi.org/10.1038/s41592-019-0582-9.

Berman, H.M., Westbrook, J., Feng, Z., Gilliland, G., Bhat, Weissig, H., et al., 2000. The protein data bank. Nucleic Acids Research 28 (1), 235–242. Available at https://doi.org/10.1093/nar/28.1.235.

Buchholz, T.-O., Jordan, M., Pigino, G., Jug, F., 2019. Cryo-CARE: content-aware image restoration for cryo-transmission electron microscopy data. Proceedings - International Symposium on Biomedical Imaging 2019-April, 502–506. Available at https://doi.org/10.1109/ISBI.2019.8759519.

Carugo, O., Djinović-Carugo, K., 2023. Structural biology: a golden era. PLoS Biology 21 (6), e3002187. Available at https://doi.org/10.1371/journal.pbio.3002187.

Chojnowski, G., Simpkin, A.J., Leonardo, D.A., Seifert-Davila, W., Vivas-Ruiz, D.E., Keegan, R.M., et al., 2022. a neural network based approach for identification of unknown proteins in X ray crystallography and cryo-EM. IUCrJ 9 (Pt 1), 86–97. Available at https://doi.org/10.1107/S2052252521011088.

Coray, R., Navarro, P., Scaramuzza, S., Stahlberg, H., Castaño-Díez, D., 2024. Automated fiducial-based alignment of cryo-electron tomography tilt series in Dynamo. Structure (London, England: 1993) 32 (10).

Dauparas, J., Anishchenko, I., Bennett, N., Bai, H., Ragotte, R.J., Milles, L.F., et al., 2022. Robust deep learning–based protein sequence design using ProteinMPNN. Science 378 (6615), 49–56. Available at https://doi.org/10.1126/science.add2187.

Draizen, E.J., Readey, J., Mura, C., Bourne, P.E., 2024. Prop3D: A flexible, Python-based platform for machine learning with protein structural properties and biophysical data. BMC Bioinformatics 25 (1). Available at https://doi.org/10.1186/s12859-023-05586-5, https://bmcbioinformatics.biomedcentral.com/.

Drenth, 2007. Principles of Protein X-ray Crystallography, 3rd ed. Springer.

Eddy, S.R., 1998. Profile hidden Markov models. Bioinformatics 14 (9), 755–763. Available at https://doi.org/10.1093/bioinformatics/14.9.755.

Emsley, P., Lohkamp, B., Scott, W.G., Cowtan, K., 2010. Features and development of Coot. Acta Crystallographica. Section D, Biological Crystallography 66 (Pt 4), 486–501. Available at https://doi.org/10.1107/S0907444910007493.

Fariselli, P., Casadio, R., 1999. A neural network based predictor of residue contacts in proteins. Protein Eng 12 (1), 15–21. Available at https://doi.org/10.1093/protein/12.1.15. PMID: 10065706.

Gainza, P., Sverrisson, F., Monti, F., Rodolà, E., Boscaini, D., Bronstein, M.M., et al., 2020. Deciphering interaction fingerprints from protein molecular surfaces using geometric deep learning. Nature Methods 17 (2), 184–192. Available at https://doi.org/10.1038/s41592-019-0666-6.

Hegde, R.S., Keenan, R.J., 2022. The mechanisms of integral membrane protein biogenesis. Nature Reviews Molecular Cell Biology 23 (2), 107–124. Available at https://doi.org/10.1038/s41580-021-00413-2, http://www.nature.com/molcellbio.

Homma, F., Huang, J., van der Hoorn, R.A.L., 2023. AlphaFold-Multimer predicts cross-kingdom interactions at the plant-pathogen interface. Nature Communications 14 (1). Available at https://doi.org/10.1038/s41467-023-41721-9.

Ito, S., Ueno, G., Yamamoto, M., 2019. DeepCentering: fully automated crystal centering using deep learning for macromolecular crystallography. Journal of Synchrotron Radiation 26 (Pt 4), 1361–1366. Available at https://doi.org/10.1107/S160057751900434X.

Jumper, J., Evans, R., Pritzel, A., Green, T., Figurnov, M., Ronneberger, O., et al., 2021. Highly accurate protein structure prediction with AlphaFold. Nature 596 (7873), 583–589. Available at https://doi.org/10.1038/s41586-021-03819-2, http://www.nature.com/nature/index.html.

Kim, H., Park, H., 2003. Protein secondary structure prediction based on an improved support vector machines approach. Protein Engineering 16 (8), 553–560. Available at https://doi.org/10.1093/protein/gzg072, www3.oup.co.uk/proeng.

Kühlbrandt, W., 2014. The resolution revolution. Science 343 (6178), 1443–1444. Available at https://doi.org/10.1126/science.1251652.

Kumar, N., Srivastava, R., 2024. Deep learning in structural bioinformatics: current applications and future perspectives. Briefings in Bioinformatics 25 (3). Available at https://doi.org/10.1093/bib/bbae042.

Lin, Z., Akin, H., Rao, R., Hie, B., Zhu, Z., Lu, W., et al., 2023. Evolutionary-scale prediction of atomic-level protein structure with a language model. Science 379 (6637), 1123–1130. Available at https://doi.org/10.1126/science.ade2574.

Liu, Y.T., Zhang, H., Wang, H., Tao, C.L., Bi, G.Q., Zhou, Z.H., 2022. Isotropic reconstruction for electron tomography with deep learning. Nature Communications 13 (1). Available at https://doi.org/10.1038/s41467-022-33957-8, https://www.nature.com/ncomms/.

Meller, A., Ward, M., Borowsky, J., Kshirsagar, M., Lotthammer, J.M., Oviedo, F., et al., 2023. Predicting locations of cryptic pockets from single protein structures using the PocketMiner graph neural network. Nature Communications 14 (1). Available at https://doi.org/10.1038/s41467-023-36699-3.

Meng, E.C., Goddard, T.D., Pettersen, E.F., Couch, G.S., Pearson, Z.J., Morris, J.H., et al., 2023. UCSF ChimeraX: tools for structure building and analysis. Protein Science: A Publication of the Protein Society 32 (11). Available at https://doi.org/10.1002/pro.4792.

Mirdita, M., Steinegger, M., Söding, J., 2019. MMseqs2 desktop and local web server app for fast, interactive sequence searches. Bioinformatics (Oxford, England) 35 (16), 2856–2858. Available at https://doi.org/10.1093/bioinformatics/bty1057.

Moebel, E., Martinez-Sanchez, A., Lamm, L., Righetto, R.D., Wietrzynski, W., Albert, S., et al., 2021. Deep learning improves macromolecule identification in 3D cellular cryo-electron tomograms. Nature Methods 18 (11), 1386–1394. Available at https://doi.org/10.1038/s41592-021-01275-4.

Pennington, A., King, O.N.F., Tun, W.M., Ho, E.M.L., Luengo, I., Darrow, M.C., et al., 2022. SuRVoS 2: accelerating annotation and segmentation for large volumetric bioimage workflows across modalities and scales. Frontiers in Cell and Developmental Biology 10. Available at https://doi.org/10.3389/fcell.2022.842342.

Pfab, J., Phan, N.M., Si, D., 2021. DeepTracer for fast de novo cryo-EM protein structure modeling and special studies on cov-related complexes. Proceedings of the National Academy of Sciences of the United States of America 118 (2). Available at https://doi.org/10.1073/pnas.2017525118.

Potter, S.C., Luciani, A., Eddy, S.R., Park, Y., Lopez, R., Finn, R.D., 2018. HMMER web server: 2018 update. Nucleic Acids Research 46 (1), W200. Available at https://doi.org/10.1093/nar/gky448, https://academic.oup.com/nar/issue.

Punjani, A., Rubinstein, J.L., Fleet, D.J., Brubaker, M.A., 2017. CryoSPARC: algorithms for rapid unsupervised cryo-EM structure determination. Nature Methods 14 (3), 290–296. Available at https://doi.org/10.1038/nmeth.4169.

Rao, R., Bhattacharya, N., Thomas, N., Duan, Y., Chen, X., Canny, J., et al., 2019. Evaluating protein transfer learning with TAPE. Advances in Neural Information Processing Systems 32. Available at https://papers.nips.cc/.

Réau, M., Renaud, N., Xue, L.C., Bonvin, A.M.J.J., 2023. DeepRank-GNN: a graph neural network framework to learn patterns in protein–protein interfaces. Bioinformatics 39 (1). Available at https://doi.org/10.1093/bioinformatics/btac759.

Rohou, A., Grigorieff, N., 2015. CTFFIND4: fast and accurate defocus estimation from electron micrographs. Journal of Structural Biology 192 (2), 216–221.

Scheres, S.H.W., 2012. RELION: implementation of a Bayesian approach to cryo-EM structure determination. Journal of Structural Biology 180 (3), 519–530.

Sievers, F., Higgins, D.G., 2018. Clustal omega for making accurate alignments of many protein sequences. Protein Science: A Publication of the Protein Society 27 (1), 135–145. Available at https://doi.org/10.1002/pro.3290.

Sonnhammer, E.L.L., Eddy, S.R., Birney, E., Bateman, A., Durbin, R., 1998. Pfam: multiple sequence alignments and HMM-profiles of protein domains. Nucleic Acids Research 26 (1), 320–322. Available at https://doi.org/10.1093/nar/26.1.320.

Steinegger, M., Meier, M., Mirdita, M., Vöhringer, H., Haunsberger, S.J., Söding, J., 2019. HH-suite3 for fast remote homology detection and deep protein annotation. BMC Bioinformatics 20 (1). Available at https://doi.org/10.1186/s12859-019-3019-7.

Tegunov, D., Cramer, P., 2019. Real-time cryo-electron microscopy data preprocessing with Warp. Nature Methods 16 (11), 1146–1152. Available at https://doi.org/10.1038/s41592-019-0580-y.

The wwPDB Consortium, EMDB—the Electron Microscopy Data Bank. Nucleic Acids Research. 52 (D1) (2024), D456, Available from: https://doi.org/10.1093/nar/gkad1019.

Trivedi, R., Nagarajaram, H.A., 2022. Intrinsically disordered proteins: an overview. International Journal of Molecular Sciences 23 (22), 14050. Available at https://doi.org/10.3390/ijms232214050.

Tunyasuvunakool, K., Adler, J., Wu, Z., Green, T., Zielinski, M., Žídek, A., et al., 2021. Highly accurate protein structure prediction for the human proteome. Nature 596 (7873), 590–596. Available at https://doi.org/10.1038/s41586-021-03828-1, http://www.nature.com/nature/index.html.

Varadi, M., Anyango, S., Deshpande, M., Nair, S., Natassia, C., Yordanova, G., et al., 2022. AlphaFold protein structure database: massively expanding the structural coverage of protein-sequence space with high-accuracy models. Nucleic Acids Research 50 (D1), D439. Available at https://doi.org/10.1093/nar/gkab1061.

Vaswani, A., Shazeer, N., Parmar, N., Uszkoreit, J., Jones, L., Gomez, A.N., et al., 2017. Neural information processing systems foundation United States Attention is all you need. Advances in Neural Information Processing Systems 2017, 5999–6009.

Wallach, I., Bernard, D., Nguyen, K., Ho, G., Morrison, A., Stecula, A., et al., 2024. AI is a viable alternative to high throughput screening: a 318-target study. Scientific Reports 14 (1). Available at https://doi.org/10.1038/s41598-024-54655-z.

Watson, J.D., Crick, F.H.C., 1953. Molecular structure of nucleic acids: a structure for deoxyribose nucleic acid. Nature 171 (4356), 737–738. Available at https://doi.org/10.1038/171737a0.

Watson, J.L., Juergens, D., Bennett, N.R., Trippe, B.L., Yim, J., Eisenach, H.E., et al., 2023. De novo design of protein structure and function with RFdiffusion. Nature 620 (7976), 1089–1100. Available at https://doi.org/10.1038/s41586-023-06415-8, https://www.nature.com/nature/.

Wiedemann, S., Heckel, R., 2024. A deep learning method for simultaneous denoising and missing wedge reconstruction in cryogenic electron tomography. Nature Communications 15 (1). Available at https://doi.org/10.1038/s41467-024-51438-y.

Wüthrich, K., 2001. The way to NMR structures of proteins. Nature Structural Biology 8 (11), 923–925. Available at https://doi.org/10.1038/nsb1101-923.

Yuan, Q., Chen, S., Rao, J., Zheng, S., Zhao, H., Yang, Y., 2022. AlphaFold2-aware protein-DNA binding site prediction using graph transformer. Briefings in Bioinformatics 23 (2). Available at https://doi.org/10.1093/bib/bbab564.

Zheng, S., Palovcak, E., Armache, J.P., et al., 2017. MotionCor2: anisotropic correction of beam-induced motion for improved cryo-electron microscopy. Nat Methods 14, 331–332. Available at https://doi.org/10.1038/nmeth.4193.

Zhong, E.D., Bepler, T., Berger, B., Davis, J.H., 2021. CryoDRGN: reconstruction of heterogeneous cryo-EM structures using neural networks. Nature Methods 18 (2), 176–185. Available at https://doi.org/10.1038/s41592-020-01049-4, http://www.nature.com/nmeth/.

CHAPTER 9

Integrating structural and omics data for drug discovery

Neha[1] and Khushi Raza[2]
[1]Department of Forensic Science, School of Basic and Applied Sciences, SGT University, Gurugram, Haryana, India,
[2]Department of Toxicology, Jamia Hamdard, New Delhi, New Delhi, India

List of abbreviations

cryo-EM	Cryo-electron microscopy
DILI	Drug-induced liver injury
FAIR	Findability, Accessibility, Interoperability, and Reusability
FBDD	Fragment-based drug discovery
GNNs	Graph neural networks
GPUs	Graphics processing units
GWAS	Genome-wide association studies
HTS	High-throughput screening
LC-MS	Liquid chromatography-mass spectrometry
MD	Molecular dynamics
MS/MS	Tandem mass spectroscopy
NMR	Nuclear magnetic resonance
ODE	Ordinary differential equation
PPI	Protein–protein interaction
PTM	Posttranslational modifications
PWAS	Protein-wide association studies
RNA-seq	RNA sequencing
scRNA-seq	Single-cell transcriptomics
SBDD	Structure-based drug design
SNPs	Single-nucleotide polymorphisms

9.1 Introduction

Integrative structural biology represents an innovative strategy in drug discovery, comprising several different experimental and computational approaches that provide extensive molecular insights into biomolecular structure, dynamics, and function (Naithani and Guleria, 2024). This multifaceted strategy combines X-ray crystallography, cryo-electron microscopy (cryo-EM), nuclear magnetic resonance (NMR) spectroscopy, molecular dynamics (MD) simulations, and biophysical methods to

Advances in Structural Biology
DOI: https://doi.org/10.1016/B978-0-443-44988-8.00003-4

convert structural data into useful chemical knowledge. When high-resolution structural data, computationally predicted outcomes, machine learning algorithms, and biophysical characterization are integrated, it becomes possible to rationally design better, more selective, and safer drugs (Son et al., 2024). This chapter explores the principles, methodologies, applications, and emerging technologies in integrative structural biology, highlighting their critical role in expediting the drug discovery process from target validation to lead optimization.

9.1.1 Historical perspective on structural biology and omics convergence

The drug discovery process has undergone an important transition from accidental discovery to mechanism-based design. The classical drug discovery process depended on qualitative screening and experimental findings, without sufficient molecular insight. X-ray crystallography in the mid-20th century transformed structural biology by providing atomic-resolution imaging of biomolecules (Moser et al., 2025). In 1958 scientists determined the first protein structure, myoglobin, marking the start of structure-based drug design (SBDD). The genomic revolution in the 1990s and early 2000s gave rise to the "omics" sciences, which complemented structural biology (Mardis and Wilson, 2025). The completion of the Human Genome Project in 2003 led to the emergence of the fields of transcriptomics, proteomics, and metabolomics. For a time, these disciplines worked largely independently, with structural biologists focusing on atomic-level details, whereas systems biologists focused on network-level properties (Karim et al., 2024). From 2010 onward, integrative methods combining structural data with multiomics profiling began to demonstrate improved prediction of drug efficacy and safety (Burley et al., 2024). Recent evidence shows that almost all FDA-approved anticancer drugs from 2019 to 2023 benefited from structural biology insights, and more than 80% were developed using structure-guided drug discovery workflows (Burley et al., 2024).

9.1.2 Fundamental principles of integrative approaches

Integrative structural biology is based on several fundamental principles that differentiate it from traditional drug discovery, which relies on a single method for drug discovery and target validation. These principles provide an analytical and realistic approach towards the systems-level understanding of protein dynamics, molecular recognition, and disease mechanisms (Schwalbe et al., 2024). The following subsections elaborate on each principle for successful drug discovery approaches (Fig. 9.1).

9.1.2.1 Bridging molecular structure and multiomics

At the atomic level, the binding of a ligand to a protein target is governed by the laws of thermodynamics and kinetics. The change in Gibbs free energy (ΔG) determines the

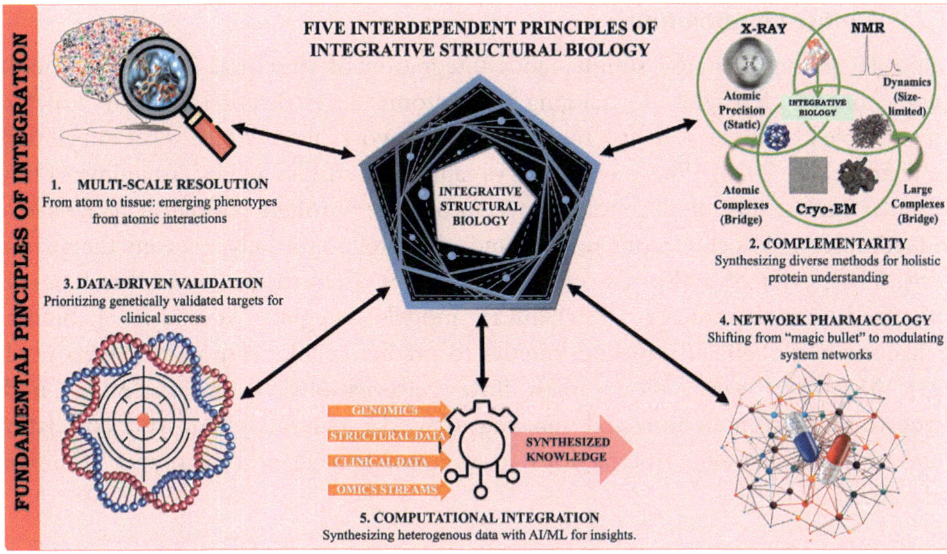

Figure 9.1 The pentagonal framework of integrative structural biology: a schematic illustration of the five foundational principles that distinguish modern systems-level discovery from traditional reductionist methodologies, as detailed in Section 9.1.2. This arrangement depicts how these areas work together and depend on each other: (1) multiscale resolution, (2) complementarity of methods, showing the functional intersections and unique domains of X-ray crystallography (atomic precision), nuclear magnetic resonance (*NMR*) spectroscopy (solution dynamics), and cryo-electron microscopy (*cryo-EM*) (macromolecular architecture); (3) data-driven validation, (4) network pharmacology, and (5) computational integration. *No permission required.*

binding affinity, whereas the association and dissociation rate constants (k_{on}, k_{off}) determine the duration of target engagement (Decherchi and Cavalli, 2020). When a drug binds to a pocket identified through X-ray crystallography or cryo-EM, it induces conformational changes that either inhibit or activate the protein's functional state. The transition to pathway-level changes occurs through the disruption of signal transduction cascades (Nada et al., 2024). For instance, the inhibition of a kinase target prevents the phosphorylation of downstream substrates. This proteomic alteration is measured via tandem mass spectrometry (MS/MS), revealing changes in the phosphorylation signatures of entire signaling modules (Lee et al., 2023). These signaling changes subsequently modulate transcription factor activity, leading to global shifts in mRNA abundance measured by RNA sequencing (RNA-seq) (Mitschka and Mayr, 2022). The integration of these molecular-level details into genome-scale models of metabolism allows researchers to predict how a mutation in a three-dimensional structure will affect the flux of metabolites throughout the metabolic network. This multiscale modeling bridge is essential for identifying "unknown unknowns" in drug toxicity and predicting whole-cell phenotypes (Seshadri et al., 2025).

9.1.2.2 Multiscale resolution

Multiscale resolution is the simultaneous integration of structural information across multiple orders of magnitude, ranging from atomic-level details (ångström scale) to more complex biomolecular structures, cellular networks, and system-level interactions (Reinhardt et al., 2023). This hierarchical organization enables an understanding of how molecular interactions at the atomic level propagate through biological pathways to produce observable cellular phenotypes and ultimately, clinically relevant therapeutic effects. The power of multiscale resolution emerges when these diverse data types are synthesized computationally. Mechanistic models integrate atomic-level binding mechanisms with cellular signaling kinetics to predict cellular responses (Myers et al., 2021). Network pharmacology maps how perturbations at one molecular node propagate through interconnected signaling networks, whereas phenotypic correlation connects specific binding interactions to measurable cellular phenotypes (Galindez et al., 2023).

9.1.2.3 Complementarity of methods

Experimental methods provide fundamentally different types of information; each method has its own strengths and weaknesses, and the weaknesses of one method are compensated for by the strengths of another (Capili and Anastasi, 2024). This principle explains why it costs a lot to use several complementary methods, such as X-ray crystallography, which provides high atomic resolution but requires crystallization. Similarly, cryo-EM reveals large macromolecular complexes in near-native states, whereas NMR spectroscopy characterizes solution-state dynamics (Zafar et al., 2025). Combining these complementary datasets provides more comprehensive biological insights than any single method.

9.1.2.4 Data-driven target validation

Modern integrative approaches prioritize genetic evidence from genome-wide association studies (GWAS) and multiomics validation over traditional biochemical screens alone. Targets with genetic support demonstrate twofold higher clinical success rates compared to those without genetic validation (King and Wu, 2025). However, target validation fundamentally changed with the emergence of human genetic approaches, providing biological confirmation that target-pathway perturbation causes disease. A landmark study by the Structural Genomics Consortium and collaborators demonstrated that drug targets with genetic evidence supporting human causal relationships show approximately twofold higher rates of clinical trial success compared to targets lacking such genetic support. They analyzed ~200 drug development programs across diverse therapeutic areas, with a ~28%–32% probability of advancing through clinical development, compared with ~14%–16% for programs lacking genetic support

(Szustakowski et al., 2021). Genetic evidence provides confidence that pathway perturbation in human disease, specifically, not just in model organisms or cell lines, produces therapeutic benefits.

9.1.2.5 Network pharmacology

Network pharmacology indicates that almost all diseases involve several dysregulated pathways rather than single-target defects. Similarly, effective therapeutics usually influence multiple interconnected targets, thereby offering greater therapeutic benefits than single-target drugs (Shah et al., 2023). It recognizes that off-target effects, once considered liabilities, often contribute to enhanced clinical efficacy. Integrative approaches embrace the concept of polypharmacology instead of the "one drug, one target" model because effective drugs often modulate interconnected targets within disease-relevant networks (Abdelsayed, 2025).

9.1.2.6 Computational integration

Modern drug discovery produces exponentially growing quantities of diverse data, including structural (atomic coordinates, NMR parameters, cryo-EM maps), genomic (sequence variants, copy number alterations, gene expression), biophysical (binding kinetics, thermodynamics, protein–ligand dynamics), proteomic (protein abundance, modifications, interactions), and phenotypic (cellular assays, organism-level responses, clinical outcomes) data. Machine learning, network analysis, and MD simulations integrate different data types into useful drug discovery insights by combining structural, chemical, and biological information. Machine learning analyzes chemical libraries to identify patterns that predict inhibitor efficacy (Ahmad et al., 2026). MD simulations elucidate the mechanisms of molecular binding and their thermodynamic stability. Network analysis examines polypharmacology and selectivity determinants by using graph-based models (Abbas et al., 2025). Integrating these diverse data types creates synergistic power; together, these computational methods bridge disciplines, transforming molecular biophysics into lead prioritization.

9.1.3 Current challenges in traditional single-method drug discovery

Despite significant advances in drug development over the last few decades and substantial investment in the field, major challenges remain. Drug development is still highly inefficient, with more than 90% of drug candidates failing during the clinical trials (Sun et al., 2022). This persistently high failure rate has remained largely unchanged over the past 2 decades, even though many optimization strategies have been implemented. This shows that traditional single-method approaches to drug discovery and development have fundamental limitations.

9.1.3.1 Target validation failure

Approximately 50% of clinical trial failures result from inadequate target validation; drug candidate advance to human trials based on compelling preclinical evidence supporting the target, yet fails to demonstrate efficacy in patients (Sun et al., 2022). This represents one of the most costly failure modes, as clinical trial expenses escalate dramatically with phase progression. A single Phase III failure in oncology costs ~$50–60 million and consumes 3–5 years of development time. Because target-related clinical attrition accounts for nearly 50% of failures, the total financial loss across the industry amounts to hundreds of millions to billions of dollars annually (Jentzsch et al., 2023).

9.1.3.2 Off-target effects and toxicity

Structural information alone cannot predict the full spectrum of drug–protein interactions across the proteome. Unexpected toxicity from off-target binding represents a major cause of late-stage attrition and postmarket withdrawals. Approximately 30% of clinical trial failures result from unmanageable off-target toxicity and side effects. Moreover, postmarket drug withdrawals frequently result from off-target interactions undetected during preclinical development (Sun et al., 2022; Amorim et al., 2024). Traditional drug discovery optimizes compounds against a single target (the "intended target"), yet typically does not systematically evaluate binding across the entire human proteome (~20,000 proteins) (Dale et al., 2021). The consequence is that unexpected off-target interactions are often discovered only after extensive human exposure.

9.1.3.3 Limited coverage of druggable space

Traditional high-throughput screening (HTS) covers only a fraction of chemical space and often identifies compounds with poor ligand efficiency. Commercial HTS libraries typically contain 500,000 to 2 million compounds. The total accessible chemical space comprises 6–8 million commercially available compounds, whereas unsynthesized virtual chemical space extends to billions of potential molecules (Yadav et al., 2023). Traditional HTS samples, of which only 7%–30% of the available chemical space, creating a risk that the best lead compounds may be missed. Compounds found through traditional HTS usually show low ligand efficiency (Shaik and Balya, 2025). Many biologically significant binding sites do not have pockets of an appropriate size in static crystal structures, which makes them "invisible" to traditional screening methods. This can lead to systematic bias in identifying compounds that bind to conserved active sites, where crystal structures clearly show binding pockets, but not allosteric sites or cryptic pockets that are inaccessible in unliganded structures (Vajda et al., 2018).

9.2 Structural biology technologies and their applications

Modern drug discovery is based on an integrated set of structural biology technologies that provide different perspectives on how biomolecules interact. These approaches go far beyond traditional static structure determination and allow the visualization of dynamic processes that are crucial for rational drug design. The integration of different structural techniques creates a comprehensive understanding of drug–target interactions that would not be possible through a single method (Sachdev and Gupta, 2019) (Table 9.1).

9.2.1 X-ray crystallography in drug design: atomic precision for small-molecule interactions

X-ray crystallography is a key component of drug design, as it provides atomic-resolution accuracy for determining how small-molecule interactions take place within protein–binding sites. The SBDD workflow uses crystal structures to facilitate the gradual optimization of compounds (Bijak et al., 2023). High-throughput crystallographic screening readily tests fragment libraries against target proteins. Crystallography exhibits some inherent limitations; for example, crystal-packing artifacts can alter binding models, static structures show time-averaged conformations that hide transient functional states, and obtaining diffraction-quality crystals is still hard to retrieve (Markiewicz et al., 2026). Computational docking, cryo-EM, and solution-phase techniques are complementary approaches that deal with these limitations and provide structural ensembles that are more physiologically relevant.

9.2.2 Cryo-electron microscopy: revolutionizing large-complex characterization

Cryo-EM has changed the characterization of large, heterogeneous protein complexes that were complicated to study with crystallography. Single-particle analysis allows direct observation of drug targets at near-atomic resolution without the need for crystallization, making cryo-EM a valuable technique for membrane proteins. (Balasco Serrão, 2025). Cryo-EM is critical because it shows conformational heterogeneity in multiple equilibrium states at the same time, providing new insights about dynamic protein states that are important for analyzing ligand-binding mechanisms and allosteric regulation (Mao, 2025). Applications of membrane protein drug discovery have been expedited as cryo-EM bypasses crystallization limitations related to integral membrane proteins, allowing structural characterization of G protein-coupled receptors, ion channels, and transporters.

Table 9.1 Comparative analysis of structural biology methodologies in drug discovery: the table compares the three primary high-resolution structural techniques (X-ray crystallography, cryo-electron microscopy [*cryo-EM*] microscopy, and nuclear magnetic resonance [*NMR*] spectroscopy), highlighting their technical capabilities, including resolution limits, sample-state requirements, and applicable molecular weight ranges.

Feature	X-ray crystallography	Cryo-EM microscopy	NMR spectroscopy	References
Primary resolution	Atomic (<1.5 Å possible)	Near-atomic (2–4 Å typical)	Atomic (local geometry)	Le Marchand et al. (2022), Balasco Serrão (2025)
Sample state	Crystalline (solid)	Vitrified (frozen-hydrated)	Solution (liquid)	Mao (2025), Zafar et al. (2025), Markiewicz et al. (2026)
Molecular weight	No strict limit (crystallization dependent)	Ideal for large complexes (>100 kDa)	Limited to small proteins (<30–50 kDa)	Moser et al. (2025)
Dynamics captured	Static snapshot (frozen in lattice)	Discrete conformational states	Continuous dynamics (ps to s timescale)	Le Marchand et al. (2022), Mao (2025)
Drug discovery role	High-throughput fragment screening; bond-level lead optimization	Mapping large, hard-to-crystallize targets (e.g., membrane proteins)	Fragment-based drug discovery: studying disordered regions	Dalvit et al. (2023), Bijak et al. (2023), Balasco Serrão (2025)
Key limitation	Requirement for diffraction-quality crystals	Computational cost; lower resolution for small molecules	Size limitations; requirement for isotopic labeling	Bijak et al. (2023)

Source: No permission required.

9.2.3 Nuclear magnetic resonance spectroscopy: solution-state insights

NMR spectroscopy detects solution-state dynamics, measuring the motion of proteins and kinetic exchange from millisecond-to-microsecond timescales. NMR directly measures changes in chemical shifts when a ligand binds, which helps to differentiate between conformational selection and induced-fit mechanisms and to describe allosteric networks using residue-specific dynamics (Le Marchand et al., 2022). NMR has improved fragment-based drug discovery (FBDD). For instance, fast-exchange NMR can detect low-affinity fragments (Kd >100 μM) that regular screening cannot detect, and chemical shift mapping can show where fragments bind without the need for labeling compounds (Dalvit et al., 2023). Ligand-detected NMR can also perform HTS of large compound libraries by using one-dimensional spectral analysis to find hits from a pool of thousands of compounds at once.

9.3 Omics technologies: comprehensive molecular profiling through integrated data layers

Modern drug discovery combines five synergistic omics modalities, that is, genomics, transcriptomics, proteomics, metabolomics, and epigenomics. Each method captures unique aspects of biological information, collectively allowing comprehensive molecular profiling that is not possible with any single technique alone. Modern pharmaceutical research indicates that omics work together in a hierarchy (Ivanisevic and Sewduth, 2023) (Fig. 9.2).

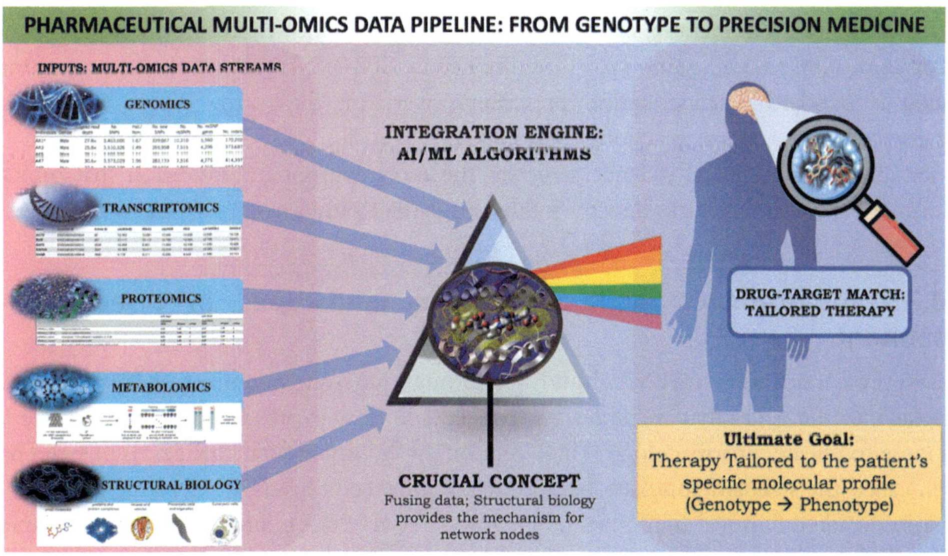

Figure 9.2 The "Structure-in-Network" integration workflow: the figure illustrates the process flow diagram showing how high-dimensional omics and structural data come together to create useful therapeutic insights. *No permission required.*

For instance, genomic variants set the baseline for how likely an individual is to develop a disease, epigenomic modifications control which genes can be transcribed, transcriptomics measures the amount of mRNA produced, proteomics measures the levels of functional proteins and posttranslational modifications (PTMs), and metabolomics measures the biochemical effects that occur afterwards (Sa et al., 2024). To understand how drug interventions move through this cascade from genetic predisposition to protein function to cellular metabolic states, all approaches must be combined; this makes precision medicine possible, where treatment plans are based on the unique molecular signatures of each patient (Table 9.2).

9.3.1 Genomics and pharmacogenomics in drug discovery

Genomics gives us the basic genetic information needed to determine how likely someone is to develop a disease and how they will respond to the intended drug. GWAS systematically identifies genetic variants linked to disease by examining millions of single-nucleotide polymorphisms across extensive patient populations, therefore locating loci that identify drug targets and biomarkers indicative of treatment response. GWAS has efficiently identified risk variants in kinases, receptors, and metabolic enzymes that can be directly targeted by drug therapy (Pereira Ciochetti et al., 2023). GWAS not only locates disease-relevant targets but also identifies genetic polymorphisms in drug-metabolizing enzymes that determine person-to-person variation in drug efficacy and clearance. Translation of genetic information to clinical predictions in pharmacogenomic biomarkers determines patient responses to a therapy, allowing personalized medicine where treatments are given according to the patient's genotype (Omidiran et al., 2024). Nowadays, pharmacogenomic testing is standard practice before prescribing warfarin, thiopurine drugs, and HLA-associated severe adverse reactions (Turongkaravee et al., 2021). This incorporation of genomic data into clinical practice serves as the most appropriate indication that omics-informed precision medicine directly influences patient outcomes.

9.3.2 Transcriptomics: gene expression landscapes in disease and treatment

This approach quantifies mRNA abundance throughout entire genomes, offering dynamic insights into the genes that are actively expressed and how expression patterns fluctuate in response to disease, pharmacological intervention, or environmental changes. RNA-seq has been called the standard approach because of its enhanced sensitivity, dynamic range, and capacity to identify novel transcripts (Antoszewski et al., 2026). Single-cell transcriptomics adds crucial cellular resolution, revealing that apparently homogeneous tissues actually comprise multiple cell types with distinct transcriptional programs (Molla Desta and Birhanu, 2025). Connectivity mapping: comparing transcriptomic signatures induced by

Table 9.2 Classification of the multiomics toolkit: tabular classification of the five primary omics modalities (genomics, transcriptomics, proteomics, metabolomics, and epigenomics) that are used in systems pharmacology. The table also highlights the key analytical technologies and primary analytes for each modality.

Omics layer	Key technology	Primary analyte	Drug discovery application	References
Genomics	GWAS, WGS	single-nucleotide polymorphisms, CNVs, indels	Target identification; pharmacogenomic stratification (e.g., *CYP2C9* variants)	Szustakowski et al. (2021), Pereira Ciochetti et al. (2023), Omidiran et al. (2024)
Transcriptomics	RNA-seq, scRNA-seq	mRNA abundance	Connectivity mapping (repurposing); mechanism-of-action elucidation	Knoll et al. (2024), Antoszewski et al. (2026), Samih et al. (2026)
Proteomics	LC-MS/MS	Protein abundance, posttranslational modifications	Target engagement assays; off-target toxicity profiling and signaling pathway mapping	Dale et al. (2021), Zhong et al. (2023), Ivanisevic and Sewduth (2023)
Metabolomics	NMR, mass spectrometry	Small molecules (<1.5 kDa)	Toxicity biomarkers (drug-induced liver injury); phenotypic readout of drug efficacy	Marcos et al. (2023), Moreno-Torres et al. (2022), Zandl-Lang et al. (2023)
Epigenomics	ChIP-seq, ATAC-seq	DNA methylation, histone modifications	Understanding drug resistance mechanisms; identifying "druggable" chromatin regulators	Igolkina et al. (2019), Woodward et al. (2022)

Source: No permission required.

experimental drugs against databases containing signatures from approved drugs and genetic perturbations enables rapid drug repurposing by identifying cases where novel compounds produce signatures phenotypically similar to successful therapeutics (Lim and Pavlidis, 2021). This method found that dexamethasone helped people with severe COVID-19 by matching inflammatory signatures (Knoll et al., 2024). It also found that kinase inhibitors worked unexpectedly against different types of cancer by matching signatures similarly (Shyam Sunder et al., 2023). This signature-based method bypasses the long process of target validation, making it possible to rapidly test the repurposing drug in clinical trials.

9.3.3 Proteomics: direct functional readouts of biological systems

Proteomics directly measures functional protein products using mass spectrometry techniques, giving an enhanced understanding of actual cellular effectors. mRNA levels do not always match up with protein levels very well because of mRNA stability, translational regulation, and protein degradation rates (Samih et al., 2026). Quantitative proteomics, comprising LC-MS/MS, measures absolute protein concentration, indicating that various disease-associated proteins exist at very low levels; on the other hand, transcriptomic alterations do not accurately indicate functional outcomes. Phosphorylation, ubiquitination, SUMOylation, and acetylation are some examples of PTMs that add on regulatory layers that genomics and transcriptomics cannot (Zhong et al., 2023). Protein-wide association studies (PWAS) strengthen GWAS by analyzing the impact of genetic variants on protein abundance or modification, therefore establishing direct mechanistic connections between genetic risk factors and functional protein alterations (Brandes et al., 2020). Merging GWAS and PWAS enables us to find targets based on pathways in which genetic variants are linked to disease through measured changes in proteins instead of statistical links.

9.3.4 Metabolomics: biochemical fingerprints of cellular states

This approach quantifies small molecules that accumulate in cells or tissues, giving direct assessments of the biochemical effects of genetic variations and pharmacological interventions. Untargeted metabolomics with liquid chromatography-mass spectrometry (LC-MS) quantifies 100s–1000s of metabolites at once (Marcos et al., 2023). This makes complete metabolic signatures that show the difference between disease states and normal physiology. Metabolic pathway analysis shows which biochemical pathways are not working properly in the disease. Metabolomics directly measures these changes and finds therapeutic weaknesses. Metabolomics-based biomarkers for toxicity, measuring metabolite signatures in preclinical studies and predicting which compounds will cause drug-induced liver injury or renal toxicity have substantially improved compound safety prediction, reducing late-stage failures from unexpected toxicity (Moreno-Torres et al., 2022). Lipid profiling within metabolomics has emerged as particularly important: altered lipid metabolism contributes to cancer, neurodegeneration, and metabolic disease; drugs targeting lipid biosynthesis produce distinct lipid signatures measurable through lipidomics; and individual genetic polymorphisms affecting lipid metabolism predict drug response and adverse effects, enabling personalized lipid-modifying therapy selection (Zandl-Lang et al., 2023).

9.3.5 Epigenomics: regulatory architecture controlling gene accessibility

Epigenomics measures chemical modifications in DNA and histones, including methylation, acetylation, and phosphorylation, which change the gene accessibility without altering the DNA sequence. DNA methylation present at gene promoters stops transcription. DNA methylation patterns show the difference between normal tissue and diseased tissue, revealing epigenetic changes that cause the disease (Younesian et al., 2024). Histone modifications, especially H3K4me3, H3K9me3, and H3K27ac, show which chromatin domains are active and which are not. Epigenomics shows that the same genetic mutations can cause different phenotypes in different cellular environments. The epigenetic state decides whether disease-causing mutations are expressed (Igolkina et al., 2019). This explains the variable penetrance of genetic variants and therapeutic heterogeneity: identical mutations in different patients give different disease manifestations and responses depending on their epigenetic landscapes (Woodward et al., 2022). Combination therapy that targets both genetic drivers and epigenetic changes solves this problem by removing driver mutations and turning on tumor suppressors that have been turned off by epigenetic changes (Coyle et al., 2017).

9.4 Computational frameworks for data integration

Modern drug discovery produces extensive heterogeneous datasets that comprise structural coordinates, genomic variants, transcriptomic signatures, proteomic measurements, and metabolic fingerprints (Singh et al., 2026). To turn this ample amount of data into useful mechanistic insights, we need advanced computational frameworks that combine different layers of information into clear biological models (Lac et al., 2024). This section brings together different computational methods, such as network analysis, machine learning, MD, and systems biology modeling. These approaches are crucial for linking structural biology and omics data into single frameworks that can predict drug efficacy, selectivity, and patient response.

9.4.1 Network-based integration approaches

A network-based approach provides a framework for integrating multiomics data by representing biological systems as graphs, where nodes signify molecular entities (proteins, genes, metabolites) and edges show functional relationships (Kumar et al., 2025). Protein-protein interaction networks provide direct physical representations of interaction among proteins, indicating that disease-associated mutations typically cluster in specific network regions rather than being randomly distributed throughout proteomes (García-Vaquero et al., 2022). Biological pathway reconstruction merges data from different omics sources into mechanistic pathway models that connect genetic variants to disease phenotypes through proteins and metabolites. Rather than relying on manually curated pathways that become rapidly outdated, modern approaches reconstruct pathways de novo from patient omics data: genomic variants are connected to disease through identified proteins and metabolites whose abundance correlates with disease phenotypes.

Multilayer network analysis simultaneously integrates networks from various omics modalities, linking genomic variants to epigenomic modifications, transcriptomic changes, proteomic alterations, and metabolomic consequences (Subramanian et al., 2020). This makes detailed maps that show how genetic variants that are higher up in the chain move down to create measurable cellular phenotypes. This multilayered method shows that the same genetic changes can lead to different phenotypes based on the epigenetic and transcriptomic environment. This is why some diseases are worse than others and why treatments work differently for different people. Drugs that target specific nodes have different effects depending on the network layer. For example, epigenetic drugs turn on transcriptionally silent pathways, kinase inhibitors stop phosphorylation cascades, and metabolic inhibitors change biochemical states directly (Marei, 2025). By using multilayer networks to understand how drugs work, we can predict which patients will respond based on the specific network architectures they have.

9.4.2 Machine learning and artificial intelligence methods

Machine learning and deep learning architectures analyze omics and structural data to find patterns that can predict drug efficacy toxicity. Deep learning models trained on historical compound properties can predict the efficacy of new compounds based solely on their chemical structure (Zhang et al., 2025). This speeds up lead optimization by allowing the virtual screening of millions of compounds before experimental synthesis. Graph neural networks (GNNs) process molecular structures and protein three-dimensional coordinates as graphs, where atoms/residues are nodes and chemical bonds/spatial proximity define edges; GNNs extract geometric and chemical features unavailable to traditional machine learning, enabling superior predictions of binding affinity, selectivity, and off-target engagement (Zhang et al., 2025). Combining omics data with machine learning increases the accuracy of predictions. Predictive modeling of drug–target interactions integrates structural features (binding pocket properties, conformational dynamics), chemical structure (molecular weight, logP, hydrogen bond donors/acceptors), omics context (target expression in relevant tissues, off-target expression patterns), and clinical outcomes to predict which drug–target pairs will succeed clinically (Oliveira et al., 2025). These integrated models significantly surpass predictions derived from structural or chemical data, illustrating that biological context obtained through omics data and network analysis is essential for forecasting real-world efficacy.

9.4.3 Molecular dynamics simulations: bridging atomic detail and biological function

This approach illustrates the atomic-level motion of proteins and ligands, uncovering dynamic mechanisms that are concealed in static structures. Enhanced sampling techniques address the constraint that traditional MD simulations are restricted to nanosecond-to-microsecond timescales, whereas drug binding and unbinding transpire on millisecond-to-second timescales (Salo-Ahen et al., 2021). Accelerated MD and replica exchange MD investigate expansive conformational spaces, revealing cryptic binding pockets that temporarily surface during protein dynamics, known as "allosteric pockets," where allosteric inhibitors bind (Castelli et al., 2024). This allows for selectivity by targeting conformations that competitors cannot reach. MD-based allosteric pathway identification shows how ligand binding at one site changes the shape of other functional sites that are far away. MD has shown how type II inhibitors bind to the DFG-out pocket and disrupt the hydrophobic spine network, transmitting effects more than 30 Å to the ATP-binding site for kinase inhibitors (Yueh et al., 2019). This knowledge allows for the logical design of inhibitors that stabilize certain conformational states, achieving selectivity that is not possible with traditional type I inhibitors that affect many conformational states.

9.4.4 Systems biology modeling approaches

Kinetic modeling of biological pathways constructs ordinary differential equation models describing how pathway components (kinases, phosphatases, transcription factors) change over time in response to perturbations. They quantitatively predict the modifications in phosphorylation cascades due to kinase inhibitors, allowing the prediction of alterations in pathways upon the inhibition of specific kinases (Feng et al., 2024). Multiscale modeling frameworks merge molecular simulations, cellular-level models, and tissue-level simulations into one framework, which predicts drugs efficacy based on molecular interactions and whole-organism effects (Ingólfsson et al., 2023). This merged approach addresses the well-known translation gap wherein compounds exhibit promising molecular and cellular activity but still fail in clinical settings due to pharmacokinetic constraints. The power of these computational frameworks emerges when applied iteratively with experimental validation. Computational predictions help decide which experiments to do first, experimental results improve computational models, and improved models help decide which analogs will show better selectivity in the next round of predictions (Ingólfsson et al., 2023). This computational-experimental loop speeds up drug development by getting rid of low-value experimental work and focusing on the most promising paths.

Contemporary pharmaceutical companies have adopted this integrated computational approach: MD simulations predict cryptic binding pockets for allosteric inhibitor design; machine learning prioritizes compounds from virtual libraries; network analysis identifies combination therapy candidates; systems biology models predict patient stratification biomarkers; FEP calculations guide lead optimization (Ahmad et al., 2026). This computational-first approach has reduced preclinical timelines by 1–2 years and substantially improved phase I candidate quality through reduced off-target engagement and toxicity.

9.5 Network-guided target selection and iterative lead optimization

Modern drug discoveries do not happen through separate, sequential phases; instead, it happens in a continuous cycle where target identification, lead discovery and optimization, and drug repurposing are all interlinked and are based on the same structural and omics data (Paul et al., 2021). This section brings together these separate processes in a clear workflow that shows how structural biology, computational modeling, multiomics profiling, and systems biology can work together to advance the process from disease biology to validated leads and clinical candidates while lowering the chances of failure by validating mechanisms at each step of the process (Fig. 9.3).

9.5.1 Network-based target discovery and simultaneous structural-functional validation

Drug discovery begins with constructing disease-relevant protein networks, densely interconnected clusters of proteins whose dysfunction directly contributes to pathologic phenotypes, by integrating multiomics data (genomics, transcriptomics, and proteomics) from diseased and healthy tissues (d'Andrea et al., 2026). Network medicine approaches find hub proteins (nodes that connect other nodes) and bottleneck proteins (nodes that control how information moves through disease pathways); these are the most important targets. Drugability filtering is very important because it quickly gets rid of targets that do not have any actionable vulnerabilities, which saves time and effort on proteins that cannot be drugged (Yu et al., 2007). This integrated assessment combines structural bioinformatics predicting small-molecule binding pockets from protein

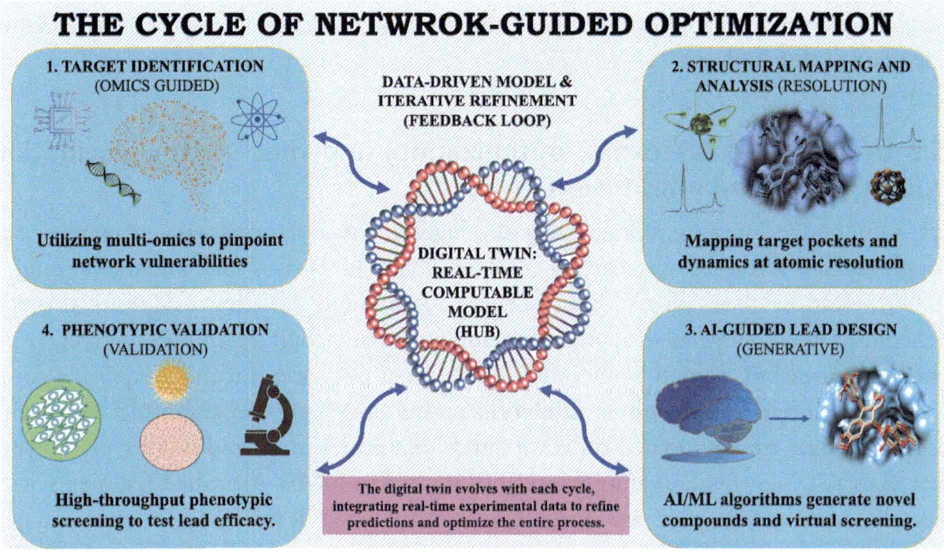

Figure 9.3 The contingent cycle of network-guided lead optimization: the figure depicts a circular model that showing the continuous feedback loop between computational predictions and experimental validation in modern drug discovery. *No permission required.*

sequences and AlphaFold models with functional genomics screening (CRISPR loss-of-function coupled with transcriptomic phenotyping) and toxicology databases. The defining innovation of contemporary target discovery is that structural characterization and target validation occur simultaneously rather than sequentially.

Traditional workflows generated crystal structures first, then validated targets later through cellular assays, a process taking 12–18 months. Modern parallel approaches employ homology modeling and AlphaFold predictions, providing rapid initial binding pocket architectures; experimental structural determination validating predictions and revealing conformational heterogeneity; CRISPR screens combined with multiomics profiling directly measuring phenotypic consequences of target modulation; chemical proteomics mapping target engagement in the cellular context; and systems-level pathway validation predicting how target inhibition propagates through disease-relevant networks (Read et al., 2023). This parallel integration eliminates the discovery of targets that prove structurally tractable but biologically irrelevant. CRISPR screening directly answers whether target knockdown produces desired phenotypes in patient-derived cells; chemical proteomics reveals whether available compounds actually engage targets in cellular context (many compounds identified as kinase inhibitors in biochemical assays fail to engage target kinases in cells due to off-target engagement); phosphoproteomics confirms that target inhibition produces predicted pathway disruption signatures (Reinecke et al., 2024). Only targets passing this integrated validation gauntlet advance to lead optimization, substantially reducing late-stage failures from inadequate target biology.

9.5.2 Unified lead discovery, optimization, and mechanism elucidation through structure-omics fusion

Lead compound discovery immediately leverages integrated structural and omics data through unified screening approaches combining structure-based and phenotype-guided selection. Rather than purely ligand-centric compound libraries evaluated through biochemical assays, integrated approaches employ pharmacophore models derived from crystal structures but refined with omics-derived pathway information (requiring compounds that engage kinases whose inhibition produces desired transcriptomic signatures); machine learning models trained on compounds that achieved desired cellular phenotypes rather than just high binding affinity; and FBDD where NMR-identified fragments with validated target engagement are expanded into full-sized leads using structure-guided elaboration with concurrent phenotypic validation (Vázquez et al., 2020).

Phenotypic screening approaches directly measure disease-relevant phenotypes in patient-derived cells or disease models rather than biochemical target engagement alone. High-content screening combining automated microscopy with multiparameter measurement (cell viability, pathway-specific phosphorylation states, gene expression signatures, and morphologic changes), coupled with computational analysis, generates comprehensive pharmacologic profiles for each compound (Vincent et al., 2022). Critical innovation: phenotypic hits are immediately profiled with rapid multiomics measurement to reveal the mechanism-of-action before advancing to lead optimization. This "phenotype-first, target-second, mechanism-simultaneously" approach has identified successful leads targeting completely novel mechanisms missed by conventional target-centric screening (Pak et al., 2024). Lead optimization progresses through iterative structure-activity relationship analysis, where each chemical modification is simultaneously evaluated through structural effects, cellular phenotypic effects, omics-guided mechanism validation, and ADMET predictions using omics-derived biomarkers.

This continuous feedback cycle from structural modification through cellular testing to omics profiling back to structural refinement ensures compounds progress only after demonstrating integrated evidence of efficacy, selectivity, and a tractable mechanism. Selectivity optimization combines multiple complementary strategies: structural selectivity exploits conformational state preferences (type II kinase inhibitors stabilizing rarely populated DFG-out conformations achieve >100-fold selectivity over off-target kinases unable to access this conformation); functional selectivity leverages omics data to preferentially engage targets whose inhibition produces desired transcriptomic changes while avoiding off-targets whose inhibition produces undesired effects (Xin et al., 2025). Mechanism validation during optimization rather than later clinical stages substantially reduces late-stage failures. For example, kinase inhibitors showing cellular potency through phosphoproteomics-confirmed target engagement are validated to produce desired pathway disruption and avoid unintended compensatory pathway activation (Rüegger et al., 2025).

9.5.3 Drug repurposing and combination therapy as extension of lead optimization

Drug repurposing and combination therapy design represent natural extensions of the unified workflow, applying identical integrated structural–omics frameworks to known drugs. Network pharmacology approaches construct drug–target–disease networks integrating known drug–target interactions, disease modules identified through patient omics profiling, and transcriptomic/proteomic signatures induced by drug treatment in relevant cell types or tissues (Federico et al., 2022). These networks reveal that clinically efficacious drugs frequently engage multiple targets whose combined inhibition produces synergistic disease pathway disruption, a "polypharmacology" principle systematically exploitable for both repurposing and combination identification (Nematzadeh and Karaul, 2025). Combination therapy identification employs network models predicting synergy more accurately than empirical screening of all possible combinations (computationally prohibitive for large drug spaces). Predictions combine biochemical knowledge of target interactions, omics data showing which drugs produce complementary phenotypes, and systems-level mathematical models simulating how dual-target inhibition disrupts disease-relevant network dynamics (Jin et al., 2025).

9.6 Personalized medicine and precision drug discovery: from population-level therapeutics to patient-specific treatment

The primary objective of integrating structural biology, computational modeling, and extensive omics profiling is to develop authentic precision medicine therapeutic strategies tailored to the molecular characteristics of individual patients, rather than applying uniform methods across all patients (Jamalinia and Weiskirchen, 2025; Chen et al., 2025). This section brings together multiomics patient stratification, pharmacogenomics-structure integration, and dynamic biomarker approaches into a single framework that shows how precision drug discovery leads to better patient outcomes.

9.6.1 Multiomics patient stratification: molecular subtyping with clinical utility

Precision medicine starts with the idea that patients with the same disease often have different molecular drivers that need different treatments. When multiomics patient stratification is used, genomic, transcriptomic, proteomic, and metabolomic data from each patient are combined simultaneously. This shows that "single diseases" often include several biologically distinct entities (Molla and Bitew, 2024). For example, nonsmall cell lung cancer (NSCLC): approximately 15% of NSCLC harbor EGFR mutations; another 5% harbor ALK fusions; 3% harbor ROS1 fusions; the remaining ~77% appear genetically undefined. Historically, these undefined NSCLCs received identical chemotherapy despite harboring distinct driver mutations (KRAS, BRAF, HER2, MET) (de Jong et al., 2023). Contemporary multiomics stratification reveals previously unrecognized molecular subtypes: comprehensive genomic profiling identifies previously unknown driver mutations in ~20% of "undefined" NSCLCs; transcriptomic profiling identifies immune-activating versus immune-suppressed subtypes, predicting immunotherapy response; and proteomic analysis measures kinase activation states, guiding targeted inhibitor selection (Wang et al., 2025).

This molecular stratification enables precision target selection matching individual patient driver mutations to specific inhibitors; immunotherapy patient selection based on transcriptomic immune signatures; biomarker-enriched clinical trials limiting enrollment to patients most likely to benefit, substantially increasing trial success rates (Pirouzbakht et al., 2025). Clinical implementation requires establishing standardized multiomics profiling in routine clinical practice (currently limited to academic centers but increasingly available through commercial laboratories); developing clinical decision support systems translating omics data into actionable treatment recommendations; and creating reimbursement mechanisms supporting biomarker testing costs.

9.6.2 Pharmacogenomics integration with structural data: predicting individual drug response

Pharmacogenomics evaluates individual drug efficacy and toxicity through genetic variants that affect drug metabolism, target identification, and toxin sensitivity. Earlier, pharmacogenomics relied on empirical databases of different effects. Now, precision approaches combine pharmacogenomic variants with structural biology to assess, mechanistically, how genetic variations alter drug–target interactions and pharmacokinetics (Asghar et al., 2025). A clinically relevant example of warfarin dosing shows a 20–40-fold difference in the doses that different people need. VKORC1 and CYP2C9 genetic variants predict ~50% of warfarin dose variation; structural characterization of variant effects enables mechanistic prediction of remaining variance. Patients with the VKORC1 rs9923231 1173 C>T variant necessitate a dose reduction of approximately 30%; those with CYP2C92 and CYP2C93 variants metabolize warfarin at a rate 40%–50% slower. The combination of genomic variants with population-specific pharmacokinetic models enables the prediction of personalized warfarin dosing with more than 80% accuracy. This reduces the risk of bleeding complications (Shafique et al., 2022).

9.6.3 Dynamic biomarkers: treatment monitoring and adaptive dosing

Pretreatment biomarkers, like genomic mutations and baseline transcriptomic signatures, allow baseline risk stratification; these are static in nature. Dynamic biomarkers, on the other hand, are measured during treatment, which helps in real-time treatment optimization by adaptive clinical trial designs (Elshimy et al., 2025). Longitudinal omics profiling elucidates therapeutic mechanisms and patterns of resistance emergence. Cancer patients treated with kinase inhibitors display distinct phosphoproteomic signatures that validate on-target engagement. The disappearance of these signatures indicates impending resistance before clinical progression (Labots et al., 2020). For instance, lung cancer patients treated with EGFR inhibitors who have sustained EGFR pathway suppression (as measured by phospho-ERK1/2 and phospho-AKT) continue to respond; the emergence of phospho-ERK1/2 reactivation allows identification weeks before clinical progression, facilitating the addition of proactive combination therapy (Schanbacher et al., 2025).

Monitoring treatment response with circulating biomarkers (circulating tumor DNA, neurofilament light chain, and troponin) allows for a quicker assessment of how well the treatment is working than traditional imaging. Serial ctDNA measurements can indicate treatment response within days; rising ctDNA levels can indicate disease progression weeks before standard imaging can detect it. Adaptive clinical trial designs use dynamic biomarkers to improve treatments in real time (Bartolomucci et al., 2025). Patients without biomarker-verified target engagement at week 2 are administered dose escalation or combination therapy; responders maintain their existing treatment; patients exhibiting early signs of resistance are given preemptive combinations. This approach reduces trial duration by 20%–30% while improving efficacy outcomes compared to fixed-dosing trials. Practical clinical implementation requires rapid (~48–72 hour) biomarker turnaround, electronic health record integration with clinical decision support, and protocols specifying treatment modifications by biomarker patterns (Liu et al., 2025). Leading academic medical centers now integrate dynamic biomarker monitoring into routine cancer care, with circulating tumor DNA monitoring standard for targeted therapy patients.

9.7 Technical challenges and limitations

It is very challenging to combine different types of structural biology, omics, and computational data. This section covers four important limitations, that is, data standardization, computational resource requirements, statistical accuracy, and reproducibility. These limitations determine whether integrated approaches produce valuable, dependable results or artifacts that can lead to incorrect decision-making.

9.7.1 Data standardization and harmonization

In order to carry out multiomics integration, it is important to combine data from various analytical platforms, such as different MS platforms, different X-ray crystallography systems, and RNA-seq data from various sequencing facilities. These platforms can have "batch effects," or systematic biases that are specific to each platform (Dai and Shen, 2022). The correlations are not accurate; they simply reflect differences in how measurements were taken if batch effects are not taken into account. For example, proteomic data from LC-MS exhibit systematic intensity variations of about 20%–40% between instruments, even with similar samples. If batch effects are not properly addressed, these differences can make it seem like there is a link between protein levels and drug efficacy (Chen et al., 2025). The FAIR data principles (Findability, Accessibility, Interoperability, and Reusability) tell you how to write down data, use standard words, and put it in public places so that computers can fix batch effects and other people can check them. Quality control procedures include using reference materials across batches to reduce systematic variations and including replicates within batches to ensure measurement accuracy. For strict quality control, 10%–20% of the experimental budget goes to control samples and replicates (de Groot et al., 2024). This is very important for credibility.

9.7.2 Computational resource requirements and implementation barriers

Integrating multiomics datasets from thousands of patients with MD simulations and machine learning modeling demands computational resources exceeding typical academic departmental budgets. MD simulations of protein–ligand binding require graphics processing units; genome-scale machine learning models demand high-performance computing clusters; cloud-based platforms (Amazon AWS, Google Cloud, Microsoft Azure) provide scalable computing but introduce data privacy concerns and substantial costs (~$10,000–50,000 annually for modest-scale projects) (Kutzner et al., 2022). Addressing resource constraints requires strategic decisions: (1) distributed computing strategies where computations are partitioned across multiple computers, reducing per-machine requirements; (2) cloud-based platforms with secure data enclaves enabling sensitive patient data processing within regulated environments; (3) open-source software (Nextflow, Snakemake) enabling workflow reproducibility across computational environments (Dai and Shen, 2022). Many academic institutions now invest in shared high-performance computing cores, providing access to institutional researchers at substantially lower cost than commercial platforms.

9.7.3 Statistical and methodological rigor: preventing spurious associations

Integration of thousands of variables (genomic variants, gene expression measurements, and protein abundances) dramatically increases multiple-testing problems: if 1000 statistical tests are performed with a $p < 0.05$ significance threshold, ~50 tests will produce false-positive associations by chance (Chen et al., 2021). Bonferroni correction and false discovery rate control are two methods for correcting for multiple testing. They reduce false positives, but they also require stringent significance thresholds, which could result in missing true effects (García-Pérez, 2023). To make causal inferences from integrated datasets, it is necessary to distinguish between correlation and causation. For example, recognizing that specific transcriptomic signatures correlate with drug response does not establish whether these signatures elicit the response or are simply artifacts of confounding variables (tumor burden, immune activation, prior treatments) that influence both drug response and transcriptomic profiles (Bhattacharjee et al., 2025).

9.7.4 Reproducibility and validation

In biomedical research, reproducibility remains a major problem, which tells us that statistical significance does not mean biological viability. Cross-validation methods (k-fold cross-validation and leave-one-out cross-validation) analyze machine learning methods that can be used on data that were not used in their training. Validation across datasets from different patient populations and ethnic groups assesses the consistency of known associations (Lumumba et al., 2024). For reliability, institutions should have strict methods, provide resources for quality control and validation, and acknowledge their limitations. Integrated structural–omics methods can be considered reliable for clinical translation only if these fundamental requirements are met.

9.8 Conclusion and future directions

Integration of structural biology, multiomics profiling, and computational frameworks represents a remarkable shift in drug discovery, from the trial-and-error method to systematic, molecularly guided precision medicine. This approach has shown a highly significant clinical effect by improving treatment results. Using high-resolution structural characterization, a wide range of molecular techniques, advanced computational methods, and dynamic biomarker monitoring together makes it possible to choose the right targets, optimize leads based on their mechanisms of action, and tailor treatment to each patient. The integrated computational-structural–omics approach is now used by modern pharmaceutical companies. It cuts the time it takes to develop drugs before they are tested in people by one to two years and improves clinical candidates by lowering off-target toxicity. Further directions must take into account the current technical challenges to guarantee therapeutic benefits for all. New opportunities include using artificial intelligence to speed up target validation by predicting protein structures; expanding dynamic biomarker monitoring into real-time adaptive clinical trial designs that allow for treatment optimization weeks earlier; and making omics more accessible by lowering costs and standardizing clinical implementation, moving precision medicine from specialized academic centers to routine community clinical practice. The ultimate goal is fully personalized medicine, where pretreatment multiomics profiling stratifies patients to therapies that are the best fit for them, dynamic biomarker monitoring enables real-time adaptive optimization, and computational frameworks predict treatment outcomes before they happen, fundamentally changing medicine from reactive, population-averaged methods to proactive, individually optimized precision therapeutics.

Acknowledgment

Dr. Neha, SGT University, gratefully acknowledges the financial support received from the Anusandhan National Research Foundation (ANRF) under the Early Career Research Grant (ECRG), File No. ANRF/ECRG/2024/001727/LS. The support of the DHR under the scheme of HRD (File No. SUG 2024-1158-reg), awarded to Dr. Neha, is gratefully acknowledged. Dr. Neha has also received the Intramural Research Grant (IMRG) from SGT University (File No. SGTU/ORG/IMRG-II/2024/1).

Conflict of interest

The authors do not have conflicts of interest to declare.

Compliance with ethical standards

This article does not contain any studies with human and/or animal participants performed by any of the authors.

References

Abbas, K., Hao, C., Yong, X., Hasan, M.K., Islam, S., Rahman, A.H.A., 2025. Graph neural network-based drug-drug interaction prediction. Scientific Reports 15 (1). Available at https://doi.org/10.1038/s41598-025-12936-1.

Abdelsayed, M., 2025. AI-driven polypharmacology in small-molecule drug discovery. International Journal of Molecular Sciences 26 (14), 6996. Available at https://doi.org/10.3390/ijms26146996.

Ahmad, B., Ouahada, K., Hamam, H., 2026. Machine learning for drug-target interaction prediction: a comprehensive review of models, challenges, and computational strategies. Computational and Structural Biotechnology Journal 31, 316–345. Available at https://doi.org/10.1016/j.csbj.2025.12.033.

Amorim, A.M.B., Piochi, L.F., Gaspar, A.T., Preto, A.J., Rosário-Ferreira, N., Moreira, I.S., 2024. Advancing drug safety in drug development: bridging computational predictions for enhanced toxicity prediction. Chemical Research in Toxicology 37 (6), 827–849. Available at https://doi.org/10.1021/acs.chemrestox.3c00352, http://pubs.acs.org/journal/crtoec.

d'Andrea, V., Loscalzo, J., De Domenico, M., 2026. Challenges and opportunities in the network medicine of complex diseases. Med 7 (2), 100920. Available at https://doi.org/10.1016/j.medj.2025.100920.

Antoszewski, K., Chmielewska, K., Jagiello, K., Puzyn, T., 2026. Designing RNA sequencing experiments: a practical guide to reproducible gene expression analysis. Computational and Structural Biotechnology Journal 31, 101–119. Available at https://doi.org/10.1016/j.csbj.2025.12.015.

Asghar, S., Shahzad Virk, M.K., Taskeen, B., Fatima, M., Hasnain, Wandala, A., 2025. The role of pharmacogenomics in advancing personalised medicine: implications for drug efficacy safety and clinical practice integration. Indus Journal of Bioscience Research 3 (5), 637–645. Available at https://doi.org/10.70749/ijbr.v3i5.1407.

Balasco Serrão, V.H., 2025. Biological breakthroughs and drug discovery revolution via cryo-electron microscopy of membrane proteins. Membranes 15 (12), 368. Available at https://doi.org/10.3390/membranes15120368.

Bartolomucci, A., Nobrega, M., Ferrier, T., Dickinson, K., Kaorey, N., Nadeau, A., et al., 2025. Circulating tumor DNA to monitor treatment response in solid tumors and advance precision oncology. npj Precision Oncology 9 (1). Available at https://doi.org/10.1038/s41698-025-00876-y.

Bhattacharjee, N., Peela, S.C.M., Halder, A., Mathew, B., Samanta, S. Gujral, S., et al., 2025. Mechanism-aware inference of response to targeted cancer therapies. bioRxiv, India bioRxiv. Available from: https://www.biorxiv.org. doi: https://doi.org/10.1101/2025.09.04.674143.

Bijak, V., Szczygiel, M., Lenkiewicz, J., Gucwa, M., Cooper, D.R., Murzyn, K., et al., 2023. The current role and evolution of X-ray crystallography in drug discovery and development. Expert Opinion on Drug Discovery 18 (11), 1221–1230. Available at https://doi.org/10.1080/17460441.2023.2246881.

Brandes, N., Linial, N., Linial, M., 2020. PWAS: proteome-wide association study—linking genes and phenotypes by functional variation in proteins. Genome Biology 21 (1). Available at https://doi.org/10.1186/s13059-020-02089-x.

Burley, S.K., Wu-Wu, A., Dutta, S., Ganesan, S., Zheng, S.X.F., 2024. Impact of structural biology and the protein data bank on us fda new drug approvals of low molecular weight antineoplastic agents 2019–2023. Oncogene 43 (29), 2229–2243. Available at https://doi.org/10.1038/s41388-024-03077-2, https://www.nature.com/onc/.

Capili, B., Anastasi, J.K., 2024. An introduction to types of quasi-experimental designs. AJN, American Journal of Nursing 124 (11), 50–52. Available at https://doi.org/10.1097/01.naj.0001081740.74815.20.

Castelli, M., Marchetti, F., Osuna, S., Oliveira, A.S.F., Mulholland, A.J., Serapian, S.A., et al., 2024. Decrypting allostery in membrane-bound K-Ras4B using complementary in silico approaches based on unbiased molecular dynamics simulations. Journal of the American Chemical Society 146 (1), 901–919. Available at https://doi.org/10.1021/jacs.3c11396.

Chen, Z., Boehnke, M., Wen, X., Mukherjee, B., De Koning, D.-J., 2021. Revisiting the genome-wide significance threshold for common variant GWAS. G3: Genes|Genomes|Genetics 11 (2). Available at https://doi.org/10.1093/g3journal/jkaa056.

Chen, Q., Cao, Z., Liu, Y., Zhang, N., Xie, Y., Chen, H., et al., 2025. Protein-level batch-effect correction enhances robustness in MS-based proteomics. Nature Communications 16 (1). Available at https://doi.org/10.1038/s41467-025-64718-y.

Coyle, K.M., Boudreau, J.E., Marcato, P., 2017. Genetic mutations and epigenetic modifications: driving cancer and informing precision medicine. BioMed Research International 2017, 1–18. Available at https://doi.org/10.1155/2017/9620870.

Dai, X., Shen, L., 2022. Advances and trends in omics technology development. Frontiers in Medicine 9. Available at https://doi.org/10.3389/fmed.2022.911861.

Dale, B., Cheng, M., Park, K.S., Kaniskan, H.Ü., Xiong, Y., Jin, J., 2021. Advancing targeted protein degradation for cancer therapy. Nature Reviews. Cancer 21 (10), 638–654. Available at https://doi.org/10.1038/s41568-021-00365-x, http://www.nature.com/cancer/.

Dalvit, C., Gmür, I., Rößler, P., Gossert, A.D., 2023. Affinity measurement of strong ligands with NMR spectroscopy: limitations and ways to overcome them. Progress in Nuclear Magnetic Resonance Spectroscopy 138-139, 52–69. Available at https://doi.org/10.1016/j.pnmrs.2023.07.001, http://www.sciencedirect.com/science/journal/00796565.

Decherchi, S., Cavalli, A., 2020. Thermodynamics and kinetics of drug-target binding by molecular simulation. Chemical Reviews 120 (23), 12788–12833. Available at https://doi.org/10.1021/acs.chemrev.0c00534.

Elshimy, Y., Alkhatib, A.R., Atassi, B., Mohammad, K.S., 2025. Biomarker-driven approaches to bone metastases: from molecular mechanisms to clinical applications. Biomedicines 13 (5). Available at https://doi.org/10.3390/biomedicines13051160, www.mdpi.com/journal/biomedicines.

Federico, A., Fratello, M., Scala, G., Möbus, L., Pavel, A., del Giudice, G., et al., 2022. Integrated network pharmacology approach for drug combination discovery: a multi-cancer case study. Cancers 14 (8), 2043. Available at https://doi.org/10.3390/cancers14082043.

Feng, J., Zhang, X., Tian, T., 2024. Mathematical modeling and inference of epidermal growth factor-induced mitogen-activated protein kinase cell signaling pathways. International Journal of Molecular Sciences 25 (18), 10204. Available at https://doi.org/10.3390/ijms251810204.

Galindez, G., Sadegh, S., Baumbach, J., Kacprowski, T., List, M., 2023. Network-based approaches for modeling disease regulation and progression. Computational and Structural Biotechnology Journal 21, 780–795. Available at https://doi.org/10.1016/j.csbj.2022.12.022.

García-Pérez, M.A., 2023. Use and misuse of corrections for multiple testing. Methods in Psychology 8, 100120. Available at https://doi.org/10.1016/j.metip.2023.100120.

García-Vaquero, M.L., Gama-Carvalho, M., Pinto, F.R., De Las Rivas, J., 2022. Biological interacting units identified in human protein networks reveal tissue-functional diversification and its impact on disease. Computational and Structural Biotechnology Journal 20, 3764–3778. Available at https://doi.org/10.1016/j.csbj.2022.07.006, www.csbj.org.

de Groot, R., van der Graaff, F., van der Doelen, D., Luijten, M., De Meyer, R., Alrouh, H., et al., 2024. Implementing findable, accessible, interoperable, reusable (FAIR) principles in child and adolescent mental health research: mixed methods approach. JMIR Mental Health 11, e59113. Available at https://doi.org/10.2196/59113.

Igolkina, A.A., Zinkevich, A., Karandasheva, K.O., Popov, A.A., Selifanova, M.V., Nikolaeva, D., et al., 2019. H3K4me3, H3K9ac, H3K27ac, H3K27me3 and H3K9me3 histone tags suggest distinct regulatory evolution of open and condensed chromatin landmarks. Cells 8 (9), 1034. Available at https://doi.org/10.3390/cells8091034.

Ingólfsson, H.I., Bhatia, H., Aydin, F., Oppelstrup, T., López, C.A., Stanton, L.G., et al., 2023. Machine learning-driven multiscale modeling: bridging the scales with a next-generation simulation infrastructure. Journal of Chemical Theory and Computation 19 (9), 2658–2675. Available at https://doi.org/10.1021/acs.jctc.2c01018.

Ivanisevic, T., Sewduth, R.N., 2023. Multi-omics integration for the design of novel therapies and the identification of novel biomarkers. Proteomes 11 (4), 34. Available at https://doi.org/10.3390/proteomes11040034.

Jamalinia, M., Weiskirchen, R., 2025. Advances in personalized medicine: translating genomic insights into targeted therapies for cancer treatment. Annals of Translational Medicine 13 (2), 18. Available at https://doi.org/10.21037/atm-25-34.

Jentzsch, V., Osipenko, L., Scannell, J.W., Hickman, J.A., 2023. Costs and causes of oncology drug attrition with the example of insulin-like growth factor-1 receptor inhibitors. JAMA Network Open 6 (7), E2324977. Available at https://doi.org/10.1001/jamanetworkopen.2023.24977, https://jamanetwork.com/journals/jamanetworkopen.

Jin, S., Long, H., Huang, A., Wang, J., Yu, X., Xu, Z., et al., 2025. Accurate prediction of synergistic drug combination using a multi-source information fusion framework. BMC Biology 23 (1). Available at https://doi.org/10.1186/s12915-025-02302-y.

de Jong, D., Das, J.P., Ma, H., Pailey Valiplackal, J., Prendergast, C., Roa, T., et al., 2023. Novel targets, novel treatments: the changing landscape of non-small cell lung cancer. Cancers 15 (10), 2855. Available at https://doi.org/10.3390/cancers15102855.

Karim, A.S., Brown, D.M., Archuleta, C.M., Grannan, S., Aristilde, L., Goyal, Y., Leonard, J.N., Mangan, N.M., Prindle, A., Rocklin, G.J., Tyo, K.J., Zoloth, L., Jewett, M.C., Calkins, S., Kamat, N.P., Tullman-Ercek, D., Lucks, J.B., 2024. Deconstructing synthetic biology across scales: a conceptual approach for training synthetic biologists. Nature Communications 15. Available at https://doi.org/10.1038/s41467-024-49626-x.

King, A., Wu, C., 2025. Integrative multi-omics approach for improving causal gene identification. Genetic Epidemiology 49 (1). Available at https://doi.org/10.1002/gepi.22601.

Knoll, R., Helbig, E.T., Dahm, K., Bolaji, O., Hamm, F., Dietrich, O., et al., 2024. The life-saving benefit of dexamethasone in severe COVID-19 is linked to a reversal of monocyte dysregulation. Cell 187 (16), 4318. Available at https://doi.org/10.1016/j.cell.2024.06.014, https://www.sciencedirect.com/science/journal/00928674.

Kumar, R., Romano, J.D., Ritchie, M.D., 2025. Network-based analyses of multiomics data in biomedicine. BioData Mining 18 (1). Available at https://doi.org/10.1186/s13040-025-00452-x, https://biodatamining.biomedcentral.com/.

Kutzner, C., Kniep, C., Cherian, A., Nordstrom, L., Grubmüller, H., de Groot, B.L., et al., 2022. GROMACS in the cloud: a global supercomputer to speed up alchemical drug design. Journal of Chemical Information and Modeling 62 (7), 1691–1711. Available at https://doi.org/10.1021/acs.jcim.2c00044.

Labots, M., Pham, T.V., Honeywell, R.J., Knol, J.C., Beekhof, R., de Goeij-De Haas, R., et al., 2020. Kinase inhibitor treatment of patients with advanced cancer results in high tumor drug concentrations and in specific alterations of the tumor phosphoproteome. Cancers 12 (2). Available at https://doi.org/10.3390/cancers12020330, https://www.mdpi.com/2072-6694/12/2/330/pdf.

Lac, L., Leung, C.K., Hu, P., 2024. Computational frameworks integrating deep learning and statistical models in mining multimodal omics data. Journal of Biomedical Informatics 152, 104629. Available at https://doi.org/10.1016/j.jbi.2024.104629.

Le Marchand, T., Schubeis, T., Bonaccorsi, M., Paluch, P., Lalli, D., Pell, A.J., et al., 2022. 1 H-Detected Biomolecular NMR under Fast Magic-Angle Spinning. Chemical Reviews 122 (10), 9943–10018. Available at https://doi.org/10.1021/acs.chemrev.1c00918.

Lee, P.Y., Yeoh, Y., Low, T.Y., 2023. A recent update on small-molecule kinase inhibitors for targeted cancer therapy and their therapeutic insights from mass spectrometry-based proteomic analysis. FEBS Journal 290 (11), 2845–2864. Available at https://doi.org/10.1111/febs.16442, http://onlinelibrary.wiley.com/journal/10.1111/(ISSN)1742-4658.

Lim, N., Pavlidis, P., 2021. Evaluation of connectivity map shows limited reproducibility in drug repositioning. Scientific Reports 11 (1). Available at https://doi.org/10.1038/s41598-021-97005-z.

Liu, H., Karsidag, I., Golin, R., Wu, G., 2025. Bridging discovery and treatment: cancer biomarker. Cancers 17 (22), 3720. Available at https://doi.org/10.3390/cancers17223720.

Lumumba, V., Kiprotich, D., Mpaine, M., Makena, N., Kavita, M., 2024. Comparative analysis of cross-validation techniques: LOOCV, K-folds cross-validation, and repeated K-folds cross-validation in machine learning models. American Journal of Theoretical and Applied Statistics 13 (5), 127–137. Available at https://doi.org/10.11648/j.ajtas.20241305.13.

Mao, Y., 2025. Dynamics-based drug discovery by time-resolved cryo-EM. Current Opinion in Structural Biology 91, 103001. Available at https://doi.org/10.1016/j.sbi.2025.103001.

Marcos, A., León, C., Moreno-Fernández, M., Castro-Rubio, F., Garrido-Matilla, L., Nozal, L., et al., 2023. Untargeted metabolomic study by liquid chromatography–mass spectrometry in brain tissues on the effects of combined cocaine and ethanol self-administration in male and female young rats. Journal of Chromatography. A 1700, 464047. Available at https://doi.org/10.1016/j.chroma.2023.464047.

Mardis, E.R., Wilson, R.K., 2025. Tracing the evolution of sequencing into the era of genomic medicine. Nature Reviews Genetics 26, 719–734. Available at https://doi.org/10.1038/s41576-025-00884-5.

Marei, H.E., 2025. Epigenetic regulators in cancer therapy and progression. npj Precision Oncology 9 (1). Available at https://doi.org/10.1038/s41698-025-01003-7, https://www.nature.com/npjprecisiononcology/.

Markiewicz, M., Gucwa, M., Bazak, J., Dec, W., Minor, W., Murzyn, K., 2026. Enhancing structural insights for advanced drug discovery by mitigating protein crystal damage. Expert Opinion on Drug Discovery 21 (1), 5–17. Available at https://doi.org/10.1080/17460441.2025.2605354.

Mitschka, S., Mayr, C., 2022. Context-specific regulation and function of mRNA alternative polyadenylation. Nature Reviews Molecular Cell Biology 23 (12), 779–796. Available at https://doi.org/10.1038/s41580-022-00507-5.

Molla Desta, G., Birhanu, A.G., 2025. Advancements in single-cell RNA sequencing and spatial transcriptomics: transforming biomedical research. Acta Biochimica Polonica 72. Available at https://doi.org/10.3389/abp.2025.13922, https://www.frontierspartnerships.org/journals/acta-biochimica-polonica.

Molla, G., Bitew, M., 2024. Revolutionizing personalized medicine: synergy with multi-omics data generation, main hurdles, and future perspectives. Biomedicines 12 (12), 2750. Available at https://doi.org/10.3390/biomedicines12122750.

Moreno-Torres, M., Quintás, G., Castell, J.V., 2022. The potential role of metabolomics in drug-induced liver injury (DILI) assessment. Metabolites 12 (6). Available at https://doi.org/10.3390/metabo12060564, https://www.mdpi.com/2218-1989/12/6/564/pdf?version=1655622578.

Moser, D., Lang, K., Birtasu, A.N., Grahammer, F., Helmstädter, M., Scheffer, M.P., et al., 2025. The slit diaphragm in Drosophila exhibits a bilayered, fishnet architecture. Nature Communications 16 (1). Available at https://doi.org/10.1038/s41467-025-64347-5, https://www.nature.com/ncomms/.

Myers, P.J., Lee, S.H., Lazzara, M.J., 2021. Mechanistic and data-driven models of cell signaling: tools for fundamental discovery and rational design of therapy. Current Opinion in Systems Biology 28. Available at https://doi.org/10.1016/j.coisb.2021.05.010, https://www.journals.elsevier.com/current-opinion-in-systems-biology/.

Nada, H., Choi, Y., Kim, S., Jeong, K.S., Meanwell, N.A., Lee, K., 2024. New insights into protein–protein interaction modulators in drug discovery and therapeutic advance. Signal Transduction and Targeted Therapy 9 (1). Available at https://doi.org/10.1038/s41392-024-02036-3.

Naithani, U., Guleria, V., 2024. Integrative computational approaches for discovery and evaluation of lead compound for drug design. Frontiers in Drug Discovery 4. Available at https://doi.org/10.3389/fddsv.2024.1362456.

Nematzadeh, S., Karaul, A., 2025. Advances in computational drug repurposing, driver genes, and therapeutics in lung adenocarcinoma. Biomolecules 15 (10), 1373. Available at https://doi.org/10.3390/biom15101373.

Oliveira, R.I., Pereira, T.O., Abbasi, M., Salvador, J.A.R., Arrais, J.P., 2025. Deep learning for small-molecule drug discovery: From molecular design to clinical translation. Journal of Pharmaceutical Analysis, 101533. Available at https://doi.org/10.1016/j.jpha.2025.101533.

Omidiran, O., Patel, A., Usman, S., Mhatre, I., Abdelhalim, H., DeGroat, W., et al., 2024. GWAS advancements to investigate disease associations and biological mechanisms. Clinical and Translational Discovery 4 (3). Available at https://doi.org/10.1002/ctd2.296.

Pak, C., Simpson, K.J., Weston, A.D., Cvijic, M.E., Evans, K., Napper, A.D., 2024. Perspectives on phenotypic screening—screen design and assay technology special interest group. SLAS Discovery 29 (2), 100146. Available at https://doi.org/10.1016/j.slasd.2024.02.001.

Paul, D., Sanap, G., Shenoy, S., Kalyane, D., Kalia, K., Tekade, R.K., 2021. Artificial intelligence in drug discovery and development. Drug Discovery Today 26 (1), 80–93. Available at https://doi.org/10.1016/j.drudis.2020.10.010.

Pereira Ciochetti, N., Lugli-Moraes, B., Santos da Silva, B., Rovaris, D.L., 2023. Genome-wide association studies: utility and limitations for research in physiology. The Journal of Physiology 601 (14), 2771–2799. Available at https://doi.org/10.1113/jp284241.

Pirouzbakht, M., Hamzeh, S., Soleimani Samarkhazan, H., 2025. Beyond single biomarkers: multi-omics strategies to predict immunotherapy outcomes in blood cancers. Clinical and Experimental Medicine 25 (1). Available at https://doi.org/10.1007/s10238-025-01902-w.

Read, R.J., Baker, E.N., Bond, C.S., Garman, E.F., Van Raaij, M.J., 2023. AlphaFold and the future of structural biology. IUCrJ 10, 377–379. Available at https://doi.org/10.1107/S2052252523004943, http://www.iucrj.org/m/.

Reinecke, M., Brear, P., Vornholz, L., Berger, B.-T., Seefried, F., Wilhelm, S., et al., 2024. Chemical proteomics reveals the target landscape of 1,000 kinase inhibitors. Nature Chemical Biology 20 (5), 577–585. Available at https://doi.org/10.1038/s41589-023-01459-3.

Reinhardt, S.C.M., Masullo, L.A., Baudrexel, I., Steen, P.R., Kowalewski, R., Eklund, A.S., et al., 2023. Ångström-resolution fluorescence microscopy. Nature 617 (7962), 711–716. Available at https://doi.org/10.1038/s41586-023-05925-9, https://www.nature.com/nature/.
Rüegger, J., Gagestein, B., Janssen, A.P.A., Valeanu, A., Mori, A.L., van der Peet, M., et al., 2025. CellEKT: a robust chemical proteomics workflow to profile cellular target engagement of kinase inhibitors. Molecular & Cellular Proteomics 24 (6), 100961. Available at https://doi.org/10.1016/j.mcpro.2025.100961.
Sachdev, K., Gupta, M.K., 2019. A comprehensive review of feature based methods for drug target interaction prediction. Journal of Biomedical Informatics 93, 103159. Available at https://doi.org/10.1016/j.jbi.2019.103159.
Salo-Ahen, O.M.H., Alanko, I., Bhadane, R., Bonvin, A.M.J.J., Honorato, R.V., Hossain, S., et al., 2021. Molecular dynamics simulations in drug discovery and pharmaceutical development. Processes 9 (1), 71. Available at https://doi.org/10.3390/pr9010071.
Samih, A., de Moura Ferreira, M.A., Nikoloski, Z., 2026. Gene expression and protein abundance: Just how associated are these molecular traits? Biotechnology Advances 86, 108720. Available at https://doi.org/10.1016/j.biotechadv.2025.108720.
Sa, P., Gòdia, M., Lewis, N., Lian, Y., Clop, A., 2024. Genomic, transcriptomic and epigenomic analysis towards the understanding of porcine semen quality traits. Past, current and future trends. Animal Reproduction Science 269, 107543. Available at https://doi.org/10.1016/j.anireprosci.2024.107543.
Schanbacher, C., Goebeler, M.-E., Gerull, B., Lorenz, K., 2025. Targeting pathological ERK1/2 signaling in cancer and beyond. Trends in Molecular Medicine 32 (3). Available at https://doi.org/10.1016/j.molmed.2025.08.001.
Schwalbe, H., Audergon, P., Haley, N., Amaro, C.A., Agirre, J., Baldus, M., Banci, L., Baumeister, W., Blackledge, M., Carazo, J.M., Carugo, K.D., Celie, P., Felli, I., Hart, D.J., Hauß, T., Lehtiö, L., Lindorff-Larsen, K., Márquez, J., Matagne, A., Pierattelli, R., Rosato, A., Sobott, F., Sreeramulu, S., Steyaert, J., Sussman, J.L., Trantirek, L., Weiss, M.S., Wilmanns, M., 2024. The future of integrated structural biology. Structure 32, 1563–1580. Available at https://doi.org/10.1016/j.str.2024.08.014.
Seshadri, K., Abad, A.N.D., Nagasawa, K.K., Yost, K.M., Johnson, C.W., Dror, M.J., et al., 2025. Synthetic biology in natural product biosynthesis. Chemical Reviews 125 (7), 3814–3931. Available at https://doi.org/10.1021/acs.chemrev.4c00567, http://pubs.acs.org/journal/chreay.
Shafique, H., Ashraf, N.M., Rashid, A., Majeed, A., Afsar, T., Daly, A.K., et al., 2022. Determination of pleiotropic effect of warfarin in VKORC1 and CYP2C9 genotypes in patients with heart valve replacement. Frontiers in Cardiovascular Medicine 9. Available at https://doi.org/10.3389/fcvm.2022.895169, www.frontiersin.org/journals/cardiovascular-medicine.
Shah, A., Patel, V., Jain, M., Parmar, G., 2023. Network Pharmacology and Systems Biology in Drug Discovery. Springer Science and Business Media LLC, pp. 231–252. Available at https://doi.org/10.1007/978-981-99-1316-9_10.
Shaik, N.S., Balya, H., 2025. High-throughput computational screening for lead discovery and development. Advances in Pharmacology 103, 185–207. Available at https://doi.org/10.1016/bs.apha.2025.01.001, https://www.sciencedirect.com/science/bookseries/10543589.
Shyam Sunder, S., Sharma, U.C., Pokharel, S., 2023. Adverse effects of tyrosine kinase inhibitors in cancer therapy: pathophysiology, mechanisms and clinical management. Signal Transduction and Targeted Therapy 8 (1). Available at https://doi.org/10.1038/s41392-023-01469-6.
Singh, V., Sinha, S., Verma, J., 2026. Bioinformatics models in drug delivery: predicting biomaterial-biological interactions for targeted therapies. Next Nanotechnology 9, 100335. Available at https://doi.org/10.1016/j.nxnano.2025.100335.
Son, A., Kim, W., Park, J., Lee, W., Lee, Y., Choi, S., et al., 2024. Utilizing molecular dynamics simulations, machine learning, Cryo-EM, and NMR spectroscopy to predict and validate protein dynamics. International Journal of Molecular Sciences 25 (17), 9725. Available at https://doi.org/10.3390/ijms25179725.

Subramanian, I., Verma, S., Kumar, S., Jere, A., Anamika, K., 2020. Multi-omics data integration, interpretation, and its application. Bioinformatics and Biology Insights 14. Available at https://doi.org/10.1177/1177932219899051, 117793221989905.

Sun, D., Gao, W., Hu, H., Zhou, S., 2022. Why 90% of clinical drug development fails and how to improve it? Acta Pharmaceutica Sinica B 12 (7), 3049–3062. Available at https://doi.org/10.1016/j.apsb.2022.02.002.

Szustakowski, J.D., Balasubramanian, S., Kvikstad, E., Khalid, S., Bronson, P.G., Sasson, A., et al., 2021. Advancing human genetics research and drug discovery through exome sequencing of the UK Biobank. Nature Genetics 53 (7), 942–948. Available at https://doi.org/10.1038/s41588-021-00885-0, http://www.nature.com/ng/index.html.

Turongkaravee, S., Jittikoon, J., Rochanathimoke, O., Boyd, K., Wu, O., Chaikledkaew, U., 2021. Pharmacogenetic testing for adverse drug reaction prevention: systematic review of economic evaluations and the appraisal of quality matters for clinical practice and implementation. BMC Health Services Research 21 (1). Available at https://doi.org/10.1186/s12913-021-07025-8.

Vajda, S., Beglov, D., Wakefield, A.E., Egbert, M., Whitty, A., 2018. Cryptic binding sites on proteins: definition, detection, and druggability. Current Opinion in Chemical Biology 44, 1–8. Available at https://doi.org/10.1016/j.cbpa.2018.05.003, http://www.elsevier.com/locate/cbi.

Vázquez, J., López, M., Gibert, E., Herrero, E., Luque, F.J., 2020. Merging ligand-based and structure-based methods in drug discovery: an overview of combined virtual screening approaches. Molecules 25 (20), 4723. Available at https://doi.org/10.3390/molecules25204723.

Vincent, F., Nueda, A., Lee, J., Schenone, M., Prunotto, M., Mercola, M., 2022. Phenotypic drug discovery: recent successes, lessons learned and new directions. Nature Reviews. Drug discovery 21 (12), 899–914. Available at https://doi.org/10.1038/s41573-022-00472-w.

Wang, C., Li, J., Chen, J., Wang, Z., Zhu, G., Song, L., et al., 2025. Multi-omics analyses reveal biological and clinical insights in recurrent stage I non-small cell lung cancer. Nature Communications 16 (1). Available at https://doi.org/10.1038/s41467-024-55068-2.

Woodward, A.A., Urbanowicz, R.J., Naj, A.C., Moore, J.H., 2022. Genetic heterogeneity: challenges, impacts, and methods through an associative lens. Genetic Epidemiology 46 (8), 555–571. Available at https://doi.org/10.1002/gepi.22497, http://onlinelibrary.wiley.com/journal/10.1002/(ISSN)1098-2272.

Xin, G., Wang, T., Zhou, Q., Zhang, S., Wei, Z., Li, K., et al., 2025. Innovative strategies for anti-fibrotic drugs discovery from traditional Chinese medicine. Chinese Herbal Medicines 16746384. Available at https://doi.org/10.1016/j.chmed.2025.10.005.

Yadav, M.R., Murumkar, P.R., Yadav, R., Joshi, K., 2023. Structure-based virtual screening in drug discovery. Cheminformatics, QSAR and Machine Learning Applications for Novel Drug Development. Elsevier, India, pp. 69–88. Available at https://www.sciencedirect.com/book/9780443186387, https://doi.org/10.1016/B978-0-443-18638-7.00006-2.

Younesian, S., Mohammadi, M.H., Younesian, O., Momeny, M., Ghaffari, S.H., Bashash, D., 2024. DNA methylation in human diseases. Heliyon 10 (11), e32366. Available at https://doi.org/10.1016/j.heliyon.2024.e32366.

Yueh, C., Rettenmaier, J., Xia, B., Hall, D.R., Alekseenko, A., Porter, K.A., et al., 2019. Kinase atlas: druggability analysis of potential allosteric sites in kinases. Journal of Medicinal Chemistry 62 (14), 6512–6524. Available at https://doi.org/10.1021/acs.jmedchem.9b00089.

Yu, H., Kim, P.M., Sprecher, E., Trifonov, V., Gerstein, M., Murray, D., 2007. The importance of bottlenecks in protein networks: correlation with gene essentiality and expression dynamics. PLoS Computational Biology 3 (4), e59. Available at https://doi.org/10.1371/journal.pcbi.0030059.

Zafar, H., Malone, K.L., Singh, A.K., Cianfrocco, M.A., Glass, K.C., 2025. Breaking barriers: transitioning from X-ray crystallography to cryo-EM for structural studies. bioRxiv, United States bioRxiv. Available from: https://www.biorxiv.org. doi: https://doi.org/10.1101/2025.10.13.682238.

Zandl-Lang, M., Plecko, B., Köfeler, H., 2023. Lipidomics—paving the road towards better insight and precision medicine in rare metabolic diseases. International Journal of Molecular Sciences 24 (2), 1709. Available at https://doi.org/10.3390/ijms24021709.

Zhang, O., Lin, H., Zhang, X., Wang, X., Wu, Z., Ye, Q., et al., 2025. Graph neural networks in modern AI-aided drug discovery. Chemical Reviews 125 (20), 10001–10103. Available at https://doi.org/10.1021/acs.chemrev.5c00461.

Zhang, R., Wen, H., Lin, Z., Li, B., Zhou, X., 2025. Artificial intelligence-driven drug toxicity prediction: advances, challenges, and future directions. Toxics 13 (7), 525. Available at https://doi.org/10.3390/toxics13070525.

Zhong, Q., Xiao, X., Qiu, Y., Xu, Z., Chen, C., Chong, B., et al., 2023. Protein posttranslational modifications in health and diseases: functions, regulatory mechanisms, and therapeutic implications. MedComm 4 (3). Available at https://doi.org/10.1002/mco2.261.

SECTION III

Applications

CHAPTER 10

Applications of structural biology in drug discovery

Saira Hamid, Sarah Siddiquah, Hanzla Shaz, and Imtaiyaz Hassan
Center for Interdisciplinary Research in Basic Sciences, Jamia Millia Islamia, New Delhi, India

10.1 Introduction

Proteins are the primary mediators of cellular function that represent the central targets of most therapeutic agents. Their activity is closely tied to their structure, making structural information important for understanding disease mechanisms and for guiding intervention strategies. The process of drug discovery has long been recognized as complex, costly, and inherently uncertain. Traditional approaches have largely relied on empirical screening methods where large libraries of compounds are tested against biological targets with limited prior knowledge of their molecular interactions. However, the inefficiency and high failure rates of these methods have created a need for more rational and targeted approaches. A major shift has been the emergence of structural biology, which helps to understand biological systems at the molecular level.

Structure-based drug design (SBDD) is based on the principle that detailed information about proteins helps to design molecules that can bind to a specific site on a target protein. This method uses molecular interactions for drug development to improve specificity and success rates (Mehmood, 2024).

Progress in this field has been driven by advances in experimental techniques capable of resolving biomolecular structures with increasing accuracy. X-ray crystallography, nuclear magnetic resonance (NMR) spectroscopy, and cryo-electron microscopy (cryo-EM) have each contributed unique strengths to the study of proteins ranging from small soluble systems to large, complex assemblies. Moreover, the complementary techniques, such as small-angle X-ray scattering, further extend this capability by providing information regarding the molecular shape and flexibility in solution (Brosey and Tainer, 2019). Together, these methods have made it possible not only to visualize static structures but also to begin capturing the dynamic nature of biomolecules, which is important for understanding the function.

Alongside experimental advances, computational methods have become integral to modern structural biology. Approaches such as homology modeling, threading, and ab initio prediction have addressed gaps where experimental structures are unavailable. More recently, artificial intelligence (AI) has transformed the field, with deep learning

Advances in Structural Biology
DOI: https://doi.org/10.1016/B978-0-443-44988-8.00013-7

models achieving near-experimental accuracy in protein structure prediction (Venkatesan et al., 2013). In parallel, molecular docking, molecular dynamics (MD) simulations, and machine learning techniques have enhanced our ability to explore protein-ligand interactions, predict binding affinities, and refine candidate molecules with greater efficiency.

Now, the field of structural biology aids in nearly every stage of the drug discovery pipeline. It supports the identification of viable drug targets and helps in the discovery of initial hit compounds and their progression into optimized leads. The detailed structure of proteins helps to characterize interaction networks and understand conformational changes that influence drug efficacy. This information is used to interpret key mechanisms such as molecular recognition, allosteric regulation, and conformational flexibility, all of which play a critical role in determining the interaction of drugs with their targets (Spassov, 2024).

Despite its importance, structural biology has certain limitations. The determination of structures of some proteins, especially membrane proteins and highly flexible systems, remains challenging. In addition, most techniques provide static snapshots, which may not fully represent the dynamic nature of proteins in biological conditions (Zagotto and Bortoli, 2021).

This chapter explains the role of structural biology in protein structures and their role in drug discovery. In addition, it shows the importance of experimental and computational techniques in designing effective drugs. Overall, it demonstrates the value of combining these approaches for improving drug development and future research.

10.2 Fundamental techniques of structural biology used in drug discovery

SBDD is an approach that uses the three-dimensional (3D) structure of a biological target, usually a protein, to design and develop of small-molecule agents that can bind to the target and modulate its function. This 3D structure is deduced by various techniques that have changed the field of drug design by providing information on the structure and function of biological macromolecules. The techniques provide a detailed understanding of protein structures and their interactions with other molecules, which allows researchers to design and develop targeted therapeutics with better accuracy and efficacy (Mehmood, 2024). The fundamental techniques of structural biology used in drug discovery are X-ray crystallography, NMR spectroscopy, cryo-EM, SAXS, and various computational structure prediction methods (Table 10.1).

10.2.1 X-ray crystallography

X-ray crystallography is one of the basic techniques in structural biology used to determine the 3D structure of biological macromolecules at atomic resolution. The field

Table 10.1 Overview of key experimental techniques used in structure-based drug design, highlighting their primary functions, typical applications in drug discovery, and representative references.

S. No.	Technique	Function in structure-based drug design	Example application	References (review/key study)
1.	X-ray crystallography	Provides high-resolution (atomic-level) structures of protein-ligand complexes; helps identify active/binding sites and key amino acid interactions, guiding rational drug design	HIV protease inhibitors design; kinase inhibitor development	Bijak et al. (2023)
2.	NMR spectroscopy	Determines structures in solution; useful for studying protein dynamics, conformational changes, flexibility, and interactions with ligands	Identification of weak-binding fragments in early drug discovery	Chen et al. (2024)
3.	Cryo-electron microscopy	Enables structure determination of large macromolecular complexes and membrane proteins by flash-freezing samples in vitreous ice	GPCR and ribosome-targeting drug development	Lees et al. (2021)
4.	Small-angle X-ray scattering	Solution-based technique; helps in 3D structural modeling and monitoring biophysical states of macromolecules	Studying flexible proteins and protein-ligand conformational changes	Brosey and Tainer (2019)

of SBDD needs detailed atomic-level information provided by X-ray crystallography better to understand the structural and chemical features of the target and thus help in designing drugs that could interact with high specificity and affinity (Schmidt et al., 2014). This technique involves the crystallization of the target molecule, and subsequently, X-rays are diffracted through the crystal lattice. The electron density and thus the 3D structure of the molecule are determined by analyzing the resulting diffraction pattern from X-ray diffraction (Chen et al., 2024).

These atomic-level structures of target proteins help to identify the active/binding sites and key amino acid interactions. This helps in the rational design and optimization of drug molecules with improved specificity and binding affinity (Bijak et al., 2023).

10.2.2 Nuclear magnetic resonance spectroscopy

NMR spectroscopy is another essential technique in structural biology, which provides information about the structure, dynamics, and interactions of biomolecules in solution. NMR relies on the magnetic properties of atomic nuclei to probe the chemical environment of atoms within a molecule. The NMR spectra of proteins provide information regarding conformational changes, flexibility, and interactions with ligands. It is well-suited for studying transient interactions and dynamic processes, which are quite challenging to record with X-ray crystallography (Chen et al., 2024).

NMR applications in drug discovery are classified into two major categories: ligand-based and protein-based. The ligand-based method has features such as one-dimensional (1D) hydrogen (1H) or fluorine (19F) NMR experiments; these approaches focus on the differences in the NMR signals derived from ligands, which makes it unnecessary to use isotopically labeled target protein samples. In the early stage of SBDD, a ligand-based approach is useful for the screening of hit ligands (Sugiki et al., 2018). For "protein-based methods," isotopic labeling with suitable stable isotopes (13C, 15N, or both) is necessary, followed by HSQC measurements. Protein-based methods are useful for hit validation, using affinity and binding modes, and for the selection of hit ligands from false-positive and nonspecific binders (Wälti et al., 2017).

10.2.3 Cryo-electron microscopy

Cryo-EM has emerged as an effective technique for analyzing large macromolecular complexes that are difficult to examine using X-ray crystallography or NMR. Cryo-EM helps in revealing the structure of large and dynamic macromolecular assemblies like membrane proteins in their native state in solution without the need to obtain crystals. Additionally, various conformational states can be resolved in a single experiment (Renaud et al., 2018). In cryo-EM, proteins are flash-frozen in a thin layer of vitreous ice, preserving their native conformations and allowing 3D reconstruction of flexible, large, and membrane-embedded complexes at atomic resolution, such as membrane proteins, particularly ion channels, viral capsids, and multisubunit assemblies (Lees et al., 2021). The benefit of cryo-EM for the structural elucidation of membrane proteins and highly dynamic proteins has led to the development of drugs targeting protein-coupled receptors (GPCRs) (Zhang et al., 2021). Cryo-EM has already proven effective in target identification, validation, and characterization, which includes the analysis of the structure of the target for a better understanding of the mechanism of action and druggability.

10.2.4 Small-angle X-ray scattering

SAXS is a technique based on the laws of biophysics used for structural analysis. It provides information on biomolecular shape, pathway interactions, and assembly states, conformational ensembles, dynamics of disordered systems, and structurally characterizes kinetic processes (Brosey and Tainer, 2019). It is a solute-based technique that helps in structural modeling of the 3D structure of a solute particle based on its measured 1D scattering profile (Tuukkanen et al., 2017). SAXS, coupled with size-exclusion chromatography and time-resolved methods, has also become a powerful tool for monitoring biophysical states related to folding, unfolding, and aggregation, analyzing biochemically relevant ligand interactions, and facilitating high-throughput library screening to assess macromolecular properties and to create opportunities for drug discovery (Zhang et al., 2021). The HT scale of SAXS can identify small molecules that modify the protein's function, facilitating researchers in targeting the most promising compounds in drug screens (Brosey et al., 2024).

10.2.5 Computational structure prediction

Apart from experimental techniques, computational structure prediction has become an important aspect of modern structural biology. Traditionally, the process of protein structure determination required methods such as X-ray crystallography and NMR spectroscopy that were often time-consuming, expensive, and technically challenging. For example, X-ray crystallography requires the formation of high-quality crystals, which is not feasible for many proteins, especially membrane proteins or large complexes. Similarly, NMR spectroscopy is limited by the size of the protein and requires significant experimental effort. Because of these constraints, many biologically important proteins remained structurally uncharacterized for years. This limitation is addressed by three major computational structure prediction methods, which estimate the 3D structure using the amino acid sequence of a protein (Venkatesan et al., 2013) (Table 10.2).

Table 10.2 Overview of computational protein structure prediction methods, including ab initio modeling, threading (fold recognition), and homology modeling.

Method	Principle	Key features/ approach	Applications in drug discovery	Tools/software	Key references
Ab initio (De Novo) structure prediction	Predicts protein structure directly from amino acid sequence without templates; based on the lowest free energy principle	Monte Carlo sampling, molecular dynamics, fragment assembly; energy-based scoring of bonded/ nonbonded interactions	Prediction of novel protein structures, useful for targets lacking homologs	Rosetta; QUARK; I-TASSER	Venkatesan et al. (2013), Pearce et al. (2022), Filgueiras et al. (2023), Leman et al. (2020), Mortuza et al. (2021), Yang and Zhang (2015)
Threading (fold recognition)	Aligns sequence to known structural folds even with low sequence similarity	Uses structural databases; evaluates compatibility via residue environment, secondary structure, solvent accessibility	Identification of distant homologs and functional domains	PHYRE2; RaptorX; HHpred	Kelley et al. (2015), Källberg et al. (2012)
Homology modelling (comparative modelling)	Builds a 3D structure based on evolutionary similarity between the target and template	Template selection, sequence alignment, backbone modeling, loop refinement, energy minimization, validation	Protein-ligand interaction studies, docking, structure-based drug design	MODELLER; SWISS-MODEL; Prime (Schrödinger)	Hameduh et al. (2020), Plonski and Reed (2023), Waterhouse et al. (2018), Jacobson et al. (2004a))

10.2.5.1 Ab-initio (De Novo) structure prediction

Ab initio protein structure prediction, also known as de novo modeling, is a method used to determine the 3D structure of a protein directly from its amino acid sequence without relying on homologous templates or experimentally resolved structures. It is based on Anfinsen's experiment, which states that the native state of a protein corresponds to its lowest free energy state. The process involves generating all possible conformations of proteins through various techniques such as Monte Carlo sampling, MD simulations, and fragment-based assembly. After this, the evaluation of each structure is carried out using physicochemical energy functions that include bonded and nonbonded interactions, hydrophobicity, and electrostatics (Pearce et al., 2022).

The most probable structure is then identified by selecting low-energy conformations, often used by clustering approaches to detect stable structural families (Filgueiras et al., 2023). Several tools have been developed to implement this approach, including Rosetta (fragment-based assembly and energy scoring) (Leman et al., 2020), QUARK (replica-exchange Monte Carlo simulations) (Mortuza et al., 2021), and I-TASSER, which combines ab initio modeling with threading for improved accuracy (Yang and Zhang, 2015). These tools can predict structures for proteins lacking homologous templates, thereby making ab initio methods important in novel drug-target identification.

10.2.5.2 Threading (fold recognition)

Threading, also known as fold recognition, is another method to detect the 3D structure of a protein by identifying the most compatible structural fold for a sequence. This method is useful when sequence similarity to known proteins is low. The principle is that protein structures are more conserved than their sequences, which means that proteins with different sequences may still adopt similar folds. This method involves the "threading" or alignment of the target sequence through a database of known structural templates. The alignment is evaluated using scoring functions that consider factors such as residue environment, secondary structure compatibility, and solvent accessibility. The alignment results are then ranked to find the best-fitting fold. Once a suitable template is identified, a structural model is constructed and refined. This method is mostly used when homology modeling fails due to low sequence identity, but structural similarity exists. It identifies distant evolutionary relationships and functional domains that are useful for drug targeting. Several tools include PHYRE2 (advanced homology detection and hidden Markov models) (Kelley et al., 2015), RaptorX (probabilistic models and deep learning) (Källberg et al., 2012), and HHpred (remote homolog detection) (Kelley et al., 2015).

10.2.5.3 Homology modelling

Homology modeling is one of the most reliable and widely used methods for protein structure prediction when a suitable template structure is available. It is based on the principle that proteins with similar amino acid sequences tend to adopt similar 3D structures. The process starts with the identification of one or more template proteins with known structures that share sequence similarity with the target protein. After the identification of the template, the target and the template are aligned to identify conserved regions. Based on this alignment, a model is constructed by aligning the backbone coordinates of the template to the target, while loops and other variable regions are modeled separately. The model is then refined through energy minimization and validated using structural assessment tools for accuracy (Hameduh et al., 2020). Homology modeling is important in the field of drug discovery because it provides sufficiently accurate models for studying protein-ligand interactions and performing molecular docking. However, the accuracy depends heavily on the level of sequence identity between the target and template, with higher similarity yielding better models (Plonski and Reed, 2023). Several well-established tools are used for homology modeling, including MODELLER (comparative modeling using spatial restraints) (Webb and Sali, 2017), SWISS-MODEL (automated model generation and evaluation) (Schwede et al., 2003), and Prime Schrödinger (advanced refinement and loop modeling capabilities) (Jacobson et al., 2004b).

10.3 Role of structural biology across drug design

The discovery of new pharmaceutical compounds is a complex, costly, and time-consuming process traditionally dependent on high-throughput screening. The field of structural biology has transformed this process through SBDD, which uses 3D structural information of biological targets to guide rational drug development. It provides information about the atomic-level details of proteins and their binding sites. This leads to the prediction of ligand-target interactions for improved specificity and potency (Biala et al., 2023). Thus, structural biology now plays a central role in identifying novel drug targets and optimizing therapeutic molecules (Table 10.3, Fig. 10.1).

10.3.1 Target identification and validation

Structure-based target identification relies on high-resolution structural data to recognize biologically relevant and druggable proteins. SBDD utilizes 3d protein structures to design small molecules that bind selectively and modulate target activity. Experimental techniques such as X-ray crystallography and cryo-EM provide detailed information on active sites for rational drug design and improving target validation accuracy (Batool et al., 2019). Identification of druggable pockets is essential at this

Table 10.3 Summary of structure-based drug design stages, outlining key structural biology roles, techniques used, resulting outcomes, and supporting references.

S. No.	Stage of drug design	Key structural biology role	Techniques used	Outcome	Key references
1.	Target identification and validation	Structure-based target selection using 3D protein structures	X-ray crystallography, cryo-EM	Accurate identification of disease-relevant targets	Schmidt et al. (2014), Anderson (2003), Verlinde and Hol (1994)
		Identification of druggable binding pockets	Q-SiteFinder, energy-based mapping	Detection of ligand-binding cavities	Kalyaanamoorthy and Chen (2011)
		Protein–protein interaction targeting	Structural interface analysis	Targeting cancer-related pathways	Cheng et al. (2020)
2.	Hit identification	Structure-based virtual screening	Molecular docking, scoring functions	Identification of hit compounds	Lavecchia and Giovanni (2013), Phatak et al. (2009)
		Ligand-based virtual screening	Pharmacophore modeling, similarity search	Scaffold identification	Pedretti et al. (2019)
		Fragment-based drug discovery	Fragment screening, structure-guided design	Initial hit generation	Bijak et al. (2023)
3.	Hit-to-lead optimization	Structural analysis of ligand-binding modes	Docking, molecular dynamics simulations	Improved binding understanding	Li et al. (2025)
		Identification of allosteric sites	Conformational dynamics studies	Development of selective modulators	Li et al. (2025)

(Continued)

Table 10.3 (Continued)

S. No.	Stage of drug design	Key structural biology role	Techniques used	Outcome	Key references
4.	Lead optimization	Structure-activity relationship (QSAR)	ML models (SVM, RF, ANN), regression	Activity prediction and optimization	Mendenhall and Meiler (2016), Hansch and Fujita (1964)
		Binding affinity and selectivity optimization	Scoring functions (Alloscore)	Improved potency and specificity	Li et al. (2023)
		ADMET and pharmacokinetics optimization	PK/PD studies, in silico ADMET	Drug safety and efficacy	Tepap Zemnou (2025), Hodgson (2001)

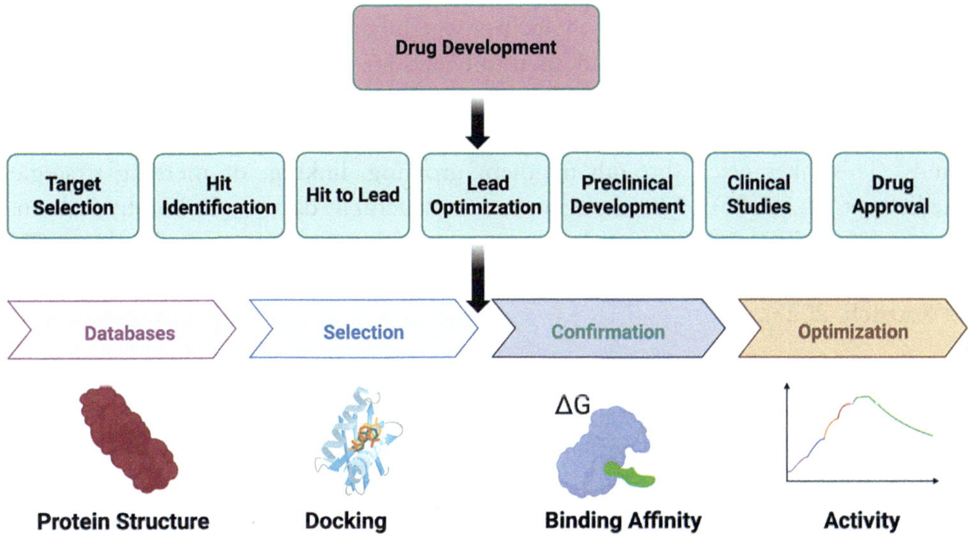

Figure 10.1 Application of structural biology in the process of drug development. It includes major steps like template selection, hit identification, hit-to-lead selection, lead optimization, and trials, which are aided by the structure.

stage, as these structural cavities determine ligand-binding feasibility. Computational tools such as Q-SiteFinder and FTMap predict energetically favorable binding sites by analyzing protein-ligand interaction energies. Furthermore, protein–protein interactions (PPIs) have emerged as important therapeutic targets, particularly in cancer, where their dysregulation contributes to disease progression (Lin et al., 2026; Comajuncosa-Creus et al., 2024; Vajda et al., 2018).

10.3.2 Hit identification

Hit identification is a critical phase in drug discovery where initial compounds with measurable activity against a biological target are discovered. The structure of the protein helps to identify the compounds that can bind effectively. Structure-based virtual screening uses experimentally determined or computationally predicted protein structures to screen large chemical libraries and identify potential hits. Furthermore, the molecular docking algorithms predict ligand-binding orientations and estimate binding affinities based on structural complementarity between the ligand and the active site. The availability of high-resolution structures from X-ray crystallography and cryo-EM significantly improves the accuracy of these predictions (Ashraf et al., 2024; Xie et al., 2022).

In addition to virtual screening, fragment-based drug discovery (FBDD) has emerged as a highly effective strategy that relies heavily on structural biology techniques. In FBDD, small, low-molecular-weight fragments are screened for weak binding to the target protein, and their interactions are characterized using X-ray crystallography or NMR. These structural features help in the precise mapping of fragment binding modes and guide their optimization into high-affinity lead compounds. This takes place through fragment growing, linking, or merging strategies (Khedkar et al., 2025). The integration of structural data with biophysical and computational methods increases the efficiency of hit identification and reduces false positives. This ultimately accelerates the discovery of viable drug candidates (Sadybekov and Katritch, 2023).

10.3.3 Hit-to-lead optimization

Hit-to-lead optimization is another stage in drug discovery in which initial hit compounds identified from screening are systematically refined into more potent, selective, and drug-like lead candidates (Nippa et al., 2025). High-resolution X-ray crystallography, NMR spectroscopy, and cryo-EM allow visualization of binding modes, key interaction residues, and conformational changes within the target protein, guiding structure-activity relationship (SAR) studies (Venien-Bryan et al., 2017). Additionally, structural analysis facilitates the identification of allosteric sites and helps optimize physicochemical and pharmacokinetic properties while minimizing off-target effects (Xie et al., 2022).

10.3.4 Lead optimization using structural information

SAR and quantitative structure-activity relationship (QSAR) analyzes are central to lead optimization. QSAR models establish relationships between chemical structure and biological activity using descriptors and machine learning methods such as support vector machines, random forests, and neural networks (Tropsha et al., 2024). These approaches enable the prediction of activity, the identification of key structural features, and the systematic optimization of lead compounds (Arya and Coumar, 2021). In parallel, optimization of pharmacokinetic and toxicity profiles is crucial for successful drug development. ADMET evaluation ensures that compounds possess favorable absorption, distribution, metabolism, excretion, and toxicity characteristics. Integrating structural insights with pharmacokinetic modeling improves drug-likeness and increases the likelihood of clinical success (Ferreira and Andricopulo, 2019).

10.4 Understanding drug-target interactions

Proteins have critical roles in energy transport, immunity, biosynthesis, degradation, and cellular communication. The link between protein 3D structure and function continues to be an important part of drug creation (Li et al., 2025). Molecular recognition is the process by which biological macromolecules join with each other or with several tiny molecules through noncovalent interactions to produce a specific complex (Tesei et al., 2024). The purpose of rational drug design is to employ structural data and protein-ligand-binding mechanisms to improve the process of discovering new medications. A thorough grasp of the nature of molecular recognition/interaction is also critical for drug discovery, design, and development (Susa et al., 2024).

10.4.1 Binding modes and interaction mapping

The binding modes and interaction mapping between a ligand and its target protein represent a central component of SBDD. This is because it directly links structural knowledge to rational drug optimization. High-resolution structural techniques such as X-ray crystallography and cryo-EM, along with computational approaches including molecular docking and MD simulations, help in the identification of ligand orientation within binding pockets and interactions with specific amino acid residues (Tran-Nguyen and Camproux, 2025). These interactions, including hydrophobic contacts, hydrogen bonds, π-stacking, salt bridges, amide–π interactions, cation–π interactions, and halogen bonding, collectively determine binding affinity, specificity, and overall drug efficacy (Ferreira De Freitas and Schapira, 2017).

Hydrophobic interactions represent the most dominant contributors to protein-ligand binding and are primarily driven by the burial of nonpolar surfaces, leading to favorable entropy changes. The frequent presence of aromatic rings in drug molecules (approximately 76% of marketed drugs) reflects their importance in forming stable hydrophobic contacts with residues such as leucine, valine, and isoleucine (Ferenczy and Kellermayer, 2022). Hydrogen bonds provide directionality and specificity, stabilizing ligands within binding sites; proteins often act as hydrogen bond donors, while residues such as glycine, serine, and aspartate play critical roles (Chen et al., 2016). Aromatic π–π stacking interactions, involving residues such as phenylalanine, tyrosine, and tryptophan, contribute significantly to molecular recognition and binding stability. However, excessive aromaticity may negatively influence solubility and pharmacokinetic properties (Shao et al., 2022).

Electrostatic interactions such as salt bridges and cation–π interactions further influence ligand binding. Salt bridges arise from interactions between oppositely charged groups but may contribute modestly to stability due to desolvation penalties. In contrast, cation–π interactions, commonly involving arginine residues and aromatic rings in ligands, play a significant role in stabilizing complexes and are frequently exploited in medicinal chemistry. Amide–π stacking interactions, involving the interaction between peptide backbone amides and aromatic systems, expand the range of noncovalent interactions relevant to ligand design. Halogen bonding, particularly involving chlorine, bromine, and iodine atoms, has emerged as an important interaction due to its directional nature and ability to enhance binding affinity, selectivity, and membrane permeability (Hazra and Levy, 2023; Gallivan and Dougherty, 2000). Thus detailed interaction mapping supports rational modification of lead compounds to enhance binding affinity, improve selectivity, and optimize pharmacokinetic properties.

10.4.2 Allosteric modulation

It is an inherent feature of all flexible biomolecules, mostly proteins, that orthosteric sites (functional sites) demonstrate a functional connection over long distances with topologically dissimilar areas known as allosteric sites. Allosteric regulation is emerging as a game-changing strategy for therapeutic development, particularly when it comes to addressing seemingly impossible protein targets. Historically, many critical protein targets have been thought to be undruggable or difficult to target using traditional orthosteric drug design. Allosteric treatments provide a potential alternative approach to overcome these obstacles, with significant advantages over their orthosteric counterparts (Fig. 10.2) (Li et al., 2023).

10.4.3 Induced fit and conformational dynamics

Three theories have been proposed to describe protein-ligand-binding mechanisms: "lock-and-key," "induced fit," and "conformational selection." The lock-and-key

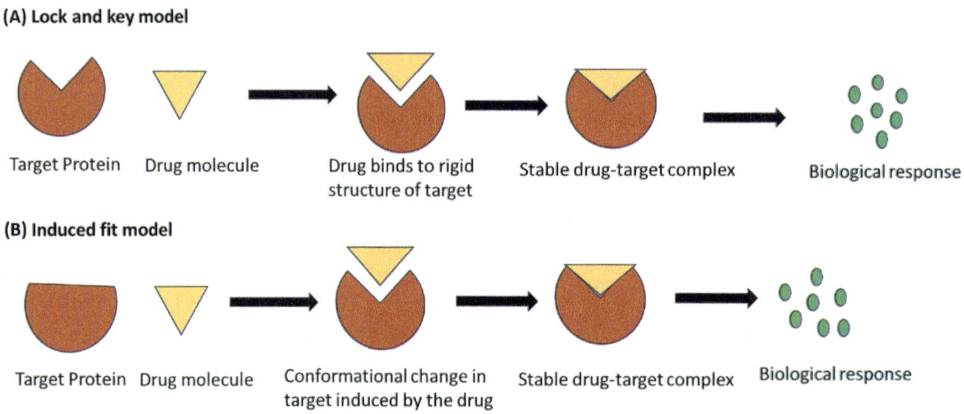

Figure 10.2 (A) Lock-and-key model, the drug binds to a rigid and structurally complementary binding site on the target protein, forming a stable complex that leads to a biological response. (B) Induced fit model, drug-binding induces conformational changes in the target protein, resulting in an optimized fit and formation of a stable drug-target complex, ultimately triggering a biological response.

concept requires that both the protein and the ligand be stiff, with binding interfaces that are ideally matched. As a result, only the suitably sized ligand (the key) can enter the binding pocket (keyhole) of the protein (the lock) (Holyoak, 2013). However, when the initial forms of a protein and ligand do not match, the model cannot explain the experimental evidence that they bind. This leads to the induced fit model, which assumes that the binding site in the protein is flexible and that the interacting ligand causes a conformational change at the binding site. Because the induced fit mechanism considers only the conformational flexibility of the ligand-binding site, this model appears to be appropriate for proteins that undergo relatively small conformational changes after ligand binding. Furthermore, both the lock-and-key and induced fit models treat the protein as a single, stable shape under certain experimental conditions. However, most proteins are fundamentally dynamic, and the conformational selection model accounts for this flexibility. The conformational selection model is based on the free energy landscape hypothesis of protein structure and dynamics, which postulates that a protein's native state is not a single, rigid conformation, but rather a vast ensemble of conformational states/substates that coexist in equilibrium with different population distributions (Holyoak, 2013). The ligand can bind selectively to the most suitable conformational state/substate, eventually shifting the equilibrium towards this state/substate. In other words, the unbound protein has a good chance of sampling the same configuration as the ligand-bound state (Sedov and Zuev, 2025). Understanding these binding mechanisms is essential in drug design, as they guide the rational development of ligands that can precisely fit rigid targets, induce favorable conformational changes, or selectively stabilize specific protein states to achieve higher affinity, selectivity, and efficacy (Spassov, 2024).

10.5 Structural biology in studying drug resistance

Drug resistance refers to the ability of pathogens (bacteria, viruses, fungi, and parasites) or cancer cells to survive and proliferate despite the presence of drugs that were previously effective against them. According to the World Health Organization, drug resistance is the capacity of microorganisms to withstand the effects of drugs designed to kill or inhibit them (Blair et al., 2015). This process has become one of the biggest challenges in drug discovery and development and is posing a serious threat to public health, affecting humans worldwide. With advances in computer science and structural biology, SBDD has become a common strategy for drug discovery to understand the role of pathogens and cancer cells in evading drug action. This helps to provide atomic-scale information on resistance mechanisms, guiding rational inhibitor design (Fig. 10.3) (Munita and Arias, 2016).

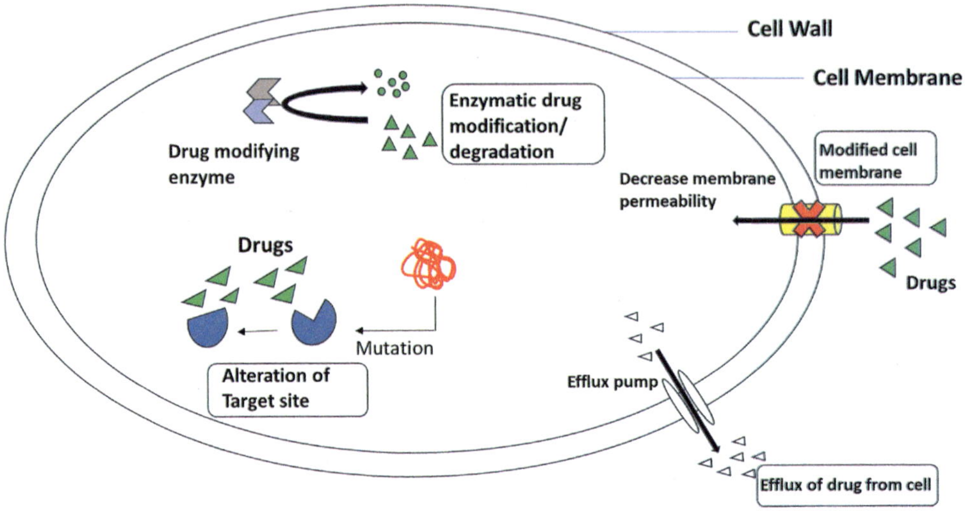

Figure 10.3 Mechanism of drug resistance. Efflux of drug from cell, removal of antibiotics from bacterial cell through protein pumps which reduces effective intracellular drug concentration; alteration of target site, mutation causes modification in the drug-target site reducing binding; enzymatic drug modification causes drug modification or degradation resulting in the inactivation of drug; cell membrane modification, cells can alter membrane lipid composition, fluidity, and permeability making it more difficult for the drugs to diffuse across.

10.5.1 Mechanisms of drug resistance

Drug resistance usually arises by several major mechanisms, namely, efflux of the drug from the cell, enzymatic drug modification, alteration of the drug target that reduces the affinity of the drug, and changes in DNA-repair or stress response pathways. Efflux of drugs is the active removal of antibiotics from bacterial cells through protein pumps, which reduces the effective intracellular drug concentration and enables bacteria to survive longer and mutate, as observed in the case of fluoroquinolone resistance (Reeve et al., 2015). The mechanism of enzymatic drug modification involves two specific classes of enzymes: one group catalyzes chemical modifications, and the second one causes drug degradation. β-lactam antibiotics, such as carbapenems, penicillins, and cephalosporins, which have a broad spectrum of activity, have become a major barrier to effective treatment due to drug modification via β-lactamases. The β-lactamases encoded by plasmids and transposons provide adaptive resistance (Reeve et al., 2015). By the mechanism of modification of the drug target, different species within a single genus of bacteria can modify the binding site of the drug or drug target by presenting various structural motifs for the same target, resulting in resistance. In the target mutation mechanism, strains with mutations in the target protein alter the conformation and dynamics of the drug-binding site (Anderson, 2012).

Extrinsic stresses such as UV radiation, oxidative stress, pH changes, and anoxia can trigger cellular responses that confer resistance to drugs. Prokaryotes regulate stress through the SOS response, the heat-shock response, and the adaptive response to alkylating agents. These responses promote resistance by mechanisms such as decreased drug absorption, increased efflux, metabolic inactivation, and drug sequestration (Ray et al., 2017).

10.5.2 Structural basis of mutation

Mutations causing antimicrobial resistance can alter the antibiotic action through the following mechanisms, including modifications of the antimicrobial target, which decrease the affinity of the drug, decrease in drug uptake, activation of efflux mechanisms, and global changes in important metabolic pathways via modulation of regulatory networks (Munita and Arias, 2016). The alterations can occur anywhere throughout the gene expression and regulatory pathways. Various molecular processes are involved in drug resistance, such as DNA mutation, deletion, or amplification, and altered transcriptional or posttranscriptional control of ribonucleic acid (RNA) levels and altered posttranslational modifications of proteins.

10.5.3 Designing resistance-overcoming drugs

Structural biology plays an important role in the development of next-generation therapeutics capable of overcoming drug resistance. Researchers can design drugs with improved binding affinity, adaptability, and specificity by analyzing and studying altered target sites. Crystal structures of targets with high resolution are important for the development of new antibiotics (Reeve et al., 2015).

By targeting the protein backbone and highly conserved residues, multiple targeting strategies have been identified as useful strategies for the rational design of novel inhibitors with a low risk of resistance. SBDD could eventually lead to the discovery and development of novel, more potent, and safer drugs with potentially different resistance profiles compared to the existing drugs (Hao et al., 2012).

10.6 Applications in biologics and macromolecular drugs

Biologics and macromolecular therapeutics form an important component of modern medicine due to the incorporation of specific targets of disease-associated molecules that are often inaccessible to conventional small-molecule drugs. These therapies include antibodies, peptides, proteins, and RNA-based drugs, all of which benefit significantly from structural biology. Structure-guided approaches, combined with advances in computational modeling and bioengineering, have accelerated the development of next-generation biologics with improved therapeutic efficacy and reduced immunogenicity (Carter and Lazar, 2018).

10.6.1 Antibody and antibody-drug conjugate design

Monoclonal antibodies and antibody-drug conjugates (ADCs) represent a major class of biologics used in oncology and autoimmune diseases. The field provides information on antigen–antibody interactions for the rational design of antibodies with enhanced binding affinity and specificity. In ADCs, antibodies are conjugated with cytotoxic payloads via specialized linkers for targeted delivery of drugs to diseased cells while minimizing systemic toxicity. The structural characterization of antibody–antigen complexes and linker-payload stability is critical for optimizing therapeutic performance and safety. Advances in cryo-EM and X-ray crystallography have led to the design of bispecific antibodies and next-generation ADCs with improved targeting capabilities (Beck et al., 2017).

10.6.2 Peptide-based therapeutics

Peptide-based therapeutics have gained increasing attention due to their high specificity, low toxicity, and ability to modulate PPIs. Structural features of peptide-receptor interactions help the rational design of peptides with improved stability, bioavailability, and receptor selectivity. Techniques such as structure-guided optimization, cyclization, and incorporation of nonnatural amino acids are commonly used to enhance therapeutic properties. Peptides are widely used in metabolic disorders, cancer, and infectious diseases for targeting receptors or enzymes that are challenging for small molecules to modulate effectively. Structural biology plays a crucial role in identifying binding conformations and guiding modifications that improve pharmacokinetic profiles (Flemming, 2016).

10.6.3 Protein and RNA-based drugs

Protein therapeutics, including enzymes and recombinant proteins, and RNA-based drugs such as small interfering RNA and messenger RNA (mRNA), represent a rapidly expanding area of drug development. Structural biology provides essential information on protein folding, stability, and interaction networks for the design of engineered proteins with enhanced activity and reduced immunogenicity. In RNA therapeutics, structural understanding of RNA folding and RNA-PPIs supports the development of stable and efficient delivery systems. Advances in lipid nanoparticle technology and nucleic acid engineering have further enabled the clinical success of RNA-based drugs, including mRNA vaccines and gene-silencing therapies. Structural approaches continue to guide the optimization of these therapeutics by improving target specificity and delivery efficiency (Pardi et al., 2018).

10.7 Integration with computational and AI-based approaches

The integration of drug discovery with computational and AI-based approaches has led to the efficient analysis of complex biological systems and large-scale datasets. These approaches provide information regarding biomolecular structure, interactions, and dynamics, which in turn accelerate target identification, validation, and optimization. This integration increases the accuracy, efficiency, and scope of structure-guided drug design.

10.7.1 Molecular dynamics simulations

MD simulation is another computational technique used to study the time-dependent behavior of biomolecules at atomic resolution by simulating their physical movements under defined conditions. In this method, atoms are treated as classical particles whose movement depends on Newton's laws of motion. Here, the forces acting on each atom are derived from a predefined potential energy function or force field. These force fields include bonded interactions, including bond stretching, angle bending, and torsional rotations, as well as nonbonded interactions such as van der Waals forces and electrostatic interactions (Tiemann et al., 2024).

The process begins with an initial structure, which is obtained from experimental data or computational prediction. The assignment of initial velocities and environmental parameters such as temperature, pressure, and solvent conditions follows this. The system then undergoes energy minimization to remove steric clashes and achieve a stable starting configuration. Moreover, equilibration steps are performed to bring the system to stability under controlled ensembles such as NVT or NPT. After equilibration, the positions and velocities of atoms are updated iteratively over very small time steps, usually in the range of femtoseconds. This generates a trajectory that captures the evolution of a molecular system over time to show various conformational changes, folding pathways, and interaction dynamics (Lüking et al., 2023).

MD simulations help elucidate protein flexibility, which is not captured by static experimental structures. It shows protein-ligand interactions, binding stability, and conformational transitions such as induced fit and allosteric regulation. Additionally, MD can be used to calculate important thermodynamic properties, including binding free energy, diffusion coefficients, and structural stability parameters. Techniques such as root mean square deviation, root mean square fluctuation, and radius of gyration are commonly employed to analyze simulation trajectories and assess structural changes over time (Ghahremanian et al., 2022).

In drug discovery, MD simulations help in refining protein structures, validating docking results, and evaluating the stability of protein-ligand complexes under dynamic conditions. It helps to identify key binding interactions, predict binding affinities, and understand mechanisms such as induced fit and allosteric modulation, which are essential for rational drug design (Kato et al., 2021). Despite their advantages, MD simulations are computationally intensive and are often limited by the accessible simulation timescales. The accuracy of results also depends heavily on the quality of the force field and the initial model. Thus recent advancements in high-performance computing, GPU acceleration, and sampling methods have significantly improved the efficiency and reliability of MD simulations (Puleva et al., 2025).

10.7.2 Machine learning in structure-guided drug discovery

Machine learning refers to a field that helps systems learn patterns and relationships from data and use this knowledge to make predictions or decisions without being explicitly programmed for each task (Vamathevan et al., 2019). It is mostly used in the field of drug discovery for analyzing complex biological and chemical data, including protein structures, ligand conformations, and experimental bioactivity profiles. Various machine learning tools are used to understand the relationship between molecular features and biological outcomes. Thus, the models can predict binding affinities, classify inactive and active compounds, and identify key determinants of protein-ligand interactions (Gawehn et al., 2016). These functions significantly increase the process of virtual screening by prioritizing potential candidates from large chemical libraries. This reduces the time and cost associated with early-stage drug discovery (Vamathevan et al., 2019). Moreover, advanced approaches such as deep learning process 3D structural information and molecular graphs for more accurate modeling of interaction mechanisms and improve the reliability of computational predictions (Gawehn et al., 2016).

Furthermore, advanced approaches such as deep learning are capable of directly processing 3D structural information and molecular graphs, enabling more accurate modeling of interaction mechanisms and improving the reliability of computational predictions (Segler et al., 2018). In addition to screening, machine learning helps in lead optimization by modifying the compounds to improve potency, selectivity, and pharmacokinetic properties. It helps to explore the chemical space through generative modeling techniques with design of novel molecules and desired characteristics.

A wide range of tools and platforms have been developed to support these applications, including DeepChem and Chemprop (Heid et al., 2024) for predicting molecular properties and bioactivity, AutoDock Vina (Trott and Olson, 2010) when integrated with machine learning for improved scoring and screening, and AlphaFold (Jumper et al., 2021) for accurate protein structure prediction. Additionally, tools such as TensorFlow (Abadi et al., 2016) and PyTorch (Paszke et al., 2019) are widely used to develop customized predictive models. Despite challenges related to data quality, interpretability, and model generalization, ongoing advancements and integration with physics-based approaches continue to enhance the impact of machine learning on the field of drug design.

10.7.3 Predictive structural models

Predictive structural models have become a fundamental component of modern structural biology. These models are designed to infer spatial arrangements of residues by integrating sequence information, physicochemical principles, and statistical patterns derived from known structures. Unlike traditional experimental techniques, these models provide rapid and cost-effective access to structural features, especially for proteins that are difficult to crystallize or characterize using laboratory methods. These models mostly identify functional domains and map potential ligand-binding sites, thereby supporting a wide range of applications in biomedical research and drug discovery (Baker and Sali, 2001).

The most advanced development in this field is AlphaFold, which uses deep learning architectures to predict protein structures with high accuracy. It is based on the analysis of multiple sequence alignments to capture evolutionary relationships among protein structures. It uses neural networks to predict interresidue distances and angles, which are then assembled into a consistent 3D model. This helps the system learn complex structural patterns and generate highly reliable predictions even in the absence of closely related templates (Jumper et al., 2021; Varadi et al., 2022). In drug design, AlphaFold has accelerated target identification and validation by providing detailed structural information about proteins, including active and binding sites. It can be integrated with molecular docking and simulation techniques to study protein-ligand interactions, optimize lead compounds, and explore mechanisms such as induced fit (Varadi et al., 2022).

10.8 Case studies in structure-guided drug discovery

In recent years, the integration of computational approaches such as molecular docking, virtual screening, and MD simulations with experimental structural data has accelerated drug discovery pipelines. This synergy enables the rapid identification of lead compounds and their optimization based on atomic-level interactions with biological targets. Structure-guided strategies have been particularly successful in targeting complex and previously "undruggable" proteins by revealing transient or allosteric binding sites. The following case studies show the role of structural biology in the development of kinase inhibitors, GPCR-targeted drugs, and antiviral and anticancer therapeutics and demonstrate their importance in modern precision medicine and targeted therapy development.

10.8.1 Kinase inhibitors

The large-scale genomic and proteomic studies have significantly advanced the therapeutic application of kinase inhibitors by identifying specific oncogenic drivers across diverse malignancies. Most small-molecule kinase inhibitors (SMKIs) target the conserved ATP-binding site within the kinase catalytic domain, enabling effective inhibition of aberrant signaling pathways that promote tumor growth and survival. Despite the structural conservation of this site across the kinome, selective targeting has been achieved through structure-guided drug design approaches that exploit subtle conformational differences and allosteric pockets. As a result, dozens of kinase inhibitors have been approved, with many more in clinical development (Li and et al., 2023). The clinical impact of these agents is particularly evident in cancers such as chronic myeloid leukemia (CML), gastrointestinal stromal tumors, melanoma, and non-small cell lung cancer, where kinase inhibitors have largely replaced traditional treatments. For example, in CML, BCR-ABL inhibitors have superseded allogeneic bone marrow transplantation as first-line therapy, demonstrating how structure-based targeting of oncogenic kinases can transform disease management. Furthermore, the ability of certain kinase inhibitors to penetrate the blood–brain barrier has expanded their use in treating central nervous system metastases, an increasingly common clinical challenge as patient survival improves (Cohen et al., 2021).

From a structural biology perspective, kinase inhibitors represent one of the most successful examples of structure-guided drug discovery. High-resolution crystallographic data have enabled detailed characterization of kinase active and inactive conformations, facilitating the rational design of inhibitors with improved specificity, potency, and pharmacokinetic properties. As of January 2026, 94 SMKIs have been approved, targeting receptor tyrosine kinases, nonreceptor tyrosine kinases, serine/threonine kinases, and dual-specificity kinases such as MEK1/2, with the majority indicated for cancer treatment (Roskoski, 2026). Notably, structural insights have also supported the development of inhibitors with diverse administration routes, including oral, intravenous, topical, and ophthalmic formulations, exemplified by agents such as ruxolitinib. Although only a subset of the ~500 human kinases has been therapeutically exploited to date, kinase-targeted drug discovery is expected to remain a dominant area in oncology, driven by ongoing advances in structural biology, precision medicine, and rational drug design (Cohen et al., 2021).

10.8.2 GPCR-targeted drug

GPCRs are one of the most common pharmacological targets, owing to their participation in a wide range of pathophysiological processes. There are 516 authorized medications targeting GPCRs, which account for 36% of all approved pharmaceuticals. These medications target 121 GPCRs, representing one-third of all nonsensory GPCRs. Furthermore, 337 medicines targeting 133 GPCRs, including 30 new targets, are being tested in clinical studies. Notably, 165 of these compounds are licensed medications under investigation for other applications, whereas innovative agents are increasingly allosteric modulators and biologics. In 2023 diabetes and obesity medications targeting GPCRs generated approximately $30 billion in sales. Clinical trials for GPCR modulators in metabolic disorders, cancer, and immunology are also on the rise (Cohen et al., 2021).

GPCR structural data provide detailed information on ligand-binding pockets, receptor conformational states, and the molecular mechanisms of receptor activation. This structural information helps in the identification of orthosteric and allosteric binding sites. This facilitates the design of selective ligands with improved efficacy and reduced off-target effects (Yang et al., 2021). Furthermore, knowledge of GPCR-ligand interactions supports virtual screening, molecular docking, and lead optimization by revealing key residues involved in binding and signaling. Collectively, structural biology has transformed GPCR drug discovery from empirical screening approaches to more rational and targeted strategies (Zhang et al., 2024).

10.8.3 Antiviral and anticancer drugs

Antiviral and anticancer drugs represent two major pillars of modern therapeutics that target infectious diseases and malignancies, which pose significant global health burdens. Antiviral drugs function by inhibiting key stages of the viral life cycle, including viral entry, replication, and assembly. Over the past decades, substantial progress has been achieved, particularly with the development of direct-acting antivirals for diseases such as HIV and hepatitis C, which have significantly improved patient outcomes (Menéndez-Arias and Gago, 2024). In the same way, anticancer drugs target dysregulated cellular pathways responsible for uncontrolled cell proliferation, survival, and metastasis. Small-molecule inhibitors, particularly kinase inhibitors, have become a cornerstone of targeted cancer therapy. These drugs primarily act by binding to ATP-binding pockets or allosteric sites of kinases. This inhibits aberrant signaling pathways that are important for tumor growth and progression (Zhang et al., 2009).

The high-resolution structural data of viral proteins and cancer-associated targets provide detailed information on binding pockets, conformational flexibility, and molecular mechanisms of action. This helps in the identification of key residues involved in ligand binding and target specificity (Bassetto et al., 2016). A notable example includes the structure-based design of inhibitors targeting the SARS-CoV-2 main protease, which demonstrated the role of structural information in directly guiding the development of potent antiviral agents (Dai et al., 2020). Furthermore, advances in structural techniques such as X-ray crystallography and cryo-EM have enabled the visualization of complex biomolecular assemblies, including virus-host interactions and tumor signaling complexes (Plavec et al., 2021).

10.9 Challenges and limitations

Despite significant progress in structural biology and its applications in drug discovery, several challenges and limitations exist. These challenges arise from both experimental and computational aspects, as well as from difficulties in translating structural features into clinically effective therapies. Most limitations exist in obtaining high-quality structural data; the dynamic nature of biomolecules and the complexity of biological systems often hinder accurate interpretation and application of structural information. Additionally, the dependency of computational tools on approximations and available data can affect reliability.

10.9.1 Experimental constraints

One of the primary challenges in structural biology is the difficulty associated with the experimental determination of biomolecular structures. Techniques such as X-ray crystallography, cryo-EM, and NMR spectroscopy require highly purified samples and specific conditions, which are not always achievable for all proteins. Certain proteins, such as membrane proteins, intrinsically disordered proteins, and large multiprotein complexes, are particularly challenging due to their instability, flexibility, and low expression levels. In crystallography, obtaining well-ordered crystals remains a major bottleneck, while cryo-EM may still face resolution limitations for smaller proteins. Additionally, these experimental methods are often time-consuming, expensive, and technically demanding for large-scale structural studies (Nogales, 2016).

10.9.2 Structural flexibility and dynamics

Proteins are dynamic molecules that undergo conformational changes essential for their biological function. However, many structural determination techniques provide static snapshots that may not fully capture this dynamic behavior. This can lead to an incomplete understanding of processes such as ligand binding, enzyme catalysis, and allosteric regulation, which often depend on transient conformational states. Computational methods like MD simulations address this issue by modeling protein motion over time, but they are constrained by computational cost and timescale limitations. Thus, the full range of biologically relevant conformations remains a significant challenge in structural biology (Moradi et al., 2024).

10.9.3 Translational challenges

The conversion of structural information into real-world effective treatment is still a major challenge in drug discovery. Although structural data can reveal potential drug targets and define ligand-binding sites, not all targets undergo pharmacological modulation. Moreover, the interactions observed in computational models or in vitro experiments often fail to demonstrate efficacy in vivo. This arises from multiple factors, including poor bioavailability, unfavorable pharmacokinetics, toxicity, and unintended off-target effects. In addition, the inherent complexity of biological systems, such as genetic diversity among patients, disease heterogeneity, and adaptive cellular responses, also complicates the successful translation of structural findings into therapeutic outcomes (Zagotto and Bortoli, 2021).

10.10 Future perspectives

The future of structural biology in drug discovery will be defined by its ability to move beyond static representations of biomolecules toward a more dynamic understanding of biological systems. Advances in high-resolution techniques are reducing long-standing experimental barriers. This leads to the characterization of increasingly complex targets, including membrane proteins, transient complexes, and flexible regions that were once considered intractable. At the same time, the growing connection between experimental methods and computational approaches helps in the generation and interpretation of structural information. Techniques such as MD simulations and predictive modeling complement experimental data by modeling conformational variability and functional motions that are essential for rational drug design. The expanding scope of structural biology in addressing therapeutic challenges continues to grow. The focus is shifting from conventional small-molecule targets to more complex systems, including PPIs, allosteric sites, and macromolecular assemblies. Despite these advances, the translation of structural knowledge into clinically effective therapies remains a significant challenge. Overcoming this gap will require improvements in data quality as well as computational accuracy and a closer integration of structural biology with pharmacology, systems biology, and clinical research. Future progress will depend on this multidisciplinary convergence, where structural features are continuously validated and refined within biological and clinical contexts (Arrowsmith, 2024).

AI disclosure

During the preparation of this work, the author(s) used ChatGPT to refine the language. After using this tool/service, the author(s) reviewed and edited the content as needed and take(s) full responsibility for the content of the publication.

References

Abadi, M., et al., 2016. {TensorFlow}: a system for {Large-Scale} machine learning. in 12th USENIX symposium on operating systems design and implementation (OSDI 16).

Anderson, A.C., 2012. Winning the arms race by improving drug discovery against mutating targets. ACS Chemical Biology 7 (2), 278–288. Available at https://doi.org/10.1021/cb200394t.
Anderson, A.C., 2003. The process of structure-based drug design. Chemistry & Biology 10 (9), 787–797. Available at https://doi.org/10.1016/j.chembiol.2003.09.002.
Arrowsmith, C.H., 2024. Structure-guided drug discovery: back to the future. Nature Structural and Molecular Biology 31 (3), 395–396. Available from. Available at https://doi.org/10.1038/s41594-024-01244-3, https://www.nature.com/nsmb/.
Arya, H., Coumar, M.S., 2021. Lead Identification and Optimization, in the Design & Development of Novel Drugs and Vaccines. Elsevier, pp. 31–63.
Ashraf, S.N., Blackwell, J.H., Holdgate, G.A., Lucas, S.C.C., Solovyeva, A., Storer, R.I., et al., 2024. Hit me with your best shot: Integrated hit discovery for the next generation of drug targets. Drug Discovery Today 29 (10), 104143. Available at https://doi.org/10.1016/j.drudis.2024.104143.
Baker, D., Sali, A., 2001. Protein structure prediction and structural genomics. Science (New York, N.Y.) 294 (5540), 93–96. Available at https://doi.org/10.1126/science.1065659.
Bassetto, M., Massarotti, A., Coluccia, A., Brancale, A., 2016. Structural biology in antiviral drug discovery. Current Opinion in Pharmacology 30, 116–130. Available at https://doi.org/10.1016/j.coph.2016.08.014.
Batool, M., Ahmad, B., Choi, S., 2019. A structure-based drug discovery paradigm. International Journal of Molecular Sciences 20 (11), 2783. Available at https://doi.org/10.3390/ijms20112783.
Beck, A., Goetsch, L., Dumontet, C., Corvaïa, N., 2017. Strategies and challenges for the next generation of antibody–drug conjugates. Nature Reviews Drug Discovery 16 (5), 315–337. Available at https://doi.org/10.1038/nrd.2016.268.
Biala, G., Kedzierska, E., Kruk-Slomka, M., Orzelska-Gorka, J., Hmaidan, S., Skrok, A., et al., 2023. Research in the field of drug design and development. Pharmaceuticals 16 (9), 1283. Available at https://doi.org/10.3390/ph16091283.
Bijak, V., Szczygiel, M., Lenkiewicz, J., Gucwa, M., Cooper, D.R., Murzyn, K., et al., 2023. The current role and evolution of X-ray crystallography in drug discovery and development. Expert Opinion on Drug Discovery 18 (11), 1221–1230. Available at https://doi.org/10.1080/17460441.2023.2246881.
Blair, J.M.A., Webber, M.A., Baylay, A.J., Ogbolu, D.O., Piddock, L.J.V., 2015. Molecular mechanisms of antibiotic resistance. Nature Reviews Microbiology 13 (1), 42–51. Available at https://doi.org/10.1038/nrmicro3380.
Brosey, C.A., Link, T.M., Shen, R., Moiani, D., Burnett, K., Hura, G.L., et al., 2024. Chemical screening by time-resolved X-ray scattering to discover allosteric probes. Nature Chemical Biology 20 (9), 1199–1209. Available from. Available at https://doi.org/10.1038/s41589-024-01609-1, https://www.nature.com/nchembio.
Brosey, C.A., Tainer, J.A., 2019. Evolving SAXS versatility: solution X-ray scattering for macromolecular architecture, functional landscapes, and integrative structural biology. Current Opinion in Structural Biology 58, 197–213. Available from. Available at https://doi.org/10.1016/j.sbi.2019.04.004, http://www.elsevier.com/locate/sbi.
Carter, P.J., Lazar, G.A., 2018. Next generation antibody drugs: pursuit of the 'high-hanging fruit' Nature Reviews. Drug Discovery 17 (3), 197–223. Available from. Available at https://doi.org/10.1038/nrd.2017.227, http://www.nature.com/nrd/index.html.
Chen, D., Oezguen, N., Urvil, P., Ferguson, C., Dann, S.M., Savidge, T.C., 2016. Regulation of protein-ligand binding affinity by hydrogen bond pairing. Science Advances 2 (3). Available at https://doi.org/10.1126/sciadv.1501240.
Chen, L., Ruan, X., Li, X., Fu, H., 2024. Molecular interactions in biological systems: technological applications and innovations. Computational Molecular Biology. Available at https://doi.org/10.5376/cmb.2024.14.0021.
Cheng, S.S., Yang, G.J., Wang, W., Leung, C.H., Ma, D.L., 2020. The design and development of covalent protein-protein interaction inhibitors for cancer treatment. Journal of Hematology and Oncology 13 (1) Available from. Available at https://doi.org/10.1186/s13045-020-00850-0, http://www.jhoonline.org/.

Cohen, P., Cross, D., Jänne, P.A., 2021. Kinase drug discovery 20 years after imatinib: progress and future directions. Nature Reviews. Drug Discovery 20 (7), 551–569. Available from. Available at https://doi.org/10.1038/s41573-021-00195-4, http://www.nature.com/nrd/index.html.

Comajuncosa-Creus, A., Jorba, G., Barril, X., Aloy, P., 2024. Comprehensive detection and characterization of human druggable pockets through binding site descriptors. Nature Communications 15 (1). Available at https://doi.org/10.1038/s41467-024-52146-3.

Dai, W., Zhang, B., Jiang, X.-M., Su, H., Li, J., Zhao, Y., et al., 2020. Structure-based design of antiviral drug candidates targeting the SARS-CoV-2 main protease. Science (New York, N.Y.) 368 (6497), 1331–1335. Available at https://doi.org/10.1126/science.abb4489.

Ferenczy, G.G., Kellermayer, M., 2022. Contribution of hydrophobic interactions to protein mechanical stability. Computational and Structural Biotechnology Journal 20, 1946–1956. Available from. Available at https://doi.org/10.1016/j.csbj.2022.04.025, www.csbj.org.

Ferreira, L.L.G., Andricopulo, A.D., 2019. ADMET modeling approaches in drug discovery. Drug Discovery Today 24 (5), 1157–1165. Available from. Available at https://doi.org/10.1016/j.drudis.2019.03.015, www.elsevier.com/locate/drugdiscov.

Ferreira De Freitas, R., Schapira, M., 2017. A systematic analysis of atomic protein-ligand interactions in the PDB. MedChemComm 8 (10), 1970–1981. Available from. Available at https://doi.org/10.1039/c7md00381a, http://www.rsc.org/Publishing/Journals/MD/About.asp.

Filgueiras, J.L., Varela, D., Santos, J., 2023. Protein structure prediction with energy minimization and deep learning approaches. Natural Computing 22 (4), 659–670. Available from. Available at https://doi.org/10.1007/s11047-023-09943-4, https://www.springer.com/journal/11047.

Flemming, A., 2016. Bivalent mTOR inhibitors — the next generation. Nature Reviews Drug Discovery 15 (7), 455. Available at https://doi.org/10.1038/nrd.2016.134 455.

Gallivan, J.P., Dougherty, D.A., 2000. A computational study of cation-π interactions vs salt bridges in aqueous media: implications for protein engineering. Journal of the American Chemical Society 122 (5), 870–874. Available at https://doi.org/10.1021/ja991755c.

Gawehn, E., Hiss, J.A., Schneider, G., 2016. Deep learning in drug discovery. Molecular Informatics 35 (1), 3–14. Available from. Available at https://doi.org/10.1002/minf.201501008, http://onlinelibrary.wiley.com/journal/10.1002/(ISSN)1868-1751.

Ghahremanian, S., Rashidi, M.M., Raeisi, K., Toghraie, D., 2022. Molecular dynamics simulation approach for discovering potential inhibitors against SARS-CoV-2: a structural review. Journal of Molecular Liquids 354, 118901. Available at https://doi.org/10.1016/j.molliq.2022.118901.

Hameduh, T., Haddad, Y., Adam, V., Heger, Z., 2020. Homology modeling in the time of collective and artificial intelligence. Computational and Structural Biotechnology Journal 18, 3494–3506. Available at https://doi.org/10.1016/j.csbj.2020.11.007.

Hansch, C., Fujita, T., 1964. ρ-σ-π Analysis. a method for the correlation of biological activity and chemical structure. Journal of the American Chemical Society 86 (8), 1616–1626. Available at https://doi.org/10.1021/ja01062a035.

Hao, G.F., Yang, G.F., Zhan, C.G., 2012. Structure-based methods for predicting target mutation-induced drug resistance and rational drug design to overcome the problem. Drug Discovery Today 17 (19-20), 1121–1126. Available at https://doi.org/10.1016/j.drudis.2012.06.018.

Hazra, M.K., Levy, Y., 2023. Cross-Talk of Cation−π interactions with electrostatic and aromatic interactions: a salt-dependent trade-off in biomolecular condensates. Journal of Physical Chemistry Letters 14 (38), 8460–8469. Available from. Available at https://doi.org/10.1021/acs.jpclett.3c01642, http://pubs.acs.org/journal/jpclcd.

Heid, E., Greenman, K.P., Chung, Y., Li, S.-C., Graff, D.E., Vermeire, F.H., et al., 2024. Chemprop: a machine learning package for chemical property prediction. Journal of Chemical Information and Modeling 64 (1), 9–17. Available at https://doi.org/10.1021/acs.jcim.3c01250.

Hodgson, J., 2001. ADMET—turning chemicals into drugs. Nature Biotechnology 19 (8), 722–726. Available at https://doi.org/10.1038/90761.

Holyoak, T., 2013. Molecular recognition: lock-and-key, induced fit, and conformational selection. Encyclopedia of Biophysics. Springer Science+Business Media, United States, pp. 1584–1588.

Available from. Available at https://link.springer.com/referencework/10.1007/978-3-642-16712-6, 10.1007/978-3-642-16712-6_468.

Jacobson, M.P., Pincus, D.L., Rapp, C.S., Day, T.J.F., Honig, B., Shaw, D.E., et al., 2004a. A hierarchical approach to all-atom protein loop prediction. Proteins: Structure, Function, and Bioinformatics 55 (2), 351–367. Available at https://doi.org/10.1002/prot.10613.

Jacobson, M.P., Pincus, D.L., Rapp, C.S., Day, T.J.F., Honig, B., Shaw, D.E., et al., 2004b. A hierarchical approach to all-atom protein loop prediction. Proteins: Structure, Function and Genetics 55 (2), 351–367. Available at https://doi.org/10.1002/prot.10613.

Jumper, J., Evans, R., Pritzel, A., Green, T., Figurnov, M., Ronneberger, O., et al., 2021. Highly accurate protein structure prediction with AlphaFold. Nature 596 (7873), 583–589. Available at https://doi.org/10.1038/s41586-021-03819-2.

Kalyaanamoorthy, S., Chen, Y.P.P., 2011. Structure-based drug design to augment hit discovery. Drug Discovery Today 16 (17-18), 831–839. Available at https://doi.org/10.1016/j.drudis.2011.07.006.

Kato, K., Nakayoshi, T., Kurimoto, E., Oda, A., 2021. Molecular dynamics simulations for the protein–ligand complex structures obtained by computational docking studies using implicit or explicit solvents. Chemical Physics Letters 781, 139022. Available at https://doi.org/10.1016/j.cplett.2021.139022.

Kelley, L.A., Mezulis, S., Yates, C.M., Wass, M.N., Sternberg, M.J.E., 2015. The Phyre2 web portal for protein modeling, prediction and analysis. Nature Protocols 10 (6), 845–858. Available from. Available at https://doi.org/10.1038/nprot.2015.053, http://www.natureprotocols.com/.

Khedkar, N.R., Sindkhedkar, M., Joseph, A., 2025. Fragment-based drug discovery: small fragments, big impact – success stories of approved oncology therapeutics. Bioorganic Chemistry 156 Available from. Available at https://doi.org/10.1016/j.bioorg.2025.108197, https://www.sciencedirect.com/science/journal/.

Källberg, M., Wang, H., Wang, S., Peng, J., Wang, Z., Lu, H., et al., 2012. Template-based protein structure modeling using the RaptorX web server. Nature Protocols 7 (8), 1511–1522. Available at https://doi.org/10.1038/nprot.2012.085.

Lavecchia, A., Giovanni, C.D., 2013. Virtual screening strategies in drug discovery: a critical review. Current Medicinal Chemistry 20 (23), 2839–2860. Available at https://doi.org/10.2174/09298673113209990001.

Lees, J.A., Dias, J.M., Han, S., 2021. Applications of Cryo-EM in small molecule and biologics drug design. Biochemical Society Transactions 49 (6), 2627–2638. Available from. Available at https://doi.org/10.1042/BST20210444, https://portlandpress.com/biochemsoctrans/article/49/6/2627/230276/Applications-of-Cryo-EM-in-small-molecule-and.

Leman, J.K., Weitzner, B.D., Lewis, S.M., Adolf-Bryfogle, J., Alam, N., Alford, R.F., et al., 2020. Macromolecular modeling and design in Rosetta: recent methods and frameworks. Nature Methods 17 (7), 665–680. Available from. Available at https://doi.org/10.1038/s41592-020-0848-2, http://www.nature.com/nmeth/.

Li, et al., 2023. An updated review on developing small molecule kinase inhibitors using computer-aided drug design approaches. International Journal of Molecular Sciences 24 (18).

Li, M., Lan, X., Lu, X., Zhang, J., 2023. A structure-based allosteric modulator design paradigm. Health Data Science 3. Available at https://doi.org/10.34133/hds.0094.

Li, R., He, X., Wu, C., Li, M., Zhang, J., 2025. Advances in structure-based allosteric drug design. Current Opinion in Structural Biology 90, 102974. Available at https://doi.org/10.1016/j.sbi.2024.102974.

Lin, A., Zhang, Z., Jiang, A., Li, K., Shi, Y., Yang, H., et al., 2026. Computational approaches to druggable site identification: current status and future perspective. Acta Pharmaceutica Sinica B 16 (1), 62–92. Available at https://doi.org/10.1016/j.apsb.2025.10.032.

Lüking, M., van der Spoel, D., Elf, J., Tribello, G.A., 2023. Can molecular dynamics be used to simulate biomolecular recognition? The Journal of Chemical Physics 158 (18). Available at https://doi.org/10.1063/5.0146899.

Mehmood, U., 2024. Structural biology techniques in drug discovery: from crystal structures to therapeutic applications. Multidisciplinary Journal of Biochemistry Technology 1 (1), 16–23.

Mendenhall, J., Meiler, J., 2016. Improving quantitative structure-activity relationship models using artificial neural networks trained with dropout. Journal of Computer-Aided Molecular Design 30 (2), 177–189. Available from. Available at https://doi.org/10.1007/s10822-016-9895-2, https://www.springer.com/journal/10822.

Menéndez-Arias, L., Gago, F., 2024. Antiviral agents: structural basis of action and rational design. Structure and Physics of Viruses: An Integrated Guide 745–784.

Moradi, S., Nowroozi, A., Aryaei Nezhad, M., Jalali, P., Khosravi, R., Shahlaei, M., 2024. A review on description dynamics and conformational changes of proteins using combination of principal component analysis and molecular dynamics simulation. Computers in Biology and Medicine 183, 109245. Available at https://doi.org/10.1016/j.compbiomed.2024.109245.

Mortuza, S.M., Zheng, W., Zhang, C., Li, Y., Pearce, R., Zhang, Y., 2021. Improving fragment-based ab initio protein structure assembly using low-accuracy contact-map predictions. Nature Communications 12 (1). Available at https://doi.org/10.1038/s41467-021-25316-w.

Munita, J.M., Arias, C.A., 2016. Mechanisms of antibiotic resistance. Virulence Mechanisms of Bacterial Pathogens wiley, United States, pp. 481–511. Available from. Available at https://onlinelibrary.wiley.com/doi/book/10.1128/97815, 10.1128/9781555819286.ch17.

Nippa, D.F., Atz, K., Stenzhorn, Y., Müller, A.T., Tosstorff, A., Benz, J., et al., 2025. Expediting hit-to-lead progression in drug discovery through reaction prediction and multi-dimensional optimization. Nature Communications 16 (1). Available at https://doi.org/10.1038/s41467-025-66324-4.

Nogales, E., 2016. The development of cryo-EM into a mainstream structural biology technique. Nature Methods 13 (1), 24–27. Available at https://doi.org/10.1038/nmeth.3694.

Pardi, N., Hogan, M.J., Porter, F.W., Weissman, D., 2018. mRNA vaccines — a new era in vaccinology. Nature Reviews. Drug Discovery 17 (4), 261–279. Available at https://doi.org/10.1038/nrd.2017.243.

Paszke, A., Gross, S., Massa, F., Lerer, A., Bradbury, J., Chanan, G., et al., 2019. PyTorch: an imperative style, high-performance deep learning library. Advances in Neural Information Processing Systems 32. Available at https://papers.nips.cc/.

Pearce, R., Li, Y., Omenn, G.S., Zhang, Y., Ouzounis, C.A., 2022. Fast and accurate Ab initio protein structure prediction using deep learning potentials. PLoS Computational Biology 18 (9), e1010539. Available at https://doi.org/10.1371/journal.pcbi.1010539.

Pedretti, A., Mazzolari, A., Gervasoni, S., Vistoli, G., 2019. Rescoring and linearly combining: a highly effective consensus strategy for virtual screening campaigns. International Journal of Molecular Sciences 20 (9), 2060. Available at https://doi.org/10.3390/ijms20092060.

Phatak, S.S., Stephan, C.C., Cavasotto, C.N., 2009. High-throughput and in silico screenings in drug discovery. Expert Opinion on Drug Discovery 4 (9), 947–959. Available from. Available at https://doi.org/10.1517/17460440903190961, http://www.informahealthcare.com/doi/pdf/10.1517/17460440903190961.

Plavec, Z., Pöhner, I., Poso, A., Butcher, S.J., 2021. Virus structure and structure-based antivirals. Current Opinion in Virology 51, 16–24. Available at https://doi.org/10.1016/j.coviro.2021.09.005.

Plonski, A.P., Reed, S.M., 2023. Assessing protein homology models with docking reproducibility. Journal of Molecular Graphics and Modelling 121, 108430. Available at https://doi.org/10.1016/j.jmgm.2023.108430.

Puleva, M., Medrano Sandonas, L., Lőrincz, B.D., Charry, J., Rogers, D.M., Nagy, P.R., et al., 2025. Extending quantum-mechanical benchmark accuracy to biological ligand-pocket interactions. Nature Communications 16 (1). Available at https://doi.org/10.1038/s41467-025-63587-9.

Ray, S., Das, S., Suar, M., 2017. Molecular mechanism of drug resistance. Drug Resistance in Bacteria, Fungi, Malaria, and Cancer. Springer International Publishing, India, pp. 47–110. Available from. Available at http://doi.org/10.1007/978-3-319-48683-3, 10.1007/978-3-319-48683-3_3.

Reeve, S.M., Lombardo, M.N., Anderson, A.C., 2015. Understanding the structural mechanisms of antibiotic resistance sets the platform for new discovery. Future Microbiology 10 (11), 1727–1733. Available from. Available at https://doi.org/10.2217/fmb.15.78, http://www.futuremedicine.com/loi/fmb.

Renaud, J.P., Chari, A., Ciferri, C., Liu, W.T., Rémigy, H.W., Stark, H., et al., 2018. Cryo-EM in drug discovery: achievements, limitations and prospects. Nature Reviews. Drug Discovery 17 (7), 471–492. Available from. Available at https://doi.org/10.1038/nrd.2018.77, http://www.nature.com/nrd/index.html.

Roskoski, R., 2026. Properties of FDA-approved small molecule protein kinase inhibitors: a 2026 update. Pharmacological Research 224, 108107. Available at https://doi.org/10.1016/j.phrs.2026.108107.

Sadybekov, A.V., Katritch, V., 2023. Computational approaches streamlining drug discovery. Nature Research, United States Nature 616 (7958), 673–685. Available from. Available at https://doi.org/10.1038/s41586-023-05905-z, https://www.nature.com/nature/.

Schmidt, T., Bergner, A., Schwede, T., 2014. Modelling three-dimensional protein structures for applications in drug design. Drug Discovery Today 19 (7), 890–897. Available from. Available at https://doi.org/10.1016/j.drudis.2013.10.027, www.elsevier.com/locate/drugdiscov.

Schwede, T., Kopp, J., Guex, N., Peitsch, M.C., 2003. SWISS-MODEL: an automated protein homology-modeling server. Nucleic Acids Research 31 (13), 3381–3385. Available at https://doi.org/10.1093/nar/gkg520.

Sedov, I.A., Zuev, Y.F., 2025. Protein–ligand interactions: recent advances in biophysics, biochemistry, and bioinformatics. International Journal of Molecular Sciences 26 (19), 9576. Available at https://doi.org/10.3390/ijms26199576.

Segler, M.H.S., Kogej, T., Tyrchan, C., Waller, M.P., 2018. Generating focused molecule libraries for drug discovery with recurrent neural networks. ACS Central Science 4 (1), 120–131. Available at https://doi.org/10.1021/acscentsci.7b00512.

Shao, J., Kuiper, B.P., Thunnissen, A.-M.W.H., Cool, R.H., Zhou, L., Huang, C., et al., 2022. The role of tryptophan in π interactions in proteins: an experimental approach. Journal of the American Chemical Society 144 (30), 13815–13822. Available at https://doi.org/10.1021/jacs.2c04986.

Spassov, D.S., 2024. Binding affinity determination in drug design: insights from lock and key, induced fit, conformational selection, and inhibitor trapping models. International Journal of Molecular Sciences 25 (13), 7124. Available at https://doi.org/10.3390/ijms25137124.

Sugiki, T., Furuita, K., Fujiwara, T., Kojima, C., 2018. Current NMR techniques for structure-based drug discovery. Molecules (Basel, Switzerland) 23 (1), 148. Available at https://doi.org/10.3390/molecules23010148.

Susa, K.J., Kruse, A.C., Blacklow, S.C., 2024. Tetraspanins: structure, dynamics, and principles of partner-protein recognition. Trends in Cell Biology 34 (6), 509–522. Available from. Available at https://doi.org/10.1016/j.tcb.2023.09.003, https://www.sciencedirect.com/science/journal/09628924.

Tepap Zemnou, C., 2025. In silico comparative study of the anti-cancer potential of inhibitors of glucose-6-phosphate dehydrogenase enzyme using ADMET analysis, molecular docking, and molecular dynamic simulation. Advanced Theory and Simulations 8 (5). Available at https://doi.org/10.1002/adts.202400757.

Tesei, G., Trolle, A.I., Jonsson, N., Betz, J., Knudsen, F.E., Pesce, F., et al., 2024. Conformational ensembles of the human intrinsically disordered proteome. Nature 626 (8000), 897–904. Available at https://doi.org/10.1038/s41586-023-07004-5.

Tiemann, J.K.S., Szczuka, M., Bouarroudj, L., Oussaren, M., Garcia, S., Howard, R.J., et al., 2024. MDverse, shedding light on the dark matter of molecular dynamics simulations. eLife 12. Available at https://doi.org/10.7554/elife.90061.

Tran-Nguyen, V.K., Camproux, A.C., 2025. Computational modeling of protein–ligand interactions: from binding site identification to pose prediction and beyond. Current Opinion in Structural Biology 95 Available from. Available at https://doi.org/10.1016/j.sbi.2025.103152, https://www.sciencedirect.com/science/journal/0959440X.

Tropsha, A., Isayev, O., Varnek, A., Schneider, G., Cherkasov, A., 2024. Integrating QSAR modelling and deep learning in drug discovery: the emergence of deep QSAR. Nature Reviews. Drug Discovery 23 (2), 141–155. Available at https://doi.org/10.1038/s41573-023-00832-0.

Trott, O., Olson, A.J., 2010. AutoDock Vina: Improving the speed and accuracy of docking with a new scoring function, efficient optimization, and multithreading. Journal of Computational Chemistry 31 (2), 455–461. Available at https://doi.org/10.1002/jcc.21334.

Tuukkanen, A.T., Spilotros, A., Svergun, D.I., 2017. Progress in small-angle scattering from biological solutions at high-brilliance synchrotrons. International Union of Crystallography, Germany IUCrJ 4, 518–528. Available from. Available at https://doi.org/10.1107/S2052252517008740, http://www.iucrj.org/m/.

Vajda, S., Beglov, D., Wakefield, A.E., Egbert, M., Whitty, A., 2018. Cryptic binding sites on proteins: definition, detection, and druggability. Current Opinion in Chemical Biology 44, 1–8. Available at https://doi.org/10.1016/j.cbpa.2018.05.003.

Vamathevan, J., Clark, D., Czodrowski, P., Dunham, I., Ferran, E., Lee, G., et al., 2019. Applications of machine learning in drug discovery and development. Nature Reviews. Drug Discovery 18 (6), 463–477. Available at https://doi.org/10.1038/s41573-019-0024-5.

Varadi, M., Anyango, S., Deshpande, M., Nair, S., Natassia, C., Yordanova, G., et al., 2022. AlphaFold protein structure database: massively expanding the structural coverage of protein-sequence space with high-accuracy models. Nucleic Acids Research 50 (D1), D439. Available at https://doi.org/10.1093/nar/gkab1061.

Venien-Bryan, C., et al., 2017. Cryo-electron microscopy and X-ray crystallography: complementary approaches to structural biology and drug discoveryThis paper is an extension of Boutin et al. [(2016), Med. Sci. (Paris), 32, 758-767]. Acta Crystallographica Section F 73 (4), 174–183.

Venkatesan, A., Gopal, J., Candavelou, M., Gollapalli, S., Karthikeyan, K., 2013. Computational approach for protein structure prediction. Healthcare Informatics Research 19 (2), 137–147. Available from. Available at https://doi.org/10.4258/hir.2013.19.2.137India, http://pdf.medrang.co.kr/Hir/2013/019/Hir019-02-10.pdf.

Verlinde, C.L., Hol, W.G., 1994. Structure-based drug design: progress, results and challenges. Structure (London, England 1993) 2 (7), 577–587. Available at https://doi.org/10.1016/S0969-2126(00)00060-5.

Waterhouse, A., Bertoni, M., Bienert, S., Studer, G., Tauriello, G., Gumienny, R., et al., 2018. SWISS-MODEL: homology modelling of protein structures and complexes. Nucleic Acids Research 46 (W1), W296. Available at https://doi.org/10.1093/nar/gky427.

Webb, B., Sali, A., 2017. Protein structure modeling with MODELLER. Methods in Molecular Biology 1654, 39–54. Available from. Available at https://doi.org/10.1007/978-1-4939-7231-9_4, http://www.springer.com/series/7651.

Wälti, M.A., Riek, R., Orts, J., 2017. Fast NMR-based determination of the 3D structure of the binding site of protein–ligand complexes with weak affinity binders. Angewandte Chemie - International Edition 56 (19), 5208–5211. Available from. Available at https://doi.org/10.1002/anie.201612304, http://onlinelibrary.wiley.com/journal/10.1002/(ISSN)1521-3773.

Xie, J., Wang, S., Xu, Y., Deng, M., Lai, L., 2022. Uncovering the dominant motion modes of allosteric regulation improves allosteric site prediction. Journal of Chemical Information and Modeling 62 (1), 187–195. Available at https://doi.org/10.1021/acs.jcim.1c01267.

Yang, D., Zhou, Q., Labroska, V., Qin, S., Darbalaei, S., Wu, Y., et al., 2021. G protein-coupled receptors: structure- and function-based drug discovery. Signal Transduction and Targeted Therapy 6 (1). Available at https://doi.org/10.1038/s41392-020-00435-w.

Yang, J., Zhang, Y., 2015. Protein structure and function prediction using I-TASSER. Current Protocols in Bioinformatics 52 (1) 5.8.1, Available from. Available at https://doi.org/10.1002/0471250953.bi0508s52, http://onlinelibrary.wiley.com/book/10.1002/0471250953;jsessionid=B64454BDEFDCDA6CFEF0F14AA47B6F02.d02t01.

Zagotto, G., Bortoli, M., 2021. Drug design: where we are and future prospects. Molecules (Basel, Switzerland) 26 (22), 7061. Available at https://doi.org/10.3390/molecules26227061.

Zhang, J., Yang, P.L., Gray, N.S., 2009. Targeting cancer with small molecule kinase inhibitors. Nature Reviews. Cancer 9 (1), 28–39. Available at https://doi.org/10.1038/nrc2559.

Zhang, M., Chen, T., Lu, X., Lan, X., Chen, Z., Lu, S., 2024. G protein-coupled receptors (GPCRs): advances in structures, mechanisms and drug discovery. Signal Transduction and Targeted Therapy 9 (1). Available at https://doi.org/10.1038/s41392-024-01803-6.

Zhang, X., Johnson, R.M., Drulyte, I., Yu, L., Kotecha, A., Danev, R., et al., 2021. Evolving cryo-EM structural approaches for GPCR drug discovery. Structure (London, England 1993) 29 (9), 963. Available at https://doi.org/10.1016/j.str.2021.04.008.

CHAPTER 11

Structural biology approaches in vaccine development

Aaliya Taiyab and Imtaiyaz Hassan
Center for Interdisciplinary Research in Basic Sciences, Jamia Millia Islamia, New Delhi, India

11.1 Introduction

The development of vaccines stands as a critical juncture in the history of biomedical science, symbolizing a pivotal triumph that contributes to long-term and economically efficient protection against infectious diseases. Throughout history, vaccines were developed through empirical methods, which included diminishing or inactivating pathogens accompanied by the observation of immune response in experimental systems (Plotkin, 2014). Despite their success, it has been observed that these approaches required years of trial-and-error experimentation, associated with significant biosafety concerns, and lacked reliability in cases of complex, intricate, and fast-evolving structural pathogens (Fig. 11.1A). Notably, over the course of two decades, it is pertinent to highlight the transformative shift in the paradigm of vaccine development that is primarily influenced by advances in structural biology (Plotkin, 2014; Ward and Wilson, 2020).

Structural biology is the discipline that seeks to illustrate the three-dimensional (3D) configuration of biological macromolecules such as proteins, nucleic acids, and their complexes at atomic- or near-atomic resolution (Fig. 11.1C). Through the analysis of the precise shape, conformational dynamics, and functional surfaces of these molecules, the researchers can decipher the process by how pathogens invade host cells, evade immune surveillance, and stimulate protective immunity. This detailed molecular information constitutes the foundation of structural vaccinology, which is a nascent field comprising the design of immunogens, epitopes, and delivery platforms governed by atomic-level insights (Byrne and McLellan, 2022; Cozzi et al., 2013).

The favorable outcomes of structural vaccinology can be attributed to various technological breakthroughs in X-ray crystallography, nuclear magnetic resonance (NMR) spectroscopy, and cryo-electron microscopy (cryo-EM). Particularly, the advent of high-resolution cryo-EM has stimulated the structural determination of viral glycoproteins, membrane proteins, and large macromolecular assemblies that were once considered unmanageable. The analogous growth, which encompasses computational modeling, bioinformatics, molecular dynamics (MD) simulations, and artificial

Advances in Structural Biology
DOI: https://doi.org/10.1016/B978-0-443-44988-8.00011-3

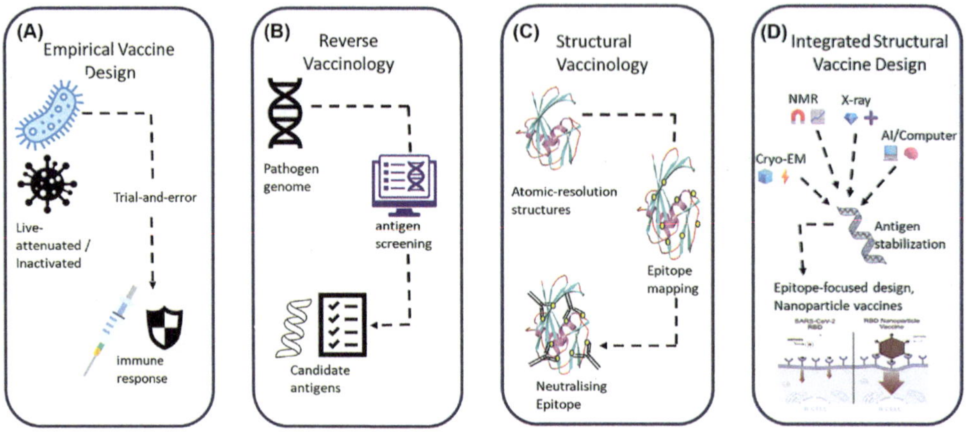

Figure 11.1 Evolution of vaccine design approaches from empirical methods to structural vaccinology. Early vaccine development relied primarily on empirical strategies, including live-attenuated and inactivated whole-pathogen vaccines, developed through trial-and-error experimentation (Panel A). The advent of reverse vaccinology introduced genome-based antigen discovery, enabling systematic identification of vaccine candidates without prior structural knowledge (Panel B). Structural vaccinology further advanced the field by incorporating high-resolution structural information to define protective epitopes, stabilize antigens in immunogenic conformations, and guide rational immunogen design (Panel C). In the current era, structural vaccinology is increasingly integrated with computational tools, molecular dynamics simulations, and artificial intelligence–based structure prediction to enable precision vaccine design, including epitope-focused and nanoparticle-based vaccines (Panel D).

intelligence (AI) has further enlarged the ability to predict, refine, and engineer vaccine antigens without relying solely on experimental architecture (Fig. 11.1D) (Thomas and Abraham, 2022).

The core objective of vaccine design is basically to present the immune system with an antigen that accurately replicates the natural pathogen while directing the immune response towards the most protective epitopes. Structural biology becomes a facilitator of this process in many ways. Firstly, it enables the identification of the functionally important surface regions, such as receptor-binding domains (RBDs), fusion peptides, loop regions, or conserved epitopes targeted by neutralizing antibodies. Second, the systematic comprehension allows the researchers to stabilize these antigens in their most immunogenic conformations. Many viral proteins, such as the respiratory syncytial virus (RSV) F glycoprotein or coronavirus spike proteins, transition through metastable conformations; stabilizing them in the prefusion state significantly enhances the immune response. Third, atomic-level structures allow for the rational design of epitope-focused immunogens, eliminating nonneutralizing regions that may distract or dilute protective immunity (McLellan et al., 2013; Wrapp et al., 2020).

Structural biology has contributed to the diversification of vaccine development for bacterial, parasitic, and even cancer antigens beyond viral pathogens. High-resolution structures of bacterial toxins, adhesins, and outer membrane proteins have shaped toxoid design and subunit vaccine engineering. Structural epitope mapping using neutralizing antibodies has directed the design of malaria and tuberculosis vaccine candidates (Koff et al., 2023). Furthermore, recent progress in structural immunology has enabled the possible visualization of interactions between antigens, antibodies, and immune receptors (such as T-cell receptors [TCRs] and major histocompatibility complex [MHC] molecules). Also, new prospects have emerged for T-cell-based vaccine design, epitope hierarchy prediction, and the development of personalized immunotherapies (Wilson and Stanfield, 2021).

The revolutionary changes that have greatly impacted the vaccine landscape should be attributed to new innovations in structural biology, like self-assembling nanoparticles, virus-like particles (VLPs), and ferritin-based platforms that are being structurally optimized to display antigens in multivalent arrangements that mimic natural viral geometry (Cappelli et al., 2022). AI models, such as AlphaFold and RosettaFold, enable accurate prediction of antigen structures even in the absence of experimental data. These computational predictions, when combined with MD simulations, help evaluate antigen stability, flexibility, and epitope exposure, thereby accelerating rational vaccine design (Seok and Tiwary, 2025). These big changes highlight how new innovations are speeding up drug discovery and helping us better understand the processes behind it.

High-resolution structure determination is resource-intensive and also requires specialized expertise. Besides this, some antigens naturally adopt multiple conformations, making stabilization complex. Additionally, structural information must be integrated with immunological knowledge to ensure that designed immunogens elicit the desired immune response in vivo.

Nonetheless, the advantages are also immense and undeniable, marking the shift from traditional experimental approaches to those of precision engineering in vaccine development. Structural biology has significantly enhanced the probability of success, reduced development timelines, and enabled the design of vaccines that can target highly variable pathogens such as HIV, influenza, and coronaviruses (Trovato and Sartorius, 2020). This can lead to long-term protection, faster response to outbreaks, and fewer chances of severe illnesses with effective preparedness.

This chapter discusses major advances in structural biology techniques used in vaccine development along with their applications in identifying and engineering vaccine antigens, and throws light on computational approaches in modern vaccinology, with case studies demonstrating how structural insights have revolutionized vaccine design. It is essential to understand the atomic-level knowledge that furnishes immunogen design and provides us with a conceptual framework of how such insights are necessary to equip us with the harmony between structural biology and immunology, further shaping the next-generation vaccine development.

11.2 Structural biology tools used in vaccine development

Structural biology is a platform that utilizes the functionality of antigens to be recognized by the immune system at a molecular level. There are several tools used in structural biology that visualize very minute details, enhancing the accuracy of vaccine design (Fig. 11.1). This section deals with such major experimental and computational techniques shaping the field of structural vaccinology (Liljeroos and Malito, 2015; Cozzi et al., 2013).

11.2.1 X-ray crystallography

X-ray crystallography has been a primary tool of structural biology. It relies on X-ray diffraction patterns produced by crystallized molecules made up of proteins, nucleic acids, etc., to reveal atomic-level structural details. It has provided several mainstay vaccines like influenza hemagglutinin (HA), HIV gp120, and bacterial toxins. The mechanism involves purification of the desired protein, crystallization, X-ray irradiation, and studying the diffraction patterns to determine the 3D structure. This technique provides resolutions better than 2 Å, enabling visualization of conformational features including side-chain interactions, glycan orientations critical for antigenicity (Malito et al., 2015).

The advantages of X-ray crystallography in vaccine development involve atomic resolution at an exceptional level, well suited for identifying neutralizing epitopes, facilitating rational antigen engineering, and detailed visualization of antigen–antibody complexes (Table 11.1). However, there are certain limitation like it requires crystallized molecules, which are difficult to achieve for membrane proteins and flexible antigens. Also, it captures the static conformations, missing out on dynamic changes (Toride King and Brooks, 2018).

Even with these challenges, X-ray crystallography remains a fundamental structural biology tool for its ability to reveal conserved structural motifs at atomic resolution that provide critical insights into pathogen biology and play a key role in the development of epitope-focused vaccine design.

11.2.2 Cryo-electron microscopy

Cryo-EM has transformed the era of structural biology by introducing electron detectors directly and exceptionally enhancing image analysis. It eliminates the need for crystallization to obtain minute structural details, making it better than

Table 11.1 Summary of experimental and computational structural biology approaches used across different stages of vaccine development, from antigen discovery to immunogen design.

S. No.	Technique	Resolution	Findings	Application in vaccinology
1.	X-ray crystallography	Atomic	Epitope mapping	Antibody–antigen complexes
2.	Cryo-EM	Near-atomic	Conformational states	Viral glycoproteins
3.	NMR	Dynamic	Molecular flexibility	Epitope dynamics
4.	Small-angle X-ray scattering	~10–30 Å (low-resolution, solution-state)	Solution-state shape and flexibility	Antigen stability and conformational validation
5.	Mass spectrometry–based structural proteomics (HDX-MS, XL-MS)	Nonspatial, residue/peptide-level structural information	Protein dynamics and interaction sites	Epitope mapping and antigen–antibody interface analysis
6.	MD simulations	Computational	Stability and motion	Antigen optimization
7.	Docking	Computational	Binding prediction	Epitope screening
8.	AlphaFold	Computational	Structure prediction	Novel antigen modeling
9.	Immunoinformatics	Computational	Epitope prediction	Vaccine prioritization

crystallography and most appropriate for large, flexible, or complex proteins like viral glycoproteins (Table 11.1). The mechanism involves rapid freezing of biological molecules as a vitreous ice thin layer followed by electron beam imaging. Computational techniques are used to deliver high-resolution 3D structures of large molecular complexes (Earl and Subramaniam, 2016; Li and Li, 2020).

The advantages of Cryo-EM involve visualization of large and complex molecules without the need for crystallization. It is ideal for membrane-bound viral proteins such as SARS-CoV-2 Spike, RSV F, and Ebola GP. It has greatly eased the process of structural characterization of multiple conformational states and is highly useful VLPs, and nanoparticle vaccines. This technique provides near-atomic resolutions of less than 3 Å. Cryo-EM has been instrumental in the design of highly effective vaccines by stabilizing antigens like the prefusion RSV F protein and the SARS-CoV-2 spike protein (Li and Li, 2020; Ma and Su, 2021).

11.2.3 Nuclear magnetic resonance spectroscopy

NMR spectroscopy opens up another domain for studying protein structures and dynamics in their native solution state. It provides details about the arrangement of atoms, flexibility of different regions, and their interactions without the need for crystallization. This technique is well suited to study conformational flexibility and dynamics, since it deciphers the high-resolution details about biomolecules in solution, making it a critical technique for analyzing antigenic epitopes (Table 11.1). The mechanism involves detecting the magnetic properties of atomic nuclei in an applied magnetic field and assessing the developed resonance patterns. This gives the information about the distances and angles between atoms, which helps in establishing the structural details about the protein (~40 kDa) (Valente and Manzano-Rendeiro, 2021; Humpierre et al., 2022).

The advantages of NMR spectroscopy involve capturing protein dynamics (multiple conformations) in solution. This is beneficial for flexible regions, loops, epitopes, and intrinsically disordered antigens. This technique is effective for studying antigen–antibody binding kinetics. However, there are certain limitations, like size constraints and large quantity requirements of isotopically labeled samples. Overall, NMR is highly effective for mapping mobile epitopes and assessing the conformational dynamics of antigens without the need for crystallization (Humpierre et al., 2022).

11.2.4 Small-angle X-ray scattering

Small-angle X-ray scattering (SAXS) is a structural biology tool for studying the overall shape, compactness, and flexibility of proteins in solution. It validates antigen conformations by studying the conformational changes over the period of infection. It is highly effective in evaluating antigen stability, which is crucial for vaccine formulation. Though it is a low-resolution method, its ability to capture structural variability and real-time conformational transitions marks its relevance as a structural biology technique (Fantini et al., 2022; Krueger et al., 2021).

11.2.5 Mass spectrometry-based structural proteomics

Mass spectrometry (MS) deals with posttranslational modifications, glycan patterns, crosslinking interactions, and protein folding states, thereby expanding the capabilities of structural biology. This technique has contributed to the field of vaccine development in several ways. It allows mapping of conformational epitopes by utilizing crosslinking MS. It provides glycoprotein profiling, which is especially important for viruses such as HIV, Ebola, and coronaviruses, where glycosylation heavily influences immune recognition, antigen stability, and vaccine effectiveness. It helps in deciphering protein conformational dynamics via hydrogen–deuterium exchange MS. It plays a crucial role in vaccine formulations by verifying antigen integrity (Ovsyannikova et al., 2007; Downard, 2000).

11.2.6 Computational structural biology

Computational structural biology is a modern approach to speed up vaccine design by initiating a rapid response of prediction, visualization, and optimization of antigen structures. It's an amalgamation of several computational techniques like homology modeling, molecular docking, MD simulations, AI, and AlphaFold, etc. (Table 11.1). Homology modeling deals with predicting antigen structures based on similar known structures in the database. It is most beneficial when there is no experimentally established protein structure. Molecular docking helps in evaluating the antigens–antibodies interaction. It is instrumental in the design of antigens with modified epitopes by recognizing key neutralizing epitopes (Liljeroos and Malito, 2015; Sunita et al., 2020).

MD simulations examine how flexible atoms are in a protein structure, thereby predicting the stability of engineered immunogens. It also assesses the structural transitions via examining the surface exposure of epitopes. AI and AlphaFold is an AI-driven protein structure prediction tool that has exceptionally enhanced antigen discovery. There are algorithms like AlphaFold, RosettaFold, etc. that predict 3D structures of viral and bacterial proteins with high accuracy, helping in recognizing conserved epitopes rapidly. It can also provide antigen models even when the structural data is not present. It has impacted vaccinology in several ways, such as shortening the duration involved in structure-based vaccine design, promoting epitope mapping and immunogen design, and facilitating massive screening of antigen candidates. The involvement of AI models has been extensively utilized in vaccine designs like nanoparticle vaccines, cancer neoantigen vaccines, and universal influenza immunogens (Sunita et al., 2020; Liljeroos and Malito, 2015).

11.2.7 Integrative structural biology

Integrative structural biology deals with combining data from multiple techniques in order to produce the most accurate and dynamic antigen models. Modern structural vaccinology uses techniques like Cryo-EM and crystallography for a refined hybrid structure. SAXS and MD simulation to establish conformational structural changes. MS and cryo-EM to decode glycosylation, especially on viral spike proteins. NMR and AI predictions to stabilize flexible epitopes. This integrative model approach results in more accurate antigen representation, providing extensive insights and details crucial for designing vaccine constructs that are highly stable, immunogenic, and safe (Liljeroos and Malito, 2015; Russo et al., 2020).

11.3 Structural biology in identifying vaccine targets

Structural Biology plays a pivotal role in the rational identification of vaccine targets by disclosing the 3D structure, antigenic determinants, and functional motifs of pathogenic proteins. Researchers use this structural information to select antigens which are surface-exposed, immunogenic, and capable of eliciting neutralizing immune responses. This integrative approach of experimental structural techniques with computational prediction tools approved structural biology for a systematic and mechanism-driven pathway to vaccine target discovery (Liljeroos and Malito, 2015; Rawal et al., 2021).

11.3.1 Structural analysis of pathogen proteins

The foremost step in the recognition of vaccine candidates includes the understanding of the structural organization of pathogen proteins. Techniques, such as X-Ray crystallography, cryo-EM, and NMR spectroscopy, allow high-resolution visualization of viral surface glycoproteins, bacterial adhesins, toxin domains, and parasitic invasion machinery. The results of structural insights have found the accessible exposed surface regions to the immune system. Conformational States that are involved in host entry or immune evasion. The shared conserved domains across strains or species, and the flexible or disordered segments, which are conceivably not ideal for vaccine targets. Such structural characterization provides a comprehensive foundation for the selection of antigens that have stable, well-defined epitopes that can induce a strong immune response (Liljeroos and Malito, 2015; Graham et al., 2019).

11.3.2 Antigenicity prediction

Structural data notably helps in rapid prediction by enabling the identification of regions that are likely to be recognized by the immune system. Structural biology helps in analyzing accessible surfaces used to identify solvent-exposed residues and mapping electrostatic and hydrophobic residues used in the evaluation of antibody-binding potential. It also assesses the stability of antigenic regions under physiological conditions and predicts conformational epitopes that rely on 3D folding. Structural Antigenicity prediction, when integrated with immuno-informatics tools, allows the prioritization of proteins and domains that are capable of providing strong immune responses with high durability (Graham et al., 2019; Liljeroos and Malito, 2015).

11.3.3 Epitope mapping: structural epitopes versus linear epitopes

Structural biology is crucial for identifying differences between linear epitopes and conformational (structural) epitopes. Linear epitopes represent continuous stretches of amino acids and can, time and again, be predicted from sequence alone, whereas structural (conformational) epitopes are formed by amino acids that are consolidated in the 3D structure but distantly separated in the initial primary sequence. These epitopes are functionally important for neutralizing antibody interactions (Gershoni et al., 2007).

The researchers can easily do the precise mapping using antigen–antibody co-crystal structures or cryo-EM reconstructions, in different explorations like neutralization of antibody binding sites, receptor-binding regions which are essential for infection, conserved epitopes which are hidden beneath variable loops, and glycan-dependent epitopes on viral glycoproteins. This mapping facilitates a rational blueprint of vaccine candidates that focuses upon the most protective epitopes (Parvizpour and Pourseif, 2020).

11.3.4 Application in viral, bacterial, and parasitic vaccines

The targeted identification driven by structural biology has been successfully implemented across major pathogen groups, like viral vaccines (which includes structures of viral fusion proteins like influenza HA, RSV F protein, SARS-CoV-2 spike that have announced prefusion confirmations essential for eliciting neutralizing antibodies), bacterial vaccines, which includes structural analysis of adhesins, toxins, secretion system proteins, and capsular polysaccharides useful for identifying stable antigenic domains for subunit vaccine and parasitic vaccines such as complex, multistage parasites that requires identification of invasion proteins, surface antigens, and conserved epitopes involved in host-cell recognition. In each of the aforementioned cases, structural insights greatly enable the selection of targets involved in virulence, cell entry, or immune evasion, all of which are the features that are paramount for effective vaccine design (Anasir and Poh, 2019; Byrne and McLellan, 2022).

11.3.5 Structural vaccinology concept

Structural vaccinology is a new frontier and a field of study that integrates high-resolution structural information into every step of vaccine design. The central principles comprise discovering the neutralizing epitopes via structural mapping and stabilizing the antigens into their biologically relevant conformations. Further immunogens are engineered to enhance immune responses to conserved epitopes and lastly, in the prediction and modulation of antigen flexibility and stability. This approach has been indispensable in the formation of several next-generation vaccines, like stabilized prefusion RSV F proteins and methodically constructed SARS-CoV-2 spike antigens (Van Regenmortel, 2016).

11.4 Structure-based antigen design/reverse vaccinology 2.0

Structure-based antigen design (SAD) or reverse vaccinology 2.0 represents a major transformation from traditional vaccine development to a systematic, structure-based design strategy. SAD is very useful in formulating highly stable, highly immunogenic, and showcasing ideal structural conformations of conserved neutralizing epitopes by utilizing high-resolution structural detail of pathogenic proteins. This framework allows the design of highly specific immunogens responsible for eliciting strong and targeted immune responses against infectious agent which were difficult to trace from conventional vaccine approaches (Rappuoli et al., 2016; Kanampalliwar, 2020).

11.4.1 Stabilizing antigens in biologically active conformations

Several viral antigens exhibit different conformations while undergoing infection, and generally, the conformation responsible for causing the strongest immune response is metastable and transient. Since it is difficult to use such transient conformations directly, structural biology helps stabilize these functional conformations. The key principles involved locking antigens in the most immunogenic prefusion conformation, unveiling helix-stabilizing mutations, proline substitutions, or disulfide (DS) bonds, preventing unnecessary conformational state shifts that expose nonneutralizing epitopes or those that divert immune responses, and ensuring optimal thermal and biochemical stability at large-scale vaccine manufacturing. Structural biology facilitates systematic stabilization approaches via unmasking the accurate arrangement of atoms of antigen states (Burton, 2017; Moxon et al., 2019).

11.4.2 Designing highly stable vaccine immunogens

Along with stabilizing the correct conformation, SAD plays a crucial role in the design of immunogens that are highly thermostable and can be easily stored and transported. It enhances the structural aspects of immunogens so as to preserve the epitope integrity. It promotes ideal epitope exposure to emphasize immune responses. It masks the unnecessary regions in order to reduce nonneutralizing immunodominance. The resultant engineered immunogens are highly potent as compared to their native form, resulting in stronger and more diverse immune responses (Rappuoli et al., 2016; Burton, 2017).

11.4.3 Key case studies in modern vaccine antigen engineering

RSV F protein prefusion design: Researchers observed that neutralizing antibodies were primarily targeting the prefusion form of RSV F protein, also called the RSV F protein. This form was highly unstable in nature. In order to achieve stability, they introduced DS bonds, cavity-filling mutations, and proline substitutions leading to the formation of DS-Cav1, which was highly stable and highly immunogenic. This major breakthrough was a milestone of reverse vaccinology 2.0 and promoted the development of several other vaccines with similar antigen designs (Alnajran et al., 2025; Anasir and Poh, 2019).

Influenza HA stem immunogens: Influenza HA is composed of a head region, which is highly variable, and a stem region that is highly conserved. Empirical existing vaccines are generally focused on head-directed antibodies that limit the protection against cross-strains. Here comes the application of structural biology that helps in the isolation of conserved HA stem domain and the design of immunogens with stem-only region that contains epitopes, that remain universal in different strains and subtypes. Further, it aids in the formation of stable engineered scaffolds that exhibit the ideal structural conformation to elicit a strong immune response. It also utilizes a nanoparticle framework to elevate exposure of stem epitopes. These structurally engineered immunogens enhance the immune response and are promisingly used for developing universal influenza vaccines that can provide protection against continuous, year-to-year antigenic changes of the virus (Anasir and Poh, 2019).

COVID-19 spike protein stabilization (2 P and 6 P mutations): Viruses like SARS-CoV-2 undergo conformational changes from prefusion state to postfusion state. Structural analyses have shown that the prefusion state is more immunogenic and triggers strong neutralizing antibody responses. Vaccines are highly effective if they are able to neutralize the antigen in this prefusion state. Structural biology has made this achievable by introducing 2 P mutations (K986P and V987P) and resulting in stabilization of the prefusion conformational state. Later, thermal stability was also enhanced by introducing 6 P (HexaPro) mutations. These resultant stabilized spike proteins were the benchmark for the design of several COVID-19 vaccines. This also highlight the power of structural vaccinology in rapid pandemic response (Zhu et al., 2024).

11.4.4 Role of structural insights in antigen engineering

Structural biology provides details that are beneficial at every step taken towards modern antigen engineering. Structural Insights are important in various processes like mapping of neutralizing epitopes, identification of structurally dynamic or unstable regions, assessing mutational impact on stability, unmasking glycan regions and masked epitopes, designing a scaffold or nanoparticles for ideal presentation of antigens, etc. The structural details at the atomic level, when combined with computational design, provide accuracy in antigen conformation, stability, and immunogenicity. Structural vaccinology is an important approach in designing vaccines that are highly effective towards complex and rapidly evolving pathogens (Rappuoli et al., 2016; Burton, 2017).

11.5 Epitope-focused vaccine design

Epitope-focused vaccine design is a targeted strategy that deals with the selective optimization of reactive regions of specific antibodies (T cells or B cells) in order to enhance the immune response. This approach is much focused on the immune responses caused by defined protective epitopes and not the complete protein, which also contains nonneutralizing or variable regions (Correia and Bates, 2014; Xu et al., 2022).

11.5.1 B-cell epitope design

Structural biology provides minute details about antigen–antibody interactions leading to the identification of neutralizing antibody epitopes. These epitopes represent the regions that are crucially involved in blocking infection or vital pathogenic processes (Fig. 11.2). The techniques utilized are X-ray crystallography (antigen–antibody complexes), Cryo-EM reconstructions (viral glycoproteins attached to neutralizing antibodies), and structural comparison (to identify conserved epitope regions in different pathogenic variants). The resultant structural information discovered were conformational (discontinuous) epitopes made via 3D protein folding, mutation-resistant epitopes, antibody-binding hotspots associated with strong immune response, and antigenic sites that were earlier masked by glycan shields or highly flexible loop structures. The information about the spatial arrangement of these epitopes and the extent of their exposure enables the researchers to design vaccines in a way that focuses immune response on defined protective epitopes (Correia and Bates, 2014; Ahmad et al., 2016).

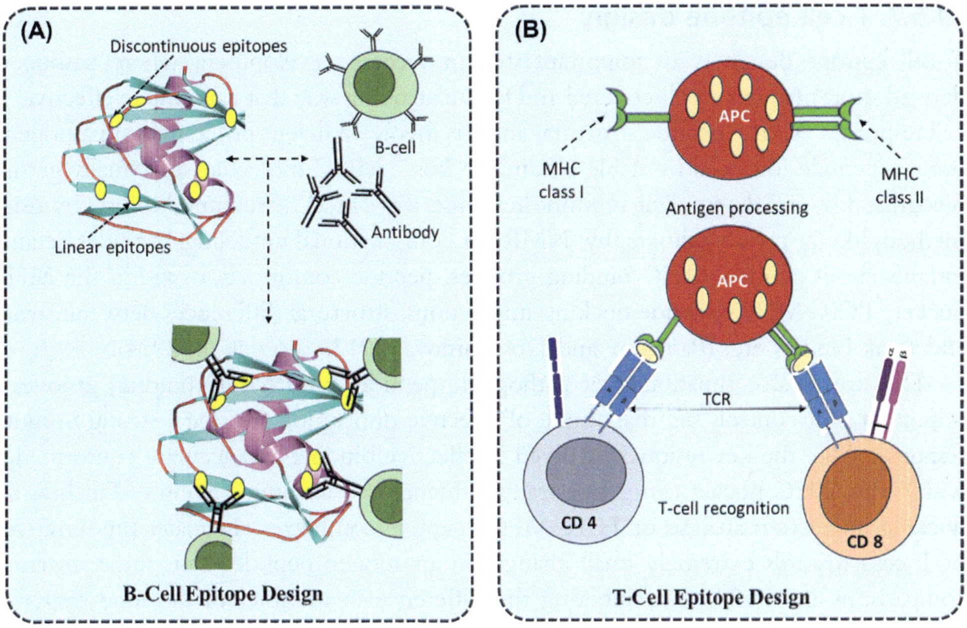

Figure 11.2 Structural distinction between B-cell and T-cell epitope design in rational vaccine development. B cell epitope design focuses on conformational and surface-exposed regions of antigens that are directly recognized by antibodies, emphasizing native three-dimensional structure, epitope accessibility, and antigen stabilization. In contrast, T-cell epitope design targets linear peptide fragments generated by antigen processing and presented by major histocompatibility complex (*MHC*) molecules to T-cell receptors (*TCRs*), highlighting peptide–MHC binding, intracellular processing, and immune repertoire specificity.

Immunofocusing is a strategy that creates immunogens that direct the selective enhancement of immune responses toward targeted epitopes while diminishing responses to off-target regions. Structural biology allows immunofocusing at various processes like epitope scaffolding where targeted epitopes are engineered in a stable, nonviral protein framework, antigen resurfacing where nondesirable epitopes are shielded, masked or mutated, glycan shielding where glycan sugars are intentionally added to shield the nonneutralizing epitopes so as to focus the immune response toward the targeted sites improving the quality and precision of the vaccine-induced immunity, and nanoparticle display where multiple copies of specific epitopes are presented for stronger and more focused antibody response. This framework is typically important for highly evolving antigens like HIV and influenza where focus is on directing the immune response towards conserved epitopes for broad and durable immune protection (Xu et al., 2022; Ren and Zhou, 2016).

11.5.2 T-cell epitope design

T-cell epitope design is an important step in vaccine development where pathogen-derived short peptides are discovered and formulated in a way that can induce effective T-cell responses. MHC–peptide structural analysis involve efficient processing of pathogenic protein peptide followed by stable binding to host's MHC molecules and finally getting recognized by TCRs to elicit immune response (Fig. 11.2). Structural biology by using methods like X-ray crystallography, NMR, or computational modeling, provides detailed insights about peptide–MHC binding grooves, peptide conformation within the MHC pocket, TCR–MHC–peptide docking interactions, structural differences between strong and weak binders etc. (Patronov and Doytchinova, 2013; Song et al., 2024).

The structural compatibility of pathogenic peptides with MHC binding grooves is important as it controls the magnitude of effective display for eliciting a strong immune response. Also, the key residues involved in effective binding and accurate conformation within the MHC pocket can dramatically influence the immune response. Further, the docking interaction analysis of TCR–MHC–peptide complexes elucidates the sensitivity of T cells towards extremely small changes in an antigen peptide. Last, these structural comparisons are useful in deciphering the difference in intensity of immune responses caused by strong binders from weak ones. These structural features collectively determine the ability of a peptide to induce strong CD4+ or CD8+ T-cell responses. Such deep insights are extremely useful in a systematic selection of naturally occurring epitopes or even engineered peptide variants that exhibit improved MHC affinity, enhanced complex stability, and more efficient T-cell activation. This structure-guided approach ultimately supports the design of vaccines that induce stronger, more targeted cellular immune responses (Patronov and Doytchinova, 2013; Song et al., 2024).

Advances in immunoinformatics combined with structural biology promote rapid, high-throughput prediction and screening of potential T-cell epitopes. This results in the selection of candidates with strong MHC-binding potential and strong immunogenicity. There are several predictive algorithms that are useful in predicting the binding affinities of MHC Class I and II molecules, patterns of antigen processing and fragmentation, extent of being immunogenic in nature, HLA allele distribution coverage, epitope conservation across strains and variants. Techniques like NetMHC, IEDB, and AI-driven models combine insightful details about sequence, structure, and binding affinity to discover ideal T-cell epitopes for effective vaccine development (Patronov and Doytchinova, 2013; Song et al., 2024).

11.6 Nanoparticle-based vaccines designed using structural biology

Nanoparticle-based vaccines are the highlights of all the applications of structural biology in enhancing the field of modern vaccinology. The minute structural details about the conformation of antigen, orientation of epitopes, and presentation of multiple copies of epitope, etc., provide the framework for systematic designing of nanoparticles that imitate viral surfaces, elevate B-cell activation, and stimulate strong immune responses. This class includes protein-based nanoparticles like ferritin, VLPs, and computationally modeled self-assembled nanostructures. The accurate geometry with which antigens are presented is ideal for combating pathogens that require potent and focused immune responses (Diaz-Arévalo and Zeng, 2020).

11.6.1 Ferritin nanoparticles

Ferritin is an endogenous protein with a characteristic of self-assembling, forming a 24-subunit spherical nanoparticle (Shahwan et al., 2022). Structural biology has led to the fusion of pathogenic antigen with subunits of ferritin without disturbing the antigen conformation and particle symmetry. It is very advantageous as it allows the multivalent epitope display, leading to the enhancement of B-cell activation. It provides stability to the structure when coming into contact with thermal or biochemical stress. It ensures the precision of spatial arrangement of antigenic domains. The antigenic presentation in multimeric or nanoparticle form (rather than as single monomers) has elevated the germinal center response. There are several vaccine designs that have utilized ferritin nanoparticle theory, like influenza (HA stem and HA head), Epstein–Barr virus, and SARS-CoV-2 antigens (Rodrigues and Alves, 2021; Tai and Chai, 2022).

11.6.2 Virus-like particles

VLPs are proteins with a multimeric form of arrangement that mimic native virus but without genetic material, which makes them the safest and highly immunogenic in nature. Structural biology plays a pivotal role in the design of these VLPs. It preserves the native state of the antigen, which includes its 3D conformation, folding, and spatial arrangement of its atomic determinants or its critical regions. It ensures proper spacing or arrangement of epitopes to resemble native virion surfaces. It also ensures that the multimeric assembly form should be stable in nature. Further, it plays an important role in establishing high compatibility between antigen domains and VLP scaffolds. There are several successful vaccines in this pipeline, like HPV and HBV vaccines (Chroboczek et al., 2014).

11.6.3 Self-assembling nanostructures

Recent innovations in computational structural biology have advanced vaccine design by providing de novo self-assembling nanostructures. These structures are capable of presenting antigens with very high accuracy at the atomic level. These assemblies include icosahedral nanoparticles, which are 20-faced symmetric structures and can present multiple antigenic copies in a consistent, repetitive framework significant for activating B cells at the germinal center. Tetrahedral and octahedral assemblies represent shorter, geometrically defined antigenic scaffolds significant for a systematic placement of epitopes, resulting in an elevated targeted immune response. The third category deals with the integration of two-component nanoparticles such as I53-50. This modular system represents self-assembly of complementary protein subunits, which is capable of facilitating the insertion of a flexible antigen and regulating multivalent presentation. This approach is important in ensuring that each antigen has an ideal orientation, maintaines prefusion or native fold, proper space alignment for B-cell receptor cross-linking, and prevents steric clashes or destabilization of the nanoparticle scaffold (López-Sagaseta et al., 2016).

11.6.4 Case studies in nanoparticle-based vaccine development

SARS-CoV-2 RBD nanoparticle vaccines: Structural biology has a major role in the design of SARS-CoV-2 RBD nanoparticles by providing an exceptionally high-resolution structural detail about spike RBD. Depending on this information, conserved antigenic sites are selected, stable monomers or dimers of RBD are modified, and accurate multivalent display on nanoparticles (ferritin, RBD-I53-50, and mosaic platforms). These nanoparticles designing leads to exceptionally higher neutralizing antibody titers when compared to their monomeric RBD and this also led to them entering into clinical trials (Ma and Zou, 2020; Chen et al., 2022).

HIV mosaic nanoparticles (I53-50 platform): Since there is diversity in HIV strains, it becomes challenging to generate antibodies that could be effective against a wide range of viral strains. This complexity has been resolved by engineering mosaic antigen nanoparticles via utilizing a computationally derived I53-50 scaffold. Structural vaccinology allows researchers to display multiple HIV variants on a single nanoparticle. Also, the conserved epitopes are exposed at their maximum limit. The optimization of multivalent geometry to enhance structural stability and immune receptor engagement. Further, it ensures the enhanced breadth of antibody responses across clades. These nanoparticles are highlights of the most advanced applications of structure-guided mosaic antigen design for a highly evolving pathogen (Brinkkemper and Kerster, 2024).

11.7 Computational integration in structural vaccine design

Computational approaches provide next-generation vaccine platforms that allow the prediction and validation of vaccine design that can complement experimental methods. It's an amalgamation of structural biology with traditional vaccine design to enable rapid optimization of antigen structure, its stability, and immunogenic nature. These computational strategies are integrated at different steps into the sequential vaccine developmental pipelines, and they play a significant role while dealing with complex pathogens and emerging infectious diseases (Ananya et al., 2024).

11.7.1 Molecular dynamics guided antigen stabilization

MD simulations are initially employed for screening at a broader range or at an early stage of antigen design in order to evaluate conformational stability and flexibility. It assesses the stability of prefusion-stabilized antigens under physiological conditions, the consequences of structural-guided mutations or truncations, etc. MD simulations also help in identifying flexible regions that may transiently expose destabilizing or nonneutralizing epitopes responsible for diverting the immune responses. Further, it allows a fair comparison between antigen alone and antigen with nanoparticle or scaffold to assess its immunogenic potential. MD simulation examines conformational changes in a time-dependent manner, and by doing so, it allows researchers to prioritize antigen constructs with retained required immunogenic structures in vivo (Mallik and Morikis, 2006; Vijayakumar et al., 2024).

11.7.2 Docking-guided refinement of antigen–antibody Interfaces

Docking methods are coordinated with the vaccine design framework to refine antigen surfaces on the basis of reported and predicted antibody interactions. Structural docking allows identification of neutralizing antibody binding sites on immunogen candidates, comparative analysis of different binding modes along the antigen variants or mutants, recognition of critical residues of antigens responsible for causing high-affinity interactions, and modification of antigen structure to increase the exposure of the most important immune-protective regions to the immune system. These insight details direct the immune response toward the most protective antigenic parts by eliminating, masking, or modifying regions that do not contribute to effective immunity (Ananya et al., 2024).

11.7.3 Structure-guided epitope selection

Computational epitope prediction works best when it is integrated with structural analysis. Structure filtering via immunoinformatic tools strengthens the accuracy and biological significance of predicted immune targets. Immunoinformatics is the amalgamation of computational and bioinformatics tools to study immune system components and responses. These tools are employed after structural analysis to make sure that predicted epitopes should have their surface exposed and accessible structurally. They should be located within the conformationally stabilized region and should be able to provide broad HLA coverage across diverse populations. This integrated approach eliminates the false positives and improves the accuracy of epitope selection with ideal immunogenic potential, especially in the case of T-cell–based vaccine components (Ananya et al., 2024; Basmenj et al., 2025).

11.7.4 AI-assisted optimization of structure-based immunogens

In modern vaccinology, artificial intelligence (AI) and machine learning (ML) are progressively integrated into the vaccine design pipeline for rapid optimization of antigen design. These combined approaches are valuable in many aspects, such as anticipating the mutational effects on expression and stability of antigen, recognizing the conserved structural motifs for providing broad-spectrum vaccines, comparing different types of biological information and further combining sequence, structural, and experimental immune response datasets. Overall, it speeds up the repeated process of designing, testing, and refining antigens to improve their stability, immunogenicity, and effectiveness (Black, n.d.; Bhardhwaj et al., 2026).

11.7.5 AlphaFold in structural vaccine design

AlphaFold and related platforms help fill critical gaps where insufficient or absence of experimental structural data exist. These models act as a supportive element in integrated vaccine design workflows. They provide initial structural assessment of novel or mutant antigens and recognition of conserved and conformationally stable epitopes. They are also helpful in designing initial vaccine construct before getting validated by crystallographic or cryo-EM, etc. While experimental validation is necessary, these predictions considerably minimize the initial stage uncertainties and facilitate faster immunogen design (Black, n.d.).

11.8 Challenges and future perspectives

Although structural vaccinology has contributed immensely to vaccine design, it has a few major challenges to overcome. One of these is the incomplete understanding of host immune responses when infected with virus, on structural and immunological level. In various cases, it has been observed that the available structural information about viral antigen does not correspond fully to the actual conformations formed during natural infection. For example, when dengue virus epitope was mapped with respect to the already available dengue envelope (E) protein structures, it was noted that the epitope was inaccessible in all existing conformational states (Li et al., 2018). This indicates that the specific conformation (E protein) on which the antibody is bound has not yet been identified experimentally and is not available in the existing structural data. This finding highlights a critical gap between structures solved experimentally and their actual biological form.

Another major pitfall lies with the static nature of the X-ray crystallography technique. Although this technique is the most widely used technique for determining the antigen–antibody complexes by providing high-resolution structural details (Toride King and Brooks, 2018), its incapability to resolve the dynamic states and flexible nature of protein-ligand interactions makes it less reliable. Structural plasticity is important for antigen identification, but static crystal structures are uncertain of displaying the actual conformational changes taking place while the antibody is binding (Van Regenmortel, 2016). At times, the binding interface of an unbound molecule differs from that of binding molecules, illustrating induced-fit or conformational selection mechanisms. In consequence of which, a crystal structure may not necessarily represent the true immunogenic structure recognized by B-cell receptors during antibody maturation.

Most research and vaccine design have focused on B-cell responses in structural vaccinology and have neglected T-cell immunity. This is an important drawback as T-cell–mediated immune responses are crucial for viral clearance and long-term immune protection (Rosendahl Huber et al., 2014). Also, T-cell responses target conserved viral peptides, which are less prone to mutation than antibody-binding sites. This is beneficial as it nullifies the viral mutation effects, thereby limiting viral escape and maintaining immune protection. The limited incorporation of T-cell–focused design strategies represents a significant gap in current structure-based vaccine development.

Structural vaccinology majorly fails when applied in the case of HIV vaccine development. Many researchers and review articles have discussed the challenges encountered while applying structure-based approaches to HIV, given its extensive antigenic diversity, glycan shielding, and conformational masking of conserved epitopes (Pejchal and Wilson, 2010; Van Regenmortel, 2016). These limitations emphasize the need for deeper and more dynamic structural and functional studies to better understand humoral immune responses, and relying only on static structures can be misleading

These challenges can be resolved by integrating some complementary structural techniques. For example, NMR spectroscopy is one such advanced technique that provides structural insights into protein dynamics, conformational flexibility, and transient interactions, which are generally not captured by crystallography alone (Moraes et al., 2016). Henceforth, the integrated approach involving cryo-EM, X-ray crystallography, and NMR can lead to a more thorough understanding of antigen–antibody interactions.

Till date, major applications of structural vaccinology have been explored with recombinant protein vaccines, peptide-based vaccines, and VLPs. Although these approaches have given significant results and are safe, they are often limited with immunogenicity. Hence, this poses a need for a combined approach of structural vaccinology with advanced vaccine delivery systems and immune-promoting platforms. For example, the DS-Cav1, a prefusion-stabilized RSV F protein, can be inserted in a single-round replication recombinant vesicular stomatitis virus (rVSV) vector to enhance immune responses. This rVSV vector efficiently delivers and presents the stabilized antigen to the immune system, enhancing antigen expression and stimulation of both antibody and cellular immune responses while maintaining safety due to its limited replication (Johnson et al., 2013).

Overall, structural vaccinology is a powerful framework for rational vaccine design with few limitations, like incomplete structural knowledge, protein dynamics, immune complexity, and translational application. Also, the future perspective involves the integration of T-cell-oriented strategies extended beyond antibody-centric designs, involving dynamic structural analyses and integrated technologies to utilize full potential of structure-based vaccine development.

11.9 Conclusions

Structural vaccinology has provided a widely used platform for a systematic, knowledge-driven approach to developing vaccines by combining high-resolution structural biology with computational and immunological tools. Experimental techniques like X-ray crystallography, cryo-EM, and NMR are powerful tools enabling detailed visualization of antigen structures and antigen–antibody complexes. These tools provide valuable insights into epitope accessibility, conformational stability, and immune recognition.

These experimental tools are further complemented with computational tools like molecular docking, MD simulations, and immune-informatics platforms to deduce epitopes, optimize antigens, and assess immune responsiveness. Recently, AI-driven structure prediction tools like AlphaFold have widened the structural horizon by allowing rapid modeling of novel unresolved antigens and escalating vaccine design at an early stage.

Although structural vaccinology is an advanced field, it must include focussing the dynamic structural analyses and not just static structural snapshots. It should move beyond antibody-centric strategies and integrate T-cell–focused immunological tools, including viral vectors and nanoparticle-based delivery systems. This collective integrative approach involving structural, computational, and immunological tools makes structural vaccinology as a foundation for next-generation vaccine development with compelling implications in combating novel and resurging infectious diseases.

References

Ahmad, T.A., Eweida, A.E., Sheweita, S.A., 2016. B-cell epitope mapping for the design of vaccines and effective diagnostics. Trials in Vaccinology 5, 71–83. Available at https://doi.org/10.1016/j.trivac.2016.04.003, http://www.journals.elsevier.com/trials-in-vaccinology/.

Alnajran, H., Awadalla, M., Aldakheel, F.M., Alam, I., Momin, A.A., Alturaiki, W., et al., 2025. Design of a peptide-based vaccine against human respiratory syncytial virus using a reverse vaccinology approach: evaluation of immunogenicity, antigenicity, allergenicity, and toxicity. Frontiers in Immunology 16. Available at https://doi.org/10.3389/fimmu.2025.1546254, https://www.frontiersin.org/journals/immunology#.

Ananya, Panchariya, D.C., Karthic, A., Singh, S.P., Mani, A., Chawade, A., et al., 2024. Vaccine design and development: exploring the interface with computational biology and AI. International Reviews of Immunology 43 (6), 361–380. Available at https://doi.org/10.1080/08830185.2024.2374546, http://www.tandfonline.com/loi/iiri20.

Anasir, M.I., Poh, C.L., 2019. Structural vaccinology for viral vaccine design. Frontiers in Microbiology 10. Available at https://doi.org/10.3389/fmicb.2019.00738, https://www.frontiersin.org/journals/microbiology#.

Basmenj, E.R., Pajhouh, S.R., Ebrahimi Fallah, A., Naijian, R., Rahimi, E., Atighy, H., et al., 2025. Computational epitope-based vaccine design with bioinformatics approach; a review. Heliyon 11 (1). Available at https://doi.org/10.1016/j.heliyon.2025.e41714, https://www.sciencedirect.com/science/journal/24058440.

Bhardhwaj, H.S., Tutika, S., Siddharth Kaushik, L.S., Gupta, P.K., Thippa Reddy, K.S., 2026. 9 - Artificial intelligence in vaccine development. In: Zarrintaj, P., Yazdi, M.K., Bencherif, S.A., Saeb, M.R., Mozafari, M. (Eds.), Artificial Intelligence in Biomaterials Design and Development. Woodhead Publishing, pp. 185–223.

Black, S., Madhi, S.A., Rappuoli, R., n.d. Artificial intelligence in the development of vaccines for infectious diseases. *The Lancet Infectious Diseases*.

Brinkkemper, M., Kerster, G., 2024. Mosaic HIV-1 glycoprotein nanoparticles elicit antibody responses. PLoS Pathogens 20, e1012558.

Burton, D.R., 2017. What are the most powerful immunogen design vaccine strategies?: Reverse vaccinology 2.0 shows great promise. Cold Spring Harbor Perspectives in Biology 9 (11). Available at https://doi.org/10.1101/cshperspect.a030262, http://cshperspectives.cshlp.org/content/9/11/a030262.full.pdf.

Byrne, P.O., McLellan, J.S., 2022. Principles and practical applications of structure-based vaccine design. Current Opinion in Immunology 77. Available at https://doi.org/10.1016/j.coi.2022.102209, http://www.journals.elsevier.com/current-opinion-in-immunology/.

Cappelli, L., Cinelli, P., Giusti, F., 2022. Self-assembling protein nanoparticles and virus like particles display β-barrel antigen. PLoS One 17, e0273322.

Chen, R., Zhang, X., Yuan, Y., Deng, X., Wu, B., Xi, Z., et al., 2022. Development of receptor binding domain (RBD)-conjugated nanoparticle vaccines with broad neutralization against SARS-CoV-2 delta and other variants. Advanced Science 9 (11). Available at https://doi.org/10.1002/advs.202105378, http://onlinelibrary.wiley.com/journal/10.1002/(ISSN)2198-3844.

Chroboczek, J., Szurgot, I., Szolajska, E., 2014. Virus-like particles as vaccine. Acta Biochimica Polonica 61 (3), 531–539. Available at https://doi.org/10.18388/abp.2014_1875, http://www.actabp.pl/pdf/3_2014/531.pdf.

Correia, B.E., Bates, J.T., 2014. Proof of principle for epitope-focused vaccine design. Nature 507, 201–206.

Cozzi, R., Scarselli, M., Ferlenghi, I., 2013. Structural vaccinology: a three-dimensional view for vaccine development. Current Topics in Medicinal Chemistry 13, 2629–2637.

Diaz-Arévalo, D., Zeng, M., 2020. Nanoparticle-based vaccines: opportunities and limitations. Nanopharmaceuticals. Elsevier, pp. 135–150.

Downard, K.M., 2000. Contributions of mass spectrometry to structural immunology. Journal of Mass Spectrometry JMS 35, 493–503.

Earl, L.A., Subramaniam, S., 2016. Cryo-EM of viruses and vaccine design. Proceedings of the National Academy of Sciences 113 (32), 8903–8905. Available at https://doi.org/10.1073/pnas.1609721113.

Fantini, M.C.De. A., Oliveira, C.L.P., Lopes, J.L.de S., Martins, T.D.S., Akamatsu, M.A., Trezena, A.G., et al., 2022. Using crystallography tools to improve vaccine formulations. IUCrJ 9 (1), 11–20. Available at https://doi.org/10.1107/s205225252101071x.

Gershoni, J.M., Roitburd-Berman, A., Siman-Tov, D.D., Freund, N.T., Weiss, Y., 2007. Epitope mapping: the first step in developing epitope-based vaccines. BioDrugs Clinical Immunotherapeutics,

Biopharmaceuticals and Gene Therapy 21 (3), 145–156. Available at https://doi.org/10.2165/00063030-200721030-00002Israel, http://biodrugs.adisonline.com/pt/re/cim/pdfhandler.00063030–200721030-00002.pdf;jsessionid=GWqd6WQL63nrGQQvyxmTWxfrhvJyzGQ8Qs1V8gxp2RBmFFHZGXpQ!959335381!-949856145!8091!-1.

Graham, B.S., Gilman, M.S., McLellan, J.S., 2019. Structure-based vaccine antigen design. Annual Review of Medicine 70, 91–104.

Humpierre, A.R., Zanuy, A., Saenz, M., Vasco, A.V., Méndez, Y., Westermann, B., et al., 2022. Quantitative NMR for the structural analysis of novel bivalent glycoconjugates as vaccine candidates. Journal of Pharmaceutical and Biomedical Analysis 214. Available at https://doi.org/10.1016/j.jpba.2022.114721, www.elsevier.com/locate/jpba.

Johnson, J.E., McNeil, L.K., Megati, S., Witko, S.E., Roopchand, V.S., Obregon, J.H., et al., 2013. Non-propagating, recombinant vesicular stomatitis virus vectors encoding respiratory syncytial virus proteins generate potent humoral and cellular immunity against RSV and are protective in mice. Immunology Letters 150 (1–2), 134–144. Available at https://doi.org/10.1016/j.imlet.2012.12.005.

Kanampalliwar, A.M., 2020. Reverse vaccinology and its applications. Immunoinformatics. Springer, pp. 1–16.

Koff, W.C., Rappuoli, R., Plotkin, S.A., 2023. Historical advances in structural and molecular biology and how they impacted vaccine development. Journal of Molecular Biology 435, 168113.

Krueger, S., Curtis, J.E., Scott, D.R., Grishaev, A., Glenn, G., Smith, G., et al., 2021. Structural characterization and modeling of a respiratory syncytial virus fusion glycoprotein nanoparticle vaccine in solution. Molecular Pharmaceutics 18 (1), 359–376. Available at https://doi.org/10.1021/acs.molpharmaceut.0c00986, http://pubs.acs.org/journal/mpohbp.

Li, J., Watterson, D., Chang, C.W., Che, X.Y., Li, X.Q., Ericsson, D.J., et al., 2018. Structural and functional characterization of a cross-reactive dengue virus neutralizing antibody that recognizes a cryptic epitope. Structure (London, England 1993) 26 (1), 51. Available at https://doi.org/10.1016/j.str.2017.11.017, http://www.journals.elsevier.com/structure/.

Li, N., Li, Z., 2020. Cryo-EM studies of virus-antibody immune complexes. Virologica Sinica 35, 1–13.

Liljeroos, L., Malito, E., 2015. Structural and computational biology in the design of immunogenic vaccine antigens. Journal of Immunology Research, 156241.

López-Sagaseta, J., Malito, E., Rappuoli, R., Bottomley, M.J., 2016. Self-assembling protein nanoparticles in the design of vaccines. Computational and Structural Biotechnology Journal 14, 58–68. Available at https://doi.org/10.1016/j.csbj.2015.11.001, www.csbj.org.

Ma, J., Su, D., 2021. Cryo-electron microscopy structure of S-trimer, a subunit vaccine candidate for COVID-19. Journal of Virology 95.

Ma, X., Zou, F., 2020. Nanoparticle vaccines based on the RBD of SARS-CoV-2. Immunity 53, 1315.

Malito, E., Carfi, A., Bottomley, M.J., 2015. Protein crystallography in vaccine research and development. International Journal of Molecular Sciences 16 (6), 13106–13140. Available at https://doi.org/10.3390/ijms160613106, http://www.mdpi.com/1422-0067/16/6/13106/pdf.

Mallik, B., Morikis, D., 2006. Applications of molecular dynamics simulations in immunology: a useful computational method in aiding vaccine design. Current Proteomics 3 (4), 259–270. Available at https://doi.org/10.2174/157016406780655568, http://docstore.ingenta.com/cgi-bin/ds_deliver/1/u/d/ISIS/37418791.1/ben/cp/2006/00000003/00000004/art00004/48CB985FEC371F80117869503441 21EA69E131EE9A.pdf?link=http://www.ingentaconnect.com/error/delivery&format=pdf.

McLellan, J.S., Chen, M., Joyce, M.G., 2013. Structure-based design of a fusion glycoprotein vaccine for respiratory syncytial virus. Science (New York, N.Y.) 342, 592–598.

Moraes, A.H., Simonelli, L., Pedotti, M., Almeida, F.C.L., Varani, L., Valente, A.P., 2016. Antibody binding modulates conformational exchange in domain III of dengue virus e protein. Journal of Virology 90 (4), 1802–1811. Available at https://doi.org/10.1128/JVI.02314–15, http://jvi.asm.org/content/90/4/1802.full.pdf.

Moxon, R., Reche, P.A., Rappuoli, R., 2019. Reverse Vaccinology. Frontiers Media SA, pp. 2776.

Ovsyannikova, I.G., Johnson, K.L., Bergen, H.R., Poland, G.A., 2007. Mass spectrometry and peptide-based vaccine development. Clinical Pharmacology & Therapeutics 82 (6), 644–652. Available at https://doi.org/10.1038/sj.clpt.6100389.

Parvizpour, S., Pourseif, M.M., 2020. Epitope-based vaccine design: a comprehensive overview of bioinformatics approaches. Drug Discovery Today 25, 1034–1042.

Patronov, A., Doytchinova, I., 2013. T-cell epitope vaccine design by immunoinformatics. Open Biology 3, 120139.

Pejchal, R., Wilson, I.A., 2010. Structure-based vaccine design in HIV: Blind men and the elephant? Current Pharmaceutical Design 16 (33), 3744–3753. Available at https://doi.org/10.2174/138161210794079173.

Plotkin, S., 2014. History of vaccination. Proceedings of the National Academy of Sciences 111 (34), 12283–12287. Available at https://doi.org/10.1073/pnas.1400472111.

Rappuoli, R., Bottomley, M.J., D'Oro, U., 2016. Reverse vaccinology 2.0: Human immunology instructs vaccine antigen design. The Journal of Experimental Medicine 213, 469–481.

Rawal, K., Sinha, R., Abbasi, B.A., Chaudhary, A., Nath, S.K., Kumari, P., et al., 2021. Identification of vaccine targets in pathogens and design of a vaccine using computational approaches. Scientific Reports 11 (1). Available at https://doi.org/10.1038/s41598-021-96863-x, www.nature.com/srep/index.html.

Ren, H., Zhou, P., 2016. Epitope-focused vaccine design against influenza A and B viruses. Current Opinion in Immunology 42, 83–90. Available at https://doi.org/10.1016/j.coi.2016.06.002, http://www.journals.elsevier.com/current-opinion-in-immunology/.

Rodrigues, M.Q., Alves, P.M., 2021. Functionalizing ferritin nanoparticles for vaccine development. Pharmaceutics 13, 1621.

Rosendahl Huber, S., van Beek, J., de Jonge, J., Luytjes, W., van Baarle, D., 2014. T cell responses to viral infections â€" opportunities for peptide vaccination. Frontiers in Immunology 5, 1664–3224. Available at https://doi.org/10.3389/fimmu.2014.00171.

Russo, G., Reche, P., Pappalardo, F., 2020. The combination of artificial intelligence and systems biology for intelligent vaccine design. Expert Opinion on Drug Discovery 15, 1267–1281.

Seok, C., Tiwary, P., 2025. Artificial intelligence methodologies in structural biology. Current Opinion in Structural Biology 95, 103156.

Shahwan, M., Khan, M.S., Husain, F.M., Shamsi, A., 2022. Understanding binding between donepezil and human ferritin: molecular docking and molecular dynamics simulation approach. Journal of Biomolecular Structure & Dynamics 40 (9), 3871–3879. Available at https://doi.org/10.1080/07391102.2020.1851302.

Song, X., Li, Y., Wu, H., Qiu, H., Sun, Y., 2024. T-Cell epitope-based vaccines: a promising strategy for prevention of infectious diseases. Vaccines 12 (10). Available at https://doi.org/10.3390/vaccines12101181, http://www.mdpi.com/journal/vaccines.

Sunita, A., Sajid, Y., Singh, P., Shukla, 2020. Computational tools for modern vaccine development. Human Vaccines & Immunotherapeutics 16 (3), 723–735. Available at https://doi.org/10.1080/21645515.2019.1670035.

Tai, W., Chai, B., 2022. Development of a ferritin-based nanoparticle vaccine against SARS-CoV-2 Omicron variant. Signal Transduction and Targeted Therapy 7, 173.

Thomas, S., Abraham, A., 2022. Progress in the development of structure-based vaccines. Methods in Molecular Biology 2412, 15–33. Available at https://doi.org/10.1007/978-1-0716–1892-9_2, http://www.springer.com/series/7651.

Toride King, M., Brooks, C.L., 2018. Epitope mapping of antibody-antigen interactions with X-ray crystallography. Methods in Molecular Biology 1785, 13–27. Available at https://doi.org/10.1007/978-1-4939–7841-0_2, http://www.springer.com/series/7651.

Trovato, M., Sartorius, R., 2020. Viral emerging diseases: challenges in developing vaccination strategies. Frontiers in Immunology 11, 2130.

Valente, A.P., Manzano-Rendeiro, M., 2021. Mapping conformational epitopes by NMR spectroscopy. Current Opinion in Virology 49, 1–6. Available at https://doi.org/10.1016/j.coviro.2021.04.001, http://www.elsevier.com/wps/find/journaldescription.cws_home/724700/description#description.

Van Regenmortel, M., 2016. Structure-based reverse vaccinology failed in the case of HIV because it disregarded accepted immunological theory. International Journal of Molecular Sciences 17 (9), 1591. Available at https://doi.org/10.3390/ijms17091591.

Vijayakumar, S., Kumar, L.L., Borkotoky, S., Murali, A., 2024. The application of MD simulation to lead identification, vaccine design, and structural studies in combat against leishmaniasis - a review. Mini-Reviews in Medicinal Chemistry 24 (11), 1089–1111. Available at https://doi.org/10.2174/1389557523666230901105231, https://www.eurekaselect.com/article/134205.

Ward, A.B., Wilson, I.A., 2020. Innovations in structure-based antigen design and immune monitoring for next generation vaccines. Current Opinion in Immunology 65, 50–56.

Wilson, I.A., Stanfield, R.L., 2021. 50 Years of structural immunology. The Journal of Biological Chemistry 296, 100745.

Wrapp, D., Wang, N., Corbett, K.S., 2020. Cryo-EM structure of the 2019-nCoV spike in the prefusion conformation. Science (New York, N.Y.) 367, 1260–1263.

Xu, D., Li, C., Utz, A., Weidenbacher, P.A.B., Tang, S., Sanyal, M., et al., 2022. Designing epitope-focused vaccines via antigen reorientation. bioRxiv 26928205. Available at https://doi.org/10.1101/2022.12.20.521291, https://www.biorxiv.org.

Zhu, Z., Zhang, L., Li, S., Gao, Y., Wang, Y., Ma, X., et al., 2024. S–6 P exhibits better immunogenicity than S–2 P at lower doses of COVID-19 mRNA vaccines. Decoding Infection and Transmission 2, 100017.

CHAPTER 12

Structural biology in antiviral drug design

Sana Tanweer[1], Sanchit Varma[1], and Malika Adam Abeeda[2]
[1]Department of Molecular Medicine, Jamia Hamdard University, New Delhi, Delhi, India,
[2]Centre for Virology, Jamia Hamdard University, New Delhi, Delhi, India

12.1 Introduction

Structural biology allows us to study high-resolution structures of critical viral targets, including proteases, polymerases, and envelope glycoproteins that may have an important contribution in designing specific therapeutics and targeted antiviral drugs for viral infections. Having prior knowledge about the three-dimensional structures of proteins can be a key asset in drug discovery, as it reveals target geometry, surface properties, and molecular interactions that guide rational drug design (Suguna, 2014).

As the availability of protein structures is fairly limited and there is a high cost and time demand associated with crystallization (Bassetto et al., 2016), combinatorial chemistry and high-throughput screening were used in place of structure-based approaches. Advances in science and technology have made structural biology faster and more cost-effective by accelerating structure determination, improving resolution, and reducing protein and crystal requirements (Congreve et al., 2005).

Viruses are microscopic infectious agents termed obligate intracellular parasites, as they can only survive and reproduce inside a host cell. Over the course of evolution, viruses have developed key mechanisms to evade and escape the immune response of the host. Antiviral drugs are hence needed to fight stubborn infections and resistant pathogens. Antiviral drugs disrupt essential viral components, mainly key proteins and viral nucleic acids. All major viral proteins can serve as potential drug targets. These include both structural and functionally critical nonstructural proteins.

Protein structure determination has thus become an early and critical step in many antiviral drug discovery programs. The development of protease inhibitors for HIV (Pearl and Taylor, 1987) and Relenza, which is an antiviral for influenza (Varghese, 1999), reinvigorated the scientific community's interest in structure-based drug discovery (Bassetto et al., 2016).

Although chemical fragments typically show low target affinity and are unsuitable for conventional high-throughput screening, advances in high-throughput nuclear magnetic resonance (NMR) and crystallography have enabled the direct structural characterization of protein–fragment complexes (Russell and Eggleston, 2000). These

Advances in Structural Biology
DOI: https://doi.org/10.1016/B978-0-443-44988-8.00004-6

insights reliably identify binding pockets and binding modes, guiding the optimization of fragments into potent lead compounds (Hardy and Malikayil, 2003).

Thus structural biology alongside modeling and high-throughput screening allows the identification of specific conserved binding pockets and allosteric sites for targeted therapeutic intervention. The successful development of multiple marketed drugs through structure-based design highlights the essential role of structural biology in targeted drug design (Scapin, 2006; Wahid, 2024).

12.2 Fundamentals of structural biology in drug design

The main goal of structural biology, which is a subfield of molecular biology, is to map the 3D structure of macromolecules such as nucleic acids, proteins, and so on. The basic tenet of structural biology is that a biomolecule's structure and function are closely linked. Researchers can comprehend how biomolecules interact with one another and with tiny molecules, such as medications, at the atomic level because of their detailed structural information (Rhodes, 2006; Branden and Tooze, 1999).

A critical component of drug discovery and development is the connection between biological function and protein structure (Kashaw et al., 2026). In order to carry out specialized biological functions, including enzyme catalysis, chemical recognition, and signal transduction, proteins take on three-dimensional conformations. Protein activity is largely determined by structural characteristics, including active sites, binding pockets, and conformational flexibility. Protein structural changes brought on by mutations or ligand binding can have a substantial impact on function and may result in illness or medication resistance (Alberts et al., 2002; Petsko and Ringe, 2004).

For logical drug design, three-dimensional structure knowledge is crucial. Finding important interaction sites for drug binding is made possible by understanding the spatial organization of amino acid residues inside a target protein. Designing compounds with increased potency, selectivity, and fewer off-target effects is supported by this information. Lead optimization is also guided by structural data, which enables compounds to be systematically modified according to how they interact with the target protein (Blundell et al., 2002; Anderson, 2003).

The two main categories of drug discovery techniques are ligand-based drug design (LBDD) and structure-based drug design (SBDD). By directly modeling protein-ligand interactions, SBDD uses high-resolution structural information of the target protein to generate or optimize therapeutic candidates (Roy, 2018). On the other hand, LBDD is used in situations when structural data are not available and relies on the examination of known ligands to pinpoint the chemical characteristics that cause biological activity. Both approaches are useful, but when trustworthy structural data are available, SBDD offers a more logical and accurate foundation for the creation of antiviral drugs (Lionta et al., 2014; Leach and Gillet, 2007).

12.3 Viral entry and immune response

Viruses enter the host cell by binding to specific host-cell factors, and the process of viral entry involves certain host cellular endocytic pathways. The ultimate goal for the virus is to release as little of its replication machinery as possible inside the host to initiate infection (Knipe and Howley, 2013).

The host innate immune system is the first barrier of defense that acts to stop the virus early. The first step is the recognition of DAMPs and PAMPs, which include viral DNA, RNA, and viral proteins (Rai et al., 2021). These are recognized by pattern recognising receptors (PRRs) such as retinoic acid-inducible gene I (RIG-I)-like receptors (RLRs), nod-like receptors (NLRs), and toll-like receptors (TLRs) (Silenzi, 2015).

This causes induction of different immune signaling cascades, leading to secretion of interferons and pro-inflammatory cytokines and chemokines (Wang et al., 2025), activation of the inflammasome and NF-κB pathway (Murphy and Whitehead, 2011). Natural Killer cells are also activated, which can kill virus-infected cells and release chemicals to promote inflammation.

12.3.1 Toll-like receptors signalling

TLR in the cytoplasm and endosomal compartments of host cells recognize RNA and DNA of viruses, causing the activation of both NF-κB and IFN pathways. Both TLR4 plus TLR3 activate the TRIF (TLR adapter protein) dependent pathway leading to "pro-inflammatory cytokine and Interferon production" (Latanova et al., 2022). All TLRS except 3 and 4, on recognizing viral genomic material, activate the pathway that is Myd88 dependent, causing the release of proinflammatory chemokines and cytokines (Li et al., 2005; Latanova et al., 2022)

TLR activation recruits the IKK complex. This then causes "proteasome-mediated degradation of IκBα by phosphorylation" (Latanova et al., 2022). Ultimately, NF-κB is caused to translocate to the nucleus, and it binds to specific DNA-binding sites of genes whose expression triggers the release of proinflammatory cytokines (Latanova et al., 2022).

12.3.2 RLR signalling

These PRRs sense viral RNA in the cytoplasm or misprocesses/mis-localize cellular RNA, which causes production of type I interferons alongside proinflammatory cytokines and chemokines (van der Lee, 2016). The two important RLRS are RIG-1 (retinoic-acid inducible gene I [RIG-I]) and melanoma differentiation-associated 5 (MDA5) (Best et al., 2006). RIG-1/MDA5 signaling depends on the function of mitochondrial antiviral-signaling protein (MAVS), which is an adapter protein, and

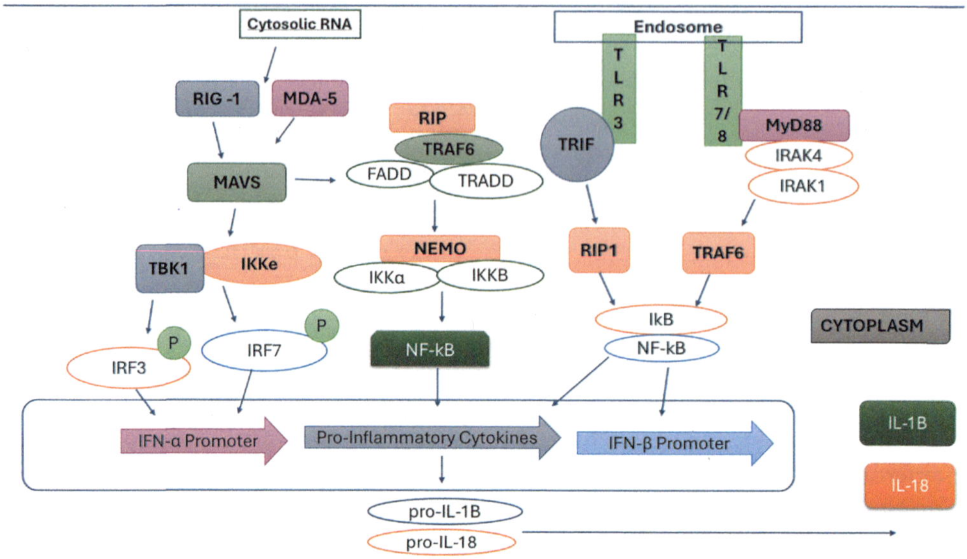

Figure 12.1 Key steps involved in toll-Like Receptors and RLR immune signalling pathways.

MAVS signaling is essential for the production of type 1 interferons (Silenzi, 2015; Rehwinkel and Gack, 2020).

The RLR signaling cascade also results in the activation of NF-κB due to the binding of the adapter protein MAVS with RIP1 and FAS-associated death domain-containing protein (FADD) (Latanova et al., 2022). This binding induces the formation of the FADD/RIP1 complex, which leads to the IKK complex being recruited, ultimately causing the translocation of NF-κB to the nucleus (Latanova et al., 2022; Wu and Hur, 2015).

Fig. 12.1 demonstrates the key steps involved in toll-Like Receptors (TLR) and RLR immune signalling pathways

12.3.3 Janus kinase-signal transducer and activator of transcription signalling

Activation of TLRs and RIG-I receptors by viral RNA leads to production of type I interferons (Koirala et al., 2025). The Interferons (IFNs) bind to their cognate IFN receptors (IFNAR1 and IFNAR2), causing the activation of the Janus kinase (JAK) and signal transducer and activator of transcription (STAT) pathway (Li, 2022; Best et al., 2005).

Activation of the JAK-STAT pathway contributes to the activation of Jak1 and Tyk2 kinases. This leads to the downstream formation of the ISGF3 transcription factor complex, which translocates to the nucleus (Koirala et al., 2025). Inside the nucleus, the complex promotes the transcription of antiviral IFN-stimulated genes (ISGs) that lead to an antiviral state inside the cell (Lin et al., 2006; Koirala et al., 2025). Key steps involved in the JAK-STAT immune signaling pathway are demonstrated in Fig. 12.2.

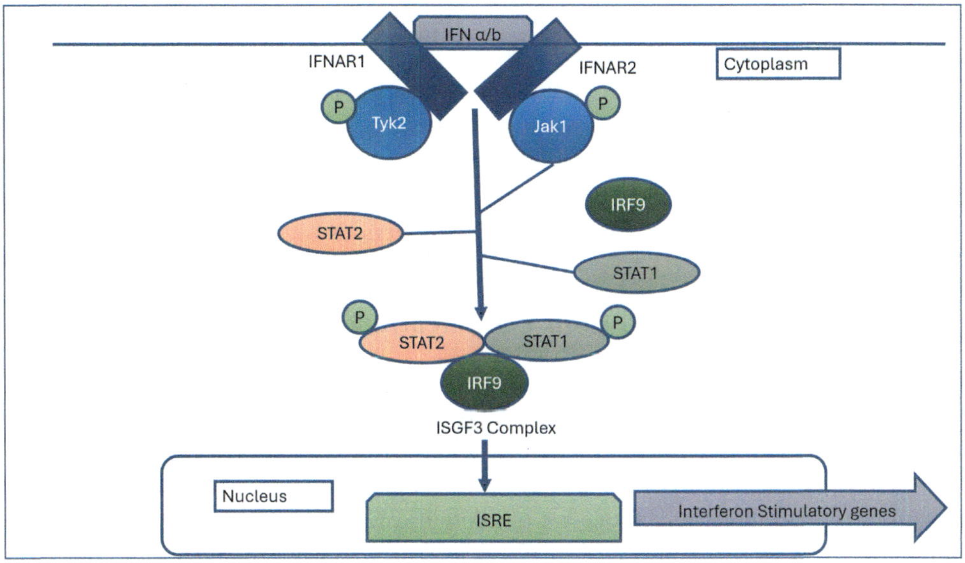

Figure 12.2 Key steps involved in Janus kinase-signal transducer and activator of transcription immune signalling pathway.

12.4 Immune evasion by viruses

With the help of structural and nonstructural proteins, many viruses hijack the host system for various processes crucial for viral replication, morphogenesis, and viral pathogenesis. They modulate host cellular behavior to make the host environment conducive for viral survival and propagation (Mueller and Rouse, 2008).

Virus nonstructural proteins, on interacting with PRRS such as TLRs, RLRs, C-type lectin receptors, NLRs, and cytoplasmic DNA/RNA sensors, cause immunomodulation of the host antiviral response (Latanova et al., 2022). They interfere with the downstream immune cascades, leading to activation or inhibition of immune signaling pathways, causing a reduction or increase in inflammatory and antiviral responses (Muñoz-Jordán et al., 2003).

Viral nonstructural proteins inhibit RLR and TLR downstream signaling and NF-κB, and type I IFN responses (Latanova et al., 2022). By inhibiting NF-κB, viruses reduce the proinflammatory immune response of the host. Also, by activation of NF-κB, they can cause increased viral spread and enhanced viral replication by exploiting apoptotic functions (Latanova et al., 2022). These nonstructural proteins interact with and suppress the primary sensors of the RLR induction pathway—RIG-1 and MDA5, and suppress the functioning of the adapter protein MAVS (Li et al., 2019).

Studies show that type I interferons are the main antiviral genes produced during most virus infections, and thus many viruses have evolved mechanisms to prevent type I IFN production (Muñoz-Jordán et al., 2003), thereby delaying host immune response.

As the host innate immune response is the first barrier of defense against controlling viral entry and replication, viruses have evolved various methods to evade host immune detection at the site of entry and replication (Owen et al., 2016). This immune evasion could thus lead to enhanced viral load and pathogenesis in the host system.

12.5 Antiviral drugs and their classification

Despite our body's strong immune system's ability to combat and withstand viral infections, viruses have evolved a number of strategies to elude this defence. Thus we need the help of antiviral drugs to fight stubborn infections and resistant pathogens.

A family of medications known as antiviral drugs is specifically intended to treat viral infections, which target key processes or proteins essential for viral survival in the host system. These drugs disrupt the viral life cycle by acting on essential viral components, mainly key proteins and viral nucleic acids.

The most common classes of antiviral drugs are:

(1) inhibitors of virus attachment, (2) inhibitors of virus entry, (3) uncoating inhibitors, (4) viral polymerase inhibitors (Ex: acyclovir, tenofovir, valganciclovir, and valacyclovir), (5) protease inhibitors, (Ex: ritonavir, atazanavir, and darunavir), (6) inhibitors of nucleoside and nucleotide reverse transcriptase and (7) integrase inhibitors (Ex: raltegravir) (Cocohoba and Dong, 2008).

12.6 Antiviral drug development and target identification

Developing effective antiviral drugs involves a multistep process that combines target selection, logical medication development, and high-throughput screening methods to assess their antiviral activity and selectivity (Barakat et al., 2017).

12.7 Antiviral drug targets

The most important step is to identify and validate viral targets, which can be essential viral proteins or enzymes that are necessary for virus replication and survival. All major viral proteins, including structural proteins and functionally critical nonstructural proteins, can serve as potential drug targets (Segal-Maurer et al., 2022). Directly inhibiting viral nucleic acids can also block replication (Lee, 2019; Wang et al., 2024). Most authorized virus-targeted medications actually employ a targeted approach, focusing on viral polymerases, proteases, capsid proteins, and a few other essential viral proteins (Table 12.1) (Wang et al., 2024).

Table 12.1 Promising viral targets against which Antiv ral drugs have been developed.

S. No.	Virus	Virus target	Therapy action	Drugs	References
Viral Surface Glycoprotein Inhibitors					
1.	Influenza	Hemagglutinin (HA)	Binds to fusion initiation region (FIR) of HA to prevent viral membrane fusion	Flufirvitide-3	Badani, Hussain et al
2.	Influenza	Neuraminidase (NA)	Inhibits binding of NA to sialic acid receptors on the host surface	Oseltamivir, Zanamivir, Laninamivir	McClellan K, et al
3.	SARS-CoV-2	S Protein	The spike fusion glycoprotein is prevented from binding the ACE 2 host receptor to inhibit viral cell entry (Chen, 2025)	Abidor (Tumifenovir)	Shuster A, et al
4.	HIV-1	gp120 or gp41 glycoprotein	Conformational change, which is required for membrane fusion, is prevented due to binding to gp41 or gp120 (Tyring et al., 2021)	Fostemsavir (gp120) and Enfuvirtide T20 (gp41)	Chen RY, et al Kuduk SD, et al
Viral Capsid Protein Inhibitors					
5.	HIV-1	p24	Inhibits p24 capsid protein, with N-terminal (NTD) and C-terminal (CTD) domains at NTD, CTD, or NTD-CTD interfaces	Lenacapavir	Segal-Maurer S, et al
6.	HBV	Viral capsid protein (Cp)	Dihydropyrimidine (DHP) derivatives are HBV capsid assembly modulators (CAMs) that block capsid formation	Heteroaryldihydropyrimidines (HAPs)	Toyama M, et al

(Continued)

Table 12.1 (Continued)

S. No.	Virus	Virus target	Therapy action	Drugs	References
Viral Protease Inhibitors					
7.	HIV Hepatitis C	Protease	Inhibits HIV protease, certain studies also exhibit inhibition of cytochrome P450-3A4	Ritonavir, Atazanavir, and Darunavir	Talha B, et al Paton NI, et al
8.	SARS-COV 2	M^{pro} Pl^{pro} (proteases)	Inhibit key proteases Mpro or Plpro, which aid in releasing nonstructural proteins and aid viral replication	Ebselen, Tideglusib, Carmofur, VIR250, Vir251	Sarkar A, et al Gunther S, et al
9.	HCV	NS3/4A Protease	It targets the serine protease and inhibits cleavage of required polyprotein	Grazoprevir	Wang SJ, et al
Viral Polymerase Inhibitors					
10.	RSV Influenza Hepatitis C	Viral Polymerase	As nucleoside analogs lack a 3'-hydroxyl group on the sugar, the elongation of the DNA chain by the polymerase is prematurely terminated (Kuriyan et al., 2012)	Ribavirin, Baloxavir (Influenza), Sofosbuvir (HCV)	Fernandez H, et al
11.	HIV	Reverse transcriptase	Thymidine analogs inhibit HIV replication by incorporating into the growing viral DNA chain. This leads to termination of viral reverse transcriptase (Elzagheid, 2021)	Stavudine, Azidothymidine (zidovudine)	Gisolf EH, et al

12.	Hepatitis B	DNA Polymerase	Competitive inhibitors block replication of HBV's DNA polymerase that combines RT and RNase H functions	Lamivudine (HBV)	Soriano V, et al
13.	SARS-CoV-2	RNA-dependent RNA polymerases (RdRp)	Binds to SARS-CoV-2 polymerase complex -(NSPs) NSP12, NSP7, and NSP8, with NSP12 acting as RdRp core to guide antiviral design via its crystal structure (Hangyu et al., 2024)	Remdesivir, Favipravir	Gao Y, et al

12.7.1 Viral surface proteins as targets

Viral surface proteins on the virion's outer envelope are essential for facilitating the adsorption and entrance of viruses into host cells and represent an important antiviral drug target. Blocking these early infection steps can halt the viral process at its outset.

In influenza, the spike proteins of neuraminidase (NA) and hemagglutinin (HA) are key surface proteins that bind inside host respiratory cells on sialic acid receptors (Hangyu et al., 2024), allowing for cytosolic invasion (Bai et al., 2021). By attaching the spike (S) protein to the host cell surface's ACE2 receptor, SARS-CoV-2 enters the host cell (Kabeerdoss et al., 2021). The serine protease TMPRSS2 on the cell surface activates this glycoprotein, making the S protein a crucial therapeutic target (Yang and Rao, 2021).

Drugs that bind to HIV gp140 and gp121 envelope glycoproteins and block them from engaging the host receptor prevent HIV viral entry and replication (Xiao et al., 2021).

12.7.2 Viral capsid protein as target

Drugs targeting viral capsid proteins disrupt the viral life cycle by preventing capsid construction or impairing capsid integrity (Hangyu et al., 2024). Examples include drugs targeting the p24 capsid protein of HIV-1 and Influenza M2 inhibitors like Amantadine and Rimantadine (Segal-Maurer et al., 2022).

HIV-1's p24 capsid protein, with N-terminal (NTD) and C-terminal (CTD) domains, is inhibited at NTD, CTD, or NTD-CTD interfaces (Aiken and Rousso, 2021). HBV capsid inhibitors like Dihydropyrimidine inhibitors primarily prevent capsid complex formation and include diverse molecular scaffolds. (Kuduk et al., 2021; Toyama et al., 2019).

12.7.3 Viral protease as target

Viral proteases cleave and process precursor proteins to produce mature, functional variants vital to the life cycle of the virus, making them important targets for antiviral medications. Proteases are critical mediators of viral maturation and replication and are thus an important tool for antiviral drug development as they enable effective pharmacological disruption of infection (Paton et al., 2021).

The aspartyl protease of HIV is responsible for cleaving the immature polypeptide chains and formation of functional proteins from the precursor proteins, which are responsible for the assembly of the infectious viral particles (Hangyu et al., 2024). This aspartyl protease is a target for HIV protease inhibitors like Ritonavir and Lopinavir (Paton et al., 2021; Talha and Dhamoon, 2023).

SARS-CoV-2 M^{pro} is a prime target for antiviral drug development because it cleaves replicase polyproteins to liberate nonstructural proteins (NSPs) (Hangyu et al., 2024; Kabeerdoss et al., 2021). SARS-CoV-2 PL^{pro}, which aids viral replication and suppresses host immunity, is another key protease target for drug development strategies

(Sarkar and Mandal, 2021; Günther et al., 2021). HCV's NS3/4A protein, a noncovalently bound dimer of NS3 serine protease and cofactor NS4A, is a key antiHCV drug target for drugs like grazoprevir, as it hydrolyzes the viral polyprotein precursor (Wang et al., 2021).

12.7.4 Viral polymerase as target

There are three types of viral polymerases: RNA-dependent RNA polymerases (RdRp), DNA-dependent DNA polymerases, and RNA-dependent DNA polymerases (Gao et al., 2020; Hangyu et al., 2024). Some viruses, like HIV and Hepatitis B, have polymerases with dual functions, including reverse transcriptase (RT) and nuclease activity (Soriano et al., 2020).

Drugs like ribavirin and lamivudine that act against RSV and Hepatitis C are Nucleoside analogs that lack a sugar's 3'-hydroxyl group. The elongation of the DNA chain by the polymerase is prematurely terminated because of the absence of the 3'-hydroxyl group (Gisolf et al., 2000).

Thymidine analogs like Stavudine and Azidothymidine inhibit HIV replication by joining the expanding viral DNA chain and inhibiting viral reverse transcriptase (Fernandez et al., 1986; Elzagheid, 2021).

12.7.5 Viral genome as target

Viruses contain nucleic acids alongside proteins, making direct targeting of viral genomes a powerful antiviral approach. RNA interference (RNAi), a natural eukaryotic mechanism, silences genes posttranscriptionally by degrading viral RNA in infected cells (Alexopoulou et al., 2020).

The bacterial CRISPR/Cas system cleaves invading viral or plasmid DNA; notably, CRISPR/Cas13d degrades HIV-1 transcripts to block replication in cell lines (Nguyen et al., 2021), while CRISPR/Cas9 directly targets the HIV genome and shows strong antiHIV effects (Hussein et al., 2023). CRISPR/Cas9 also effectively inhibits HBV and HCV, highlighting their broad therapeutic promise against viruses (Lee, 2019).

12.8 Target identification and selection

This involves identifying essential viral proteins or enzymes that are critical for the virus's replication, life cycle, or pathogenesis (Chen, 2025).

Several methods are used, including:

1. **Bioinformatics analysis:**
   In this process, viral genomes and protein sequences are analyzed to identify potential targets based on their function, homology to known targets, or presence of conserved domains (Martin et al., 2016).

2. **Reverse genetics:**
   Here, genetically modified viruses are created to assess the function of specific genes and identify their role in viral replication (Zhang et al., 2018).
3. **High-throughput screening**:
   In High-throughput screening, viral proteins or cells are tested against several large libraries of compounds to identify potential inhibitors (Aiken and Rousso, 2021).
4. **Expression profiling:**
   This involves analyzing changes in protein and gene expression in infected cells in order to identify potential targets (Lesk and Chothia, 1986).

12.9 In silico techniques for synthesis of candidate drugs

Once a target is identified, scientists can design and synthesize novel antiviral compounds that specifically interact with and inhibit the target.

This can be done through various methods, including:

12.9.1 High-throughput virtual screening

This computational approach screens large libraries of existing or novel compounds in order to identify the few that have the potential to bind to the target (Ul-Haq and Wilson, 2020). Virtual screening (VS) comprises computational methods that prioritize compounds in large chemical databases based on their predicted biological activity against a specific molecular target (Martin et al., 2016).

Acting as an efficient filtering process, VS reduces vast libraries to a small set of promising candidates by selecting molecules with favorable drug-like properties while eliminating those associated with toxicity or poor pharmacokinetic and pharmacodynamic profiles. This prioritization enables experimental testing to focus on the most viable compounds, thereby reducing cost and development time (Zhang et al., 2018).

12.9.1.1 Similarity search in virtual screening

In the virtual screening workflow, Similarity searching is of critical importance. In this method, the features of the target compound are compared with the features of every available compound in the database (Yusuf et al., 2008).

There are three main steps in similarity searching:

Firstly, the target compound and the database compounds are represented with an equivalent – these are a key collection of binary elements that describe whether compound attributes (called fingerprints) are absent or present (Kabeerdoss et al., 2021).

Secondly, the score of similarity is calculated between two compounds representation using similarity coefficients. And lastly, ranking of the compounds in the database is done in the order of the similarity score. This is done to determine which are the active compounds (Kristensen et al., 2013).

12.9.2 Structure-based drug design

This approach uses the known three-dimensional structure of the target in order to design small molecules that fit into its active site and prevent it from functioning (Gautam et al., 2025). Structure-based virtual screening (SBVS), which is also called target-based virtual screening, relies on the three-dimensional structure of the biological target to predict optimal ligand binding orientations and complex stability (Ylikangas, 2024). This approach is most effective when the target structure has been experimentally resolved (Berrhail and Belhadef, 2019).

12.9.3 Ligand-based drug design

This approach uses information about existing drugs that have activity against the target to design new compounds with improved potency and selectivity (Fig. 12.3). Ligand-based virtual screening (LBVS), in contrast, does not require target structural information and instead identifies candidate molecules based on whether they are similar to already known active ligands. This method follows the principle that structurally related compounds tend to exhibit comparable biological activity (Banegas-Luna et al., 2018).

The field of structural biology involves the mapping of molecular structures of biological macromolecules alongside studying the dynamics of their interactions (Suguna, 2014). Structural Biology allows us to study high-resolution structures of critical viral targets, including proteases, polymerases, and envelope glycoproteins, that may have a critical role in designing antiviral drugs and therapeutics for viral infections (Suguna, 2014; Mtambo, 2022).

This method lets us study the three-dimensional structure of viral proteins using advanced structural techniques such as X-ray crystallography, cryo-electron microscopy (cryo-EM), and NMR spectroscopy, and study their interactions with host components (Creative Biostructure, 2024). These proteins could be used as potential sites for the development of novel target-specific drugs and vaccines.

Structural Biology combined with computer modeling and high-throughput screening could allow identification of conserved binding pockets and allosteric sites for therapeutic intervention and optimization of lead compounds (Scapin, 2006).

12.10 Computer aided structure based drug design

After target identification, computational models are created of the target protein's three-dimensional structure and possible therapeutic possibilities to predict how

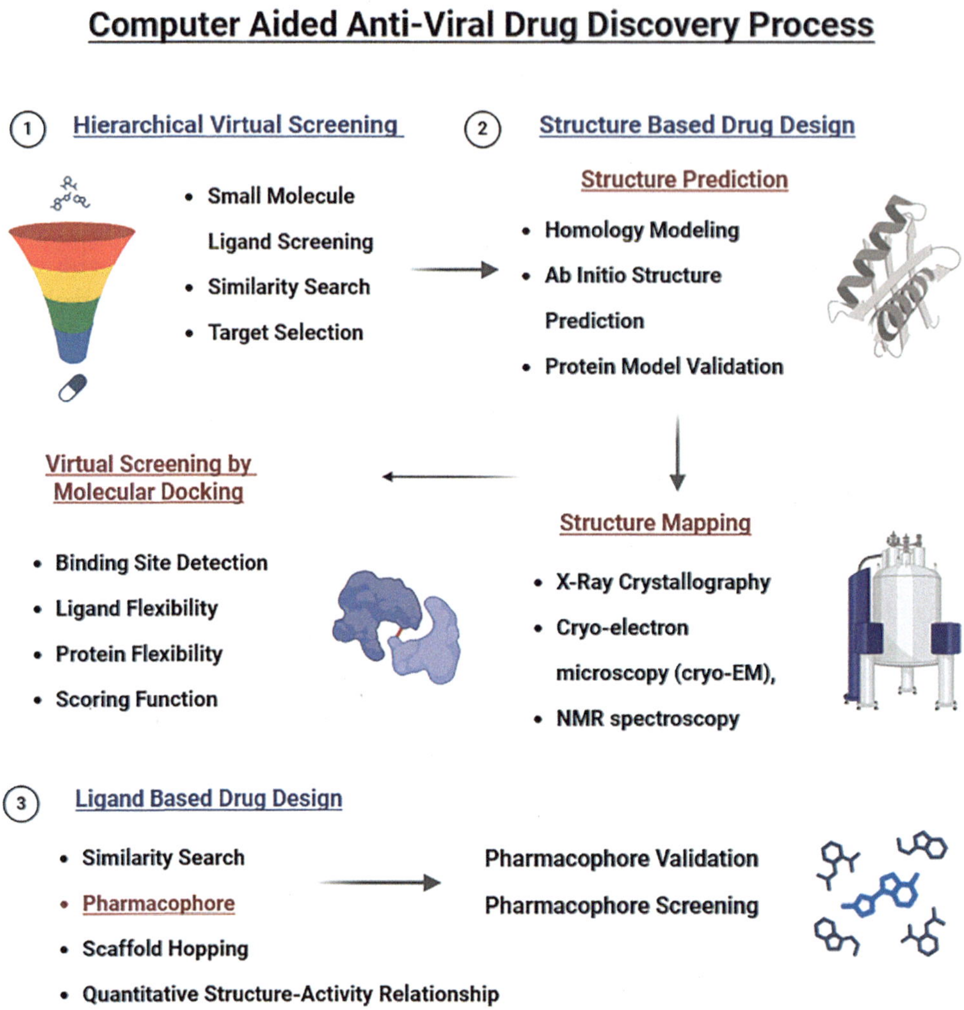

Figure 12.3 In silico techniques used in computer aided antiviral drug discovery.

they might interact (Elzagheid, 2021). Structure-based drug design uses the known 3D structure of the target to design small molecules that fit into its active site and prevent it from functioning (Berrhail and Belhadef, 2019; Gautam et al., 2025).

Novel antiviral compounds can be synthesized that specifically interact with and inhibit the target. High-throughput virtual screening is used to screen large libraries of existing or novel compounds to identify those that have the potential to bind to the target (Zhang et al., 2018; Ul-Haq and Wilson, 2020).

12.10.1 Homology modelling

The 3D structure of a related protein with a comparable amino acid sequence can be predicted using homology modeling, which makes use of existing experimental protein structures (Hillisch and Hilgenfeld, 2003). It relies on nature's principle that the structure of proteins is more conserved than their sequence, so modest sequence changes typically cause only minor structural shifts (Murzin et al., 1995; Hillisch and Hilgenfeld, 2003).

Homology modeling relies on high-resolution experimental protein structures, enabled by improved protein expression systems and advances in X-ray crystallography and NMR spectroscopy, which have exponentially increased available 3D structures (Hillisch and Hilgenfeld, 2003).

Protein structural data—and thus homology models—are useful throughout drug discovery, including lead optimization and target identification (Renaud, 2020). Homology models have significantly supported finding leads, identifying and validating targets, and then optimizing those leads (Hardy and Malikayil, 2003; Roy, 2018).

12.10.2 Molecular docking and dynamics

SBDD uses molecular docking as a computer technique to VS thousands of drug-like compounds against a given target protein structure (Wang and Zhu, 2016; Jia et al., 2023). Docking scores help pinpoint potential leads for further development by estimating binding affinity. Its key strength lies in flexible pose prediction, where algorithms iteratively test ligand orientations, then rank the best protein-ligand complex using scoring functions to gauge binding accuracy and strength (Meng et al., 2011).

Molecular dynamics simulations (MDS) enhance docking results by offering a powerful tool for identifying strong lead candidates. MDS reveals ligand–target protein interactions via energy calculations and verifies the stability of the resulting complex (Ganesan et al., 2017). In drug discovery, MD tools are vital for developing candidates against specific diseases, as they deliver dynamic insights into drug-target interactions across various protein conformations (Saurabh et al., 2020).

12.10.3 Quantitative structure-activity relationship

Artificial intelligence-based quantitative structure-activity relationship (QSAR) modeling is driving drug discovery increasingly, replacing traditional trial-and-error approaches with data-driven computational strategies (Cherkasov et al., 2014; Koirala et al., 2025). This has resulted in a reduction of costs, speedier drug research, and the creation of safer medications.

From straightforward linear models to sophisticated machine learning and deep learning strategies that capture intricate structure–activity connections across vast chemical domains, modern QSAR methodologies have seen significant development. The predictive capacity of QSAR can be used for virtual screening of large chemical databases with billions of chemicals, lead optimization for particular targets, and de novo molecular design, significantly reducing preclinical costs and development time (Ren, 2002; Koirala et al., 2025).

Preclinical drug discovery is greatly accelerated by computational methods, which lower expenses, minimize attrition, and find promising candidates more quickly (Ou-Yang et al., 2012; Koirala et al., 2025). Moreover, Virtual screening and QSAR techniques are becoming essential tools for ranking compounds prior to synthesis, which immediately lowers the cost of experiments (Lavecchia, 2015; Koirala et al., 2025). In pharmaceutical R&D, computational filtering and optimization techniques demonstrate measurable advantages by cutting down on the time needed to move leads into the preclinical pipeline (Paul et al., 2021; Koirala et al., 2025).

QSAR models rely on molecular descriptors that encode chemical and physicochemical properties, typically categorized as one-, two-, or three-dimensional (De et al., 2022). These correspond, respectively, to the molecular weight, topological indices, molecular shape, and electrostatic potential maps of the chemical (Roy et al., 2015; Koirala et al., 2025). Careful descriptor selection is critical for building robust, predictive, and interpretable models that support informed decision-making in medicinal chemistry.

12.10.4 Lead optimization

One crucial phase in the drug development process is lead optimization in which promising lead compounds are systematically altered to enhance their biological performance and developability (Bhat and Ahmed, 2025). Following initial evidence of activity, optimization efforts aim to enhance potency, selectivity, pharmacokinetic behavior, and chemical stability while reducing toxicity and adverse effects (Venkatesh and Lipper, 2000).

These refinements are guided by structure-activity relationship (SAR) analyses, where targeted structural changes are evaluated for their impact on biological response. QSAR models and molecular docking are examples of contemporary computational techniques, which have greatly streamlined this process by elucidating ligand-target interactions and predicting property improvements (Hansch et al., 2002).

The ultimate objective of lead optimization is to generate a candidate with an optimal balance of efficacy and safety suitable for clinical development.

Fig. 12.4 demonstrates the structure-based antiviral drug design pipeline.

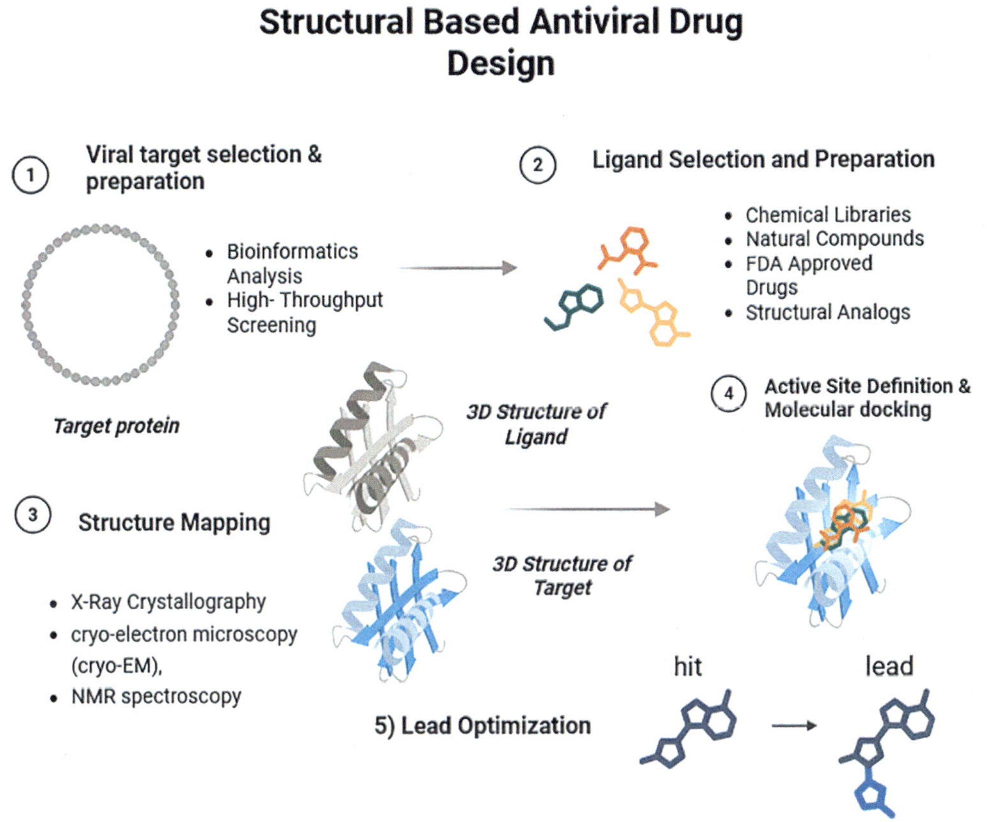

Figure 12.4 Structure-based antiviral drug design pipeline.

12.11 Structural mapping methods

There are several methods, each having unique principles, which are employed for structural mapping, with the most prolific ones being NMR spectroscopy, X-ray crystallography, and cryo-EM (Jakubowski, n.d.). All these techniques are used to map the spatial relationship and distances between all the atoms of a molecule and to dynamically track the changes in their relative location (Egli, 2010).

12.12 X-ray crystallography

It is widely believed that X-Ray Crystallography is the most powerful structural mapping technique available. This method is based on the basic principle of employing X-rays in order to spatially map the arrangement of all the given atoms in any crystal (Wang and Wang, 2017).

12.12.1 Principle employed

There is a fundamental law in physics which governs how crystals of any molecule diffract X-rays incident on them. This law, called Bragg's law, states that when irradiated by X-Ray beams, crystals tend to diffract light at various intensities and at random angles. This pattern of difference in intensities at different angles is referred to as a specific "diffraction pattern" which most of the time has the appearance of a collection of focused spots and is wholly indicative of the structural arrangement of the crystal (Egli, 2010; Wang and Wang, 2017).

Additionally, phase information of the spots is computationally calculated and both the intensity and phase information of all the diffraction patterns, which are resolved by X-ray crystallography, are refined and fitted into the atomic model to create a final 3D virtual structure of the molecule under study (Muench et al., 2019; Wang and Wang, 2017).

12.12.2 Workflow of the technique

The steps involved in this technique are as follows:

Fig. 12.5 shows the principle and workflow of X-Ray Crystallography.

1. An appropriately large, chemically pure crystal is generated of the molecule whose structure has to be resolved. Certain key points need to be kept in mind, such as the fact that the crystal should not have cracks, holes, or any irregularities in structure or any other imperfections and blemishes which may produce irregular diffraction patterns.
2. The crystal is kept in the path of a highly focused and intense beam of X-ray light, and the intensities and angles of the produced diffraction pattern are measured and recorded at every spot. The X-ray light used is usually of a single wavelength (Vénien-Bryan et al., 2017; Baianu, 2010).
3. The phase information of all the spots is also recorded, and all the data is combined and computed electronically to generate a refined 3D model corresponding to the unique arrangements of the atoms within the crystal. This crystal structure can then be stored in a database and made available publicly or privately (Vénien-Bryan et al., 2017; Baianu, 2010).

12.12.3 Advantages of X-ray crystallography

X-ray crystallography is considered the gold standard for high-resolution 3D structural mapping of macromolecules and is thus invaluable in mapping molecular structures for drug designing (Fig. 12.5).

The technique offers high atomic resolution and is not highly molecule-specific. This is because X-ray crystallography is not hindered by the molecular weight of the

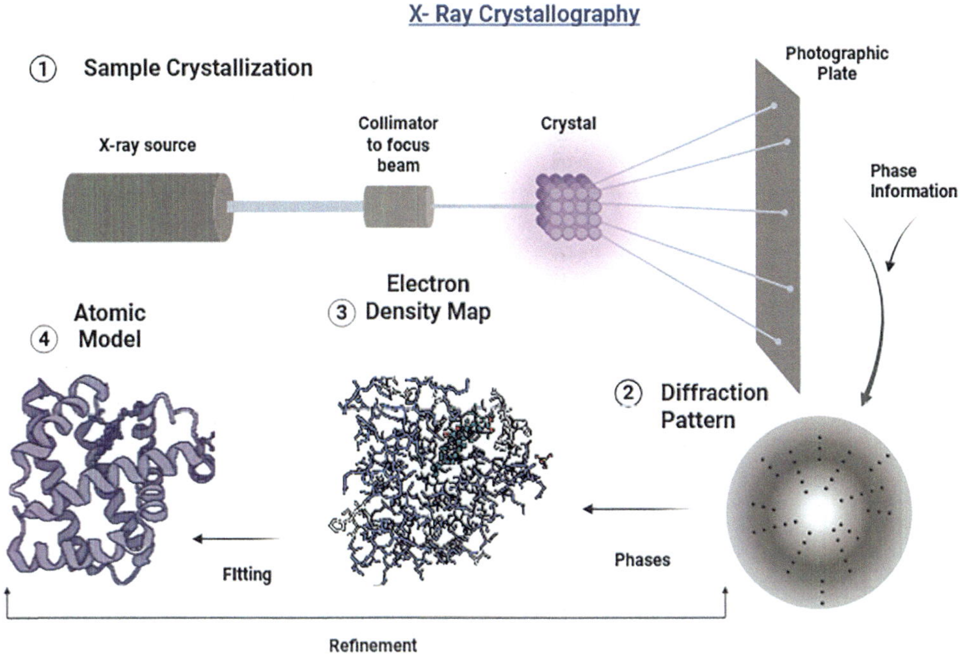

Figure 12.5 Principle and workflow of X-ray crystallography.

macromolecule. Hence, it can be used for mapping a broad range of drug targets and ligands, including high molecular weight proteins and complexes of macromolecules Table 12.2 (Egli, 2010; Creative Biostructure, 2024).

Besides structural information, the technique can also help us determine information about spatial relationships and homogeneity of protein structures (Egli, 2010; Creative Biostructure, 2024).

12.13 Nuclear magnetic resonance Spectroscopy

12.13.1 Principle employed

The basic principle that governs NMR Spectroscopy is the spinning of charged nuclei, which leads to the formation of magnetic dipoles inside every atom. Under normal circumstances, there is zero overall magnetic field due to these dipoles as they are oriented in a random direction. Under specific conditions, such as a strong outside magnetic field, some random nulcei align with the magnetic field and a measurable magnetic moment or resonance is thus produced (Chatham and Blackband, 2001) (Fig. 12.6).

As nuclei of different molecules produce unique moments or signals at specific frequencies, an NMR spectrum is generated when complex samples with

Table 12.2 Advantages and disadvantages of structural mapping techniques.

Technique	Advantages	Disadvantages	Target molecules	Resolution
X-Ray crystallography	High atomic resolution	Difficult for crystallization	Soluble proteins	High
	The molecular weight of molecule is not a limitation (Creative Biostructure, 2024)	Difficult for diffraction	Membrane proteins	
	Easy for model building	Samples must be crystallizable	Ribosomes	
	Reveals information about heterogeneity of protein structures	Obtained structures represent only static form rather than dynamic	DNA/RNA	
NMR Spectroscopy	Good for dynamic and intramolecular interaction study	Sample preparation complex	Water soluble samples	High
	We can map the 3D structures of samples in their native state	In order to get an acceptable signal to noise ratio, there is a need for highly pure samples	Samples with molecular weight below 40–50 kDa	
	Highly flexible, nondestructive and noninvasive	Highly sensitive to motion		
Cryo-EM	Small amounts of sample (0.1 mg) needed	The size of samples observed here is limited by the low signal to noise ratio of Captured Images (Tsegaye et al., 2021).	>250 kDa	Approaching high resolution (typically sub 4Å)
	Sample doesn't require crystallization	Relatively low resolution	Virions, membrane proteins, large proteins	
	Structure resolved in native state	EM techniques are costly	Ribosomes, complex compound	

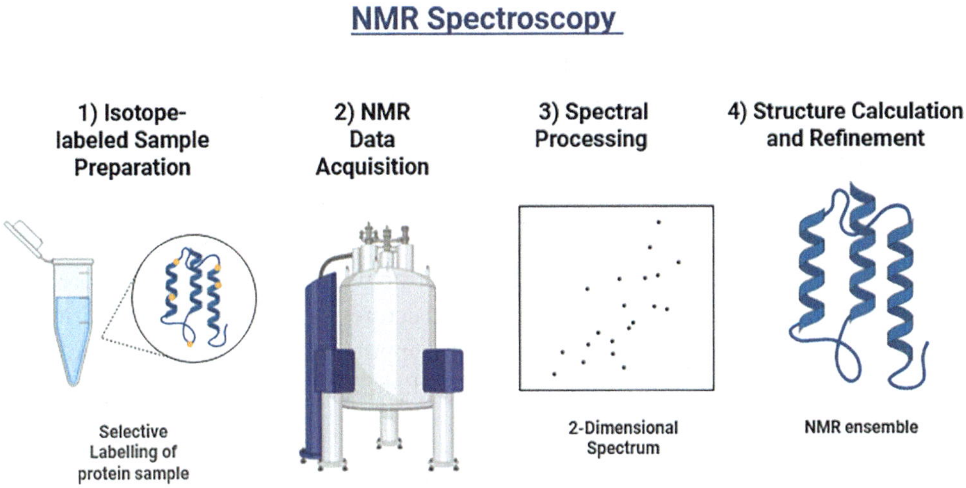

Figure 12.6 Principle and workflow of nuclear magnetic resonance (*NMR*) spectroscopy.

different molecules produce multiple NMR signals, which is indicative of the atomic structure and chemical composition of the sample (Chatham and Blackband, 2001).

12.13.2 Workflow of the technique

The steps involved in this technique are as follows:

1. Sample preparation
   For this method of structure mapping, pure, concentrated, and stable aqueous samples are used. Proteins and amino acids are labeled selectively with very stable isotopes like 2H, 13C, and 15N (Chatham and Blackband, 2001).
2. Data acquisition
   The protein whose structure is to be mapped is subjected to multiple rounds of a series of multidimensional NMR analysises to study and acquire as much structural data as possible, including NOE and J coupling constants (Chatham and Blackband, 2001).
3. Spectral processing
   The NMR spectra that are obtained from multidimensional experiments are then processed and computationally analyzed to map the atoms that correspond to every specific peak on the spectrum (Chatham and Blackband, 2001).
4. Structural analysis
   Molecular dynamics and geometric techniques are employed to map the entire 3D spatial structure (Chatham and Blackband, 2001).

12.13.3 Advantages of nuclear magnetic resonance spectroscopy

This technique may not be as widely used as X-ray crystallography, but it still provides some specific advantages, making it a critical asset of structural mapping (Liu et al., 2004). NMR can give insights into the dynamics and kinetics of a wide range of biological processes and intramolecular interactions. It can be used to dynamically track protein location and movement inside cells and study the mechanism of ligand and protein binding (Creative Biostructure, 2024).

NMR is inherently noninvasive and nondestructive in nature, and therefore it can be used to measure the native and natural 3D structures of macromolecules. Moreover, the technique can be employed for structural mapping of a broad range of molecules, having a resolution of as low as subnanometer (Egli, 2010).

12.14 Cryo-electron microscopy

12.14.1 Principle employed

This method is used for the imaging of fully preserved samples by first freezing the samples in vitreous ice, also called amorphous ice. This vitreous ice exists as an extremely thin layer of water in a noncrystalline form and is usually kept at temperatures that are the same as those of liquid nitrogen. As this technique utilizes electron microscopy for the observation of these low-temperature specimens, it is termed cryo-EM (Bai et al., 2015)

The principle behind electron microscopy involves the use of electrons inplace of light alongside electromagnetic lenses in order to magnify images. The electron beam used can either pass through the sample and thus get scattered by the atoms of the molecules (transmission electron microscopy), or the electrons, on interacting with the atoms of the sample, can be reflected back from the sample (scanning electron microscopy) (Savva, 2019). Before using Cryo-EM, the samples can first be subjected to negative staining EM. This step is used in the case of samples that are homogeneous to ensure rapid screening (Bai et al., 2015).

12.14.2 Workflow of the technique

A typical Cryo-EM workflow includes three steps:

1. Preparation of the sample
   In this step, vitrification of the sample is done, where the solution of the protein sample is subjected to instant cooling to form an amorphous solid layer. The cooling needs to be quick to ensure ice crystals are not formed inside the cells (Muench et al., 2019; Creative Biostructure, 2024).
2. Low dosage data collection
   A series of two-dimensional images is taken during screening of the frozen sample, and all the acquired data is electronically recorded and stored (Creative Biostructure, 2024).

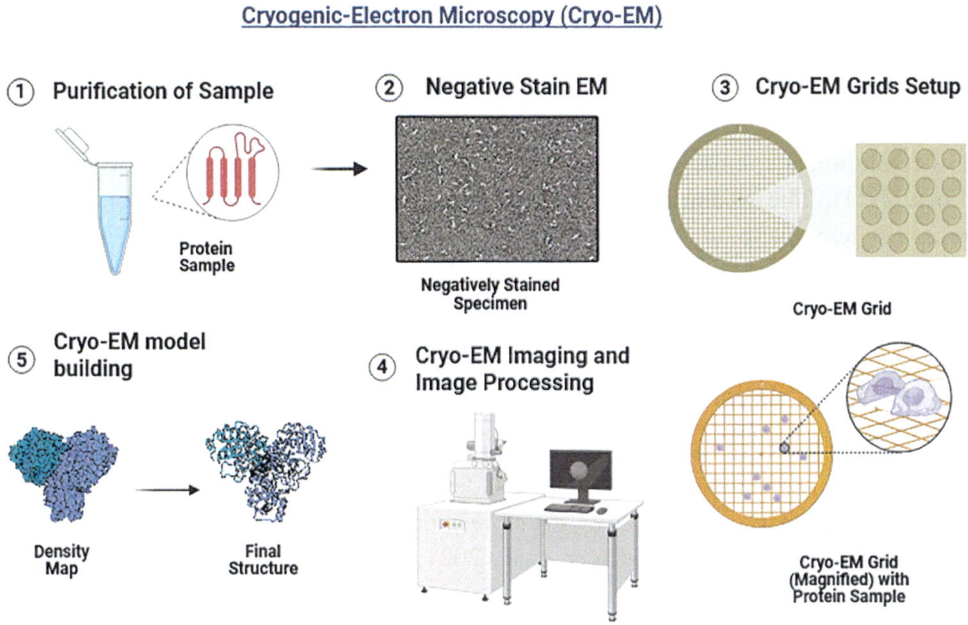

Figure 12.7 Principle and workflow of cryo-electron microscopy (*cryo-EM*).

3. Building of models
   Particle alignment and classification are performed based on the two-dimensional images acquired in the previous step are used in order to aligna nd classify the particles of the sample. Lastly, a three-dimensional model of the atomic structure is constructed after the data is processed by software, which reconstructs all the acquired information (Muench et al., 2019).

   Fig. 12.7 shows the principle and workflow of cryo-electron microscopy.

12.14.3 Advantages of cryo-electron microscopy

Cryo-electron microscopy is used for mapping the 3D structure of drug targets and ligands in structure based drug designing as this technique offers some unique advantages over other techniques such as X-Ray crystallography and NMR spectroscopy. As all samples in cryo-EM undergo rapid freeze treatment, the structure mapped of the frozen sample resembles that of its native state. Thus this technique imparts the least restrictions on sample purity (Bai et al., 2015).

Moreover, the sample preparation is much easier here as compared to X-ray crystallography as in this method, the sample does not need to undergo crystallization in advance. Also, a relatively small amount of sample (0.1 mg) is required for structure mapping using this method. Cryo-EM is helping us understand various complex systems and 3D structures of biological molecules, which allows us to study mechanisms of diseases, helping in designing novel diagnostic and therapeutic agents (Vénien-Bryan et al., 2017).

The biggest disadvantage of cryo-EM for structural mapping was its comparatively low resolution as compared to X-ray crystallography and NMR Spectroscopy (Voet and Voet, 2011). The resolution of the technique has seen drastic improvements due to recent developments of advances such as direct electron detectors (DED) in electron microscopy alongside advancements of computational and microprocessor technologies and image processing techniques (Bai et al., 2015; Vega and Fernández, 2024).

12.15 Drugs developed by structure-based drug design

Structural biology can have a significant impact on current and future antiviral therapies (Table 12.3). During the 2019 COVID pandemic, the antiviral drugs developed against SARS-CoV-2 have shown how structure-guided drug development leads to the discovery of novel antivirals (Plavec et al., 2021; Chen, 2024).

12.16 Structure-based drug design for antivirals against HIV

Recent development of antiviral drugs using structural-based computational methods includes saquinavir and ritonavir, which specifically target the viral protease and are effectively used to treat HIV/AIDS (Talha and Dhamoon, 2023; Gisolf et al., 2000). These drugs were one of the pioneer antivirals developed using structural-based drug design with their target and ligand structure mapped using X-ray crystallography, and are still used in first-line treatment of HIV (Talha and Dhamoon, 2023).

Raltegravir is an HIV integrase inhibitor that exhibits poor chemical stability and aqueous solubility. Hence, studies were done to search for an alternate form with better physical properties. The crystalline form of the drug, categorized as impurity C, had its structure mapped using X-ray diffraction, thermal analysis, and vibrational spectroscopy, and it was found that this form of the drug offered higher aqueous and thermal stability (Fayaz et al., 2024).

12.17 Structure-based drug design for antivirals against hepatitis B

Lamivudine is a nucleoside reverse transcriptase inhibitor (NRTI) used in the treatment of HIV and chronic hepatitis B (Hangyu et al., 2024). Crystals of lamivudine salts such as lamivudine hydrochloride monohydrate and lamivudine hydrochloride were prepared and their structures mapped using X-ray crystallography (Perumalla and Sun, 2014). It was observed that these salts demonstrated significantly improved aqueous

Table 12.3 Promising antivirals developed by structural-based drug design.

S. No.	Name of Drug	Virus	Viral Target	Structural Method	References
1.	Saquinavir	HIV	Protease	X-ray crystallography	Talha B, et al
2.	Ritonavir	HIV	Protease	X-ray crystallography	Talha B, et al
3.	Raltegravir impurity C	HIV	Integrase	X-ray diffraction and vibrational spectroscopy	Fayaz, T. et al
4.	Lamivudine	HIV, Hepatitis B	Reverse transcriptase	X-ray crystallography and crystal engineering	Felipe T. et al.
5.	Oseltamivir derivatives (43b)	Influenza (H1N1, H5N1 and H5N8)	Neuraminidase 150 cavity	X-ray crystallography and molecular docking	Ruifang Jia, et al
6.	Zanamivir derivatives	Influenza	Neuraminidase 150 cavity	Molecular docking and in-silico ADME evaluation	McClellan K, et al
7.	BBH-1 inhibitor	SARS-CoV-2	Main protease (Mpro) (Yevsieieva et al., 2023)	Neutron diffraction and X-ray crystallography	Kneller, D.W., et al
8.	PLpro inhibitor Jun9722 and Jun9843.	SARS-CoV-2	Papain-like protease (PLpro) (Wlodawer et al., 2020)	X-ray crystallography	Prakash Jadhav et al

solubility and drug solubility profiles compared to therapeutic formulations (Martins et al., 2012).

12.18 Structure-based drug design for antivirals against influenza

Oseltamivir and zanamivir, which are second-generation influenza drugs with much fewer systemic side effects, as they specifically target viral neuraminidase. Oseltamivir was approved as an oral first-line drug against Influenza in 1999, but because of the extensive use of both Oseltamivir and zanamivir, multiple mutants of the influenza virus have evolved to have resistance to inhibitors of neuraminidase (Jia et al., 2023).

Novel derivatives of oseltamivir that target the neuraminidase 150 cavity were developed that exhibited more favorable neuraminidase inhibition alongside improved drug (McClellan and Perry, 2001).

Oseltamivir Derivative 43b targets the neuraminidase 150 Cavity in influenza strains H1N1, H5N1 and H5N8, and the structure of the macromolecule was mapped using X-crystallography followed by molecular docking with the target cavity (Jia et al., 2023; Fahim, 2026). Molecular docking and in-silico ADME evaluation were used to elucidate the structure and dynamics of Zanamivir derivatives specifically targeting the 150-cavity of Neuraminidase (McClellan and Perry, 2001).

12.19 Structure-based drug design for SARS-CoV-2 antivirals

The COVID-19 pandemic caused a rush for the development of more potent and novel drugs and vaccines targeting SARS-CoV-2. Structure-based drug discovery and development led to the discovery of more effective antivirals. The surface spike glycoprotein is an effective antiviral target as it plays critical roles in viral entry and pathogenicity. Several studies identified and deposited the three-dimensional crystal structures of the SARS-CoV-2 spike protein in the Protein Data Bank during the pandemic (Yang and Rao, 2021). This accelerated the research in computer-aided drug designing, especially in the field of SBDD (Macalino et al., 2015).

Different approaches based on structural biology have been used for the development of various antivirals against SARS-COV-2, such as spike inhibitors, allosteric inhibitors, and spike-ACE-2 interface inhibitors. The various approaches, like high-throughput virtual screening, molecular docking, and molecular dynamics simulation, are used on various ligands and datasets, including natural compounds, small molecular chemical compounds, chemical databases, structural analogs, and FDA-approved drugs (Shanmugam et al., 2022; Macalino et al., 2015).

SARS-CoV-2 encodes two cysteine proteases that are critical for the replication of the virus. Papain-like protease (PL^{pro}) is one of those two cysteine proteases, and the mechanism of PL^{pro} is by cleaving the viral polyproteins at three distinct sites (Jadhav et al., 2024). Crystal Structures of these proteases were elucidated using X-ray crystallography, and structure-activity relationships were explored, leading to the discovery of potent PL^{pro} inhibitors showing antiviral activity against SARS-CoV-2 (Jadhav et al., 2024).

BBH-1 is a promising SARS-CoV-2 inhibitor that was developed using structural-based methods by the US Department of Energy by combining promising features of existing approved drugs for hepatitis C and similar viruses like SARS-CoV-1. Structural methods such as Neutron diffraction and X-Ray Crystallography were used to study the structure of the BBH-1 Inhibitor at an atomic level to find the exact conserved pocket where it binds to the SARS-CoV-2 protein to prevent virus replication (Chen, 2025). A hybrid novel BBH-1 inhibitor was created, which combined atomic features of preexisting approved antivirals, including Boceprevir, which is a hepatitis C inhibitor and GC-376, a SARS-CoV-1 inhibitor (Kneller et al., 2022).

12.20 Advantages of structural-based drug design

Viruses can mutate quickly and are rapidly evolving, thus lead to the development of resistance of drugs. Thus established antiviral drugs are no longer effective in combating persistent viral infections, and novel viral targets need to be identified.

But there is a distinct lack of validated targets, and identifying new viral targets that don't have erroneous side effects in the host poses to be a very difficult and time-consuming process. Structural-based approaches can find a "desirable target structure complexed with a specific ligand even if there are no experimental data available" (Dronina et al., 2021; Hangyu et al., 2024).

Identifying and validating potential targets using structural-based methods can shorten the time taken for drug discovery and development. Moreover, we also don't need to physically screen the compounds in the lab, as it can be done in silico (Plavec et al., 2021).

Combining structural biology, computational data, and modeling helps predict the efficacy and any undesirable off-target side effects posed by the various drug candidates before costly animal and human trials (Berg et al., 2012). We can also target specific viral proteins and pathways that can lead to the development of more effective drugs with fewer off-target effects (Congreve et al., 2005; Rao and Alavala, 2025).

12.21 Disadvantages of structural-based drug design

Despite the many advantages of structural-ased drug design and screening, there are many limitations and disadvantages also. The most critical one being that many drugs identified using structural-based drug discovery tool may not be applicable in actual clinical cases. That is why so many leads identified through this process fail in animal or clinical trials (Lionta et al., 2014).

Most lead compounds that are identified by Structural Screening are usually less bioactive, and very potent leads are rarely identified. There is increasingly high computational cost and high complexity in structural-based drug discovery, making this process resource- and money-intensive (Lionta et al., 2014).

One of the most demanding challenges of structural-based screening is that it is difficult to successfully identify the complexity of the ligand-receptor binding interactions. Because of this, prediction of the accurate binding poses is challenging and hence, scoring is not that accurate, leading to the leads identified not being extremely potent (Hangyu et al., 2024). Predicted protein structures from homology modeling and predicted protein-ligand complexes have increasingly resulted in false positive or negative results in actual clinical trials (Anderson, 2003; Roy, 2018).

12.22 Challenges and limitations

Despite its significant contributions to antiviral drug discovery, structural biology faces several challenges and limitations that can affect the translation of structural insights into effective therapies.

Methodological and technical difficulties in producing high-quality viral protein structures are one of the main obstacles in structural biology. It is challenging to synthesize, purify, and crystallize many viral proteins because they are unstable, membrane-associated, or form transitory complexes. Even if the reliance on crystallization has decreased because of developments in cryo-electron microscopy, problems such as sample heterogeneity and poor resolution for tiny proteins still exist. Furthermore, structural determination frequently requires a lot of time and resources, which can hinder its quick implementation during new outbreaks. (Chayen, 2004; Kühlbrandt, 2014).

Conformational Dynamics and Structural Flexibility of Proteins exist in a variety of conformational states that are necessary for their biological activity, making them dynamic entities. The dynamic character of viral proteins under physiological settings may not be properly represented by traditional structural methods, which sometimes only capture one or a small number of conformations. Because molecules optimized for one conformation could be less effective against another, this constraint can make medication creation more difficult. Therefore to gain a better understanding of protein flexibility and ligand binding behavior, structural data must be integrated with methods like molecular dynamics simulations and NMR spectroscopy (Henzler-Wildman and Kern, 2007; Dror et al., 2012).

Problems with drug resistance and viral mutation, drug-resistant versions can evolve quickly due to the high mutation rates found in viral genomes, especially those of RNA viruses. Drug target protein mutations may change binding locations and make structure-based inhibitors less effective. Resistance is still a major problem, despite structural biology's assistance in identifying conserved areas that are less vulnerable to mutation. To maintain treatment efficacy and modify drug design techniques, ongoing structural monitoring of developing variations is necessary (Domingo and Perales, 2019; Irwin et al., 2016).

12.23 Prospects for the future and global health readiness

It is believed that structural biology will become more crucial in combating upcoming viral threats and bolstering international health readiness. A more thorough understanding of viral proteins and multiprotein complexes, including large, flexible, and dynamic systems that were previously challenging to characterize, has been made possible by recent developments in integrative structural biology, which combine X-ray crystallography, cryo-electron microscopy, NMR spectroscopy, computational modeling, and complementary biophysical techniques. It is expected that these advancements will greatly improve the precision and effectiveness of structure-based antiviral medication discovery (Ward et al., 2013; Alber et al., 2007).

The COVID-19 pandemic further illustrated the crucial role structural biology plays in quick reaction plans, as the quick identification of high-resolution structures of important viral proteins sped up the development of medications and vaccines. By facilitating quick data sharing and coordinated actions during upcoming pandemics, the creation of international structural biology platforms, open-access databases, and cooperative research networks will further enhance preparedness (Wrapp et al., 2020; Burley et al., 2021).

Furthermore, structural biology supports the rational design of therapeutic medicines that can be quickly optimised against newly developing viruses and enables the identification of conserved viral targets appropriate for the creation of broad-spectrum antivirals, both of which contribute to long-term readiness. Strategies to avoid, mitigate, and manage future viral epidemics will continue to rely heavily on structural biology as part of a multidisciplinary framework (Bekerman and Einav, 2015).

12.24 Conclusion

By highlighting how three-dimensional structural data improves knowledge of viral protein function, virus–host interactions, and therapeutic targeting, this chapter emphasizes the crucial role that structural biology plays in the development of antiviral drugs. The rational design and optimization of antiviral drugs have been influenced by advances in structural methods, which have made it possible to identify important viral therapeutic targets. By moving drug discovery from conventional empirical screening techniques to precise, mechanism-based methodologies, structural biology has revolutionized antiviral research. This has enhanced medication specificity, efficacy, and hastened development, especially during public health emergencies. In the future, it is anticipated that integrated structural techniques and ongoing technical advancements will increase structural biology's contribution to the development of antiviral drugs. Structural biology will continue to play a crucial role in defending public health against current and new viral threats by bolstering quick reaction plans and improving worldwide readiness.

References

Aiken, C., Rousso, I., 2021. The HIV-1 capsid and reverse transcription. Retrovirology 18 (1). Available at https://doi.org/10.1186/s12977-021-00566-0.

Alber, F., Dokudovskaya, S., Veenhoff, L.M., Zhang, W., Kipper, J., Devos, D., et al., 2007. The molecular architecture of the nuclear pore complex. Nature 450 (7170), 695–701. Available at https://doi.org/10.1038/nature06405.

Alberts, B., Johnson, A., Lewis, J., Raff, M., Roberts, K., Walter, P., et al., 2002. Molecular Biology of the Cell, 4th ed. Garland Science, New York, ISBN 0-8153-3218-1.

Alexopoulou, A., Vasilieva, L., Karayiannis, P., 2020. New approaches to the treatment of chronic hepatitis B. Journal of Clinical Medicine 9 (10), 3187. Available at https://doi.org/10.3390/jcm9103187.

Anderson, A.C., 2003. The process of structure-based drug design. Chemistry and Biology 10 (9), 787–797. Available at https://doi.org/10.1016/j.chembiol.2003.09.002, http://www.cell.com.

Bai, Xc, McMullan, G., Scheres, S.H.W., 2015. How cryo-EM is revolutionizing structural biology. Trends in Biochemical Sciences 40 (1), 49–57. Available at https://doi.org/10.1016/j.tibs.2014.10.005, www.elsevier.com/locate/tibs.

Bai, Y., Jones, J.C., Wong, S.-S., Zanin, M., 2021. Antivirals targeting the surface glycoproteins of influenza virus: mechanisms of action and resistance. Viruses 13 (4), 624. Available at https://doi.org/10.3390/v13040624.

Baianu, I.C., editor. 2010 November 27. Techniques and instrumentation for structure determination. [Planet Physics].

Banegas-Luna, A.J., Cerón-Carrasco, J.P., Pérez-Sánchez, H., 2018. A review of ligand-based virtual screening web tools and screening algorithms in large molecular databases in the age of big data. Future Medicinal Chemistry 10 (22), 2641–2658. Available at https://doi.org/10.4155/fmc-2018-0076, http://www.future-science.com/loi/fmc.

Barakat, K.H., Houghton, M., Tyrrel, D.L., Tuszynski, J.A., 2017. Rational drug design: one target, many paths to it. Pharmaceutical Sciences: Breakthroughs in Research and Practice. IGI Global, Hershey [PA], pp. 31. Available at https://doi.org/10.4018/978-1-5225-1762-7.ch044.

Bassetto, M., Massarotti, A., Coluccia, A., Brancale, A., 2016. Structural biology in antiviral drug discovery. Current Opinion in Pharmacology 30, 116–130. Available at https://doi.org/10.1016/j.coph.2016.08.014.

Bekerman, E., Einav, S., 2015. Combating emerging viral threats. Science (New York, N.Y.) 348 (6232), 282–283. Available at https://doi.org/10.1126/science.aaa3778.

Berg, J.M., Tymoczko, J.L., Stryer, L., Gatto Jr., G.J., 2012. Biochemistry, 7th ed. W.H. Freeman and Company, New York [NY], ISBN: 9781429229364.

Berrhail, F., Belhadef, H., Hentabli, H., Saeed, F., 2019. International Journal of Intelligent Systems Technologies and Applications 18 (4), 405–425.

Best, S.M., Morris, K.L., Shannon, J.G., Robertson, S.J., Mitzel, D.N., Park, G.S., et al., 2005. Inhibition of interferon-stimulated JAK-STAT signaling by a tick-borne flavivirus and identification of NS5 as an interferon antagonist. Journal of Virology 79 (20), 12828–12839. Available at https://doi.org/10.1128/JVI.79.20.12828-12839.2005.

Best, S.M., Mitzel, D.N., Bloom, M.E., 2006. Action and reaction: the arthropod-borne flaviviruses and host interferon responses. Future Virology 1 (4), 447–459. Available at https://doi.org/10.2217/17460794.1.4.447.

Bhat, A.R., Ahmed, S., 2025. Artificial intelligence (AI) in drug design and discovery: a comprehensive review. In Silico Research in Biomedicine 1, 100049. Available at https://doi.org/10.1016/j.insi.2025.100049.

Blundell, T.L., Jhoti, H., Abell, C., 2002. High-throughput crystallography for lead discovery in drug design. Nature Reviews. Drug Discovery 1 (1), 45–54. Available at https://doi.org/10.1038/nrd706, http://www.nature.com/nrd/index.html.

Branden, C., Tooze, J., 1999. Introduction to Protein Structure, 2nd ed. Garland Science, New York, ISBN 9780815323051.

Burley, S.K., et al., 2021. RCSB protein data bank: powerful new tools for exploring 3D structures. Nucleic Acids Research 49 (D1), D437–D451.

Chatham, J.C., Blackband, S.J., 2001. Nuclear magnetic resonance spectroscopy and imaging in animal research. ILAR Journal 42 (3), 189–208. Available at https://doi.org/10.1093/ilar.42.3.189.

Chayen, N.E., 2004. Turning protein crystallisation from an art into a science. Current Opinion in Structural Biology 14 (5), 577–583. Available at https://doi.org/10.1016/j.sbi.2004.08.002.

Chen, J.T., 2024. Anti-SARS-CoV-2 activity of flavonoids Anti-SARS-CoV-2. Activity of Flavonoids. CRC Press, Taiwan, pp. 1–256. Available at https://doi.org/10.1201/9781003433200, https://www.routledge.com/Anti-SARS-CoV-2-Activity-of-Flavonoids/Chen/p/book/9781032556215.

Chen, J.T., 2025. Integrative Omics and Systems Pharmacology for COVID-19 Integrative Omics and Systems Pharmacology for COVID-19. Apple Academic Press, Taiwan, pp. 1–442. Available at https://doi.org/10.1201/9781998511211, https://www.taylorfrancis.com/books/edit/10.1201/9781998511211.

Cherkasov, A., Muratov, E.N., Fourches, D., Varnek, A., Baskin, I.I., Cronin, M., et al., 2014. QSAR modeling: where have you been? Where are you going to? Journal of Medicinal Chemistry 57 (12), 4977–5010. Available at https://doi.org/10.1021/jm4004285, http://pubs.acs.org/jmc.

Cocohoba, J., Dong, B.J., 2008. Raltegravir: the first HIV integrase inhibitor. Clinical Therapeutics 30 (10), 1747–1765. Available at https://doi.org/10.1016/j.clinthera.2008.10.012.

Congreve, Murray, Blundell, 2005. Structural biology and drug discovery. Drug Discovery Today 10, 895–907. Available at https://doi.org/10.1016/S1359-6446[05]03484-7.

Creative Biostructure, 2024. Comparison of Crystallography, NMR and EM. Creative Biostructure Sep 3 [cited 2026 Feb 20]. Available at https://www.creative-biostructure.com/comparison-of-crystallography-nmr-and-em_6.htm.

De, P., Kar, S., Ambure, P., Roy, K., 2022. Prediction reliability of QSAR models: an overview of various validation tools. Archives of Toxicology 96 (5), 1279–1295. Available at https://doi.org/10.1007/s00204-022-03252-y.

Domingo, E., Perales, C., 2019. Viral quasispecies. PLoS Genetics 15 (10), e1008271. Available at https://doi.org/10.1371/journal.pgen.1008271.

Dronina, J., Samukaite-Bubniene, U., Ramanavicius, A., 2021. Advances and insights in the diagnosis of viral infections. Journal of Nanobiotechnology 19 (1). Available at https://doi.org/10.1186/s12951-021-01081-2.

Dror, R.O., Dirks, R.M., Grossman, J.P., Xu, H., Shaw, D.E., 2012. Biomolecular simulation: A computational microscope for molecular biology. Annual Review of Biophysics 41 (1), 429–452. Available at https://doi.org/10.1146/annurev-biophys-042910-155245, http://www.annualreviews.org/journal/biophys.

Egli, M., 2010. Diffraction techniques in structural biology. Current Protocols in Nucleic Acid Chemistry / Edited by Serge L. Beaucage… [et al.], Chapter 7:Unit 7.13.

Elzagheid, M.I., 2021. Nucleosides and nucleoside analogues as emerging antiviral drugs. Mini-Reviews in Organic Chemistry 18 (6), 672–679. Available at https://doi.org/10.2174/1570193x17999200820163234.

Fahim, A.M., 2026. Structure-based drug design; Computational strategies in drug discovery; antihypertensive agents; Antiviral drugs; molecular docking; QSAR; pharmacological insights.

Computational Biology and Chemistry 120 (Pt 1), 108663. Available at https://doi.org/10.1016/j.compbiolchem.2025.108663.

Fayaz, T.K.S., Chanduluru, H.K., Lal, P., Ghosh, A., Chernyshev, V., Sanphui, P., 2024. Structural analysis of anti-retroviral drug raltegravir and its potential impurity C: investigation of solubility and stability. CrystEngComm 26 (4), 517–531. Available at https://doi.org/10.1039/d3ce01071f.

Fernandez, H., Banks, G., Smith, R., 1986. Ribavirin: a clinical overview. European Journal of Epidemiology 2 (1), 1–14. Available at https://doi.org/10.1007/BF00152711.

Ganesan, A., Coote, M.L., Barakat, K., 2017. Molecular dynamics-driven drug discovery: leaping forward with confidence. Drug Discovery Today 22 (2), 249–269. Available at https://doi.org/10.1016/j.drudis.2016.11.001, www.elsevier.com/locate/drugdiscov.

Gao, Y., Yan, L., Huang, Y., Liu, F., Zhao, Y., Cao, L., et al., 2020. Structure of the RNA-dependent RNA polymerase from COVID-19 virus. Science (New York, N.Y.) 368 (6492), 779–782. Available at https://doi.org/10.1126/science.abb7498.

Gautam, V., Gautam, H.K., Pathak, R., Kumar, L., 2025. Medicinal Compounds in Modern Healthcare: from Tradition to Innovation. CRC Press. Available at https://doi.org/10.1201/9781003588832.

Gisolf, E.H., Dreezen, C., Danner, S.A., Weel, J.L.F., Weverling, G.J., Reiss, P., et al., 2000. Risk factors for hepatotoxicity in HIV-1-infected patients receiving ritonavir and saquinavir with or without stavudine. Clinical Infectious Diseases 31 (5), 1234–1239. Available at https://doi.org/10.1086/317449.

Günther, S., Reinke, P.Y.A., Fernández-Garciá, Y., Lieske, J., Lane, T.J., Ginn, H.M., et al., 2021. X-ray screening identifies active site and allosteric inhibitors of SARS-CoV-2 main protease. Science (New York, N.Y.) 372 (6542), 642–646. Available at https://doi.org/10.1126/science.abf7945, https://science.sciencemag.org/content/372/6542/642.full.

Hangyu, W., Panpan, L., Jie, S., Hongyan, W., Linmiao, W., Kangning, H., et al., 2024. Advancements in antiviral drug development: comprehensive insights into design strategies and mechanisms targeting key viral proteins. Journal of Microbiology and Biotechnology 34 (7), 1376–1384. Available at https://doi.org/10.4014/jmb.2403.03008, https://www.jmb.or.kr/journal/view.html?uid=6482&vmd=Full.

Hansch, C., Hoekman, D., Leo, A., Weininger, D., Selassie, C.D., 2002. Chem-bioinformatics: comparative QSAR at the interface between chemistry and biology. Chemical Reviews 102 (3), 783–812. Available at https://doi.org/10.1021/cr0102009.

Hardy, L.W., Malikayil, A., 2003. The impact of structure-guided drug design on clinical agents. Current Drug Discovery 15–19.

Henzler-Wildman, K., Kern, D., 2007. Dynamic personalities of proteins. Nature 450 (7172), 964–972. Available at https://doi.org/10.1038/nature06522.

Hillisch, A., Hilgenfeld, R., 2003. The role of protein 3D-structures in the drug discovery process.EXS. In: In: Hillisch, A., Hilgenfeld, R. (Eds.), Modern Methods of Drug Discovery, vol. 93. Birkhäuser, Basel [CH], pp. 157–181.. EXS. Available at https://doi.org/10.1007/978-3-0348-7997-2_8. EXS.

Hussein, M., Molina, M.A., Berkhout, B., Herrera-Carrillo, E., 2023. A CRISPR-cas cure for HIV/AIDS. International Journal of Molecular Sciences 24 (2), 1563. Available at https://doi.org/10.3390/ijms24021563.

Irwin, K.K., Renzette, N., Kowalik, T.F., Jensen, J.D., 2016. Antiviral drug resistance as an adaptive process. Virus Evolution 2 (1), vew014. Available at https://doi.org/10.1093/ve/vew014.

Jadhav, P., Huang, B., Osipiuk, J., Zhang, X., Tan, H., Tesar, C., et al., 2024. Structure-based design of SARS-CoV-2 papain-like protease inhibitors. European Journal of Medicinal Chemistry 264, 116011. Available at https://doi.org/10.1016/j.ejmech.2023.116011.

Jakubowski, no date. Biochemistry Online – Protein Structure: Composition, Sequence and Conformational Analysis of Proteins: Analysis of Protein Tertiary Structure. Muovetevikant.

Jia, R., Zhang, J., Shi, F., Bonomini, A., Lucca, C., Bertagnin, C., et al., 2023. Discovery of N-substituted oseltamivir derivatives as novel neuraminidase inhibitors with improved drug resistance profiles and favorable drug-like properties. European Journal of Medicinal Chemistry 252, 115275. Available at https://doi.org/10.1016/j.ejmech.2023.115275.

Kabeerdoss, J., Pilania, R.K., Karkhele, R., Kumar, T.S., Danda, D., Singh, S., 2021. Severe COVID-19, multisystem inflammatory syndrome in children, and Kawasaki disease: immunological mechanisms, clinical manifestations and management. Rheumatology International 41 (1), 19–32. Available at https://doi.org/10.1007/s00296-020-04749-4, link.springer.de/link/service/journals/00296/index.htm.

Kashaw, S.K., Dixit, A., Agarwal, S., Nema, P., 2026. Bioinformatics, Computational Chemistry, and AI in Drug Innovation: Advances and Applications Bioinformatics, Computational Chemistry, and AI in Drug Innovation: Advances and Applications. CRC Press, India, pp. 1–250. Available at https://doi.org/10.1201/9781003464679, https://www.taylorfrancis.com/books/10.1201/9781003464679.

Kneller, D.W., Li, H., Phillips, G., Weiss, K.L., Zhang, Q., Arnould, M.A., et al., 2022. Covalent narlaprevir- and boceprevir-derived hybrid inhibitors of SARS-CoV-2 main protease. Nature Communications 13 (1). Available at https://doi.org/10.1038/s41467-022-29915-z, https://www.nature.com/ncomms/.

Knipe, D.M., Howley, P.M. (Eds.), 2013. Fields Virology, 6th Ed., Vol. 1, Lippincott Williams & Wilkins, Philadelphia [PA].

Koirala, M., Yan, L., Mohamed, Z., DiPaola, M., 2025. AI-integrated QSAR modeling for enhanced drug discovery: from classical approaches to deep learning and structural insight. International Journal of Molecular Sciences 26 (19), 9384. Available at https://doi.org/10.3390/ijms26199384.

Kristensen, T.G., Nielsen, J., Pedersen, C.N.S., 2013. Methods for similarity-based virtual screening. Computational and Structural Biotechnology Journal 5 (6), e201302009. Available at https://doi.org/10.5936/csbj.201302009, http://csbj.org/articles/e201302009.pdf.

Kuduk, S.D., Stoops, B., Lam, A.M., Espiritu, C., Vogel, R., Lau, V., et al., 2021. Oxadiazepinone HBV capsid assembly modulators. Bioorganic & Medicinal Chemistry Letters 52, 128353. Available at https://doi.org/10.1016/j.bmcl.2021.128353.

Kühlbrandt, W., 2014. The resolution revolution. Science (New York, N.Y.) 343 (6178), 1443–1444. Available at https://doi.org/10.1126/science.1251652.

Kuriyan, J., Konforti, B., Wemmer, D., 2012. The Molecules of Life: Physical and Chemical Principles. Garland Science, New York [NY].

Latanova, A., Starodubova, E., Karpov, V., 2022. Flaviviridae nonstructural proteins: the role in molecular mechanisms of triggering inflammation. Viruses 14 (8), 1808. Available at https://doi.org/10.3390/v14081808.

Lavecchia, A., 2015. Machine-learning approaches in drug discovery: methods and applications. Drug Discovery Today 20 (3), 318–331. Available at https://doi.org/10.1016/j.drudis.2014.10.012.

Leach, A.R., Gillet, V.J., 2007. An Introduction to Chemoinformatics an Introduction to Chemoinformatics. Springer Netherlands, United Kingdom, pp. 1–255. Available at https://doi.org/10.1007/978-1-4020-6291-9, http://www.springerlink.com/openurl.asp?genre=book&isbn=978-1-4020-6290-2.

Lee, C., 2019. CRISPR/Cas9-based antiviral strategy: current status and the potential challenge. Molecules (Basel, Switzerland) 24 (7), 1349. Available at https://doi.org/10.3390/molecules24071349.

Lesk, A.M., Chothia, C., 1986. The response of protein structures to amino-acid sequence changes. Philosophical Transactions of the Royal Society of London. Series B, Biological Sciences 317 (1188), 345–356.

Li, Y., 2022. Virus-Host Interplay: Insights from Interferon and Inflammasome Responses. [Doctoral Thesis, Erasmus, University Rotterdam].

Li, K., Foy, E., Ferreon, J.C., Nakamura, M., Ferreon, A.C.M., Ikeda, M., et al., 2005. Immune evasion by hepatitis C virus NS3/4A protease-mediated cleavage of the Toll-like receptor 3 adaptor protein TRIF. Proceedings of the National Academy of Sciences of the United States of America 102 (8), 2992–2997. Available at https://doi.org/10.1073/pnas.0408824102.

Li, W., Li, N., Dai, S., Hou, G., Guo, K., Chen, X., et al., 2019. Zika virus circumvents host innate immunity by targeting the adaptor proteins MAVS and MITA. The FASEB Journal 33 (9), 9929–9944. Available at https://doi.org/10.1096/fj.201900260r.

Lin, R.J., Chang, B.L., Yu, H.P., Liao, C.L., Lin, Y.L., 2006. Blocking of interferon-induced Jak-Stat signaling by Japanese encephalitis virus NS5 through a protein tyrosine phosphatase-mediated mechanism. Journal of Virology 80 (12), 5908–5918. Available at https://doi.org/10.1128/JVI.02714-05.

Lionta, E., Spyrou, G., Vassilatis, D., Cournia, Z., 2014. Structure-based virtual screening for drug discovery: principles, applications and recent advances. Current Topics in Medicinal Chemistry 14 (16), 1923–1938. Available at https://doi.org/10.2174/1568026614666140929124445.

Liu, R., Hsieh, C.Y., Lam, K.S., 2004. New approaches in identifying drugs to inactivate oncogene products. Seminars in Cancer Biology 14 (1), 13–21. Available at https://doi.org/10.1016/j.semcancer.2003.11.003, http://www.elsevier.com/inca/publications/store/6/2/2/9/4/3/index.htt.

Macalino, S.J.Y., Gosu, V., Hong, S., Choi, S., 2015. Role of computer-aided drug design in modern drug discovery. Archives of Pharmacal Research 38 (9), 1686–1701. Available at https://doi.org/10.1007/s12272-015-0640-5.

Martin, Y.C., Abagyan, R., Ferenczy, G.G., Gillet, V.J., Oprea, T.I., Ulander, J., et al., 2016. Glossary of terms used in computational drug design, part II (IUPAC Recommendations 2015). Pure and Applied Chemistry 88 (3), 239–264. Available at https://doi.org/10.1515/pac-2012-1204, http://www.degruyter.com/view/j/pac.

Martins, F.T., Bonfilio, R., De Araújo, M.B., Ellena, J., 2012. Lamivudine salts with improved solubilities. Journal of Pharmaceutical Sciences 101 (6), 2143–2154. Available at https://doi.org/10.1002/jps.23117, www.interscience.wiley.com/jpages/0022-3549.

McClellan, K., Perry, C.M., 2001. Oseltamivir: a review of its use in influenza. Drugs 61 (2), 263–283. Available at https://doi.org/10.2165/00003495-200161020-00011, http://rd.springer.com/journal/40265.

Meng, X.Y., Zhang, H.X., Mezei, M., Cui, M., 2011. Molecular docking: a powerful approach for structure-based drug discovery. Current Computer-Aided Drug Design 7 (2), 146–157. Available at https://doi.org/10.2174/157340911795677602, www.eurekaselect.com/582/journal/current-computer-aided-drug-design.

Mtambo, S.E., 2022. In silico insights into the effect of mutation on peramivir resistance against influenza H7N9 virus and the development of potential inhibitors [dissertation]. KwaZulu-Natal [ZA]: University of KwaZulu-Natal.

Mueller, S.N., Rouse, B.T., 2008. Immune responses to viruses. Clinical Immunology (Orlando, Fla.) 421–431. Available at https://doi.org/10.1016/B978-0-323-04404-2.10027-2, Epub 2009 May 15. PMCID:PMC7151814.

Muench, S.P., Antonyuk, S.V., Hasnain, S.S., 2019. The expanding toolkit for structural biology: synchrotrons, X-ray lasers and cryoEM. IUCrJ 6 (2), 167–177. Available at https://doi.org/10.1107/s2052252519002422.

Muñoz-Jordán, J.L., Sánchez-Burgos, G.G., Laurent-Rolle, M., García-Sastre, A., 2003. Inhibition of interferon signaling by dengue virus. Proceedings of the National Academy of Sciences of the United States of America 100 (2), 14333–14338. Available at https://doi.org/10.1073/pnas.2335168100.

Murphy, B.R., Whitehead, S.S., 2011. Immune response to dengue virus and prospects for a vaccine. Annual Review of Immunology 29, 587–619. Available at https://doi.org/10.1146/annurev-immunol-031210-101315.

Murzin, A.G., Brenner, S.E., Hubbard, T., Chothia, C., 1995. SCOP: A structural classification of proteins database for the investigation of sequences and structures. Journal of Molecular Biology 247 (4), 536–540. Available at https://doi.org/10.1016/S0022-2836(05)80134-2.

Nguyen, H., Wilson, H., Jayakumar, S., Kulkarni, V., Kulkarni, S., 2021. Efficient inhibition of HIV using CRISPR/Cas13d nuclease system. Viruses 13 (9), 1850. Available at https://doi.org/10.3390/v13091850.

Ou-Yang, S.S., Lu, J.Y., Kong, X.Q., Liang, Z.J., Luo, C., Jiang, H., 2012. Computational drug discovery. Acta Pharmacologica Sinica 33 (9), 1131–1140. Available at https://doi.org/10.1038/aps.2012.109.

Owen, K.A., Anderson, C.J., Casanova, J.E., 2016. Salmonella suppresses the TRIF-dependent type I interferon response in macrophages. mBio 7 (1). Available at https://doi.org/10.1128/mBio.02051-15, http://mbio.asm.org/content/7/1/e02051-15.full.pdf.

Paton, N.I., Musaazi, J., Kityo, C., Walimbwa, S., Hoppe, A., Balyegisawa, A., et al., 2021. Dolutegravir or darunavir in combination with zidovudine or tenofovir to treat HIV. New England Journal of

Medicine 385 (4), 330–341. Available at https://doi.org/10.1056/NEJMoa2101609, http://www.nejm.org/medical-index.

Paul, D., Sanap, G., Shenoy, S., Kalyane, D., Kalia, K., Tekade, R.K., 2021. Artificial intelligence in drug discovery and development. Drug Discovery Today 26 (1), 80–93. Available at https://doi.org/10.1016/j.drudis.2020.10.010, www.elsevier.com/locate/drugdiscov.

Pearl, L.H., Taylor, W.R., 1987. A structural model for the retroviral proteases. Nature 329 (6137), 351–354. Available at https://doi.org/10.1038/329351a0.

Perumalla, S.R., Sun, C.C., 2014. Design and preparation of a 4:1 lamivudine–oxalic acid CAB cocrystal for improving the lamivudine purification process. Crystal Growth & Design 14 (8), 3990–3995. Available at https://doi.org/10.1021/cg5005825.

Petsko, G.A., Ringe, D., 2004. Protein Structure and Function. New Science Press, London, ISBN 9781405119221.

Plavec, Z., Pöhner, I., Poso, A., Butcher, S.J., 2021. Virus structure and structure-based antivirals. Current Opinion in Virology 51, 16–24. Available at https://doi.org/10.1016/j.coviro.2021.09.005, http://www.elsevier.com/wps/find/journaldescription.cws home/724700/description#description.

Rai, K.R., Shrestha, P., Yang, B., Chen, Y., Liu, S., Maarouf, M., et al., 2021. Acute infection of viral pathogens and their innate immune escape. Frontiers in Microbiology 12. Available at https://doi.org/10.3389/fmicb.2021.672026, https://www.frontiersin.org/journals/microbiology#.

Rao, G.S.N.K., Alavala, R.R., 2025. Applications of Computational Tools in Drug Design and Development Applications of Computational Tools in Drug Design and Development. Springer Science+Business Media, India, pp. 1–1022. Available at https://doi.org/10.1007/978-981-96-4154-3.

Rehwinkel, J., Gack, M.U., 2020. RIG-I-like receptors: their regulation and roles in RNA sensing. Nature Reviews. Immunology 20 (9), 537–551. Available at https://doi.org/10.1038/s41577-020-0288-3, http://www.nature.com/nri/index.html.

Ren, B., 2002. Novel atomic-level-based ai topological descriptors: application to QSPR/QSAR modeling. Journal of Chemical Information and Computer Sciences 42 (4), 858–868. Available at https://doi.org/10.1021/ci020362l.

Renaud, J.P., 2020. Structural Biology in Drug Discovery: Methods, Techniques, and Practices Structural Biology in Drug Discovery: Methods, Techniques, and Practices. wiley, France, pp. 1–660. Available at https://doi.org/10.1002/9781118681121, https://onlinelibrary.wiley.com/doi/book/10.1002/9781118681121.

Rhodes, G., 2006. Crystallography Made Crystal Clear: A Guide for Users of Macromolecular Models, Third Edition. Elsevier, United States, pp. 1–306. Available at https://doi.org/10.1016/B978-0-12-587073-3.X5000-4, https://www.sciencedirect.com/book/9780125870733.

Roy, K. (Ed.), 2018. Computational Modeling of Drugs Against Alzheimer's Disease. Neuromethods. Humana Press, New York [NY]. Available at https://doi.org/10.1007/978-1-4939-7404-7.

Roy, K., Kar, S., Das, R.N., 2015. Understanding the Basics of QSAR for Applications in Pharmaceutical Sciences and Risk Assessment. Elsevier Inc., India, pp. 1–479. Available at https://doi.org/10.1016/C2014-0-00286-9, http://www.sciencedirect.com/science/book/9780128015056.

Russell, R.B., Eggleston, D.S., 2000. New roles for structure in biology and drug discovery. Nature Structural Biology 7, 928–930. Available at https://doi.org/10.1038/80691.

Sarkar, A., Mandal, K., 2021. Repurposing an antiviral drug against SARS-CoV-2 main protease. Angewandte Chemie International Edition 60 (44), 23492–23494. Available at https://doi.org/10.1002/anie.202107481.

Saurabh, S., Sivakumar, P.M., Perumal, V., Khosravi, A., Sugumaran, A., Prabhawathi, V., 2020. Molecular dynamics simulations in drug discovery and drug delivery. In: Krishnan, A., Chuturgoon, A. (Eds.), Integrative nanomedicine for new therapies. Springer Nature, Cham, pp. 275–301. Available at https://doi.org/10.1007/978-3-030-36260-7_10.

Savva, C., 2019. A beginner's guide to cryogenic electron microscopy. The Biochemist 41 (2), 46–52. Available at https://doi.org/10.1042/bio04102046.

Scapin, G., 2006. Structural biology and drug discovery. Current Pharmaceutical Design 12 (17), 2087–2097. Available at https://doi.org/10.2174/138161206777585201.

Segal-Maurer, S., DeJesus, E., Stellbrink, H.J., Castagna, A., Richmond, G.J., Sinclair, G.I., et al., 2022. Capsid inhibition with lenacapavir in multidrug-resistant HIV-1 infection. New England Journal of Medicine 386 (19), 1793–1803. Available at https://doi.org/10.1056/NEJMoa2115542, http://www.nejm.org/medical-index.

Shanmugam, A., Venkattappan, A., Gromiha, M.M., 2022. Structure based drug designing approaches in SARS-CoV-2 spike inhibi-tor design. Current Topics in Medicinal Chemistry 22 (29), 2396–2409. Available at https://doi.org/10.2174/1568026623666221103091658, https://www.eurekaselect.com/article/127347.

Silenzi, S., 2015. Characterization of West Nile virus strains isolated in Italy [thesis]. Bologna [IT]: Alma Mater Studiorum–Università di Bologna.

Soriano, V., Barreiro, P., Cachay, E., Kottilil, S., Fernandez-Montero, J.V., de Mendoza, C., 2020. Advances in hepatitis B therapeutics. Therapeutic Advances in Infectious Disease 7. Available at https://doi.org/10.1177/2049936120965027, http://tai.sagepub.com/.

Suguna, K., 2014. Crystallography and drug design. Resonance 19 (12), 1093–1103. Available at https://doi.org/10.1007/s12045-014-0135-6.

Talha, B., Dhamoon, A.S., 2023. Ritonavir. StatPearls [Internet]. StatPearls Publishing, Treasure Island [FL], 2025. Updated 2023 Aug 8.

Toyama, M., Sakakibara, N., Takeda, M., Okamoto, M., Watashi, K., Wakita, T., et al., 2019. Pyrimidotriazine derivatives as selective inhibitors of HBV capsid assembly. Virus Research 271, 197677. Available at https://doi.org/10.1016/j.virusres.2019.197677.

Tsegaye, S., Dedefo, G., Mehdi, M., 2021. Biophysical applications in structural and molecular biology. Biological Chemistry 402 (10), 1155–1177. Available at https://doi.org/10.1515/hsz-2021-0232.

Tyring, S.K., Moore, S.A., Moore, A.Y., Lupi, O., 2021. Overcoming antimicrobial resistance of the skin. Updates in Clinical Dermatology. Springer International Publishing. Available at https://doi.org/10.1007/978-3-030-68321-4.

Ul-Haq, Z., Wilson, A.K. (Eds.), 2020. Frontiers in Computational Chemistry Vol. 8 Bentham Science Publishers, Sharjah [UAE].

van der Lee, R., 2016. Integrative genomics of complex molecular systems: a systems biology approach to chronic inflammatory diseases [doctoral thesis]. Nijmegen [NL]: Radboud University Medical Center, 252 p.

Varghese, J.N., 1999. Development of neuraminidase inhibitors as anti-influenza virus drugs. Drug Development Research 46 (3-4), 176–196. Available at https://doi.org/10.1002/(SICI)1098-2299(199903/04)46:3/4<176::AID-DDR4>3.0.CO;2-6.

Vega, M.C., Fernández, F.J., 2024. Advanced technologies for protein complex production and characterization: volume II. Advances in Experimental Medicine and Biology 1453. Available at https://doi.org/10.1007/978-3-031-52193-5.

Vénien-Bryan, C., Li, Z., Vuillard, L., Boutin, J.A., 2017. Cryo-electron microscopy and X-ray crystallography: complementary approaches to structural biology and drug discovery. Acta Crystallographica Section:F Structural Biology Communications 73 (4), 174–183. Available at https://doi.org/10.1107/S2053230X17003740, http://journals.iucr.org/f/journalhomepage.html.

Venkatesh, S., Lipper, R.A., 2000. Role of the development scientist in compound lead selection and optimization. Journal of Pharmaceutical Sciences 89 (2), 145–154. Available at https://doi.org/10.1002/(SICI)1520-6017(200002)89:2<145::AID-JPS2>3.0.CO;2-6, www.interscience.wiley.com/jpages/0022-3549.

Voet, D., Voet, J.G., 2011. Biochemistry, 4th ed. John Wiley & Sons, Hoboken [NJ], ISBN: 978-0470-57095-1; 978-0470-91745-9.

Wahid, M.F., 2024. A review on crystallography and its role on drug design. Texas Journal of Multidisciplinary Studies 38.

Wang, H.W., Wang, J.W., 2017. How cryo-electron microscopy and X-ray crystallography complement each other. Protein Science 26 (1), 32–39. Available at https://doi.org/10.1002/pro.3022, http://onlinelibrary.wiley.com/journal/10.1002/(ISSN)1469-896X.

Wang, G., Zhu, W., 2016. Molecular docking for drug discovery and development: a widely used approach but far from perfect. Future Medicinal Chemistry 8 (14), 1707–1710. Available at https://doi.org/10.4155/fmc-2016-0143.

Wang, S.J., Huang, C.F., Yu, M.L., 2021. Elbasvir and grazoprevir for the treatment of hepatitis C. Expert Review of Anti-Infective Therapy 19 (9), 1071–1081. Available at https://doi.org/10.1080/14787210.2021.1874351, http://www.tandfonline.com/loi/ierz20.

Wang, H., Li, P., Sun, J., Wang, H., Wang, L., Huang, K., et al., 2024. Advancements in antiviral drug development: comprehensive insights into design strategies and mechanisms targeting key viral proteins. Journal of Microbiology and Biotechnology 34 (7), 1376–1388.

Wang, X., Wang, Q., Zheng, C., Wang, L., 2025. MAVS: the next STING in cancers and other diseases. Critical Reviews in Oncology/Hematology 207, 104610. Available at https://doi.org/10.1016/j.critrevonc.2024.104610.

Ward, A.B., Sali, A., Wilson, I.A., 2013. Integrative structural biology. Science (New York, N.Y.) 339 (6122), 913–915. Available at https://doi.org/10.1126/science.1228565, http://www.sciencemag.org/content/339/6122/913.full.pdf.

Wlodawer, A., Dauter, Z., Shabalin, I.G., Gilski, M., Brzezinski, D., Kowiel, M., et al., 2020. Ligand-centered assessment of SARS-CoV-2 drug target models in the Protein Data Bank. The FEBS Journal 287 (17), 3703–3718. Available at https://doi.org/10.1111/febs.15366.

Wrapp, D., et al., 2020. Cryo-EM structure of the SARS-CoV-2 spike protein. Science (New York, N.Y.) 367 (6483), 1260–1263.

Wu, B., Hur, S., 2015. How RIG-I like receptors activate MAVS. Current Opinion in Virology 12, 91–98. Available at https://doi.org/10.1016/j.coviro.2015.04.004, http://www.elsevier.com/wps/find/journaldescription.cws_home/724700/description#description.

Xiao, T., Cai, Y., Chen, B., 2021. HIV-1 entry and membrane fusion inhibitors. Viruses 13 (5), 735. Available at https://doi.org/10.3390/v13050735.

Yang, H., Rao, Z., 2021. Structural biology of SARS-CoV-2 and implications for therapeutic development. Nature Reviews. Microbiology 19 (11), 685–700. Available at https://doi.org/10.1038/s41579-021-00630-8.

Yevsieieva, L.V., Lohachova, K.O., Kyrychenko, A., Kovalenko, S.M., Ivanov, V.V., Kalugin, O.N., 2023. Main and papain-like proteases as prospective targets for pharmacological treatment of coronavirus SARS-CoV-2. RSC Advances 13 (50), 35500–35524. Available at https://doi.org/10.1039/d3ra06479d, http://pubs.rsc.org/en/journals/journal/ra.

Ylikangas, H., 2024. Virtual screening of ligands for kallikrein-related peptidase 3 and the L-type amino acid transporter 1 [dissertation]. Kuopio [FI]: University of Eastern Finland, 137 p. [Publications of the University of Eastern Finland. Dissertations in Health Sciences, No. 869].

Yusuf, D., Davis, A.M., Kleywegt, G.J., Schmitt, S., 2008. An alternative method for the evaluation of docking performance: RSR vs RMSD. Journal of Chemical Information and Modeling 48 (7), 1411–1422. Available at https://doi.org/10.1021/ci800084x, http://pubs.acs.org/journal/jcisd8

Zhang, G., Guo, S., Cui, H., Qi, J., 2018. Virtual screening of small molecular inhibitors against DprE1. Molecules (Basel, Switzerland) 23 (3), 524. Available at https://doi.org/10.3390/molecules23030524.

CHAPTER 13

Structural bioinformatics in personalized medicine

Sarika Bano[1], Aaliya Javed[1], Anas Shamsi[2], and Asimul Islam[1]
[1]Centre for Interdisciplinary Research in Basic Sciences, Jamia Millia Islamia, New Delhi, India,
[2]Centre of Medical and Bio-allied Health Sciences Research, Ajman University, Ajman, United Arab Emirates

13.1 Introduction

The landscape of medicinal practice has undergone a fundamental transformation over the past 2 decades, shifting from a traditional paradigm based on population-level evidence to an increasingly sophisticated, patient-centric approach. Traditional medicine, rooted in evidence-based practice guidelines developed from randomized controlled trials, treats diseases based on clinical phenotypes and population-level efficacy data. However, this conventional approach inherently overlooks significant individual variability in disease susceptibility, progression, treatment response, and adverse effects. The concept of personalized medicine, also termed "precision medicine" or "genomic medicine," advocates for healthcare delivery that incorporates individual variation in genes, environments, and lifestyles to inform diagnosis, prognosis, prevention, and treatment strategies (Delpierre and Lefèvre, 2023) (Fig. 13.1). This paradigm recognizes the genetic heterogeneity observed in complex diseases and that molecular profiling can identify patient subpopulations likely to respond to specific interventions (Woodward et al., 2022).

Structural bioinformatics has emerged as a discipline that leverages computational methods to analyze, predict, and manipulate 3D protein structures, providing mechanistic insights into biological processes and disease pathogenesis. The discipline has emerged as a critical enabler of personalized medicine by enabling precise molecular characterization of disease mechanisms at the atomic level. The integration of structural knowledge with genomics, proteomics, and clinical data creates a comprehensive foundation for understanding individual patient biology and tailoring therapeutic interventions accordingly (Patel and Mani, 2024). This approach transcends the traditional one-size-fits-all medical paradigm by incorporating structural insights that directly govern how genetic variation manifests as functional changes in proteins and affects disease phenotype and drug responses (Molla and Bitew, 2024).

The human genome contains over 4 million missense variants, yet fewer than 2% have been clinically classified, leaving the majority of variants with unknown functional consequences. Structural bioinformatics addresses this critical gap by predicting how

Advances in Structural Biology
DOI: https://doi.org/10.1016/B978-0-443-44988-8.00001-0

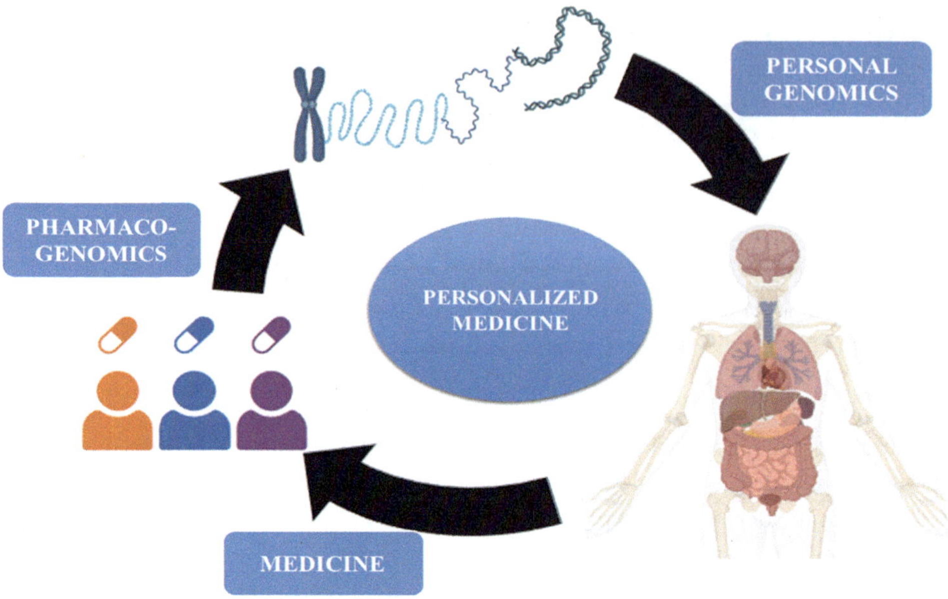

Figure 13.1 Schematic representation of personalized medicine, highlighting the continuous interaction between genomic insights, pharmacogenomics, and tailored medical treatment.

sequence changes alter protein architecture and biochemical function, enabling more accurate disease risk assessment and treatment selection. Furthermore, the field leverages structural information to identify druggable targets and predict how individual variations in drug-metabolizing enzymes will affect pharmaceutical efficacy and safety (Li et al., 2026).

13.2 Personalized medicine and structural medicine: introduction

PM is fundamentally built upon the molecular understanding of diseases and the therapeutic targeting of disease-specific alterations. At its core lies the integration of structural biology, the 3D characterization of biological macromolecules and their interactions, with genomic data and clinical information. Structural biology has traditionally relied on experimental techniques such as X-ray crystallography, nuclear magnetic resonance spectroscopy, and cry-EM to elucidate protein structures at atomic resolution. However, the exponential growth of genomic sequence data, combined with revolutionary advances in computational protein structure prediction, has transformed structural biology from a primarily experimental discipline into an increasingly computational source (Mishra et al., 2025).

The fundamental principle underlying PM in structural biology is that genetic variations—primarily single nucleotide polymorphism (SNPs), insertion/deletion, and structural variants—alter protein sequences, which in turn impact protein structure, stability, and function. By understanding how genetic variants affect protein 3D structure and function, clinicians can predict disease risk, anticipate treatment response, and design targeted therapies. This structure–function relationship is particularly crucial for understanding how nonsynonymous SNPs affect protein stability and drug binding, information essential for personalizing pharmacotherapies and identifying novel drug targets (Hossain et al., 2025).

13.3 Significance of structural bioinformatics and PM

Proteins are the primary molecular executors of biological function, and their 3D structure determines their ability to perform enzymatic reactions, bind regulatory molecules, interact with other proteins, and respond to cellular signals. Mutations that alter protein structural bioinformatics and PM derive from its capacity to predict and characterize how genetic variants affect protein structure and, consequently, disease phenotype. Recent advances in AI-driven protein structure prediction, exemplified by AlphaFold 2, have achieved near-experimental accuracy in predicting protein 3D structure from amino acid sequence alone (Peng et al., 2024; Jumper et al., 2021). This breakthrough has profound implications for PM, enabling rapid computational assessment of how patient-specific mutations affect protein structure and function without requiring time-consuming experimental validation.

Structural bioinformatics provides a mechanistic bridge between genotype and phenotype. While traditional genome-wide association studies identify statistical associations between genetic variants and disease susceptibility, they typically provide limited insight into the biological mechanisms underlying these associations. Structural analysis can reveal how specific variants disrupt critical functional elements—such as active sites, allosteric sites, protein-protein interaction interfaces, or protein stability—therefore explaining the molecular basis of disease and identifying pathways amenable to therapeutic interventions (Pande et al., 2025). This mechanistic understanding is essential for the rational design of patient-specific therapeutics and for predicting both therapeutic efficacy and adverse effects based on individual molecular profiles.

Structural bioinformatics enhances PM by providing tools to interpret the functional consequences of patient-specific molecular alterations. Structural analysis can predict whether a missense mutation will disrupt protein folding, alter active site geometry, impair protein–protein interactions, or modulate protein–drug binding (Pandurangan and Blundell, 2020). These predictions enable clinicians to identify actionable mutations likely to drive disease and amenable to targeted therapy, thereby enabling personalized treatment selection based on molecular evidence rather than population-level data.

13.3.1 Integration with genomics

The impact of PM is maximized through the integration of structural bioinformatics with genomic analysis. Genomics provides comprehensive inventories of genetic variants and alterations in individual patients, while structural bioinformatics enables the functional interpretation of these variants. The combination yields powerful predictive systems capable of anticipating disease risk, disease progression, treatment response, and adverse effects. Multiomics integration, that is, combining genomic, transcriptomic, proteomic, metabolomic, and phenotypic data, further improves personalization by capturing the dynamic complexity of biological systems (Zheng et al., 2024). SB tools can integrate these diverse data types by predicting how genetic variants affect protein structure and, consequently, transcriptional regulation, protein function, and metabolic fluxes.

This integrated approach has demonstrated clinical utility across multiple disease domains; in cancer, genomic profiling combined with structural analysis of driver mutations enables subtyping and prediction of treatment response to targeted therapies (Chakravarty and Solit, 2021; Sadee et al., 2023). In rare genetic disorders, structural analysis of disease-causing mutations enables functional annotation of variants, while analyzing genetic polymorphisms affecting drug-metabolizing enzymes and drug targets enables the prediction of individual drug response phenotypes and support dose optimization (Umair and Waqas, 2023).

The workflow involves the systematic progression from raw genomic data through functional annotation to clinical interpretation and therapeutic decision-making. The foundational step involves high-quality variant discovery from patient genomic data, obtained via next-generation sequencing (NGS). Modern MGS workflows employ multiple variants-calling algorithms to maximize sensitivity and specificity. Tools such as SIFT, PolyPhen-2, PROVEAN, and FATHMM evaluate sequence-level conservation and physicochemical properties of amino acid substitutions to predict deletion mutations (Choudhury et al., 2022). However, structure-based variant annotation tools provide predictive power by analysing how mutations affect protein folding, stability, and function in 3D space (David and Sternberg, 2023).

13.4 Bioinformatics steps in genomic data processing for PM

The translation of raw sequencing data into clinically actionable insights requires a sophisticated, mutistep bioinformatics workflow. Raw sequencing data consist of millions to billions of short DNA sequences, typically 50–300 bp in length, that have to be aligned to a reference genome and processed for quality and variant analysis. The standard NGS data processing pipeline includes the raw sequence quality assessment with trimming and adapter removal, then read alignment to the reference genome, followed by alignment quality assessment, variant calling, variant annotation, variant filtering, and clinical interpretation (Beura et al., 2025).

13.4.1 Raw sequence quality assessment

Raw sequence quality assessment employs tools to evaluate read quality score, GC content distribution, and metrics that might indicate sequencing biases. Trimming and adapter removal eliminate low-quality bases and sequencing that could introduce false positives in subsequent variant calling. Read alignment matches sequences to position in the reference genome that employs algorithms that balance alignment sensitivity and computational efficiency (Alser et al., 2021). Alignment quality metrics assess the percentage of reads successfully aligned, uniformity of alignment coverage, and evidence of technical artifacts.

13.4.2 Variant calling, annotation, and functional prediction

Variant calling identifies genomic positions where patient sequences differ from the reference genome, and it typically employs multiple algorithms to balance sensitivity and specificity. Variants with multiple algorithms and accepting only consistent calls across approaches provide superior sensitivity and specificity, reducing both false positives and false negatives (Zverinova and Guryev, 2022).

Identified variants must be annotated with information regarding their genomic context, predicted functional impact, and clinical significance. Annotation tools assign variants to genomic features, predict coding consequences, and query databases of known pathogenic variants and population frequencies. Functional impact prediction employs diverse computational approaches to assess whether variants are likely to be pathogenic (Hebbar and Sowmya, 2021)

Phenotype prediction models integrate genomic and phenotypic data to predict disease characteristics and treatment response. Machine learning approaches, including random forests, support vector machines, and neural networks, have proven particularly powerful for this prediction, often outperforming traditional statistical approaches (Martinović et al., 2025). Deep learning models, particularly when trained on large datasets, can identify complex, nonlinear relationships between genetic features and phenotypes (Ahmed et al., 2023).

13.4.3 Data integration and clinical interpretation

The culminating step in bioinformatics data processing is the integration of genomic findings with clinical data and their translation into clinical recommendations. Electronic health record systems, patient registries, and outcome databases contain rich clinical information, including comorbidities, medical history, treatment response, adverse effects, and outcomes. Integration of complex structure, function, and phenotype outcome relationships (de Oliveira et al., 2021).

Clinical interpretation frameworks provide standardized approaches to assigning pathogenicity classifications to identify variants and integrating information into clinical reports, bioinformatics systems increasingly provide structured reports that translate complex genomic data into actionable clinical recommendations, enabling direct translation of genomic findings into treatment planning.

13.5 Application of structural bioinformatics in PM and vaccine development

Bioinformatics tools may provide better diagnosis at the genomic level with earlier detection of disease and better-targeted therapy through efficient PM development. Genomics analysis provides great assistance in the development of PM. Bioinformatics tools help in diagnosis, intervention, drug development, therapy, and personalized vaccines. Personalized vaccine means for an optimized prevention of disease with minimum reactions and side effects. Personalized vaccine is developed to take care of haplotypes and polymorphisms that can become risk of an adverse vaccine reaction. Variation in gender, racial, ethnic, gene polymorphism, and other environmental factors affects the immune response to a particular vaccine, where we require a personalized vaccine and drug (Thitame and Aher, 2025).

13.5.1 Structure-based precision oncology and targeted therapeutics

Cancer represents the most advanced domain of PM, with multiple FDA-approved therapeutics guided by genomic profiling and molecular diagnostics. Structural bioinformatics has been instrumental in translating cancer genomics into precision therapeutics. In nonsmall cell lung cancer, identification of specific EGFR mutations predicts response to tyrosine kinase inhibitors. Structural analysis has revealed how specific mutations affect kinase domain structure and inhibitor binding, enabling the design of inhibitors with activity against specific mutant variants (Robichaux et al., 2021).

Structural bioinformatics has been crucial for understanding acquired resistance to targeted therapies. Secondary mutations arising during treatment can alter drug-binding sites, impacting therapeutic efficacy. Structural analysis of resistance-causing mutation enables the design of next-generation inhibitors with activity against resistant variants (Tunstall et al., 2020).

13.5.2 Personalized cancer vaccine development

Vaccines are designed from each patient's individual tumor mutations. Neoantigen vaccines represent a paradigm shift in cancer immunotherapy. The discovery of immunogenic neoantigens requires the identification of patient-specific mutations through tumor genomic sequencing, followed by the prediction of which mutations encode immunogenic neoantigens, vaccine manufacturing, then vaccination; this is followed by immune monitoring.

13.5.3 Infectious disease and preventive vaccine design

Structural bioinformatics similarly transforms infectious disease prevention through structure-guided vaccine design. Understanding pathogen protein structures and their interaction with the host immune systems enables the rational design of more immunogenic vaccines (Sarvmeili et al., 2024). SARS-CoV-2 vaccine development benefited tremendously from rapid determination of the spike protein structure, which guided the design of optimized immunogens for COVID-19 vaccines.

Reverse vaccinology increasingly incorporates structural bioinformatics to identify conserved, immunogenic, surface-accessible proteins amenable to vaccine targeting (da Silva et al., 2023). Structural analysis identifies epitopes for immunogen design to enhanced the immune response (Fig. 13.2).

13.6 Genomic analysis in PM

Structural bioinformatics contributes to pharmacogenomics by predicting how genetic variants in drug targets affect during-binding affinity. A patient with a variant in a drug

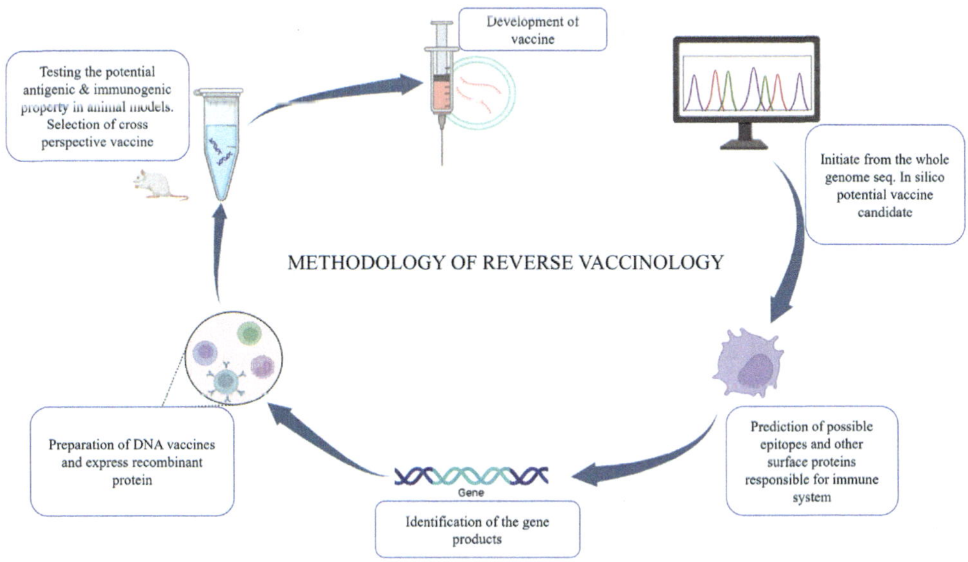

Figure 13.2 Methodology of reverse vaccinology.

target might exhibit diminished drug response despite appropriate drug levels, due to reduced drug-target binding. Structure-based drug design by patient-specific target mutations enables optimization of drugs for individual patient genotypes (Mozaffari et al., 2025).

Phylogenetic risk scores integrate information from multiple genetic variants across the genome to predict disease risk or treatment response, increasingly leveraging structural bioinformatics. Risk prediction models trained on multiomics data, including genomic, structural, transcriptomic, and proteomic information, have demonstrated superior predictive accuracy compared to sequence-level prediction (Pasat et al., 2024).

13.6.1 Multiomics integration for complex disease understanding

Complex diseases result from intricate interactions among multiple genetic variants, epigenetic modifications, environmental factors, and age-related changes. Comprehensive understanding requires integration of multiomics layers. Structural bioinformatics bridges these diverse omics layers by predicting how genetic variants affect protein structure, which consequently impacts protein function and metabolic consequences (Pasat et al., 2024).

Multiomics integration has proven ideal in cancer research, where comprehensive characterization of tumor molecular landscapes enables identification of cancer subtypes with distinct molecular drivers and treatment sensitivities (Thapliyal et al., 2025).

13.6.2 Systems biology and network medicine approaches

Network medicine is the application of network analysis to understand disease mechanisms and identify therapeutic targets, representing an emerging frontier integrating structural bioinformatics with systems biology. Biological networks capture protein–protein interaction, metabolic pathways, signaling cascades, and transcriptional regulatory relationships; structural bioinformatics informs network analysis by predicting how mutations alter protein–protein interactions or modify regulatory protein binding.

13.7 Challenges in incorporating structural bioinformatics into PM design

13.7.1 Technical and computational challenges

Despite remarkable progress, substantial technical challenges persist in implementing structural bioinformatics to PM. Protein structure prediction accuracy, while vastly improved by deep learning, remains incomplete for complex scenarios, including protein complexes, disordered regions, and conformational ensembles. AlphaFold 2 excels at predicting single protein conformations but may miss alternative conformations crucial for understanding protein function and drug binding (Chakravarty et al., 2025).

The blind spots of structure prediction methods can lead to mispredictions with clinical consequences (Chakravarty et al., 2025).

Protein dynamics is the temporal fluctuation in protein conformation that proves essential for understanding drug binding and protein–protein interactions but remains incompletely captured by static structures. Molecular dynamics simulations can sample conformational ensembles but require substantial computational resources (Sil et al., 2025). Data integration challenges arise from the heterogeneity of omics technologies and analysis methods (Sibilio et al., 2025).

13.7.2 Clinical translation and regulatory challenges

Translating structural bioinformatics predictions into clinical practice requires rigorous validation that predictions translate into improved patient outcomes. Many publications report excellent prediction accuracy in retrospective validation using existing datasets, but prospective validity and clinical utility versus clinical validity remain substantial for bioinformatics applications.

13.7.3 Equity, access, and implementation challenges

PM risks exacerbating healthcare inequities if it is accessible only to affluent populations in developed nations. Genomic databases reflect a population skewing toward European ancestry, creating reduced prediction accuracy for non-European populations (Green et al., 2023). The economic burden of genomic testing and personalized therapeutics renders them inaccessible to many patients globally, particularly in poor countries.

Integration of PM into routine clinical practice requires not merely technological development but substantial clinical and educational infrastructure changes. Clinicians require training in genomic interpretation; bioinformatics platforms must integrate seamlessly with electronic health records (Abul-Husn and Kenny, 2019).

13.8 Advantages and disadvantages of PM

13.8.1 Advantages of PM

- The potential advantages of PM are substantial and increasingly validated in clinical practice. Improved therapeutic efficacy represents perhaps the most immediate benefit; matching patients to therapies predicted to be effective for their specific molecular profiles yields a higher response rate compared to standard approaches (Thitame and Aher, 2025).
- Reduced adverse effects constitute another advantage, avoiding therapies predicted to be ineffective for individual patients prevents unnecessary toxicity exposure and associated morbidity and mortality.

- Improved disease prevention represents another transformative potential of PM. Identifying individuals at elevated genetic risk enables targeted prevention strategies (Dong et al., 2025).
- Advancement in drug discovery efficiency represents an additional benefit. Structural bioinformatics and PM approaches enable drug development targeting specific patient subpopulations rather than assuming uniform efficacy across disease subtypes (Batool et al., 2019).

13.8.2 Disadvantages and limitation of PM

- Despite immense success, substantial limitations and challenges constrain PM implementation and clinical utility. High cost of genomic testing and personalized treatment renders them inaccessible to many patients, particularly in limited resource areas.
- Clinical complexity and uncertain clinical utility represent additional limitations. Phenotypic heterogeneity reveals that genetic information is insufficient for complete phenotype prediction, as epigenetic modifications, environmental factors, and other influences affect disease manifestation.
- Incidental findings raise ethical challenges regarding informed consent, genetic counseling requirements, and patient preference regarding disclosure.
- Data privacy, genetic discrimination, and potential misuse of genomic information create legitimate patient concerns.

13.9 Advancement in methods for PM development

13.9.1 AI and machine learning innovation

AI and machine learning have revolutionized personalized medicine by enabling analysis of high-dimensional data and discovery of complex patterns invisible to traditional statistical approaches. Deep learning models excel at learning nonlinear relationships in multiomics data (Alum and Ugwu, 2025). AI-driven protein structure prediction, exemplified by AlphaFold 2, has transformed structural biology and drug discovery. Generative AI models enable de novo design of novel molecules with desired properties, accelerating drug discovery (Qiu et al., 2024).

13.9.2 Advanced sequencing and genomic technologies

Third-generation sequencing technologies enable the detection of structural variants, epigenetic modifications, and transcript isoforms that are impossible with second-generation short-read sequencing (Searle et al., 2023). Single-cell sequencing enables cellular-resolution analysis, revealing intra-tumor heterogeneity and transcriptional states influencing treatment response (Jia et al., 2022). Circulating cell-free DNA and circulating tumor DNA enable noninvasive tumor monitoring and early detection (Bruhm et al., 2025).

13.9.3 High-performance computing and cloud infrastructure

PM requires processing substantial volumes of sequence data and integrating results with clinical decision systems. Cloud computing platforms provide scalable, flexible computing infrastructure, enabling rapid analysis of large cohorts. Workflow management systems enable the construction of reproducible, scalable bioinformatics systems (Iglesias et al., 2024).

13.10 Future perspective

13.10.1 Integration of multimodal data and holistic patient understanding

The future of PM will increasingly integrate diverse data modalities, including genomics, imaging, clinical notes, wearable device data, and environmental data into unified models of individual patient biology and health status. Multimodal AI systems are capable of integrating heterogeneous data types that will enable more accurate disease prediction and treatment response forecasting (Acosta et al., 2022).

Real-time wearable monitoring integrated with genomic data will enable the detection of disease signatures before clinical analysis. Cellular and subcellular understanding of disease will shift from the tissue level to the single-cell and spatial level. Integration of spatial transcriptomics, proteomics, and imaging will create comprehensive maps of tissue composition (Tu et al., 2025).

13.10.2 Precision therapeutics and engineered therapeutics

Gene therapy using CRISPR-based approaches will enable correction of patient-specific disease-causing mutations (Far et al., 2025). CAR-T therapies and TCR-engineered T cells will become increasingly personalized, with T cells engineered to recognize patient-specific tumor neoantigens. Personalized RNA therapeutics will enable treatment of rare genetic disorders.

13.10.3 Precision prevention and wellness

Future healthcare systems will increasingly shift from reactive disease treatment toward proactive disease prevention enabled by personalized risk prediction. Genomic prediction combined with longitudinal health monitoring will identify individuals at elevated risk years before clinical trials. Personalized cancer vaccines developed from individual patient tumor mutations may prevent cancer recurrence (Lu et al., 2025).

13.11 Conclusion

Structural bioinformatics has emerged as an indispensable discipline enabling the translation of genomic discoveries into personalized therapeutic strategies. By bridging molecular understanding of disease mechanisms with patient-specific genomic profiles, structural bioinformatics enables rational identification of therapeutic targets, design of patient-specific drugs, prediction of treatment response, and development of individualized prevention strategies.

The revolutionary advances in protein structure prediction through deep learning, coupled with expanding genomic data resources and increasingly sophisticated computational methods, have created unprecedented opportunities for precision medicine (Jänes and Beltrao, 2024). Current evidence demonstrates substantial clinical utility of PM in defined domains, including oncology, pharmacogenomics, and rare genetic disease diagnosis (Cecchin and Stocco, 2020). However, realizing the full potential of PM requires addressing substantial technical, clinical, regulatory, and equity challenges.

Integration of diverse omics layers through machine learning and network medicine approaches promises a deeper mechanistic understanding of complex disorders. Future directions, including cellular analysis, longitudinal monitoring, and engineered therapeutics, will further advance PM capabilities (Dai et al., 2021).

The convergence of structural bioinformatics with genomics, AI, and clinical medicine represents one of the most consequential developments in biomedical science, offering transformative potential to improve individual health (Olawade et al., 2025). However, this potential can only be realized through continued technical innovation and advances, rigorous clinical validation, equitable access strategies, and interdisciplinary collaboration. PM shows structural understanding of individual patient biology, promises a future where healthcare is truly tailored to each individual's unique molecular characteristics, enabling optimal prevention, diagnosis, and treatment while advancing medical science toward the longstanding goal of precision therapeutic intervention.

References

Abul-Husn, N.S., Kenny, E.E., 2019. Personalized medicine and the power of electronic health records. Cell 177 (1), 58–69. Available at https://doi.org/10.1016/j.cell.2019.02.039, https://www.sciencedirect.com/journal/cell.

Acosta, J.N., Falcone, G.J., Rajpurkar, P., Topol, E.J., 2022. Multimodal biomedical AI. Nature Medicine 28 (9), 1773–1784. Available at https://doi.org/10.1038/s41591-022-01981-2, http://www.nature.com/nm/index.html.

Ahmed, S.F., Alam, M.S.B., Hassan, M., Rozbu, M.R., Ishtiak, T., Rafa, N., et al., 2023. Deep learning modelling techniques: current progress, applications, advantages, and challenges. Artificial Intelligence Review 56 (11), 13521–13617. Available at https://doi.org/10.1007/s10462-023-10466-8, https://www.springer.com/journal/10462.

Alser, M., Rotman, J., Deshpande, D., Taraszka, K., Shi, H., Baykal, P.I., et al., 2021. Technology dictates algorithms: recent developments in read alignment. Genome Biology 22 (1). Available at https://doi.org/10.1186/s13059-021-02443-7.

Alum, E.U., Ugwu, O.P.C., 2025. Artificial intelligence in personalized medicine: transforming diagnosis and treatment. Discover Applied Sciences 7 (3). Available at https://doi.org/10.1007/s42452-025-06625-x, https://link.springer.com/journal/42452.

Batool, M., Ahmad, B., Choi, S., 2019. A structure-based drug discovery paradigm. International Journal of Molecular Sciences 20 (11). Available at https://doi.org/10.3390/ijms20112783, https://www.mdpi.com/1422-0067/20/11/2783/pdf.

Beura, A., Manjunath, G.K., Mahalingam, S., Rajguru, M.S., Dakal, T.C., Kumar, A., 2025. Next generation DNA sequencing data analysis and its application in clinical genomics. Pathology - Research and Practice 276, 156280. Available at https://doi.org/10.1016/j.prp.2025.156280.

Bruhm, D.C., Vulpescu, N.A., Foda, Z.H., Phallen, J., Scharpf, R.B., Velculescu, V.E., 2025. Genomic and fragmentomic landscapes of cell-free DNA for early cancer detection. Nature Reviews Cancer 25 (5), 341–358. Available at https://doi.org/10.1038/s41568-025-00795-x, https://www.nature.com/cancer/.

Cecchin, E., Stocco, G., 2020. Pharmacogenomics and personalized medicine. Genes 11 (6), 1–5. Available at https://doi.org/10.3390/genes11060679, https://www.mdpi.com/2073-4425/11/6/679/pdf.

Chakravarty, D., Solit, D.B., 2021. Clinical cancer genomic profiling. Nature Reviews. Genetics 22 (8), 483–501. Available at https://doi.org/10.1038/s41576-021-00338-8, http://www.nature.com/reviews/genetics.

Chakravarty, D., Lee, M., Porter, L.L., 2025. Proteins with alternative folds reveal blind spots in AlphaFold-based protein structure prediction. Current Opinion in Structural Biology 90. Available at https://doi.org/10.1016/j.sbi.2024.102973, https://www.sciencedirect.com/science/journal/0959440X.

Choudhury, A., Mohammad, T., Anjum, F., Shafie, A., Indrakant, K., Singh, B., et al., 2022. Tripathi, Comparative analysis of web-based programs for single amino acid substitutions in proteins. PLoS One 17 (5), e0267084. Available at https://doi.org/10.1371/journal.pone.0267084

Dai, G., Feinberg, A.W., Wan, L.Q., 2021. Recent advances in cellular and molecular bioengineering for building and translation of biological systems. Cellular and Molecular Bioengineering 14 (4), 293–308. Available at https://doi.org/10.1007/s12195-021-00676-x, http://www.springer.com/engineering/biomedical+eng/journal/12195.

David, A., Sternberg, M.J.E., 2023. Protein structure-based evaluation of missense variants: resources, challenges and future directions. Current Opinion in Structural Biology 80, 102600. Available at https://doi.org/10.1016/j.sbi.2023.102600.

Delpierre, C., Lefèvre, T., 2023. Precision and personalized medicine: what their current definition says and silences about the model of health they promote. Implication for the development of personalized health. Frontiers in Sociology 8. Available at https://doi.org/10.3389/fsoc.2023.1112159, www.frontiersin.org/journals/sociology.

Dong, M., Wang, L., Hu, N., Rao, Y., Wang, Z., Zhang, Y., 2025. Integration of multi-omics approaches in exploring intra-tumoral heterogeneity. Cancer Cell International 25 (1). Available at https://doi.org/10.1186/s12935-025-03944-2.

Far, B.F., Akbari, M., Habibi, M.A., Katavand, M., Nasseri, S., 2025. CRISPR technology in disease management: an updated review of clinical translation and therapeutic potential. Cell Proliferation 58 (11). Available at https://doi.org/10.1111/cpr.70099, http://onlinelibrary.wiley.com/journal/10.1111/(ISSN)1365-2184.

Green, S., Prainsack, B., Sabatello, M., 2023. Precision medicine and the problem of structural injustice. Medicine, Health Care, and Philosophy 26 (3), 433–450. Available at https://doi.org/10.1007/s11019-023-10158-8, https://www.springer.com/journal/11019.

Hebbar, P., Sowmya, S.K., 2021. Genomic variant annotation: a comprehensive review of tools and techniques. In: International Conference on Intelligent Systems Design and Applications. Springer.

Hossain, T., Islam, M.N., Hossain, M.A., Rahman, M.M., Islam, M.M., Gupta, S.D., 2025. Unraveling the structural and functional consequences of non-synonymous single-nucleotide polymorphisms (nsSNPs) in human SOCS2: an in silico approach. Egyptian Journal of Medical Human Genetics 26 (1). Available at https://doi.org/10.1186/s43042-025-00669-9.

Iglesias, N.E., et al., 2024. Revolutionizing Next-Generation Sequencing: Advanced Analysis Pipeline and Integrated Solutions. In: International Conference on Practical Applications of Computational Biology & Bioinformatics. Springer.

Jia, Q., Chu, H., Jin, Z., Long, H., Zhu, B., 2022. High-throughput single-cell sequencing in cancer research. Signal Transduction and Targeted Therapy 7 (1). Available at https://doi.org/10.1038/s41392-022-00990-4.

Jumper, J., Evans, R., Pritzel, A., Green, T., Figurnov, M., Ronneberger, O., et al., 2021. Highly accurate protein structure prediction with AlphaFold. Nature 596 (7873), 583–589. Available at https://doi.org/10.1038/s41586-021-03819-2.

Jänes, J., Beltrao, P., 2024. Deep learning for protein structure prediction and design—progress and applications. Molecular Systems Biology 20 (3), 162–169. Available at https://doi.org/10.1038/s44320-024-00016-x, https://www.embopress.org/page/journal/17444292/about.

Li, D., Shi, S., Yu, Z., Xu, P., Zhang, C., 2026. AI accelerate the identification of druggable targets by 3D structures of proteins and compounds. npj Precision Oncology 10 (1). Available at https://doi.org/10.1038/s41698-026-01310-7.

Lu, L., Lu, X., Luo, W., 2025. Personalized cancer vaccines: current advances and emerging horizons. Vaccines 13 (12), 1231. Available at https://doi.org/10.3390/vaccines13121231.

Martinović, M., Dokic, K., Pudić, D., 2025. Comparative analysis of machine learning models for predicting innovation outcomes: an applied AI approach. Applied Sciences 15 (7), 3636. Available at https://doi.org/10.3390/app15073636.

Mishra, A., Khan, K., Kumar, D., 2025. Structural Biology: Current Trends and Future Perspectives Biochemical and Biophysical Methods in Molecular and Cellular Biology. Springer Nature, India, pp. 285–303. Available at http://doi.org/10.1007/9789819620883_11, https://link.springer.com/book/9789819620883.

Molla, G., Bitew, M., 2024. Revolutionizing personalized medicine: synergy with multi-omics data generation, main hurdles, and future perspectives. Biomedicines 12 (12). Available at https://doi.org/10.3390/biomedicines12122750, www.mdpi.com/journal/biomedicines.

Mozaffari, S., Moen, A., Ng, C.Y., Nicolaes, G.A.F., Wichapong, K., 2025. Structural bioinformatics for rational drug design. Research and Practice in Thrombosis and Haemostasis 9 (1), 102691. Available at https://doi.org/10.1016/j.rpth.2025.102691.

Olawade, D.B., Kade, A., Egbon, E., Usman, S.O., Fapohunda, O., Ijiwade, J., et al., 2025. Bioinformatics and artificial intelligence in genomic data analysis: current advances and future directions. Molecular Genetics and Genomics 300 (1). Available at https://doi.org/10.1007/s00438-025-02314-x, https://www.springer.com/journal/438.

de Oliveira, J.M., da Costa, C.A., Antunes, R.S., 2021. Data structuring of electronic health records: a systematic review. Health and Technology 11 (6), 1219–1235. Available at https://doi.org/10.1007/s12553-021-00607-w, http://www.springer.com/engineering/biomedical+eng/journal/12553.

Pande, S., Dawood, M., Grochowski, C.M., 2025. Structural variants: mechanisms, mapping, and interpretation in human genetics. Genes 16 (8). Available at https://doi.org/10.3390/genes16080905, http://www.mdpi.com/journal/genes/.

Pandurangan, A.P., Blundell, T.L., 2020. Prediction of impacts of mutations on protein structure and interactions: SDM, a statistical approach, and mCSM, using machine learning. Protein Science 29 (1), 247–257. Available at https://doi.org/10.1002/pro.3774, http://onlinelibrary.wiley.com/journal/10.1002/(ISSN)1469-896X.

Pasat, B.A., Pilalis, E., Mnich, K., Samali, A., Chatziioannou, A., Gorman, A.M., 2024. MultiOmicsIntegrator: a nextflow pipeline for integrated omics analyses. Bioinformatics

Advances 4 (1). Available at https://doi.org/10.1093/bioadv/vbae175, https://academic.oup.com/bioinformaticsadvances.

Patel, K., Mani, A., 2024. Structural bioinformatics and protein structure prediction. Unraveling New Frontiers and Advances in Bioinformatics. Springer Nature, India, pp. 143–162. Available at http://doi.org/10.1007/978-981-97-7123-3_8, https://link.springer.com/book/9789819771233.

Peng, C.X., Liang, F., Xia, Y.H., Zhao, K.L., Hou, M.H., Zhang, G.J., 2024. Recent advances and challenges in protein structure prediction. Journal of Chemical Information and Modeling 64 (1), 76–95. Available at https://doi.org/10.1021/acs.jcim.3c01324, http://pubs.acs.org/journal/jcisd8.

Qiu, X., Li, H., Ver Steeg, G., Godzik, A., 2024. Advances in AI for protein structure prediction: implications for cancer drug discovery and development. Biomolecules 14 (3), 339. Available at https://doi.org/10.3390/biom14030339.

Robichaux, J.P., Le, X., Vijayan, R.S.K., Hicks, J.K., Heeke, S., Elamin, Y.Y., et al., 2021. Structure-based classification predicts drug response in EGFR-mutant NSCLC. Nature 597 (7878), 732–737. Available at https://doi.org/10.1038/s41586-021-03898-1, http://www.nature.com/nature/index.html.

Sadee, W., Wang, D., Hartmann, K., Toland, A.E., 2023. Pharmacogenomics: driving personalized medicine. Pharmacological Reviews 75 (4), 789–814. Available at https://doi.org/10.1124/pharmrev.122.000810.

Sarvmeili, J., Baghban Kohnehrouz, B., Gholizadeh, A., Shanehbandi, D., Ofoghi, H., 2024. Immunoinformatics design of a structural proteins driven multi-epitope candidate vaccine against different SARS-CoV-2 variants based on fynomer. Scientific Reports 14 (1). Available at https://doi.org/10.1038/s41598-024-61025-2.

Searle, et al., 2023. Third-generation sequencing of epigenetic DNA. Angewandte Chemie (International Ed. in English) 62 (14).

Sibilio, P., De Smaele, E., Paci, P., Conte, F., 2025. Integrating multi-omics data: methods and applications in human complex diseases. Biotechnology Reports 48, e00938. Available at https://doi.org/10.1016/j.btre.2025.e00938.

da Silva, M.K., Campos, D.Md.O., Akash, S., Akter, S., Yee, L.C., Fulco, U.L., et al., 2023. Advances of reverse vaccinology for mRNA vaccine design against SARS-CoV-2: a review of methods and tools. Viruses 15 (10). Available at https://doi.org/10.3390/v15102130, http://www.mdpi.com/journal/viruses/.

Sil, S., Datta, I., Basu, S., 2025. Use of AI-methods over MD simulations in the sampling of conformational ensembles in IDPs. Frontiers in Molecular Biosciences 12. Available at https://doi.org/10.3389/fmolb.2025.1542267, journal.frontiersin.org/journal/molecular-biosciences.

Thapliyal, A.P., et al., 2025. Advancements in multi-omics integration for comprehensive cancer understanding and treatment. Advancing Biotechnology: From Science to Therapeutics and Informatics: Technological Advancements in Biosciences and Biotechnology. Springer, pp. 195–202.

Thitame, S.N., Aher, A.A., 2025. AI-driven advancements in bioinformatics: transforming healthcare and science. Journal of Pharmacy and Bioallied Sciences 17 (Suppl 1), S24. Available at https://doi.org/10.4103/jpbs.jpbs_389_25.

Tunstall, T., Portelli, S., Phelan, J., Clark, T.G., Ascher, D.B., Furnham, N., 2020. Combining structure and genomics to understand antimicrobial resistance. Computational and Structural Biotechnology Journal 18, 3377–3394. Available at https://doi.org/10.1016/j.csbj.2020.10.017.

Tu, J., Flynn, C.D., Yeom, J., Wu, Z., Kelley, S.O., Gao, W., 2025. Wearable biomolecular sensing nanotechnologies in chronic disease management. Nature Nanotechnology 20 (10), 1388–1404. Available at https://doi.org/10.1038/s41565-025-02010-2, https://www.nature.com/nnano/.

Umair, M., Waqas, A., 2023. Undiagnosed rare genetic disorders: importance of functional characterization of variants. Genes 14 (7). Available at https://doi.org/10.3390/genes14071469, http://www.mdpi.com/journal/genes/.

Woodward, A.A., Urbanowicz, R.J., Naj, A.C., Moore, J.H., 2022. Genetic heterogeneity: challenges, impacts, and methods through an associative lens. Genetic Epidemiology 46 (8), 555–571. Available at https://doi.org/10.1002/gepi.22497, http://onlinelibrary.wiley.com/journal/10.1002/(ISSN)1098-2272.

Zheng, Y., Liu, Y., Yang, J., Dong, L., Zhang, R., Tian, S., et al., 2024. Multi-omics data integration using ratio-based quantitative profiling with Quartet reference materials. Nature Biotechnology 42 (7), 1133–1149. Available at https://doi.org/10.1038/s41587-023-01934-1.

Zverinova, S., Guryev, V., 2022. Variant calling: considerations, practices, and developments. Human Mutation 43 (8), 976–985. Available at https://doi.org/10.1002/humu.24311, http://onlinelibrary.wiley.com/journal/10.1002/(ISSN)1098-1004.

CHAPTER 14

Structural biology in understanding bacterial resistance

Deepali Gupta[1], Mukesh Kumar[2], Dominic Skaria Mathew[3], Ishaan Bhardwaj[3], Shivani Dixit[1], and Punit Kaur[1]
[1]Department of Biophysics, All India Institute of Medical Sciences, New Delhi, India,
[2]Department of Optometry, School of Allied Health Sciences, Galgotias University, Noida, Uttar Pradesh, India,
[3]Department of Computer Science, Jamia Millia Islamia, South Delhi, Delhi, India

14.1 Introduction

Antimicrobial resistance (AMR) has become one of the most serious global health issues of the 21st century. It has the potential to destroy the years of advancement that have been made so far for controlling infectious diseases (Larsson and Flach, 2022; Van Boeckel et al., 2014; Browne et al., 2021; Murray et al., 2022). Bacterial resistance patterns are changing so quickly that they have made almost all major groups of antibiotics ineffective, which has led to prolonged illness, an increase in mortality rate, and growing healthcare expenses (Larsson and Flach, 2022; Van Boeckel et al., 2014; Browne et al., 2021; Murray et al., 2022; Dadgostar, 2019). Millions of people die every year because of AMR, and the number is expected to increase dramatically if effective countermeasures are not taken soon (Larsson and Flach, 2022; Van Boeckel et al., 2014; Dadgostar, 2019; Prestinaci et al., 2015).

The World Health Organization (WHO) has recognized AMR as a significant global threat, as it can challenge modern medical interventions. Almost all infectious diseases are heavily dependent on antibiotics to prevent and control the disease, and bacterial resistance may lead to fail the treatment, including surgical procedures, organ transplantation, or even cancer chemotherapy.

The impact of AMR is particularly severe in many developing countries, as infectious diseases are highly prevalent in such settings. Additionally, due to their insufficient healthcare system and lack of surveillance infrastructure, antibiotics are often used without proper regulation or medical supervision, which further complicates the monitoring and control of resistance pathogen and contributes to the development of bacterial resistance (Browne et al., 2021; Murray et al., 2022; Laxminarayan et al., 2016; Pokharel et al., 2019).

India is a very important epicenter of this crisis, as it is one of the largest consumers of antibiotics in the world, and the number of multidrug-resistant bacteria, such as carbapenem-resistant *Enterobacteriaceae*, methicillin-resistant *Staphylococcus aureus* (MRSA), or extensively drug-resistant *Mycobacterium tuberculosis*, is increasing alarmingly (Nordmann et al., 2011;

Advances in Structural Biology
DOI: https://doi.org/10.1016/B978-0-443-44988-8.00021-6

Dheda et al., 2017; Klein et al., 2018). These hard-to-treat infectious diseases have now become a high socioeconomic burden on healthcare systems, as it has increased prolonged hospital stay, treatment cost, and high mortality rate. International travel further contributes to the spread of these resistant microorganisms and leads to an increase in human mortality (Murray et al., 2022; Dadgostar, 2019; Prestinaci et al., 2015; Laxminarayan et al., 2016; Pokharel et al., 2019; Nordmann et al., 2011; Dheda et al., 2017; Klein et al., 2018; Naylor et al., 2018).

Although microbiological, biochemical, and genomic studies have been effective in identifying the major determinants of resistance, including resistance genes, mutations, and mobile genetic elements, such methods in isolation are usually inadequate to explain the molecular mechanisms of resistance (Partridge et al., 2018; Blair et al., 2015b). This is one of the knowledge gaps that is critical. The knowledge of resistance at the sequence level or the phenotypic level does not provide the ability to design effective and long-term antimicrobial interventions. To fulfill this lacuna, structural biology provides the molecular mechanism of AMR at the atomic level. Structural studies have shown that small changes in bacterial structure can reduce antibiotic binding, promote drug degradation, upregulate the efflux mechanism, or limit the intracellular drug accumulation (Blair et al., 2015b). Structural biologists have revealed the key mechanisms of bacterial resistance by resolving the three-dimensional structures of antibiotic targets such as resistance enzymes, efflux transporters, and membrane-bound proteins. This approach has also supported the development of structure-guided drug design and therapeutic interventions (Wilson, 2014; Fischbach and Walsh, 2009).

This chapter explores the structural basis underlying major AMR pathways, including target modification, enzyme drug inhibition, efflux pump activation, and membrane permeability alteration. It also highlights how structural techniques such as X-ray crystallography, cryo-electron microscopy (cryo-EM), and nuclear magnetic resonance (NMR), along with molecular dynamics simulations (MDS), have advanced the understanding of molecular interactions and conformational changes that lead to antibiotic resistance.

Lastly, this chapter discusses how structure-based drug discovery with machine-learning-based predictions is transforming the future of antimicrobial discovery. Collectively, these strategies provide new avenues to improve the effectiveness of antibiotics and intensify efforts to tackle AMR worldwide (Stokes et al., 2020).

14.2 Molecular mechanism underlying antimicrobial resistance

AMR has become a major global health concern as several bacteria have developed the ability to escape the effects of many antibiotics. This growing resistance makes the treatment of several diseases ineffective that were once considered treatable (Murray et al., 2022; Prestinaci et al., 2015).

Resistance is usually described in genetic or phenotypic terms, but the real basis of resistance is fundamentally molecular in nature (Fig. 14.1a) (Blair et al., 2015b). Antibiotics exert their effects by binding to the essential bacterial macromolecules, such as enzymes involved in the synthesis of the cell wall, nucleic acids, or proteins, and inhibit bacterial survival (Draper et al., 2015). Alterations in bacterial structure components lead to the disruption of drug binding and then the drug is no longer effective to inhibit bacterial growth (Partridge et al., 2018). This is how resistance towards a drug is developed (Blair et al., 2015b).

Microbiological, biochemical, and genetic methods have played a significant role in the identification of resistance determinants, such as resistance genes, mutations, and regulatory pathways. However, these methods describe only the associated changes rather than the mechanism of how these changes are altering the interaction between the drug and its molecular targets (Fig. 14.1b). This limitation highlights a crucial gap in knowledge that can be filled with structural understanding. The efficiency of an antibiotic depends on the three-dimensional complementarity between the drug and its target molecule (Fig. 14.1c). This interaction is mediated through hydrogen bonds, hydrophobic contacts, electrostatic forces, and steric compatibility. Even a small change in such an interaction, like reorientation of the amino acid side chain, minor backbone movements, or increased conformational

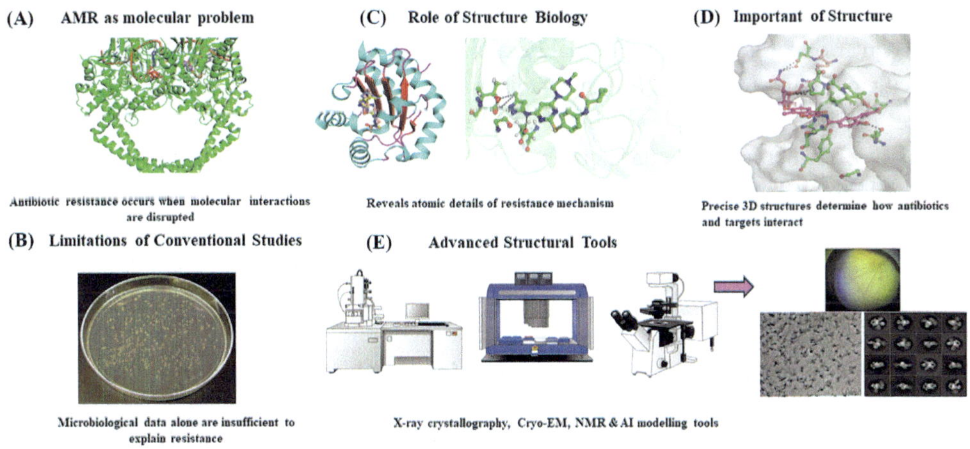

Figure 14.1 Structural basis of antimicrobial resistance (*AMR*) in *Salmonella Typhi*. (A) Overall structure of the target protein (e.g., DNA gyrase) showing domain organization and active-site location. (B) Bacterial colony growth on agar plates showing antimicrobial susceptibility and resistance phenotype. (C) Secondary structure representation highlighting α-helices, β-sheets, and ligand-binding region. (D) Close-up view of inhibitor binding within the active pocket, depicting key interactions, such as hydrogen bonds and hydrophobic contacts. (E) Advanced structural tools, including X-ray crystallography, cryo-electron microscopy (*cryo-EM*), nuclear magnetic resonance (*NMR*), and artificial intelligence (*AI*) modeling.

flexibility, may weaken the binding efficacy of the antibiotic while maintaining the normal biological activity of the target (Fig. 14.1d).

The effectiveness of an antibiotic, as well as the emergence of resistance, depends on the molecular interaction between the drug and its bacterial target. Therefore structural biology plays a very important role in understanding the mechanism behind the mechanistic and predictive studies of bacterial resistance.

14.3 Structure biology approaches to study antimicrobial resistance

Molecular Biophysics explains the three-dimensional structure of biological macromolecules and directly correlates their molecular structure with their biological function (Fig. 14.1e). It enables the elucidation of atomic or near-atomic structures of proteins, nucleic acids, as well as their complexes, that determine protein folding and their active conformations, and their biological functions to regulate cellular processes (Kühlbrandt, 2014). This method is used in the context of AMR to gain critical insight into the mechanism behind the recognition of the molecular target, their modifications in context to mutations, and the survival of bacteria (Blair et al., 2015b). Using X-ray crystallography, it is possible to visualize the target and enzyme-drug complexes in high resolution with binding pockets, catalytic residues, and small structural alterations related to resistance. Cryo-EM lifts the technical limitations of the structural method to large dynamic macromolecular assemblies like ribosomes or multidrug efflux pumps, and captures conformational heterogeneity in near-native states, which are functional (Benjin and Ling, 2020). NMR spectroscopy is the only method that has access to molecular dynamics, conformational flexibility, and transient interactions in solution and can facilitate insight into resistance mechanisms that rely on structural plasticity but not fixed changes (Lerner et al., 2018). Computational methods, such as molecular docking, MDS, and AI-based structure prediction, combine and complement experimental data on the basis of modeling antibiotic binding, mutation-induced structural effects, and possible evolutionary paths of resistance. Together, these tools make structural biology a key resource in the understanding, prediction, and overcoming of AMR (Stokes et al., 2020; Jumper et al., 2021; Ferreira et al., 2015) Fig. 14.1.

14.4 Structural basis of major antimicrobial resistance mechanism

Bacterial resistance against any drug develops through several interconnected mechanisms at the molecular level that untimely effect the binding of the drug to its bacterial target. Any structural changes in the bacterial macromolecule can reduce the binding of the drug, enhance drug modification, or restrict the accumulation of the drug inside the cell (Darby et al., 2023). The key resistance mechanisms include genetic mutation (Fig. 14.2a), horizontal gene transfer (Fig. 14.2b), target modification (Fig. 14.2c), reduced membrane permeability (Fig. 14.2d), enzymatic drug inactivation (Fig. 14.2e), metabolic bypass or target replacement (Fig. 14.2f), and active efflux (Fig. 14.2g)

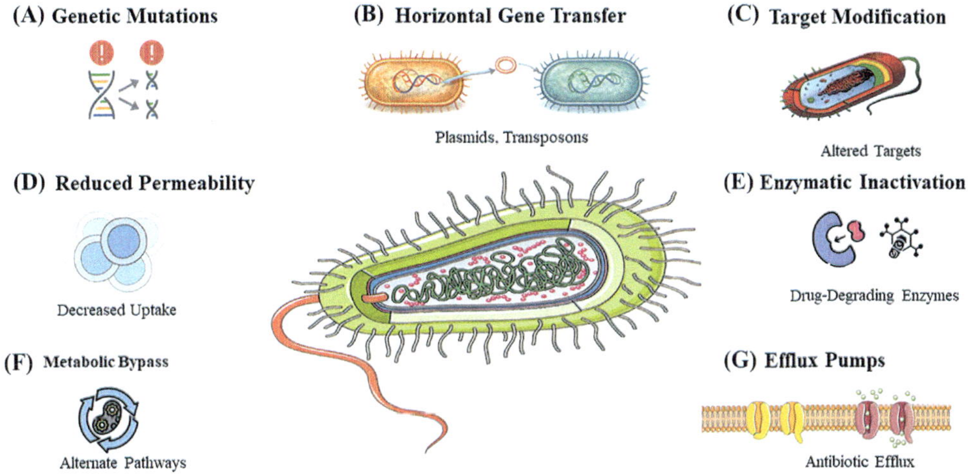

Figure 14.2 Mechanism of antibiotic resistance. (A) Genetic mutations in bacterial DNA. (B) Horizontal gene transfer via plasmids. (C) Target modification reducing antibiotic binding. (D) Reduced permeability limiting drug uptake. (E) Enzymatic inactivation of antibiotics. (F) Metabolic bypass through alternative pathways. (G) Efflux pumps actively export antibiotics.

(Fig. 14.2) (Sartori et al., 2026). Each of these mechanisms involves specific structural or conformational changes that help the bacteria to survive antibiotic exposure while maintaining normal cellular function. Structural biology has played a very important role in understanding these molecular alterations that are contributing towards antibiotic resistance. By analysing the three-dimensional structure of bacterial targets and resistance proteins, researchers are able to understand the structural alterations that can weaken antibiotic activity. The following section discusses the resistance mechanisms in detail (Darby et al., 2023; Sartori et al., 2026; Varela et al., 2021) (Fig. 14.2).

14.4.1 Target modification

Target modification is one of the most common mechanisms of antibiotic resistance, where structural alterations occur in essential bacterial targets such as penicillin-binding proteins (PBPs), ribosomes, DNA gyrase, or RNA polymerase (Darby et al., 2023). These changes occur within or near the antibiotic-binding region of the target molecule and reduce the ability of the drug to bind efficiently (Wilson, 2014). Structural studies have shown that resistance can be developed by even small modifications in the amino acid side chain or nucleotide bases at the antibiotic-binding pocket (Blair et al., 2015b). These minor structural adjustments weaken the interactions that are required for strong drug binding while maintaining the normal biological activity of the targeted protein. As a result, bacteria are able to survive even in the presence of antibiotics (Svetlov et al., 2019). All these methods have been summarized in Table 14.1.

Table 14.1 Structural insights into bacterial antimicrobial resistance mechanisms.

Resistance pathway	Conventional example	Antibiotic class	Target/system	Structural basis of resistance	Structural biology tools used
Genetic mutation	Quinolone resistance-determining region mutation in DNA gyrase (Ser83→Leu, Asp87→Asn)	Fluoroquinolones	DNA gyrase (GyrA)	Point mutations alter the coordination of Mg^{2+} ions and break hydrogen bonding, making the drug-enzyme-DNA complex unstable without affecting catalysis	X-ray crystallography, molecular dynamics (MD) simulations (Kim et al., 2022; Das et al., 2021).
Horizontal gene transfer	Acquisition of low-affinity PBP2a	β-lactams	Penicillin-binding proteins	Active site of the enzyme is being sterically shielded, reducing the accessibility and preventing the acylation of catalytic serine	X-ray crystallography, comparative structural analysis (Koch, 2000).
Target modification	Ribosomal RNA (rRNA) methylation in the ribosomal exit tunnel	Macrolides	23S rRNA	Presence of a methyl group introduces steric hindrance, altering the electrostatic surface and thus blocking antibiotic binding	Cryo-electron microscopy (cryo-EM), rRNA structural mapping (Svetlov et al., 2021).
Enzymatic inactivation	β-lactamase production	β-lactams	Antibiotic molecule	Catalytic serine residue hydrolyzes the β-lactam ring, thus destroying the structure of the drug	X-ray crystallography, QM/MM, MD simulations (Koch, 2000).

Reduced permeability	Porin loop modification	β-lactams, quinolones	Outer membrane porins	Narrowing of the channel, as well as the alteration of charge, inhibits the diffusion of the antibiotic	X-ray crystallography, cryo-EM, and electrostatic analysis (Koch, 2000; Prajapati et al., 2021; Zhou et al., 2023).
Efflux pumps	Multidrug efflux systems	Multiple classes	Membrane transporters	Large, flexible binding cavities and conformational cycling actively discharge antibiotics	Cryo-EM (multiple states), MD simulations (Eicher et al., 2014; Aborah et al., 2025; Deng et al., 2012).
Metabolic bypass	Alternative dihydrofolate reductase	Trimethoprim	Folate biosynthesis enzyme	Altered geometry of the active site inhibits the inhibitors from binding and thus maintaining catalytic function	X-ray crystallography, docking studies (Brolund et al., 2010).
Target replacement	D-Ala-D-Lac cell-wall precursor	Glycopeptides	Peptidoglycan terminus	The disruption of the key hydrogen bond reduces antibiotic affinity by ~1000-fold	X-ray crystallography, free-energy calculations (Dheda et al., 2017).

14.4.1.1 Alteration of penicillin-binding proteins

β-Lactam antibiotics, including penicillin, methicillin, and ampicillin, show their effect by inhibiting the synthesis of the bacterial cell wall through binding to the enzymes known as PBPs (Koch, 2000). These enzymes are responsible for the catalysis of the final step of peptidoglycan cross-linking, that are essential for maintaining cell-wall integrity (Koch, 2000; Turner et al., 2019). In MRSA, resistance arises through the acquisition of a modified PBP known as PBP2a, encoded by newly acquired gene mecA, which produce this altered enzyme (Koch, 2000; Turner et al., 2019).

These enzymes perform the same role as PBPs but have a slightly altered structural configuration at their antibiotic-binding region (Koch, 2000). X-ray crystallographic studies have revealed that the active site of PBP2a is relatively narrow and partially shielded by surrounding residues (Koch, 2000). This structural arrangement restricts the entry of penicillin and other β-lactam antibiotics to the catalytic center, thereby reducing their ability to bind effectively and inactivate the enzyme (Koch, 2000; Turner et al., 2019). At the same time, the enzyme is still able to perform its normal function of cell-wall synthesis. This structural feature explains how MRSA cells remain viable even in the presence of high concentrations of β-lactam antibiotics.

14.4.1.2 Resistance through mutation in ribosomal proteins

Streptomycin is an aminoglycoside antibiotic that targets the 30S subunit of the bacterial ribosome. Specifically, it binds to ribosomal protein S12 and interferes with the decoding process during protein synthesis (Carter et al., 2000). In several bacterial species, including *Mycobacterium*, resistance arises through point mutation in the gene encoding ribosome protein S12. Mutations occur at residues such as Lys42 and Lys87, leading to amino acid substitutions including Lys42Arg (K42R), Lys42Asn (K42N), or Lys87Arg (K87R). Although a mutations occur only in a single amino acid, the mutation is small enough to alter the local structure of the ribosome-binding site.

Structural studies of ribosome–streptomycin complexes revealed that these substitutions interfere with the hydrogen bonds required for stable antibiotic binding (Carter et al., 2000). As a result, streptomycin can no longer bind effectively to the ribosome, while the ribosome itself maintains its protein synthesis functioning actively. This structural evidence proved the ability of a one-letter genetic variation to prevent the binding of antibiotics while conserving bacterial survival (Carter et al., 2000).

14.4.1.3 Resistance through RNA polymerase mutation

Rifampicin and other rifamycin antibiotics, including rifabutin, rifaximin, and rifapentine, are a group of antibiotics primarily used for the treatment of mycobacterial infections like tuberculosis and specific bacterial infections (Campbell et al., 2001). They inhibit the transcription mechanism of bacteria by targeting the RNA polymerase enzyme responsible for the synthesis of RNA (Campbell et al., 2001). The antibiotics act by binding to the β-subunit of the RNA polymerase and blocking the elongation of the RNA chain (Campbell et al., 2001). Resistance against these antibiotics arises through point mutations in a small part of the RNA polymerase gene encoding the β-subunit (Zaw et al., 2018). Structural studies of RNA polymerase complexes with rifampicin showed that mutations occur in a specific region called the rifampicin resistance-determining region (Zaw et al., 2018). Substitutions like Ser531Leu (S531L), His526Tyr/Asp (H526Y or H526D), and Asp516Val (D516V) slightly alter the structural changes in the antibiotic-binding site, resulting in reduced affinity of rifamycin to the RNA polymerase while maintaining the normal bacterial enzymatic activity (Molodtsov et al., 2017).

14.4.1.4 Resistance through mutation in DNA gyrase

Fluoroquinolone antibiotics bind to DNA gyrase, an essential bacterial enzyme involved in DNA replication and supercoiling (Aldred et al., 2014b). The antibiotics inhibit bacterial growth by binding to the DNA gyrase-DNA complex and trapping the enzyme in the DNA-bound form (Aldred et al., 2014b). Thereby inhibiting the DNA replication process. Point mutations occur in the quinolone resistance-determining region of the DNA gyrase gene (Aldred et al., 2014b). Substitutions including Ser83Leu, Ser83Phe, Asp87Asn, Asp87Gly, or Asp87Tyr are commonly observed in resistant bacterial strains such as *Escherichia coli* and *Salmonella Typhi* (Aldred et al., 2014b). Structural studies have shown that mutations in the GyrA (Ser83Ala) (Fig. 14.3a, b, c) and the GyrB subunit (R136H) introduce small changes in the enzyme that disrupt the magnesium-mediated interactions required for stabilizing the antibiotic–enzyme–DNA complex (Fig. 14.3d, e, f) (Aldred et al., 2014a). As a result, the fluoroquinolones are no longer able to bind to the DNA gyrase enzyme while the enzyme remains catalytically active, allowing DNA replication to continue (Kumar et al., 2015). Hence, the bacteria are continuing to survive in the presence of antibiotics Fig. 14.3.

14.4.1.5 Macrolide resistance via ribosomal RNA methylation

The macrolide antibiotics bind to the bacterial ribosomal RNA (rRNA) and inhibit protein synthesis machinery (Svetlov et al., 2021). In resistant bacteria, the methyltransferase enzyme chemically modifies the structure of RNA by adding a methyl group close to the macrolide antibiotic-binding site. This modification alters the structure of the ribosome. Cryo-EM studies of the ribosome in the presence and absence of a methyl group (cryo-EM structure) revealed that the introduced methyl group caused steric hindrance that physically prevented the binding of macrolide (Svetlov et al., 2021). Despite this modification, the normal functional activity of the ribosome is maintained (Svetlov et al., 2021).

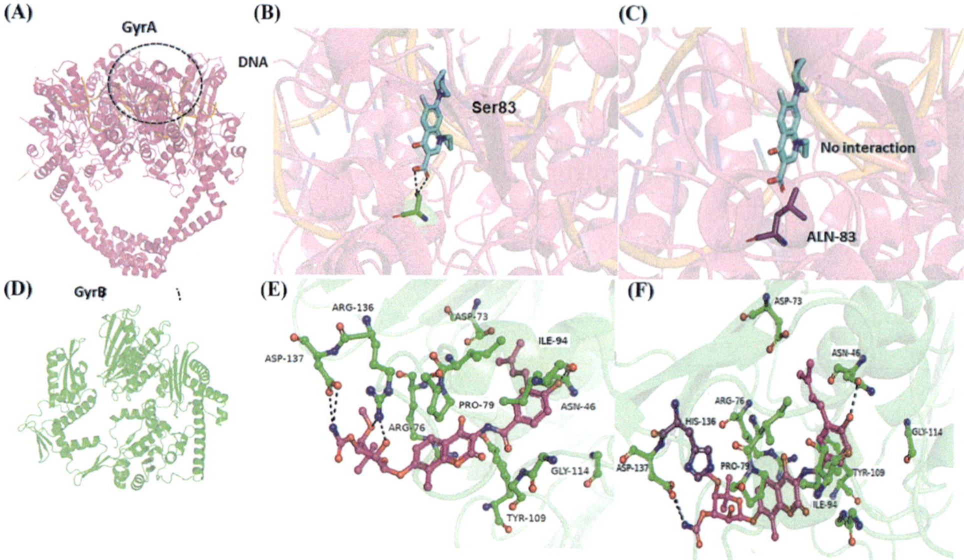

Figure 14.3 Structural analysis of DNA gyrase mutations affecting ligand (Cipro) binding. (A) Overall structure of DNA gyrase A–DNA complex (PDB–2XCT), highlighting the active-site region. (B) Zoomed wild-type interaction showing hydrogen bonding between ligand and catalytic residue Ser83 of GyrA. (C) Mutant GyrA (Ser83→Ala) showing loss of key hydrogen-bonding interactions. (D) Overall structure of the GyrB subunit having ATPase domain (PDB–5ZXM). (E) Wild-type GyrB interaction showing novobiocin stabilization via Arg136. (F) Mutant GyrB (Arg136→His) showing altered interaction pattern and reduced binding affinity.

14.4.1.6 Tetracycline resistance via ribosome protection proteins

Tetracycline antibiotics act by inhibiting the protein synthesis machinery of the bacteria (Lin et al., 2018). The antibiotics bind to the ribosome and block the attachment of aminoacyl-tRNA to the aminoacyl site responsible for the addition of an amino acid to the growing peptide chain (Lin et al., 2018). Some resistance bacteria escape from the antibiotic effect by expressing ribosome protection proteins (RPPs) rather than through mutation in ribosomal components (Li et al., 2013). These RPPs bind to the ribosome and alter its structures, thereby weakening the interaction between tetracycline and its target protein (Li et al., 2013). Structural studies by cryo-EM revealed how these RPPs attach to the ribosome, remodel the tetracycline-binding site, and promote antibiotic dissociation (Arenz et al., 2015). Once tetracycline dissociates from the complex, the ribosome returns to its functional conformation and resumes to synthesize proteins (Lin et al., 2018; Arenz et al., 2015). This mechanism demonstrates that resistance could be developed by reversible structural remodeling of the ribosome instead of permanent genetic mutation (Lin et al., 2018; Arenz et al., 2015).

14.4.2 Enzymatic inactivation of antibiotics

Enzymatic inactivation is one of the most potent mechanisms by which bacteria survive antibiotic therapy (Blair et al., 2015b; Darby et al., 2023). Here, bacterial enzymes destroy or alter the antibiotics chemically before they reach their target (Blair et al., 2015b; Darby et al., 2023). Structural biology has played a key role in understanding the mechanism through which these enzymes detect antibiotics, orientation in their active sites and catalyze chemical reactions, which nullify antibacterial activity (Varela et al., 2021).

14.4.2.1 β-Lactamase-mediated inactivation of β-lactam antibiotics

Many bacteria develop resistance against β-lactam antibiotics, including penicillins, cephalosporins, carbapenems, and monobactams, by producing β-lactamase enzymes (Fisher et al., 2005). These enzymes hydrolyze β-lactam ring, a structural component that is crucial for the antibacterial activity of these drugs (Fisher et al., 2005). Structural analysis using X-ray crystallography elucidated that the β-lactam antibiotic binds within a deep active-site pocket of the enzyme (Fig. 14.4) (Matagne et al., 1998). In class A β-lactamases, the key catalytic residues, such as Ser70, Lys73, Ser130, Glu166, and Asn170, surround the binding site and play crucial roles in catalysis process (Matagne et al., 1998). The Ser70 residue behaves like a nucleophile that attacks the β-lactam ring of the antibiotic and forms a covalent acyl-enzyme intermediate (Fig. 14.4a, b, c) (Matagne et al., 1998). This intermediate is subsequently hydrolyzed by a water molecule activated by Glu166, resulting in the opening of the β-lactam ring and inactivation of the antibiotics (Fig. 14.4 d, e) (Fisher et al., 2005; Matagne et al., 1998) Fig. 14.4.

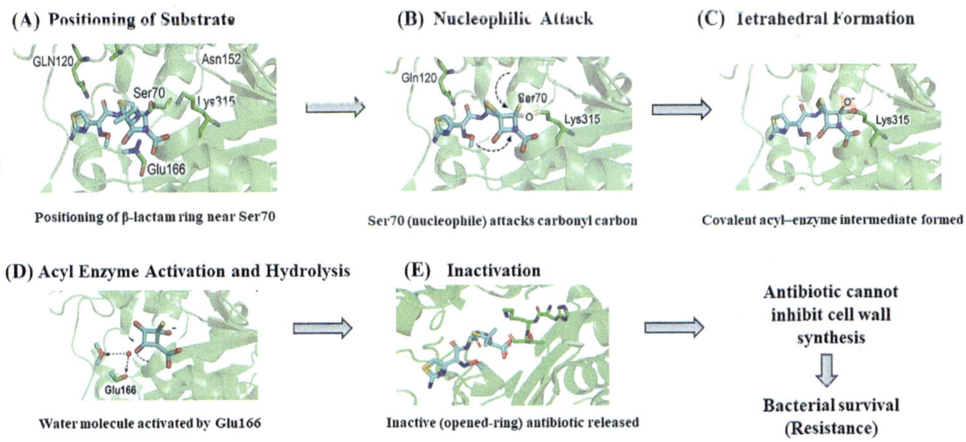

Figure 14.4 Catalytic mechanism of β-lactamase-mediated antibiotic hydrolysis. (A) Substrate (penicillin) positioning in the active site (PDB- 1LLB). (B) Nucleophilic attack by Ser70 on the β-lactam ring. (C) Formation of a tetrahedral acyl-enzyme intermediate. (D) Activation of water by Glu166 and hydrolysis of the intermediate. (E) Release of inactive (opened-ring) antibiotic and regeneration of free enzyme, leading to bacterial resistance.

14.4.2.2 Chloramphenicol acetyltransferase mediated inactivation of chloramphenicol

Chloramphenicol blocks the synthesis of bacterial proteins by binding to the peptidyltransferase enzyme center of ribosomes and blocking peptide bond formation. Many bacteria develop resistance to this antibiotic by synthesizing the chloramphenicol acetyltransferase (CAT) enzyme that inactivates chloramphenicol through acetylation. This enzyme was one of the first resistance proteins to have been solved by X-ray crystallography. The structural investigations showed that CAT forms a symmetrical trimeric enzyme having clearly defined antibiotic-binding regions (Leslie et al., 1988). Within the active site, the enzyme transfers one acetyl group from acetyl-coenzyme A to chloramphenicol. This acetylation modification changes the antibiotic structure and prevents the formation of the hydrogen-bonding interaction required for the binding of chloramphenicol to rRNA, making the drug inactive (Leslie et al., 1988).

14.4.2.3 Macrolide inactivation by macrolide esterases

Macrolide antibiotics like erythromycin act upon bacteria by binding to the ribosome (Alvarez-Elcoro and Enzler, 1999). These antibiotics contain large lactone rings that are needed for their binding to the rRNA (Alvarez-Elcoro and Enzler, 1999). Some bacteria have developed resistance against the drug by producing an enzyme, macrolide esterase, that hydrolyzes this ring, thereby inactivating the drug (Varela et al., 2021). X-ray crystallographic analysis revealed that the macrolide esterase enzyme active site can accommodate the large macrolide molecule and position the ester bond in a proper orientation to undergo hydrolysis. Cleavage of this lactone ring disrupts the structural entity of the drug needed for ribosome binding. As a result, the modified antibiotic is no longer effective in interacting with the ribosome and hence loses its antibacterial activity.

14.4.3 Active efflux mechanism

Active efflux is a major antibiotic resistance strategy in which bacteria pump antibiotics out of the cell faster than they can act, keeping intracellular drug levels below thresholds (Gaurav et al., 2023). Unlike mechanisms such as target modification or enzymatic inactivation, the antibiotic molecule in the active efflux mechanism remains chemically unchanged, but the resistance arises from the insufficient drug accumulation within the bacterial cell (Belay et al., 2024). Structural biology has provided valuable insights into explaining the architecture and function of these efflux pumps (Gaurav et al., 2023). Structural studies have revealed how efflux pumps recognize a range of antibiotics, how energy sources such as proton gradients drive drug transport, and how mutations or overexpression of efflux pumps contribute to multidrug resistance (Gaurav et al., 2023).

14.4.3.1 Gram-negative bacterial resistance-nodulation-division family efflux pumps

Among the various efflux systems, the most widely studied are the resistance-nodulation-division (RND) family in Gram-negative bacteria (Toone et al., 2011). A well-characterized example is the AcrAB-TolC efflux pump, which forms a tripartite complex spanning the inner membrane, the periplasmic space, and the outer membrane, creating a continuous channel that leads from the cytoplasm to the extracellular space (Miller and Salama, 2018). The architecture of the AcrAB-TolC has been resolved by cryo-EM, and provided deep insight into its transport mechanism (Du et al., 2014). The snapshots of the structure revealed that the pump operates through a "rotating access" mechanism, in which transporters pass through different conformational cycle switches between drug entry, binding, and extrusion (Toone et al., 2011). As a result of this dynamic process, a single pump can expel a wide range of structurally diverse drugs, including antibiotics, detergents, and dyes, and ultimately contribute to multidrug resistance (Toone et al., 2011). The inner membrane transporter (AcrB) plays a very important role in substrate recognition. The structural studies of AcrB revealed that they contain large and flexible binding pockets, enriched with hydrophobic and aromatic residues that help in recognizing multiple substrates (Ababou et al., 2016). They present overlapping pockets that are responsible for the polyspecific binding of the pump (Ababou et al., 2016). Resistance-associated mutations can cause minute remodeling of these binding pockets, thereby reducing the drug residence time within the transporter without disrupting pump activity (Blair et al., 2015a). Efflux is an energy-dependent process in RND systems, as conformational changes across the membrane are driven by the proton-motive force (Miller and Salama, 2018). MDS together with cryo-EM and X-ray structural data revealed how protonation-coupled conformational changes propagate through the protein to power drug extrusion (Eicher et al., 2014). These findings also explained that mutations that occur at distant locations from the binding pocket might also be capable of strengthening efflux by maximizing the energy conversion instead of directly modifying the drug binding (Ababou et al., 2016; Blair et al., 2015a; Eicher et al., 2014).

14.4.3.2 Efflux pumps in gram-positive bacteria

Gram-positive bacteria lack an outer membrane, but they depend on an efflux pump to reduce intracellular antibiotic concentrations (Aborah et al., 2025). A well-established example is the NorA efflux pump of S. aureus, which pumps out fluoroquinolones and a range of other antimicrobial drugs (Deng et al., 2012). NorA is a membrane protein that belongs to the family of major facilitator superfamily (Aborah et al., 2025). It has 12 transmembrane helices that form a central cavity to actively transport drugs across the membrane (Aborah et al., 2025). Besides NorA, there are some other efflux systems that also contribute to multidrug resistance in Gram-positive bacteria, such as NorB, NorC,

MepA, and LmrS (Aborah et al., 2025). Structural studies have provided important information about the working mechanism of these MFS transporters. For instance, X-ray crystallography and homology modeling studies of NorA showed that these transporters operate through an alternating access or rocker-switch mechanism in which the protein switches between inward- and outward-facing conformations, allowing the substrate to bind at the intracellular side and be expelled to the extracellular environment (Deng et al., 2012). NMR studies further revealed the role of flexible loops that support these conformational changes (Lerner et al., 2018). In addition to this, resistance-associated mutations can stabilize this conformational change, thereby enhancing the efficacy of the pump to export the drug (Deng et al., 2012).

14.4.4 Reduced membrane permeability in antibiotic resistance

Reduced membrane permeability is a significant mechanism of antibiotic resistance whereby bacteria reduce the uptake of antibiotics by the cell. Bacteria, rather than altering the drug or pumping it out, complicate the process of the antibiotic getting inside the bacterium in the first place. Structural biology has played a critical role in the understanding of how simple architectural modifications in bacterial membranes and porin channels have severe impacts on drug uptake and still allow cell survival (Prajapati et al., 2021).

14.4.4.1 Loss or reduced porin expression

The outer membrane of Gram-negative bacteria is a strong barrier to antibiotics (Zhou et al., 2023). The hydrophilic antibiotics normally gain access to the cell through porins that are protein channels embedded in the outer membrane (Zhou et al., 2023). Structural studies using X-ray crystallography determined that the porins are composed of a β-barrel-shaped structure, with water-filled channels that permit the passive diffusion of small molecules (Koebnik et al., 2000). These channels are very selective in size, charge, and shape (Koebnik et al., 2000). Even slight modifications in the porin architecture can cause great modifications in permeability (Prajapati et al., 2021). Downregulation or total loss of porin proteins is one of the usual strategies of resistance that reduces the number of entry pathways of antibiotics into bacterial cells (Ghai and Ghai, 2018). As a result, the intracellular concentration of the antibiotic/drug is reduced inside the cell to a sublethal level, limiting the effectiveness of the drug (Mammeri et al., 2025). Structural and functional studies have further validated that bacteria can tolerate loss of porins, as alternative nutrient uptake systems partially compensate for function (Mammeri et al., 2025).

14.4.4.2 Narrowing of porin channels through loop mutations

Mutations in loop regions of proteins are another common process utilized by bacteria to reduce membrane permeability (Ghai and Ghai, 2018). These loop regions span throughout the channel interior point mutations in these regions can alter the structural geometry of the pore (Ghai and Ghai, 2018). Structure studies of mutated porins using the X-ray diffraction method revealed that these loop changes the diameter of the channels (Vasan et al., 2022). Even a small fraction of angstroms can be sufficient to prevent the passage of large molecules, such as β-lactam antibiotics and fluoroquinolones, while still permitting the passage of small molecules that are necessary (Vasan et al., 2022). This allows bacteria to control membrane permeability to an optimal extent without compromising cellular viability (Vasan et al., 2022). Additionally, it was found that resistance may occur even through the alteration of the electrostatic charge in porin channels (Zhou et al., 2023). For example, mutations that add a positively charged residue within the pore can repel positively charged antibiotics, thereby reducing their uptake (Zhou et al., 2023). These charge shifts were revealed by high-resolution X-ray structures and electrostatic mapping as they lead to an adverse environment for the diffusion of the antibiotics, and the uptake efficiency decreases without physical obstructions in the pore (Zhou et al., 2023). Although X-ray crystallography gives more detailed structural information about porins, cryo-EM has enabled us to visualize the porins in a near-native membrane environment (Mammeri et al., 2025). These studies reveal the mechanism of interaction between porins with the surrounding lipopolysaccharides, and the membrane organization further influences the entry of drugs (Arunmanee et al., 2016). MDS, combined with experimental structure, explain the dynamic nature of porins (Vasan et al., 2022). These experiments demonstrate temporary constrictions, hydration alterations, and specific interaction hotspots that can retard or arrest antibiotic diffusion in resistant strains (Arunmanee et al., 2016; Phale et al., 1997). These dynamic features help us in understanding how antibiotics of similar size can exhibit markedly different permeability in resistant bacterial strains (Tran et al., 2013).

14.4.5 Metabolic bypass mechanism

Bacteria do not always resist antibiotics by directly modifying the drug or its target, but they evade the inhibitory effect by bypassing the affected pathways (Elbediwi and Rolff, 2024). One such ingenious survival technique is metabolic bypass, in which bacteria divert their biochemical pathway silently around the step that is blocked by the antibiotics (Passalacqua et al., 2016). By doing so, the bacteria use an alternative enzyme that performs the same function but is not affected by the antibiotics (Passalacqua et al., 2016). As a result, bacteria continue to survive in the presence of antibiotics (Passalacqua et al., 2016).

14.4.5.1 Resistance to trimethoprim

Trimethoprim is an antimicrobial agent that kills bacteria by targeting the activity of the enzyme dihydrofolate reductase (DHFR) (Brolund et al., 2010). DHFR catalyzes the conversion of dihydrofolate into tetrahydrofolate, a key cofactor required for the synthesis of nucleotides and other important amino acids (Brolund et al., 2010). Inhibition of DHFR disrupts this folate metabolism and blocks the DNA synthesis mechanism (Brolund et al., 2010). Resistance against trimethoprim arises through the mutation in DHFR or alternative DHFR variants, which continue to catalyzes folate metabolism and reduce the affinity of the drug effectively (Brolund et al., 2010). Structural studies, including X-ray crystallography of sensitive and resistant DHFR, showed that resistance is due to point mutations that cause only minimal rearrangement of the drug-binding pocket (Krucinska et al., 2022; Cammarata et al., 2017). These mutations disrupt the hydrogen-bonding pattern and change the geometry of the pockets such that trimethoprim is not able to bind, but the enzyme binds its natural substrate (Krucinska et al., 2022; Cammarata et al., 2017; Then, 1982). This allows the folate metabolism to continue in resistant bacteria even in the presence of antibiotics (Krucinska et al., 2022; Cammarata et al., 2017; Then, 1982).

14.4.5.2 Sulfonamide resistance

Sulfonamides act as structural analogs of para-aminobenzoic acid (PABA) and inhibit the activity of the dihydropteroate synthase (DHPS), an enzyme that is necessary for folate biosynthesis (Sköld, 2000; Yun et al., 2012). Sulfonamide resistance is primarily achieved through mutations or alternate DHPS enzymes encoded by genes like sul1, sul2, or sul3, which exhibit low affinity for the drug (Yun et al., 2012; Venkatesan et al., 2023). In addition, some bacteria increase the production of PABA or reduce the intracellular drug concentration by diminished permeability or active efflux (Yun et al., 2012). These modifications enable the folate biosynthesis process to proceed even in the presence of sulfonamide antibiotics, thereby maintaining bacterial survival (Sköld, 2000; Yun et al., 2012; Venkatesan et al., 2023).

14.4.6 Target replacement

Target replacement is another resistance mechanism in which bacteria replace the original antibiotic target with an alternative protein that performs the same function but is not recognized by the drug (Prestinaci et al., 2015; Darby et al., 2023). A well-known example is MRSA (Turner et al., 2019). β-Lactam antibiotics normally kill bacteria by inhibiting PBPs involved in cell-wall synthesis (Fisher et al., 2005). However, MRSA acquires a modified enzyme, PBP2a, which substitutes for the original PBPs and carries out cell-wall synthesis without being affected by β-lactam antibiotics (Turner et al., 2019). Structural studies using X-ray crystallography have shown that the active site of PBP2a is more restricted, which limits the access of β-lactam antibiotics to the catalytic center while still allowing the natural substrate to bind efficiently (Turner et al., 2019;

Fisher et al., 2005). This structural modification enables continued cell wall synthesis, allowing the bacterium to survive in the presence of β-lactam antibiotics (Fisher et al., 2005).

14.5 Advanced structural and computational approaches for understanding antimicrobial resistance

Structural biology, together with computational modeling, helps us to "see" the resistance at the molecular and atomic levels (Stokes et al., 2020; Jumper et al., 2021). This integrated approach helps the scientists to transform drug discovery from trial and error into more precise mechanism-derived strategies (Ferreira et al., 2015). Structural biology offers a blueprint of the mechanisms that bacteria use to modify drug targets, inactivate antibiotics, alter permeability, or enhance drug efflux (Reeve et al., 2015). High-resolution techniques such as X-ray crystallography, cryo-EM, and NMR spectroscopy provide detailed information about the structure and behavior of antibiotic targets (Kühlbrandt, 2014; Benjin and Ling, 2020; Lerner et al., 2018). These approaches can explain that even minute structural changes, including point mutations, conformational changes, or altered interactions, can reduce drug binding and promote resistance. In addition to experimental methods, computational approaches have further expanded our knowledge to understand AMR at the molecular level (Ferreira et al., 2015). MDS provide an understanding of the dynamic behavior of proteins, revealing transient conformational states, flexible regions, and allosteric sites that cannot be understood through static structures alone. These approaches help us to understand certain mutations that lead to structural destabilization or antibiotic resistance. With the help of these studies, scientists can measure the level of protein flexibility or overall structural stability due to mutation. A combination of structural biology, along with artificial intelligence (AI), has opened new insights to analyzing resistance mechanisms (Zhan et al., 2025). AI-based tools can predict protein structures, map resistance-associated mutations, and simulate drug-target interactions at high-throughput levels (Jumper et al., 2021; Zhan et al., 2025). In recent years, AI-assisted virtual screening has been used to efficiently screen large chemical libraries in a short time, focusing on the compounds that are most likely to defeat resistance (Roy et al., 2025). In addition to this, machine learning algorithms are being used for predicting novel resistance mechanisms, enabling resistance-resilient drugs to be designed in advance prior to clinical failure (Sakagianni et al., 2023; Kim et al., 2022). Structural biology also plays an important role in the development of new antibiotics against bacterial resistance (Reeve et al., 2015). It helps the researchers to understand how drugs interact with their target at the molecular level and can optimize properties like binding strength, specificity, and stability. Table 14.2 lists the structural and computational strategies in understanding and combating antibiotic resistance.

Table 14.2 Structural and computational strategies in understanding and combating antibiotic resistance.

Outcome	Description	Tools/methods	Impact
Mechanistic understanding	Explains how mutations or modifications disrupt drug binding	X-ray crystallography, nuclear magnetic resonance (NMR), cryo-electron microscopy (cryo-EM)	Translates resistance from descriptive to mechanistic (Dadgostar, 2019; Prestinaci et al., 2015).
Prediction of resistance evolution	Identifies mutation-tolerant versus intolerant regions	Structural modeling, artificial intelligence (AI) prediction	Predicts likely resistance pathways (Zhan et al., 2025).
Structure-guided drug design	Remodeling antibiotics for modified binding pockets	X-ray, cryo-EM, molecular dynamics (MD) simulations, docking	Accelerates next-generation drug discovery (De Vivo et al., 2016; Lin et al., 2018; Arenz et al., 2015).
Resistance-blocking strategies	Targets resistance enzymes and efflux pumps	X-ray, cryo-EM, inhibitor binding studies	Restores potency of existing antibiotics (Dadgostar, 2019; Prestinaci et al., 2015).
On-target specificity	Confirms precise drug-binding sites	Crystallography, docking, NMR	Reduces off-target effects and toxicity.
Understanding dynamics	Reveals flexibility and reparatory mechanisms	NMR, MD simulations	Explains how bacteria maintain function while resisting drugs.
AI and computational integration	Predicts structure, binding, and mutation effects	AlphaFold, ML algorithms, docking	Accelerates screening and rational compound design (Zhan et al., 2025; Roy et al., 2025).
Clinical application	Guides therapy selection, dosing, and combinations	Integrated structural and computational insights	Enhances treatment outcomes and resistance management (Browne et al., 2021; Murray et al., 2022).
Accelerated discovery	Integrates all structural and AI approaches	Crystallography, cryo-EM, NMR, MD, AI	Predictive, mechanism-driven drug development (Zhan et al., 2025; Roy et al., 2025).

14.5.1 Insights into aminoglycoside resistance through nuclear magnetic resonance dynamics

A group of aminoglycoside antibiotics, including streptomycin, gentamicin, and kanamycin, inhibit bacterial growth by binding to the 30S ribosomal subunit, particularly the 16S ribosomal subunit (Carter et al., 2000; Fourmy et al., 1996). They interrupt accurate translation and cause misreading of mRNA (Fourmy et al., 1996). In some resistance strains, mutations in genes encoding ribosomal components do not produce a substantial impact on the static structure of the ribosome (Carter et al., 2000). Instead, these mutations affect the dynamic behavior of the ribosomal binding site (Carter et al., 2000). NMR spectroscopy studies have shown that resistant ribosomal regions become more flexible, resulting in a shortened period of attachment between aminoglycosides and the ribosome (Carter et al., 2000). As a result, antibiotics dissociate more rapidly from the ribosome, decreasing their inhibitory effect (Carter et al., 2000). These findings highlight the key role of structural approaches such as NMR in uncovering the resistance mechanism that cannot be understood through static structural methods alone (Carter et al., 2000).

14.5.2 Prediction of resistance mutation using AlphaFold

AlphaFold is an AI-based program that is used to predict the probable structure of a protein with high accuracy, in cases where the macromolecular structures are not known experimentally (Jumper et al., 2021, 2021). Through the modeling of resistant and nonresistant variants of proteins, researchers can identify positions where mutations are structurally tolerated without significantly affecting protein functions (Varadi et al., 2022). For example, mutations in proteins targeted by antibiotics such as RNA polymerase, ribosomal proteins, or DNA gyrase can be modeled using AlphaFold to visualize alterations in the active site or ligand-binding region (Akdel et al., 2022). This approach helps predict potential resistance pathways and enhances our understanding of mutation-driven adaptation (Jumper et al., 2021; Varadi et al., 2022; Akdel et al., 2022).

14.5.3 Structure-guided strategies for developing resistance-resilient antibiotics

Structural biology has played a very significant role in the development of new antibiotics that are more effective against resistant bacteria (Reeve et al., 2015). It provides a framework for designing molecules that can overcome common resistance mechanisms. Structural insights reveal how a drug interacts with its target at the molecular level and make us understand why certain antibiotics fail and how they can be improved (Reeve et al., 2015). With the help of this information, it is possible to design molecules that can effectively bind to their target bacteria that are even slightly altered. Structural biology helped in designing this molecule to bypass common resistance mechanisms such as efflux and evading ribosomal protection (Gaurav et al., 2023; Lin et al., 2018). Structural biology has also given us information about how antibiotics are expelled by the efflux pumps or degraded by the resistance enzymes (Gaurav et al., 2023; Varela et al., 2021). These important insights help scientists to design drugs that can avoid or overcome these hurdles (Reeve et al., 2015). A well-known example is tigecycline, a broad-spectrum antibiotic that was developed through structural modification of earlier tetracyclines (Lin et al., 2018). In addition to this, computational tools and AI in combination now help in predicting mutations and how they might affect drug binding, making it easier to develop and design more effective compounds in advance (Zhan et al., 2025; Roy et al., 2025). Overall, these approaches make drug development more focused and efficient and improve the chances of developing antibiotics that remain effective even in resistant bacteria (Sakagianni et al., 2023; Kim et al., 2022).

14.6 Future perspectives

Future strategies to combat AMR will rely on an integrated approach that combines structural biology with AI, genomics, and systems biology (Sakagianni et al., 2023; Kim et al., 2022; Das et al., 2021). Such interdisciplinary effort will allow a more in-depth understanding of bacterial adaptation and help in the development of new, more effective therapeutic interventions (Sakagianni et al., 2023). The combination therapies are likely to play an important role in simultaneously addressing multiple bacterial pathways (Sartori et al., 2026). For instance, efflux pump inhibitors may be combined with traditional antibiotics, and the mechanisms of metabolic bypass may be predicted with the help of structural predictions (Sartori et al., 2026). In addition, the combination of high-throughput structural techniques and AI-directed models will result in the acceleration of the discovery of new antimicrobial compounds (Zhan et al., 2025; Das et al., 2021). Ultimately, structural biology is transforming the nature of drug discovery from more of an empirical effort to a more predictive, rational, and precision-directed activity (Reeve et al., 2015). Since there is knowledge about bacterial resistance, the level of atoms and the use of computational improvement, scientists can create the next-generation of antibiotics, which are not only effective but will offer a sustainable solution to the worldwide menace of AMR (Sakagianni et al., 2023; Kim et al., 2022; Das et al., 2021).

14.7 Conclusion

AMR remains a major global challenge and has been declared by the WHO as one of the top 10 threats to public health. Bacteria develop resistance to effective treatment of infectious disease through multiple molecular mechanisms (Blair et al., 2015b; Darby et al., 2023). These include mutation of drug targets, replacement of targets, metabolic bypass pathways, and reduced membrane permeability involving antibiotic targets, resistance enzymes, efflux pumps, and membrane proteins (Blair et al., 2015b; Darby et al., 2023; Zhou et al., 2023). Structural studies have shown that even small changes in these molecules can weaken drug binding while allowing normal cellular functions to continue (Campbell et al., 2001; Molodtsov et al., 2017; Kumar et al., 2015; Svetlov et al., 2021; Fourmy et al., 1996; Leslie et al., 1988; Koebnik et al., 2000). Structural techniques such as X-ray crystallography, cryo-EM, NMR spectroscopy, and molecular simulations have provided detailed views of protein structures and their interactions with antibiotics (Kühlbrandt, 2014; Benjin and Ling, 2020; Lerner et al., 2018). These approaches have helped researchers identify resistance mechanisms, conserved structural regions, and bacterial adaptation strategies (Reeve et al., 2015). Structural knowledge also supports the design of new antibiotics, resistance inhibitors, and effective combination therapies (Reeve et al., 2015). Understanding the structural basis of resistance is therefore essential for developing improved treatments against the growing problem of AMR (Blair et al., 2015b).

AI disclosure

During the preparation of this work, the author(s) used Wordtune in order to present this manuscript in a formal scientific tone. After using this tool/service, the author(s) reviewed and edited the content as needed and take(s) full responsibility for the content of the publication.

References

Ababou, A., Koronakis, V., Cascales, E., 2016. Structures of gate loop variants of the AcrB drug efflux pump bound by erythromycin substrate. PLoS One 11 (7), e0159154. Available at https://doi.org/10.1371/journal.pone.0159154.

Aborah, E., Ayitah, M., Boafo, K.F., Ortiz-Alegria, A., Lekshmi, M., Dhanush, C.K., et al., 2025. Multidrug resistance and major facilitator superfamily antimicrobial efflux pumps of the ESKAPEE pathogen *Staphylococcus aureus*. Exploration of Drug Science. Available at https://doi.org/10.37349/eds.2025.100897.

Akdel, M., Pires, D.E.V., Pardo, E.P., Jänes, J., Zalevsky, A.O., Mészáros, B., et al., 2022. A structural biology community assessment of AlphaFold2 applications. Nature Structural and Molecular Biology 29 (11), 1056–1067. Available at https://doi.org/10.1038/s41594-022-00849-w, https://www.nature.com/nsmb/.

Aldred, K.J., Breland, E.J., Vlčková, V., Strub, M.P., Neuman, K.C., Kerns, R.J., et al., 2014a. Role of the water-metal ion bridge in mediating interactions between quinolones and *Escherichia coli* topoisomerase IV. Biochemistry 53 (34), 5558–5567. Available at https://doi.org/10.1021/bi500682e, http://pubs.acs.org/journal/bichaw.

Aldred, K.J., Kerns, R.J., Osheroff, N., 2014b. Mechanism of quinolone action and resistance. Biochemistry 53 (10), 1565–1574. Available at https://doi.org/10.1021/bi5000564, http://pubs.acs.org/journal/bichaw.

Alvarez-Elcoro, S., Enzler, M.J., 1999. The macrolides: erythromycin, clarithromycin, and azithromycin. Mayo Clinic Proceedings 74 (6), 613–634. Available at https://doi.org/10.4065/74.6.613, http://www.journals.elsevier.com/mayo-clinic-proceedings.

Arenz, S., Nguyen, F., Beckmann, R., Wilson, D.N., 2015. Cryo-EM structure of the tetracycline resistance protein TetM in complex with a translating ribosome at 3.9-Å resolution. Proceedings of the National Academy of Sciences of the United States of America 112 (17), 5401–5406. Available at https://doi.org/10.1073/pnas.1501775112, http://www.pnas.org/content/112/17/5401.full.pdf.

Arunmanee, W., Pathania, M., Solovyova, A.S., Le Brun, A.P., Ridley, H., Baslé, A., et al., 2016. Gram-negative trimeric porins have specific LPS binding sites that are essential for porin biogenesis. Proceedings of the National Academy of Sciences of the United States of America 113 (34), E5034. Available at https://doi.org/10.1073/pnas.1602382113, http://www.pnas.org/content/113/34/E5034.full.pdf.

Belay, W.Y., Getachew, M., Tegegne, B.A., Teffera, Z.H., Dagne, A., Zeleke, T.K., et al., 2024. Mechanism of antibacterial resistance, strategies and next-generation antimicrobials to contain antimicrobial resistance: a review. Frontiers in Pharmacology 15. Available at https://doi.org/10.3389/fphar.2024.1444781, http://www.frontiersin.org/Pharmacology.

Benjin, X., Ling, L., 2020. Developments, applications, and prospects of cryo-electron microscopy. Protein Science 29 (4), 872–882. Available at https://doi.org/10.1002/pro.3805.

Blair, J.M.A., Bavro, V.N., Ricci, V., Modi, N., Cacciotto, P., Kleinekathöfer, U., et al., 2015a. AcrB drug-binding pocket substitution confers clinically relevant resistance and altered substrate specificity. Proceedings of the National Academy of Sciences 112 (11), 3511–3516. Available at https://doi.org/10.1073/pnas.1419939112.

Blair, J.M.A., Webber, M.A., Baylay, A.J., Ogbolu, D.O., Piddock, L.J.V., 2015b. Molecular mechanisms of antibiotic resistance. Nature Reviews. Microbiology 13 (1), 42–51. Available at https://doi.org/10.1038/nrmicro3380, http://www.nature.com/nrmicro/index.html.

Brolund, A., Sundqvist, M., Kahlmeter, G., Grape, M., Mylonakis, E., 2010. Molecular characterisation of trimethoprim resistance in *Escherichia coli* and *Klebsiella pneumoniae* during a two year intervention on trimethoprim use. PLoS One 5 (2), e9233. Available at https://doi.org/10.1371/journal.pone.0009233.

Browne, A.J., Chipeta, M.G., Haines-Woodhouse, G., Kumaran, E.P.A., Hamadani, B.H.K., Zaraa, S., et al., 2021. Global antibiotic consumption and usage in humans, 2000–18: a spatial modelling study. The Lancet Planetary Health 5 (12), e893. Available at https://doi.org/10.1016/S2542-5196(21)00280-1, https://www.journals.elsevier.com/the-lancet-planetary-health.

Cammarata, M., Thyer, R., Lombardo, M., Anderson, A., Wright, D., Ellington, A., et al., 2017. Characterization of trimethoprim resistant *E. coli* dihydrofolate reductase mutants by mass spectrometry and inhibition by propargyl-linked antifolates. Chemical Science 8 (5), 4062–4072. Available at https://doi.org/10.1039/c6sc05235e.

Campbell, E.A., Korzheva, N., Mustaev, A., Murakami, K., Nair, S., Goldfarb, A., et al., 2001. Structural mechanism for rifampicin inhibition of bacterial RNA polymerase. Cell 104 (6), 901–912. Available at https://doi.org/10.1016/S0092-8674(01)00286-0, https://www.sciencedirect.com/journal/cell.

Carter, A.P., Clemons, W.M., Brodersen, D.E., Morgan-Warren, R.J., Wimberly, B.T., Ramakrishnan, V., 2000. Functional insights from the structure of the 30S ribosomal subunit and its interactions with antibiotics. Nature 407 (6802), 340–348. Available at https://doi.org/10.1038/35030019.

Dadgostar, P., 2019. <p>Antimicrobial resistance: implications and costs</p> Infection and Drug Resistance Volume 12, 3903–3910. Available at https://doi.org/10.2147/idr.s234610.

Darby, E.M., Trampari, E., Siasat, P., Gaya, M.S., Alav, I., Webber, M.A., et al., 2023. Molecular mechanisms of antibiotic resistance revisited. Nature Reviews Microbiology 21 (5), 280–295. Available at https://doi.org/10.1038/s41579-022-00820-y, https://www.nature.com/nrmicro/.

Das, P., Sercu, T., Wadhawan, K., Padhi, I., Gehrmann, S., Cipcigan, F., et al., 2021. Accelerated antimicrobial discovery via deep generative models and molecular dynamics simulations. Nature Biomedical Engineering 5 (6), 613–623. Available at https://doi.org/10.1038/s41551-021-00689-x.

Deng, X., Sun, F., Ji, Q., Liang, H., Missiakas, D., Lan, L., et al., 2012. Expression of multidrug resistance efflux pump gene nora is iron responsive in *Staphylococcus aureus*. Journal of Bacteriology 194 (7), 1753–1762. Available at https://doi.org/10.1128/jb.06582-11.

De Vivo, M, asetti M, M, Bottegoni, G, Cavalli, A, 2016. Role of Molecular Dynamics and Related Methods in Drug Discovery. J Med Chem 59 (9), 4035–4061. Available at https://doi.org/10.1021/acs.jmedchem.5b01684.

Dheda, K., Gumbo, T., Maartens, G., Dooley, K.E., McNerney, R., Murray, M., et al., 2017. The epidemiology, pathogenesis, transmission, diagnosis, and management of multidrug-resistant, extensively drug-resistant, and incurable tuberculosis. The Lancet Respiratory Medicine 5 (4), 291–360. Available at https://doi.org/10.1016/S2213-2600(17)30079-6, http://www.elsevier.com/journals/the-lancet-respiratory-medicine/2213-2600.

Draper, L.A., Cotter, P.D., Hill, C., Ross, R.P., 2015. Lantibiotic resistance. Microbiology and Molecular Biology Reviews 79 (2), 171–191. Available at https://doi.org/10.1128/MMBR.00051-14, http://mmbr.asm.org/content/79/2/171.full.pdf+html.

Du, D., Wang, Z., James, N.R., Voss, J.E., Klimont, E., Ohene-Agyei, T., et al., 2014. Structure of the AcrAB-TolC multidrug efflux pump. Nature 509 (7501), 512–515. Available at https://doi.org/10.1038/nature13205, http://www.nature.com/nature/index.html.

Eicher, T., Seeger, M.A., Anselmi, C., Zhou, W., Brandstätter, L., Verrey, F., et al., 2014. Coupling of remote alternating-access transport mechanisms for protons and substrates in the multidrug efflux pump AcrB. eLife 3. Available at https://doi.org/10.7554/eLife.03145.

Elbediwi, M., Rolff, J., 2024. Metabolic pathways and antimicrobial peptide resistance in bacteria. Journal of Antimicrobial Chemotherapy 79 (7), 1473–1483. Available at https://doi.org/10.1093/jac/dkae128.

Ferreira, L., Dos Santos, R., Oliva, G., Andricopulo, A., 2015. Molecular docking and structure-based drug design strategies. Molecules (Basel, Switzerland) 20 (7), 13384–13421. Available at https://doi.org/10.3390/molecules200713384.

Fischbach, M.A., Walsh, C.T., 2009. Antibiotics for emerging pathogens. Science (New York, N.Y.) 325 (5944), 1089–1093. Available at https://doi.org/10.1126/science.1176667.

Fisher, J.F., Meroueh, S.O., Mobashery, S., 2005. Bacterial resistance to β-lactam antibiotics: compelling opportunism, compelling opportunity. Chemical Reviews 105 (2), 395–424. Available at https://doi.org/10.1021/cr030102i.

Fourmy, D., Recht, M.I., Blanchard, S.C., Puglisi, J.D., 1996. Structure of the A site of *Escherichia coli* 16S ribosomal RNA complexed with an aminoglycoside antibiotic. Science (New York, N.Y.) 274 (5291), 1367–1371. Available at https://doi.org/10.1126/science.274.5291.1367.

Gaurav, A., Bakht, P., Saini, M., Pandey, S., Pathania, R., 2023. Role of bacterial efflux pumps in antibiotic resistance, virulence, and strategies to discover novel efflux pump inhibitors. Microbiology (Reading, England) 169 (5). Available at https://doi.org/10.1099/mic.0.001333.

Ghai, I., Ghai, S., 2018. Understanding antibiotic resistance via outer membrane permeability. Infection and Drug Resistance 11, 523–530. Available at https://doi.org/10.2147/IDR.S156995, https://www.dovepress.com/getfile.php?fileID=41466.

Jumper, J., Evans, R., Pritzel, A., Green, T., Figurnov, M., Ronneberger, O., et al., 2021. Highly accurate protein structure prediction with AlphaFold. Nature 596 (7873), 583–589. Available at https://doi.org/10.1038/s41586-021-03819-2.

Kim, J.I., Maguire, F., Tsang, K.K., Gouliouris, T., Peacock, S.J., McAllister, T.A., et al., 2022. Machine learning for antimicrobial resistance prediction: current practice, limitations, and clinical perspective. Clinical Microbiology Reviews 35 (3). Available at https://doi.org/10.1128/cmr.00179-21, https://journals.asm.org/doi/10.1128/cmr.00179-21.

Klein, E.Y., Van Boeckel, T.P., Martinez, E.M., Pant, S., Gandra, S., Levin, S.A., et al., 2018. Global increase and geographic convergence in antibiotic consumption between 2000 and 2015. Proceedings of the National Academy of Sciences of the United States of America 115 (15), E3463. Available at https://doi.org/10.1073/pnas.1717295115, http://www.pnas.org/content/pnas/115/15/E3463.full.pdf.

Koch, A.L., 2000. Penicillin binding proteins, β-lactams, and lactamases: offensives, attacks, and defensive countermeasures. Critical Reviews in Microbiology 26 (4), 205–220. Available at https://doi.org/10.1080/10408410091154228.

Koebnik, R., Locher, K.P., Van Gelder, P., 2000. Structure and function of bacterial outer membrane proteins: Barrels in a nutshell. Molecular Microbiology 37 (2), 239–253. Available at https://doi.org/10.1046/j.1365-2958.2000.01983.x.

Krucinska, J., Lombardo, M.N., Erlandsen, H., Estrada, A., Si, D., Viswanathan, K., et al., 2022. Structure-guided functional studies of plasmid-encoded dihydrofolate reductases reveal a common mechanism of trimethoprim resistance in Gram-negative pathogens. Communications Biology 5 (1). Available at https://doi.org/10.1038/s42003-022-03384-y.

Kumar, M., Dahiya, S., Sharma, P., Sharma, S., Singh, T.P., Kapil, A., et al., 2015. Structure based in silico analysis of quinolone resistance in clinical isolates of *Salmonella* Typhi from India. PLoS One 10 (5), e0126560. Available at https://doi.org/10.1371/journal.pone.0126560.

Kühlbrandt, W., 2014. The resolution revolution. Science (New York, N.Y.) 343 (6178), 1443–1444. Available at https://doi.org/10.1126/science.1251652.

Larsson, D.G.J., Flach, C.F., 2022. Antibiotic resistance in the environment. Nature Reviews Microbiology 20 (5), 257–269. Available at https://doi.org/10.1038/s41579-021-00649-x, http://www.nature.com/nrmicro/index.html.

Laxminarayan, R., Matsoso, P., Pant, S., Brower, C., Røttingen, J.A., Klugman, K., et al., 2016. Access to effective antimicrobials: a worldwide challenge. The Lancet 387 (10014), 168–175. Available at https://doi.org/10.1016/S0140-6736(15)00474-2, http://www.journals.elsevier.com/the-lancet/.

Lerner, E., Cordes, T., Ingargiola, A., Alhadid, Y., Chung, S.Y., Michalet, X., et al., 2018. Toward dynamic structural biology: two decades of single-molecule Förster resonance energy transfer. Science (New York, N.Y.) 359 (6373). Available at https://doi.org/10.1126/science.aan1133.

Leslie, A.G., Moody, P.C., Shaw, W.V., 1988. Structure of chloramphenicol acetyltransferase at 1.75-A resolution. Proceedings of the National Academy of Sciences 85 (12), 4133–4137. Available at https://doi.org/10.1073/pnas.85.12.4133.

Li, W., Atkinson, G.C., Thakor, N.S., Allas, Ü., Lu, C.-chao, Chan, K.-Y., et al., 2013. Mechanism of tetracycline resistance by ribosomal protection protein Tet(O). Nature Communications 4 (1). Available at https://doi.org/10.1038/ncomms2470.

Lin, J., Zhou, D., Steitz, T.A., Polikanov, Y.S., Gagnon, M.G., 2018. Ribosome-targeting antibiotics: modes of action, mechanisms of resistance, and implications for drug design. Annual Review of Biochemistry 87, 451–478. Available at https://doi.org/10.1146/annurev-biochem-062917-011942, http://www.annualreviews.org/journal/biochem.

Mammeri, H., Sereme, Y., Toumi, E., Faury, H., Skurnik, D., 2025. Daria Van Tyne, Interplay between porin deficiency, fitness, and virulence in carbapenem-non-susceptible *Pseudomonas aeruginosa* and Enterobacteriaceae. PLoS Pathogens 21 (2), e1012902. Available at https://doi.org/10.1371/journal.ppat.1012902.

Matagne, Lamotte-Brasseur, Frère, J.M., 1998. Catalytic properties of class A beta-lactamases: efficiency and diversity. The Biochemical Journal 330 (Pt 2), 581–598. Available at https://doi.org/10.1042/bj3300581.

Miller, S.I., Salama, N.R., 2018. The gram-negative bacterial periplasm: size matters. PLoS Biology 16 (1), e2004935. Available at https://doi.org/10.1371/journal.pbio.2004935.

Molodtsov, V., Scharf, N.T., Stefan, M.A., Garcia, G.A., Murakami, K.S., 2017. Structural basis for rifamycin resistance of bacterial RNA polymerase by the three most clinically important RpoB mutations found in *Mycobacterium tuberculosis*. Molecular Microbiology 103 (6), 1034–1045. Available at https://doi.org/10.1111/mmi.13606, http://www.blackwellpublishing.com/journals/MMI.

Murray, C.J., Ikuta, K.S., Sharara, F., Swetschinski, L., Robles Aguilar, G., Gray, A., et al., 2022. Global burden of bacterial antimicrobial resistance in 2019: a systematic analysis. The Lancet 399 (10325), 629–655. Available at https://doi.org/10.1016/S0140-6736(21)02724-0, http://www.journals.elsevier.com/the-lancet/.

Naylor, N.R., Atun, R., Zhu, N., Kulasabanathan, K., Silva, S., Chatterjee, A., et al., 2018. Estimating the burden of antimicrobial resistance: a systematic literature review. Antimicrobial Resistance & Infection Control 7 (1). Available at https://doi.org/10.1186/s13756-018-0336-y.

Nordmann, P., Naas, T., Poirel, L., 2011. Global spread of carbapenemase-producing enterobacteriaceae. Emerging Infectious Diseases 17 (10), 1791–1798. Available at https://doi.org/10.3201/eid1710.110655.

Partridge, S.R., Kwong, S.M., Firth, N., Jensen, S.O., 2018. Mobile genetic elements associated with antimicrobial resistance. Clinical Microbiology Reviews 31 (4). Available at https://doi.org/10.1128/CMR.00088-17, http://cmr.asm.org/.

Passalacqua, K.D., Charbonneau, M.E., O'Riordan, M.X.D., 2016. Bacterial metabolism shapes the host-pathogen interface. Microbiology Spectrum 4 (3). Available at https://doi.org/10.1128/microbiolspec.VMBF-0027-2015, http://www.asmscience.org/docserver/fulltext/microbiolspec/4/3/VMBF-0027-2015.pdf?expires=1486113500&id=id&accname=45052&checksum=FB5E459AA97E6F093990CA5EC5753F5C.

Phale, P.S., Schirmer, T., Prilipov, A., Lou, K.L., Hardmeyer, A., Rosenbusch, J.P., 1997. Voltage gating of *Escherichia coli* porin channels: role of the constriction loop. Proceedings of the National Academy of Sciences of the United States of America 94 (13), 6741–6745. Available at https://doi.org/10.1073/pnas.94.13.6741.

Pokharel, S., Raut, S., Adhikari, B., 2019. Tackling antimicrobial resistance in low-income and middle-income countries. BMJ Global Health 4 (6), e002104. Available at https://doi.org/10.1136/bmjgh-2019-002104.

Prajapati, J.D., Kleinekathöfer, U., Winterhalter, M., 2021. How to enter a bacterium: bacterial porins and the permeation of antibiotics. Chemical Reviews 121 (9), 5158–5192. Available at https://doi.org/10.1021/acs.chemrev.0c01213, http://pubs.acs.org/journal/chreay.

Prestinaci, F., Pezzotti, P., Pantosti, A., 2015. Antimicrobial resistance: a global multifaceted phenomenon. Pathogens and Global Health 109 (7), 309–318. Available at https://doi.org/10.1179/2047773215y.0000000030.

Reeve, S.M., Lombardo, M.N., Anderson, A.C., 2015. Understanding the structural mechanisms of antibiotic resistance sets the platform for new discovery. Future Microbiology 10 (11), 1727–1733. Available at https://doi.org/10.2217/fmb.15.78, http://www.futuremedicine.com/loi/fmb.

Roy, S., Nagaraj, K., Mittal, A., Shah, F.C., Raja, K., 2025. Artificial intelligence in virtual screening: transforming drug research and discovery—a review. Journal of Bio-X Research 8. Available at https://doi.org/10.34133/jbioxresearch.0041, https://spj.science.org/toc/jbioxresearch/8.

Sakagianni, A., Koufopoulou, C., Feretzakis, G., Kalles, D., Verykios, V.S., Myrianthefs, P., et al., 2023. Using machine learning to predict antimicrobial resistance—a literature review. Antibiotics 12 (3). Available at https://doi.org/10.3390/antibiotics12030452, http://www.mdpi.com/journal/antibiotics.

Sartori, M., Toppo, S., Lavezzo, E., 2026. Molecular resistance mechanisms to newly approved antibiotics (2017–2025) in WHO priority pathogens. Frontiers in Microbiology 16. Available at https://doi.org/10.3389/fmicb.2025.1719798.

Sköld, O., 2000. Sulfonamide resistance: mechanisms and trends. Drug Resistance Updates 3 (3), 155–160. Available at https://doi.org/10.1054/drup.2000.0146.

Stokes, J.M., Yang, K., Swanson, K., Jin, W., Cubillos-Ruiz, A., Donghia, N.M., et al., 2020. A deep learning approach to antibiotic discovery. Cell 180 (4), 688. Available at https://doi.org/10.1016/j.cell.2020.01.021, https://www.sciencedirect.com/journal/cell.

Svetlov, M.S., Plessa, E., Chen, C.W., Bougas, A., Krokidis, M.G., Dinos, G.P., et al., 2019. High-resolution crystal structures of ribosome-bound chloramphenicol and erythromycin provide the ultimate basis for their competition. RNA (New York, N.Y.) 25 (5), 600–606. Available at https://doi.org/10.1261/rna.069260.118, https://rnajournal.cshlp.org/content/25/5/600.full.pdf+html.

Svetlov, M.S., Syroegin, E.A., Aleksandrova, E.V., Atkinson, G.C., Gregory, S.T., Mankin, A.S., et al., 2021. Structure of Erm-modified 70S ribosome reveals the mechanism of macrolide resistance. Nature Chemical Biology 17 (4), 412–420. Available at https://doi.org/10.1038/s41589-020-00715-0, http://www.nature.com/nchembio.

Then, R.L., 1982. Mechanisms of resistance to trimethoprim, the sulfonamides, and trimethoprim-sulfamethoxazole. Reviews of Infectious Diseases 4 (2), 261–269. Available at https://doi.org/10.1093/clinids/4.2.261.

Toone, E.J., Zalucki, Y., Su, C.-C., Long, F., Zhang, Q., Shafer, W.M., et al., 2011. Efflux pumps of the resistance–nodulation–division family: a perspective of their structure, function, and regulation in gram-negative bacteria. wiley. Advances in Enzymology and Related Areas of Molecular Biology 77, 109–146. Available at https://doi.org/10.1002/9780470920541.ch3.

Tran, Q.T., Williams, S., Farid, R., Erdemli, G., Pearlstein, R., 2013. The translocation kinetics of antibiotics through porin OmpC: insights from structure-based solvation mapping using WaterMap. Proteins: Structure, Function and Bioinformatics 81 (2), 291–299. Available at https://doi.org/10.1002/prot.24185.

Turner, N.A., Sharma-Kuinkel, B.K., Maskarinec, S.A., Eichenberger, E.M., Shah, P.P., Carugati, M., et al., 2019. Methicillin-resistant *Staphylococcus aureus*: an overview of basic and clinical research. Nature Reviews. Microbiology 17 (4), 203–218. Available at https://doi.org/10.1038/s41579-018-0147-4, http://www.nature.com/nrmicro/index.html.

Van Boeckel, T.P., Gandra, S., Ashok, A., Caudron, Q., Grenfell, B.T., Levin, S.A., et al., 2014. Global antibiotic consumption 2000 to 2010: an analysis of national pharmaceutical sales data. The Lancet Infectious Diseases 14 (8), 742–750. Available at https://doi.org/10.1016/S1473-3099(14)70780-7, http://www.journals.elsevier.com/the-lancet-infectious-diseases.

Varadi, M., Anyango, S., Deshpande, M., Nair, S., Natassia, C., Yordanova, G., et al., 2022. AlphaFold protein structure database: massively expanding the structural coverage of protein-sequence space with high-accuracy models. Nucleic Acids Research 50 (D1), D439. Available at https://doi.org/10.1093/nar/gkab1061.

Varela, M.F., Stephen, J., Lekshmi, M., Ojha, M., Wenzel, N., Sanford, L.M., et al., 2021. Bacterial resistance to antimicrobial agents. Antibiotics 10 (5). Available at https://doi.org/10.3390/antibiotics10050593, https://www.mdpi.com/2079-6382/10/5/593/pdf.

Vasan, A.K., Haloi, N., Ulrich, R.J., Metcalf, M.E., Wen, P.C., Metcalf, W.W., et al., 2022. Role of internal loop dynamics in antibiotic permeability of outer membrane porins. Proceedings of the National Academy of Sciences of the United States of America 119 (8). Available at https://doi.org/10.1073/pnas.2117009119, http://www.pnas.org/lookup/doi/10.1073/pnas.2117009119.

Venkatesan, M., Fruci, M., Verellen, L.A., Skarina, T., Mesa, N., Flick, R., et al., 2023. Molecular mechanism of plasmid-borne resistance to sulfonamide antibiotics. Nature Communications 14 (1). Available at https://doi.org/10.1038/s41467-023-39778-7, https://www.nature.com/ncomms/.

Wilson, D.N., 2014. Ribosome-targeting antibiotics and mechanisms of bacterial resistance. Nature Reviews. Microbiology 12 (1), 35–48. Available at https://doi.org/10.1038/nrmicro3155.

Yun, M.K., Wu, Y., Li, Z., Zhao, Y., Waddell, M.B., Ferreira, A.M., et al., 2012. Catalysis and sulfa drug resistance in dihydropteroate synthase. Science (New York, N.Y.) 335 (6072), 1110–1114. Available at https://doi.org/10.1126/science.1214641, http://www.sciencemag.org/content/335/6072/1110.full.pdf.

Zaw, M.T., Emran, N.A., Lin, Z., 2018. Mutations inside rifampicin-resistance determining region of rpoB gene associated with rifampicin-resistance in *Mycobacterium tuberculosis*. Journal of Infection and Public Health 11 (5), 605–610. Available at https://doi.org/10.1016/j.jiph.2018.04.005.

Zhan, Z.H., Hong, J., Li, J.Y., Wang, C., He, L., Xu, Z., et al., 2025. Artificial intelligence-based methods for protein structure prediction: a survey. Artificial Intelligence Review 58 (10). Available at https://doi.org/10.1007/s10462-025-11325-4, https://www.springer.com/journal/10462.

Zhou, G., Wang, Q., Wang, Y., Wen, X., Peng, H., Peng, R., et al., 2023. Outer membrane porins contribute to antimicrobial resistance in gram-negative bacteria. Microorganisms 11 (7), 1690. Available at https://doi.org/10.3390/microorganisms11071690.

CHAPTER 15

Structural biology approaches in neurodegeneration and aging

Minhal Abidi[1], Sana Iram[2], Jamal-E-Fatima[3], Hamda Khan[1], Shaheda Perween[1], Safia Habib[1], and Moinuddin[1]
[1]Department of Biochemistry, Jawaharlal Nehru Medical College, Aligarh Muslim University, Aligarh, Uttar Pradesh, India,
[2]Department of Medical Biotechnology, Yeungnam University, Gyeongsan, Gyeongsangbuk-do, South Korea,
[3]BMI-EP Lab, School of Electronic Science and Engineering, University of Electronic Science and Technology of Chine, Chengdu, Sichuan, P.R. China

15.1 Introduction: aging and neurodegeneration–a structural view

Biological aging is a multifactorial and progressive process that leads to the deterioration of molecular, physiological, and cellular integrity (Aunan et al., 2016; Passos and Kirkwood, 2015). On the molecular level, the process of aging is regulated by a set of hallmarks, including proteostasis disturbances, the accumulation of genomic damage, mitochondrial dysfunction, cellular senescence, autophagy dysfunction, and disturbances in cellular communication (Aunan et al., 2016; Terracina et al., 2023). The aforementioned hallmarks of cellular aging processes function in a complex feedback loop that compromises neuronal resilience (Guo et al., 2022). Neurons are particularly vulnerable to these disturbances due to their long lifespan, high metabolic demands, and limited regenerative capacity (Mattson and Magnus, 2006). Thus the most important risk factor for a wide range of neurodegenerative diseases, including Alzheimer's disease (AD), Parkinson's disease (PD), Huntington's disease (HD), and amyotrophic lateral sclerosis (ALS), is the process of aging (Hou et al., 2019). Even though the diseases have different clinical and pathological features, one can converge on the view that they share common molecular mechanisms, including protein misfolding, synaptic dysfunction, mitochondrial dysfunction, and chronic neuroinflammation (Fig. 15.1).

Deterioration of proteostasis that occurs with age is one of the earliest and most convergent physiological markers associated with neurodegeneration (Krisko and Radman, 2019). The detection of misfolded proteins, such as amyloid-β (Aβ), tau, and α-synuclein, previously described as hallmarks of neurodegenerative diseases, was first reported in early biochemical studies (Sweeney et al., 2017). Structural biology techniques, including cryo-electron microscopy and solid-state nuclear magnetic resonance (NMR) spectroscopy, have now revealed the atomic structures of these pathogenic protein aggregates, decades after their initial detection. These studies have revealed that different polymorphs of protein fibrils exhibit disease- or patient-specific

Advances in Structural Biology
DOI: https://doi.org/10.1016/B978-0-443-44988-8.00008-3

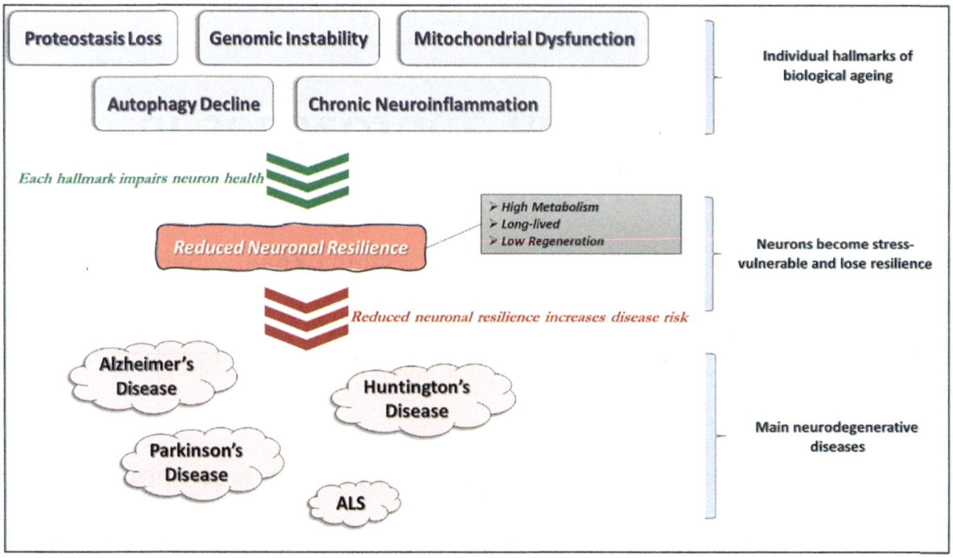

Figure 15.1 Biological aging hallmarks compromise neuronal resilience, increasing neurodegeneration.

phenotypes (Tycko, 2025; Creekmore et al., 2021). The results of these studies have challenged long-held assumptions that protein aggregates are merely disordered structures. Instead, protein aggregates are highly ordered structures composed of β-sheets with strain-like polymorphisms that can transmit their toxicity in a prion-like manner. The results of these studies now provide a link between age-related malfunction of proteostasis, which results from a loss of chaperone function, destruction of ubiquitin–proteasome function, and enhancement of autophagic flux, and a cellular environment that is permissive for the nucleation, maturation, and spread of toxic oligomeric species (Hipp et al., 2019; Frankowska et al., 2022).

Mitochondrial degradation, a major characteristic of aging, has also undergone a conceptual shift. Classical research has placed a lot of emphasis on age-related increases in ROS levels and mitochondrial DNA (mtDNA) mutations that may cause neurodegenerative diseases (Alexeyev, 2009). Although this research provided a backbone for the structural analysis, further levels of complexity have been revealed through the new structural exploration. The structure of respiratory chain complexes that are purified in the presence of cryogenic electron microscopy (cryo-EM) images of tissues collected from animals at old age shows that there are changes in conformation that interrupt the efficiency of electron transfer (Yaikhomba, 2022). In addition, molecular kinetograms of the PINK1 and Parkin proteins, which are implicated in PD have used to explain the mechanism of mitophagy, which explains the mechanism of age-related mitochondrial quality surveillance failure that leads to neuronal degeneration (Hou et al., 2018; Mishra and Thakur, 2023). These morphological details emphasize that mitochondrial dysfunction is not merely a side effect of a progressive oxidative damage process, but is a correct reflection of architectural and regulatory failure of the organelle (Khan et al., 2022).

Another hallmark closely related to neuronal aging is genomic instability. This is characterized by the occurrence of DNA double-strand breaks, the inability to repair them, and epigenetic drift (López-Gil et al., 2023). Previous cytological and biochemical reports had correlated genomic instability with neurodegenerative pathology. However, these reports lacked a mechanistic approach. Recent advances in structural biology enable visualization of protein complexes involved in DNA repair at near-atomic resolution. Indicatively, the structural descriptions of ATM, DNA-PK, and PARP1 have revealed how age-related conformational changes compromise their catalytic activity (Edogbanya, 2023). Moreover, structural studies at a chromatin level have revealed that aging is associated with disturbed nucleosome positioning and histone modifications that compromise transcription of synaptic and metabolic genes. From a comparison of the above structural implications and initial genetic findings, it is evident that genomic instability increases with age and destabilizes neuronal identity and function through complex chromatin restructuring (López-Gil et al., 2023).

It is the combination of these three features, proteostasis, mitochondrial, and genomic instability, that provides a favorable framework for the emergence of a disease, leading to its progression in a neurodegenerative manner (Höhn et al., 2020). While genetic susceptibility and target cell types vary among AD, PD, HD, and ALS, a striking degree of mechanistic overlap among these neurodegenerative diseases has been revealed by structural biology approaches (Toader et al., 2024). The amyloid fibrils in AD, α-synuclein fibrils in PD, mutant huntingtin in HD, and TDP-43 in ALS, although different in three-dimensional structures, follow a common mechanism for nucleation, elongation, and propagation (Candelise et al., 2021). While neuropathological studies carried out in the early days could not go beyond a superficial level, describing only the gross pathological features, today, with atomic-resolution images, the entire maps are available for researchers, who can correlate different structure variants with clinical heterogeneity and progression rates. For example, different conformers of tau filaments are linked with AD, corticobasal degeneration, or Pick's disease, contradicting earlier suppositions of a homogeneous tau pathology in various disorders (Goedert et al., 2018; Zhang et al., 2020).

The advent of structural biology, beginning with the original X-ray crystallography, continuing to the more modern techniques of cryo-EM, and now integrative modeling, has given a whole new perspective on the study of aging and neurodegeneration (Creekmore et al., 2021). Previously, the study of aging under the category of aging studied was carried out through the alterations observed at the cellular and biochemical levels. This provided an enormous amount of information at low resolution. However, the advent of atomic-resolution techniques has now enabled the visualization of the effects of aging at the molecular level. This includes the effects of the derailment of enzymes, receptors, and protein complexes over the decades to destabilize the neuron (Georgiev et al., 2004). This is especially true of the cryo-EM revolution, which has opened up structures of large assemblies of molecules, allowing the creation of mechanistic structures that were not accessible previously (Egelman, 2016).

Given the increasing burden of age-related neurodegenerative diseases worldwide, structural biology is playing a major role in understanding the molecular mechanisms of aging in neurons (Jiang et al., 2025). High-resolution structures are being employed to design drugs that can target specific fibril interfaces, correct misfolded proteins, stabilize mitochondrial complexes, and enhance the DNA repair machinery (Galkin et al., 2024). The integration of structural information with other approaches in biochemical, genetic, and cellular studies has provided this field with the opportunity to develop "revolutionary therapeutic approaches."

The current chapter follows a structural approach, dividing the molecular processes that lead to aging and ultimately to neurodegeneration. Neurodegeneration is discussed based on the background and the latest findings to offer a comprehensive and in-depth view of the topic.

15.2 Molecular pathways of protein misfolding and aggregation during aging of the brain

Protein aggregation and misfolding are significant pathological processes associated with age-related neurodegenerative diseases (Johnson et al., 2025). Despite the brain's use of the most efficient quality control mechanisms to maintain the integrity of the proteome, the aging brain undergoes continuous loss of protein homeostasis (proteostasis) and thus becomes more susceptible to the formation of toxic protein species (Koga et al., 2011). Over the last three decades, there have been dramatic advances in the conceptual understanding of protein aggregation. From the biochemical observations of insoluble plaques to the atomic-resolution structures that have revealed the exact molecular structures of the pathogenic fibrils. In addition to this, the mechanistic principles of the protein misfolding associated with the aging brain are discussed. A comparison of current and historical paradigms provides a comprehensive understanding of the progression of neurodegenerative diseases (Fig. 15.2).

Protein homeostasis and aging: There is a decline in protein homeostasis with age, which involves chaperones, the proteasome, and autophagy (Kaushik and Cuervo, 2015). The proteostasis process is controlled by a highly regulated network, which involves molecular chaperones, the ubiquitin-protein aspartamer system (UPS), and the autophagy-lysosomal pathway (ALP). This system is essential for the correct folding,

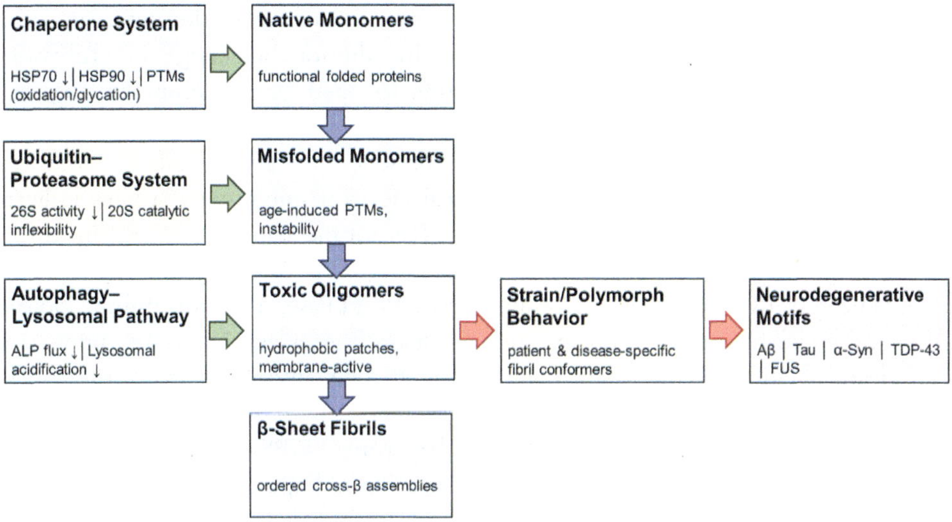

Figure 15.2 Proteostasis decline and structural aggregation cascade in aging brain.

trafficking, and degradation of thousands of proteins in the neurons of young individuals. With age, many of the nodes of the proteostasis network are impaired, leading to the accumulation of damaged organelles, aggregated proteins, and the subsequent disruption of cellular functions. As one gets older, several nodes of this network get impaired (Wang, 2015; Wang and Le, 2019).

Chaperone dysfunction: Previous studies indicate that the levels of heat shock proteins (HSPs) decline during aging, leading to a loss of folding capacity (Calderwood et al., 2009). This hypothesis was primarily backed by the key evidence obtained from transcriptional and biochemical tests that indicated that the level of HSP70 and HSP90 in an aged tissue in the brain is reduced (Macario and Macario, 2002). The latest structural studies have further supported these theories by indicating that some age-specific changes in proteins, such as oxidation and glycation, affect the substrate-binding sites of chaperones (Asomugha, 2011; Lee and Hsu, 2017). The results from experiments using cryo-EM techniques on chaperonin complexes indicated that minor conformational destabilization in aged neurons results in a loss of the ability to interact with misfolded polypeptides. The reduction in the chaperone capacity is therefore not only a quantitative but also a structural reduction (Creekmore et al., 2021).

Proteasomal impairment: In addition, previous experiments have shown that with age, ubiquitinated proteins were accumulated, which indicates a deficiency in proteasomal functioning (Blasiak et al., 2019). Firstly, a decrease in proteasomal activity with age was observed, resulting from reduced proteasome concentration. However, recent structural studies carried out on the 26S proteasome from aged rat brain tissue revealed that proteasomal dysfunction in aging results from oxidative stress on its catalytic core, the 20S proteasome, which leads to a decrease in the flexibility of the active sites, as well as in substrate capacity. In addition, a change in proteasomal structure with age, from closed-gate competent to open-gate incompetent, has been revealed using single-particle cryo-electron microscopy (Ding et al., 2006).

Autophagy–lysosomal dysfunction: The presence of autophagic vacuoles that accumulated in aged neurons through classical electron microscopy studies indicated degradation impairment (Terman et al., 2007). Recent mechanistic studies confirm that lysosomal acidification is decreased with age, thereby reducing the efficiency of autophagic flux (Moore, 2020). The analysis of V-ATPase, a proton pump, indicates that lipofuscin accumulation during aging impairs the assembly of this pump in a mechanistic relationship between aging and inefficient autophagy that could not be explained until now (Eaton et al., 2021).

The three components of proteostasis, chaperones, UPS, and ALP, constitute a proteostasis system, which is structurally impaired in old age and renders the neurons susceptible to the continuous accumulation of misfolded proteins. The proteostasis system is crucial in the maintenance of cellular homeostasis (Rajak, 2025).

The aggregation of these proteins involves a series of conformational changes, beginning with the misfolding of the native monomers, followed by the oligomerization of these misfolded structures, and culminating in the deposition of amyloid plaques or the formation of these structures (Almeida and Brito, 2020). Among the previously known types of amyloid structures, amorphous deposition was included, but breakthroughs from the application of X-ray crystallography and then cryo-EM revealed the ordered structures of amyloid fibrils, consisting of β-sheets (Fitzpatrick and Saibil, 2019).

It has undergone an enormous shift since the first atomic structures of Aβ and tau filaments were resolved in solid-state NMR and cryo-EM studies, to the dismay of earlier suggestions based on low-resolution Fourier transform infrared spectroscopy (FTIR) and electron diffraction (Lippens and Gigant, 2019; Ramachandran, 2016). Previous studies had proposed generic cross-β structures, but recent data suggest that a particular polymorph contains disease-specific structures. What is particularly interesting is that oligomers, and not the complete oligomer structure known as fibrils, are now thought to be the main cytotoxic species (Grasso et al., 2019). The reason for this is that, according to the recent structure, oligomers have hydrophobic patches that are accessible to solvent and can cause membrane disruption and synaptic signaling, something that had not been taken into account in previous studies that had focused on the fibril structure (Teplow, 2013).

A comparison with the more recent and the preceding data reveals that a paradigm shift has occurred: instead of the passive accumulation of proteins, the aggregation process is now discussed as an active process, a structurally directed conversion into highly selective pathological conformers (Aguzzi and O'Connor, 2010).

15.2.1 Structural basis of Alzheimer's disease: tau and amyloid-β protein aggregates

Tau and Aβ are the two most common proteins in AD and among the best-studied pathological aggregates.

Tau: Initial immunohistochemical investigations found paired helical filaments (PHFs) to be the prime constituents of neurofibrillary tangles. The structure was proposed in the electron micrographs of a twisted ribbon (Mena et al., 1991), but they did not include any atomic detail. The atomic structures of PHFs and straight filaments are now understood due to recent cryo-EM studies, which reveal that tau adopts a C-shaped conformation with interlocking β-strands (Zhu and Qian, 2022).

These high-resolution structures demonstrate that tau fibrils differ in AD, Pick's disease, and chronic traumatic encephalopathy, and thus the assumption that tau aggregates would be identical across tauopathies is refuted (Cieplak, 2019).

Aβ: Historically, Aβ plaques have been termed insoluble extracellular deposits. Although the first models suggested that the core should be a uniform 42-strand β-sheet, recent cryo-EM and NMR studies indicate several Aβ polymorphs with different protofilament structures (Takahashi et al., 2017; Yang et al., 2022). The oligomeric Aβ subunits, determined by single-molecule FRET and structural ensemble modeling, are more dynamic and structurally heterogeneous than previously believed (Lyubchenko et al., 2006). Notably, the mechanism by which Aβ oligomers destabilize synaptic membranes by forming β-barrel-like holes has not been recognized by plaque-centric theories.

15.2.2 α-Synuclein in Parkinson's disease

The main constituent of Lewy bodies is α-synuclein, and it has undergone equivalent advances in structural knowledge (Hauser et al., 2011). Past biochemical research considered Lewy bodies to be amorphous aggregates composed of multiple proteins and lipids (Shahmoradian et al., 2019). Cryo-EM reconstructions have, however, shown that β-sheet motifs of α-synuclein fibrils are tightly packed and are available in various polymorphs depending on disease subtype or mutation (e.g., A53T, E46K) (Mehra et al., 2021). The structure of α-synuclein fibrils is not random, as there are highly specific steric zippers and protofilament interfaces. This structural specificity is why different fibril strains have distinct pathological consequences, similar to prion strains (Holec et al., 2022).

15.2.3 TDP-43 and FUS aggregates in amyotrophic lateral sclerosis

TDP-43 and FUS are cytoplasmic inclusions associated with neuronal death in ALS. The initial research determined these aggregates through immunostaining, but it was not mechanistic. Recent structural studies have demonstrated that the low-complexity domains of TDP-43 and FUS have liquid–liquid phase separation (LLPS), and these states gradually form solid aggregates during cellular stress or aging (Aksoy et al., 2020; Alberti and Hyman, 2021). The crystallized forms of TDP-43 fibrils using cryo-EM are used to show that pathogenic fragments do not take the form of the loose and disordered forms of their physiological counterparts (Li et al., 2021a). These observations refute the previous hypotheses that TDP-43 aggregation was a loss-of-function event; rather, it is a gain-of-toxicity structure that impairs RNA processing and nucleocytoplasmic transport. Similar transitions of FUS have been observed, where aging-linked phosphorylation changes FUS LLPS behavior (Brandi, 2019). Previous studies were of the opinion that FUS mutations only disturbed nuclear positioning, but current findings show that the mutations reform the structural environment of the LLPS, increasing the rate of conversion into pathogenic solid aggregates (Rhoads et al., 2018).

15.2.4 The posttranslational modifications and structural transitions in roles

PTMs play a crucial role in protein misfolding and aggregation. Previous studies identified aberrant PTMs, hyperphosphorylated tau, oxidized Aβ, and ubiquitinated α-synuclein, yet they were unable to mechanistically link these findings to structural transitions. Modern structural and mass spectrometry-based analyses give the following gaps:

Oxidation adds carbonyl groups that destabilize native folds, favoring aggregation-prone intermediates (Rajan et al., 2021).

Phosphorylation changes the charge distribution. Phosphorylation of specific residues removes the protein-microtubule interaction in tau and promotes aggregation-prone conformations (Akoury, 2013).

Ubiquitination is a tag on misfolded proteins that targets them to be degraded; ubiquitination of proteins increases during aging and functions as a nucleation seed (Hagai and Levy, 2010).

These processes explain why aging, with an increase in oxidative stress, reduced proteasome function, and a change in the kinase/phosphatase ratio, provides an environment in which pathological aggregation is accelerated.

In short, aggregation and misfolding of proteins in the aging brain are related to a complicated interrelation between impaired proteostasis, structurally guided conformational changes, and disease-specific aggregation pathways (Hipp et al., 2019). The groundwork for historical studies has been laid, and contemporary structural biology has transformed mechanistic understanding, showing that pathogenic aggregates are heterogeneous and structurally diverse, with different toxicities and biological actions. These results have provided a point of contact for rational therapy design based on select aggregation processes and reestablishing proteome homeostasis in the aging brain.

15.3 Mitochondrial dysfunction and oxidative stress structural biology

Mitochondria serve as key regulators of neuronal health, bioenergetic potential, calcium homeostasis, and programmed cell death. Their unique structure, including two lipid layers, a compartmentalized proteome, and morphology that can be dynamically changed, makes them particularly susceptible to age-associated damage (Srivastava, 2017). During the last two decades, the impairment of mitochondrial function has become one of the classic signs of neurodegenerative diseases, such as AD, PD, ALS, and HD (Alqahtani et al., 2023). Initial biochemical and cellular studies associated aging with decreased ATP production, an increase in the levels of ROS, and an increase in the number of mutations in the mtDNA (Wei and Lee, 2002). Another development in high-resolution structural biology revealed conformational remodeling of specific proteins and complexes that mediates mitochondrial competence loss in neurons as they age. This part synthesizes these structural explanations and compares them with historical discoveries to highlight the development of the science (Fig. 15.3).

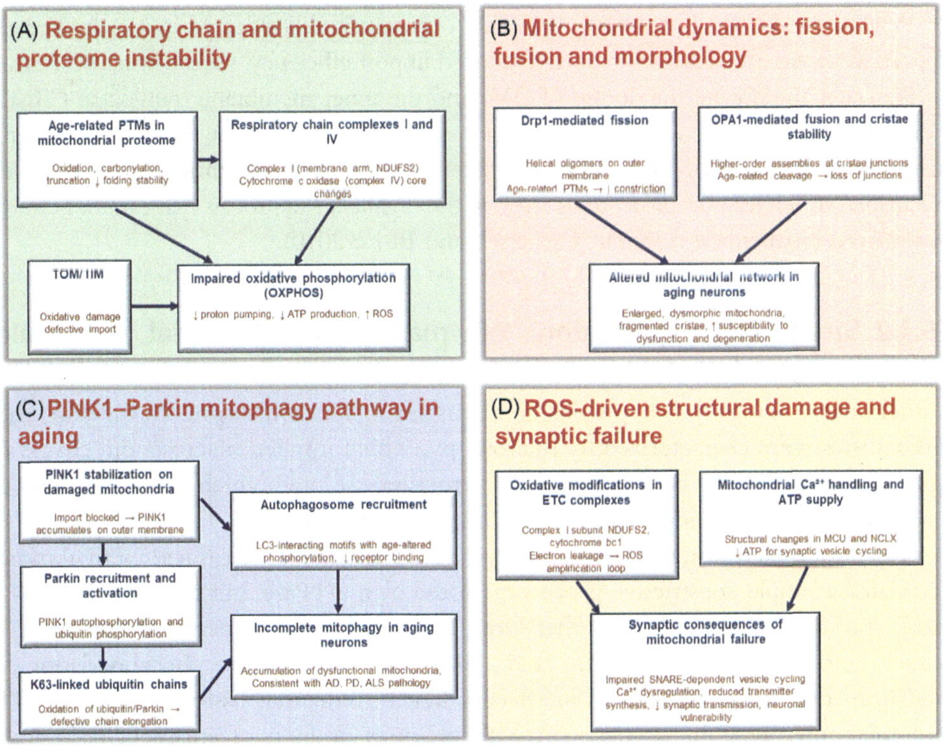

Figure 15.3 Mitochondrial dysfunction and oxidative stress in aging neurons.

15.3.1 Mitochondrial proteomics dysregulation during aging and neurodegeneration

Initial research hypothesized that mitochondrial dysfunction during senescence was primarily due to the accumulation of mitochondrial damage, overproduction of ROS, and mitochondrial dysfunction (Zia et al., 2022). Although these are contributory factors, modern proteomic and structural studies have shown that an unstable mitochondrial proteome is a major cause of neurodegeneration (Table 15.1).

Posttranslational modifications, including oxidation, carbonylation, and truncation, accumulate on mitochondrial proteins, reducing their folding stability. An example is the report of attenuated activity of respiratory chain complexes I and IV in aged tissues by early biochemical assays, but the mechanism involved could not be identified (Fernández Sanz, 2015). Cryo-EM structures of aged complex I also reveal new fine changes in the conformational change of its membrane arm, which compromise proton-pumping performance (Grba et al., 2023). Similarly, young and aged cytochrome c oxidases exhibit structural differences that reveal age-related changes in the catalytic core, impairing electron transfer. Such findings improve the antecedent models by providing a molecular explanation of the impaired oxidative phosphorylation capacity that causes aging neurons to exist (Yoshikawa and Shimada, 2015).

When mitochondrial proteomic instability occurs, it is also reflected in impaired protein import. Whereas early research reported reduced import efficiency, recent structural research on the outer membrane translocase (TOM) and the inner membrane translocase (TIM23) indicates that oxidative alterations of channel constituents disrupt their pore structure (Chaudhuri et al., 2021). These structural damages result in the accumulation of misfolded precursors, which leads to proteotoxic stress in the organelle—a process that cannot be solved by previous biochemical research (Gregersen and Bross, 2010).

15.3.2 Structural explanation of major mitochondrial molecules: dynamin-related protein 1, optic atrophy protein 1, PINK1 and Parkin

Mitochondrial integrity is dependent on constant fission and fusion. In the past, these mechanisms were characterized by microscopy, which implied macroscopic changes in aging neurons. This has been enhanced by structural biology, which has shown how the machinery of mitochondrial dynamics is impaired with age.

Mitochondrial fission takes place via dynamin-related protein 1 (Drp1). Initial models postulated a simple constrictive process mediated by a GTPase, but nowadays, cryo-EM and X-ray crystallography show that Drp1 forms helical oligomers that surround the mitochondrial membrane (Rochon et al., 2024). Age effects cause phosphorylation and oxidation of these oligomeric interfaces, hence reducing fission efficiency. This molecular interpretation is consistent with previous findings of enlarged, dysmorphic mitochondria in old neurons, with accurate structural breakdowns (Zorov et al., 2019).

Table 15.1 Structural biology impistences to the mitochondrial dysfunction and oxidative stress in aging and neurodegeneration.

Aspect	Important molecular processes/players	Traditional view (historical findings)	Modern structural biology insights	Aging/neurodegeneration: implication
Mitochondrial proteome instability	Mitochondrial matrix proteins, respiratory chain complexes (complex I and IV)	↓ ATP production, ↑ ROS, mitochondrial DNA mutations; unknown molecular basis for enzymatic decline	Cryo-EM: complex I membrane arm conformational remodeling; disturbed catalytic core in complex IV	Hampered oxidative phosphorylation activity, energy deficit in the aging neurons
Protein import machinery decline	TIM23/TOM translocase complexes	Reduced import efficiency biochemically observed, mechanisms unknown	Oxidative changes alter pore structure; cumulative degradation of aggregated proteins	Intramitochondrial proteotoxic stress and neuronal degeneration
Failure of mitochondrial dynamics	Dynamin-related protein 1 (Drp1) (fission), optic atrophy protein 1 (fusion/cristae integrity)	Macroscopic mitochondrial enlargement and fragmentation observed on EM	Drp1: age-related phosphorylation/oxidation impairs the formation of oligomers; OPA1: higher-order assembly cleavage destabilizes cristae	There are dysmorphic mitochondria, impaired bioenergetics, and the collapse of the cristae associated with AD
Mitophagy defects	PINK1, Parkin, LC3-interacting receptors, ubiquitin chains	Evidence of age-related degradation of mitochondrial clearance; etiology unclear	There is evidence of PINK1 autophosphorylation-dependent dimerization formation; Parkin large-scale activation changes; failure to form K63-linked ubiquitin chains during aging	Damaged mitochondria buildup stage effectors-enhanced neuronal stress in AD/PD/ALS

(*Continued*)

Table 15.1 (Continued)

Aspect	Important molecular processes/players	Traditional view (historical findings)	Modern structural biology insights	Aging/ neurodegeneration: implication
Respiratory chain oxidative vulnerability	NDUFS2, bc1 complex	ROS was regarded as nonspecific undesirable byproducts	Oxidation site structural mapping reveals that there is a particular folding perturbation -> self-amplifying loop of ROS	Additional damage to mitochondria -> compensatory metabolic acidosis
Synaptic energy failure	Mitochondrial ATP-dependent SNARE machinery; MCU/NCLX calcium exchange	Electrophysiology: age-related decrease in synaptic activity, mechanism unknown	The direct effects of the oxidative misfolding of mitochondrial and synaptic proteins are the disruption of vesicle cycling and Ca^{2+} buffering	Synaptic loss → loss of memory and breakdown of neural circuit
Overall conceptual shift	Entire mitochondrial structural proteome	Correlations between biochemical details and mechanism	Structural alterations at the atomic resolution bring together mitochondrial aging processes	Proteostasis failure-neuronal death cascade

The protein optic atrophy protein 1 (OPA1) is responsible for inner membrane fusion as well as cristae maintenance (Yuan and Zhang, 2023). Genetic analyses have associated OPA1 mutations with optic atrophy, but it was not until recent structural studies that the mechanism by which OPA1 forms dimeric and higher-order structures that stabilize cristae junctions was understood. The age-related proteolytic cleavage of OPA1 breaks these assemblies, resulting in the fragmentation of the cristae, a pathology present in aging brain and AD tissue (Erchova et al., 2021). This mechanistic understanding connects previous electron microscopic findings of cristae destruction to the molecular-based disease pathways.

The PINK1-Parkin pathway coordinates mitophagy, the selective degradation of damaged mitochondria (Springer and Kahle, 2011). Before the structural understanding, PINK1 was simply considered a depolarization sensor. However, crystallographic information shows that PINK1 autophosphorylation triggers conformational activation, allowing ubiquitin and Parkin phosphorylation (Rasool and Trempe, 2018). Later cryo-EM analyses then show that Parkin undergoes radical structural reorganizations and becomes an active E3 ligase, as demonstrated by its ability to ubiquitinate mitochondrial targets (Panicker et al., 2017). These conformational transitions during aging explain the inhibition of mitophagy in aging neurons. Past researchers have observed a decrease in the clearance of impaired mitochondria as people age; recent structural evidence now helps identify the origin of this malfunction.

15.3.3 Mitochondrial basis of mitophagy and aging: neural failure

Mitochondrial failure is one of the characteristics of neurodegenerative disease. The structural studies have made our understanding deeper as they define the stepwise architecture of this pathway:

PINK1 stabilization on the outer membrane: Cryo-EM structures reveal that healthy mitochondria quickly import and degrade PINK1; damaged mitochondria keep PINK1 on the outer membrane, and it forms active dimers (Osman, 2022).

Parkin recruitment and activation: Structural studies indicate that phosphorylated ubiquitin is a scaffold during Parkin activation. It is impaired by aging, which interferes with ubiquitin maintenance and leads to loss of PINK1 kinase activity (Rasool and Trempe, 2018).

Ubiquitin chain elongation: NMR and cryo-EM results explain the formation of K63-linked ubiquitin chains by Parkin to promote autophagy receptors. Ubiquitin or Parkin oxidation disrupts chain formation, explaining previous findings of partial mitophagic flux in aging neurons (Liang et al., 2023).

Autophagosome formation: Recent structural studies of LC3-interacting motifs (LIR motifs) show that phosphorylation alterations during aging undermine receptor binding (Suzuki et al., 2014).

This has the cumulative effect of slowing mitochondrial clearance and stimulating the development of dysfunctional organelles, consistent with classical neuropathological findings in aging and AD, PD, and ALS.

15.3.4 Structural changes and synaptic consequences caused by ROS

Oxidative stress is an established cause of aging, although it was initially thought to be a diffuse, nonspecific destructive agent. This has been improved by structural biology, which has localized oxidative modifications capable of altering protein conformation (Krishnamurthy et al., 2024).

Local changes in tertiary structure occur because of oxidation of cysteine and methionine residues in complexes of electron transport. Recent structural investigations based on mass-spectrometry signatures indicate that ROS disrupts the complex I subunit NDUFS2 and the cytochrome bc1 complex, resulting in electron leakage and subsequent ROS production , generating additional ROS, which is a feed-forward loop not predicted in earlier models (Li et al., 2021b).

Oxidative structural damage to mitochondrial proteins influences synapses, which are extremely energy-dependent, thus affecting the production of ATP and disrupting:

synaptic vesicle cycling (through ATP-dependent SNARE machineries),

- modified MCU and NCLX structures (buffering of calcium), and
- neurotransmitter precursors are produced.

The previous electrophysiological literature reported decreased synaptic transmission with age, but with little mechanistic understanding; structural studies demonstrated that ROS-induced misfolding of key mitochondrial and synaptic proteins led to this degradation (Nesci et al., 2021).

Overall, structural biology has had a significant influence on understanding mitochondrial dysfunction in aging and neurodegeneration. Although the previous models were based on biochemical measurements and microscopy, which offer initial insight, atomic-level information reveals that definite conformational changes are dominant. All these structural changes, including fission/fusion apparatus, mitophagy regulators, respiratory chain complexes, and ROS-sensitive proteins, are the reasons why aging promotes synaptic loss and neurodegeneration. All these results together create a mechanistic continuum between age-dependent proteome instability and neuronal death and disease development.

15.4 Cell and structural mechanisms of synaptic degeneration

Synapses are the basic units of neuronal communication and plasticity, and their structural integrity has been proposed to be crucial for learning, memory, and executive functioning (Selemon, 2013). The common pathological feature of neurodegenerative disorders and normal aging of the brain is a progressive deterioration of synapses. Although some initial studies described synapse loss using histological stains and electrophysiological recordings, the development of high-resolution structural technologies has radically changed our understanding (Kleinfeld et al., 2011). Recent advances in cryo-EM, super-resolution imaging, and integrative structural biology have provided an unprecedented view of the synaptic molecular structure, thereby offering a mechanistic explanation of how age-dependent changes in key synaptic proteins lead to functional loss (Table 15.2 (Barrantes, 2024).

15.4.1 Core synaptic protein core synaptic proteins are organized structurally

The transmission at the synaptic level requires a very complex set of proteins to mediate the processes of vesicle docking, fusion, receptor organization, and postsynaptic signaling. The heart of this architecture comprises three protein systems: SNAREs, postsynaptic density (PSD-95), and synaptotagmin (Iasevoli et al., 2013).

15.4.1.1 The SNARE complex

SNARE machinery, which is a complex of syntaxin-1, SNAP-25, and synapotbrevin/VAMP2, facilitates the fusion of synaptic vesicles (Walch-Solimena et al., 1995). The first biochemical studies identified these proteins through coimmunoprecipitation experiments and proposed a zipper model of membrane fusion. Later crystallographic studies, however, defined a four α-helix bundle structure that triggers fusion. Investigations into the SNARE complex using more recent cryo-EM experiments and single-molecule force spectroscopy techniques provide dynamic data: the SNARE complex undergoes tightening of its coils, which precipitates membranes into the immediate vicinity (Jahn et al., 2024). This structural accuracy was not available in previous models, which treated fusion mostly as a stochastic process.

15.4.1.2 Postsynaptic density 95 and the postsynaptic density

PSD-95 is a scaffold protein present in the PSD, and it arranges glutamate receptors, kinases, and signaling complexes. Traditionally, the PSD was referred to as an electron-dense thickening under the postsynaptic membrane (Rutter, 2001). Super-resolution microscopy currently depicts a modular, nano-compartmentalized scaffold, wherein receptors and signaling proteins are bound in spatially ordered clusters by PSD-95 PDZ, SH3, and GK domains. The ability of PSD-95 to form mesh-like lattices is further shown by cryo-EM tomography and provides structural stability to postsynaptic signaling centers—a feature that was unavailable using previous EM (Tao et al., 2018).

Table 15.2 Structural mechanisms of age-associated synaptic degeneration.

Synaptic component	Normal structural role	Structural changes during aging	Functional consequences	Key technologies informing insights
SNARE complex (syntaxin-1, SNAP-25, synaptobrevin/ VAMP2)	Four-helix coiled package mediating vesicle docking and membrane fusion	Weakening of helix–helix interactions by oxidation; loss of syntaxin clustering by glycation	Partial zippering of the membranes leads to low release probability of the vesicles, slower fusion kinetics	Cryo-EM, single-molecule force spectroscopy
Postsynaptic density (PSD-95) and PSD	Scaffold AMPA/NMDA receptors, kinases/ signaling complexes in nano-clusters	Phosphorylation/ oxidation caused by age interferes with PDZ folding; destabilization and thinning of PSD lattice	Reduction in receptor nano-clusters stability leads to damaged long-term potentiation and memory mechanisms	Cryo-EM tomography, super-resolution microscopy
Synaptotagmin-1	Ca^{2+} sensor links Ca^{2+} to fast neurotransmitter release	C_2 domain oxidation decreases Ca^{2+} affinity and membrane insertion	Delayed synaptic fusion of vesicles and temporal accuracy of neurotransmission	Cryo-EM, NMR
Synaptic architecture	Extremely synchronized coordination between release sites of vesicles and receptor nanodomains	Nano-domain dispersion with age; loss in active zone structural integrity	Premature synaptic dysfunction before cell death and cognitive impairment	Cryo-EM, STORM/ PALM imaging

15.4.1.3 Synaptotagmin

The major calcium sensor of neurotransmitter release is synaptotagmin-1. The first biochemical research described the Ca^{2+}-binding C_2 domains, yet modern structural research has clarified the underlying mechanism by which calcium binding induces conformational changes that result in rapid vesicle fusion (Radhakrishnan et al., 2021). Recent NMR and cryo-EM evidence has mapped the mechanism of synaptotagmin's interactions with all three SNAREs and phospholipids, coordinating vesicle exocytosis with calcium spikes in the presynaptic cell. Combined, these structural systems are examples of the accuracy and coordination that is needed in neurotransmission (Jaczynska, 2024).

15.4.2 Effects of age on conformational change and influence on neurotransmission

The problem lies in aging, which creates structural instability throughout synaptic proteins as a result of which they change their folding, interactions, and responsiveness.

15.4.2.1 SNARE complex instability

The initial electrophysiological experiments revealed reduced vesicle release probability in aged synapses, but the mechanism underlying this phenomenon remained unclear. Recent structural studies show oxidative changes in SNAP-25 and syntaxin, which undermine helix–helix interactions in the SNARE bundle, leading to incomplete zippering and slower fusion kinetics (Van Dinh, 2022). Glycation also induces age-related impairment of syntaxin clustering at active zones that reduces docking performance (Uceda et al., 2024).

15.4.2.2 PSD-95 and postsynaptic vulnerability

The aging process decreases PSD thickness, as reported in EM experiments since the 1970s. Contemporary cryo-EM tomograms and proteomic evidence can now be used to describe this decrease: age-related phosphorylation and oxidation of PSD-95 impair its PDZ domain folding and, in turn, diminish its ability to tether N-methyl-D-aspartate (NMDA) and α-amino-3-hydroxy-5-methyl-4-isoxazolepropionic acid (AMPA) receptors (Villmann and Becker, 2007). Weakening of the PSD leads to the destabilization of receptor nano-clusters, which disrupts long-term potentiation (LTP)—a connection that could be proposed by previous physiological studies but not structurally demonstrated before (Clavet-Fournier et al., 2023).

15.4.2.3 Synaptotagmin dysregulation

Previous structural studies confirming the lack of calcium-sensitivity loss in aged neurons are supported by new structural evidence showing that oxidation of the C_2

domains of synaptotagmin reduces Ca^{2+} affinity and hinders membrane insertion, thereby slowing synaptic vesicle release. These results interconnect biochemical evidence of depressed Ca^{2+} reactivity and specific structural changes in the sensor apparatus (Südhof and Rizo, 1996).

Overall, these protein-level changes in aging lead to direct effects on synaptic communication and refute the previous paradigm of a generalized decrease in synaptic strength.

15.4.3 Cryogenic electron microscopy and super-resolution microscopy of synaptic architecture mapping

Structural imaging techniques have provided a clearer understanding of synapse architecture at a scale never before seen.

15.4.3.1 Cryo-electron microscopy

Cryo-EM captures presynaptic active regions, vesicle pools, and postsynaptic densities with nanometer-scale accuracy (Held et al., 2024). Previously unexplored conventional EM provided snapshots of the functional state, but the improved contrast and preservation provided by cryo-EM can reveal dynamic functional states, for example, priming of vesicles and the conformation of SNAREs and synaptotagmin (Zabeo and Davies, 2022). The mechanisms of organizing release sites and the molecular basis of previous electrophysiological findings on release probability have been explained by cryo-EM tomography of synaptic vesicle-tethering proteins such as RIM and Munc13.

Super-resolution methods have challenged long-held ideas about the arrangement of synapses. Fluorescence imaging earlier indicated uniform distributions of receptors, but STORM/PALM revealed receptor nanotrails of the AMPA and NMDA receptors, which perfectly match the release apparatus of synapses (Wang, 2015). Dispersal of these nanodomains with age, as observed by super-resolution imaging, is associated with reduced LTP and impaired cognition. These imaging advances connect cellular phenotypes to molecular structure and bring synaptic biology into a mechanistically accurate field.

Synaptic depletion, but not plaque load or neurodegeneration, has been shown by decades of neuroscience studies to be the best predictor of cognitive impairment in old age and in dementia (Lista and Hampel, 2017). Initial research has focused on the number of synapses in postmortem brains, but structural biology provides a mechanistic explanation for this relationship (Fig. 15.4).

- Decreased SNARE complex function leads to defective vesicle release and, consequently, suppresses synaptic transmission.
- PSD destabilization reduces the receptor density and interferes with plasticity processes involved in memory.

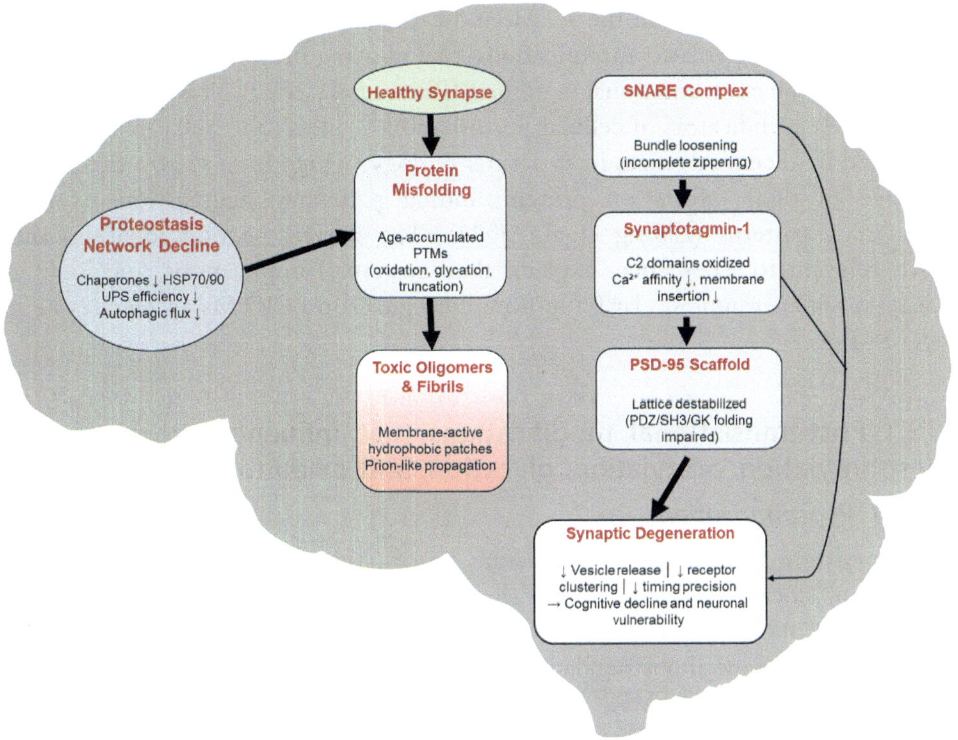

Figure 15.4 Aging triggers the decline of neuronal proteostasis, allowing the misfolding of synapse proteins into pathogenic oligomeric and β-sheet fibrillar aggregates with prion-like propagation. Conformational impairments of SNARE complexes, synaptotagmin-1, the Ca^{2+} sensors, and PSD-95, the scaffold, impair vesicular fusion, disrupt receptor nanocluster organization, and induce temporal fidelity disruption in neurotransmission. These anomalies in early structure thus trigger a progressive synaptic degeneration, leading to measurable cognitive deterioration

- Synaptotagmin failure retards neurotransmission, compromising the temporal accuracy of network synchrony.

Super-resolution and cryo-EM images are unanimous in showing that synaptic degeneration occurs before neurons are lost in AD, PD, and normal aging, thereby validating previous neuropathological findings with greater molecular detail. Structural degradation at the synapse, therefore forms a mechanistic spectrum—from protein misfolding to defective neurotransmission to cognitive impairment (Bogdanovic, 2024).

15.5 Role of posttranslational modifications in structural alterations

PTMs are critical factors that affect protein structure, stability, and function. Despite the fact that PTMs refine canonical cellular processes, the aging environment promotes radical changes in the PTM landscape, thereby increasing the number of aberrant or nonenzymatic modifications that disrupt protein folding and trigger disease-inducing conformational equilibria (Ebert et al., 2022). Previous conceptions of the PTMs as indicators of cellular perturbation or metabolic imbalance have been confirmed by recent advances in structural biology, which have shown that they are causative of their effects on protein folding pathways, aggregation rates, and cytotoxicity patterns (Zhong et al., 2023). The current section identifies major PTMs relevant to senescence, explains their structural implications, and highlights disease-specific examples, clarifying how dysregulation of PTMs accelerates neurodegeneration.

15.5.1 Post-translational modifications that influence protein folding: phosphorylation, acetylation, glycation, and oxidation

15.5.1.1 Phosphorylation

Phosphorylation is an enzyme-modulating process that changes electrostatic charge, precipitating local or global conformational changes. Classical biochemical assays characterized phosphorylation using a few structural measurements based on kinase activity or enzyme metabolic regulation (Johnson and Lewis, 2001). On the other hand, recent high-resolution NMR and cryo-electron microscopy reveal the specific mechanistic effects of phosphate groups on the stabilization of secondary structure, thereby explaining the phosphorylation effect on aggregation propensity in neurodegenerative proteins, including tau and α-synuclein.

15.5.1.2 Acetylation

Acetylation reverses the positive charge on lysine side chains, thereby affecting protein–protein interactions and folding. Initial studies had focused on histone acetylation as a transcriptional regulator; subsequent structural studies show that cytosolic protein acetylation (tau, mitochondrial enzymes) can change conformational ensembles and aggregate kinetics (Kabir et al., 2023).

15.5.1.3 Glycation

Advanced aging and hyperglycaemic conditions enhance nonenzymatic glycation, leading to advanced glycation end-products (Arivazhagan et al., 2022). Previously, diabetic sequelae were associated primarily with glycation. Recent structural studies have shown that glycation significantly stabilizes the Aβ protein (Aβ) and α-synuclein, promoting their transition to β-sheet-rich conformations and the formation of toxic oligomers (Prosad Banik, 2024).

15.5.1.4 Oxidation

Oxidation is caused by ROS and targets methionine, cysteine, and aromatic residues. Past research has generalized the concept of oxidation to mean the destruction of proteins. Recent structural methods demonstrate that oxidation triggers discrete conformational changes, for example, helix-to-coil transitions or destablization of β-sheet formations, with far-reaching effects on synaptic and mitochondrial components (Griffiths, 2000).

15.5.2 Structural consequences of post-translational modifications in aging-related proteins

The aging process results in changes in the enzymatic milieu, increases pro-modification enzymes (e.g., kinases), and decreases clearance pathways (e.g., autophagy, proteasome). At the same time, nonenzymatic PTMs, including glycation and nitration, are enhanced by oxidative stress. These alterations affect the architecture of proteins in a variety of ways:

15.5.2.1 Manipulation of electrostatic landscapes

Local charge changes induced by acetylation or phosphorylation near flexible domains trigger conformational rearrangement. Structural studies indicate that phosphorylation of intrinsically disordered proteins (IDPs) biases their ensembles to aggregation-prone conformations (Mukhopadhyay, 2020).

15.5.2.2 Breaking intramolecular interactions

Glycation removes lysine or arginine side chains, thereby blocking the hydrogen-bond networks necessary to maintain the native state. This destablization changes the equilibrium of folding, which increases the size of the misfolded intermediates (Milordini, 2019).

15.5.2.3 Promoting β-sheet formation

Oxidation or glycation disrupts α-helix structures and induces a β-sheet transition, a typical amyloid aggregation hallmark (Milordini, 2019). This phenomenon was first postulated in early FTIR but is now supported by modern NMR and cryo-EM techniques, indicating a specific mechanism for residue levels.

15.5.2.4 Sabotaging proteinacious vapidation

Some PTMs block proteasomal recognition or autophagic receptor binding, promoting the buildup of modified proteins. Previous biochemical data reported ubiquitinated inclusions, but without details of their structure. Recent reports show that PTMs can conceal degron signals or distort protein surfaces that are important for engaging the clearance system (Zafar et al., 2024).

In turn, PTMs are not only biomarkers but also the active factors that trigger the development of structural and functional changes.

15.5.3 Neurodegeneration structural alterations due to post-translational modifications

15.5.3.1 Tau hyperphosphorylation

One of the features of AD is massive hyperphosphorylation of tau. Previously, phosphorylation was associated with the detachment of microtubules, and new cryo-EM structures indicate residue-specific disturbances of native folding leading to the development of PHFs with an architecturally defined 2D β core (Rani and Mallajosyula, 2021). Comparison of the structural data reveals that phosphorylated tau adopts conformers that are not observed in the native form, which explains the faster aggregation during AD pathology (Zabik et al., 2017).

15.5.3.2 α-Synuclein nitration

Nitration of tyrosine residues in α-synuclein was once considered to be a collateral effect of oxidative stress. New structural data have shown that the breakdown of long-range intramolecular interactions stabilizing the monomer induced by nitration alters the folding equilibrium toward oligomeric and fibrillar forms. Fibrillographic evidence obtained through cryo-EM reveals that nitrated α-synuclein adopts polymorphs that are not similar to canonical fibrils and are associated with increased toxicity in PD (Guerrero-Ferreira et al., 2020).

15.5.3.3 Aβ glycation

In the past, Aβ glycation was linked to higher risks of AD in diabetes. Current mass spectrometry and structural studies show that glycation results in β-sheet-rich conformers that initiate aggregation at a faster rate than native Aβ (Emendato et al., 2018). Glycation-dependent protofilament interfaces are revealed by cryo-EM reconstructions, which can provide molecular explanations of observations that hyperglycemia increases Aβ toxicity (Prosad Banik, 2024).

These models highlight PTMs' ability to directly tune physiologically important structural transitions that drive neurodegenerative disease.

15.5.4 Structural biology probes of post-translational modification heterogeneity

The heterogeneity and substoichiometric nature of PTMs in the past hindered their study. Modern structural biology has overcome these challenges using integrative approaches:

15.5.4.1 Mass spectrometry

MS provides accurate localization of different PTM sites and stoichiometric quantification. Previously, the detection of PTMs relied on antibody-based methods, which had poor specificity; new MS methods can be used to explain PTM mosaics at the level of whole proteins, even for low-abundance modifications (Doll and Burlingame, 2015).

15.5.4.2 Cryo-electron microscopy

The PTM-dependent fibril polymorphs can be determined with almost atomic precision using cryo-EM. Comparing fibrils with unmodified fibrils also allows understanding the way in which PTMs reorganize β-sheet packing and the protofilament structure (Cereghetti, 2021). The ability of cryo-EM to be able to capture conformational heterogeneity is a significant methodological improvement over traditional electron microscopy (Tang et al., 2023).

15.5.4.3 Nuclear magnetic resonance

With atomic-resolution analysis of PTM-induced changes in IDPs, NMR reveals dynamic changes in the ensembles that cannot be observed by crystallography (Csizmok et al., 2016). This sensitivity was not previously available in lower-resolution NMR, and today, with high-field systems, finer mapping of the effect of PTM on transient structural states is feasible.

15.5.4.4 Mass spectrometry-cryogenic electron microscopy/nuclear magnetic resonance workflows

The current hybrid strategies incorporate PTM profiles and structural state mapping. For example, MS will localize glycation hotspots and, using cryo-EM, assess the subsequent impact on amyloid architecture, thereby closing the biochemical descriptors and structural phenotypes (Landrieu et al., 2022).

PTMs radically carve out the structural fate of aging-linked proteins. The first studies portrayed PTMs as biochemical equivalents of disease; more recent structural biology studies depict them as direct regulators of protein folding, aggregation, and cytotoxicity. By integrating MS, cryo-EM, NMR, and related methods, researchers can capture the heterogeneity of PTMs with incomparable granularity, thereby supporting the assumption that PTMs are active agents of neurodegeneration rather than byproducts (Azevedo et al., 2022). These findings are an omen of new therapeutic approaches based on PTM-specific structural changes in the aging brain.

15.6 Interplay of proteostasis, autophagy, and neuroinflammation

Proteostasis, autophagy, and neuroinflammation are three pillars that rely on each other and determine neuronal well-being (Sonninen et al., 2020). Although the mechanisms have been studied in-depth individually, aging and neurodegeneration demonstrate that these systems are interdependent at the structural and functional levels in fundamental ways. The recent developments in structural biology, such as cryo-EM, X-ray crystallography, integrative modeling, mass spectrometry, and others, show that the molecular network is tightly coordinated, and its failure leads to the spread of synaptic dysfunction, neuronal loss, and chronic inflammation (Bogdanovic, 2024). In this part, one considers the structural basis of these interactions, compares historical paradigms with modern insights, and suggests a mechanistic model of how neuroinflammation and proteostatic malfunction feed off each other, and vice versa.

15.6.1 Structural properties of proteostasis networks: heat shock protein, ubiquitin ligases and autophagy receptors

15.6.1.1 Heat shock proteins

HSP70 and HSP90 are HSPs that act as molecular chaperones, controlling protein folding and aggregation (Morán Luengo et al., 2019). Initial biochemical analyses indicated that they were increased with stress, but with no structural detail. Recent cryo-EM and NMR studies show that HSPs are based on dynamic allosteric cycles: ATP binding opens substrate-binding domains, and hydrolysis activates client release (Gates, 2017). These allosteric transitions are impaired by age-related oxidation and glycation, which lowers the affinity of the substrate. Such structural changes explain the previous findings of reduced chaperone activity in older neurons despite no change in their levels of expression (Moldogazieva et al., 2019).

15.6.1.2 Ubiquitin ligases

Misfolded proteins are also tagged by E3 ubiquitin ligases such as Parkin and CHIP. In the past, ubiquitination was inferred from immunoblotting of high-molecular weight aggregates, and this approach has not provided much mechanistic information (Labbadia et al., 2012). In modern structural studies, it has been found that E3 ligases maintain autoinhibited conformations that are deactivated upon binding phosphorylated ubiquitin or unfolded substrates (Buetow and Huang, 2016). The posttranslational changes, such as oxidation of catalytic cysteines and phosphorylation of inhibitory sites, triggered by age, hinder these conformational changes. This causes a greater accumulation of misfolded proteins in old age despite an intact ubiquitin machinery, which can currently be explained by the structural impairment underlying it (Cuanalo-Contreras et al., 2013).

15.6.1.3 *Autophagy receptors (p62, LC3)*

Selective degradation is facilitated by autophagy receptors, which bind to ubiquitinated cargo and ubiquitinated LC3 on autophagosomes (Shaid et al., 2013). Initial microscopy identified p62-positive inclusions, but was unable to resolve the molecular architecture (Sergin et al., 2016). It is now known that p62 forms helical polymers that scaffold ubiquitinated aggregates and facilitate LC3 binding via its LIR motif, as shown by cryo-EM and molecular dynamics (MD) simulations. Autophagic failure in neurodegeneration is attributed to the destabilizing effect of p62 polymers through oxidation and phosphorylation (Palara, 2021). Likewise, autophagosome curvature and cargo loading (processes now visualized in near-native conditions) are controlled by LC3 lipidation. Together, structural biology depicts proteostasis as a multidomain, age-sensitive, dynamic, and conformation-regulated network that is vulnerable to stresses with age (Ju et al., 2025).

The increase in the accumulation of misfolded proteins is a complex process involving intercellular interactions that include the activation of innate immunity (Grootjans et al., 2016).

Previously, neuroinflammation was regarded as a secondary effect of neuronal damage. Modern evidence puts it as a proponent of proteostasis collapse (Sonninen et al., 2020).

15.6.1.4 *Microglial activation*

Pattern-recognition receptors such as TLR2 and TLR4 directly interact with misfolded proteins, including α-synuclein oligomers, α-synuclein fibrils, and TDP-43 fragments (Prasad et al., 2019). Initial experimental findings on microglial activation have measured cytokine release as a surrogate endpoint of microglial activation; more recent structural studies have demonstrated that aggregates bind receptors and cause ectodomain changes leading to propagation of inflammatory signaling (Colton and Wilcock, 2010). TLR4–MD2 complexes cryo-EM structures reveal that during amyloid binding, asymmetric dimerization, a requirement of downstream MyD88 recruitment, is stimulated. These observations enhance previous models, which relied on pharmacological inhibition but were not molecular in detail (Fu et al., 2025).

15.6.1.5 *Astrocytic response*

Misfolded proteins are internalized by astrocytes through endocytosis, whereas lysosomal overload caused by aggregates causes membrane rupture. X-ray structural analyses of lysosomal membrane proteins indicate that oxidative changes destabilize luminal domains, which shows previous findings of astrocytic dysfunction during aging and AD (Fanlo-Ucar et al., 2024). The accumulation of misfolded proteins triggers inflammation, which is further exacerbated by proteostatic impairment, creating a vicious cycle (Agyemang et al., 2015).

The NLRP3 inflammasome is a key mediator of neuroinflammation. Initial models of inflammasome activation described a binary on/off switch that was activated by reactive oxygen species or mitochondrial injury. The more complex mechanism has been revealed through structural biology (Sebastian-Valverde and Pasinetti, 2020).

15.6.1.6 NLRP3 activation

Cryo-EM experiments indicate that NLRP3 transitions from a compact, autoinhibited state to an oligomeric, ring-shaped state when mitochondrial dysfunction is detected (Liu et al., 2023). This conformational change is caused by age-related changes in mitochondrial lipid composition and ROS levels. Biochemical evidence from previous age studies of NLRP3 upregulation in elderly microglia lacked a specific mechanistic focus; structural models currently reveal the exact domain rotations and rearrangements of the ATP-binding pocket, initiating assembly (Hanslik and Ulland, 2020).

15.6.1.7 ASC filament formation

The adapter ASC polymerizes into prion-like filaments that attract caspase-1. Solid-state NMR shows that the pyrin domain in ASC forms helical assemblies composed of distinct polymorphs, which explains the different responses to inflammation in the aging disease and acute injury (Diaz-parga, 2022). Previous immunofluorescence experiments had visualized ASC speckles but failed to resolve their filament structure.

15.6.1.8 Caspase-1 activation

The cryo-EM structures of caspase-1 indicate that ASC filaments cause dimerization and stabilization of the active site. Oxidative stress is correlated with neuroinflammation through age-associated oxidation of the caspase-1 catalytic cysteine, thereby enhancing its activity (Wang et al., 2022). These structural data indicate that inflammasome activation in aging neurons is a compartmentalized, multistep assembly driven by age-sensitive conformational changes, rather than an accidental damage response.

Alternative perspective on neuroinflammation and proteostatic failure, showing that one molecular interface mediates the pathophysiological connections between the conditions (Jurcău et al., 2022).

Some of the problematic interaction points between misfolded proteins and inflammatory signaling include:

- Proteins of proteolysis as immuno-regulators:
  HSP70 and HSP90, in addition to refolding misfolded proteins, bind toll-like receptors and regulate NF-κB signaling. Structural studies indicate that HSP90 stabilizes components of the IKK complex, thereby associating proteostasis

malfunction with excessive inflammatory response (Vahid et al., 2017). The previous examples of immunohistochemistry of HSP upregulation in glial cells gain a mechanistic background in this way.

- Receptors of autophagy and inflammation:
  p62 directly interacts with TRAF6 by its ubiquitin-associated domain (Lin et al., 2013). Recent structural evidence indicates that glycation enhances the structural stability of amyloid-β (Aβ) and α-synuclein, facilitating their conversion into β-sheet-rich aggregates and promoting the formation of neurotoxic oligomers (Kawai et al., 2008). This is the reason why there are paradoxical results in which the p62 deficiency reduces the aggregate load but amplifies inflammatory signaling.
- Autophagy destruction by inflammasomes:
  The active caspase-1 cleaves proteins involved in autophagy, such as Beclin-1 and Parkin (Tsapras and Nezis, 2017). Structural mapping reveals the effects of cleavage on central domains required for mitophagy and autophagosome formation, supporting previous functional studies linking inflammasome activation to autophagy suppression.
- Proteotoxic stress caused by cytokines:

TNF-α and interleukin-1 signaling increase ROS and inhibit chaperones (Brigelius-Flohé et al., 2004). HSP70 oxidation structural studies reveal impaired nucleotide-binding pockets, rendering the proteostasis machinery ineffective during chronic inflammation (Galli et al., 2024).

Proteostasis, autophagy, and neuroinflammation are interrelated systems that regulate their interactions through precise structural interfaces (Sonninen et al., 2020). The previous models treated these processes as separate determinants of aging and disease; modern structural biology shows functional cross-talk, in which changes in the conformation of key proteins lead to dysfunction across the entire system (Tompa, 2014). Abnormal proteins trigger the immune system; inflammation interferes with proteostasis and autophagy; and clearance failure enhances aggregate deposition, creating a vicious cycle that drives neurodegeneration (Sonninen et al., 2020). The mechanistic clarity provided by understanding these structural interdependencies and the therapeutic opportunities identified by interfaces between proteostatic failure and inflammatory signaling.

15.7 Computational approaches and integrative structural biology

Aging and neurodegeneration would not be comprehensible without analytical methods that can render visualization of dynamic, heterogeneous, and usually transient protein conformations (Bashir et al., 2024). Traditional structural biology methods, such as X-ray crystallography, NMR spectroscopy, and, more recently, cryo-EM, have provided essential information on the structures of proteins involved in neurodegenerative diseases (Arar et al., 2025). However, even these methods often cannot detect intrinsically disordered regions (IDRs), infrequent aggregation intermediates, or multi-protein assemblies in the aging brain. This scientific landscape has thus been changed by integrative structural biology, which has been enhanced with complex computational modeling and artificial intelligence (AI).

This part explores how integrated experimental and computational approaches can address complex conformational changes in pathological proteins, and compares past perspectives with recent breakthroughs to demonstrate the influence of integrative approaches on the discipline.

15.7.1 This mixture of cryogenic electron microscopy, X-ray crystallography, nuclear magnetic resonance, and molecular dynamics simulations is combined

Traditionally, X-ray crystallography has provided the first atomic structures of a large fraction of neuronal proteins, such as synaptic enzymes and chaperones (Shi, 2014). NMR has addressed this knowledge gap by elucidating the intrinsically disordered regions (IDRs) and dynamic conformational states of proteins such as α-synuclein and TDP-43 (Haider, 2024). However, its applicability was curtailed by constraints on system size.

Cryo-electron microscopy (cryo-EM) has revolutionized structural studies of amyloids and large macromolecular assemblies. Early low-resolution electron microscopy depicted amyloid plaques as amorphous deposits with limited structural detail. In contrast, cryo-EM now resolves amyloid-β (Aβ) and tau fibril polymorphs at near-atomic resolution, enabling the identification of disease-specific structural conformations and establishing correlations between fibril architecture and clinical phenotypes (Scheres et al., 2020).

However, even cryo-EM snapshots represent only a subset of biologically relevant conformations. As a result, MD simulations are no longer dispensable when it comes to the modeling of aggregation pathways, energetics, and transient oligomer species that could not be directly visualized (Ghosh et al., 2024).

Integrative methods—combining cryo-EM density with NMR constraints and MD refinement—have enlightened:

- tau filament nucleation processes;
- Aβ-to-fibril changes in Aβ oligomers; and
- toxic pore formation in membranes of α-synuclein.

Previous theories were based on bulk biochemical kinetics and indirect aggregation reagents. Currently, techniques can directly relate the folding landscapes of atomic-resolution structural folds with previously inaccessible mechanistic precision.

15.7.2 Artificial intelligence-driven prediction and structural modeling of age-associated protein variants

Structural biology has been radically changed by AI. The AlphaFold and RoseTTAFold systems can predict protein structures from sequence alone, with unprecedented accuracy (Orr, 2025). In the past, computational prediction had a weakness in disease variants, particularly in disordered and aggregation-prone proteins (Ghosh et al., 2024). To model a range of structural states, AI tools now integrate evolutionary coupling and machine learning.

Prediction of structural effects of age-related posttranslational modifications:

- Mapping risk variants of genes: Age-related mutations (such as LRRK2 in PD, APP and PSEN1 in AD) may be predicted in silico to stabilize changes, perturbed folding, or rearrangement of surfaces (Perrone et al., 2021)
- Anticipating structural changes by posttranslational modification: The AI-enhanced modeling assesses the effects of phosphorylation or oxidation on tau and TDP-43 assemblies, which standard crystallography was not able to factor out (Recharla, n.d.)
- IDR conformation ensembles: AlphaFold has also made modeling of proteins that have been considered unstructured, including initial geometries to be refined with MD (Omidi et al., 2024).

Transition from traditional variant analysis to AI-guided structural hypothesis generation:

- Previous studies had associated mutations with pathology by statistical or biochemical tests.
- Experimental confirmation of specific structural perturbations due to variants with aging before generation of hypothesis: now it is possible to visualize specific structural perturbations due to aging and generate hypotheses more quickly (Medina-Carmona et al., 2019).

15.7.3 Conformational landscape mapping of neurotoxic aggregates in *Z. burgdorferi*

Complex metastable intermediates of folding occupy neurodegenerative disease proteins (Horwich, 2002). Previously, such species were not visible; the researchers inferred their existence from toxicity test results or antibody capture. Modern computerized devices allow direct mapping of such landscapes:

- Enhanced-sampling MD techniques can show early oligomer formation events, which identify hydrophobic patches and aggregation (Williams-Noonan et al., 2023).
- Free-energy landscape methods define the important transition states at which therapeutic treatment can halt misfolding (Plotkin and Onuchic, 2002).

- Computational docking simulations of interactions between misfolded proteins and membranes, which clarify the pathways of toxicity that are supported by electrophysiological data (Gonzalez-Garcia et al., 2021).

With cutting-edge computation and AI, integrative structural biology has changed the nature of neurodegeneration studies into precise mechanisms rather than descriptive pathology (Rao et al., 2024). Through experimentation and in silico modeling, scientists now have access to complete conformational continua, both between monomers and fibrils and also between multiprotein complexes, and have the ability to predict how aging alters these landscapes (LiWang, n.d.). This methodology not only fills gaps in our knowledge that have persisted since ancient times, but also brings disparate datasets into coherent frameworks and accelerates therapeutic development by analyzing structural weaknesses.

Given the ongoing complexity of neurodegenerative diseases, integrative structural biology is the solution, offering predictive biology and individualized intervention in the aging brain (Rao et al., 2024).

15.8 Translational implications: structural insights to therapeutic development

Neurodegenerative diseases such as AD, PD, HD, and ALS are among the most challenging diseases to treat (Pathak et al., 2022). Conventional approaches to treatment have been mainly focused on alleviating symptoms, that is, improving cognitive impairments, enhancing dopaminergic stimulation, or reducing behavioral disruption, and have largely ignored the underlying molecular pathways of neuronal death. Recently, due to advancements in structural biology, the therapeutic focus has shifted towards the major pathogenic processes: protein misfolding, aggregation, and dysregulated proteostasis (Lindquist and Kelly, 2011). The visualization of pathogenic structures can now be performed with atomic precision with technologies including cryo-EM, NMR, AI-assisted prediction of protein structures, and integrative modeling (Jeyaraj et al., 2025). As a result, scientists can develop interventions that prevent aggregation, break up propagative fibril cores, or increase clearance pathways. The chapter discusses how structural knowledge is used in drug development, compares modern approaches with earlier ones, and outlines current translational issues.

15.8.1 Structure-based drug design to misfolded proteins

15.8.1.1 Targeting Aβ

The initial antiamyloid treatments, including 2-secretase inhibitors, have been developed with little structural background and have failed because of their toxicity or lack of target interaction (Giorgetti et al., 2018). As crystal-EM has demonstrated polymorphic structures of Aβ fibrils in AD, structure-based design now targets certain fibril interfaces that stabilize β-sheet stacking. Retrospective binding of small molecules such as tramiprosate and EGCG to prefibrillar species was identified, with new Aβ structures optimized using structural templates (Attar et al., 2013).

15.8.1.2 Targeting tau

Tau-directed therapies have previously been based on kinase inhibition or on mechanisms that decrease phosphorylation, which are biochemical intuitions rather than structural suggestions. The structural explanation of tau PHFs has identified core segments (e.g., the PHF6 motif) that act as aggregation nuclei (Di Lorenzo, 2023). Small molecules and peptides are currently designed to interlock with these β-sheet cores, thereby interfering with fibril formation. These approaches have greater specificity and reduced off-target effects compared to older nonspecific kinase inhibitors (Francioso et al., 2015).

15.8.1.3 Targeting α-synuclein

Historically, general hydrophobic interactions were inhibited by α-synuclein aggregation inhibitors with low success (Galkin et al., 2024). Conformer-selective ligand design has been driven by the identification of distinct fibril polymorphs in PD and multiple system atrophy (Mehra et al., 2021). Drugs are now targeting protofilament interfaces characteristic of disease-specific strains, a precision strategy that was not used in early therapeutic programs (Villalón-García et al., 2020). All these examples demonstrate a significant paradigm shift: structural knowledge enables the targeted inhibition of toxic conformers without inhibiting the physiological activity of proteins.

15.8.2 Molecular chaperone mimetics small-molecule stabilizers

A potential treatment option does not seek to decompose aggregates but to stabilize native protein structures, preventing misfolding at its origin.

15.8.2.1 Pharmacological chaperones

Small molecules can stabilize and bind partially folded intermediates, allowing proper protein folding (Kim and Baldwin, 1982). This has been effective in peripheral proteinopathies, for example, tafamidis in transthyretin amyloidosis (Cruz, 2019). Understanding the transient structural intermediates in neurodegenerative diseases has been challenging. Nevertheless, new cryo-EM and MD experiments are now uncovering transient conformations of tau, TDP-43, and α-synuclein, suggesting the rational development of stabilizing therapeutics.

15.8.2.2 Oscillating pathways of proteostasis

The compounds promoting the activity of HSP70/HSP90 or proteasome activity were first screened empirically (Nitzsche et al., 2023). Structural mapping of chaperone-substrate interactions can now be used to engineer cochaperone mimetics that specifically rescue aggregation-prone proteins (Karunanayake and Page, 2021). These developments address the limitations of early chaperone therapies, which were not selective for their targets and caused undesirable stress responses.

15.8.3 Designing the antibodies and proteolysis-targeting chimeras with structures to achieve selective clearance

15.8.3.1 Antibody-based therapeutics

The most advanced structure-based therapies in the pipeline for neurodegeneration are monoclonal antibodies (mAbs) (Masters et al., 2006).

Early antibodies had low clinical efficacy and were directed to monomeric A4 (e.g., solanezumab) (Sperling et al., 2023).

New mAbs, including lecanemab and donanemab, specifically target toxic oligomers/N-terminally truncated fibrils, according to structurally mapped epitopes (Ameen et al., 2025).

In the case of tau and α-synuclein, epitope mapping using cryo-EM reveals areas of exposure in pathological forms but are buried in physiological forms, making it possible to use conformer-specific immunotherapy (Scheres et al., 2023).

15.8.3.2 Therapeutic peptides

Peptides are designed using:

- breaking of amyloid cores, β-sheet breaker motifs, and
- binding sequences which block propagation surfaces in a competitive way (Ghosh et al., 2024).

Previous peptide therapies were unstable and poorly penetrative to the brain. Present-day structural engineering enhances the delivery of protease-resistant and blood-brain barrier-permeable agents using cyclization or nanoparticle carriers.

15.8.3.3 Proteolysis-targeting chimeras

The proteolysis-targeting chimeras (PROTACs) identify misfolded proteins and recruit them to ubiquitin ligases to target their degradation (Neklesa et al., 2017). Structural modeling is used to design linkers and to make ternary complexes. Even though PROTACs are not yet in the neurodegeneration preclinical stage, they represent a potent means to eliminate intracellular aggregates (e.g., TDP-43 and tau)

(Kong et al., 2025). These modalities demonstrate that structural specificity allows selective clearing without affecting normal protein activity, a property that contrasts with the limitations of previous broad-spectrum approaches.

The paradigm shift in the development of neurodegenerative disease therapies has been driven by structural biology, which has shifted the focus beyond symptomatic treatment to the specific targeting of pathogenic conformer aggregates. By combining sophisticated imaging modalities, complex computational models, and protein engineering, researchers have developed molecular agents that stabilize native protein folds, disaggregate toxic assemblies, or induce their specific elimination. It is pointed out that the classification, in contrast to the former broad and nonspecific methods, proves the tremendous improvements that atomic-resolution knowledge can achieve. Recent progress in imaging, biomarker discovery, and personalization of protein polymorphic diversity can increase the pace of the next generation of structure-based therapeutics for age-related neurodegenerative diseases.

15.9 Novel horizons: structural biomarkers and interventions to aging

Biological aging is a process of ordered molecular events that change the structural integrity of cells, tissues, and organs (DiLoreto and Murphy, 2015). Traditional physiological biomarkers have limited predictive value for age-related reduction; hence, structural biology offers a revolutionary prism for finding early molecular markers and developing rational interventions (Furrer and Handschin, 2025). Newer developments in structural proteomics, metabolomics, and advanced imaging modalities can currently capture dynamic structural degenerative processes and adaptive remodeling by aging cellular populations, especially in the nervous system. These advances have the potential to revolutionize biomarkers of aging in the brain, clarify predictors of exceptional longevity, and inform the future of neurotherapeutics.

15.9.1 Biological proteomics, structural proteomics and metabolomics in biomarker discovery

One of the key processes in aging is proteostasis failure, which results in protein misfolding, aberrant aggregation, and reduced turnover (Hipp et al., 2019). Structural proteomics (including cross-linking mass spectrometry, hydrogen-deuterium exchange, and cryo-EM techniques) enables proteins to be mapped with respect to their conformational states at a proteome-scale level (Lu et al., 2024). Research on aging brain tissue has identified structural weaknesses in synaptic proteins, including impairments in the structure of AMPA and NMDA receptor complexes and disruptions in chaperone networks (Michaelis, 1998). Such structural fingerprints can serve as predictors of neurodegeneration and as targets for structure-based therapy.

In addition, metabolomics captures age-related changes in cellular metabolites that influence macromolecular stability (Hoffman et al., 2017). The buildup of reactive lipid species, disrupted redox balance, and cross-links generated during glycation damage the structure of proteins and membranes. Integrative structural metabolomics has recently enabled the correlation of metabolite changes with structural damage, such as mapping glycation-mediated β-sheet stabilization during early cognitive impairment in amyloidogenic proteins (Sirangelo and Iannuzzi, 2021). These types of multimodal structural signatures can be used to presymptomatically identify disease and to detect therapeutic response with greater accuracy.

15.9.2 Longevity and neuronal resilience structural correlates

Among the most interesting findings in comparative structural aging is that structural deterioration is not experienced equally across organisms or individuals. Variants in genes related to longevity, such as FOXO3A, APOE2, and insulin/IGF-1 pathway regulators, maintain the structural integrity of neuronal membranes and organelles (Ni et al., 2021). Super-aging phenotypes have been maintained in terms of synaptic ultrastructure, an increase in mitochondrial cristae density, and anti-protein aggregation ability despite chronological age (Shigemoto et al., 2010).

Exceptional longevity in animal models, such as the naked mole rat, is associated with enhanced proteome stability, efficient clearance of misfolded proteins, and structural adaptations that confer resistance to oxidative stress (Pérez et al., 2009). These observations highlight that sustained neuronal functioning results from structural maintenance rather than the deceleration of metabolic processes. Structural biomarkers of resilience, including mitochondrial structural fidelity, synaptic density retention, and amyloid resistance, are new biological age predictors and indicators of cognitive health (Yang et al., 2025).

15.9.3 Centenarian proteomics structural adaptation

The centenarians offer a special biological template of healthy aging. Proteomic studies show a preference for proteins with highly optimized folding kinetics, increased thermodynamic stability, and resistance to post-translational damage (Qiu et al., 2024). Structural adaptations are:

- Strengthened chaperone networks that inhibit premature misfolding pathways.
- Decreased aggregation-prone protein intrinsic disorder.
- Structural amyloidogenicity-reducing evolutionarily conserved structural variants.
- Enhanced proteasomal and autophagic clearance, which restrains toxic assemblies.

In addition, centenarian brain tissue has fewer conformational changes in vital synaptic proteins and a more stable lipid structure that forms membrane microdomains (Ganz et al., 2023). Such structural optimizations are evolutionary measures to slow proteostasis collapse and neurodegeneration and, as such, represent models of biomimetic anti-aging intervention.

The future development of aging research will be based on sophisticated imaging technologies that can visualize structural degradation in living tissues and throughout the lifespan. New technologies, such as in situ cryo-electron tomography, are now enabling nanometer-scale mapping of aging neurons in their natural environment, revealing dynamic processes such as cytoskeletal disassembly, mitochondrial fragmentation, and synaptic collapse in real time. The mechanistic insights provided by complementary time-resolved structural biology methods, such as rapid-freeze electron microscopy and ultrafast X-ray scattering, into transient, pathogenic events, such as prion-like tau propagation and oxidative membrane damage, are based on the observation of molecular transitions between initiation events and endpoints. A combination of these novel methodologies will be able to elucidate the structural pathways underlying neuronal aging and neurodegeneration.

A combination of these structural datasets with machine-learning-based aging clocks will provide multiscale predictive indices that connect atomic-level variations with circuit dysfunction and behavioral aging. This type of predictive modeling will hasten interventions to preserve structure: pharmacological stabilizers of weak folds, lipid therapeutics for membrane rigidity, and gene therapies for mitochondrial structure.

15.10 Conclusions and perspectives

The process of aging of the human brain is essentially a molecular and structural challenge. Neurodegenerative diseases are not the result of cell loss or an overload of proteins, but rather of specific conformational changes that alter the normal functioning of the synapses, mitochondria, and neuronal identity. Traditionally, aging studies have relied on general cellular phenotypes, for example, oxidative stress, neuronal death, and synaptic thinning, providing little mechanistic rigor. Modern developments in structural biology have overhauled our understanding of how atomic-level changes can be translated into network dysfunction and clinical impairment.

Structural studies have disproved the belief that aggregates are amorphous waste products by revealing the fibril cores of amyloid-β, tau, α-synuclein, and TDP-43. Rather, they are ordered polymorphic assemblies with disease-specific strain properties that propagate, are toxic, and exhibit heterogeneous phenotypes. Equally, the molecular structures of mitochondrial complexes, the SNARE machinery, chaperones, and autophagy receptors depict how small, gradual changes in age degrading neuronal resilience. These observations are useful for interpolating decades of biochemical observations on memory loss, motor impairment, and cognitive decline to clinical manifestations along a direct molecular-to-symptom continuum.

A paradigm shift in biomarker discovery has also been driven by structural biology. Diagnostic methods have taken advantage of conformation-selective antibodies, structural proteomics, and metabolomic signatures, which have the potential to identify pathology many years before the disease manifests. Different structure-directed antibody reengineering, aggregation-interfering molecules, chaperone mimetics, and targeted degradation methods are replacing previous symptomatic or trial-and-error methods. The current achievements of new precision therapies highlight the need to incorporate structural reasoning into drug design, patient stratification, and clinical testing.

In the future, a number of converging innovations are likely to transform the field in even more radical ways. Structural Systems Biology notes that the next generation of research will not focus on single proteins but rather on interacting structural networks. The insight into the collective failure of proteostasis machinery, immune sensors, and synaptic complexes during aging will provide the means to intervene to restore system-level stability. Time-resolved structuralomics—dynamic structuralomics such as time-resolved cryo-EM, cryo-tomography, high-field NMR, and real-time molecular simulations—will capture the mechanics of transient intermediates, which are the real determinants of misfolding toxicity and were long before classical techniques. The previous static models of protein aggregation will evolve into dynamic groups associated with disease progression. Personalized neurotherapeutics posits that intraprofessional disparities in fibril polymorphs, post-translational modification patterns, and resilience features will lead to more accurate structural medicine. Treatment based on AI-improved prediction and multiomics integration will be tailored to a patient's conformational landscape, turning prevention and therapy into a patient-focused molecular approach.

To sum up, structural biology has transformed descriptive pathology into precision and mechanism, providing the tools to predict, detect, and counter age-related neurodegeneration. The coming generation of brain-aging studies can not only extend life expectancy, but also cognitive health, due to the combination of molecular structures and systems-level modeling with individualized treatment design.

References

Aguzzi, A., O'Connor, T., 2010. Protein aggregation diseases: pathogenicity and therapeutic perspectives. Nature Reviews Drug Discovery 9 (3), 237–248. Available at https://doi.org/10.1038/nrd3050.

Agyemang, A.F., Harrison, S.R., Siegel, R.M., McDermott, M.F., 2015. Protein misfolding and dysregulated protein homeostasis in autoinflammatory diseases and beyond. Seminars in Immunopathology 37 (4), 335–347. Available at https://doi.org/10.1007/s00281-015-0496-2, http://www.springer.com/east/home/generic/search/results?SGWID=5-40109-70-1042641-0.

Akoury, E., 2013. Molecular Mechanisms of Tau Protein Aggregation Inhibition. Niedersächsische Staats- und Universitätsbibliothek Göttingen.

Aksoy, Y.A., Deng, W., Stoddart, J., Chung, R., Guillemin, G., Cole, N.J., et al., 2020. "STRESSED OUT": the role of FUS and TDP-43 in amyotrophic lateral sclerosis. The International Journal of Biochemistry & Cell Biology 126, 105821.

Alberti, S., Hyman, A.A., 2021. Biomolecular condensates at the nexus of cellular stress, protein aggregation disease and ageing. Nature Reviews Molecular Cell Biology 22 (3), 196–213. Available at https://doi.org/10.1038/s41580-020-00326-6, http://www.nature.com/molcellbio.

Alexeyev, M.F., 2009. Is there more to aging than mitochondrial DNA and reactive oxygen species? FEBS Journal 276 (20), 5768–5787. Available at https://doi.org/10.1111/j.1742-4658.2009.07269.x.

Almeida, Z.L., Brito, R.M.M., 2020. Structure and aggregation mechanisms in amyloids. Molecules 25 (5), 1195. Available at https://doi.org/10.3390/molecules25051195.

Alqahtani, T., Deore, S.L., Kide, A.A., Shende, B.A., Sharma, R., Dadarao Chakole, R., et al., 2023. Mitochondrial dysfunction and oxidative stress in Alzheimer's disease, and Parkinson's disease, Huntington's disease and Amyotrophic Lateral Sclerosis -An updated review. Mitochondrion 71, 83–92. Available at https://doi.org/10.1016/j.mito.2023.05.007, https://www.journals.elsevier.com/mitochondrion.

Ameen, Tb, Ali, U., Salma, O., Abdul Samee, M., Iraj Abbas, S.M., Naveera Kashif, S., et al., 2025. Amyloid solutions: lecanemab, gantenerumab, and donanemab in the treatment of Alzheimer's disease. Egyptian Journal of Neurology, Psychiatry and Neurosurgery 61 (1). Available at https://doi.org/10.1186/s41983-025-00968-3, https://www.springer.com/journal/41983.

Arar, S., Haque, M.A., Kayed, R., 2025. Protein aggregation and neurodegenerative disease: Structural outlook for the novel therapeutics. Proteins: Structure, Function and Bioinformatics 93 (8), 1314–1329. Available at https://doi.org/10.1002/prot.26561, http://onlinelibrary.wiley.com/journal/10.1002/(ISSN)1097-0134.

Arivazhagan, L., López-Díez, R., Shekhtman, A., Ramasamy, R., Schmidt, A.M., 2022. Glycation and a spark of ALEs (advanced lipoxidation end products) – igniting RAGE/diaphanous-1 and cardiometabolic disease. Frontiers in Cardiovascular Medicine 9. Available at https://doi.org/10.3389/fcvm.2022.937071, www.frontiersin.org/journals/cardiovascular-medicine.

Asomugha, C., 2011. Post-translational modifications, structure and function of human lens crystallins.

Attar, A., Rahimi, F., Bitan, G., 2013. Modulators of amyloid protein aggregation and toxicity: EGCG and CLR01. Translational Neuroscience 4 (4), 385–409. Available at https://doi.org/10.2478/s13380-013-0137-y.

Aunan, J.R., Watson, M.M., Hagland, H.R., Søreide, K., 2016. Molecular and biological hallmarks of ageing. British Journal of Surgery 103 (2), e29. Available at https://doi.org/10.1002/bjs.10053

Azevedo, R., Jacquemin, C., Villain, N., Fenaille, F., Lamari, F., Becher, F., 2022. Mass spectrometry for neurobiomarker discovery: the relevance of post-translational modifications. Cells 11 (8), 1279. Available at https://doi.org/10.3390/cells11081279.

Barrantes, F.J., 2024. Cognitive synaptopathy: synaptic and dendritic spine dysfunction in age-related cognitive disorders. Frontiers in Aging Neuroscience 16. Available at https://doi.org/10.3389/fnagi.2024.1476909, https://www.frontiersin.org/journals/aging-neuroscience#.

Bashir, S., Aiman, A., Chaudhary, A.A., Khan, N., Ahanger, I.A., Sami, N., et al., 2024. Probing protein aggregation through spectroscopic insights and multimodal approaches: a comprehensive review for counteracting neurodegenerative disorders. Heliyon 10 (7). Available at https://doi.org/10.1016/j.heliyon.2024.e27949, https://www.sciencedirect.com/science/journal/24058440.

Blasiak, J., Pawlowska, E., Szczepanska, J., Kaarniranta, K., 2019. Interplay between autophagy and the ubiquitin-proteasome system and its role in the pathogenesis of age-related macular degeneration. International Journal of Molecular Sciences 20 (1), 210. Available at https://doi.org/10.3390/ijms20010210.

Bogdanovic, N., 2024. Cryo-EM structural features of glutamate receptors involved in synaptic neurotransmission or brain pathology. Cryo-Electron Microscopy in Structural Biology. CRC Press, pp. 349–366.

Brandi, O., 2019. Role of the ALS-linked RNA binding protein TDP-43 in the DNA damage response.

Brigelius-Flohé, R., Banning, A., Kny, M., Böl, G.F., 2004. Redox events in interleukin-1 signaling. Archives of Biochemistry and Biophysics 423 (1), 66–73. Available at https://doi.org/10.1016/j.abb.2003.12.008, http://www.elsevier.com/inca/publications/store/6/2/2/7/8/7/index.htt.

Buetow, L., Huang, D.T., 2016. Structural insights into the catalysis and regulation of E3 ubiquitin ligases. Nature Reviews Molecular Cell Biology 17 (10), 626–642. Available at https://doi.org/10.1038/nrm.2016.91, http://www.nature.com/molcellbio.

Calderwood, S.K., Murshid, A., Prince, T., 2009. The shock of aging: molecular chaperones and the heat shock response in longevity and aging - a mini-review. Gerontology 55 (5), 550–558. Available at https://doi.org/10.1159/000225957.

Candelise, N., Scaricamazza, S., Salvatori, I., Ferri, A., Valle, C., Manganelli, V., et al., 2021. Protein aggregation landscape in neurodegenerative diseases: clinical relevance and future applications. International Journal of Molecular Sciences 22 (11), 6016. Available at https://doi.org/10.3390/ijms22116016.

Cereghetti, G., 2021. Disorder, Order, and Life: Reversible Protein Aggregation as a Fundamental Process in Cell Physiology. ETH Zurich.

Chaudhuri, M., Tripathi, A., Gonzalez, F.S., 2021. Diverse functions of Tim50, a component of the mitochondrial inner membrane protein translocase. International Journal of Molecular Sciences 22 (15), 7779. Available at https://doi.org/10.3390/ijms22157779.

Cieplak, A.S., 2019. Tau inclusions in Alzheimer's, chronic traumatic encephalopathy and Pick's disease. A speculation on how differences in backbone polarization underlie divergent pathways of tau aggregation. Frontiers in Neuroscience 13 (MAY). Available at https://doi.org/10.3389/fnins.2019.00488, https://www.frontiersin.org/journals/neuroscience#.

Clavet-Fournier, V., Lee, C., Wegner, W., Brose, N., Rhee, J., Willig, K.I., 2023. Pre-and postsynaptic nanostructures increase in size and complexity after LTP induction. bioRxiv 2023.02.09.527812.

Colton, C.A., Wilcock, D.M., 2010. Assessing activation states in microglia. CNS & Neurological Disorders-Drug Targets (Formerly Current Drug Targets-CNS & Neurological Disorders), 9 (2), 174-191.

Creekmore, B.C., Chang, Y.W., Lee, E.B., 2021. The cryo-EM effect: structural biology of neurodegenerative disease aggregates. Journal of Neuropathology and Experimental Neurology 80 (6), 514–529. Available at https://doi.org/10.1093/jnen/nlab039, http://journals.lww.com/jneuropath.

Cruz, M.W., 2019. Tafamidis for autonomic neuropathy in hereditary transthyretin (ATTR) amyloidosis: a review. Clinical Autonomic Research 29, 19–24. Available at https://doi.org/10.1007/s10286-019-00625-9, http://www.springerlink.com/content/0959-9851.

Csizmok, V., Follis, A.V., Kriwacki, R.W., Forman-Kay, J.D., 2016. Dynamic protein interaction networks and new structural paradigms in signaling. Chemical Reviews 116 (11), 6424–6462. Available at https://doi.org/10.1021/acs.chemrev.5b00548, http://pubs.acs.org/journal/chreay.

Cuanalo-Contreras, K., Mukherjee, A., Soto, C., 2013. Role of protein misfolding and proteostasis deficiency in protein misfolding diseases and aging. International Journal of Cell Biology 2013, 1–10. Available at https://doi.org/10.1155/2013/638083.

Diaz-parga, P., 2022. Self-assembly and Interdomain Dynamics of Natural and Designed Inflammasome Adaptor Isoforms of ASC: Implications for Inflammasome Regulation. University of California, Merced.

Di Lorenzo, D., 2023. Using β-strand and β-hairpin peptidomimetics induced by constrained amino acids to modulate the aggregation of the microtubule associated protein Tau: design, synthesis and evaluation. Université Paris-Saclay; Università degli studi (Milan, Italie). Convegno di ....

DiLoreto, R., Murphy, C.T., 2015. The cell biology of aging. Molecular Biology of the Cell 26 (25), 4524–4531. Available at https://doi.org/10.1091/mbc.E14-06-1084, http://www.molbiolcell.org/content/26/25/4524.full.pdf+html.

Ding, Q., Dimayuga, E., Keller, J.N., 2006. Proteasome regulation of oxidative stress in aging and age-related diseases of the CNS. Antioxidants and Redox Signaling 8 (1-2), 163–172. Available at https://doi.org/10.1089/ars.2006.8.163.

Doll, S., Burlingame, A.L., 2015. Mass spectrometry-based detection and assignment of protein posttranslational modifications. ACS Chemical Biology 10 (1), 63–71. Available at https://doi.org/10.1021/cb500904b, http://pubs.acs.org/journal/acbcct.

Eaton, A.F., Merkulova, M., Brown, D., 2021. The H+-ATPase (V-ATPase): from proton pump to signaling complex in health and disease. American Journal of Physiology - Cell Physiology 320 (3), C392. Available at https://doi.org/10.1152/ajpcell.00442.2020, https://journals.physiology.org/doi/pdf/10.1152/ajpcell.00442.2020.

Ebert, T., Tran, N., Schurgers, L., Stenvinkel, P., Shiels, P.G., 2022. Ageing-oxidative stress, PTMs and disease. Molecular Aspects of Medicine 86, 101099.

Edogbanya, U.-O.J., 2023. Understanding the Role of Novel Cancer-related Gene C1ORF112 in Relation to DNA Damage and Repair. University of Liverpool.

Egelman, E.H., 2016. The current revolution in cryo-EM. Biophysical Journal 110 (5), 1008–1012. Available at https://doi.org/10.1016/j.bpj.2016.02.001, http://www.sciencedirect.com/science/journal/00063495.

Emendato, A., Milordini, G., Zacco, E., Sicorello, A., Dal Piaz, F., Guerrini, R., et al., 2018. Glycation affects fibril formation of Aβ peptides. Journal of Biological Chemistry 293 (34), 13100–13111. Available at https://doi.org/10.1074/jbc.ra118.002275.

Erchova, I., Sun, S., Votruba, M., 2021. A perspective on accelerated aging caused by the genetic deficiency of the metabolic protein, OPA1. Frontiers in Neurology 12. Available at https://doi.org/10.3389/fneur.2021.641259.

Fanlo-Ucar, H., Picón-Pagès, P., Herrera-Fernández, V., ILL-Raga, G., Muñoz, F.J., 2024. The dual role of amyloid beta-peptide in oxidative stress and inflammation: unveiling their connections in Alzheimer's disease etiopathology. Antioxidants 13 (10), 1208. Available at https://doi.org/10.3390/antiox13101208.

Fernández Sanz, C., 2015. Defective Sarcoplasmic Reticulum-mitochondria Communication in Aged Heart and its Effect on Ischemia and Reperfusion Injury. Universitat Autònoma de Barcelona.

Fitzpatrick, A.W., Saibil, H.R., 2019. Cryo-EM of amyloid fibrils and cellular aggregates. Current Opinion in Structural Biology 58, 34–42. Available at https://doi.org/10.1016/j.sbi.2019.05.003, http://www.elsevier.com/locate/sbi.

Francioso, A., Punzi, P., Boffi, A., Lori, C., Martire, S., Giordano, C., et al., 2015. β-Sheet interfering molecules acting against β-amyloid aggregation and fibrillogenesis. Bioorganic & Medicinal Chemistry 23 (8), 1671–1683. Available at https://doi.org/10.1016/j.bmc.2015.02.041.

Frankowska, N., Lisowska, K., Witkowski, J.M., 2022. Proteolysis dysfunction in the process of aging and age-related diseases. Frontiers in Aging 3. Available at https://doi.org/10.3389/fragi.2022.927630, www.frontiersin.org/journals/aging.

Fu, Y., Kim, H., Lee, D.S., Han, A.R., Heine, H., Zamyatina, A., et al., 2025. Structural insight into TLR4/MD-2 activation by synthetic LPS mimetics with distinct binding modes. Nature Communications 16 (1). Available at https://doi.org/10.1038/s41467-025-59550-3, https://www.nature.com/ncomms/.

Furrer, R., Handschin, C., 2025. Biomarkers of aging: from molecules and surrogates to physiology and function. Physiological Reviews 105 (3), 1609–1694. Available at https://doi.org/10.1152/physrev.00045.2024.

Galkin, M., Priss, A., Kyriukha, Y., Shvadchak, V., 2024. Navigating α-synuclein aggregation inhibition: methods, mechanisms, and molecular targets. The Chemical Record 24 (2). Available at https://doi.org/10.1002/tcr.202300282.

Galli, F., Bartolini, D., Ronco, C., 2024. Oxidative stress, defective proteostasis and immunometabolic complications in critically ill patients. European Journal of Clinical Investigation 54 (9). Available at https://doi.org/10.1111/eci.14229.

Ganz, A.B., Zhang, M., Koopmans, F., Li, K.W., Miedema, S.S.M., Rozemuller, A.J.M., et al., 2023. Proteomic profiling of aging brains identifies key proteins by which cognitively healthy centenarians defy their age by decades. medRxiv. Available at https://doi.org/10.1101/2023.11.30.23299224, https://www.medrxiv.org/.

Gates, S., 2017. Understanding protein folding mechanisms of chaperone proteins using cryo-EM.

Georgiev, D.D., Papaioanou, S.N., Glazebrook, J.F., 2004. Neuronic system inside neurons: molecular biology and biophysics of neuronal microtubules. Biomedical Reviews 15 (1), 67–75. Available at https://doi.org/10.14748/bmr.v15.103, http://press.mu-varna.bg/ojs/index.php/bmr/issue/view/171.

Ghosh, D., Biswas, A., Radhakrishna, M., 2024. Advanced computational approaches to understand protein aggregation. Biophysics Reviews 5 (2). Available at https://doi.org/10.1063/5.0180691.

Giorgetti, S., Greco, C., Tortora, P., Aprile, F.A., 2018. Targeting amyloid aggregation: an overview of strategies and mechanisms. International Journal of Molecular Sciences 19 (9), 2677. Available at https://doi.org/10.3390/ijms19092677.

Goedert, M., Falcon, B., Zhang, W., Ghetti, B., Scheres, S.H.W., 2018. Distinct conformers of assembled tau in Alzheimer's and Pick's diseases. Cold Spring Harbor Symposia on Quantitative Biology 83, 163–171. Available at https://doi.org/10.1101/sqb.2018.83.037580, http://www.cshl-symposium.org.

Gonzalez-Garcia, M., Fusco, G., De Simone, A., 2021. Membrane interactions and toxicity by misfolded protein oligomers. Frontiers in Cell and Developmental Biology 9. Available at https://doi.org/10.3389/fcell.2021.642623.

Grasso, G., Rebella, M., Morbiducci, U., Tuszynski, J.A., Danani, A., Deriu, M.A., 2019. The role of structural polymorphism in driving the mechanical performance of the Alzheimer's beta amyloid fibrils. Frontiers in Bioengineering and Biotechnology 7. Available at https://doi.org/10.3389/fbioe.2019.00083, https://www.frontiersin.org/articles/10.3389/fbioe.2019.00083/full.

Grba, D.N., Chung, I., Bridges, H.R., Agip, A.N.A., Hirst, J., 2023. Investigation of hydrated channels and proton pathways in a high-resolution cryo-EM structure of mammalian complex I. Science Advances 9 (31). Available at https://doi.org/10.1126/sciadv.adi1359, https://www.science.org/toc/sciadv/9/31.

Gregersen, N., Bross, P., 2010. Protein misfolding and cellular stress: an overview. Protein Misfolding and Cellular Stress in Disease and Aging: Concepts and Protocols, 3-23.

Griffiths, H.R., 2000. Antioxidants and protein oxidation. Free Radical Research 33 S47-S58S47 United Kingdom.

Grootjans, J., Kaser, A., Kaufman, R.J., Blumberg, R.S., 2016. The unfolded protein response in immunity and inflammation. Nature Reviews Immunology 16 (8), 469–484. Available at https://doi.org/10.1038/nri.2016.62, http://www.nature.com/nri/index.html.

Guerrero-Ferreira, R., Kovacik, L., Ni, D., Stahlberg, H., 2020. New insights on the structure of alpha-synuclein fibrils using cryo-electron microscopy. Current Opinion in Neurobiology 61, 89–95. Available at https://doi.org/10.1016/j.conb.2020.01.014.

Guo, J., Huang, X., Dou, L., Yan, M., Shen, T., Tang, W., et al., 2022. Aging and aging-related diseases: from molecular mechanisms to interventions and treatments. Signal Transduction and Targeted Therapy 7 (1). Available at https://doi.org/10.1038/s41392-022-01251-0.

Hagai, T., Levy, Y., 2010. Ubiquitin not only serves as a tag but also assists degradation by inducing protein unfolding. Proceedings of the National Academy of Sciences 107 (5), 2001–2006. Available at https://doi.org/10.1073/pnas.0912335107.

Haider, R., 2024. The Role of Phosphorylation as a Regulatory Mechanism in Liquid-Liquid Phase Separation of TDP-43 Low Complexity Domain. Case Western Reserve University.

Hanslik, K.L., Ulland, T.K., 2020. The Role of Microglia and the Nlrp3 Inflammasome in Alzheimer's Disease. Frontiers in Neurology 11. Available at https://doi.org/10.3389/fneur.2020.570711, http://www.frontiersin.org/Neurology.

Hauser, P.S., Narayanaswami, V., Ryan, R.O., 2011. Apolipoprotein E: from lipid transport to neurobiology. Progress in Lipid Research 50 (1), 62–74. Available at https://doi.org/10.1016/j.plipres.2010.09.001.

Held, R.G., Liang, J., Brunger, A.T., 2024. Nanoscale architecture of synaptic vesicles and scaffolding complexes revealed by cryo-electron tomography. Proceedings of the National Academy of Sciences 121 (27). Available at https://doi.org/10.1073/pnas.2403136121.

Hipp, M.S., Kasturi, P., Hartl, F.U., 2019. The proteostasis network and its decline in ageing. Nature Reviews Molecular Cell Biology 20 (7), 421–435. Available at https://doi.org/10.1038/s41580-019-0101-y, http://www.nature.com/molcellbio.

Hoffman, J.M., Lyu, Y., Pletcher, S.D., Promislow, D.E.L., 2017. Proteomics and metabolomics in ageing research: from biomarkers to systems biology. Essays in Biochemistry 61 (3), 379–388. Available at https://doi.org/10.1042/EBC20160083, http://essays.biochemistry.org/content/61/3/379.full-text.pdf.

Höhn, A., Tramutola, A., Cascella, R., 2020. Proteostasis failure in neurodegenerative diseases: focus on oxidative stress. Oxidative Medicine and Cellular Longevity 2020, 1–21. Available at https://doi.org/10.1155/2020/5497046.

Holec, S.A.M., Liu, S.L., Woerman, A.L., 2022. Consequences of variability in α-synuclein fibril structure on strain biology. Acta Neuropathologica 143 (3), 311–330. Available at https://doi.org/10.1007/s00401-022-02403-w, link.springer.de/link/service/journals/00401/index.htm.

Horwich, A., 2002. Protein aggregation in disease: a role for folding intermediates forming specific multimeric interactions. Journal of Clinical Investigation 110 (9), 1221–1232. Available at https://doi.org/10.1172/jci0216781.

Hou, X., Fiesel, F.C., Truban, D., Castanedes Casey, M., Lin, Wl, Soto, A.I., et al., 2018. Age- and disease-dependent increase of the mitophagy marker phospho-ubiquitin in normal aging and Lewy body disease. Autophagy 14 (8), 1404–1418. Available at https://doi.org/10.1080/15548627.2018.1461294, http://www.tandfonline.com/toc/kaup20/current.

Hou, Y., Dan, X., Babbar, M., Wei, Y., Hasselbalch, S.G., Croteau, D.L., et al., 2019. Ageing as a risk factor for neurodegenerative disease. Nature Reviews Neurology 15 (10), 565–581. Available at https://doi.org/10.1038/s41582-019-0244-7, http://www.nature.com/nrneurol/archive/index.html.

Iasevoli, F., Tomasetti, C., de Bartolomeis, A., 2013. Scaffolding proteins of the post-synaptic density contribute to synaptic plasticity by regulating receptor localization and distribution: relevance for neuropsychiatric diseases. Neurochemical Research 38 (1), 1–22. Available at https://doi.org/10.1007/s11064-012-0886-y.

Jaczynska, K.M., 2024. Molecular Basis of Coupling Calcium Sensing to Fast Membrane Fusion by Synaptotagmin-1 in Neurotransmitter Release.

Jahn, R., Cafiso, D.C., Tamm, L.K., 2024. Mechanisms of SNARE proteins in membrane fusion. Nature Reviews Molecular Cell Biology 25 (2), 101–118. Available at https://doi.org/10.1038/s41580-023-00668-x, https://www.nature.com/molcellbio.

Jeyaraj, G., Rajendran, A.K., Sathishkumar, K., Almutairi, B.O., Vadivelu, A., Chokkakula, S., et al., 2025. High-resolution protein modeling through Cryo-EM and AI: current trends and future perspectives – a review. Frontiers in Molecular Biosciences 12. Available at https://doi.org/10.3389/fmolb.2025.1688455, journal.frontiersin.org/journal/molecular-biosciences.

Jiang, Q., Liu, J., Huang, S., Wang, X.Y., Chen, X., Liu, G.H., et al., 2025. Antiageing strategy for neurodegenerative diseases: from mechanisms to clinical advances. Signal Transduction and Targeted Therapy 10 (1). Available at https://doi.org/10.1038/s41392-025-02145-7, https://www.nature.com/sigtrans.

Johnson, L.N., Lewis, R.J., 2001. Structural basis for control by phosphorylation. Chemical Reviews 101 (8), 2209–2242. Available at https://doi.org/10.1021/cr000225s.

Johnson, J.T., Awosiminiala, F.W., Anumudu, C.K., 2025. Exploring protein misfolding and aggregate pathology in neurodegenerative diseases: from molecular mechanisms to clinical interventions. Applied Sciences 15 (18), 10285. Available at https://doi.org/10.3390/app151810285.

Ju, J., Neuen, S.M.L., Zandvoort, Mv, Keulers, T.G.H., Rouschop, K.M.A., 2025. Extracellular vesicles: cargo loading, degradation and secretory pathways, and their intersection with autophagy. Extracellular Vesicles and Circulating Nucleic Acids 6 (3), 355–380. Available at https://doi.org/10.20517/evcna.2025.21, https://f.oaes.cc/xmlpdf/published/article/2bbba4cc60735f497cfe6bec3ee2c1b8/evcna6021.pdf.

Jurcău, M.C., Andronie-Cioara, F.L., Jurcău, A., Marcu, F., Ţiţ, D.M., Paşcalău, N., et al., 2022. The link between oxidative stress, mitochondrial dysfunction and neuroinflammation in the pathophysiology of Alzheimer's disease: therapeutic implications and future perspectives. Antioxidants 11 (11). Available at https://doi.org/10.3390/antiox11112167, www.mdpi.com/journal/antioxidants.

Kabir, F., Atkinson, R., Cook, A.L., Phipps, A.J., King, A.E., 2023. The role of altered protein acetylation in neurodegenerative disease. Frontiers in Aging Neuroscience 14. Available at https://doi.org/10.3389/fnagi.2022.1025473, https://www.frontiersin.org/journals/aging-neuroscience#.

Karunanayake, C., Page, R.C., 2021. Cytosolic protein quality control machinery: Interactions of Hsp70 with a network of co-chaperones and substrates. Experimental Biology and Medicine 246 (12), 1419–1434. Available at https://doi.org/10.1177/1535370221999812, http://www.uk.sagepub.com/journals/Journal202180.

Kaushik, S., Cuervo, A.M., 2015. Proteostasis and aging. Nature Medicine 21 (12), 1406–1415. Available at https://doi.org/10.1038/nm.4001, http://www.nature.com/nm/index.html.

Kawai, K., Saito, A., Sudo, T., Osada, H., 2008. Specific regulation of cytokine-dependent p38 MAP kinase activation by p62/SQSTM1. Journal of Biochemistry 143 (6), 765–772. Available at https://doi.org/10.1093/jb/mvn027.

Khan, T., Waseem, R., Zehra, Z., Aiman, A., Bhardwaj, P., Ansari, J., et al., 2022. Mitochondrial dysfunction: pathophysiology and mitochondria-targeted drug delivery approaches. Pharmaceutics 14 (12), 2657. Available at https://doi.org/10.3390/pharmaceutics14122657.

Kim, P.S., Baldwin, R.L., 1982. Specific intermediates in the folding reactions of small proteins and the mechanism of protein folding. Annual Review of Biochemistry 51, 459–489. Available at https://doi.org/10.1146/annurev.bi.51.070182.002331.

Kleinfeld, D., Bharioke, A., Blinder, P., Bock, D.D., Briggman, K.L., Chklovskii, D.B., et al., 2011. Large-scale automated histology in the pursuit ofconnectomes. Journal of Neuroscience 31 (45), 16125–16138. Available at https://doi.org/10.1523/JNEUROSCI.4077-11.2011, http://www.jneurosci.org/content/31/45/16125.full.pdf#page=1&view=FitH.

Koga, H., Kaushik, S., Cuervo, A.M., 2011. Protein homeostasis and aging: the importance of exquisite quality control. Ageing Research Reviews 10 (2), 205–215. Available at https://doi.org/10.1016/j.arr.2010.02.001.

Kong, D., Meng, L., Lin, P., Wu, G., 2025. Advancements in PROTAC-based therapies for neurodegenerative diseases. Future Medicinal Chemistry 17 (5), 591–605. Available at https://doi.org/10.1080/17568919.2025.2463310.

Krishnamurthy, H.K., Pereira, M., Rajavelu, I., Jayaraman, V., Krishna, K., Wang, T., et al., 2024. Oxidative stress: fundamentals and advances in quantification techniques. Frontiers in Chemistry 12. Available at https://doi.org/10.3389/fchem.2024.1470458, http://journal.frontiersin.org/journal/chemistry.

Krisko, A., Radman, M., 2019. Protein damage, ageing and age-related diseases. Open Biology 9 (3). Available at https://doi.org/10.1098/rsob.180249.

Labbadia, J., Novoselov, S.S., Bett, J.S., Weiss, A., Paganetti, P., Bates, G.P., et al., 2012. Suppression of protein aggregation by chaperone modification of high molecular weight complexes. Brain 135 (4), 1180–1186. Available at https://doi.org/10.1093/brain/aws022, http://brain.oxfordjournals.org/.

Landrieu, I., Dupré, E., Sinnaeve, D., El Hajjar, L., Smet-Nocca, C., 2022. Deciphering the structure and formation of amyloids in neurodegenerative diseases with chemical biology tools. Frontiers in Chemistry 10. Available at https://doi.org/10.3389/fchem.2022.886382.

Lee, Y.-T.C., Hsu, S.-T.D., 2017. Familial mutations and post-translational modifications of UCH-L1 in Parkinson's disease and neurodegenerative disorders. Current Protein & Peptide Science 18 (7), 733–745. Available at https://doi.org/10.2174/1389203717666160217143721.

Li, Q., Babinchak, W.M., Surewicz, W.K., 2021a. Cryo-EM structure of amyloid fibrils formed by the entire low complexity domain of TDP-43. Nature Communications 12 (1). Available at https://doi.org/10.1038/s41467-021-21912-y, https://www.nature.com/ncomms/.

Li, H., Uittenbogaard, M., Hao, L., Chiaramello, A., 2021b. Clinical insights into mitochondrial neurodevelopmental and neurodegenerative disorders: their biosignatures from mass spectrometry-based metabolomics. Metabolites 11 (4), 233. Available at https://doi.org/10.3390/metabo11040233.

Liang, Y., Zhong, G., Ren, M., Sun, T., Li, Y., Ye, M., et al., 2023. The role of ubiquitin–proteasome system and mitophagy in the pathogenesis of Parkinson's disease. NeuroMolecular Medicine 25 (4), 471–488. Available at https://doi.org/10.1007/s12017-023-08755-0.

Lin, X., Li, S., Zhao, Y., Ma, X., Zhang, K., He, X., et al., 2013. Interaction domains of p62: a bridge between p62 and selective autophagy. DNA and Cell Biology 32 (5), 220–227. Available at https://doi.org/10.1089/dna.2012.1915.

Lindquist, S.L., Kelly, J.W., 2011. Chemical and biological approaches for adapting proteostasis to ameliorate protein misfolding and aggregation diseases-progress and prognosis. Cold Spring Harbor Perspectives in Biology 3 (12), a004507. Available at https://doi.org/10.1101/cshperspect.a004507.

Lippens, G., Gigant, B., 2019. Elucidating Tau function and dysfunction in the era of cryo-EM. Journal of Biological Chemistry 294 (24), 9316–9325. Available at https://doi.org/10.1074/jbc.rev119.008031.

Lista, S., Hampel, H., 2017. Synaptic degeneration and neurogranin in the pathophysiology of Alzheimer's disease. Expert Review of Neurotherapeutics 17 (1), 47–57. Available at https://doi.org/10.1080/14737175.2016.1204234.

Liu, Y., Zhai, H., Alemayehu, H., Boulanger, J., Hopkins, L.J., Borgeaud, A.C., et al., 2023. Cryo-electron tomography of NLRP3-activated ASC complexes reveals organelle co-localization. Nature Communications 14 (1). Available at https://doi.org/10.1038/s41467-023-43180-8, https://www.nature.com/ncomms/.

LiWang, A., n.d. Track: Structure and Dynamics Perspectives on Enzyme Function.

López-Gil, L., Pascual-Ahuir, A., Proft, M., 2023. Genomic instability and epigenetic changes during aging. International Journal of Molecular Sciences 24 (18), 14279. Available at https://doi.org/10.3390/ijms241814279.

Lu, H., Zhu, Z., Fields, L., Zhang, H., Li, L., 2024. Mass spectrometry structural proteomics enabled by limited proteolysis and cross-linking. Mass Spectrometry Reviews. Available at https://doi.org/10.1002/mas.21908.

Lyubchenko, Y.L., Sherman, S., Shlyakhtenko, L.S., Uversky, V.N., 2006. Nanoimaging for protein misfolding and related diseases. Journal of Cellular Biochemistry 99 (1), 52–70. Available at https://doi.org/10.1002/jcb.20989, http://onlinelibrary.wiley.com/journal/10.1002/(ISSN)1097-4644.

Macario, A.J.L., De Macario, E.C., 2002. Sick chaperones and ageing: a perspective. Ageing Research Reviews 1 (2), 295–311. Available at https://doi.org/10.1016/S1568-1637(01)00005-8.

Masters, C.L., Cappai, R., Barnham, K.J., Villemagne, V.L., 2006. Molecular mechanisms for Alzheimer's disease: implications for neuroimaging and therapeutics. Journal of Neurochemistry 97 (6), 1700–1725. Available at https://doi.org/10.1111/j.1471-4159.2006.03989.x.

Mattson, M.P., Magnus, T., 2006. Ageing and neuronal vulnerability. Nature Reviews Neuroscience 7 (4), 278–294. Available at https://doi.org/10.1038/nrn1886.

Medina-Carmona, E., Betancor-Fernández, I., Santos, J., Mesa-Torres, N., Grottelli, S., Batlle, C., et al., 2019. Insight into the specificity and severity of pathogenic mechanisms associated with missense mutations through experimental and structural perturbation analyses. Human Molecular Genetics 28 (1), 1–15. Available at https://doi.org/10.1093/hmg/ddy323, http://hmg.oxfordjournals.org/.

Mehra, S., Gadhe, L., Bera, R., Sawner, A.S., Maji, S.K., 2021. Structural and functional insights into α-synuclein fibril polymorphism. Biomolecules 11 (10). Available at https://doi.org/10.3390/biom11101419, https://www.mdpi.com/2218-273X/11/10/1419/pdf.

Mena, R., Wischik, C.M., Novak, M., Milstein, C., Cuello, A.C., 1991. A progressive deposition of paired helical filaments (PHF) in the brain characterizes the evolution of dementia in alzheimer's disease: an immunocytochemical study with a monoclonal antibody against the PHF core. Journal of Neuropathology and Experimental Neurology 50 (4), 474–490. Available at https://doi.org/10.1097/00005072-199107000-00008.

Michaelis, E.K., 1998. Molecular biology of glutamate receptors in the central nervous system and their role in excitotoxicity, oxidative stress and aging. Progress in Neurobiology 54 (4), 369–415. Available at https://doi.org/10.1016/S0301-0082(97)00055-5, www.elsevier.com/locate/pneurobio.

Milordini, G., 2019. A Structural Approach to Investigate Glycation Effect on Protein Aggregation. King's College, London.

Mishra, E., Thakur, M.K., 2023. Mitophagy: a promising therapeutic target for neuroprotection during ageing and age-related diseases. British Journal of Pharmacology 180 (12), 1542–1561. Available at https://doi.org/10.1111/bph.16062, http://onlinelibrary.wiley.com/journal/10.1111/(ISSN)1476-5381.

Moldogazieva, N.T., Mokhosoev, I.M., Mel'Nikova, T.I., Porozov, Y.B., Terentiev, A.A., 2019. Oxidative stress and advanced lipoxidation and glycation end products (ALEs and AGEs) in aging and age-related diseases. Oxidative Medicine and Cellular Longevity 2019. Available at https://doi.org/10.1155/2019/3085756, http://www.hindawi.com/journals/oximed/.

Moore, M.N., 2020. Lysosomes, autophagy, and hormesis in cell physiology, pathology, and age-related disease. Dose-Response 18 (3). Available at https://doi.org/10.1177/1559325820934227, http://dos.sagepub.com/.

Morán Luengo, T., Mayer, M.P., Rüdiger, S.G.D., 2019. The Hsp70–Hsp90 chaperone cascade in protein folding. Trends in Cell Biology 29 (2), 164–177. Available at https://doi.org/10.1016/j.tcb.2018.10.004, www.elsevier.com/locate/tcb.

Mukhopadhyay, S., 2020. The dynamism of intrinsically disordered proteins: binding-induced folding, amyloid formation, and phase separation. The Journal of Physical Chemistry B 124 (51), 11541–11560. Available at https://doi.org/10.1021/acs.jpcb.0c07598.

Neklesa, T.K., Winkler, J.D., Crews, C.M., 2017. Targeted protein degradation by PROTACs. Pharmacology & Therapeutics 174, 138–144. Available at https://doi.org/10.1016/j.pharmthera.2017.02.027.

Nesci, S., Trombetti, F., Pagliarani, A., Ventrella, V., Algieri, C., Tioli, G., et al., 2021. Molecular and supramolecular structure of the mitochondrial oxidative phosphorylation system: implications for pathology. Life 11 (3), 242. Available at https://doi.org/10.3390/life11030242.

Ni, X., Wang, Z., Gao, D., Yuan, H., Sun, L., Zhu, X., et al., 2021. A description of the relationship in healthy longevity and aging-related disease: from gene to protein. Immunity & Ageing 18 (1). Available at https://doi.org/10.1186/s12979-021-00241-0.

Nitzsche, B., Höpfner, M., Biersack, B., 2023. Synthetic small molecule modulators of Hsp70 and Hsp40 chaperones as promising anticancer agents. International Journal of Molecular Sciences 24 (4), 4083. Available at https://doi.org/10.3390/ijms24044083.

Omidi, A., Møller, M.H., Malhis, N., Bui, J.M., Gsponer, J., 2024. AlphaFold-multimer accurately captures interactions and dynamics of intrinsically disordered protein regions. Proceedings of the National Academy of Sciences 121 (44). Available at https://doi.org/10.1073/pnas.2406407121.

Orr, B., 2025. Improved Prediction of Structurally Diverse De Novo Proteins. University of California, San Francisco.

Osman, S., 2022. PINK spots: diseased mitochondria prepare for mitophagy. Nature Structural & Molecular Biology 29 (2), 82. Available at https://doi.org/10.1038/s41594-022-00733-7 82.

Palara, A., 2021. Mechanism and structural requirements for formation of p62 bodies and degradation of p62 by selective autophagy.

Panicker, N., Dawson, V.L., Dawson, T.M., 2017. Activation mechanisms of the E3 ubiquitin ligase parkin. Biochemical Journal 474 (18), 3075–3086. Available at https://doi.org/10.1042/BCJ20170476, http://www.biochemj.org/content/ppbiochemj/474/18/3075.full.pdf.

Passos, J.F., Kirkwood, T., 2015. Biological and physiological aspects of aging. Textbook of Geriatric Dentistry 7, 29–36.

Pathak, N., Vimal, S.K., Tandon, I., Agrawal, L., Hongyi, C., Bhattacharyya, S., 2022. Neurodegenerative disorders of Alzheimer, Parkinsonism, amyotrophic lateral sclerosis and multiple sclerosis: an early diagnostic approach for precision treatment. Metabolic Brain Disease 37 (1), 67–104. Available at https://doi.org/10.1007/s11011-021-00800-w, www.wkap.nl/journalhome.htm/0885-7490.

Pérez, V.I., Buffenstein, R., Masamsetti, V., Leonard, S., Salmon, A.B., Mele, J., et al., 2009. Protein stability and resistance to oxidative stress are determinants of longevity in the longest-living rodent, the naked mole-rat. Proceedings of the National Academy of Sciences 106 (9), 3059–3064. Available at https://doi.org/10.1073/pnas.0809620106.

Perrone, F., Cacace, R., van der Zee, J., Van Broeckhoven, C., 2021. Emerging genetic complexity and rare genetic variants in neurodegenerative brain diseases. Genome Medicine 13 (1). Available at https://doi.org/10.1186/s13073-021-00878-y.

Plotkin, S.S., Onuchic, J.N., 2002. Understanding protein folding with energy landscape theory. Part I: basic concepts. Quarterly Reviews of Biophysics 35 (2), 111–167. Available at https://doi.org/10.1017/S0033583502003761.

Prasad, A., Bharathi, V., Sivalingam, V., Girdhar, A., Patel, B.K., 2019. Molecular mechanisms of TDP-43 misfolding and pathology in amyotrophic lateral sclerosis. Frontiers in Molecular Neuroscience 12. Available at https://doi.org/10.3389/fnmol.2019.00025, https://www.frontiersin.org/journals/molecular-neuroscience.

Prosad Banik, S., 2024. Glycation-induced amyloid formation in proteins: an emerging perspective to explore diabetes associated onset of neurodegenerative symptoms. Current Nutrition & Food Science 20 (1), 2–7. Available at https://doi.org/10.2174/1573401319666230224094812.

Qiu, X., Lu, Y., Mu, C., Tang, P., Liu, Y., Huang, Y., et al., 2024. The biomarkers in extreme longevity: insights gained from metabolomics and proteomics. International Journal of Medical Sciences 21 (14), 2725–2744. Available at https://doi.org/10.7150/ijms.98778, https://www.medsci.org/v21p2725.pdf.

Radhakrishnan, A., Li, X., Grushin, K., Krishnakumar, S.S., Liu, J., Rothman, J.E., 2021. Symmetrical arrangement of proteins under release-ready vesicles in presynaptic terminals. Proceedings of the National Academy of Sciences 118 (5). Available at https://doi.org/10.1073/pnas.2024029118.

Rajak, A., 2025. Degradation of misfolded proteins in neurodegenerative diseases: therapeutic targets and strategies. Proteostasis: Investigating Molecular Dynamics in Neurodegenerative Disorders. Springer Science+Business Media, India, pp. 371–406. Available at http://doi.org/10.1007/978-981-96-6202-9_17.

Rajan, R., Ahmed, S., Sharma, N., Kumar, N., Debas, A., Matsumura, K., 2021. Review of the current state of protein aggregation inhibition from a materials chemistry perspective: special focus on polymeric materials. Materials Advances 2 (4), 1139–1176. Available at https://doi.org/10.1039/d0ma00760a.

Ramachandran, G., 2016. Fourier Transform Infrared (FTIR) Spectroscopy, Ultraviolet Resonance Raman (UVRR) Spectroscopy, and Atomic Force Microscopy (AFM) for Study of the Kinetics of Formation and Structural Characterization of Tau Fibrils1523. Springer Science and Business Media LLC, pp. 113–128. Available at http://doi.org/10.1007/978-1-4939-6598-4_7.

Rani, L., Mallajosyula, S.S., 2021. Phosphorylation-induced structural reorganization in tau-paired helical filaments. ACS Chemical Neuroscience 12 (9), 1621–1631. Available at https://doi.org/10.1021/acschemneuro.1c00084, http://pubs.acs.org/journal/acncdm.

Rao, N.G.R., Singh, G., Patil, A.R.B., Aparna, T.N., Vippamakula, S., Dharmalingam, S., et al., 2024. Advances in computational biology for diagnosing neurodegenerative diseases: a comprehensive review. Chinese Journal of Applied Physiology 40, e20240008. Available at https://doi.org/10.62958/j.cjap.2024.008.

Rasool, S., Trempe, J.F., 2018. New insights into the structure of PINK1 and the mechanism of ubiquitin phosphorylation. Critical Reviews in Biochemistry and Molecular Biology 53 (5), 515–534. Available at https://doi.org/10.1080/10409238.2018.1491525, http://www.tandfonline.com/loi/ibmg20.

Recharla, M., n.d. Translational Advances in Neurological and Renal Therapeutics. Global Pen Press UK PUBLICATION.

Rhoads, S.N., Monahan, Z.T., Yee, D.S., Shewmaker, F.P., 2018. The role of post-translational modifications on prion-like aggregation and liquid-phase separation of FUS. International Journal of Molecular Sciences 19 (3). Available at https://doi.org/10.3390/ijms19030886.

Rochon, K., Bauer, B.L., Roethler, N.A., Buckley, Y., Su, C.C., Huang, W., et al., 2024. Structural basis for regulated assembly of the mitochondrial fission GTPase Drp1. Nature Communications 15 (1). Available at https://doi.org/10.1038/s41467-024-45524-4, https://www.nature.com/ncomms/.

Rutter, A.R., 2001. Biochemical and pharmacological characterisation of the interaction between nMDA receptors and the scaffolding protein, PSDd-95. University of London, University College London (United Kingdom).

Scheres, S.H., Zhang, W., Falcon, B., Goedert, M., 2020. Cryo-EM structures of tau filaments. Current Opinion in Structural Biology 64, 17–25. Available at https://doi.org/10.1016/j.sbi.2020.05.011, http://www.elsevier.com/locate/sbi.

Scheres, S.H.W., Ryskeldi-Falcon, B., Goedert, M., 2023. Molecular pathology of neurodegenerative diseases by cryo-EM of amyloids. Nature 621 (7980), 701–710. Available at https://doi.org/10.1038/s41586-023-06437-2.

Sebastian-Valverde, M., Pasinetti, G.M., 2020. The NLRP3 inflammasome as a critical actor in the inflammaging process. Cells 9 (6), 1552. Available at https://doi.org/10.3390/cells9061552.

Selemon, L.D., 2013. A role for synaptic plasticity in the adolescent development of executive function. Translational Psychiatry 3 (3), e238. Available at https://doi.org/10.1038/tp.2013.7.

Sergin, I., Bhattacharya, S., Emanuel, R., Esen, E., Stokes, C.J., Evans, T.D., et al., 2016. Inclusion bodies enriched for p62 and polyubiquitinated proteins in macrophages protect against atherosclerosis. Science Signaling 9 (409). Available at https://doi.org/10.1126/scisignal.aad5614, http://stke.sciencemag.org/content/sigtrans/9/409/ra2.full.pdf.

Shahmoradian, S.H., Lewis, A.J., Genoud, C., Hench, J., Moors, T.E., Navarro, P.P., et al., 2019. Lewy pathology in Parkinson's disease consists of crowded organelles and lipid membranes. Nature Neuroscience 22 (7), 1099–1109. Available at https://doi.org/10.1038/s41593-019-0423-2, http://www.nature.com/neuro/index.html.

Shaid, S., Brandts, C.H., Serve, H., Dikic, I., 2013. Ubiquitination and selective autophagy. Cell Death & Differentiation 20 (1), 21–30. Available at https://doi.org/10.1038/cdd.2012.72.

Shi, Y., 2014. A glimpse of structural biology through X-ray crystallography. Cell 159 (5), 995–1014. Available at https://doi.org/10.1016/j.cell.2014.10.051.

Shigemoto, K., Kubo, S., Mori, S., Yamada, S., Akiyoshi, T., Miyazaki, T., 2010. Muscle weakness and neuromuscular junctions in aging and disease. Geriatrics and Gerontology International 10 (1), S137. Available at https://doi.org/10.1111/j.1447-0594.2010.00608.x.

Sirangelo, I., Iannuzzi, C., 2021. Understanding the role of protein glycation in the amyloid aggregation process. International Journal of Molecular Sciences 22 (12), 6609. Available at https://doi.org/10.3390/ijms22126609.

Sonninen, T.-M., Goldsteins, G., Laham-Karam, N., Koistinaho, J., Lehtonen, Š., 2020. Proteostasis disturbances and inflammation in neurodegenerative diseases. Cells 9 (10), 2183. Available at https://doi.org/10.3390/cells9102183.

Sperling, R.A., Donohue, M.C., Raman, R., Rafii, M.S., Johnson, K., Masters, C.L., et al., 2023. Trial of solanezumab in preclinical Alzheimer's disease. New England Journal of Medicine 389 (12), 1096–1107. Available at https://doi.org/10.1056/NEJMoa2305032, http://www.nejm.org/medical-index.

Springer, W., Kahle, P.J., 2011. Regulation of PINK1-Parkin-mediated mitophagy. Autophagy 7 (3), 266–278. Available at https://doi.org/10.4161/auto.7.3.14348, http://www.landesbioscience.com/journals/autophagy/SpringerAUTO7-3.pdf.

Srivastava, S., 2017. The mitochondrial basis of aging and age-related disorders. Genes 8 (12), 398. Available at https://doi.org/10.3390/genes8120398.

Südhof, T.C., Rizo, J., 1996. Synaptotagmins: C2-domain proteins that regulate membrane traffic. Neuron 17 (3), 379–388. Available at https://doi.org/10.1016/S0896-6273(00)80171-3, www.neuron.org.

Suzuki, H., Tabata, K., Morita, E., Kawasaki, M., Kato, R., Dobson, R.C.J., et al., 2014. Structural basis of the autophagy-related LC3/Atg13 LIR complex: recognition and interaction mechanism. Structure 22 (1), 47–58. Available at https://doi.org/10.1016/j.str.2013.09.023.

Sweeney, P., Park, H., Baumann, M., Dunlop, J., Frydman, J., Kopito, R., et al., 2017. Protein misfolding in neurodegenerative diseases: implications and strategies. Translational Neurodegeneration 6 (1). Available at https://doi.org/10.1186/s40035-017-0077-5.

Takahashi, R.H., Nagao, T., Gouras, G.K., 2017. Plaque formation and the intraneuronal accumulation of β-amyloid in Alzheimer's disease. Pathology International 67 (4), 185–193. Available at https://doi.org/10.1111/pin.12520, http://onlinelibrary.wiley.com/journal/10.1111/(ISSN)1440-1827.

Tang, W.S., Zhong, E.D., Hanson, S.M., Thiede, E.H., Cossio, P., 2023. Conformational heterogeneity and probability distributions from single-particle cryo-electron microscopy. Current Opinion in Structural Biology 81. Available at https://doi.org/10.1016/j.sbi.2023.102626, http://www.elsevier.com/locate/sbi.

Tao, C.L., Liu, Y.T., Sun, R., Zhang, B., Qi, L., Shivakoti, S., et al., 2018. Differentiation and characterization of excitatory and inhibitory synapses by cryo-electron tomography and correlative microscopy. Journal of Neuroscience 38 (6), 1493–1510. Available at https://doi.org/10.1523/JNEUROSCI.1548-17.2017, http://www.jneurosci.org/content/38/6/1493.full.pdf.

Teplow, D.B., 2013. On the subject of rigor in the study of amyloid β-protein assembly. Alzheimer's Research and Therapy 5 (4). Available at https://doi.org/10.1186/alzrt203.

Terman, A., Gustafsson, B., Brunk, U., 2007. Autophagy, organelles and ageing. The Journal of Pathology: A Journal of the Pathological Society of Great Britain and Ireland 211 (2), 134–143.

Terracina, S., Ferraguti, G., Petrella, C., Bruno, S.M., Blaconà, G., Di Certo, M.G., et al., 2023. Characteristic hallmarks of aging and the impact on carcinogenesis. Current Cancer Drug Targets 23 (2), 87–102. Available at https://doi.org/10.2174/1568009622666220816120353, https://www.eurekaselect.com/issue/12055.

Toader, C., Tataru, C.P., Munteanu, O., Serban, M., Covache-Busuioc, R.A., Ciurea, A.V., et al., 2024. Decoding neurodegeneration: a review of molecular mechanisms and therapeutic advances in Alzheimer's, Parkinson's, and ALS. International Journal of Molecular Sciences 25 (23). Available at https://doi.org/10.3390/ijms252312613, http://www.mdpi.com/journal/ijms.

Tompa, P., 2014. Multisteric regulation by structural disorder in modular signaling proteins: an extension of the concept of allostery. Chemical Reviews 114 (13), 6715–6732. Available at https://doi.org/10.1021/cr4005082.

Tsapras, P., Nezis, I.P., 2017. Caspase involvement in autophagy. Cell Death and Differentiation 24 (8), 1369–1379. Available at https://doi.org/10.1038/cdd.2017.43, http://www.nature.com/cdd/index.html.

Tycko, R., 2025. The evolving role of solid state nuclear magnetic resonance methods in studies of amyloid fibrils. Current Opinion in Structural Biology 92, 103043. Available at https://doi.org/10.1016/j.sbi.2025.103043.

Uceda, A.B., Mariño, L., Casasnovas, R., Adrover, M., 2024. An overview on glycation: molecular mechanisms, impact on proteins, pathogenesis, and inhibition. Biophysical Reviews 16 (2), 189–218. Available at https://doi.org/10.1007/s12551-024-01188-4, https://www.springer.com/journal/12551.

Vahid, S., Thaper, D., Zoubeidi, A., 2017. Chaperoning the cancer: the proteostatic functions of the heat shock proteins in cancer. Recent Patents on Anti Cancer Drug Discovery 12 (1), 35–47. Available at https://doi.org/10.2174/1574892811666161102125252.

Van Dinh, Q., 2022. Coarse-Grained Molecular Simulations for Conformational Changes of Transmembrane and Membrane Trafficking Proteins. Washington State University.

Villalón-García, I., Álvarez-Córdoba, M., Suárez-Rivero, J.M., Povea-Cabello, S., Talaverón-Rey, M., Suárez-Carrillo, A., et al., 2020. Precision medicine in rare diseases. Diseases 8 (4). Available at https://doi.org/10.3390/diseases8040042, https://www.mdpi.com/journal/diseases.

Villmann, C., Becker, C.-M., 2007. On the hypes and falls in neuroprotection: targeting the NMDA receptor. The Neuroscientist 13 (6), 594–615. Available at https://doi.org/10.1177/1073858406296259.

Walch-Solimena, C., Blasi, J., Edelmann, L., Chapman, E.R., Von Mollard, G.F., Jahn, R., 1995. The t-SNAREs syntaxin 1 and SNAP-25 are present on organelles that participate in synaptic vesicle recycling. Journal of Cell Biology 128 (4), 637–645. Available at https://doi.org/10.1083/jcb.128.4.637, rupress.org/jcb.

Wang, L., 2015. The role of synaptopodin for the diffusion of membrane protein in the dendritic spine neck. Ecole normale supérieure-ENS PARIS.

Wang, Y., Le, W.-D., 2019. Autophagy and ubiquitin-proteasome system. Autophagy: biology and diseases: basic science, 527-550.

Wang, T., Ruan, B., Wang, J., Zhou, Z., Zhang, X., Zhang, C., et al., 2022. Activation of NLRP3-Caspase-1 pathway contributes to age-related impairments in cognitive function and synaptic plasticity. Neurochemistry International 152, 105220. Available at https://doi.org/10.1016/j.neuint.2021.105220.

Wei, Y.H., Lee, H.C., 2002. Oxidative stress, mitochondrial DNA mutation, and impairment of antioxidant enzymes in aging. Experimental Biology and Medicine 227 (9), 671–682. Available at https://doi.org/10.1177/153537020222700901, http://www.uk.sagepub.com/journals/Journal202180.

Williams-Noonan, B.J., Kamboukos, A., Todorova, N., Yarovsky, I., 2023. Self-assembling peptide biomaterials: Insights from spontaneous and enhanced sampling molecular dynamics simulations. Chemical Physics Reviews 4 (2). Available at https://doi.org/10.1063/5.0142302.

Yaikhomba, M., 2022. Electron cryomicroscopy structures of respiratory supercomplexes from alphaproteobacteria suggest mechanisms to enhance catalysis and to prevent deactivation of respiratory complex I..

Yang, Y., Arseni, D., Zhang, W., Huang, M., Lövestam, S., Schweighauser, M., et al., 2022. Cryo-EM structures of amyloid-b 42 filaments from human brains. Science 375 (6577), 167–172. Available at https://doi.org/10.1126/science.abm7285, https://www.science.org/doi/10.1126/science.abm7285.

Yang, Z., Sheng, J., Zhang, Q., Zhao, X., Song, Y., Dong, G., et al., 2025. Multimodal imaging markers of cognitive resilience and molecular mechanisms of brain resilience in Alzheimer's disease. Journal of Integrative Neuroscience 24 (5). Available at https://doi.org/10.31083/jin26584.

Yoshikawa, S., Shimada, A., 2015. Reaction mechanism of cytochrome c oxidase. Chemical Reviews 115 (4), 1936–1989. Available at https://doi.org/10.1021/cr500266a.

Yuan, Y., Zhang, Xm, 2023. Mechanistic study of optic atrophy 1 in ischemia-reperfusion disease. Journal of Molecular Medicine 101 (1-2), 1–8. Available at https://doi.org/10.1007/s00109-022-02271-7, https://www.springer.com/journal/109.

Zabeo, D., Davies, K.M., 2022. Studying membrane modulation mechanisms by electron cryo-tomography. Current Opinion in Structural Biology 77, 102464. Available at https://doi.org/10.1016/j.sbi.2022.102464.

Zabik, N.L., Imhof, M.M., Martic-Milne, S., 2017. Structural evaluations of tau protein conformation: methodologies and approaches. Biochemistry and Cell Biology 95 (3), 338–349. Available at https://doi.org/10.1139/bcb-2016-0227, www.nrc.ca/cgi-bin/cisti/journals/rp/rp_desy_e?bcb.

Zafar, S., Fatima, S.I., Schmitz, M., Zerr, I., 2024. Current technologies unraveling the significance of post-translational modifications (PTMs) as crucial players in neurodegeneration. Biomolecules 14 (1), 118. Available at https://doi.org/10.3390/biom14010118.

Zhang, W., Tarutani, A., Newell, K.L., Murzin, A.G., Matsubara, T., Falcon, B., et al., 2020. Novel tau filament fold in corticobasal degeneration. Nature 580 (7802), 283–287. Available at https://doi.org/10.1038/s41586-020-2043-0, http://www.nature.com/nature/index.html.

Zhong, Q., Xiao, X., Qiu, Y., Xu, Z., Chen, C., Chong, B., et al., 2023. Protein posttranslational modifications in health and diseases: functions, regulatory mechanisms, and therapeutic implications. MedComm 4 (3). Available at https://doi.org/10.1002/mco2.261.

Zhu, L., Qian, Z., 2022. Recent studies of atomic-resolution structures of tau protein and structure-based inhibitors. Quantitative Biology 10 (1), 17–34. Available at https://doi.org/10.15302/j-qb-021-0271.

Zia, A., Farkhondeh, T., Pourbagher-Shahri, A.M., Samarghandian, S., 2022. The roles of mitochondrial dysfunction and reactive oxygen species in aging and senescence. Current Molecular Medicine 22 (1), 37–49. Available at https://doi.org/10.2174/1566524021666210218112616, http://www.eurekaselect.com/600/journal/current-molecular-medicine.

Zorov, D.B., Vorobjev, I.A., Popkov, V.A., Babenko, V.A., Zorova, L.D., Pevzner, I.B., et al., 2019. Lessons from the discovery of mitochondrial fragmentation (fission): a review and update. Cells 8 (2). Available at https://doi.org/10.3390/cells8020175, https://www.mdpi.com/2073-4409/8/2/175/pdf.

CHAPTER 16

Structural biology of protein misfolding and aggregation: from disease to therapeutic innovations

Neha Kausar Ansari and Aabgeena Naeem
Department of Biochemistry, Faculty of Life Sciences, Aligarh Muslim University, Aligarh, Uttar Pradesh, India

16.1 Introduction

Proteins must fold into an accurate three-dimensional shape, known as the native state, to accomplish biological activities such as the transportation of materials, the catalysis of reactions, and skeletal support. The folding of the naïve polypeptide chain is guided by the amino acid sequence, the so-called primary structure. The primary structure is further stabilized by intramolecular interactions, mainly hydrogen bonds, into secondary conformations such as α-helices and β-sheets. In addition, hydrophobic interactions, electrostatic contacts, and van der Waals forces determine the formation of tertiary structure and quaternary structure (Anfinsen, 1972). To keep a check on the folding pathway under physiological conditions, several machineries, such as the ubiquitin-proteasome system, molecular chaperones, quality control pathways, and the autophagy-mediated degradation pathway, operate within the cell to ensure proteins are correctly translated, localized within the cells, and cleared when damaged (Kuzu et al., 2025). The deviation of proteins from normal folding pathways resulted in the formation of misfolded conformations as a result of genetic mutations, posttranslational modifications (PTMs), and/or cellular stress. Overwhelming of the cellular proteostasis machinery results in the accumulation of misfolded proteins. These entities expose hydrophobic as well as aggregation-prone regions (APRs), which are normally shielded in the core of the proteins. Exposed regions form aberrant intra/intermolecular interactions with one another and assemble into aggregated species, either soluble oligomers or insoluble amyloid fibrils (Grad et al., 2011). Amyloid fibrils are associated with numerous diseases collectively called proteinopathies or protein-conformational anomalies. Aggregation of proteins is not only linked to neurodegenerative conditions (Alzheimer's, Parkinson's, amyotrophic lateral sclerosis [ALS], and Huntington's disease) but also gives rise to systemic disorders like type 2 diabetes (T2D), lysozyme-associated systemic amyloidosis, cancer, cystic fibrosis, and corneal dystrophies (Wang et al., 2025a). A typical structural attribute of aggregated proteins is the presence of nonnative cross-β-sheet structures. In this architecture, the different β-strands are joined together

Advances in Structural Biology
DOI: https://doi.org/10.1016/B978-0-443-44988-8.00016-2

perpendicularly to the fibril axis via intermolecular hydrogen bonding, forming a long, stack-like fibrillar structure. These fibrils have extensive stability and tensile strength, being the most determining cause of the endurance of amyloids in the living system (Maji et al., 2009). Moreover, the prion-like propagation of misfolded proteins is the crucial step that leads to the assembly of normally folded proteins into misfolded forms, driving them towards aggregation.

16.2 Structural determinants of protein misfolding and aggregation

Structural determinants are protein sequence elements that modulate folding forces, making them prone to aggregation. These determinants are primarily determined by intrinsic factors, such as those encoded within the sequences; that is, these sequences maintain the three-dimensional structure and overall conformational states. Fig. 16.1 shows the diagrammatic representation of the various protein folding pathways that make proteins prone to aggregation.

16.2.1 Amino acid sequence and composition

The sequence of amino acids, the so-called primary structure, affects the functional conformation of proteins. Modifications to these sequences destabilize the native

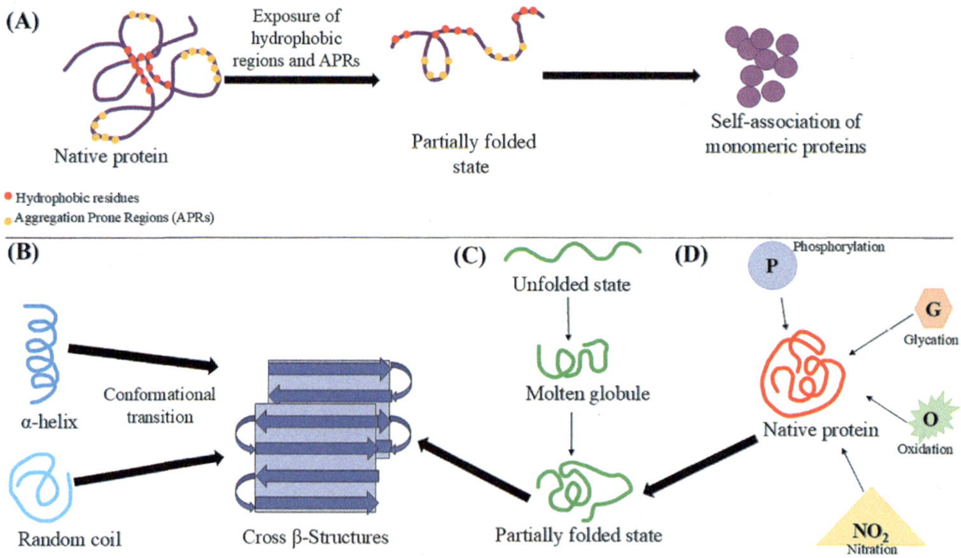

Figure 16.1 Diagrammatic representation of various pathways of protein aggregation (A) exposure of hydrophobic and aggregation-prone regions (*APR*), (B) conformational transition, (C) molten globule state formation, (D) posttranslational modifications.

structure, leading to misfolding and consequently aggregation (Dobson, 2003). Various factors governed by amino acid sequences are hydrophobicity, APRs or hotspots, net charge, and gatekeeper residues. For proper protein folding, the hydrophobic residues are normally shielded in the protein's core from the surrounding water molecules. During misfolding, these residues are exposed to the solvent, allowing the protein molecules to associate with one another, resulting in aggregation (van Gils et al., 2020). Certain small sequences in the polypeptide chain are susceptible to aggregation. These regions, specifically known as APRs or hotspots, are rich in hydrophobic amino acids with a greater propensity to form β-sheets, thereby driving the native proteins to aggregated forms (Chiti and Dobson, 2006). The net charge on proteins also helps maintain the native contacts. As the net charge decreases, the electrostatic repulsion between the protein molecules also decreases. This destabilizes the protein structure globally, allowing the self-association of the molecules more readily (Calamai et al., 2003). Moreover, certain amino acids, such as proline, are known as gatekeeper residues because they hinder the formation of cross-β-sheets due to their distinctive structure, which does not allow hydrogen bonding with neighboring molecules and efficiently inhibits protein aggregation (Pal and Udgaonkar, 2022).

16.2.2 Secondary structure propensity

One of the most distinctive features that assists in the formation of protein aggregates is the conformational transformation from a predominantly α-helical or disordered native form to an increased proportion of β-sheets. These β-sheets, via intermolecular hydrogen bonding, form a stable stack-like extended β-sheet network. As the extended β-sheets stabilize, they pave the way for cross-β-structures, termed amyloid fibrils (Chiti and Dobson, 2006). The cross-β-conformations have β-strands running perpendicular to the fibril axis, and the hydrogen bonds are extended parallel to the axis (Eanes and Glenner, 1968). This arrangement generates a highly ordered, repetitive, fibrous backbone that makes the fibrils thermodynamically as well as chemically stable, resistant to degradation, and therefore persistent in the living system for a longer time (Nelson et al., 2005). The structural proneness of a protein to adopt β-sheets under destabilizing conditions, irrespective of its original structure, is the determining factor for the aggregation of that protein. The α-helical, nonpathogenic proteins, such as insulin, prion protein, and apolipoprotein A1, as well as intrinsically disordered pathogenic forms, that is, tau, huntingtin (HTT) protein, and α-synuclein, can be transformed into β-rich structures upon disruption of the native conformation (Lee et al., 2020). The so-formed cross-β-architecture can act as a template that assists misfolded proteins in forming early oligomeric species, eventually assembling into mature fibrils.

16.2.3 Protein stability and folding intermediates

During the folding process, the naïve polypeptide chain goes through several intermediate states to achieve the native form. In addition, under normal physiological conditions, cellular machinery prevents the accumulation of misfolded proteins by either correctly folding them with the help of molecular chaperones or targeting them for degradation by the ubiquitin-ligase proteasome system (Naeem and Fazili, 2011). These cellular systems monitor protein homeostasis and suppress aggregation to a greater extent. But as this machinery is compromised by oxidative stress, genetic mutations, aging, or the accumulating burden of misfolded proteins, the cells' quality control mechanism saturates. Gradually, this imbalance led to the formation of soluble oligomers, protofibrils, and, eventually, insoluble amyloid fibrils. The less stable or partially folded intermediate states, such as the molten globule or expanded conformational entities, are more liable to form amyloids, as these forms have exposed hydrophobic residues that favor nonnative intermolecular interactions with other exposed residues instead of solvent (Iram et al., 2013). As a result of these interactions, a metastable nucleus is formed in the form of oligomers, which subsequently assemble into larger aggregates. Marginal protein stability, partially unfolded intermediates, and/or breakdown of cellular quality-control pathways collectively lead to protein aggregation. Therefore aggregation is a cooperative phenomenon of protein biophysics and cellular stress responses.

16.2.4 Posttranslational modifications

The newly synthesized polypeptides undergo a wide range of PTMs or chemical alterations that regulate the protein's structure, interactions, dynamics, and functions. In some cases, PTMs are crucial for physiological roles, but on the other hand, they can also unusually affect the folding pathway, making them susceptible to misfolding or aggregation. Anomalous PTMs facilitate the transformation of a biologically functional conformation into a misfolded or aggregated form, paving the way for proteinopathies. For instance, hyperphosphorylation of the tau protein promotes structural rearrangements that accelerate the formation of helical filaments and neurofibrillary tangles (NFTs). Likewise, the phosphorylation of α-synuclein is linked to Parkinson's disease (PD). Glycation is a nonenzymatic form of glycosylation that triggers the crosslinking of proteins, leading to structural destabilization and ultimately reinforcing the formation of amyloid fibrils in various proteins. The oxidation of methionine and cysteine residues disrupts the disulfide bonds in the protein's tertiary structure, consequently increasing structural flexibility and accelerating aggregation. In Alzheimer's disease (AD) and PD, oxidative stress is linked to Aβ aggregation and α-synuclein oligomer formation, respectively (Fu et al., 2022; Kawai et al., 2025). The addition of bulky anionic nitro groups to tyrosine residues can also disrupt the original protein interactions and lead to the formation of oligomers and fibrils, resulting in increased cytotoxicity, as observed in the case of α-synuclein. Protein aggregation resulting from PTMs does not take place alone in the living system. Rather, they piled up under cellular stress, aging, or inflammation, overburdening the cellular response machinery. As a result, these modified, misfolded entities accumulate in the cell, forming toxic fibrillar assemblies. Briefly, chemical and PTMs can efficiently influence protein energy landscapes, affect domain stability, and expose aggregation hotspots. Therefore PTMs can either govern or initiate the misfolding and aggregation of various proteins.

16.3 Mechanism of misfolding and aggregation

Protein misfolding and aggregation refer to the formation of nonnative and functionally inactive structures as a result of deviations from normal folding pathways. The misfolding and aggregation are not a single event; instead, they represent a multistep process involving a combination of structural destabilization, failure of the proteostasis machinery, and/or self-assembly mechanisms.

16.3.1 Mechanism of misfolding

Fig. 16.2 depicts the mechanism of protein misfolding governed by various factors as stated below.

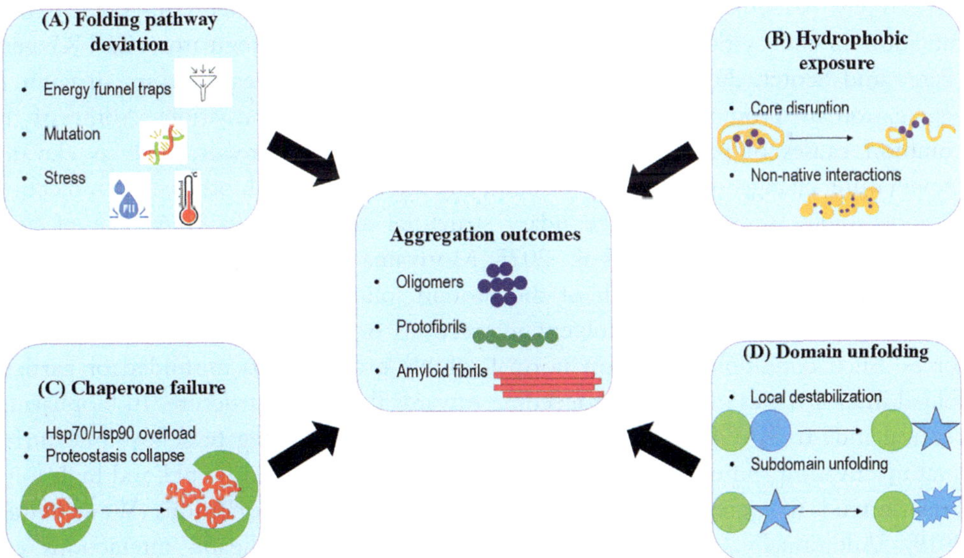

Figure 16.2 Mechanism of protein misfolding (A) deviation from protein folding pathways, (B) protein misfolding leading to hydrophobic exposure, (C) Incompetence of chaperones, and (D) destabilization of domains.

16.3.1.1 Deviation of proteins from folding pathways

The folding of proteins in the crowded environment of cells is a tightly controlled and highly coordinated procedure in which the primary structure of the protein folds into a specific three-dimensional structure that is not only biologically active but also thermodynamically stable. During the normal course of folding, the newly synthesized polypeptide chain proceeds along a defined pathway known as the energy landscape or folding funnel (Dobson, 2003). At each folding step, the energy of the protein molecule decreases; therefore the fully unfolded form has high energy and high entropy, and the folded native state is the global free-energy minimum. This landscape is not smooth and also has some rugged features or traps, termed as local energy minima, where proteins become kinetically trapped, and the folding slows down (Onuchic and Wolynes, 2004). The disruption of the balance between the forces that stabilize the native structure is mainly due to the discrepancies in the folding pathways, which compel polypeptide chains to adopt nonnative conformations (Onuchic and Wolynes, 2004). However, efficient folding depends on various factors such as the sequence of amino acids, environmental stress, and molecular chaperones. Mutations play a crucial role in deviating proteins from their native folding pathways. Even a single amino acid substitution can alter the folding kinetics and stability by decreasing the energy barrier for misfolding or producing APRs. For example, fibrinogen Aα-chain and transthyretin (TTR) amyloidosis are the result of mutations in the fibrinogen and TTR genes, respectively (Lachmann et al., 2002). Cystic fibrosis is another consequence of a mutation in the cystic fibrosis transmembrane conductance regulator (CFTR) gene (Férec and Scotet, 2020). These mutations shift the folding equilibrium in favor of aggregation by promoting the formation of nonnative conformations. Some of the common causes of protein misfolding are environmental stresses, such as elevated temperature, changes in pH, ionic strength, and interactions with other small molecules. Various studies showed that secondary structural alterations in BSA take place at elevated temperatures (Ansari et al., 2025; Moriyama et al., 2008). In a similar way, a change in pH or ionic strength of the protein solutions also destabilizes the native structure by disrupting either solvent interactions or intramolecular and electrostatic forces. Such conditions transform normal soluble proteins into misfolded or partially folded intermediates instead of proceeding towards the native structure. In biopharmaceutical industries, monoclonal antibodies like IgG4-N1 aggregate at low pH during viral inactivation (Jin et al., 2019). Hen egg white lysozyme at low pH and high ionic strength tends to aggregate more readily than at physiological conditions (Wawer et al., 2019). Additionally, anomalous protein-protein or protein-ligand interactions can augment undesirable intermolecular associations that stabilize misfolded conformations or facilitate polypeptide cross-linking, thereby perturbing the normal folding process and favoring alternative aggregation pathways in the crowded microenvironment of the living system. The association of the amyloid β-peptide with tau protein eventually leads to the formation of the most toxic oligomeric species (Do et al., 2014). Moreover, the binding of metal ions such as Fe^{3+}, Cu^{2+}, and Zn^{2+} accelerates the formation of toxic Aβ-complexes (Chen et al., 2011), and the interaction of α-synuclein with anionic lipid membrane induces conformational changes that facilitate its aggregation (Gaspar et al., 2018).

16.3.1.2 Exposure of hydrophobic patches

The amino acids of globular proteins are arranged in highly ordered three-dimensional structures to adopt biologically active native conformations. In these conformations, the protein shields the hydrophobic residues from the aqueous solvent within the core of the protein and minimizes free energy. The hydrophobic core reduces inappropriate molecular interactions between nonpolar amino acids and the aqueous environment, thereby stabilizing the overall protein structure. The hydrophobic effect is the primary force that stabilizes the folded state during protein folding. Therefore to maintain protein solubility, conformational stability, and biological function, hydrophobic residues should be located in the protein's core (Lins and Brasseur, 1995). Misfolding disrupts the exquisite balance between the hydrophobic core and the structural integrity of proteins, leading to partial or complete unfolding. Factors such as genetic mutations, PTMs, damage by oxidative stress, changes in pH or temperature, or proteostasis machinery failure result in the exposure of hydrophobic residues and aid in the anomalous intermolecular interactions with the nonpolar amino acids of the surrounding protein molecules to minimize solvent exposure (Naeem and Fazili, 2011). This type of molecular interaction remarkably increases the tendency of misfolded entities to self-associate and form unordered amorphous aggregates or oligomers that transform into highly ordered fibrillar assemblies. In living systems, molecular chaperones, heat shock proteins (Hsp70 and Hsp90), recognize the exposed hydrophobic core and bind to it, thus preventing aggregation. In addition, these chaperones assist in refolding or degradation as needed by the cell (Tanu et al., 2025). However, when the system is overwhelmed by these misfolded moieties, the hydrophobic residues facilitate excessive aggregation and accompany the pathophysiological conditions of protein-conformational anomalies, thereby highlighting the primary role of hydrophobic residues in disease progression.

16.3.1.3 Chaperones failure

Molecular chaperones, also known as heat-shock proteins (Hsp), help other proteins to assemble correctly, preventing their misfolding and aggregation. These proteins play a predominant role in maintaining cellular proteostasis by aiding the naïve polypeptide chain in attaining functionally active three-dimensional conformations. They facilitate systematic folding pathways, stabilize partially folded intermediates, and hinder undesirable molecular interactions without actually being part of the final functional protein themselves (Beckmann et al., 1990). Chaperones are widely classified into well-known families that perform specific roles in a protein's life cycle. ATP-dependent key chaperones such as Hsp70, Hsp90, and Hsp60 (chaperonins), together with small heat shock proteins, work coherently to maintain cellular protein homeostasis from cotranslational folding to posttranslational recovery. Hsp70 binds to hydrophobic residues of the newly synthesized polypeptide chain and prevents aggregation (Beckmann et al., 1990). Hsp90 helps fold proteins that are mainly involved in signaling pathways, such as kinases (Sima et al., 2021). Hsp60 has a barrel-like structure that allows the naïve polypeptide to fold correctly without interference from the surrounding environment. It also sometimes helps fold proteins in association with Hsp70. ATP-independent small heat shock proteins bind to misfolded proteins in the cellular system, and consequently obstruct their aggregation till Hsp70 and Hsp60 come into play and promote proper refolding. Typically, chaperones are competent at managing steady-state protein folding pathways and eliminating rare misfolded species. These proteins either facilitate the refolding of client proteins or direct misfolded species to the ubiquitin ligase degradation pathway. Yet, chaperones fail when the living system is impaired, dysregulated, or overwhelmed due to aging, mutations that reduce the expression or function of chaperones, environmental stress, etc. When the cellular system failed to meet the demand for chaperones, the misfolded proteins with exposed hydrophobic patches and aggregation hotspots accumulated by escaping the quality control machineries. These nonnative protein species tend to associate with one another, promoting aggregate formation. The abundance of chaperones is noticeable, especially during stress conditions with extensive protein unfolding. For example, heat shock initiates upregulation of chaperone expression, but this dynamic process is either deficient or delayed, leading to the accumulation of misfolded proteins. Therefore chronic stress can shift the equilibrium toward aggregation pathways. Additionally, various misfolded moieties also hinder chaperones within aggregates, thereby depleting their pools and exacerbating the failure of the protein homeostasis cycle on its own (Sima et al., 2021). Consequently, chaperone failure leads to compromised proteostasis, allowing the living system to harbor misfolded proteins, expose aggregation-prone hotspots, and from toxic aggregates, contributing to proteotoxicity.

16.3.1.4 Specific domain unfolding

The initial trigger for the misfolding of proteins is not always the global unfolding of the whole polypeptide chain, but rather, many proteins go through destabilization of certain local domains, resulting in aggregation. Structural entities such as α-helices, β-sheets, and interdomain boundaries within the multidomain protein exhibited different conformational stabilities. The interaction between these domains plays a significant role in maintaining the conformational integrity of the proteins by concealing the aggregation-specific sequences. Interferences in these interactions lead to the formation of partially folded states with exposed aggregation hotspots without affecting the overall protein structure. The partially folded structures are liable for misfolding and thus aggregation. A well-established example of this is the cellular prion protein (Prusiner, 2001). It contains a domain having three α-helices with a short antiparallel β-sheet. These secondary structures are held together by hydrophobic interactions and hydrogen bonding, making this domain minimally stable in the cellular microenvironment and sensitive to aberrant environmental conditions, such as mutations and nonspecific plasma membrane interactions, which can lead to prion protein misfolding and aggregation into β-sheet-rich conformations. Analogous mechanisms of aggregation have been reported in several proteins, such as superoxide dismutase (SOD) (Nordlund and Oliveberg, 2006), TTR (Zhou et al., 2019), and β2-microglobulin (McParland et al., 2002), among others. These conclusions favor the idea that misfolding originates from local unfolding of a marginally stable subdomain rather than global unfolding of the polypeptide chain.

16.3.2 Mechanism of aggregation

Aggregation is a multifactorial process facilitated by nonnative interactions that accompany the destabilization of native conformations. These nonnative interactions can result from the formation of partially folded or locally unfolded intermediates rather than complete unfolding, thereby exposing aggregation hotspots. These intermediate structures are trapped in local energy minima of the energy landscape and serve as fundamental precursors to oligomer and fibril formation. The various pathways followed by the proteins to form fibrils are described below:

16.3.2.1 Reversible native state self-association

The native monomers have some self-complementary regions on their surface that bind to each other through noncovalent interactions to form small oligomers. These assemblies are in dynamic equilibrium with the native monomers and can promptly dissociate upon dilution. However, factors such as pH, temperature, and ionic strength can critically influence the reversible association of these oligomers, leading to larger, more stable irreversible aggregates with covalent disulfide bonds. Insulin monomers naturally associate to form a reversible hexamer (Brems et al., 1992), reducing the bioavailability and efficacy, slowing absorption in blood, and/or leading to immunological responses (Grudzielanek et al., 2007). Interleukin-1 receptor antagonist undergoes reversible dimerization to form irreversible dimers and trimers at higher concentrations (Krishnan and Raibekas, 2009). Monoclonal antibodies also show reversible self-association, leading to physical instability as well as potentially triggering immune responses (Esfandiary et al., 2015). Fig. 16.3 A outlines the reversible native state self-association mechanism of aggregation.

16.3.2.2 Aggregation via misfolded monomer self-association

The first step in this mechanism is the formation of a competent structure that exposes hydrophobic regions, along with altered protein conformations. Intermolecular hydrogen bonding and hydrophobic interactions are thermodynamically favored by these structures in the cellular system, leading to the formation of highly ordered fibrils or disordered amorphous aggregates. The initial trigger for this mechanism can be change in temperature or pH, exposed APRs, mutations, etc. Fig. 16.3 B illustrates the diagrammatic representation of aggregation via conformationally altered monomer self-association. Proteins such as TTR, amyloid-β peptide, and α-synuclein form aggregates via this mechanism. TTR monomers associate into a tetramer in the native state. Under certain conditions, these monomers dissociate, adopt conformationally altered non-native structures, and assemble into fibrils (Sun and Dyson, 2018). Under cellular stress, normally soluble α-synuclein misfolds into toxic oligomers and insoluble fibrils (Zhou and Kurouski, 2020). The amyloid-β peptide monomers, specifically Aβ42, adopt misfolded conformations due to their hydrophobic C-terminus and consequently form amyloid fibrils (Barnes et al., 2019).

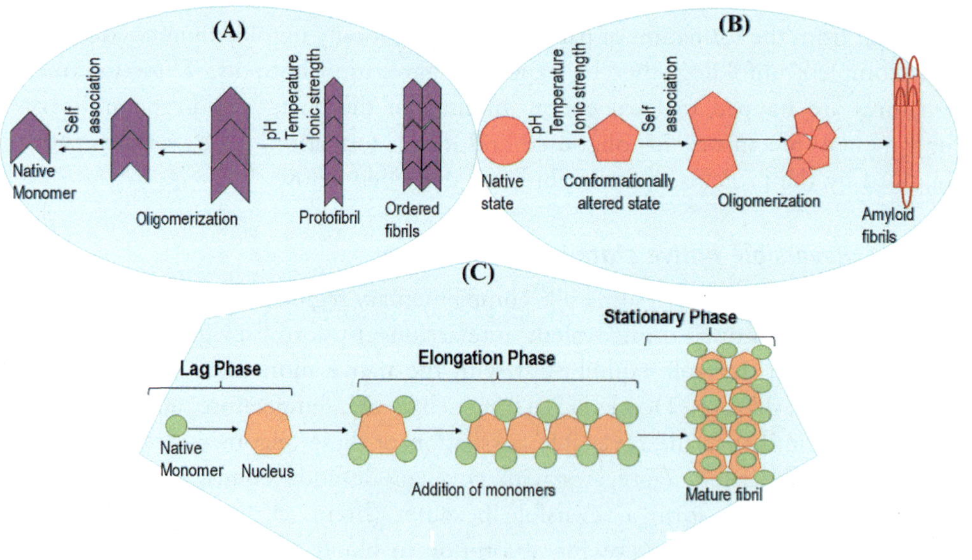

Figure 16.3 Mechanism of protein aggregation (A) native monomers self-associate, forming irreversible ordered fibrils; (B) monomers undergo altered conformation, forming self-associated oligomers; and (C) nucleation-dependent aggregation.

16.3.2.3 Nucleation-driven protein aggregation

It is an extensively observed pathway of protein aggregation. The nucleation-driven protein aggregation shows sigmoidal kinetics, with three major phases: the lag, elongation, and saturation/stationary phases (Fig. 16.3C). Native monomers undergo structural alterations, adopting misfolded or unfolded states. These monomers gradually interact with each other, forming small oligomers. Oligomerization is thermodynamically unfavorable, as the energy barrier is high and the system remains dominated by monomers. This phase is termed the lag phase. As soon as a stable critical nucleus, also termed a seed, is formed, the energy barrier has been overcome; further addition of monomers to the nucleus is thermodynamically favorable and results in the formation of fibrillar species. This phase is termed the elongation phase, and the rapid growth surges the concentration of the aggregates. This phase is followed by the stationary or saturation phase. As all the monomers are incorporated into the fibrils, the process slows down, as no more new fibrils can be formed, thus exhibiting the saturation point. Proteins such as tau (Congdon et al., 2008), prion proteins (Sneideris et al., 2020), β2-microglobulin (Xue et al., 2008), insulin (Foderà et al., 2008), islet amyloid polypeptide (Ruschak and Miranker, 2007), amyloid-β peptide (Cohen et al., 2013), and α-synuclein (Gaspar et al., 2017) follow this mechanism and contribute to various pathological conditions.

16.4 Molecular and cellular consequences

Protein aggregation not only disrupts cellular functions but also has distinct effects at the molecular level. These molecular events drive cellular damage, influencing the function and viability of the cells, eventually leading to cell death. Fig. 16.4 shows the schematic representation of the molecular and cellular consequences of protein aggregation, as mentioned below.

16.4.1 Molecular consequences of protein aggregation

At the molecular level, protein aggregates interfere with the normal protein structure, functions, and interactions, leading to extensive disruptions of biochemical pathways. These effects arise before noticeable cellular damage and underlie various protein conformational anomalies.

16.4.1.1 Loss of normal protein function

Native three-dimensional conformations are essential for proteins' biological activity. As proteins misfold and aggregate, they become incapable of performing physiological functions, even when synthesized at normal levels. The most common example is a mutation in the CFTR gene. The mutated protein results in improper folding and, consequently, the formation of aggregates. The functional CFTR protein failed to reach the plasma membrane, resulting in defective chloride ion transport and contributing to cystic fibrosis pathogenesis (Du et al., 2015).

16.4.1.2 Gain of toxic functions

As a result of aggregation, protein aggregates can inappropriately interact with several other cellular components, such as membranes, enzymes, receptors, or nucleic acids, and can give rise to various molecular abnormalities, including loss of membrane integrity, ion dyshomeostasis, or disruption of signal transduction pathways. The best example is the transformation of the amyloid β-peptide into soluble oligomers and/or insoluble plaques that interact with neuronal membranes and synaptic receptors (Patel and Jhamandas, 2012). This gain in the toxic function of oligomers gives rise to conditions such as cognitive impairment with substantial neurodegeneration and is the root cause of AD (Chang and Chen, 2014). Amylin (islet amyloid polypeptide) is an important human protein secreted by pancreatic β-cells along with insulin, and it helps regulate blood glucose (Visa et al., 2015). Under certain cellular conditions, such as a mutation in the IAPP gene (Sakagashira et al., 2000), or physiological conditions such as hyperglycemia (Janson et al., 1996), amylin can form toxic soluble oligomeric species that can efficiently damage the pancreatic β-cells and eventually result in T2D mellitus (Huang et al., 2007).

16.4.1.3 Disruption of cellular functions by protein sequestration

The toxic protein aggregates act as sinks, trapping other functional proteins through various specific and nonspecific interactions, resulting in the loss of function of these essential components, ultimately hindering various biochemical pathways in the living system. RNA-binding protein; TDP-43 is a transcriptional repressor that gets trapped by TDP-43 fragments and mislocalized to the cytoplasm instead of the nucleus, leading to functional loss (Peinado et al., 2022). The aggregates of HTT proteins facilitated by polyglutamine expansion interact with the transcription factor CREB-binding protein, thereby disrupting the regulation of genes for neuronal survival and aiding in the progression of neurodegeneration (Arancibia-Opazo et al., 2023). Similarly, α-synuclein aggregates sequester synaptic proteins, such as vesicle-associated membrane protein 2 and synaptosomal-associated protein 25, by directly binding to them and contributing to neurotoxicity (Choi et al., 2018).

16.4.1.4 Aggregate-induced cellular proteostasis failure

The living systems maintain protein homeostasis through molecular chaperones, the ubiquitin-ligase proteasomal degradation system, and/or autophagy. The chaperones guide the protein folding and refolding (Hartl et al., 2011). Ubiquitin-proteasome system degrades individual misfolded species, and lysozyme-induced autophagy clears large protein aggregates that damage the cell organelles (Klaips et al., 2018). As this system is overwhelmed by the protein aggregates, a vicious cycle of the overwhelmed network continues and results in the disruption of the balance of vital processes, leading to aggregate toxicity, mitochondrial dysfunction, and substantial cell death (Santra et al., 2019). The saturation of chaperones, inhibition of proteasomal machinery, as well as piling up of ubiquitinated proteins in the cell, acts as positive feedback that accelerates further aggregation (Ciechanover and Kwon, 2017). In PD, the aggregation of α-synuclein leads to the formation of Lewy bodies, which hinder the proteasomal degradation machinery. As a result, the misfolded proteins accumulate further, leading to protein misfolding stress that eventually damages the neurons, making them susceptible to cell death (Sitron et al., 2025). Likewise, the proteostasis failure was found to amplify the altered excitability, neurotransmitter imbalance, and structural changes in the neuronal cells and, therefore is a potential contributing factor in the progression of AD (Morrone et al., 2023).

16.4.2 Cellular consequences of protein aggregation

Protein aggregates have severe effects at the cellular level, and one or more impaired molecular actions are exhibited as structural damage and physiological process dysfunction, which eventually leads to cell death. All these cellular effects mark the shift from fundamental biochemical disturbances to apparent pathology and are predominantly found in aggregation-related neurodegenerative disorders.

16.4.2.1 Proteotoxic stress-induced cellular responses

Cellular homeostasis is significantly affected by the accumulation of misfolded and aggregated proteins. The cells activate several stress response pathways to cope with these situations, which help obliterate the damage and thereby restore protein homeostasis. The two main responses included in this list are the unfolded protein response (UPR) and heat shock response. UPR operates in response to the piling up of misfolded proteins in the endoplasmic reticulum (ER) (Zapun et al., 1999). This inordinate accumulation leads to reduced protein synthesis, amplifies chaperone expression, and augments the degradation of misfolded proteins (Travers et al., 2000). However, the heat shock response elevates the expression of heat shock proteins (HSPs), also known as molecular chaperones, to assist in the refolding of misfolded proteins or direct them towards degradation pathways (Chen et al., 2020a). In several protein conformation-related anomalies, the continuous accumulation of misfolded proteins in the ER triggers a stress response. The repeated activation of UPR and HSR shifts cell metabolism toward pro-apoptotic signaling pathways rather than protecting against misfolding, thereby leading to functional disturbances and cell degeneration

(Zhang et al., 2010). In addition, the heat shock response and the UPR work interdependently within the cell, particularly in proteinopathies. Several studies have revealed crosstalk and coordination between them, as the activation of both pathways depends heavily on the protein quality control system and shares components, such as the proteasome machinery and eIF2α kinases (Liu and Chang, 2008). In some cases, UPR directly regulates HSR elements, as seen in maize, where an important UPR transcription factor, bZIP60, further promotes the expression of the HSR transcription factor (Li et al., 2020). Moreover, some genes of the stress response machinery harbor binding sites for transcription factors required for the UPR and HSR, such as UPRE elements and HSF1, respectively, facilitating the coordinated regulation of gene expression (Truttmann et al., 2017). Certain feedback loops, for example, Sir2 (a protein found in yeast), are upregulated by UPR, act as an important link between UPR and HSR, serving as a negative regulator to confirm that the UPR should be transient, and hence attenuating the stress responses and manipulating the protein homeostasis (Weindling and Bar-Nun, 2015).

16.4.2.2 Proteotoxic stress-induced organelle

The accumulation of protein aggregates within the cellular organelles efficiently disrupts the normal structural and functional integrity. This spatial repositioning is detrimental to cellular homeostasis. The two main organelles, mitochondria and ER, are severely affected by protein aggregate accumulation because of their crucial roles in cellular energy metabolism and protein folding, respectively (Zaffagnini et al., 2024). The misfolded/aggregated species in the mitochondria interfere with the electron transport chain, leading to impaired mitochondrial respiration and reduced ATP production. This low energy availability is toxic to cells, particularly neurons, which have high energy demands for maintaining resting potential, restoring ion gradients, and packaging neurotransmitters (Calkins and Reddy, 2011). In addition, impaired electron transport leads to mitochondrial electron leakage, which in turn produces reactive oxygen species, resulting in oxidative damage to macromolecules and further worsening mitochondrial dysfunction and facilitating the vicious cycle of organelle damage. PD is characterized by α-synuclein aggregation, during which oligomers or protofibrils induce mitochondrial depolarization and the release of cytochrome c, leading to cell death (Fredenburg et al., 2007). Studies conducted in animal models of PD have shown that α-synuclein can interact with the components of the electron transport system and trigger oxidative stress in neurons that specifically produce dopamine (Coyne et al., 2022). Overexpression of α-synuclein in in vivo models was found to be linked with mitochondrial depolarization and reduced energy production in neuronal cells, leading to oxidative stress and brain injury (Jiao et al., 2025; Vasquez et al., 2024). Another cell organelle profoundly affected by the accumulation of misfolded proteins and aggregates is the ER, the prime site of protein folding. Within the ER, conformationally altered protein species overwhelm the protein quality control system, triggering chronic ER stress that drastically alters intracellular calcium signaling, leading to calcium imbalance and further impairing numerous downstream pathways, mitochondrial function, and synaptic activity (Xu and Wang, 2024). Additionally, prolonged proteostasis imbalance also restrains the clearance of protein aggregates via ER-associated degradation pathways, leading to cellular malfunctions (Chen et al., 2023). For example, valosin-containing protein (VCP) detects aggregated and misfolded proteins and targets them to ER-associated degradation pathways. But when protein clearance machinery fails, VCP initiates apoptotic pathways, especially in response to polyglutamine-induced ER stress, leading to neuronal death (Hirabayashi et al., 2001). Numerous studies have shown that certain mutant proteins that accumulate as conformationally altered forms and evade the quality control machinery are likely to induce ER stress, leading to disease progression. Examples include the accumulation of amyloid-β-peptide oligomers, which induces ER stress in AD (Soejima et al., 2013). In PD, cytoplasmic inclusions comprising α-synuclein, parkin, and an unfolded parkin-associated endothelial receptor-like protein disrupt ER homeostasis, contributing to its pathogenesis (Imai et al., 2000). Similarly, mutant copper-zinc SOD and polyglutamine-expanded aggregates trigger ER-induced cell death in the ALS (Sundaramoorthy et al., 2013) and Huntington's (Melo et al., 2022) model systems, respectively. Therefore the cumulative organelle dysfunctions facilitate wide cellular impairments, increasing susceptibility to damage and having a major impact on the progression of various protein conformation-related disorders.

16.4.2.3 Aggregation-mediated cytoskeletal disorganization and intracellular transport defects

Protein aggregates not only damage cellular organelles but also physically interfere with the cytoskeletal framework, such as actin and microtubules, consequently hindering the transportation of various molecules within the cell. Protein aggregates can impair intracellular transport and cause cytoskeletal disorganization in several ways, including physically blocking large motor proteins such as kinesin and dynein, thereby preventing effective movement (Yang and Hu, 2016). Moreover, aggregates also influence the polymerization and depolymerization dynamics of the actin filament, thereby hindering short-distance transport in the cell and signaling pathways (Fang et al., 2025). The mislocalization or reduced mobility of cell organelles, such as mitochondria or lysosomes, can lead to energy deficits and the accumulation of toxic waste, respectively, due to impaired transport, resulting in cellular dysfunction and cell death (Kiraly et al., 2025). The consequences of protein aggregation are injurious to cells, particularly neurons, making them vulnerable to synaptic failure, energy deficits, and amplification of neuronal stress, leading to degeneration. Some physiological conditions characterized by impairment of protein aggregate-mediated transport include AD, ALS, Huntington's disease, etc. The hyperphosphorylated tau protein forms NFTs. Under normal conditions, the tau protein stabilizes microtubules, but aggregated tau detaches from microtubules and destabilizes their assembly, leading to fragmentation. This disrupts the transport of nutrients, mitochondria, and neurotransmitter-containing vesicles across the axon, triggering synaptic dysfunction and neuronal death (Liu et al., 2007; Takei et al., 2000). Similarly, the HTT protein with polyglutamine expansion in Huntington's disease forms inclusion bodies that trap the microtubule- and actin-mediated transport machinery and inhibit mitochondrial localization, triggering a biochemical signaling cascade that leads to oxidative stress and neurodegeneration (Gunawardena et al., 2003; Yadav and Tapadia, 2013). The mislocalization and aggregation of RNA-binding proteins TDP-43 and fused in sarcoma (FUS) are linked to ALS and frontotemporal dementia. The protein aggregates inhibit nucleocytoplasmic RNA transport and obstruct axonal transport across motor neurons, overwhelm the protein quality control machinery, and are partly responsible for motor neuron degeneration (Naumann et al., 2018; Wu et al., 2012).

16.4.2.4 Proteotoxic stress-induced cell death pathways and loss of tissue/organ function

The conformationally altered proteins form either toxic oligomers or larger aggresomes, disrupting cellular homeostasis by creating proteotoxic stress. As the cellular quality control machinery is overwhelmed, these aggregates can further increase oxidative stress, ER, and mitochondrial dysfunction, ultimately leading to cell death. Moreover, aggregates can also trigger autophagy, also known as aggrephagy, to clear the aggregates, thereby promoting cell survival (Yu et al., 2025). Autophagy can also promote cell death if the essential cellular components are depleted. The larger protein aggregates are either degraded by the macro-autophagy pathway or chaperone-mediated autophagy. Studies have shown that protein aggregates can induce programmed cell death via autophagy in various living systems (Ueno et al., 2019). Besides, aggregates can activate caspase signaling pathways either by directly binding to them or by facilitating conformational changes that lead to apoptosis (Pan et al., 2011). The conformational alteration of α-synuclein into Lewy bodies leads to apoptosis of dopaminergic neurons in the substantia nigra region, which contributes to the progression of PD (Choubey et al., 2011). The progressive loss of cells caused by protein aggregation eventually leads to tissue degeneration and organ dysfunction. This cell loss leads to chronic impairment, particularly in the tissues that have poor regenerative potential, namely the central nervous system. As seen in the case of AD, the extensive neuronal loss in the hippocampus and cerebral cortex leads to memory and cognitive impairments, significantly contributing to the pathophysiology of the disease (Reddy et al., 2018).

In summary, molecular consequences reflect protein-level disturbances, such as misfolding and aggregation, while cellular consequences are their downstream effects on cellular physiology, including stress responses, organelle dysfunction, and subsequent cell death. Fig. 16.4 shows the schematic representation of the molecular and cellular consequences of protein aggregation.

16.5 Structural themes for protein aggregation diseases

Protein aggregation-related disorders exhibit specific, repetitive conformational features that depend on the aggregated protein and the affected region. The predominant structural themes involve the formation of highly ordered amyloid fibrils, rich in cross-β structures, and less ordered amorphous aggregates that lack a defined repeating structure. Interestingly, it has been found that amyloid fibrils are associated with a greater number of diseases than amorphous aggregates.

16.5.1 Amyloid fibrils

Amyloid fibrils are the proteinaceous, nonsoluble fibrillar structures found in the living system, which are resistant to degradation. The formation of fibrils is associated with specific diseases, including amyloidosis, characterized by the aggregation of proteins or peptides. The fibrils are structurally stable and are abundantly composed of β-sheets

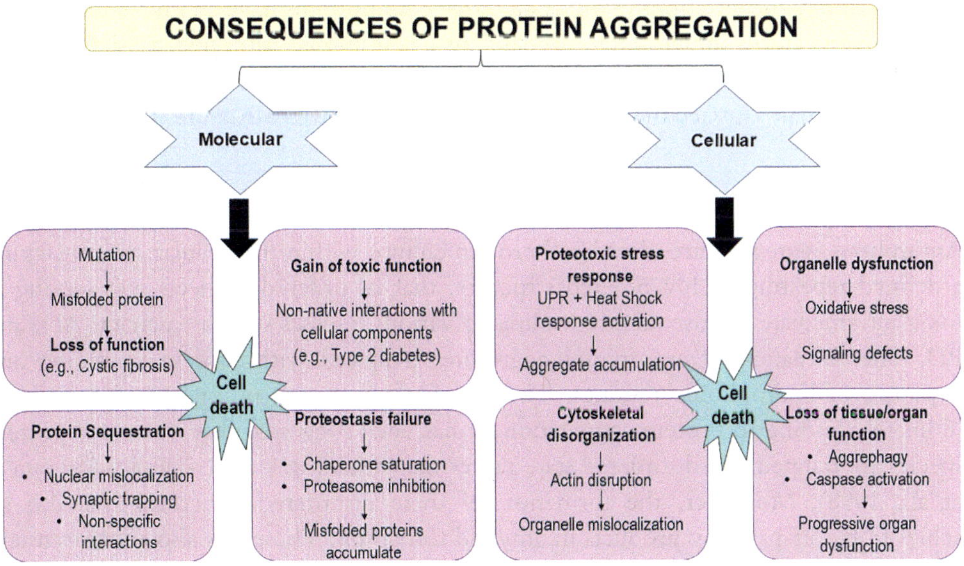

Figure 16.4 Schematic representation of the molecular and cellular consequences of protein aggregation.

arranged in a cross-β conformation, as revealed by X-ray fiber diffraction (Eanes and Glenner, 1968; Yagi et al., 2007). In this conformation, the β-strands run perpendicular to the long fibrillar axis, and the hydrogen bonds are parallel to the axis (Nelson et al., 2005). As discovered by electron and atomic force microscopy, the fibrils are 0.1–10 μm long and 10 nm wide (Yagi et al., 2007; Khurana et al., 2003). The macromolecular assembly of amyloid fibrils is made up of monomers, also known as protofibrils. Protofibrils are generally thinner, shorter, and less rigid than when joined together to form straight, unbranched mature fibrillar assemblies (Serpell et al., 2000). The fibrils formed from different proteins share common cross-β-structures, despite differing in their amino acid sequences and secondary/tertiary structures. As evident from in vitro studies, the formation of amyloid fibrils from normal soluble proteins typically entails the formation of thermodynamically unstable partially folded intermediates that stabilize into amyloidogenic forms (Jahn et al., 2006; McParland et al., 2000). The formation of fibrils involves nucleation-dependent kinetics characterized by a lag phase, which is followed by an elongation phase and a final stationary or saturation phase, depicted by a characteristic sigmoidal curve (Lee et al., 2007). Soluble monomers form micelle-like structures when the critical threshold concentration is exceeded, serving as precursors to the formation of a seed or nucleus. The nucleation is the rate-determining step. Around this nucleus, the free monomers bind irreversibly and elongate the fibril's length. Adding preformed seeds to the native protein eliminates the initial lag phase and bypasses the nucleation step, thereby leading to the formation of fibrillar assemblies more quickly. Therefore the formation of micelle-like structures is a critical step at physiologically lower protein concentrations (Šarić et al., 2014). All proteins have an innate property to form amyloid fibrils under certain destabilizing conditions. Although the tendency to form fibrils widely depends on the sequence of amino acids, the charge present on the polypeptide chain, surface hydrophobicity, and the secondary structure (Dobson, 2003).

16.5.2 Amorphous aggregates

Amorphous aggregates are disordered protein clumps within the cellular system lacking a defined repeating highly ordered structure, that is, ordered β-sheets, yet serving as toxic intermediates, consequently leading to various diseases such as cataract, AD, and PD. The formation of amorphous aggregates is linked with protein unfolding and exposure of hydrophobic residues. These residues promote nonspecific interactions, ultimately leading to protein aggregation. It has been suggested that misfolded protein species are trapped in a disordered state rather than attaining kinetic stability (Nymeyer et al., 1998). Moreover, the formation of these disordered structures can also be achieved by off-pathway products in amyloid formation. These can also act as transient intermediates in the formation of ordered fibrils, underscoring the interplay between fibrils and amorphous aggregation pathways. Environmental factors drive the protein to form either ordered fibrils or amorphous aggregates. For example, β2-microglobulin was found to form amyloid fibrils at acidic pH (2.5) and at moderate NaCl concentrations between 100 and 300 millimolar, following a nucleation-driven pathway. Contrastingly, the prompt formation of amorphous aggregates was driven by high NaCl concentration, that is, 500 millimolar, without a lag phase, which leads to the conclusion that amorphous aggregation takes place without a significant energy barrier, and solubility as well as supersaturation can play vital roles in the pathways of protein aggregation (Yoshimura et al., 2012). This aggregation pathway also gives rise to structural heterogeneity and cellular impairments, leading to disease progression. Also, the aggregation of amyloidogenic proteins with disordered nonamyloidogenic analogs can promote the formation of amorphous aggregates, further aggravating the cellular injury (Wu et al., 2022).

16.6 Disease-specific structural themes

These are the distinctive physical/molecular anomalies affecting the living system and describe a particular disorder. These themes help develop therapeutics and diagnostics. Pathological conditions related to protein misfolding and aggregation are not limited to neurological conditions such as AD but also extend to several metabolic disorders, like phenylketonuria, and to diseases associated with structural proteins such as collagen and keratin. Despite differences in the pathological conditions as a consequence of aggregation, the amyloid structures are found to be similar in all cases (Dobson, 2003). Some of the diseases are discussed below:

16.6.1 Alzheimer's disease

The most common form of dementia, that is, AD, is characterized either by the presence of extracellular plaques of amyloid beta peptide or by the formation of NFTs of the microtubule-associated tau protein in the neurons due to its hyperphosphorylation (Calhoun et al., 1999; Fitzpatrick et al., 2017). Amyloid precursor protein (APP) is a single-pass transmembrane protein that helps in neuronal stability and signaling, neuronal development, and intraneuronal transport (Banote et al., 2016). APP is the precursor of the amyloid β-peptide that is found in the brains of AD patients. Human APP can be cleaved by α- or β-secretase to produce membrane-bound C-terminal fragments (Haass et al., 1995). The product released by the action of β-secretase is amyloidogenic in nature and requires further processing by γ-secretase at the N-terminal to generate 40- or 42-amino acid-long fragments termed Aβ-40 and Aβ-42, respectively (Olsson et al., 2014). The accumulation and binding of Aβ to various cellular compartments lead to the formation of toxic fibrillar species, the generation of oxidative stress, or the disruption of mitochondrial function, causing neurotoxicity that ultimately induces neuronal cell death and memory loss (Liu et al., 2022). Oligomeric Aβ species bind to various biomolecules, such as the prion protein, low-density lipoprotein receptor-related protein, N-methyl-D-aspartic acid receptor, paired immunoglobulin-like receptor B, and others, leading to the generation of noxious signals in neurons, including a stress response in the ER and mitochondrial dysfunction (Deshpande et al., 2009; Kessels et al., 2010; Kim et al., 2013). The pathogenicity of Aβ widely depends on its aggregation, transport, and clearance. Intrinsically disordered Aβ monomers interact with each other to form toxic oligomers, which slowly form plaques and fibrils. These Aβ plaques alter the normal kinase/phosphatase activity of the receptors, disrupting the balance between them. This imbalance causes hyperphosphorylation of the tau protein and, consequently, detachment from microtubules, forming NFTs. These NFTs disrupt the neuronal transport system, causing synaptic malfunction and, in turn, underpinning the progression of AD (Mondragón-Rodríguez et al., 2008). The Aβ aggregates also cause lipid peroxidation by generating a toxic compound, 4-hydroxynonenal (an aldehyde), which disrupts the functioning of ion-motive ATPase, glucose transporters, and glutamate transporters (Keller et al., 1997). Aβ peptide also triggers an innate immune response by activating microglial-induced phagocytosis of the neuronal cell, leading to its degeneration (Weggen et al., 2001). Both forms of protein aggregates associated with AD, whether extracellular Aβ plaques or intracellular NFTs, are deposited in major brain regions, including the hippocampus, amygdala, forebrain, and cortex, affecting learning and memory behavior. Cognitive deficits, such as memory loss and impaired thinking, are associated with the formation of Aβ deposits, which consequently lead to the development of AD (Koistinaho et al., 2001).

16.6.2 Parkinson's disease

The second most common form of dementia is PD, which constitutes approximately 15% of the total cases. It is a neurodegenerative disease characterized by the degeneration of neurons that synthesize dopamine (a neurotransmitter) in the substantia nigra pars compacta region of the basal ganglia, resulting in low dopamine levels (Ramesh and Arachchige, 2023). The protein associated with PD pathology is α-synuclein. It is a 14-kDa protein with 140 amino acids, primarily found at presynaptic terminals in the central nervous system and encoded by the SCNA gene (Polymeropoulos et al., 1997). Physiologically, α-synuclein at presynaptic terminals interacts with proteins that help in the release and uptake of neurotransmitters, as well as help in the formation of the SNARE complex for vesicle anchorage and exocytosis. Lower levels of α-synuclein decrease the number of synaptic vesicles anchored to the membrane, thereby making it an important molecule for vesicle pools and trafficking (Burré et al., 2010). Cellular α-synuclein levels are highly dependent on the balance between synthesis and clearance, and thus an imbalance can lead to the formation of cytotoxic oligomers and/or fibrils. Various factors, such as oxidative stress, PTMs, and genetic mutations, can modulate the native conformation of α-synuclein and drive the protein to form oligomeric and fibrillar species (Hashimoto et al., 1999; Li et al., 2005; Paleologou et al., 2010); hence, α-synuclein aggregation is primarily associated with neurodegenerative diseases collectively known as synucleopathies. These nonnative, abnormal conformations overwhelm the cell's quality control system, disrupt autophagic pathways, and increase toxicity (Desplats et al., 2009). The fibrils of α-synuclein were found as Lewy bodies, intraneuronal aggregates associated with PD and Lewy body dementia (Spillantini et al., 1998). Once oligomers form, they can be transferred from one neuron to another or to glial cells via exosomes and can trigger further aggregation via nucleation-dependent pathways, thereby triggering neuroinflammation (Emmanouilidou et al., 2010). Since α-synuclein helps in the release and uptake of neurotransmitters. Therefore the dopaminergic neurons are severely affected by the levels as well as conformations of α-synuclein. The oligomeric α-synuclein inserts into the neuronal plasma membrane, forming pores, and binds to calcium ATPase pumps in the ER, causing unregulated calcium influx into the cell and initiating cell death pathways (Betzer et al., 2018). Moreover, they increase intracellular calcium levels in neurons by modulating voltage-gated calcium channels, which, in turn, release dopamine and alter synaptic function, thereby leading to neurotoxicity and degeneration in the central nervous system and contributing to the progression of PD (Angelova et al., 2016). The symptoms include impairment in voluntary motor control and slow movement or freezing, reduction in facial expression, difficulty in communication, insomnia, confusion, memory loss, depression, psychosis, etc. In the initial stages of PD, pathological changes occur in the medulla and olfactory bulb, which may spread to the amygdala, basal forebrain, and temporal lobes, and eventually to the neocortex (Wang et al., 2025b).

16.6.3 Amyotrophic lateral sclerosis

ALS is a pathological condition of motor neurons characterized by degeneration of upper and lower motor neurons, extensively affecting the brainstem, spinal cord, and muscles, leading to muscle weakness, amyotrophy, and paralysis, which ultimately results in respiratory failure and consequent death (Masrori and Van Damme, 2020). The progressive degeneration is associated with the accumulation of conformationally altered protein species, resulting in a physiological perturbation in the living system. The main proteins associated with ALS are SOD, transactive response DNA-binding protein (TARDBP; TDP-43), FUS, and chromosome 9 open reading frame 72 (C9ORF72). Mutations in the SOD1 gene make the protein less stable and likely cause misfolding and aggregation within motor neurons, a prime cause of familial ALS (Bosco et al., 2010). Meanwhile, the sporadic ALS is found to be linked with a different gene, that is, TDP-43. This nuclear protein, when mislocalized as a result of phosphorylation or ubiquitination, forms toxic cytoplasmic insoluble inclusion bodies, augmenting the coaggregation of other functional proteins (Shenouda et al., 2022). Mutations in the FUS gene can give rise to mislocalized cytoplasmic aggregates that impair RNA metabolism and inhibit protein translation in neurons, resulting from the gain of toxic functions (Mitchell et al., 2013; Rogelj et al., 2012). The C9ORF72 harbors a hexanucleotide repeat expansion that is either transcribed into a toxic nuclear RNA inclusion that sequesters significant RNA-binding proteins or translated into a dipeptide repeat protein, forming toxic cytoplasmic aggregates (Kortazar-Zubizarreta et al., 2023; Ryan et al., 2022). All these aggregates, irrespective of the original protein, widely affect the cellular physiology, leading to impaired nucleocytoplasmic transport and RNA metabolism, mitochondrial and ER dysfunction, hindered actin dynamics, and proteostasis disruption. These molecule-level effects result in excitotoxicity and neuroinflammation, causing widespread neuronal damage and cell death, consequently leading to the progression of ALS (Hu et al., 2024). The loss of motor neurons affects voluntary movements such as walking, speaking, and breathing, leading to coordination issues and paralysis.

16.6.4 Huntington's disease

Huntington's disease is a fatal neurological condition causing neuron degeneration in the brain, resulting in uncontrolled and impaired movements, cognitive defects, depression, and psychosis. The main culprit in this disease is a mutation in the HTT gene that leads to expansion of the CAG trinucleotide repeat. The translated HTT protein now has polyglutamine expansion, which adopts an altered conformation, making it susceptible to aggregation (MacDonald et al., 1993). The normal HTT protein helps in synaptic transmission and the transport of brain-derived neurotrophic factor. However, the mutated HTT protein is inefficient in performing these functions (Ravalia et al., 2021). Moreover, the cleaved N-terminal fragments of the mutated HTT are relocated to the nucleus and piled up as inclusion bodies (Schilling et al., 2004). These aggregates trap transcription factors (Steffan et al., 2000) and impair proteostasis (Yamanaka et al., 2008), further disrupting axonal transport and perturbing communication between neurons (Gunawardena et al., 2003). The microscopic-level synaptic disruption can lead to macroscopic degeneration of gamma-aminobutyric acid-producing medium spiny neurons, selectively in the striatum of the basal ganglia (Paraskevopoulou et al., 2021). The extensive loss of medium spiny neurons leads to fluid-filled cavities and striatal degeneration, which gradually affect the cerebral cortex and thalamus.

16.6.5 Type 2 diabetes

T2D is a metabolic disorder characterized by faults in the production, secretion, and signaling of insulin, an important hormone made by pancreatic β-cells. Insulin helps maintain blood glucose levels by allowing cells to take up glucose and utilize it for energy metabolism (Okon et al., 2007). The pancreatic β-cells release another hormone, that is, islet amyloid peptide (amylin), along with insulin, which regulates blood sugar and appetite by slowing gastric emptying, thereby preventing sharp glucose spikes in the blood. Excessive amylin production and secretion are associated with diabetes (Visa et al., 2015). Normally, amylin exists as a soluble protein with a random coil structure, which transforms into toxic β-sheet structures at high concentrations and forms insoluble fibrils (Huang et al., 2007). The fibrillar assemblies further damage pancreatic β-cells, thereby disrupting insulin production and secretion, and perpetuating the deleterious cycle. Moreover, elevated glucose levels also contribute to the formation of oligomers and amyloid fibrils of amylin through glycation, thereby forming advanced end glycation products with altered conformation (Hsu et al., 2019). The amylin aggregates can create pores in the plasma membrane, leading to ion dyshomeostasis (Demuro et al., 2005) and increased inflammatory responses (Westwell-Roper et al., 2011), leading to a gradual loss of the pancreatic β-cells, further worsening the insulin deficiency and contributing to cell death.

16.6.6 Cancer

The uncontrolled growth and abnormal spread of cells that bypass the normal stop-dividing signals and invade other tissues or organs in the form of a tumor mass lead to cancer. The hindrance to following the normal stop signal can result from various factors, such as genetic mutations, exposure to carcinogens, or viral infections. The most common cause of cancer is a mutation in the p53 gene, which encodes the tumor suppressor protein that normally detects cellular stress and arrests the cell cycle (Goncharoff et al., 2025). However, as a result of the mutation, the p53 protein misfolds, leading to the formation of amyloid-like fibrils in the nucleus (Yang-Hartwich et al., 2015). Aggregated p53 is functionally inactive, leading to loss of tumor suppression and allowing damaged cells to divide in an uncontrolled manner. Moreover, the p53 aggregates can also transform the normal p53 protein into aggregates, thus effectively hampering its tumor-suppressing activity (Ano Bom et al., 2012). Aggregation of the BRCA1-associated protein (BAP1) and the KRAS protein is also associated with cancer progression. BAP1 aggregates are retained in the cytoplasm rather than the nucleus and accumulate, leading to impairment of cell cycle control, DNA repair, and transcription (Hsiao et al., 2025). Likewise, KRAS aggregates with altered conformation are hyperactivated, leading to a sustained signal for cell proliferation (Janssen et al., 2023). These aggregated proteins thus halt the cell's apoptotic pathways by promoting cell proliferation, differentiation, and genomic instability. The cumulative outcome led to the formation of malignant tumors and eventually death.

In addition to the aforementioned proteinopathies, several other protein conformation-related anomalies are listed in Table 16.1

16.7 Therapeutic and structural modulation strategies

The therapeutic interventions against protein aggregation are intended to inhibit the formation of toxic amyloid fibrils or upregulate their clearance to treat associated disorders.

16.7.1 Therapeutic modulation strategies

This approach primarily targets various steps of protein folding and/or the proteostasis mechanism. Several approaches are listed below to eliminate the production of toxic fibrillar species and to mitigate their toxicity.

16.7.1.1 Suppressing pathogenic protein expression

Certain approaches, such as gene silencing or gene therapy, help to prevent the formation of aggregating proteins, thus inhibiting the first step in the process of aggregation. Gene silencing by RNA interference or antisense oligonucleotides helps block protein translation by binding to messenger RNA, thereby reducing the production of amyloidogenic proteins. The translation of the mutant TTR protein was reduced by RNA interference and antisense oligonucleotide (Benson et al., 2006; Coelho et al., 2013). These approaches have been approved for the treatment of hereditary TTR amyloidosis (Adams et al., 2018). Another strategy is gene therapy by organ transplantation. For instance, in familial amyloid polyneuropathy, the mutant TTR was effectively replaced by the wild-type TTR gene by transplanting the liver from healthy donors, thereby reducing the production of amyloidogenic TTR protein and hence mitigating familial amyloid polyneuropathic abnormalities (Holmgren et al., 1993).

16.7.1.2 Modulation of proteostasis and immune-mediated aggregate clearance

The clearance of toxic aggregates in the cellular system by upregulating proteostasis pathways or directly targeting the aggregates could be a possible curative approach. Drugs such as arimoclomol upregulate the expression of Hsp and enhance the degradation of conformationally altered protein species via ubiquitin-proteasomal or autophagic pathways (Kieran et al., 2004). Another drug, that is, ibudilast, an antiinflammatory medication, was found to remove SOD1 and TDP-43 aggregates by inducing autophagy and lysosomal biogenesis through mammalian target of

Table 16.1 List of diseases caused by various proteins as a result of aggregation.

Disease	Protein involved in aggregation	Affected region	References
Primary systemic amyloidosis	Immunoglobulin light and heavy chains	Heart, kidneys, liver, skin, gastrointestinal tract, peripheral and autonomic nervous system	Andrich et al. (2017)
Senile systemic amyloidosis	Wild-type transthyretin	Heart	Saelices et al. (2015)
Secondary systemic amyloidosis	Serum amyloid A	Kidney, spleen, gastrointestinal tract, and heart	Lachmann et al. (2007)
Hemodialysis-related amyloidosis	β2-microglobulin	Osteoarticular tissues	Jadoul et al. (1997)
Spongiform encephalopathy	Prion protein	Brain stem, cerebellum, and other parts of the brain	Kovács et al. (2002)
Lysozyme systemic amyloidosis	Lysozyme	Gastrointestinal tract and kidneys	Li et al. (2019)
Hypercholesterolemia	Low-density lipoprotein receptors	Liver and arteries	Brown and Goldstein (1986)
Familial hypoalphalipoproteinemia	Apolipoprotein-A1	Kidneys, heart, and liver	Petrlova et al. (2014)
Spinocerebellar ataxias	Ataxians	Brainstem, spinal cord, and cerebellum	Seki et al. (2005)
Hereditary gelsolin amyloidosis	Gelsolin	Eyes, nervous system, skin, kidneys, heart, and muscles	de la Chapelle et al. (1992)
Dentatorubral-pallidolusian atrophy	Atrophin-1	Central Nervous System and Cerebellum	Nuga et al. (2025)
Spinal and bulbar muscular atrophy	Androgen receptor	Spinal cord, brain stem, and skeletal muscles	Lieberman and Robins (2008)
Cystic fibrosis	Cystic fibrosis transmembrane conductance regulator	Pancreas, lungs, intestines, liver, and reproductive organs	Du et al. (2015)
Chronic pancreatitis	Cationic trypsinogen (PRSS1) Carboxyl ester lipase Carboxypeptidase A1 Pancreatic lipase Lithostathine (pancreatic stone protein)	Pancreas	Sahin-Tóth (2017)
Cataract	Crystallins	Eyes	Das et al. (2011)

rapamycin complex 1-transcription factor EB signaling pathways (Y. Chen et al., 2020b). The anticancer drug nilotinib has been shown to increase proteasome activity by upregulating the lysosomal degradation pathway in α-synuclein A53T transgenic mice (Hebron et al., 2013). Various monoclonal antibodies have been shown to target protein aggregates and could be promising therapeutic agents for their clearance. FDA-approved antiamyloid monoclonal antibodies, such as Lecanemab, donanemab, and aducanumab, are potential agents against protein aggregates. All these antibodies bind to amyloid β protofibrils, plaques, or extracellular aggregates and clear them from the cellular system via microglial activation-mediated phagocytosis (Budd Haeberlein et al., 2022; Mintun et al., 2021; Van Dyck et al., 2023). Prasinezumab, an antibody against aggregated α-synuclein, has shown promise as a potential therapeutic in slowing the progression of PD (Pagano et al., 2021).

16.7.1.3 Nanoconjugates-assisted drug delivery

Certain aggregate inhibitors have poor bioavailability and lower blood-brain barrier penetration. Therefore to overcome these problems, ligands were attached to the nanoparticles to enhance stability, effectiveness, and targeted delivery to the specific location. The fluorescent quantum dot-siRNA nanoplexes target β-secretase in nerve cells, thus reducing the amyloid-β levels (Li et al., 2012). Another nano-formulation made of β-sheet breaker peptide, KLVFF, showed promising effects in inhibiting amyloid-β peptide aggregation in rat hippocampal cells (Pederzoli et al., 2019). In addition, nanoconjugates not only inhibit the amyloid fibril formation but can also help in the disintegration of fibrils by converting them to soluble forms. One such instance is mApoE-PA-LIP nanoparticles that effectively clear amyloid β aggregates from the brain in APP/PS1 mice without notable adverse effects (Balducci et al., 2014).

16.7.2 Structural modulation strategies

Structural modulation strategies against protein aggregation focus on direct protein-small molecule interactions or on altering the solvent environment by adding formulation additives to downregulate aggregation.

16.7.2.1 Ligand-induced kinetic stabilization of native conformation

Pharmacological chaperones are small molecules that act as kinetic stabilizers. They bind to the native conformation of the proteins, thereby reducing their propensity to misfold or aggregate. For example, tafamidis (Bulawa et al., 2012) binds to the TTR tetramer and stabilizes it, thus preventing aggregation.

16.7.2.2 Ligand-induced inhibition of nonnative protein assembly

Natural compounds such as curcumin (Jakubczyk et al., 2020), epigallocatechin-3-gallate (Seidler et al., 2022), or synthetic compounds like doxycycline (Balducci and Forloni, 2019), bexarotene (Miziak et al., 2021), methylene blue (Soeda et al., 2019) and molecular tweezers such as CLR01 (Herrera-Vaquero et al., 2019) were found to inhibit α-synuclein oligomerization/fibrillation.

16.7.2.3 Peptide modulators

The small molecules, such as molecular glues, promote protein degradation by binding the target protein to ubiquitin ligase and directing it for degradation. For example, compound BIO-2007817 is a molecular glue degrader that activates the parkin/PINK ubiquitin ligase pathway and facilitates the degradation of misfolded proteins and damaged mitochondria, thus preventing the accumulation of aberrant proteins (Sauvé et al., 2024). Polyglutamine-binding peptide (QBP1) is an example of a stabilizing peptide modulator that can specifically bind to polyglutamine expansions of disease-causing proteins, thus blocking their interaction with other polyQ proteins that promote aggregation (Popiel et al., 2011).

16.8 Conclusions

Structural biology plays a fundamental role in transforming the comprehension of protein misfolding and aggregation, as key factors in various human pathophysiological conditions such as neurodegenerative diseases, systemic amyloidosis, and metabolic disorders. Misfolded proteins do not randomly aggregate; rather, they form systematic, highly ordered amyloid fibrils. The structural insights of the protein now elucidate the destabilization of native conformations, the formation of intermediates that readily aggregate, and correlations of particular fibril architectures with the disease. Significantly, conformational studies have revealed that oligomers are more toxic than mature fibrils, and therefore the studies should focus beyond aggregation models, which is critical for diagnosing and treating amyloid disorders. Moreover, structural biology provides insights into the relationship between atomic structures and cellular disruption, elucidating the mechanisms by which misfolded or aggregated proteins disrupt normal cellular physiology and lead to the development of pathological conditions. Broadly, insights from structural biology also help researchers develop therapeutic or diagnostic strategies for proteinopathies. therapeutic approaches to find already existing drugs to treat amyloid-related disorders.

Acknowledgments

The authors are highly thankful for the facilities provided by the Department of Biochemistry, Faculty of Life Sciences, AMU, Aligarh. N. K. A is a recipient of MANF- SRF funded by the Ministry of Minority Affairs, India.

Conflict of interest

The authors declare no conflict of interest in this work.

References

Adams, D., Gonzalez-Duarte, A., O'Riordan, W.D., Yang, C.C., Ueda, M., Kristen, A.V., et al., 2018. Patisiran, an RNAi therapeutic, for hereditary transthyretin amyloidosis. New England Journal of Medicine 379 (1), 11–21. Available at https://doi.org/10.1056/NEJMoa1716153, http://www.nejm.org/medical-index.

Andrich, K., Hegenbart, U., Kimmich, C., Kedia, N., Bergen, H.R., Schönland, S., et al., 2017. Aggregation of full-length immunoglobulin light chains from systemic light chain amyloidosis (AL) patients is remodeled by epigallocatechin-3-gallate. Journal of Biological Chemistry 292 (6), 2328–2344. Available at https://doi.org/10.1074/jbc.M116.750323, http://www.jbc.org/content/292/6/2328.full.pdf.

Anfinsen, C.B., 1972. The formation and stabilization of protein structure. The Biochemical Journal 128 (4), 737–749. Available at https://doi.org/10.1042/bj1280737.

Angelova, P.R., Ludtmann, M.H.R., Horrocks, M.H., Negoda, A., Cremades, N., Klenerman, D., et al., 2016. Ca2+ is a key factor in α-synuclein-induced neurotoxicity. Journal of Cell Science 129 (9), 1792–1801. Available at https://doi.org/10.1242/jcs.180737, http://jcs.biologists.org/content/joces/129/9/1792.full.pdf.

Ano Bom, A.P.D., Rangel, L.P., Costa, D.C.F., de Oliveira, G.A.P., Sanches, D., Braga, C.A., et al., 2012. Mutant p53 aggregates into prion-like amyloid oligomers and fibrils. Journal of Biological Chemistry 287 (33), 28152–28162. Available at https://doi.org/10.1074/jbc.m112.340638.

Ansari, N.K., Hasan, S., Siddiqui, G.A., Naeem, A., 2025. Protective effect of curcumin on thermally aggregated bovine serum albumin. Cell Biochemistry and Biophysics 83 (4), 4897–4906. Available at https://doi.org/10.1007/s12013-025-01810-6, https://www.springer.com/journal/12013.

Arancibia-Opazo, S., Contreras-Riquelme, J.S., Sánchez, M., Cisternas-Olmedo, M., Vidal, R.L., Martin, A. J.M., et al., 2023. Transcriptional and Histone acetylation changes associated with CRE elements expose key factors governing the regulatory circuit in early stage of Huntington's disease models. bioRxiv, Chile bioRxiv. Available at https://www.biorxiv.org, https://doi.org/10.1101/2023.01.19.524732.

Balducci, C., Mancini, S., Minniti, S., La Vitola, P., Zotti, M., Sancini, G., et al., 2014. Multifunctional liposomes reduce brain β-amyloid burden and ameliorate memory impairment in alzheimer's disease mouse models. Journal of Neuroscience 34 (42), 14022–14031. Available at https://doi.org/10.1523/JNEUROSCI.0284-14.2014, http://www.jneurosci.org/content/34/42/14022.full.pdf.

Balducci, C., Forloni, G., 2019. Doxycycline for Alzheimer's disease: fighting β-amyloid oligomers and neuroinflammation. Frontiers in Pharmacology 10. Available at https://doi.org/10.3389/fphar.2019.00738, http://www.frontiersin.org/Pharmacology.

Banote, R.K., Edling, M., Eliassen, F., Kettunen, P., Zetterberg, H., Abramsson, A., 2016. β-Amyloid precursor protein-b is essential for Mauthner cell development in the zebrafish in a Notch-dependent manner. Developmental Biology 413 (1), 26–38. Available at https://doi.org/10.1016/j.ydbio.2016.03.012, http://www.elsevier.com/inca/publications/store/6/2/2/8/1/6/index.htt.

Barnes, C.A., Robertson, A.J., Louis, J.M., Anfinrud, P., Bax, A., 2019. Observation of β-Amyloid peptide oligomerization by pressure-jump NMR spectroscopy. Journal of the American Chemical Society 141 (35), 13762–13766. Available at https://doi.org/10.1021/jacs.9b06970, http://pubs.acs.org/journal/jacsat.

Beckmann, R.P., Mizzen, L.A., Welch, W.J., 1990. Interaction of Hsp 70 with newly synthesized proteins: implications for protein folding and assembly. Science 248 (4957), 850–854. Available at https://doi.org/10.1126/science.2188360.

Benson, M.D., Kluve-Beckerman, B., Zeldenrust, S.R., Siesky, A.M., Bodenmiller, D.M., Showalter, A.D., et al., 2006. Targeted suppression of an amyloidogenic transthyretin with antisense oligonucleotides. Muscle and Nerve 33 (5), 609–618. Available at https://doi.org/10.1002/mus.20503.

Betzer, C., Lassen, L.B., Olsen, A., Kofoed, R.H., Reimer, L., Gregersen, E., et al., 2018. Alpha-synuclein aggregates activate calcium pump SERCA leading to calcium dysregulation. EMBO Reports 19 (5). Available at https://doi.org/10.15252/embr.201744617.

Bosco, D.A., Morfini, G., Karabacak, N.M., Song, Y., Gros-Louis, F., Pasinelli, P., et al., 2010. Wild-type and mutant SOD1 share an aberrant conformation and a common pathogenic pathway in ALS. Nature Neuroscience 13 (11), 1396–1403. Available at https://doi.org/10.1038/nn.2660.

Brems, D.N., Alter, L.A., Beckage, M.J., Chance, R.E., Dimarchi, R.D., Green, L.K., et al., 1992. Altering the association properties of insulin by amino acid replacement. Protein Engineering, Design and Selection 5 (6), 527–533. Available at https://doi.org/10.1093/protein/5.6.527.

Brown, M.S., Goldstein, J.L., 1986. A receptor-mediated pathway for cholesterol homeostasis. Science 232 (4746), 34–47. Available at https://doi.org/10.1126/science.3513311.

Budd Haeberlein, S., Aisen, P.S., Barkhof, F., Chalkias, S., Chen, T., Cohen, S., et al., 2022. Two randomized phase 3 studies of aducanumab in early Alzheimer's disease. Serdi-Editions, United States Journal of Prevention of Alzheimer's Disease 9 (2), 197–210. Available at https://doi.org/10.14283/jpad.2022.30, https://rd.springer.com/journal/42414.

Bulawa, C.E., Connelly, S., DeVit, M., Wang, L., Weigel, C., Fleming, J.A., et al., 2012. Tafamidis, a potent and selective transthyretin kinetic stabilizer that inhibits the amyloid cascade. Proceedings of the National Academy of Sciences of the United States of America 109 (24), 9629–9634. Available at https://doi.org/10.1073/pnas.1121005109, http://www.pnas.org/content/109/24/9629.full.pdf+html.

Burré, J., Sharma, M., Tsetsenis, T., Buchman, V., Etherton, M.R., Südhof, T.C., 2010. α-Synuclein promotes SNARE-complex assembly in vivo and in vitro. Science 329 (5999), 1663–1667. Available at https://doi.org/10.1126/science.1195227, http://www.sciencemag.org/cgi/reprint/329/5999/1663.pdf.

Calamai, M., Taddei, N., Stefani, M., Ramponi, G., Chiti, F., 2003. Relative influence of hydrophobicity and net charge in the aggregation of two homologous proteins. Biochemistry 42 (51), 15078–15083. Available at https://doi.org/10.1021/bi030135s.

Calhoun, M.E., Burgermeister, P., Phinney, A.L., Stalder, M., Tolnay, M., Wiederhold, K.H., et al., 1999. Neuronal overexpression of mutant amyloid precursor protein results in prominent deposition of cerebrovascular amyloid. Proceedings of the National Academy of Sciences of the United States of America 96 (24), 14088–14093. Available at https://doi.org/10.1073/pnas.96.24.14088.

Calkins, M.J., Reddy, P.H., 2011. Assessment of newly synthesized mitochondrial DNA using BrdU labeling in primary neurons from Alzheimer's disease mice: implications for impaired mitochondrial biogenesis and synaptic damage. Biochimica et Biophysica Acta - Molecular Basis of Disease 1812 (9), 1182–1189. Available at https://doi.org/10.1016/j.bbadis.2011.01.006.

Chang, Y.-J., Chen, Y.-R., 2014. The coexistence of an equal amount of Alzheimer's amyloid-β 40 and 42 forms structurally stable and toxic oligomers through a distinct pathway. The FEBS Journal 281 (11), 2674–2687. Available at https://doi.org/10.1111/febs.12813.

de la Chapelle, A., Tolvanen, R., Boysen, G., Santavy, J., Bleeker-Wagemakers, L., Maury, C.P.J., et al., 1992. Gelsolin–derived familial amyloidosis caused by asparagine or tyrosine substitution for aspartic acid at residue 187. Nature Genetics 2 (2), 157–160. Available at https://doi.org/10.1038/ng1092-157.

Chen, W.-T., Liao, Y.-H., Yu, H.-M., Cheng, I.H., Chen, Y.-R., 2011. Distinct effects of Zn^{2+}, Cu^{2+}, Fe^{3+}, and Al^{3+} on amyloid-β stability, oligomerization, and aggregation. Journal of Biological Chemistry 286 (11), 9646–9656. Available at https://doi.org/10.1074/jbc.m110.177246.

Chen, B., Yang, B., Zhu, J., Wu, J., Sha, J., Sun, J., et al., 2020a. Hsp90 relieves heat stress-induced damage in mouse kidneys: involvement of antiapoptotic PKM2-AKT and autophagic HIF-1α signaling. International Journal of Molecular Sciences 21 (5), 1646. Available at https://doi.org/10.3390/ijms21051646.

Chen, Y., Wang, H., Ying, Z., Gao, Q., 2020b. Ibudilast enhances the clearance of SOD1 and TDP-43 aggregates through TFEB-mediated autophagy and lysosomal biogenesis: the new molecular mechanism of ibudilast and its implication for neuroprotective therapy. Biochemical and Biophysical Research Communications 526 (1), 231–238. Available at https://doi.org/10.1016/j.bbrc.2020.03.051, http://www.sciencedirect.com/science/journal/0006291X.

Chen, G., Wei, T., Ju, F., Li, H., 2023. Protein quality control and aggregation in the endoplasmic reticulum: from basic to bedside. Frontiers in Cell and Developmental Biology 11. Available at https://doi.org/10.3389/fcell.2023.1156152.

Chiti, F., Dobson, C.M., 2006. Protein misfolding, functional amyloid, and human disease. Annual Review of Biochemistry 75, 333–366. Available at https://doi.org/10.1146/annurev.biochem.75.101304.123901.

Choi, M.-G., Kim, M.J., Kim, D.-G., Yu, R., Jang, Y.-N., Oh, W.-J., et al., 2018. Sequestration of synaptic proteins by alpha-synuclein aggregates leading to neurotoxicity is inhibited by small peptide. PLOS ONE 13 (4), e0195339. Available at https://doi.org/10.1371/journal.pone.0195339.

Choubey, V., Safiulina, D., Vaarmann, A., Cagalinec, M., Wareski, P., Kuum, M., et al., 2011. Mutant A53T α-Synuclein induces neuronal death by increasing mitochondrial autophagy. Journal of Biological Chemistry 286 (12), 10814–10824. Available at https://doi.org/10.1074/jbc.M110.132514Estonia, http://www.jbc.org/content/286/12/10814.full.pdf+html.

Ciechanover, A., Kwon, Y.T., 2017. Protein quality control by molecular chaperones in neurodegeneration. Frontiers in Neuroscience 11. Available at https://doi.org/10.3389/fnins.2017.00185, http://journal.frontiersin.org/article/10.3389/fnins.2017.00185/full.

Coelho, T., Adams, D., Silva, A., Lozeron, P., Hawkins, P.N., Mant, T., et al., 2013. Safety and efficacy of RNAi therapy for transthyretin amyloidosis. New England Journal of Medicine 369 (9), 819–829. Available at https://doi.org/10.1056/NEJMoa1208760, http://www.nejm.org/doi/pdf/10.1056/NEJMoa1208760.

Cohen, S.I.A., Linse, S., Luheshi, L.M., Hellstrand, E., White, D.A., Rajah, L., et al., 2013. Proliferation of amyloid-β42 aggregates occurs through a secondary nucleation mechanism. Proceedings of the National Academy of Sciences of the United States of America 110 (24), 9758–9763. Available at https://doi.org/10.1073/pnas.1218402110, http://www.pnas.org/content/110/24/9758.full.pdf+html.

Congdon, E.E., Kim, S., Bonchak, J., Songrug, T., Matzavinos, A., Kuret, J., 2008. Nucleation-dependent tau filament formation: the importance of dimerization and an estimation of elementary rate constants. Journal of Biological Chemistry 283 (20), 13806–13816. Available at https://doi.org/10.1074/jbc.M800247200, http://www.jbc.org/cgi/reprint/283/20/13806.

Coyne, L.P., Rana, A., Wang, X., Bhagwagar, S., Umino, Y., Solessio, E.C., et al., 2022. Mitochondrial protein import stress augments α-synuclein aggregation and neurodegeneration independent of bioenergetics. bioRxiv, United States bioRxiv. Available at https://www.biorxiv.org, https://doi.org/10.1101/2022.09.20.508793.

Das, P., King, J.A., Zhou, R., 2011. Aggregation of γ-crystallins associated with human cataracts via domain swapping at the C-terminal β-strands. Proceedings of the National Academy of Sciences of the United States of America 108 (26), 10514–10519. Available at https://doi.org/10.1073/pnas.1019152108, http://www.pnas.org/content/108/26/10514.full.pdf+html.

Demuro, A., Mina, E., Kayed, R., Milton, S.C., Parker, I., Glabe, C.G., 2005. Calcium dysregulation and membrane disruption as a ubiquitous neurotoxic mechanism of soluble amyloid oligomers. Journal of Biological Chemistry 280 (17), 17294–17300. Available at https://doi.org/10.1074/jbc.M500997200.

Deshpande, A., Kawai, H., Metherate, R., Glabe, C.G., Busciglio, J., 2009. A role for synaptic zinc in activity-dependent aβ oligomer formation and accumulation at excitatory synapses. Journal of Neuroscience 29 (13), 4004–4015. Available at https://doi.org/10.1523/JNEUROSCI.5980-08.2009, http://www.jneurosci.org/cgi/reprint/29/13/4004.

Desplats, P., Lee, H.J., Bae, E.J., Patrick, C., Rockenstein, E., Crews, L., et al., 2009. Inclusion formation and neuronal cell death through neuron-to-neuron transmission of α-synuclein. Proceedings of the National Academy of Sciences of the United States of America 106 (31), 13010–13015. Available at https://doi.org/10.1073/pnas.0903691106, http://www.pnas.org/content/106/31/13010.full.pdf.

Do, T.D., Economou, N.J., Chamas, A., Buratto, S.K., Shea, J.E., Bowers, M.T., 2014. Interactions between amyloid-β and tau fragments promote aberrant aggregates: implications for amyloid toxicity. Journal of Physical Chemistry B 118 (38), 11220–11230. Available at https://doi.org/10.1021/jp506258g, http://pubs.acs.org/journal/jpcbfk.

Dobson, C.M., 2003. Protein folding and misfolding. Nature 426 (6968), 884–890. Available at https://doi.org/10.1038/nature02261.

Du, K., Karp, P.H., Ackerley, C., Zabner, J., Keshavjee, S., Cutz, E., et al., 2015. Aggregates of mutant CFTR fragments in airway epithelial cells of CF lungs: new pathologic observations. Journal of Cystic Fibrosis 14 (2), 182–193. Available at https://doi.org/10.1016/j.jcf.2014.09.012, http://www.elsevier.com/locate/jcf.

Eanes, E.D., Glenner, G.G., 1968. X-ray diffraction studies on amyloid filaments. The Journal of Histochemistry and Cytochemistry: Official Journal of the Histochemistry Society 16 (11), 673–677. Available at https://doi.org/10.1177/16.11.673.

Emmanouilidou, E., Melachroinou, K., Roumeliotis, T., Garbis, S.D., Ntzouni, M., Margaritis, L.H., et al., 2010. Cell-produced α-synuclein is secreted in a calcium-dependent manner by exosomes and impacts neuronal survival. Journal of Neuroscience 30 (20), 6838–6851. Available at https://doi.org/10.1523/JNEUROSCI.5699-09.2010Greece, http://www.jneurosci.org/cgi/reprint/30/20/6838.

Esfandiary, R., Parupudi, A., Casas-Finet, J., Gadre, D., Sathish, H., 2015. Mechanism of reversible self-association of a monoclonal antibody: role of electrostatic and hydrophobic interactions. Journal of Pharmaceutical Sciences 104 (2), 577–586. Available at https://doi.org/10.1002/jps.24237.

Fang, R., Bai, L., Verheijen, B.M., Li, B., Dong, K., Paulo, J.A., et al., 2025. Episodic transport of protein aggregates achieves a positive size selectivity in aggresome formation. Nature Communications 16 (1). Available at https://doi.org/10.1038/s41467-025-62751-5, https://www.nature.com/ncomms/.

Férec, C., Scotet, V., 2020. Genetics of cystic fibrosis: basics. Archives de Pediatrie 27, eS4. Available at https://doi.org/10.1016/S0929-693X(20)30043-9, https://www.elsevier.com/journals/archives-de-pediatrie/0929-693x.

Fitzpatrick, A.W.P., Falcon, B., He, S., Murzin, A.G., Murshudov, G., Garringer, H.J., et al., 2017. Cryo-EM structures of tau filaments from Alzheimer's disease. Nature 547 (7662), 185–190. Available at https://doi.org/10.1038/nature23002, http://www.nature.com/nature/index.html.

Foderà, V., Librizzi, F., Groenning, M., Van De Weert, M., Leone, M., 2008. Secondary nucleation and accessible surface in insulin amyloid fibril formation. Journal of Physical Chemistry B 112 (12), 3853–3858. Available at https://doi.org/10.1021/jp710131u, http://pubs.acs.org/journal/jpcbfk.

Fredenburg, R.A., Rospigliosi, C., Meray, R.K., Kessler, J.C., Lashuel, H.A., Eliezer, D., et al., 2007. The impact of the E46K mutation on the properties of α-synuclein in its monomelic and oligomeric states. Biochemistry 46 (24), 7107–7118. Available at https://doi.org/10.1021/bi7000246.

Fu, Y.H., He, Y., Phan, K., Bhatia, S., Pickford, R., Wu, P., et al., 2022. Increased unsaturated lipids underlie lipid peroxidation in synucleinopathy brain. Acta Neuropathologica Communications 10 (1). Available at https://doi.org/10.1186/s40478-022-01469-7.

Gaspar, R., Meisl, G., Buell, A.K., Young, L., Kaminski, C.F., Knowles, T.P.J., et al., 2017. Secondary nucleation of monomers on fibril surface dominates α-synuclein aggregation and provides autocatalytic amyloid amplification. Quarterly Reviews of Biophysics 50. Available at https://doi.org/10.1017/S0033583516000172, http://journals.cambridge.org/action/displayJournal?jid=QRB.

Gaspar, R., Pallbo, J., Weininger, U., Linse, S., Sparr, E., 2018. Ganglioside lipids accelerate α-synuclein amyloid formation. Biochimica et Biophysica Acta - Proteins and Proteomics 1866 (10), 1062–1072. Available at https://doi.org/10.1016/j.bbapap.2018.07.004, http://www.elsevier.com/locate/bbapap.

van Gils, Juami Hermine Mariama, van Dijk, Erik, Peduzzo, A., Hofmann, A., Vettore, N., Schützmann, M.P., et al., 2020. The hydrophobic effect characterises the thermodynamic signature of amyloid fibril growth. PLOS Computational Biology 16 (5), e1007767. Available at https://doi.org/10.1371/journal.pcbi.1007767.

Goncharoff, D., Du, Z., Venkatesan, S., Cho, B., Zhao, J., Alasady, M.J., et al., 2025. Investigating the aggregation and prionogenic properties of human cancer-related proteins. Molecular and Cellular Biology 45 (4), 154–168. Available at https://doi.org/10.1080/10985549.2025.2481054, https://www.tandfonline.com/journals/tmcb20.

Grad, L.I., Guest, W.C., Yanai, A., Pokrishevsky, E., O'Neill, M.A., Gibbs, E., et al., 2011. Intermolecular transmission of superoxide dismutase 1 misfolding in living cells. Proceedings of the National Academy of Sciences of the United States of America 108 (39), 16398–16403. Available at https://doi.org/10.1073/pnas.1102645108Canada, http://www.pnas.org/content/108/39/16398.full.pdf+html.

Grudzielanek, S., Velkova, A., Shukla, A., Smirnovas, V., Tatarek-Nossol, M., Rehage, H., et al., 2007. Cytotoxicity of insulin within its self-assembly and amyloidogenic pathways. Journal of Molecular Biology 370 (2), 372–384. Available at https://doi.org/10.1016/j.jmb.2007.04.053, https://www.journals.elsevier.com/journal-of-molecular-biology.

Gunawardena, S., Her, L.S., Brusch, R.G., Laymon, R.A., Niesman, I.R., Gordesky-Gold, B., et al., 2003. Disruption of axonal transport by loss of huntingtin or expression of pathogenic polyQ proteins in *Drosophila*. Cell Press, United States Neuron 40 (1), 25–40. Available at https://doi.org/10.1016/S0896-6273(03)00594-4, www.neuron.org.

Haass, C., Lemere, C.A., Capell, A., Citron, M., Seubert, P., Schenk, D., et al., 1995. The Swedish mutation causes early-onset Alzheimer's disease by β-secretase cleavage within the secretory pathway. Nature Medicine 1 (12), 1291–1296. Available at https://doi.org/10.1038/nm1295-1291.

Hartl, F.U., Bracher, A., Hayer-Hartl, M., 2011. Molecular chaperones in protein folding and proteostasis. Nature 475 (7356), 324–332. Available at https://doi.org/10.1038/nature10317.

Hashimoto, M., Hsu, L.J., Xia, Y., Takeda, A., Sisk, A., Sundsmo, M., et al., 1999. Oxidative stress induces amyloid-like aggregate formation of NACP/α- synuclein in vitro. NeuroReport 10 (4), 717–721. Available at https://doi.org/10.1097/00001756-199903170-00011, http://journals.lww.com/neuroreport.

Hebron, M.L., Lonskaya, I., Moussa, C.E.H., 2013. Nilotinib reverses loss of dopamine neurons and improvesmotorbehavior via autophagic degradation of α-synuclein in parkinson's disease models. Human Molecular Genetics 22 (16), 3315–3328. Available at https://doi.org/10.1093/hmg/ddt192, http://hmg.oxfordjournals.org/.

Herrera-Vaquero, M., Bouquio, D., Kallab, M., Biggs, K., Nair, G., Ochoa, J., et al., 2019. The molecular tweezer CLR01 reduces aggregated, pathologic, and seeding-competent α-synuclein in experimental multiple system atrophy. Biochimica et Biophysica Acta (BBA) - Molecular Basis of Disease 1865 (11), 165513. Available at https://doi.org/10.1016/j.bbadis.2019.07.007.

Hirabayashi, M., Inoue, K., Tanaka, K., Nakadate, K., Ohsawa, Y., Kamei, Y., et al., 2001. VCP/p97 in abnormal protein aggregates, cytoplasmic vacuoles, and cell death, phenotypes relevant to neurodegeneration. Cell Death & Differentiation 8 (10), 977–984. Available at https://doi.org/10.1038/sj.cdd.4400907.

Holmgren, G., Steen, L., Suhr, O., Ericzon, B.G., Groth, C.G., Andersen, O., et al., 1993. Clinical improvement and amyloid regression after liver transplantation in hereditary transthyretin amyloidosis. The Lancet 341 (8853), 1113–1116. Available at https://doi.org/10.1016/0140-6736(93)93127-M.

Hsiao, L.-C., Puri, S., Kumar Sriramoju, M., Wang, Y.-S., Lee, T.-N., Wang, L.H.-C., et al., 2025. Hsu, characterization of BRCA1-associated protein-1 (BAP1) aggregation properties induced by cancer-associated mutations. ChemBioChem 26 (23). Available at https://doi.org/10.1002/cbic.202500372.

Hsu, Y.H., Chen, Y.W., Wu, M.H., Tu, L.H., 2019. Protein glycation by glyoxal promotes amyloid formation by islet amyloid polypeptide. Biophysical Journal 116 (12), 2304–2313. Available at https://doi.org/10.1016/j.bpj.2019.05.013, http://www.sciencedirect.com/science/journal/00063495.

Hu, Y., Chen, W., Wei, C., Jiang, S., Li, S., Wang, X., et al., 2024. Pathological mechanisms of amyotrophic lateral sclerosis. Neural Regeneration Research 19 (5), 1036–1044. Available at https://doi.org/10.4103/1673-5374.382985, http://www.nrronline.org/.

Huang, C.J., Lin, C.Y., Haataja, L., Gurlo, T., Butler, A.E., Rizza, R.A., et al., 2007. High expression rates of human islet amyloid polypeptide induce endoplasmic reticulum stress-mediated β-cell apoptosis, a characteristic of humans with type 2 but not type 1 diabetes. Diabetes 56 (8), 2016–2027. Available at https://doi.org/10.2337/db07-0197, http://diabetes.diabetesjournals.org/cgi/reprint/56/8/2016.

Imai, Y., Soda, M., Takahashi, R., 2000. Parkin suppresses unfolded protein stress-induced cell death through its E3 ubiquitin-protein ligase activity. Journal of Biological Chemistry 275 (46), 35661–35664. Available at https://doi.org/10.1074/jbc.C000447200.

Iram, A., Alam, T., Khan, J.M., Khan, T.A., Khan, R.H., Naeem, A., 2013. Molten globule of hemoglobin proceeds into aggregates and advanced glycated end products. PLoS ONE 8 (8). Available at https://doi.org/10.1371/journal.pone.0072075India, http://www.plosone.org/article/fetchObjectAttachment.action;jsessionid=3713A55ADE848F6399B1E86AE8B0873C?uri=info%3Adoi%2F10.1371%2Fjournal.pone.0072075&representation=PDF.

Jadoul, M., Garbar, C., Noël, H., Sennesael, J., Vanholder, R., Bernaert, P., et al., 1997. Histological prevalence of β2-microglobulin amyloidosis in hemodialysis: a prospective post-mortem study. Kidney International 51 (6), 1928–1932. Available at https://doi.org/10.1038/ki.1997.262.

Jahn, T.R., Parker, M.J., Homans, S.W., Radford, S.E., 2006. Amyloid formation under physiological conditions proceeds via a native-like folding intermediate. Nature Structural and Molecular Biology 13 (3), 195–201. Available at https://doi.org/10.1038/nsmb1058.

Jakubczyk, K., Drużga, A., Katarzyna, J., Skonieczna-Żydecka, K., 2020. Antioxidant potential of curcumin—a meta-analysis of randomized clinical trials. Antioxidants 9 (11), 1092. Available at https://doi.org/10.3390/antiox9111092.

Janson, J., Soeller, W.C., Roche, P.C., Nelson, R.T., Torchia, A.J., Kreutter, D.K., et al., 1996. Spontaneous diabetes mellitus in transgenic mice expressing human islet amyloid polypeptide. Proceedings of the National Academy of Sciences of the United States of America 93 (14), 7283–7288. Available at https://doi.org/10.1073/pnas.93.14.7283.

Janssen, K., Claes, F., Van de Velde, D., Wehbi, V.L., Houben, B., Lampi, Y., et al., 2023. Exploiting the intrinsic misfolding propensity of the KRAS oncoprotein. Proceedings of the National Academy of Sciences of the United States of America 120 (9). Available at https://doi.org/10.1073/pnas.2214921120, https://www.pnas.org/doi/10.1073/pnas.2214921120.

Jiao, J., Liu, W., Gao, G., Yang, H., 2025. Serine-129 phosphorylated α-synuclein drives mitochondrial dysfunction and calcium dysregulation in Parkinson's disease model. Frontiers in Aging Neuroscience 17. Available at https://doi.org/10.3389/fnagi.2025.1538166.

Jin, W., Xing, Z., Song, Y., Huang, C., Xu, X., Ghose, S., et al., 2019. Protein aggregation and mitigation strategy in low pH viral inactivation for monoclonal antibody purification. mAbs 11 (8), 1479–1491. Available at https://doi.org/10.1080/19420862.2019.1658493, http://www.tandfonline.com/loi/kmab20.

Kawai, Y., Kuriyama, M., Sohma, Y., Kanai, M., Hori, Y., Tomita, T., 2025. The dynamics of oxygenation to Aβ fibrils using an azobenzene–boron complex type photocatalyst and light energy. Scientific Reports 15 (1). Available at https://doi.org/10.1038/s41598-025-10880-8.

Keller, J.N., Mark, R.J., Bruce, A.J., Blanc, E., Rothstein, J.D., Uchida, K., et al., 1997. 4-hydroxynonenal, an aldehydic product of membrane lipid peroxidation, impairs glutamate transport and mitochondrial function in synaptosomes. Neuroscience 80 (3), 685–696. Available at https://doi.org/10.1016/S0306-4522(97)00065-1.

Kessels, H.W., Nguyen, L.N., Nabavi, S., Malinow, R., 2010. The prion protein as a receptor for amyloid γ 2. Nature 466 (7308), E3. Available at https://doi.org/10.1038/nature09217.

Khurana, R., Ionescu-Zanetti, C., Pope, M., Li, J., Nielson, L., Ramírez-Alvarado, M., et al., 2003. A general model for amyloid fibril assembly based on morphological studies using atomic force microscopy. Biophysical Journal 85 (2), 1135–1144. Available at https://doi.org/10.1016/s0006-3495(03)74550-0.

Kieran, D., Kalmar, B., Dick, J.R.T., Riddoch-Contreras, J., Burnstock, G., Greensmith, L., 2004. Treatment with arimoclomol, a coinducer of heat shock proteins, delays disease progression in ALS mice. Nature Medicine 10 (4), 402–405. Available at https://doi.org/10.1038/nm1021, http://www.nature.com/nm/index.html.

Kim, T., Vidal, G.S., Djurisic, M., William, C.M., Birnbaum, M.E., Garcia, K.C., et al., 2013. Human LilrB2 is a β-amyloid receptor and its murine homolog PirB regulates synaptic plasticity in an Alzheimer's model. American Association for the Advancement of Science, United States Science 341 (6152), 1399–1404. Available at https://doi.org/10.1126/science.1242077, http://www.sciencemag.org/content/341/6152/1399.full.pdf.

Kiraly, S., Stanley, J., Eden, E.R., 2025. Lysosome-mitochondrial crosstalk in cellular stress and disease. Antioxidants 14 (2), 125. Available at https://doi.org/10.3390/antiox14020125.

Klaips, C.L., Jayaraj, G.G., Hartl, F.U., 2018. Pathways of cellular proteostasis in aging and disease. Journal of Cell Biology 217 (1), 51–63. Available at https://doi.org/10.1083/jcb.201709072, http://jcb.rupress.org/content/217/1/51.

Koistinaho, M., Ort, M., Cimadevilla, J.M., Vondrous, R., Cordell, B., Koistinaho, J., et al., 2001. Specific spatial learning deficits become severe with age in β-amyloid precursor protein transgenic mice that harbor diffuse β-amyloid deposits but do not form plaques. Proceedings of the National Academy of Sciences of the United States of America 98 (25), 14675–14680. Available at https://doi.org/10.1073/pnas.261562998.

Kortazar-Zubizarreta, I., Manero-Azua, A., Afonso-Agüera, J., Perez de Nanclares, G., 2023. C9ORF72 gene GGGGCC hexanucleotide expansion: a high clinical variability from amyotrophic lateral sclerosis to frontotemporal dementia. Journal of Personalized Medicine 13 (9), 1396. Available at https://doi.org/10.3390/jpm13091396.

Kovács, G.G., Trabattoni, G., Hainfellner, J.A., Ironside, J.W., Knight, R.S.G., Budka, H., 2002. Mutations of the prion protein gene. Journal of Neurology 249 (11), 1567–1582. Available at https://doi.org/10.1007/s00415-002-0896-9.

Krishnan, S., Raibekas, A.A., 2009. Multistep aggregation pathway of human interleukin-1 receptor antagonist: kinetic, structural, and morphological characterization. Biophysical Journal 96 (1), 199–208. Available at https://doi.org/10.1016/j.bpj.2008.10.002, http://download.cell.com/biophysj/pdf/PIIS0006349508000192.pdf?intermediate=true.

Kuzu, O.F., Granerud, L.J.T., Saatcioglu, F., 2025. Navigating the landscape of protein folding and proteostasis: from molecular chaperones to therapeutic innovations. Signal Transduction and Targeted Therapy 10 (1). Available at https://doi.org/10.1038/s41392-025-02439-w, https://www.nature.com/sigtrans.

Lachmann, H.J., Booth, D.R., Booth, S.E., Bybee, A., Gilbertson, J.A., Gillmore, J.D., et al., 2002. Misdiagnosis of hereditary amyloidosis as AL (primary) amyloidosis. New England Journal of Medicine 346 (23), 1786–1791. Available at https://doi.org/10.1056/NEJMoa013354.

Lachmann, H.J., Goodman, H.J.B., Gilbertson, J.A., Gallimore, J.R., Sabin, C.A., Gillmore, J.D., et al., 2007. Natural history and outcome in systemic AA amyloidosis. New England Journal of Medicine 356 (23), 2361–2371. Available at https://doi.org/10.1056/NEJMoa070265, http://content.nejm.org/cgi/reprint/356/23/2361.pdf.

Lee, C.C., Nayak, A., Sethuraman, A., Belfort, G., McRae, G.J., 2007. A three-stage kinetic model of amyloid fibrillation. Biophysical Journal 92 (10), 3448–3458. Available at https://doi.org/10.1529/biophysj.106.098608, http://www.biophysj.org/cgi/reprint/92/10/3448.

Lee, S., Choi, M.C., Al Adem, K., Lukman, S., Kim, T.Y., 2020. Aggregation and cellular toxicity of pathogenic or non-pathogenic proteins. Scientific Reports 10 (1). Available at https://doi.org/10.1038/s41598-020-62062-3, www.nature.com/srep/index.html.

Li, W., West, N., Colla, E., Pletnikova, O., Troncoso, J.C., Marsh, L., et al., 2005. Aggregation promoting C-terminal truncation of α-synuclein is a normal cellular process and is enhanced by the familial Parkinson's disease-linked mutations. Proceedings of the National Academy of Sciences of the United States of America 102 (6), 2162–2167. Available at https://doi.org/10.1073/pnas.0406976102.

Li, S., Liu, Z., Ji, F., Xiao, Z., Wang, M., Peng, Y., et al., 2012. Delivery of quantum Dot-siRNA nanoplexes in SK-N-SH cells for BACE1 gene silencing and intracellular imaging. Molecular Therapy - Nucleic Acids 1 (4), e20. Available at https://doi.org/10.1038/mtna.2012.11.

Li, Z., Xu, H., Liu, D., Li, D., Liu, G., Wang, S.X., 2019. Hereditary renal amyloidosis with a variant lysozyme p.Trp82Arg in a Chinese family: case report and literature review. BMC Nephrology 20 (1). Available at https://doi.org/10.1186/s12882-019-1496-6, http://www.biomedcentral.com/bmcnephrol/.

Li, Z., Tang, J., Srivastava, R., Bassham, D.C., Howell, S.H., 2020. The transcription factor bZIP60 links the unfolded protein response to the heat stress response in maize. Plant Cell 32 (11), 3559–3575. Available at https://doi.org/10.1105/TPC.20.00260, http://www.plantcell.org/lookup/doi/10.1105/tpc.20.00260.

Lieberman, A.P., Robins, D.M., 2008. The androgen receptor's CAG/glutamine tract in mouse models of neurological disease and cancer. Journal of Alzheimer's Disease 14 (2), 247–255. Available at https://doi.org/10.3233/JAD-2008-14212, www.iospress.nl/journal/journal-of-alzheimers-disease/.

Lins, L., Brasseur, R., 1995. The hydrophobic effect in protein folding. FASEB, Belgium FASEB Journal 9 (7), 535–540. Available at https://doi.org/10.1096/fasebj.9.7.7737462, https://onlinelibrary.wiley.com/journal/15306860.

Liu, Y., Chang, A., 2008. Heat shock response relieves ER stress. EMBO Journal 27 (7), 1049–1059. Available at https://doi.org/10.1038/emboj.2008.42.

Liu, F., Li, B., Tung, E.J., Grundke-Iqbal, I., Iqbal, K., Gong, C.X., 2007. Site-specific effects of tau phosphorylation on its microtubule assembly activity and self-aggregation. European Journal of Neuroscience 26 (12), 3429–3436. Available at https://doi.org/10.1111/j.1460-9568.2007.05955.x.

Liu, F., Sun, J., Wang, X., Jin, S., Sun, F., Wang, T., et al., 2022. Focal-type, but not diffuse-type, amyloid beta plaques are correlated with Alzheimer's neuropathology, cognitive dysfunction, and neuroinflammation in the human hippocampus. Neuroscience Bulletin 38 (10), 1125–1138. Available at https://doi.org/10.1007/s12264-022-00927-5, https://www.springer.com/journal/12264.

MacDonald, M.E., Ambrose, C.M., Duyao, M.P., Myers, R.H., Lin, C., Srinidhi, L., et al., 1993. , A novel gene containing a trinucleotide repeat that is expanded and unstable on Huntington's disease chromosomes. Cell 72 (6), 971–983. Available at https://doi.org/10.1016/0092-8674(93)90585-E.

Maji, S.K., Wang, L., Greenwald, J., Riek, R., 2009. Structure-activity relationship of amyloid fibrils. FEBS Letters 583 (16), 2610–2617. Available at https://doi.org/10.1016/j.febslet.2009.07.003.

Masrori, P., Van Damme, P., 2020. Amyotrophic lateral sclerosis: a clinical review. European Journal of Neurology 27 (10), 1918–1929. Available at https://doi.org/10.1111/ene.14393, http://onlinelibrary.wiley.com/journal/10.1111/(ISSN)1468-1331.

McParland, V.J., Kad, N.M., Kalverda, A.P., Brown, A., Kirwin-Jones, P., Hunter, M.G., et al., 2000. Partially unfolded states of β-microglobulin and amyloid formation in vitro. Biochemistry 39 (30), 8735–8746. Available at https://doi.org/10.1021/bi000276j.

McParland, V.J., Kalverda, A.P., Homans, S.W., Radford, S.E., 2002. Structural properties of an amyloid precursor of β2-microglobulin. Nature Structural Biology 9 (5), 326–331. Available at https://doi.org/10.1038/nsb791.

Melo, E.P., Konno, T., Farace, I., Awadelkareem, M.A., Skov, L.R., Teodoro, F., et al., 2022. Stress-induced protein disaggregation in the endoplasmic reticulum catalysed by BiP. Nature Communications 13 (1). Available at https://doi.org/10.1038/s41467-022-30238-2, http://www.nature.com/ncomms/index.html.

Mintun, M.A., Lo, A.C., Evans, C.D., Wessels, A.M., Ardayfio, P.A., Andersen, S.W., et al., 2021. Donanemab in early Alzheimer's disease. New England Journal of Medicine 384 (18), 1691–1704. Available at https://doi.org/10.1056/NEJMoa2100708, http://www.nejm.org/medical-index.

Mitchell, J.C., McGoldrick, P., Vance, C., Hortobagyi, T., Sreedharan, J., Rogelj, B., et al., 2013. Overexpression of human wild-type FUS causes progressive motor neuron degeneration in an age- and dose-dependent fashion. Acta Neuropathologica 125 (2), 273–288. Available at https://doi.org/10.1007/s00401-012-1043-z.

Miziak, B., Błaszczyk, B., Czuczwar, S.J., 2021. Some candidate drugs for pharmacotherapy of Alzheimer's disease. Pharmaceuticals 14 (5), 458. Available at https://doi.org/10.3390/ph14050458.

Mondragón-Rodriguez, S., Basurto-Islas, G., Santa-Maria, I., Mena, R., Binder, L.I., Avila, J., et al., 2008. Cleavage and conformational changes of tau protein follow phosphorylation during Alzheimer's disease. International Journal of Experimental Pathology 89 (2), 81–90. Available at https://doi.org/10.1111/j.1365-2613.2007.00568.x.

Moriyama, Y., Watanabe, E., Kobayashi, K., Harano, H., Inui, E., Takeda, K., 2008. Secondary structural change of bovine serum albumin in thermal denaturation up to 130 °C and protective effect of sodium dodecyl sulfate on the change. The Journal of Physical Chemistry B 112 (51), 16585–16589. Available at https://doi.org/10.1021/jp8067624.

Morrone, C.D., Raghuraman, R., Hussaini, S.A., Yu, W.H., 2023. Proteostasis failure exacerbates neuronal circuit dysfunction and sleep impairments in Alzheimer's disease. Molecular Neurodegeneration 18 (1). Available at https://doi.org/10.1186/s13024-023-00617-4, https://molecularneurodegeneration.biomedcentral.com.

Naeem, A., Fazili, N.A., 2011. Defective protein folding and aggregation as the basis of neurodegenerative diseases: the darker aspect of proteins. Cell Biochemistry and Biophysics 61 (2), 237–250. Available at https://doi.org/10.1007/s12013-011-9200-x.

Naumann, M., Pal, A., Goswami, A., Lojewski, X., Japtok, J., Vehlow, A., et al., 2018. Impaired DNA damage response signaling by FUS-NLS mutations leads to neurodegeneration and FUS aggregate formation. Nature Communications 9 (1). Available at https://doi.org/10.1038/s41467-017-02299-1.

Nelson, R., Sawaya, M.R., Balbirnie, M., Madsen, A., Riekel, C., Grothe, R., et al., 2005. Structure of the cross-β spine of amyloid-like fibrils. Nature 435 (7043), 773–778. Available at https://doi.org/10.1038/nature03680.

Nordlund, A., Oliveberg, M., 2006. Folding of Cu/Zn superoxide dismutase suggests structural hotspots for gain of neurotoxic function in ALS: parallels to precursors in amyloid disease. Proceedings of the National Academy of Sciences of the United States of America 103 (27), 10218–10223. Available at https://doi.org/10.1073/pnas.0601696103.

Nuga, O., Pourhadi, M., Rausch, J.P., Todi, S.V., 2025. Disrupted transcriptional networks by mutant atrophin-1 in a cell culture model of dentatorubral-pallidoluysian atrophy [Internet]. Neuroscience [cited 2026 Jan 11]. Available at http://biorxiv.org/lookup/doi/10.1101/2025.08.08.669318.

Nymeyer, H., García, A.E., Onuchic, J.N., 1998. Folding funnels and frustration in off-lattice minimalist protein landscapes. Proceedings of the National Academy of Sciences of the United States of America 95 (11), 5921–5928. Available at https://doi.org/10.1073/pnas.95.11.5921, www.pnas.org.

Okon, E.B., Chung, A.W.Y., Zhang, H., Laher, I., Van Breemen, C., 2007. Hyperglycemia and hyperlipidemia are associated with endothelial dysfunction during the development of type 2 diabetes. Canadian Journal of Physiology and Pharmacology 85 (5), 562–567. Available at https://doi.org/10.1139/Y07-026.

Olsson, F., Schmidt, S., Althoff, V., Munter, L.M., Jin, S., Rosqvist, S., et al., 2014. Characterization of intermediate steps in amyloid beta (Aβ) production under near-native conditions. Journal of Biological Chemistry 289 (3), 1540–1550. Available at https://doi.org/10.1074/jbc.M113.498246Sweden, http://www.jbc.org/content/289/3/1540.full.pdf+html.

Onuchic, J.N., Wolynes, P.G., 2004. Theory of protein folding. Current Opinion in Structural Biology 14 (1), 70–75. Available at https://doi.org/10.1016/j.sbi.2004.01.009, http://www.elsevier.com/locate/sbi.

Pagano, G., Boess, F.G., Taylor, K.I., Ricci, B., Mollenhauer, B., Poewe, W., et al., 2021. A phase II study to evaluate the safety and efficacy of prasinezumab in early Parkinson's disease (PASADENA): rationale, design, and baseline data. Frontiers in Neurology 12. Available at https://doi.org/10.3389/fneur.2021.705407.

Pal, S., Udgaonkar, J.B., 2022. Evolutionarily conserved proline residues impede the misfolding of the mouse prion protein by destabilizing an aggregation-competent partially unfolded form. Journal of Molecular Biology 434 (23), 167854. Available at https://doi.org/10.1016/j.jmb.2022.167854.

Paleologou, K.E., Oueslati, A., Shakked, G., Rospigliosi, C.C., Kim, H.Y., Lamberto, G.R., et al., 2010. Phosphorylation at S87 is enhanced in synucleinopathies, inhibits α-synuclein oligomerization, and influences synuclein-membrane interactions. Journal of Neuroscience 30 (9), 3184–3198. Available at https://doi.org/10.1523/JNEUROSCI.5922-09.2010Switzerland, http://www.jneurosci.org/cgi/reprint/30/9/3184.

Pan, J.A., Ullman, E., Dou, Z., Zong, W.X., 2011. Inhibition of protein degradation induces apoptosis through a microtubule-associated protein 1 light chain 3-mediated activation of caspase-8 at intracellular membranes. Molecular and Cellular Biology 31 (15), 3158–3170. Available at https://doi.org/10.1128/MCB.05460-11, http://mcb.asm.org/cgi/reprint/31/15/3158.pdf.

Paraskevopoulou, F., Parvizi, P., Senger, G., Tuncbag, N., Rosenmund, C., Yildirim, F., 2021. Impaired inhibitory GABAergic synaptic transmission and transcription studied in single neurons by Patch-seq in Huntington's disease. Proceedings of the National Academy of Sciences of the United States of America 118 (19). Available at https://doi.org/10.1073/pnas.2020293118, https://www.pnas.org/content/118/19/e2020293118.

Patel, A.N., Jhamandas, J.H., 2012. Neuronal receptors as targets for the action of amyloid-beta protein (Aβ) in the brain. Expert Reviews in Molecular Medicine 14. Available at https://doi.org/10.1017/S1462399411002134.

Pederzoli, F., Ruozi, B., Duskey, J., Hagmeyer, S., Sauer, A.K., Grabrucker, S., et al., 2019. Nanomedicine against Aβ aggregation by β–sheet breaker peptide delivery: in vitro evidence. Pharmaceutics 11 (11). Available at https://doi.org/10.3390/pharmaceutics11110572, https://www.mdpi.com/1999-4923/11/11/572/pdf.

Peinado, J.R., Chaplot, K., Jarvela, T.S., Barbieri, E.M., Shorter, J., Lindberg, I., 2022. Sequestration of TDP-43216-414aggregates by cytoplasmic expression of the proSAAS chaperone. ACS Chemical Neuroscience 13 (11), 1651–1665. Available at https://doi.org/10.1021/acschemneuro.2c00156, http://pubs.acs.org/journal/acncdm.

Petrlova, J., Bhattacherjee, A., Boomsma, W., Wallin, S., Lagerstedt, J.O., Irbäck, A., 2014. Conformational and aggregation properties of the 1-93 fragment of apolipoprotein A-I. Protein Science 23 (11), 1559–1571. Available at https://doi.org/10.1002/pro.2534, http://onlinelibrary.wiley.com/journal/10.1002/(ISSN)1469-896X.

Polymeropoulos, M.H., Lavedan, C., Leroy, E., Ide, S.E., Dehejia, A., Dutra, A., et al., 1997. Mutation in the α-synuclein gene identified in families with Parkinson's disease. Science 276 (5321), 2045–2047. Available at https://doi.org/10.1126/science.276.5321.2045.

Popiel, H.A., Burke, J.R., Strittmatter, W.J., Oishi, S., Fujii, N., Takeuchi, T., et al., 2011. The aggregation inhibitor peptide qbp1 as a therapeutic molecule for the polyglutamine neurodegenerative diseases. Journal of Amino Acids 2011, 1–10. Available at https://doi.org/10.4061/2011/265084.

Prusiner, S.B., 2001. Neurodegenerative diseases and prions. New England Journal of Medicine 344 (20), 1516–1526. Available at https://doi.org/10.1056/nejm200105173442006.

Ramesh, S., Arachchige, A.S.P.M., 2023. Depletion of dopamine in Parkinson's disease and relevant therapeutic options: a review of the literature. AIMS Neuroscience 10 (3), 200–231. Available at https://doi.org/10.3934/NEUROSCIENCE.2023017, http://www.aimspress.com/journal/neuroscience.

Ravalia, A.S., Lau, J., Barron, J.C., Purchase, S.L.M., Southwell, A.L., Hayden, M.R., et al., 2021. Super-resolution imaging reveals extrastriatal synaptic dysfunction in presymptomatic Huntington disease mice. Neurobiology of Disease 152. Available at https://doi.org/10.1016/j.nbd.2021.105293, http://www.elsevier.com/inca/publications/store/6/2/2/9/2/3/index.htt.

Reddy, P.H., Yin, X.L., Manczak, M., Kumar, S., Pradeepkiran, J.A., Vijayan, M., et al., 2018. Mutant APP and amyloid beta-induced defective autophagy, mitophagy, mitochondrial structural and functional changes and synaptic damage in hippocampal neurons from Alzheimer's disease. Human Molecular Genetics 27 (14), 2502–2516. Available at https://doi.org/10.1093/hmg/ddy154, http://hmg.oxfordjournals.org/.

Rogelj, B., Easton, L.E., Bogu, G.K., Stanton, L.W., Rot, G., Curk, T., et al., 2012. Widespread binding of FUS along nascent RNA regulates alternative splicing in the brain. Scientific Reports 2. Available at https://doi.org/10.1038/srep00603.

Ruschak, A.M., Miranker, A.D., 2007. Fiber-dependent amyloid formation as catalysis of an existing reaction pathway. Proceedings of the National Academy of Sciences of the United States of America 104 (30), 12341–12346. Available at https://doi.org/10.1073/pnas.0703306104.

Ryan, S., Rollinson, S., Hobbs, E., Pickering Brown, S., 2022. C9orf72 dipeptides disrupt the nucleocytoplasmic transport machinery and cause TDP-43 mislocalisation to the cytoplasm. Scientific Reports 12 (1). Available at https://doi.org/10.1038/s41598-022-08724-w.

Saelices, L., Johnson, L.M., Liang, W.Y., Sawaya, M.R., Cascio, D., Ruchala, P., et al., 2015. Uncovering the mechanism of aggregation of human transthyretin. Journal of Biological Chemistry 290 (48), 28932–28943. Available at https://doi.org/10.1074/jbc.M115.659912, http://www.jbc.org/content/290/48/28932.full.pdf+html.

Sahin-Tóth, M., 2017. Genetic risk in chronic pancreatitis: the misfolding-dependent pathway. Current Opinion in Gastroenterology 33 (5), 390–395. Available at https://doi.org/10.1097/MOG.0000000000000380, http://journals.lww.com/co-gastroenterology/pages/default.aspx.

Sakagashira, S., Hiddinga, H.J., Tateishi, K., Sanke, T., Hanabusa, T., Nanjo, K., et al., 2000. S20G mutant amylin exhibits increased in vitro amyloidogenicity and increased intracellular cytotoxicity compared to wild-type amylin. The American Journal of Pathology 157 (6), 2101–2109. Available at https://doi.org/10.1016/s0002-9440(10)64848-1.

Santra, M., Dill, K.A., De Graff, A.M.R., 2019. Proteostasis collapse is a driver of cell aging and death. Proceedings of the National Academy of Sciences of the United States of America 116 (44), 22173–22178. Available at https://doi.org/10.1073/pnas.1906592116, https://www.pnas.org/content/pnas/116/44/22173.full.pdf.

Šarić, A., Chebaro, Y.C., Knowles, T.P.J., Frenkel, D., 2014. Crucial role of nonspecific interactions in amyloid nucleation. Proceedings of the National Academy of Sciences of the United States of America 111 (50), 17869–17874. Available at https://doi.org/10.1073/pnas.1410159111, http://www.pnas.org/content/111/50/17869.full.pdf+html.

Sauvé, V., Stefan, E., Croteau, N., Goiran, T., Fakih, R., Bansal, N., et al., 2024. Activation of parkin by a molecular glue. Nature Communications 15 (1). Available at https://doi.org/10.1038/s41467-024-51889-3.

Schilling, G., Savonenko, A.V., Klevytska, A., Morton, J.L., Tucker, S.M., Poirier, M., et al., 2004. Nuclear-targeting of mutant huntingtin fragments produces Huntington's disease-like phenotypes in transgenic mice. Human Molecular Genetics 13 (15), 1599–1610. Available at https://doi.org/10.1093/hmg/ddh175.

Seidler, P.M., Murray, K.A., Boyer, D.R., Ge, P., Sawaya, M.R., Hu, C.J., et al., 2022. Structure-based discovery of small molecules that disaggregate Alzheimer's disease tissue derived tau fibrils in vitro. Nature Communications 13 (1). Available at https://doi.org/10.1038/s41467-022-32951-4, https://www.nature.com/ncomms/.

Seki, T., Adachi, N., Ono, Y., Mochizuki, H., Hiramoto, K., Amano, T., et al., 2005. Mutant protein kinase Cγ found in spinocerebellar ataxia type 14 is susceptible to aggregation and causes cell death. Journal of Biological Chemistry 280 (32), 29096–29106. Available at https://doi.org/10.1074/jbc.M501716200.

Serpell, L.C., Sunde, M., Benson, M.D., Tennent, G.A., Pepys, M.B., Fraser, P.E., 2000. The protofilament substructure of amyloid fibrils11Edited by F. E. Cohen. Journal of Molecular Biology 300 (5), 1033–1039. Available at https://doi.org/10.1006/jmbi.2000.3908.

Shenouda, M., Xiao, S., MacNair, L., Lau, A., Robertson, J., 2022. A C-terminally truncated TDP-43 splice isoform exhibits neuronal specific cytoplasmic aggregation and contributes to TDP-43 pathology in ALS. Frontiers in Neuroscience 16. Available at https://doi.org/10.3389/fnins.2022.868556, https://www.frontiersin.org/journals/neuroscience#.

Sima, S., Barkovits, K., Marcus, K., Schmauder, L., Hacker, S.M., Hellwig, N., et al., 2021. HSP-90/kinase complexes are stabilized by the large PPIase FKB-6. Scientific Reports 11 (1). Available at https://doi.org/10.1038/s41598-021-91667-5, www.nature.com/srep/index.html.

Sitron, C.S., Trinkaus, V.A., Galesic, A., Garhammer, M., Yuste-Checa, P., Dransfeld, U., et al., 2025. α-Synuclein aggregates inhibit ESCRT-III through sequestration and collateral degradation. Molecular Cell 85 (18), 3505. Available at https://doi.org/10.1016/j.molcel.2025.08.022, https://www.sciencedirect.com/science/journal/10972765.

Sneideris, T., Ziaunys, M., Chu, B.K.Y., Chen, R.P.Y., Smirnovas, V., 2020. Self-replication of prion protein fragment 89-230 amyloid fibrils accelerated by prion protein fragment 107-143 aggregates. International Journal of Molecular Sciences 21 (19), 1–16. Available at https://doi.org/10.3390/ijms21197410, https://www.mdpi.com/1422-0067/21/19/7410/pdf.

Soeda, Y., Saito, M., Maeda, S., Ishida, K., Nakamura, A., Kojima, S., et al., 2019. Methylene blue inhibits formation of tau fibrils but not of granular tau oligomers: a plausible key to understanding failure of a clinical trial for Alzheimer's disease. Journal of Alzheimer's Disease 68 (4), 1677–1686. Available at https://doi.org/10.3233/jad-181001.

Soejima, N., Ohyagi, Y., Nakamura, N., Himeno, E., M. Iinuma, K., Sakae, N., et al., 2013. Intracellular accumulation of toxic turn amyloid-β is associated with endoplasmic reticulum stress in Alzheimer's disease. Current Alzheimer Research 10 (1), 11–20. Available at https://doi.org/10.2174/156720513804871372, www.eurekaselect.com/643/journal/current-alzheimer-research.

Spillantini, M.G., Crowther, R.A., Jakes, R., Hasegawa, M., Goedert, M., 1998. α-Synuclein in filamentous inclusions of Lewy bodies from Parkinson's disease and dementia with Lewy bodies. Proceedings of the National Academy of Sciences of the United States of America 95 (11), 6469–6473. Available at https://doi.org/10.1073/pnas.95.11.6469.

Steffan, J.S., Kazantsev, A., Spasic-Boskovic, O., Greenwald, M., Zhu, Y.Z., Gohler, H., et al., 2000. The Huntington's disease protein interacts with p53 and CREB-binding protein and represses transcription. Proceedings of the National Academy of Sciences of the United States of America 97 (12), 6763–6768. Available at https://doi.org/10.1073/pnas.100110097.

Sun, X., Dyson, H.J., Wright, P.E., 2018. Kinetic analysis of the multistep aggregation pathway of human transthyretin. Proceedings of the National Academy of Sciences of the United States of America [Internet] 115 (27) July 3 [cited 2025 Dec 27]. Available at https://www.pnas.org/doi/full/10.1073/pnas.1807024115.

Sundaramoorthy, V., Walker, A.K., Yerbury, J., Soo, K.Y., Farg, M.A., Hoang, V., et al., 2013. Extracellular wildtype and mutant SOD1 induces ER-Golgi pathology characteristic of amyotrophic lateral sclerosis in neuronal cells. Cellular and Molecular Life Sciences 70 (21), 4181–4195. Available at https://doi.org/10.1007/s00018-013-1385-2.

Takei, Y., Teng, J., Harada, A., Hirokawa, N., 2000. Defects axonal elongation and neuronal migration in mice with disrupted tau and map1b genes. Journal of Cell Biology 150 (5), 989–1000. Available at https://doi.org/10.1083/jcb.150.5.989.

Tanu, Chaturvedi, M., Fatima, S., Yadav, S.S., Padhy, P.K., Tiwari, S., et al., 2025. Expression analysis of molecular chaperones associated with disaggregation complex in rotenone-induced Parkinsonian rat model. International Journal of Biochemistry and Cell Biology 181. Available at https://doi.org/10.1016/j.biocel.2025.106752, https://www.sciencedirect.com/science/journal/13572725.

Travers, K.J., Patil, C.K., Wodicka, L., Lockhart, D.J., Weissman, J.S., Walter, P., 2000. Functional and genomic analyses reveal an essential coordination between the unfolded protein response and ER-associated degradation. Cell 101 (3), 249–258. Available at https://doi.org/10.1016/S0092-8674(00)80835-1, https://www.sciencedirect.com/journal/cell.

Truttmann, M.C., Zheng, X., Hanke, L., Damon, J.R., Grootveld, M., Krakowiak, J., et al., 2017. Unrestrained AMPylation targets cytosolic chaperones and activates the heat shock response [Internet]. Proceedings of the National Academy of Sciences 114 (2) Jan 10 [cited 2025 Dec 30]. Available at https://www.pnas.org/doi/full/10.1073/pnas.1619234114.

Ueno, N., Kashiwagi, M., Kanekatsu, M., Marubashi, W., Yamada, T., 2019. Accumulation of protein aggregates induces autolytic programmed cell death in hybrid tobacco cells expressing hybrid lethality. Scientific Reports 9 (1). Available at https://doi.org/10.1038/s41598-019-46619-5.

Van Dyck, C.H., Swanson, C.J., Aisen, P., Bateman, R.J., Chen, C., Gee, M., et al., 2023. Lecanemab in early Alzheimer's disease. New England Journal of Medicine 388 (1), 9–21. Available at https://doi.org/10.1056/NEJMoa2212948, http://www.nejm.org/medical-index.

Vasquez, V., Kodavati, M., Mitra, J., Vedula, I., Hamilton, D.J., Garruto, R.M., et al., 2024. Mitochondria-targeted oligomeric α-synuclein induces TOM40 degradation and mitochondrial dysfunction in Parkinson's disease and parkinsonism-dementia of Guam. Cell Death and Disease 15 (12). Available at https://doi.org/10.1038/s41419-024-07258-5, https://www.nature.com/cddis/.

Visa, M., Alcarraz-Vizán, G., Montane, J., Cadavez, L., Castaño, C., Villanueva-Peñacarrillo, M.L., et al., 2015. Islet amyloid polypeptide exerts a novel autocrine action in β cell signaling and proliferation. The FASEB Journal 29 (7), 2970–2979. Available at https://doi.org/10.1096/fj.15-270553.

Wang, J., Dai, L., Zhang, Z., 2025a. Protein aggregation in neurodegenerative diseases. Chinese Medical Journal 138 (21), 2753–2768. Available at https://doi.org/10.1097/CM9.0000000000003802, https://journals.lww.com/cmj/pages/default.aspx.

Wang, X., Xiong, Y., Duan, C., Hu, J., Lu, H., Yang, M., et al., 2025b. The disease-specific structural pattern in Parkinson's disease and its cortical characteristics associated with gene function: a 7-Tesla MRI study. Journal of Neurology 272 (4). Available at https://doi.org/10.1007/s00415-025-13035-x, https://www.springer.com/journal/415.

Wawer, J., Szociński, M., Olszewski, M., Piątek, R., Naczk, M., Krakowiak, J., 2019. Influence of the ionic strength on the amyloid fibrillogenesis of hen egg white lysozyme. International Journal of Biological Macromolecules 121, 63–70. Available at https://doi.org/10.1016/j.ijbiomac.2018.09.165, www.elsevier.com/locate/ijbiomac.

Weggen, S., Eriksen, J.L., Das, P., Sagi, S.A., Wang, R., Pietrzik, C.U., et al., 2001. A subset of NSAIDs lower amyloidogenic Aβ42 independently of cyclooxygenase activity. Nature 414 (6860), 212–216. Available at https://doi.org/10.1038/35102591.

Weindling, E., Bar-Nun, S., 2015. Sir2 links the unfolded protein response and the heat shock response in a stress response network. Biochemical and Biophysical Research Communications 457 (3), 473–478. Available at https://doi.org/10.1016/j.bbrc.2015.01.021, http://www.sciencedirect.com/science/journal/0006291X.

Westwell-Roper, C., Dai, D.L., Soukhatcheva, G., Potter, K.J., Van Rooijen, N., Ehses, J.A., et al., 2011. IL-1 blockade attenuates islet amyloid polypeptide-induced proinflammatory cytokine release and pancreatic islet graft dysfunction. Journal of Immunology 187 (5), 2755–2765. Available at https://doi.org/10.4049/jimmunol.1002854Canada, http://www.jimmunol.org/content/187/5/2755.full.pdf+html.

Wu, L.S., Cheng, W.C., Shen, C.K.J., 2012. Targeted depletion of TDP-43 expression in the spinal cord motor neurons leads to the development of amyotrophic lateral sclerosis-like phenotypes in mice. Journal of Biological Chemistry 287 (33), 27335–27344. Available at https://doi.org/10.1074/jbc.M112.359000Taiwan, http://www.jbc.org/content/287/33/27335.full.pdf+html.

Wu, J., Österlund, N., Wang, H., Sternke-Hoffmann, R., Pupart, H., Ilag, L.L., et al., 2022. Identifying the role of co-aggregation of Alzheimer's amyloid-β with amorphous protein aggregates of non-amyloid proteins. Cell Reports Physical Science 3 (9), 101028. Available at https://doi.org/10.1016/j.xcrp.2022.101028.

Xu, F., Wang, L., 2024. Deciphering ER stress-unfolded protein response relationship by visualizing unfolded proteins in the ER. Cell Reports 43 (6), 114358. Available at https://doi.org/10.1016/j.celrep.2024.114358.

Xue, W.F., Homans, S.W., Radford, S.E., 2008. Systematic analysis of nucleation-dependent polymerization reveals new insights into the mechanism of amyloid self-assembly. Proceedings of the National Academy of Sciences of the United States of America 105 (26), 8926–8931. Available at https://doi.org/10.1073/pnas.0711664105, http://www.pnas.org/content/105/26/8926.full.pdf.

Yadav, S., Tapadia, M.G., 2013. Neurodegeneration caused by polyglutamine expansion is regulated by P-glycoprotein in *Drosophila melanogaster*. Genetics 195 (3), 857–870. Available at https://doi.org/10.1534/genetics.113.155077India, http://www.genetics.org/content/195/3/857.full.pdf.

Yagi, H., Ban, T., Morigaki, K., Naiki, H., Goto, Y., 2007. Visualization and classification of amyloid β supramolecular assemblies. Biochemistry 46 (51), 15009–15017. Available at https://doi.org/10.1021/bi701842n.

Yamanaka, T., Miyazaki, H., Oyama, F., Kurosawa, M., Washizu, C., Doi, H., et al., 2008. Mutant Huntingtin reduces HSP70 expression through the sequestration of NF-Y transcription factor. EMBO Journal 27 (6), 827–839. Available at https://doi.org/10.1038/emboj.2008.23.

Yang, H., Hu, H.Y., 2016. Sequestration of cellular interacting partners by protein aggregates: implication in a loss-of-function pathology. FEBS Journal 283 (20), 3705–3717. Available at https://doi.org/10.1111/febs.13722, http://onlinelibrary.wiley.com/journal/10.1111/(ISSN)1742-4658.

Yang-Hartwich, Y., Bingham, J., Garofalo, F., Alvero, A.B., Mor, G., 2015. Detection of p53 protein aggregation in cancer cell lines and tumor samples. In: In: Mor, G., Alvero, A.B. (Eds.), Apoptosis and Cancer [Internet]., vol. 1219. Springer New York, New York, NY, pp. 75–86. [cited 2026 Jan 14]. Available at https://link.springer.com/10.1007/978-1-4939-1661-0_7.

Yoshimura, Y., Lin, Y., Yagi, H., Lee, Y.H., Kitayama, H., Sakurai, K., et al., 2012. Distinguishing crystal-like amyloid fibrils and glass-like amorphous aggregates from their kinetics of formation. Proceedings of the National Academy of Sciences of the United States of America 109 (36), 14446–14451. Available at https://doi.org/10.1073/pnas.1208228109Japan, http://www.pnas.org/content/109/36/14446.full.pdf+html.

Yu, X., Fang, Y., Zhang, J., Li, W., Zhang, Y., Cui, Q., et al., 2025. Glycyrrhizic acid combined with human adipose-derived MSCs synergistically alleviates the MPP+/MPTP-induced parkinson's disease by inducing autophagy through PI3K/AKT/HIF-1α pathway. Stem Cell Research & Therapy 16 (1). Available at https://doi.org/10.1186/s13287-025-04626-6.

Zaffagnini, G., Cheng, S., Salzer, M.C., Pernaute, B., Duran, J.M., Irimia, M., et al., 2024. Mouse oocytes sequester aggregated proteins in degradative super-organelles. Cell 187 (5), 1109. Available at https://doi.org/10.1016/j.cell.2024.01.031, https://www.sciencedirect.com/science/journal/00928674.

Zapun, A., Jakob, C.A., Thomas, D.Y., Bergeron, J.J.M., 1999. Protein folding in a specialized compartment: the endoplasmic reticulum. Structure 7 (8), R173. Available at https://doi.org/10.1016/S0969-2126(99)80112-9, http://www.journals.elsevier.com/structure/.

Zhang, H.M., Ye, X., Su, Y., Yuan, J., Liu, Z., Stein, D.A., et al., 2010. Coxsackievirus B3 infection activates the unfolded protein response and induces apoptosis through downregulation of p58IPK and activation of CHOP and SREBP1. Journal of Virology 84 (17), 8446–8459. Available at https://doi.org/10.1128/JVI.01416-09Canada, http://jvi.asm.org/cgi/reprint/84/17/8446.
Zhou, L., Kurouski, D., 2020. Characterization of individual α-synuclein oligomers formed at different stages of protein aggregation by atomic force microscopy-infrared spectroscopy. Analytical Chemistry 92 (10), 6806–6810. Available at https://doi.org/10.1021/acs.analchem.0c00593, http://pubs.acs.org/journal/ancham.
Zhou, S., Cheng, J., Yang, T., Ma, M., Zhang, W., Yuan, S., et al., 2019. Exploration of the misfolding mechanism of transthyretin monomer: insights from hybrid-resolution simulations and markov state model analysis. Biomolecules 9 (12), 889. Available at https://doi.org/10.3390/biom9120889.

CHAPTER 17

Integrated structural approaches in immunotherapy

Hamda Khan[1], Minhal Abidi[1], Safia Habib[1], Arbab Husain[2], Afreen Khanam[2], Rizwan Ahmad[1], Mohd Mustafa[1], and Moinuddin[1]
[1]Department of Biochemistry, Jawaharlal Nehru Medical College, Aligarh Muslim University, Aligarh, Uttar Pradesh, India,
[2]Department of Biotechnology and Life Sciences, Faculty of Sciences, Mangalayatan University, Aligarh, Uttar Pradesh, India

17.1 Immunotherapy

Immunotherapy plays a critical role in achieving effective cancer remission. Cancer immunotherapy, which includes both experimental and conventional treatments, has become a viable strategy for using the immune system to suppress tumors. A key treatment for a variety of immunological conditions, including immunodeficiencies and cancers, is immunotherapy. Numerous methods for triggering an antitumor response against tumors have been explored. Surgery, radiation, and chemotherapy are examples of conventional treatment techniques; each has its pros and cons and has not been very successful in improving clinical outcomes. Immunotherapy has successfully treated more than 10 tumors such as bladder cancer, skin cancer, lung cancer, and kidney cancer; additionally, this number is still rising. Cancer is treated using a variety of immunotherapies (Zhang et al., 2025). Among these are immune checkpoint inhibitors (ICIs), T-cell transfer therapy, monoclonal antibodies (mAbs), immune system modulators, and vaccines. ICIs, T-cell transfer treatment, mAbs, vaccinations, and immune system modulators are a few of them. Chronic and deadly autoimmune illnesses affect millions of people globally, posing a challenge to modern treatment. These conditions arise when the immune system damages and inflames its own tissues. The therapeutic care of a variety of tumors has been greatly enhanced by recent developments in cancer immunotherapy, such as ICIs and chimeric antigen receptor (CAR)-T-cell therapy (Baumeister et al., 2016). Physicians explain how immunotherapy is used to treat certain critical immune system illnesses. Future research must address issues such as production restrictions, limited efficacy in solid tumors, and toxicity associated with immunotherapy. The main focus of traditional therapy has been widespread immunosuppression, which can reduce symptoms but has serious side effects and overlooks immunological dysregulation. Significant obstacles still exist despite these advancements, particularly in the treatment of solid tumors. The variety of tumor antigens, physical barriers that hinder CAR-T-cell penetration, and the immunosuppressive tumor microenvironment (TME) all restrict the effectiveness of these

Advances in Structural Biology
DOI: https://doi.org/10.1016/B978-0-443-44988-8.00009-5

treatments (Alsayb, 2025). Millions of people worldwide suffer from systemic lupus erythematosus, type 1 diabetes, rheumatoid arthritis, and multiple sclerosis, which present significant treatment challenges in terms of efficacy, tolerability, and long-term disease control. By genetically modifying T-cells to target tumor-specific antigens, CAR-T-cell therapy has shown impressive success in hematologic malignancies, especially B-cell leukemias and lymphomas (Bangolo et al., 2025). Unlike conventional cancer therapies, immunotherapy uses the body's own immune system to identify and combat cancer cells, providing a natural way to slow the disease's growth. ICIs, cellular immunotherapy, exosome immunotherapy, and other cutting-edge treatments are the core components of tumor immunotherapy (Xu et al., 2020). Immunotherapy has been developed for hematologic malignancies due to its dramatic improvement in survival benefits when compared to cytotoxic treatment for lung tumors and melanoma. For many solid tumors, including lung cancers and melanoma, immunotherapy with ICIs has taken the potential role of the original treatment (Shahid et al., 2019). An increasing number of cancer types that were previously thought to be nonimmunogenic and to have a restricted function for immunotherapy are now exhibiting signals of transformation. Another example that has long been recognized to be resistant to immunotherapy is pancreatic cancer. Increased macrophages in the stroma and significantly higher VISTA expression are indicative of an immunotherapy resistance mechanism, as per the research available on the milieu in pancreatic cancer (Blando et al., 2019). Immunotherapy has transformed the treatment landscape for head and neck cancer. Numerous ICIs that target the programmed cell death-1/programmed cell death-ligand 1 (PD-1/PD-L1) axis have been approved for use in various disease contexts, and numerous others are presently being studied either alone or in combination (Fasano et al., 2022). There is still a need for a novel treatment for metastatic melanoma because many patients receiving immunotherapy experience disease progression despite these recent developments (Franklin et al., 2017). High mortality and morbidity are linked to advanced stages of melanoma, which entail metastatic processes. This is mostly because of the disease's quick spread and variable reactions to existing treatments, such as immunotherapy.

17.2 Mechanism of cancer immunotherapy

To maximize therapeutic effectiveness and minimize adverse effects, it is essential to comprehend the fundamental mechanisms governing immunotherapy-related responses and different host-tumor interactions. Cancer immunotherapy, which uses the body's natural immune system to detect and eradicate cancer cells, is a cutting-edge treatment. Numerous immunotherapy strategies, including ICIs, CAR-T, tumor-infiltrating lymphocytes (TILs), cancer vaccines, cytokines, and immune modulators, have been studied. It makes use of materials produced in a lab or by the body to boost or stimulate the immune system to make it more effective against cancer or adds synthetic

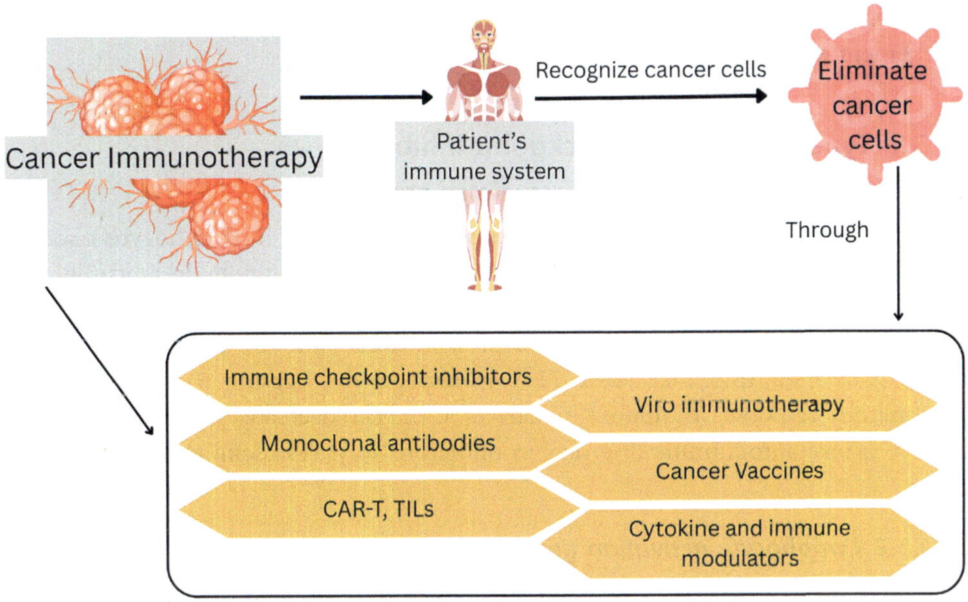

Figure 17.1 Overview of cancer Immunotherapy.

compounds that function similarly to the immune system's natural components (Disis, 2014). Using techniques such as ICIs to unblock T-cells or cell therapies (CAR-T-cells) (Fig. 17.1) that modify immune cells to more effectively target tumors, immunotherapy essentially "releases the brakes" or gives the immune system "superpowers" to overcome cancer's ability to hide. It has been administered in a variety of methods: (1) IV infusion: this is the most popular approach; in which the drug is administered over several hours via a vein; (2) tablets or pills; some immunotherapy drugs are administered as tablets; and (3) Injections: certain therapies are administered subcutaneously.

17.3 Immune checkpoint inhibitors

Immune checkpoint therapy has significantly improved cancer patients' clinical results and provided long-lasting clinical improvements, including a cure in some individuals. By disrupting the CD28-costimulatory signaling pathway, PD-1 engagement can decrease cell proliferation and cytokine release, including IFN-γ, TNF-α, and IL-2 (Shiravand et al., 2022). In fact, a more comprehensive understanding of cancer has emerged from the success of immunotherapy, which depends on activating the host immune system to destroy cancer. It now considers the cancer immunological environment in addition, to the cancer cells that need to be targeted and eliminated. On the other hand, with ICIs immunotherapy, the potential for many neoantigens emerging from highly altered tumors emerges as a favorable factor for response (Parchment et al., 2016).

17.3.1 Types of immune checkpoint inhibitors

17.3.1.1 Cytotoxic T-lymphocyte-associated protein 4

T-cells express the inhibitory receptor known as cytotoxic T-lymphocyte-associated protein 4 (CTLA-4), which is a member of the CD28 immunoglobulin subfamily. Two crucial stimulatory receptors for controlling T-cell activation are CD28 and CTLA-4. Usually located on the surface of antigen-presenting cells (APCs), its ligands, CD80 and CD86, can bind to CD28 or CTLA-4, respectively, causing a costimulatory or coinhibitory response. Inhibitory signals of CTLA-4 decrease T-cell responses and protect against autoimmune disorders in numerous ways (Hosseini et al., 2020).

17.3.1.2 Lymphocyte activation gene 3 inhibitor

Similar to PD-L1 and CTLA-4, lymphocyte activation gene-3 (LAG-3) regulates immune activity, cell proliferation, homeostasis, and cytokine release (Fig. 17.2). Exhausted T cells display a high level of the inhibitory receptor known as LAG-3. Two other prominent ligands of LAG-3 in the TME are liver sinusoidal endothelial cell lectin (LSECtin), which significantly inhibits the antitumor response, and galectin-3, which is a galactoside-binding soluble lectin that has been demonstrated to regulate antigen-specific T-cell activation. LAG-3 is a promising target for immunotherapy, with more than 20 LAG-3-targeting agents in clinical trials and a fixed-dose combination of anti-LAG-3 and anti-PD-1 now licensed to treat metastatic or incurable melanoma (Aggarwal et al., 2023).

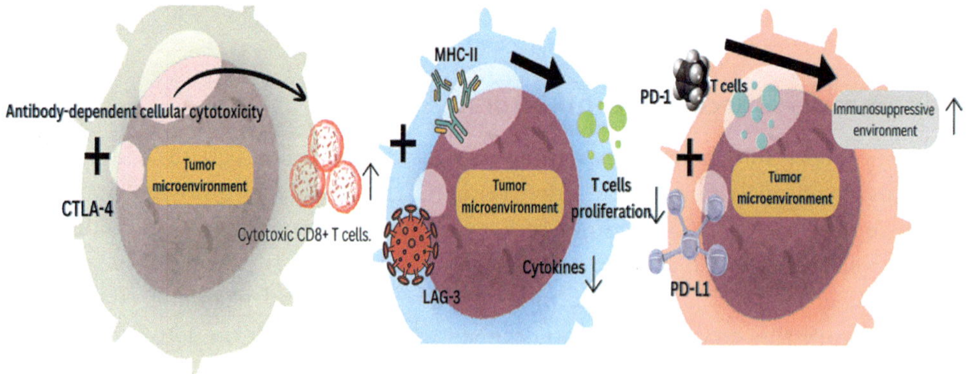

Figure 17.2 Activity of cytotoxic T-lymphocyte-associated protein 4 (*CTLA-4*), lymphocyte activation gene (*LAG3*), and programmed cell death-ligand 1 (*PD-L1*) in cancer therapy.

17.3.2 Mechanism of immune checkpoint inhibitors

The therapeutic use of cancer immunotherapy is severely hampered by both primary and acquired resistance to ICIs. During this phase, certain tumor cells use several methods, including immunological checkpoint proteins, cytokines, and antigens, to elude the antitumor immune response (Ohaegbulam et al., 2015). ICIs are mAbs that "release the brakes" of the immune system by inhibiting inhibitory mechanisms that cancer cells exploit to elude detection. ICIs reawaken T cells by obstructing these "off-switch" signals, enabling them to identify and eliminate tumor cells. T cells' PD-1 attaches to normal cells' PD-L1 to stop autoimmune damage (peripheral tolerance). Many tumors overexpress PD-L1, sending an "off" signal to attacking T cells, causing them to become "exhausted" and inactive. By blocking this interaction, medications like pembrolizumab or atezolizumab revitalize worn-out T cells within the TME.

17.3.3 Immune checkpoint inhibitors in cancer therapy

During the development of the PD-1/PD-L1 pathway, which sends (Disis, 2014) an inhibitory signal to lower T cell activity, PD-1 and PD-L1 significantly contribute to immune system suppression. Numerous malignant cancers frequently express PD-L1. In 83% of SIA cases, intratumoral and peritumoral cells expressed PD-1. In cancer-suffering patients, PD-L1 expression was seen on tumor cells and immune cells. As an immune activation inhibitor, CTLA-4 controls T-cell immunological responses. One strategy to overcome tumor-induced immunological tolerance is to use mAbs that target negative signaling receptors like CTLA-4 to enhance or prolong T-cell activation. By blocking CTLA-4, the CTLA-4 inhibitor ipilimumab enables T-cells to eliminate cancer cells. Another immunotherapy medication, or monoclonal antibody, targeting CTLA-4 is tremelimumab (MDX-010, Yervoy) (Tarhini and Iqbal, 2010). The utilization of combinations of drugs that target multiple important immune system regulatory pathways and alter the host's immune response in a controlled and synergistic manner is also anticipated to contribute to future advancements in CTLA-4 blockade therapy. In antiPD1-based neoadjuvant treatments, anti-CTLA-4 doubles substantial pathologic response rates. LAG3 expression limits the size of the memory pool and the growth of activated T cells, suggesting that LAG3 could be a useful target for immunotherapy. Prior research indicated that individuals with metastatic melanoma who had not previously received treatment could benefit from the combination of LAG-3 inhibitors, such as relatlimab and nivolumab, following the phase I clinical trial. Additionally, via binding to galectin-3 and LSECtin, LAG-3 inhibitors can impact T cell activity. Furthermore, LAG-3 inhibitors can enhance the body's antitumor immune response by blocking the FGL1-LAG-3 interaction (Aggarwal et al., 2023; Huo et al., 2022). The PD1 protein is one of the most significant immunological mechanisms. PD1 regulates immunological responses and is expressed on the surface of T cells (Fig. 17.2).

Myeloid lineage cells, such as dendritic cells (DCs), macrophages, tumor-associated suppressor cells, effector CD8+ T cells, tumor cells, and express CD274, B7–H1, or PD-L1 (Shiri Aghbash et al., 2024). PD-1 is expressed on activated T cells, B-cells, monocytes, DCs, regulatory T cells (Tregs), and natural killer T cells, just like other inhibitory coreceptors. T-cell exhaustion, which is well-defined in cancer and persistent viral infection, is characterized by PD-1 expression (Shiri Aghbash et al., 2024). As effective ICIs, anti-PD-1/PD-L1 inhibitors are quickly taking over as the mainstay treatment for many cancers. In order to produce antitumor immunological responses, tumor immunotherapy attempts to decrease the action of inhibitory immune checkpoint proteins and stimulate T-cell activation (Lin et al., 2024).

17.4 Monoclonal antibodies and cancer therapy

One promising type of targeted anticancer drug is mAbs, which improve the immune system's ability to fight and eradicate cancer cells. Monoclonal antibody-based immunotherapy is now regarded as a key element of cancer treatment in addition, to surgery, radiation, and chemotherapy. mAbs can perform a variety of tasks in breast cancer, such as identifying and delivering therapeutic medications to targets, suppressing immune system inhibitors, preventing cell development, and occasionally directly destroying cancer cells (Behl et al., 2023). Although mAbs have not been extensively studied for other forms of breast cancer, including triple-negative breast cancers (TNBCs), they are also one of the key effective therapy approaches, particularly against human epidermal growth factor receptor 2 (HER2) (Zahavi and Weiner, 2020). Therefore by preventing heterodimerization and internalization, antibodies that target HER2 disrupt signaling. The first FDA-approved anti-HER2 monoclonal antibody, trastuzumab, is still an essential part of treatment for breast cancer that has been amplified by HER2. Complement-dependent cytotoxicity, ADCC, and antibody-dependent cellular phagocytosis are examples of indirect modes of action of mAbs that necessitate the involvement of elements of the host immune system (Jin et al., 2022). Combination therapies involving multispecific antibodies and other immunomodulatory agents, such as immune checkpoint blockers, personalized neoantigen vaccines, and oncolytic viruses, are necessary to address the complex immunosuppressive TME and poor infiltration of immune effector cells in solid tumors (Patel et al., 2009). Combining normal clinical data with molecular profiling is a real attempt to improve prognosis and treatment responsiveness in patients with breast cancer. The three main molecular markers that distinguish people with breast cancer into subtypes are the estrogen receptor, progesterone receptor, and HER2. The primary hormone controlling the proliferation of breast cancer cells is estrogen (Behl et al., 2023). Eliminating cancer cells while causing the least amount of harm to healthy cells is the biggest problem in cancer treatment. mAbs are frequently employed to deliver nanoparticles and enhance therapeutic targeting to cancer cells because of the high number of antigens on the surface of these cells (Wathoni et al., 2022). Bispecific antibodies (bsAbs) have developed quickly in recent years, with numerous novel target combinations and structures being developed. Due to the surge in bsAbs, industry guidelines for their development have been issued in China and the United States in succession. The emergence of bsAbs (Sun et al., 2023) may be influenced by the high degree of similarity in target selection. mAbs are a good option for treating cancer since they may be utilized in a variety of ways and are more effective than other forms of treatment. However, there are a number of adverse effects associated with monoclonal antibody treatment, including fever, shaking, exhaustion, headache, bleeding, breathing difficulties, nausea, muscle pain, rashes, and vomiting. Targeting distinct HER2 domain, the mAbs trastuzumab and pertuzumab can be used to treat breast cancers with HER2 mutations. One monoclonal antibody drug combo that targets HER2 and delivers the deadly medication to tumor cells is trastuzumab emtansine, which is used to treat breast cancer (Shuel, 2022). According to a study, 3C11, a monoclonal antibody that targets the IgV domain of human CEACAM1, and anti-PD1 antibodies work together to create a synergistic antitumor effect on nonsmall cell lung cancer. The expression of p-SHP-1 and p-ZAP70 in T cells is controlled by antibody 3C11. It increased anti-PD-1's tumor growth inhibition value from 18% to 85% in vivo. According to these results, 3C11 may be a useful immunotherapy medication for nonsmall cell lung cancer (Zhao et al., 2024). Most significantly, the FDA has approved trastuzumab deruxtecan for previously treated advanced HER2-mutant nonsmall cell lung cancer (Yu et al., 2023). Additionally, antitumor activity against castration-resistant prostate cancer was enhanced by combined treatment with CAR NK-92 and anti-PD-L1 monoclonal antibody, which has exceptional translational potential in clinical treatment (Wang et al., 2022). A number of examples of mAbs and their beneficial effects in cancer treatment are displayed in Table 17.1.

Table 17.1 Monoclonal antibodies: positive outcomes in cancer therapy.

Monoclonal antibody	Target
Ipilimumab	Advanced gastric cancer (Pento, 2017)
Pembrolizumab	
Osimertinib	NSCLC patients (Barr et al., 2024)
Ado-trastuzumab	NSCLC (Antibody-drug conjugates: A promising novel therapeutic approach in lung cancer, n.d)
Bevacizumab	Advanced NSCLC (Araghi et al., 2023)
Trastuzumab	HER2-positive breast cancer (von Arx et al., 2023)
Paclitaxel	Triple-negative breast cancer (Liu et al., 2022)
Prostate-specific membrane antigen	Prostate cancer-associated protease (Alati et al., 2023)

17.5 Chimeric antigen receptor T cell in cancer therapy

From being a research tool, CAR T-cell treatments has developed into a game-changing treatment with remarkable results in B-cell malignancies. Patient-derived T cells that have been genetically reprogrammed to express CD19-specific CARs are commercially available for the treatment of patients with specific advancedstage B-cell malignancies and have shown promising therapeutic outcomes. In both pediatric and adult patients, the introduction of CAR T-cell treatment has led to an unparalleled long-term clearance of relapse/refractory hematological malignancies (Milone et al., 2021). The two primary functional domains of a typical CAR are an external antigen recognition domain, known as a single-chain variable fragment, and an intracellular signaling domain. CARs are divided into four generations according to the quantity of intracellular signaling molecules they contain. One promising treatment for hematologic malignancies is CAR T-cell therapy. Its clinical trials on solid tumors, however, have not produced promising outcomes. Currently, little is known about how immunogenicity may affect the function and persistence of CAR T cells. Hematopoietic cancers have been successfully treated with CAR T-cell therapy; however, its efficacy against solid tumors has not yet been established (Muhammad et al., 2017). The danger of anti-CAR immunity can be decreased by a variety of monitoring, mitigation, and management strategies (Fig. 17.3), although there is still a need for validated assays

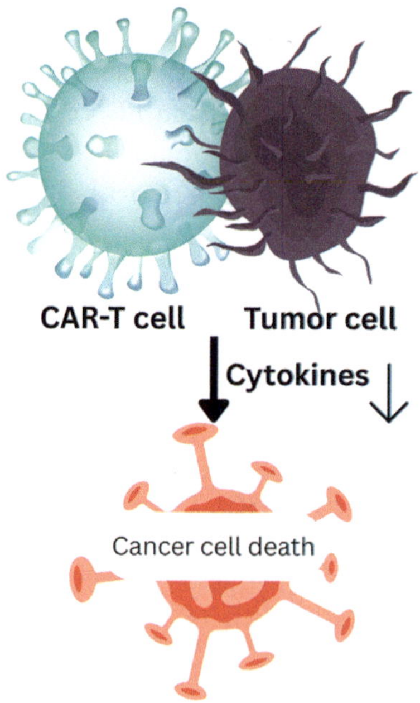

Figure 17.3 Activity and mechanism of chimeric antigen receptor T (*CAR-T*) in cancer immunotherapy.

that allow for sufficient evaluations of anti-CAR immune responses. The CD3 chain is a crucial signal transmitter from endogenous TCRs in the first-generation CAR T model. This kind of medication entered phase I clinical trials for leukemia, lymphoma, and several other cancers, such as ovarian cancer and neuroblastoma, after a favorable result in preclinical trials (Wagner et al., 2021; Mohanty et al., 2019). This novel therapeutic T cell engineering method entails modifying patient blood-derived T cells in vitro to express artificial receptors that are targeted against a specific cancer antigen. The complex and suppressive TME, tumor antigen heterogeneity, cell transport challenges, CAR-T-cell depletion, and decreased cytotoxicity at the tumor site restrict the use of CAR-T-cell therapy and emphasize the need to enhance its efficacy. Five generations of CAR-T cells have been distinguished based on the shape of their endodomains. Consequently, a single signaling domain derived from the TCR/CD3 complex forms the first generation. First-generation CARs exhibit poor antitumor cytotoxicity, durability, and proliferation due to the lack of costimulatory signals. When CAR-T cells and chemotherapy are administered together, the autoimmune and immunosuppressive cells are inhibited, increasing the CAR-T cells' in vivo durability. Hence, the effectiveness of both therapies is maximized and a healthy balance is achieved (Muranski et al., 2006). Future research efforts should focus on improving tumor-site trafficking and infiltration, preventing CAR-T-cell exhaustion and immunological synapse dysfunction, overcoming antigen loss and heterogeneity, increasing resistance to the immunosuppressive TME, enhancing CAR-T-cell efficacy, metabolic reprogramming, and most importantly limiting CAR adverse events (Al-Haideri et al., 2022).

17.5.1 Tumor-infiltrating lymphocytes in cancer therapy

Immunotherapy is a crucial component of cancer treatment. Numerous research focus on biomarkers that can predict response, such as TILs. Advances in data science and genetics, along with a growing understanding of T cell growth and function, make it possible to employ a patient's own T cells for ACT (TIL ACTs). In TIL ACT, TILs are produced and expanded ex vivo from removed tumor material before being administered back to the patient as therapeutic agents (Kumar et al., 2021; Uryvaev et al., 2018). TIL treatment is a multistep procedure that involves removing patient-specific tumor-resident T cells from tumors, growing T cells ex vivo, reinfusing the patient, and then delivering supportive IL-2 after a lymphodepleting preparation regimen. TILs in the setting of postneoadjuvant residual disease are becoming increasingly important as a stratifying marker in clinical research, especially in light of the increased interest in immunotherapeutic methods after neoadjuvant chemotherapy (Dieci et al., 2018; Matsueda et al., 2024). TIL therapy is a rapidly developing modality that is expected to replace ICIs as part of the growing immunotherapy toolset used to treat melanoma and other cancer types. Evidence is mounting that unaltered TILs have strong antitumor activity in melanoma. TIL therapy is the way forward for more individualized, efficient, and humane cancer treatments. It is impossible to overestimate its ability to completely transform the lives of patients and their families. In the face of one of the most difficult enemies, cancer, this treatment is proof of the unwavering spirit of human inventiveness and compassion (Matsueda et al., 2024; Warner et al., 2023). Enhancing the therapeutic window to realize the full potential of TIL therapy is made possible by developments in our understanding of the TME. Cell engineering, process enhancements, and combinations with other existing agents are some of the many ways to achieve this result (El Bairi et al., 2021). Scientists, doctors, and biotech companies must continue their research and work together to advance TIL therapy and expedite its entry into clinical practice, which will ultimately benefit cancer patients worldwide.

17.6 Viro immunotherapy in cancer therapy

Viro immunotherapy causes patients to develop general antitumor and antiviral immune responses. The science of cancer immunotherapy is seeing a sharp increase in interest in oncolytic viral immunotherapy. About 20% of all breast tumors are TNBCs, which have a poor prognosis and a high risk of metastasizing. A new approach to treating pancreatic cancer is needed because of the disease's dismal prognosis, which is partly caused by the limited effectiveness of conventional and targeted therapy. The ability of the immune system to specifically eliminate tumors has come to light, and when it is deliberately stimulated, a long-lasting advantage can be obtained (Brito Baleeiro et al., 2023). Oncolytic viruses may therefore be a potential way to make tumor cells more sensitive to promising immunotherapies. In order to stimulate the immune system and specifically infect and destroy particular cancer cells, oncolytic viruses expose pathogen-associated molecular patterns and damage-associated molecular patterns (DAMPs) to induce inflammatory responses. At the same time, the damaged cancer cells can release neoantigens, which prime and induce antitumor immunity (Palanivelu et al., 2023). Oncolytic viruses are significant therapeutic vectors for a highly immunosuppressed tumor like pancreatic ductal adenocarcinoma because of their capacity to induce inflammation, transfer genes and immunomodulatory drugs, and decrease tumor burden through direct cell lysis (Ibrahim and Wang, 2016). Oncolytic virus therapy is an intriguing therapeutic approach due to its low toxicity during treatment and its dual function of direct oncolysis and immune stimulation (de Graaf et al., 2020). It may therefore be a potential way to make tumor cells more sensitive to promising immunotherapies. Intravenous administration is made possible by certain oncolytic viro therapy (OVTs) ability to cross the blood-brain barrier and their central nervous system tropism. Although oncolytic viruses have been demonstrated to reverse immunosuppressive TME, virotherapy is now recognized as a type of immunotherapy. However, factors that restrict other traditional immunotherapies, like T-cell exhaustion, may also apply to virotherapy (Shin et al., 2021; Zeng et al., 2021). However, in June 2021, teserpaturev, a genetically modified OV based on the herpes simplex virus-1 (G47Δ), received conditional and time-limited approval in Japan to treat malignant gliomas. As a unique immunotherapy, OVT has the ability to both directly destroy gliomas and trigger the body's potent antitumor immunity, which may work in concert with other immunotherapeutic techniques. Additionally, there is a greater chance that oncolytic viruses made from common viruses like AdV or HSV will be neutralized by preexisting antibodies (Propper and Balkwill, 2022). There is still much to learn about OVT and combination therapy, though, such as the mechanisms underlying the immunosuppressive microenvironment of brain tumors, the best way to deliver OVs intratumorally, how to choose the best OVs and immune targets for various tumor types, and molecular markers for predicting the effectiveness of treatment.

17.7 Cytokines and immune modulators in cancer therapy

Therapeutic approaches to improve the growth-inhibiting and immunostimulatory effects of interleukins and interferons to prevent the inflammatory and tumor-promoting effects of cytokines, such as IL-6, TNF, and IL-1β are strongly supported by preclinical evidence (Propper and Balkwill, 2022). The immune system's cells can communicate across short distances, which are important regulators of both innate and adaptive immunity. The majority of trials, however, included individuals with advanced stages of the illness, which may not be the best environment for cytokine-based treatment. Activating the immune system of cancer patients by cytokine therapy has been a significant therapeutic approach and is still a major factor in clinical cancer research today. Adjuvant treatment with IFNα is approved for patients with high-risk melanoma who have undergone total resection as well as a number of refractory cancers (Conlon et al., 2019). Each can improve the efficacy of adoptive immunotherapies and has been shown to mediate strong vaccine adjuvant effects in tumor models (Fewkes and Mackall, 2010). Clinical trials are being conducted to assess antibody IL-2 fusion proteins, which contain IL-2, IL-21, IL-12, TNF, and IFN-αβγ. These proteins help direct cytokines to tumor sites, where they trigger an antitumor response without causing the systemic toxicity of free cytokine therapy. There are still many obstacles to overcome, particularly with regard to delivery strategies, context-dependent effects, and the pleiotropic, redundant, and frequently contradictory functions of several cytokines. While many cytokines' production and activity are dysregulated in cancer, certain cytokines aid in the host's antitumor responses. Due to dose-limiting toxicities, these features, along with subpar pharmacokinetics, have made it more difficult to deliver cytokines using systemic administration routes. In patients with advanced-stage malignancies, cytokines and cytokine antagonists have not been shown to be successful monotherapies, despite compelling preclinical findings. New strategies for using cytokine networks in cancer treatment have been made possible by the development of more potent immunotherapies and a better understanding of the TME. These strategies include targeting early-stage cancers and employing cytokine-based therapies to boost the effectiveness of other treatments or reduce immune-related toxicities (Parmiani et al., 2000). Recent advances in targeting on these pathways have shown promising therapeutic potential, offering novel strategies for treatment resistance, inhibit cancer progression inhibition, and immune system regulation. Given the chance to create fresh approaches and substitutes for TME immunomodulation, cytokines play a significant role in the establishment of the suppressive TME. For this reason, fourth-generation CAR vectors are being developed to include additional genes that can express pro-inflammatory cytokines in addition, to the CAR construct, which may increase the effectiveness of these treatments (Fraietta et al., 2018; Silveira et al., 2022). Cytokine and chemokine signaling pathway targeting is a new area of cancer treatment that has the potential to greatly enhance patient outcomes. The accomplishments to date offer a solid basis for upcoming studies and clinical advancements. Utilizing the expanding knowledge of cytokine biology in conjunction with developments in biotechnology and precision medicine, investigators can anticipate more potent, individualized cancer treatments.

17.8 Cancer Immunotherapy targeting glycation

Advanced glycation end-products (AGEs) are generated when reducing sugars, such as fructose or glucose, establish nonenzymatic covalent bonds with proteins, lipids, or nucleic acids. After being exposed to sugars, proteins or lipids undergo nonenzymatic changes known as AGEs. The glycation process is consistent with the idea that aging is facilitated by the build-up of metabolic waste because the human body lacks the enzymes necessary to eliminate glycated products. Reactive oxygen species (ROS) also contribute significantly to the development of physiological conditions in humans, such as atherosclerosis, aging, cancer, diabetes, and hyperglycemia (Schröter and Höhn, 2018). The development, spread, and occasionally recurrence of malignancies are caused by the ongoing accumulation of cellular and molecular changes. Glycative stress and the production of protein AGEs in the human body are primarily caused by a number of prevalent risk factors for pancreatic cancer. In addition, to producing energy, glycation invariably results in the production of AGEs and reactive dicarbonyl by-products (Dariya and Nagaraju, 2020). In order to change their structure and cause immunogenicity or genotoxicity for the development of cancer, reducing sugars, lipoproteins, or DNAs can also undergo the maillard reaction (MR). AGEs biomarker present in damaged or dysfunctional cells exhibiting aerobic glycolysis and oxidative stress specifically targeted by SIWA318H, which is a new monoclonal antibody. SIWA318H attaches to patient-derived pancreatic tumor xenografts as well as cancer-associated fibroblasts and pancreatic cancer cells. Senescent cells were significantly reduced in the TME of SIWA318H-treated animals as compared to control-treated mice, according to immunohistochemical examinations of the tumor tissues ($P<.05$). SIWA318H is a promising new treatment for pancreatic cancer, as suggested by the previous study (Rossi et al., 2023). As DAMPs, AGEs mediate a variety of pathogenic effects by activating MAPK, PI3K-Akt, P38, ERK1/2-JNK, and MyD88-induced NF-κB signaling pathways through binding to receptors for advanced glycation end-products (RAGE) or other scavenger receptors on the cell surface (Shen et al., 2024).

17.8.1 Activity as immunomodulators

Patients with poorly managed conditions are known to have higher levels of AGEs, and the accumulation of these molecules is a key process that leads to problems in conditions like diabetes and cancer. When the T-lymphocyte mitogen phytohemagglutinin is stimulated, a surface receptor for AGE-modified proteins is present on T cells (Kankia, 2015). Processing encourages the MR, a nonenzymatic interaction between proteins and carbohydrates. The glycation reaction can alter the immunomodulatory potential of feed proteins, and glycation reaction proteins interact with a variety of immunological receptors, such as the RAGE. Advanced glycation targets a number of proteins, including intracellular proteins like hemoglobin and lens crystalline, plasma proteins like immunoglobulins and albumin, and structural proteins like collagen and aggrecan (Teodorowicz et al., 2018). Additional reactions triggered by AGEs include monocytes inducing IL-1 and expressing tissue factor, and THP-1 monocytic cells treated with AGE-human serum albumin releasing TNFα. A decrease in thrombomodulin expression, a rise in endothelin-1 levels, and an increase in the procoagulant response to TNF-α are additional effects of the activation of AGE-RAGE axis (Shen et al., 2024). By scavenging ROS and reducing NADPH oxidase activity, intracellular ROS formation was suppressed, which was a key regulatory mechanism for stopping the AGE-mediated inflammatory response.

Numerous immune cells, such as monocytes/macrophages, neutrophils, DC, and lymphocytes, have been shown to express activated RAGE, and it has been detected in a broad variety of cells, including vascular endothelial cells, hepatocytes, cardiomyocytes, renal cells, etc. According to Taguchi et al., RAGE's interaction with its ligands and the signaling that results are linked to an increase in metastases and play a causal role in the growth of brain tumors (Petriv et al., 2021). According to one study, the activation of AGE-RAGE axis increases the number of cancer cells by triggering the activation of prosurvival (AKT), antiapoptotic (BCl2), cell-cycle (cyclin D1), and autophagic proteins. RAGE has also been found to play a part in the activation of myeloid derived suppressor cells and the angiogenesis in cancer cells (Malik et al., 2015). Research on RAGE's function in humans and murine knockout models points to a homeostatic function in innate immunity, particularly in the control of sepsis. It was crucial to show that there was no toxicity within the host because the bio-distribution investigations had shown that anti-RAGE was widely distributed and that some accumulation was seen in the spleen, uterus, liver, and ovary. There was no discernible toxicity linked to any of the doses employed in the initial in vivo assessment of RAGE-antibody drug conjugate toxicity (Hofer et al., 2016; Deslandes, 2014). High-mobility group box 1 protein (HMGB1), which is one of the important receptor of AGEs is involved in processes linked to immunotherapy resistance and is sensitive to exportin 1 (XPO1)-dependent nuclear export. It is anticipated that a thorough investigation of the roles of HMGB1 mRNA and protein will reveal new possible targets and serve as a foundation for the treatment of metastatic NSCLC in conjunction with ICIs (González-Cao et al., 2024).

17.9 Glycosylation and cancer immunotherapy

One important cellular mechanism controlling a number of physiological and pathological processes is glycosylation. Cancer cells frequently exhibit changes in glycoproteins, glycosphingolipids, and proteoglycans. All cancer cells have altered glycosylation, and specific glycans are recognized indicators of tumor growth. By affecting protein folding, transportation, and localization, glycosylation is a crucial method of protein modification in living things which is essential for controlling a number of biological processes (Xu et al., 2024). Overexpression of fucosylation of antibodies increased branched-glycan structures, and enhanced sialylation. It is the most common cancer associated changes in protein glycosylation. Cancerous cells can exhibit a range of glycan modifications such as increased expression of incomplete glycans, lack of expression or excessive expression of specific glycans, and the emergence of novel glycans. The biological biomarkers approved for clinical use include glycoproteins such as carcinoembryonic antigen for colon cancer, prostate-specific antigen for prostate cancer, HER2 for breast cancer, and alpha-fetoprotein for liver cancer (Varki et al., 2009). For increased anticancer effectiveness, several immunotherapies use a combination of treatments. For instance, anti-PD-1/CTLA-4, nivolumab + ipilimumab, is a commonly used commercially accessible combination therapy for a wide range of metastatic solid tumors. It targets the immunological checkpoints on T cells or their ligands on the tumor cell. It has been shown that the glycan motifs on the structures of these specific proteins, which are processed by the cell's glycosylation apparatus, affect the effectiveness of these combination therapies (Varki et al., 2009; Choi et al., 2022). A novel strategy has been suggested previously to overcome resistance to existing cancer immunotherapies and enhance patient outcomes, which is the alteration of the the glycosylation of malignant cells in the TME. To increase the effectiveness of tumor immunotherapy, targeting glycosylation on immune checkpoint proteins works in concert with ICIs (Pinho et al., 2025; Lee et al., 2025). Tumor-associated immune escape is caused by the extracellular interaction of PD-1 and PD-L1. Glycans on PD-L1 have a crucial role in modulating immunosuppression in cancer. The PD-1/PD-L1 pathway lowers T-cell activation by sending an inhibitory signal. The expression of PD-L1 is frequently seen in malignant tumors. By preventing T-cell activation, glycosylation of PD-L1 at N35, N192, N200, and N219 stabilizes PD-L1 on the surface of cancer cells, aiding in immune evasion. Only angioimmunoblastic T-cell lymphoma tissue cells and Jurkat cells under specific circumstances exhibit surface expression of PD-1 among malignant cells (Yamane et al., 2015). Protein glycosylation is involved in many aspects of homeostasis regulation, including immunological response, protein sorting, and cell migration.

Glycosylation-targeting medications have significant effects on cancer regression, diagnosis, and immune enhancement (Li et al., 2016). By preventing T-cell activation, glycosylation of PD-L1 at N200, N35, N192, and N219 stabilizes PD-L1 on the surface of cancer cells, aiding in immune evasion. As of right now, PD-L1 glycosylation is known to be regulated by at least six glycosyltransferases and four associated proteins. CTLA-4 is a protein mostly generated on the surface of activated lymphocytes and was first discovered in 1987. The surface of T (Benicky et al., 2021) cells expresses CTLA-4, which may help inhibit lymphocyte growth by binding to B7 on APCs. Three levels of analysis are provided by liquid chromatography-mass spectrometry (Zhu et al., 2014). In an allogeneic MLR, Fab fragments and anti-CTLA-4 antibodies increased T-cell proliferation. However, anti-CTLA-4 mAbs inhibited T cell growth in the presence of adequate costimulation and Fc cross-linking (Fang and Chen, 2019).

17.10 Conclusion

Alongside surgery, chemotherapy, and radiation therapy, cancer immunotherapy has become an important aspect in the treatment of contemporary oncology. It offers enhanced quality of life, success in advanced malignancies, and greater precision and specificity. Because cancer immunotherapy is such a large topic, merely analyzing a few points and coming to a conclusion won't explain why immunotherapy has been successful in treating cancer. We covered the latest immunotherapy research and its effects on treatment in this chapter. Immunotherapy can change lives, but not everyone responds to it. For many solid tumors, response rates fall between 15% and 30% on average. Hence, it is important to learn the integrated approach to cancer immunotherapy for better investigation.

References

Aggarwal, V., Workman, C.J., Vignali, D.A.A., 2023. LAG-3 as the third checkpoint inhibitor. Nature Immunology 24 (9), 1415–1422. Available at https://doi.org/10.1038/s41590-023-01569-z, https://www.nature.com/ni.

Alati, S., Singh, R., Pomper, M.G., Rowe, S.P., Banerjee, S.R., 2023. Preclinical development in radiopharmaceutical therapy for prostate cancer. Seminars in Nuclear Medicine 53 (5), 663–686. Available at https://doi.org/10.1053/j.semnuclmed.2023.06.007, http://www.elsevier.com/inca/publications/store/6/2/3/1/8/4/index.htt.

Al-Haideri, M., Tondok, S.B., Safa, S.H., Maleki, A.H., Rostami, S., Jalil, A.T., et al., 2022. CAR-T cell combination therapy: the next revolution in cancer treatment. Cancer Cell International 22 (1). Available at https://doi.org/10.1186/s12935-022-02778-6, https://cancerci.biomedcentral.com/.

Alsayb, M.A., 2025. Innovations in immunotherapy for autoimmune diseases: recent breakthroughs and future directions. Frontiers in Immunology 16. Available at https://doi.org/10.3389/fimmu.2025.1647066, https://www.frontiersin.org/journals/immunology#.

Antibody-drug conjugates: a promising novel therapeutic approach in lung cancer, n.d.

Araghi, M., Mannani, R., Heidarnejad maleki, A., Hamidi, A., Rostami, S., Safa, S.H., et al., 2023. Recent advances in non-small cell lung cancer targeted therapy; an update review. Cancer Cell International 23 (1). Available at https://doi.org/10.1186/s12935-023-02990-y, https://cancerci.biomedcentral.com/.

Bangolo, A., Amoozgar, B., Mansour, C., Zhang, L., Gill, S., Ip, A., et al., 2025. Comprehensive review of early and late toxicities in CAR T-cell therapy and bispecific antibody treatments for hematologic malignancies. Cancers 17 (2), 282. Available at https://doi.org/10.3390/cancers17020282.

Barr, T., Ma, S., Li, Z., Yu, J., 2024. Recent advances and remaining challenges in lung cancer therapy. Chinese Medical Journal 137 (5), 533–546. Available at https://doi.org/10.1097/cm9.0000000000002991.

Baumeister, S.H., Freeman, G.J., Dranoff, G., Sharpe, A.H., 2016. Coinhibitory pathways in immunotherapy for cancer. Annual Review of Immunology 34, 539–573. Available at https://doi.org/10.1146/annurev-immunol-032414-112049, http://arjournals.annualreviews.org/loi/immunol.

Behl, A., Wani, Z.A., Das, N.N., Parmar, V.S., Len, C., Malhotra, S., et al., 2023. Monoclonal antibodies in breast cancer: a critical appraisal. Critical Reviews in Oncology/Hematology 183. Available at https://doi.org/10.1016/j.critrevonc.2023.103915, www.elsevier.com/locate/critrevonc.

Benicky, J., Sanda, M., Brnakova Kennedy, Z., Grant, O.C., Woods, R.J., Zwart, A., et al., 2021. PD-L1 Glycosylation and its impact on binding to clinical antibodies. Journal of Proteome Research 20 (1), 485–497. Available at https://doi.org/10.1021/acs.jproteome.0c00521.

Blando, J., Sharma, A., Higa, M.G., Zhao, H., Vence, L., Yadav, S.S., et al., 2019. Comparison of immune infiltrates in melanoma and pancreatic cancer highlights VISTA as a potential target in pancreatic cancer. Proceedings of the National Academy of Sciences 116 (5), 1692–1697. Available at https://doi.org/10.1073/pnas.1811067116.

Brito Baleeiro, R., Liu, P., Chard Dunmall, L.S., Di Gioia, C., Nagano, A., Cutmore, L., et al., 2023. Personalized neoantigen viro-immunotherapy platform for triple-negative breast cancer. Journal for ImmunoTherapy of Cancer 11 (8), e007336. Available at https://doi.org/10.1136/jitc-2023-007336.

Choi, Y., Kim, J., Chae, J., Hong, J., Park, J., Jeong, E., et al., 2022. Surface glycan targeting for cancer nano-immunotherapy. Journal of Controlled Release 342, 321–336. Available at https://doi.org/10.1016/j.jconrel.2022.01.004.

Conlon, K.C., Miljkovic, M.D., Waldmann, T.A., 2019. Cytokines in the treatment of cancer. Journal of Interferon and Cytokine Research 39 (1), 6–21. Available at https://doi.org/10.1089/jir.2018.0019, www.liebertonline.com/jir.

Dariya, B., Nagaraju, G.P., 2020. Advanced glycation end products in diabetes, cancer and phytochemical therapy. Drug Discovery Today 25 (9), 1614–1623. Available at https://doi.org/10.1016/j.drudis.2020.07.003, www.elsevier.com/locate/drugdiscov.

de Graaf, J.F., Huberts, M., Fouchier, R.A.M., van den Hoogen, B.G., 2020. Determinants of the efficacy of viro-immunotherapy: a review. Cytokine & Growth Factor Reviews 56, 124–132. Available at https://doi.org/10.1016/j.cytogfr.2020.07.001.

Deslandes, A., 2014. Comparative clinical pharmacokinetics of antibody-drug conjugates in first-in-human Phase 1 studies. mAbs 6 (4), 859–870. Available at https://doi.org/10.4161/mabs.28965.

Dieci, M.V., Radosevic-Robin, N., Fineberg, S., van den Eynden, G., Ternes, N., Penault-Llorca, F., et al., 2018. Update on tumor-infiltrating lymphocytes (TILs) in breast cancer, including recommendations to assess TILs in residual disease after neoadjuvant therapy and in carcinoma in situ: a report of the international immuno-oncology biomarker working group on breast cancer.

Seminars in Cancer Biology 52, 16–25. Available at https://doi.org/10.1016/j.semcancer.2017.10.003, http://www.elsevier.com/inca/publications/store/6/2/2/9/4/3/index.htt.

Disis, M.L., 2014. Mechanism of action of immunotherapy. Seminars in Oncology 41 (5), S3. Available at https://doi.org/10.1053/j.seminoncol.2014.09.004, http://www.elsevier.com/inca/publications/store/6/2/3/3/7/8/index.htt.

El Bairi, K., Haynes, H.R., Blackley, E., Fineberg, S., Shear, J., Turner, S., et al., 2021. The tale of TILs in breast cancer: a report from the international immuno-oncology biomarker working group. npj Breast Cancer 7 (1). Available at https://doi.org/10.1038/s41523-021-00346-1, http://www.nature.com/npjbcancer/.

Fang, W.-M., Chen, C., 2019. Progress in research on cytotoxic T lymphocyte-associated antigen 4 (CTLA-4) and bladder cancer. Annals of Urologic Oncology 1–7. Available at https://doi.org/10.32948/auo.2019.01.22.

Fasano, M., Corte, C.M.D., Liello, R.D., Viscardi, G., Sparano, F., Iacovino, M.L., et al., 2022. Immunotherapy for head and neck cancer: present and future. Critical Reviews in Oncology/Hematology 174. Available at https://doi.org/10.1016/j.critrevonc.2022.103679, www.elsevier.com/locate/critrevonc.

Fewkes, N.M., Mackall, C.L., 2010. Novel gamma-chain cytokines as candidate immune modulators in immune therapies for cancer. Cancer Journal 16 (4), 392–398. Available at https://doi.org/10.1097/PPO.0b013e3181eacbc4.

Fraietta, J.A., Lacey, S.F., Orlando, E.J., Pruteanu-Malinici, I., Gohil, M., Lundh, S., et al., 2018. Determinants of response and resistance to CD19 chimeric antigen receptor (CAR) T cell therapy of chronic lymphocytic leukemia. Nature Medicine 24 (5), 563–571. Available at https://doi.org/10.1038/s41591-018-0010-1, http://www.nature.com/nm/index.html.

Franklin, C., Livingstone, E., Roesch, A., Schilling, B., Schadendorf, D., 2017. Immunotherapy in melanoma: recent advances and future directions. European Journal of Surgical Oncology (EJSO) 43 (3), 604–611. Available at https://doi.org/10.1016/j.ejso.2016.07.145.

González-Cao, M., Cai, X., Bracht, J.W.P., Han, X., Yang, Y., Pedraz-Valdunciel, C., et al., 2024. HMGB1 expression levels correlate with response to immunotherapy in non-small cell lung cancer. Lung Cancer: Targets and Therapy 15, 55–67. Available at https://doi.org/10.2147/LCTT.S455034, https://www.dovepress.com/getfile.php?fileID=98950.

Hofer, S., Uhle, F., Fleming, T., Hell, C., Schmoch, T., Bruckner, T., et al., 2016. RAGE-mediated inflammation in patients with septic shock. Journal of Surgical Research 202 (2), 315–327. Available at https://doi.org/10.1016/j.jss.2016.01.019.

Hosseini, A., Gharibi, T., Marofi, F., Babaloo, Z., Baradaran, B., 2020. CTLA-4: From mechanism to autoimmune therapy. International Immunopharmacology 80, 106221. Available at https://doi.org/10.1016/j.intimp.2020.106221.

Huo, J.L., Wang, Y.T., Fu, W.J., Lu, N., Liu, Z.S., 2022. The promising immune checkpoint LAG-3 in cancer immunotherapy: from basic research to clinical application. Frontiers in Immunology 13. Available at https://doi.org/10.3389/fimmu.2022.956090, https://www.frontiersin.org/journals/immunology#.

Ibrahim, A.M., Wang, Y.H., 2016. VIRO-immune therapy: a new strategy for treatment of pancreatic cancer. World Journal of Gastroenterology 22 (2), 748–763. Available at https://doi.org/10.3748/wjg.v22.i2.748, http://www.wjgnet.com/esps/DownLoadFile.aspx?Type=Digital&SubType=1&DOI=10.3748%2fwjg.v22.i2.748&FilePath=Pub%5c10.3748%5cv22%5ci2%5cWJG-22-748.pdf.

Jin, S., Sun, Y., Liang, X., Gu, X., Ning, J., Xu, Y., et al., 2022. Emerging new therapeutic antibody derivatives for cancer treatment. Signal Transduction and Targeted Therapy 7 (1). Available at https://doi.org/10.1038/s41392-021-00868-x.

Kankia, H.I., 2015. Immunomodulatory effects of carbohydrates and advanced glycation end products. Biotechnology and Molecular Biology Reviews 10 (1), 1–11. Available at https://doi.org/10.5897/bmbr2014-0237.

Kumar, A., Watkins, R., Vilgelm, A.E., 2021. Cell therapy with TILs: training and taming T cells to fight cancer. Frontiers in Immunology 12. Available at https://doi.org/10.3389/fimmu.2021.690499, https://www.frontiersin.org/journals/immunology#.

Lee, T.A., Tsai, E.Y., Liu, S.H., Chou, W.C., Hsu Hung, S.D., Chang, C.Y., et al., 2025. Regulation of PD-L1 glycosylation and advances in cancer immunotherapy. Cancer Letters 612. Available at https://doi.org/10.1016/j.canlet.2025.217498, https://www.sciencedirect.com/science/journal/03043835.

Li, C.W., Lim, S.O., Xia, W., Lee, H.H., Chan, L.C., Kuo, C.W., et al., 2016. Glycosylation and stabilization of programmed death ligand-1 suppresses T-cell activity. Nature Communications 7. Available at https://doi.org/10.1038/ncomms12632, https://www.nature.com/ncomms/.

Lin, X., Kang, K., Chen, P., Zeng, Z., Li, G., Xiong, W., et al., 2024. Regulatory mechanisms of PD-1/PD-L1 in cancers. Molecular Cancer 23 (1). Available at https://doi.org/10.1186/s12943-024-02023-w.

Liu, J., Wang, P., Huang, B., Cheng, Q., Duan, Y., Chen, L., et al., 2022. Effective suppression of triple negative breast cancer by paclitaxel nanoparticles conjugated with transmembrane TNF-α monoclonal antibody. International Journal of Pharmaceutics 624, 121969. Available at https://doi.org/10.1016/j.ijpharm.2022.121969.

Malik, P., Chaudhry, N., Mittal, R., Mukherjee, T.K., 2015. Role of receptor for advanced glycation end products in the complication and progression of various types of cancers. Biochimica et Biophysica Acta - General Subjects 1850 (9), 1898–1904. Available at https://doi.org/10.1016/j.bbagen.2015.05.020, www.elsevier.com/locate/bbagen.

Matsueda, S., Chen, L., Li, H., Yao, H., Yu, F., 2024. Recent clinical researches and technological development in TIL therapy. Cancer Immunology, Immunotherapy 73 (11). Available at https://doi.org/10.1007/s00262-024-03793-4.

Milone, M.C., Xu, J., Chen, S.J., Collins, M.K.A., Zhou, J., Powell, D.J., et al., 2021. Engineering-enhanced CAR T cells for improved cancer therapy. Nature Cancer 2 (8), 780–793. Available at https://doi.org/10.1038/s43018-021-00241-5, https://www.nature.com/natcancer/.

Mohanty, R., Chowdhury, C.R., Arega, S., Sen, P., Ganguly, P., Ganguly, N., 2019. CAR T cell therapy: a new era for cancer treatment. Oncology Reports 42 (6), 2183–2195.

Muhammad, N., Mao, Q., Xia, H., 2017. CAR T-cells for cancer therapy. Biotechnology and Genetic Engineering Reviews 33 (2), 190–226. Available at https://doi.org/10.1080/02648725.2018.1430465.

Muranski, P., Boni, A., Wrzesinski, C., Citrin, D.E., Rosenberg, S.A., Childs, R., et al., 2006. Increased intensity lymphodepletion and adoptive immunotherapy - How far can we go? Nature Clinical Practice. Oncology 3 (12), 668–681. Available at https://doi.org/10.1038/ncponc0666.

Ohaegbulam, K.C., Assal, A., Lazar-Molnar, E., Yao, Y., Zang, X., 2015. Human cancer immunotherapy with antibodies to the PD-1 and PD-L1 pathway. Trends in Molecular Medicine 21 (1), 24–33. Available at https://doi.org/10.1016/j.molmed.2014.10.009, www.elsevier.com/locate/molmed.

Palanivelu, L., Liu, C.H., Lin, L.T., 2023. Immunogenic cell death: the cornerstone of oncolytic viro-immunotherapy. Frontiers in Immunology 13. Available at https://doi.org/10.3389/fimmu.2022.1038226, https://www.frontiersin.org/journals/immunology#.

Parchment, R.E., Voth, A.R., Doroshow, J.H., Berzofsky, J.A., 2016. Immuno-pharmacodynamics for evaluating mechanism of action and developing immunotherapy combinations. Seminars in Oncology 43 (4), 501–513. Available at https://doi.org/10.1053/j.seminoncol.2016.06.008, http://www.elsevier.com/inca/publications/store/6/2/3/3/7/8/index.htt.

Parmiani, G., Rivoltini, L., Andreola, G., Carrabba, M., 2000. Cytokines in cancer therapy. Immunology Letters 74 (1), 41–44. Available at https://doi.org/10.1016/s0165-2478(00)00247-9.

Patel, D., Bassi, R., Hooper, A., Prewett, M., Hicklin, D.J., Kang, X., 2009. Anti-epidermal growth factor receptor monoclonal antibody cetuximab inhibits EGFR/HER-2 heterodimerization and activation. International Journal of Oncology 34 (1), 25–32. Available at https://doi.org/10.3892/ijo_00000125, http://www.spandidos-publications.com/serveFile/ijo_34_1_25_PDF.pdf?type=article&article_id=ijo_34_1_25&item=PDF.

Pento, J.T., 2017. Monoclonal antibodies for the treatment of cancer. Anticancer Research 37 (11), 5935–5939. Available at https://doi.org/10.21873/anticanres.12040, http://ar.iiarjournals.org/content/37/11/5935.full.pdf.

Petriv, N., Neubert, L., Vatashchuk, M., Timrott, K., Suo, H., Hochnadel, I., et al., 2021. Increase of α-dicarbonyls in liver and receptor for advanced glycation end products on immune cells are linked to nonalcoholic fatty liver disease and liver cancer. OncoImmunology 10 (1). Available at https://doi.org/10.1080/2162402x.2021.1874159.

Pinho, S.S., Macauley, M.S., Läubli, H., 2025. Tumor glyco-immunology, glyco-immune checkpoints and immunotherapy. Journal for ImmunoTherapy of Cancer 13 (6), e012391. Available at https://doi.org/10.1136/jitc-2025-012391.

Propper, D.J., Balkwill, F.R., 2022. Harnessing cytokines and chemokines for cancer therapy. Nature Reviews Clinical Oncology 19 (4), 237–253. Available at https://doi.org/10.1038/s41571-021-00588-9, https://www.nature.com/nrclinonc/archive/.

Rossi, G.R., Jensen, A., Ng, S., Yin, Z., Li, A., Misra, A., et al., 2023. Advanced glycation end product (AGE) targeting antibody SIWA318H is efficacious in preclinical models for pancreatic cancer. Scientific Reports 13 (1). Available at https://doi.org/10.1038/s41598-023-44211-6, https://www.nature.com/srep/.

Schröter, D., Höhn, A., 2018. Role of advanced glycation end products in carcinogenesis and their therapeutic implications. Current Pharmaceutical Design 24 (44), 5245–5251. Available at https://doi.org/10.2174/1381612825666190130145549, http://www.eurekaselect.com/169476/article.

Shahid, K., Khalife, M., Dabney, R., Phan, A.T., 2019. Immunotherapy and targeted therapy—the new roadmap in cancer treatment. Annals of Translational Medicine 7 (20), 595. Available at https://doi.org/10.21037/atm.2019.05.58.

Shen, C.Y., Lu, C.H., Cheng, C.F., Li, K.J., Kuo, Y.M., Wu, C.H., et al., 2024. Advanced glycation end-products acting as immunomodulators for chronic inflammation, inflammaging and carcinogenesis in patients with diabetes and immune-related diseases. Biomedicines 12 (8). Available at https://doi.org/10.3390/biomedicines12081699, www.mdpi.com/journal/biomedicines.

Shin, D.H., Nguyen, T., Ozpolat, B., Lang, F., Alonso, M., Gomez-Manzano, C., et al., 2021. Current strategies to circumvent the antiviral immunity to optimize cancer virotherapy. Journal for ImmunoTherapy of Cancer 9 (4), e002086. Available at https://doi.org/10.1136/jitc-2020-002086.

Shiravand, Y., Khodadadi, F., Kashani, S.M.A., Hosseini-Fard, S.R., Hosseini, S., Sadeghirad, H., et al., 2022. Immune checkpoint inhibitors in cancer therapy. Current Oncology 29 (5), 3044–3060. Available at https://doi.org/10.3390/curroncol29050247, https://www.mdpi.com/1718-7729/29/5/247.

Shiri Aghbash, P., Mehdizadeh, F., Pourbeiragh, G., Yazdani, Y., Bannazadeh Baghi, H., Jafari Sales, A., et al., 2024. PD-L1 and PD-1 in immune regulation and their implications in blood cancers. Advances in Cancer Biology - Metastasis 11. Available at https://doi.org/10.1016/j.adcanc.2024.100125, https://www.sciencedirect.com/science/journal/26673940.

Shuel, S.L., 2022. Targeted cancer therapies: clinical pearls for primary care. Canadian Family Physician 68 (7), 515.

Silveira, C.R.F., Corveloni, A.C., Caruso, S.R., Macêdo, N.A., Brussolo, N.M., Haddad, F., et al., 2022. Cytokines as an important player in the context of CAR-T cell therapy for cancer: their role in tumor immunomodulation, manufacture, and clinical implications. Frontiers in Immunology 13. Available at https://doi.org/10.3389/fimmu.2022.947648, https://www.frontiersin.org/journals/immunology#.

Sun, Y., Yu, X., Wang, X., Yuan, K., Wang, G., Hu, L., et al., 2023. Bispecific antibodies in cancer therapy: target selection and regulatory requirements. Acta Pharmaceutica Sinica B 13 (9), 3583–3597. Available at https://doi.org/10.1016/j.apsb.2023.05.023.

Tarhini, A.A., Iqbal, F., 2010. CTLA-4 blockade: therapeutic potential in cancer treatments. OncoTargets and Therapy 3, 15–25. Available at https://doi.org/10.2147/ott.s4833, http://www.dovepress.com/getfile.php?fileID=5833.

Teodorowicz, M., Hendriks, W.H., Wichers, H.J., Savelkoul, H.F.J., 2018. Immunomodulation by processed animal feed: the role of maillard reaction products and advanced glycation end-products (AGEs). Frontiers in Immunology 9. Available at https://doi.org/10.3389/fimmu.2018.02088, www.frontiersin.org/Immunology.

Uryvaev, A., Passhak, M., Hershkovits, D., Sabo, E., Bar-Sela, G., 2018. The role of tumor-infiltrating lymphocytes (TILs) as a predictive biomarker of response to anti-PD1 therapy in patients with metastatic non-small cell lung cancer or metastatic melanoma. Medical Oncology 35 (3). Available at https://doi.org/10.1007/s12032-018-1080-0.

Varki, A., Kannagi, R., Toole, B.P., 2009. Glycosylation changes in cancer. Essentials of Glycobiology. 2nd edition.

von Arx, C., De Placido, P., Caltavituro, A., Di Rienzo, R., Buonaiuto, R., De Laurentiis, M., et al., 2023. The evolving therapeutic landscape of trastuzumab-drug conjugates: future perspectives beyond HER2-positive breast cancer. Cancer Treatment Reviews 113, 102500. Available at https://doi.org/10.1016/j.ctrv.2022.102500.

Wagner, D.L., Fritsche, E., Pulsipher, M.A., Ahmed, N., Hamieh, M., Hegde, M., et al., 2021. Immunogenicity of CAR T cells in cancer therapy. Nature Reviews Clinical Oncology 18 (6), 379–393. Available at https://doi.org/10.1038/s41571-021-00476-2, http://www.nature.com/nrclinonc/archive/index.html.

Wang, F., Wu, L., Yin, L., Shi, H., Gu, Y., Xing, N., 2022. Combined treatment with anti-PSMA CAR NK-92 cell and anti-PD-L1 monoclonal antibody enhances the antitumour efficacy against castration-resistant prostate cancer. Clinical and Translational Medicine 12 (6). Available at https://doi.org/10.1002/ctm2.901.

Warner, A.B., Corrie, P.G., Hamid, O., 2023. Tumor-infiltrating lymphocyte therapy in melanoma: facts to the future. Clinical Cancer Research 29 (10), 1835–1854. Available at https://doi.org/10.1158/1078-0432.CCR-22-1922, https://aacrjournals.org/clincancerres/article/29/10/1835/726243/Tumor-Infiltrating-Lymphocyte-Therapy-in-Melanoma.

Wathoni, N., Puluhulawa, L.E., Joni, I.M., Muchtaridi, M., Mohammed, A.F.A., Elamin, K.M., et al., 2022. Monoclonal antibody as a targeting mediator for nanoparticle targeted delivery system for lung cancer. Drug Delivery 29 (1), 2959–2970. Available at https://doi.org/10.1080/10717544.2022.2120566, https://www.tandfonline.com/loi/idrd20.

Xu, Z., Zeng, S., Gong, Z., Yan, Y., 2020. Exosome-based immunotherapy: a promising approach for cancer treatment. Molecular Cancer 19 (1). Available at https://doi.org/10.1186/s12943-020-01278-3.

Xu, X., Peng, Q., Jiang, X., Tan, S., Yang, W., Han, Y., et al., 2024. Altered glycosylation in cancer: molecular functions and therapeutic potential. Cancer Communications 44 (11), 1316–1336. Available at https://doi.org/10.1002/cac2.12610.

Yamane, H., Isozaki, H., Takeyama, M., Ochi, N., Kudo, K., Honda, Y., et al., 2015. Programmed cell death protein 1 and programmed death-ligand 1 are expressed on the surface of some small-cell lung cancer lines. American Journal of Cancer Research 5 (4), 1553–1557. Available at http://www.ajcr.us/files/ajcr0006165.pdf.

Yu, Y., Yang, Y., Li, H., Fan, Y., 2023. Targeting HER2 alterations in non-small cell lung cancer: therapeutic breakthrough and challenges. Cancer Treatment Reviews 114, 102520. Available at https://doi.org/10.1016/j.ctrv.2023.102520.

Zahavi, D., Weiner, L., 2020. Monoclonal antibodies in cancer therapy. Antibodies 9 (3), 34. Available at https://doi.org/10.3390/antib9030034.

Zeng, J., Li, X., Sander, M., Zhang, H., Yan, G., Lin, Y., 2021. Oncolytic viro-immunotherapy: an emerging option in the treatment of gliomas. Frontiers in Immunology 12. Available at https://doi.org/10.3389/fimmu.2021.721830.

Zhang, M., Liu, C., Tu, J., Tang, M., Ashrafizadeh, M., Nabavi, N., et al., 2025. Advances in cancer immunotherapy: historical perspectives, current developments, and future directions. Molecular Cancer 24 (1). Available at https://doi.org/10.1186/s12943-025-02305-x.

Zhao, L., Li, T., Zhou, Y., Wang, P., Luo, L., 2024. Monoclonal antibody targeting CEACAM1 enhanced the response to anti-PD1 immunotherapy in non-small cell lung cancer. International Immunopharmacology 143 (Pt 2), 113395. Available at https://doi.org/10.1016/j.intimp.2024.113395.

Zhu, L., Guo, Q., Guo, H., Liu, T., Zheng, Y., Gu, P., et al., 2014. Versatile characterization of glycosylation modification in CTLA4-Ig fusion proteins by liquid chromatography-mass spectrometry. mAbs 6 (6), 1474–1485. Available at https://doi.org/10.4161/mabs.36313, http://www.tandfonline.com/doi/pdf/10.4161/mabs.36313.

SECTION IV

Translational biology

CHAPTER 18

Structural and systems biology in cancer therapy

Mohd Basheeruddin[1], Sana Qausain[1], Ashish Anjankar[1], Faez Iqbal Khan[2], Samarth Shukla[3], Shazia Jamal[4], Neesar Ahmed[4], Vaibhav Anjankar[5], Zahoor Ahmad Parray[6], and Netri Upadhyay[1]

[1]Department of Biochemistry, Jawaharlal Nehru Medical College, Datta Meghe Institute of Higher Education and Research, Wardha, Maharashtra, India,
[2]Department of Biological Sciences, School of Science, Xi'an Jiaotong-Liverpool University, Suzhou, Jiangsu, P.R. China,
[3]Department of Pathology, Jawaharlal Nehru Medical College, Datta Meghe Institute of Higher Education and Research, Wardha, Maharashtra, India,
[4]School of Life Sciences, BSA Crescent Institute of Science & Technology, Chennai, Tamil Nadu, India,
[5]Department of Anatomy, Jawaharlal Nehru Medical College, Datta Meghe Institute of Higher Education and Research, Wardha, Maharashtra, India,
[6]Department of Bio-Science and Technology, M.M College of Engineering, Maharishi Markandeshwar University (Deemed to be), Ambala, Haryana, India

18.1 Introduction

Cancer is one of the most complicated and problematic health conditions of the 21st century, which is characterized by its heterogeneity, flexibility, and capacity to resist treatment pressure. Even with impressive progress in diagnostics, therapeutics, and population-level cancer control programs, the global cancer burden is steadily increasing and requires integrative scientific approaches that do not limit themselves to conventional reductionist research models (Chakraborty and Rahman, 2012).

The essential biology of cancer is a system of molecular, structural, physiological, and environmental interactions that all interact to affect disease development and progression, metastasis, and treatment sensitivity. A new paradigm based on integrative biology, which brings together molecular, cellular, biophysical, computational, and clinical aspects, has become a disruptive concept in better cancer understanding and treatment in recent years. The present introductory chapter describes the ongoing increase in the global cancer burden, the relevance of the structural and systems biology paradigm, and the accelerating nature of the precision-oncology landscape, which currently aims to personalize diagnosis and treatment by matching interventions with patient-specific molecular signatures and tumor phenotypes (Laubenbacher et al., 2009).

18.1.1 Global cancer burden: an expanding public health challenge

Cancer burden continues to increase at an unprecedented rate in the world, due to the transformations in demographic, lifestyle, and environmental factors, in addition to the

Advances in Structural Biology
DOI: https://doi.org/10.1016/B978-0-443-44988-8.00023-X

development of diagnostic tools. Modern epidemiological statements suggest a persistent rise in cancer morbidity and mortality in almost all places on earth. The trend is not only observable in high-income countries but also in low- and middle-income countries, where the rate of growth is often even steeper. Millions of people are diagnosed with cancer each year, and a significant percentage of them die of the illness, turning cancer into one of the leading causes of morbidity and mortality on the global level. Cancer cases are projected to increase in the next decades even more than they have already in the world, as populations are aging and modifiable lifestyle-related risk factors, including tobacco use, poor diet, exposure to carcinogens, obesity, and physical inactivity, continue to increase (Kanavos, 2006).

The growing trend in cancer prevalence has also stimulated the socioeconomic burden on the healthcare systems. Contemporary oncology is resource-consuming, and the cost is increasing due to advanced treatment, supportive care, follow-up, and survivorship treatment. Extensive and lengthy treatment courses, increasing indications of targeted agents, and the increasing number of cancer survivors together constitute an inconceivable strain on the national health budgets and clinical infrastructure. These facts emphasize the necessity of creative, economically efficient, and universal methods that will be able to maximize patient results and minimize the number of unnecessary interventions, treatment-related toxicity, and financial burden. To address these accruing challenges, there has been an increase in the movement to appreciate cancer at a more biological and mechanistic level (Laviana et al., 2020). The old methods of examining tumors by only one biological dimension cannot be used anymore in relation to the heterogeneity of the disease. The overall architecture of the cancer is decipherable using integrative biology, a synthesis of findings in genomics, transcriptomics, proteomics, metabolomics, and the tumor microenvironment. This systems-level view defines how genetic mutations, epigenetic alterations, distorted signaling pathways, metabolic rewiring, and microenvironmental factors engage together to initiate tumor formation, control tumor growth, and promote therapeutic resistance. The interconnected layers that integrative biology connects support the creation of predictive, individualized, and mechanistically informed strategies that have the potential to revolutionize future cancer prevention, diagnosis, and treatment (Dong et al., 2025).

18.1.1.1 Need for integrative biology in cancer research

Cancer biology is multifactorial in nature, mediated by interactions among cellular processes, molecular pathways, structural conformations of biomolecules, metabolic processes, multiorgan cross-communication, and systemic feedback mechanisms. Traditional single-parameter or pathway-specific research models often cannot represent this complexity. Integrative biology thus comes to the rescue in filling the gap between molecular understanding and clinical reality. Integrative biology is an interdisciplinary approach that integrates genomics, transcriptomics, metabolomics, proteomics, structural biology, biophysics, and systems-level modeling to create a more consistent view of cancer behavior. It is group of individuals who recognize that cancer cannot be precisely comprehended from individual perspectives; rather, it is an emergent, dynamic network (Koutsogiannouli et al., 2013).

For example, mutations in oncogenes or tumor-suppressor genes affect not only transcriptional output but also protein structure, stability, folding, protein–protein interactions (PPIs), and intracellular localization. Likewise, signaling pathways are interconnected circuits and not linear cascades, and cross-talk mechanisms allow tumor cells to circumvent therapeutic blockade. Factors in the tumor microenvironment, such as immune cells, stromal cells, extracellular matrix molecules, and angiogenic networks, have further effects on tumor evolution, migration, and drug sensitivity (Oncogenes and genes: functions and roles in cancers - PMC [Internet], no date). Integrative biology enables a multidimensional assessment of these aspects, enabling the identification of new biomarkers, therapeutic targets, and resistance pathways. It can also be used to design interventions that cannot regard isolated molecular events, but take into account the whole biological context. Integrative approaches, by bringing together structural and systems-level perspectives, contribute to better predictive capabilities for disease pathways, classify patients on the basis of biological risk signatures, and develop combination therapy approaches that circumvent compensatory mechanisms (Ahmad et al., 2023).

18.1.2 Overview of structural and systems biology paradigms

18.1.2.1 Structural biology: lighting up the structures of molecules and processes

Structural biology provides precise insights into the three-dimensional (3D) structures, folding dynamics, and function of biomolecules that determine cellular and physiological processes. This field in cancer research is crucial for unravel the mechanism through which genetic changes can redefine protein structure, alter binding affinities, and eventually cause cellular pathways to go awry, as seen in Fig. 18.1. A large number of oncogenic mutations cause subtle or dramatic changes in conformations that affect protein stability, enzyme activity, and the formation of functional complexes by the molecules. Through these structural implications, researchers will be in a better position to comprehend the molecular pathway of malignant transformation (Ahmad et al., 2023).

The analysis methods available today offer a broad spectrum of methods that have facilitated the visualization of biological macromolecules at the atomic level and the near-atomic level. X-ray crystallography, nuclear magnetic resonance (NMR) spectroscopy, cryogenic electron microscopy (cryo-EM), small-angle X-ray scattering, and high-end molecular modeling tools can offer complementary insights into biomolecular form and function. Through these systems, scientists are able to study proteins, nucleic

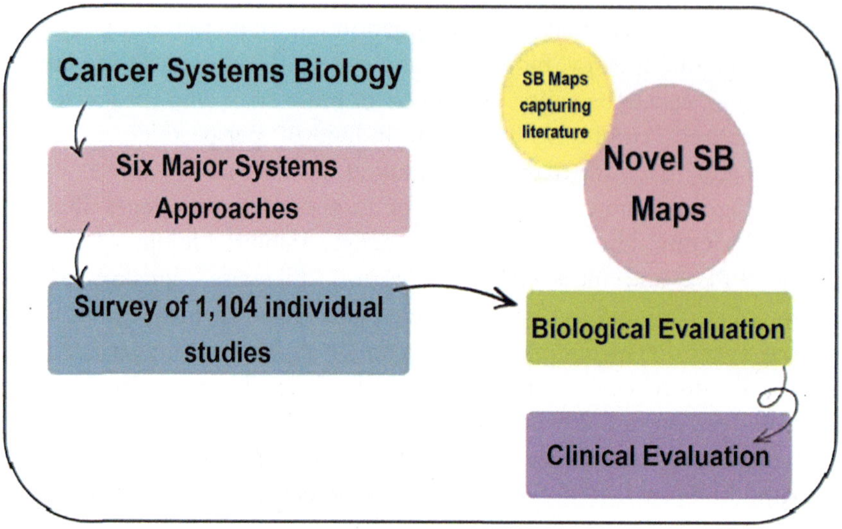

Figure 18.1 A pathway map in cancer system biology. *No Permission Required.*

acids, and multicomponent assemblies with a degree of detail never previously possible, providing a range of valuable information on involving cancer mechanisms and therapeutic innovation (High-resolution protein modeling through Cryo-EM and AI: current trends and future perspectives – a review - PMC [Internet], no date).

18.1.2.1.1 Chronological understanding of oncogenic variants

This is among the greatest contributions of structural biology. Comparing wild-type and mutated protein structures, scientists can determine the effects of individual point mutations on catalytic residues, disruptive interfaces, or constitutive activation of oncogenic proteins. These structural lessons explain the molecular etiology of tumor-favoring changes and the design of specific treatment plans (Pandurangan and Blundell, 2020).

18.1.2.1.2 Rational drug design

The other application is in rational drug design. Structural data can provide information on accessible pockets to which a ligand can bind, direct optimization of ligand binding, and even predict off-target interactions. This method will expedite the development of more discriminatory and powerful anticancer drugs (Pandurangan and Blundell, 2020).

18.1.2.1.3 Learning about protein misfolding and aggregation

Structural biology is also able to augment our knowledge of protein misfolding and aggregation, which are gradually being identified as contributors to the development of cancer. The visualization of misfolded intermediates and aberrant complexes illuminates the failure of proteostasis and the distorted interaction of chaperones (Protein Misfolding and Aggregation in Proteinopathies: Causes, Mechanism and Cellular Response - PMC [Internet], no date).

18.1.2.1.4 Mapping signaling complexes

Lastly, structural analyses allow the mapping of signaling complexes, which assist in clarifying the structure and regulation of receptor complexes, scaffolding proteins, and multiprotein networks that lie at the heart of oncogenic signaling (Prischi and Filippakopoulos, 2021).

18.1.2.2 Systems biology: cellular and network-level interactions decoding

Systems biology offers a more integrative insight, which goes beyond the molecules themselves to the interactions between cells, tissues, and all of nature that determine biological behavior. In contrast to the reductionist methods, which study one part at a time, systems biology focuses on the interlocking characteristics of biological pathways and how emergent properties can be developed out of complex networks. It can be used to simulate the behavior of dynamic biological systems in response to perturbations, stresses, or therapeutic interventions by integrating experimental data with computational tools (Systems Biology as a Comparative Approach to Understand Complex Gene Expression in Neurological Diseases - PMC [Internet], no date). One of the strengths of systems biology is the ability to combine multiomics datasets such as genomics, transcriptomics, proteomics, metabolomics, and epigenomics, as shown in Table 18.1. Those data layers are constructed by mathematical modeling, network theory, and pathway reconstruction to produce holistic images of cellular functionality. This method is especially useful in the field of cancer since tumors are immensely heterogeneous and adaptive (Chen et al., 2023).

18.1.2.3 Mapping oncogenic networks

Systems-level mapping reveals how signaling pathways communicate, compensate, or reorganize when exposed to therapeutic pressure. This information aids in the determination of the key regulatory centers that can have the potential to act as drug targets (Behar et al., 2013).

18.1.2.4 Modeling tumor evolution

Computational models are dynamic representations of cancer evolution, shown to emerge as clonal subpopulations that compete and evolve in response to the selective pressure of situations like hypoxia or drug exposure (A comprehensive review of computational cell cycle models in guiding cancer treatment strategies - PMC [Internet], no date).

Table 18.1 Major databases for multiomics and systems biology research in cancer.

Database name	Omics type	Description/key features	URL/access
The Cancer Genome Atlas	Genomics/ transcriptomics	Multicancer omics collection	https://www.cancer.gov/tcga
International Cancer Genome Consortium	Genomics	Global cancer genome collaboration	https://dcc.icgc.org
Human Protein Atlas	Proteomics	Tissue protein expression resources	https://www.proteinatlas.org
ProteomicsDB	Proteomics	Comprehensive human proteome quantification	https://www.proteomicsdb.org
MetaboLights	Metabolomics	Metabolomics data study repository	https://www.ebi.ac.uk/metabolights
Gene Expression Omnibus	Transcriptomics	Functional genomics dataset archive	https://www.ncbi.nlm.nih.gov/geo
Reactome pathway database	Pathway/systems biology	Disease-linked pathway networks	https://reactome.org

No Permission Required.

18.1.2.5 Tumor-immune interactions to be decoded

Systems immunology has offered an understanding of how immune cells sense and react to malignant cells, and how tumors find a way to evade immune monitoring by taking advantage of network imbalances (Immune Surveillance and Immune Escape in Cancer: Mechanisms and Immunotherapy - PMC [Internet], no date).

18.1.2.6 Recognizing combinations in therapy

Simulations based on networks can be useful in predicting synergistic combinations of drugs and finding pathways that mediate resistance to guide the design of more effective multiagent therapies (Jin et al., 2025).

18.1.2.7 Making microenvironmental influences

By including the extracellular matrix properties, vascular structures, immune cell invasion, and metabolic gradients, systems biology creates more realistic models of tumor behavior in real physiological settings (Dong et al., 2024).

18.1.3 Emerging precision-oncology landscape

Precision oncology is a revolution in the field of cancer treatment that has changed the direction of treatment toward a more personalized approach to treating the disease. It appreciates the fact that tumors are biologically heterogeneous, though they may have a similar origin, and thus need individually tailored diagnostic and treatment protocols. Through the combination of genomic, proteomic, metabolic, and microenvironmental data, precision oncology seeks to provide interventions that more accurately target the molecular drivers of the disease in each individual patient (Heterogeneity in precision oncology - PMC [Internet], no date).

18.1.3.1 Genomic and molecular profiling

The rapid development of high-throughput sequencing has made it possible to characterize cancer genomes without the complexity of traditional methods. Contemporary technologies can detect somatic mutations, changes in copy number, epigenetic changes, and changes in gene expression. Overall profiling offers several clinical benefits, including the identification of actionable mutations, selection of targeted therapy, risk assessment, and monitoring of minimal residual disease. In addition, therapeutic modifications can be made in a timely manner through the early identification of emerging resistance mechanisms. Molecular profiling has been the focus of the treatment of cancers such as lung, breast, colorectal, prostate, and hematological malignancies (Ghoreyshi et al., 2025).

18.1.3.2 Targeted and mechanism-driven therapies

The development of precision oncology is highly associated with mechanism-based therapeutic design. Monoclonal antibodies, antibody-drug conjugates, small-molecule inhibitors, and synthetic lethality approaches have increased the therapeutic options for patients with specific molecular changes. These treatments are directly directed towards mutant proteins, dysregulated pathways, or altered structural motifs in cancer cells. Such interventions can be quite efficacious with minimal systemic toxicity by virtue of their molecular specificity, compared to conventional chemotherapies (He et al., 2024).

18.1.3.3 Immuno-oncology and personalized immune approaches

As a part of contemporary oncology, immunotherapy has established its status. Immunomodulators, adoptive T-cell therapy, neoantigen vaccines, and engineered immune cells are engineered to match the tumor's immune and antigenic profiles. Precision immuno-oncology allows clinicians to pair the therapy with biomarkers, like PD-L1 or tumor mutational burden, or the neoantigen repertoire, leading to improved therapy response and survival (Neoantigen-driven personalized tumor therapy: An update from discovery to clinical application - PMC [Internet], no date).

18.1.3.4 Biomarker-driven clinical decisions

Circulating tumor DNA, extracellular vesicles, proteomic and metabolomic changes produce noninvasive biomarkers that are increasingly incorporated into the management of cancer. Such biomarkers facilitate early diagnosis, immediate assessment of treatment response, risk assessment for recurrence, and assessment of therapeutic efficacy. Their integration within standard clinical operations enhances customized care and supports adaptive treatment plans (Tamura et al., 2021).

18.1.3.5 Integrative multiomics approaches

Accurate oncology has now been based on the intersection of various omics. The combination of genomic, proteomic, metabolomic, epigenomic, and microbiome data will enable clinicians and researchers to define tumor subclasses more precisely, identify latent vulnerabilities, and tailor therapeutic combinations. The multiomics analysis will give a more in-depth insight into tumor evolution, heterogeneity, and treatment adaptation (A Detailed Catalogue of Multi-Omics Methodologies for Identification of Putative Biomarkers and Causal Molecular Networks in Translational Cancer Research [Internet], no date).

18.1.3.6 Structural and systems-guided interventions

Structural biology also plays a role in the identification of drug-binding sites, the prediction of resistance-inducing mutations, and rational drug design. This can be supplemented by systems biology, which produces computational models of disease progression, predicts responses to therapy, and recommends the most suitable sequence of therapy. These methods can be used together to increase the accuracy of clinical decision-making (Hao et al., 2012).

18.1.3.7 Transition toward predictive and preventive oncology

Oncology is slowly moving from treatment to prevention. Genetic risk profiling and population-based stratification can enable earlier screening, closer population observation, and more personalized prevention measures. The purpose of this prospective method is to decrease the occurrence of cancer and enhance future outcomes (Shieh et al., 2016).

18.2 Foundations of structural biology in cancer research

The 3D structures of proteins and nucleic acids can be identified using structural biology to understand the mechanisms by which cancer arises from changes in molecular structure. It connects genetic mutations to effects on protein folding, stability, and interactions, and is used to explain how oncogenes become abnormally active or how tumor suppressors become inactive. X-ray crystallography, cryo-EM, and NMR are techniques that have revolutionized cancer research by enabling visualization of key molecules, including p53, RAS, epidermal growth factor receptor (EGFR), BRAF, and HER2. These structural understandings have been used to design specific therapies by structure-based and fragment-based drug-design methods, which aid in the design of more specific and effective anticancer agents (Structural Variations in Cancer and the 3D Genome - PMC [Internet], no date).

18.2.1 Methods of structural biology: concepts and scientific significance

A number of complementary methods are relied upon in the structural analysis of cancer-associated proteins. There is no universal technique that can be used with all types of biomolecules; rather, researchers combine various techniques to obtain a complete picture of protein conformation, flexibility, dynamics, and interactions. All the techniques offer a distinct view into the molecular landscape, and all of them can be used as a toolkit, facilitating both fundamental research and therapeutic innovation (Zhou et al., 2024).

18.2.1.1 X-ray crystallography

X-ray crystallography is still considered one of the most fundamental techniques in structural biology and has been used to elucidate thousands of protein structures, including those involved in cancer. The method entails purifying a target protein, then inducing it to form an ordered crystal lattice. When an X-ray beam is incident upon the crystal, atoms scatter the radiation, producing a diffraction pattern that is distinctive of the atomic arrangement in the lattice (Smyth and Martin, 2000). This pattern can be mathematically recreated to give an electron-density map into which the protein structure can be modeled. X-ray crystallography has a number of strong features.

18.2.1.2 Atomic-scale resolution

Resolutions of less than 2 Å are often achievable in crystallography, enabling researchers to observe the orientations of side chains, hydrogen-bond networks, and even water molecules that stabilize an active site. Such accuracy is important in determining the effects of mutations on protein structure or the interactions of small molecules with catalytic sites (Wlodawer et al., 2008).

18.2.1.3 Finer perspectives on mechanisms of enzymes

The structural configuration of catalytic loops, ATP-binding residues, and activation segments of many kinases, phosphatases, proteins, and other enzymes of interest in cancer biology has been identified through crystallographic studies. It is based on these insights that selective inhibitors are designed to distinguish between an oncogenic and a wild-type form (Conformational Dynamics and Structural Transitions of Arginine Kinase: Implications for Catalysis and Allergen Control [Internet], no date).

18.2.1.4 Drug–protein interaction mapping

Crystals soaked with inhibitors or drug candidates provide direct visualization of how therapeutic molecules occupy binding pockets. This has been essential in developing inhibitors against BCR-ABL, EGFR, BRAF, and numerous other cancer targets (Leveraging artificial intelligence and machine learning in kinase inhibitor development: advances, challenges, and future prospects - PMC [Internet], no date).

18.2.2 Cryogenic electron microscopy

Over the past decade, cryo-EM has transformed structural biology and has been referred to as the start of the "resolution revolution." Unlike crystallography, cryo-EM does not require protein crystals. Instead, proteins or complexes are rapidly frozen in a thin layer of vitreous ice, preserving them in near-native conformations (Developments, applications, and prospects of cryo-electron microscopy - PMC [Internet], no date). An electron beam is then used to generate thousands to millions of particle images, which are computationally combined to reconstruct a 3D structure. Cryo-EM is particularly well suited for cancer research for several reasons, such as the following:

18.2.2.1 Visualization of large complexes

Many cancer-related pathways involve immense molecular assemblies—as DNA repair complexes, spliceosomes, chromatin remodelers, and receptor clusters—cannot be crystallized. Cryo-EM excels in capturing these megadalton-scale assemblies with high-resolution (Cryo-EM of NHEJ supercomplexes provides insights into DNA repair - PMC [Internet], no date).

18.2.2.2 Structural snapshots of dynamic states

Proteins often adopt multiple conformations during signaling or catalysis. Cryo-EM is able to record various states that are coexisting within the sample, and therefore researchers can map conformational changes that occur during activation, inhibition, or dysregulation due to mutation (Conformational changes studied by cryo-electron microscopy - PubMed [Internet], no date).

18.2.2.3 Membrane protein structures

A number of receptor tyrosine kinases, ion channels, and transporters are important in cancer cell signaling. The cryo-EM technique can be used to analyze these membrane-bound proteins in lipid conditions or nanodiscs that simulate cellular membranes (How Neuromembrane Lipids Modulate Membrane Proteins: Insights from G-Protein-Coupled Receptors [GPCRs] and Receptor Tyrosine Kinases [RTKs] - PMC [Internet], no date).

18.2.3 Nuclear magnetic resonance spectroscopy

The concept of NMR spectroscopy is radically different since the proteins are examined in solution instead of in crystal form or in frozen form. This enables the researcher to obtain conformational flexibility and temporary interactions that cannot be observed in crystallography and are frequently difficult to observe in cryo-EM. NMR derives structural data using the magnetic character of atomic nuclei. The nuclear spins produce signals that are dependent on the chemical environment when subjected to a powerful magnetic field and to radiofrequency pulses (New Horizons in Structural Biology of Membrane Proteins: Experimental Evaluation of the Role of Conformational Dynamics and Intrinsic Flexibility [Internet], no date). The study of these signals shows the interatomic distances, bond angles, and dynamic movement. NMR is especially beneficial in cancer studies in some respects, such as the following:

18.2.3.1 Study of intrinsically disordered proteins

Intrinsically disordered regions that do not have a stable secondary or tertiary structure are found in many cancer-associated proteins, including transcription factors, signaling adapters, and regulatory domains. These loose regions, which commonly mediate crucial PPIs, are best studied using NMR (Pathological Unfoldomics of Uncontrolled Chaos: Intrinsically Disordered Proteins and Human Diseases - PMC [Internet], no date).

18.2.3.2 Mapping interaction interfaces

Chemical-shift perturbation NMR experiments can accurately determine which residues contribute to interactions with proteins, DNA, RNA, or small molecules. This becomes precious in determining drug-designable surfaces (Kleckner and Foster, 2011).

18.2.3.3 Fragment-based drug discovery

The binding of small fragments is detected by NMR and is consequently regularly utilized to recognize weak-binding molecules, which may be further enhanced into effective inhibitors (Fragment-Based Drug Discovery Using NMR Spectroscopy - PMC [Internet], no date).

18.2.4 Structural insights into oncogenic proteins and tumor suppressors

Research in structural biology has provided some of the most revolutionary information to cancer biology, as it has discovered the mechanism through which mutations can modify protein structure and cell behavior. Structural studies explain how disease-related variants alter folding, binding, and catalytic activity, which can explain how oncogenic activation and tumor-suppressor loss happen, as well as how drug resistance develops (Protein structural context of cancer mutations reveals molecular mechanisms and candidate driver genes - PMC [Internet], no date).

18.2.4.1 p53 tumor-suppressor protein

The p53 protein is popularly described as the genome guardian, as it plays a key role in keeping the cells intact. As a transcription factor, p53 balances pathways involved in DNA repair, apoptosis, cell-cycle arrest, and senescence. It is, however, one of the most frequently mutated genes in human malignancies, and structural studies have played a crucial role in understanding why its loss has such a tremendous impact (Unraveling the Guardian: p53's Multifaceted Role in the DNA Damage Response and Tumor Treatment Strategies - PMC [Internet], no date).

18.2.4.1.1 DNA-binding domain

The DNA-binding domain of p53 is also a characteristic feature that includes a zinc-stabilized motif that is necessary to make the target gene promoters more accurate. A structural analysis of the domain has revealed that this domain is dependent on a network of β-sheets and loops, which must stay rigid and yet well orchestrated so as to allow efficient DNA binding. The majority of cancer-related mutations, like R175H, R248Q, and R273H, are in this region, which interferes with the necessity of maintaining DNA contacts or the framework. These modifications greatly affect the capacity of the protein to regulate the transcription (The Tumor Suppressor p53: From Structures to Drug Discovery - PMC [Internet], no date).

18.2.4.1.2 Reduced thermodynamic stability

Substantial numbers of mutant variants are also characterized by a diminished thermodynamic stability, which predisposes the protein to partial aggregation and unfolding. These aggregates are able to disrupt related proteins, such as p63 and p73, and increase oncogenic signaling. Structural considerations have been used to produce molecules that stabilize the native state of p53 or regain its ability to interact with DNA (The effect of mutation on an aggregation-prone protein: An in vivo, in vitro, and in silico analysis - PMC [Internet], no date).

18.2.4.2 RAS oncogene

RAS proteins can be considered molecular switches that can be inactive in the GDP-bound form or active in the GTP-bound form. RAS mutations, particularly KRAS, are the cause of some of the most aggressive cancers, including pancreatic, colorectal, and lung cancers. Structural studies have played an imperative role in elucidating how these mutations distort RAS regulation (Simanshu et al., 2017).

18.2.4.2.1 Switch I and switch II

Activation is associated with major conformational changes in two regions: switch I and switch II. These active conformations are frequently stabilized by cancer-related mutations, and this allows them to sustain downstream signaling via MAPK and

PI3K. Mutations such as G12V or Q61H prevent GTP hydrolysis, either by the inability to properly position essential residues or by the inability to interact with GTPase-activating proteins (KRAS Mutation Subtypes and Their Association with Other Driver Mutations in Oncogenic Pathways - PMC [Internet], no date).

18.2.4.2.2 Allosteric pockets

Notably, structural biology has revealed allosteric pockets that were previously inaccessible, especially around the switch II region. These cryptic sites were discovered, which allowed the development of covalent inhibitors against KRAS G12C—a milestone in the history of RAS-directed therapy (Biophysical and structural analysis of KRAS switch-II pocket inhibitors reveals allele-specific binding constraints[image] - PMC [Internet], no date).

18.2.4.3 Epidermal growth factor receptor

EGFR is a cell survival and growth receptor tyrosine kinase. Its kinase domain mutations are major causes of lung cancer, and structural studies have been used to guide the design of several generations of targeted inhibitors (EGFR in Cancer: Signaling Mechanisms, Drugs, and Acquired Resistance - PMC [Internet], no date).

18.2.4.3.1 Receptor dimerization

Depending on the binding of the ligands, the extracellular part of EGFR takes different folds that allow receptor dimerization, a condition that enables receptor activation. In the intracellular kinase domain, oncogenic mutations, for example, L858R or exon 19 deletions, support the active conformation, increasing the affinity for ATP and allowing better signaling. Comparisons between structures indicate that first-generation inhibitors select these mutant forms (Depending on the binding with the ligands, the extracellular part of EGFR takes different folds that allow receptor dimerization, a condition that enables receptor activation. In the intracellular kinase domain, oncogenic mutations, e.g. L858R or exon 19 deletions, support the active conformation, increasing the affinity of ATP and allowing better signaling. Comparisons between structures indicate that first-generation inhibitors select these mutant forms. - Google Search [Internet], no date).

18.2.4.3.2 T790M or C797S

Resistance frequently arises from additional mutations, such as T790M or C797S, which alter the binding site and reduce the drug's affinity. These variants have been structurally mapped to design next-generation inhibitors that can overcome resistance (Overcoming EGFR T790M and C797S resistance with mutant-selective allosteric inhibitors - PMC [Internet], no date).

18.2.4.4 BRAF kinase

BRAF is a serine/threonine kinase that is part of the MAPK pathway. The V600E mutation is one of the most common mutations related to cancer, and structural analyses have given a clear understanding of the transforming potential of the V600E mutation (Mutations in the Serine/Threonine Kinase BRAF: Oncogenic Drivers in Solid Tumors - PMC [Internet], no date).

18.2.4.4.1 Activation loop

The V600E substitution results in the rearrangement of the activation loop, placing it in an active state. Concurrently, the αC-helix is stabilized to enable ATP binding and catalysis. These modifications also increase the kinase's capacity to form active dimers. Structural biological insights were directly used to develop selective inhibitors such as vemurafenib and dabrafenib, which target this hyperactive conformation (The mechanism of activation of monomeric B-Raf V600E - PMC [Internet], no date).

18.2.4.5 HER2 receptor

HER2, a EGFR family member, is particular in the sense that it does not have a known ligand but is prone to activation. The structural analysis demonstrates that it is intrinsically in an open, dimerization-bound form in its extracellular region, and thus HER2 is specifically strong upon overexpression (Structure and dynamics of the EGFR/HER2 heterodimer - PMC [Internet], no date).

18.2.4.5.1 Trastuzumab and pertuzumab

Trastuzumab and pertuzumab are monoclonal antibodies that target distinct epitopes on HER2 to prevent dimerization and downstream signaling. Also, the heterodimers of HER2/HER3 have been found to be particularly oncogenic units, particularly in cancer. The insights are the basis of HER2-based treatments that are common in gastric and breast cancer (HER2 Dimerization Inhibitor Pertuzumab – Mode of Action and Clinical Data in Breast Cancer - PMC [Internet], no date).

18.2.5 Structural drug-design approaches in cancer biology

The knowledge of the folding, interactions, and changes in shape of the cancer-related proteins can be crucial in informing contemporary drug development. Structural knowledge can be used by scientists to identify sites of vulnerability, anticipate resistance mechanisms, and build molecules that specifically interact with modified folding (Strategies for the drug development of cancer therapeutics - PMC [Internet], no date).

18.2.5.1 Structure-based drug design

One of the key technologies in contemporary therapeutic development is the so-called structure-based drug design (SBDD) that utilizes the accurate 3D structural data of biological targets in the development of increasingly selective and effective molecules. SBDD is based on the principles of protein folding, the organization of their active or regulatory sites, and the disease-related changes in these structural elements. With the aid of high-resolution information from X-ray crystallography, cryo-EM, or NMR spectroscopy, scientists can see the precise structure of residues composing the drug-binding pocket. The structural map is a guide in the synthesis of candidate molecules that have the ability to interact with the protein and do so in a very specific way (Structure-based drug design: aiming for a perfect fit - PMC [Internet], no date).

The initial significant step in SBDD is to determine drug-binding sites or allosteric sites. After these pockets have been defined, medicinal chemists develop chemical scaffolds that supplement the shape and physicochemical characteristics of the binding region. Further refinement cycles are aimed at enhancing hydrogen bonding, hydrophobic packing, salt-bridging, and π–π stacking. These optimizations lead to increased affinity, better selectivity, and less off-target toxicity (Structure-Based Virtual Screening for Drug Discovery: Principles, Applications and Recent Advances - PMC [Internet], no date).

SBDD can also be used to predict possible resistance by modeling how mutations can be bent or blocked by other mutations. This understanding can be used to design molecules that can be active against mutant variants in advance. SBDD has had a significant effect on the creation of a number of clinically useful anticancer agents, such as EGFR, BRAF kinase, anaplastic lymphoma kinase (ALK), and poly (ADP-ribose) polymerase (PARP) inhibitors. These achievements reflect the ways structural knowledge can turn drug discovery into a process of trial and error, which could be a rational approach that is based on mechanisms (Lionta et al., 2014).

18.2.5.2 Fragment-based drug discovery

Fragment-based drug discovery (FBDD) has become an effective approach to finding new therapeutic scaffolds through screening of tiny chemical fragments instead of large and intricate molecules. The fragments are usually of low-molecular-weight and bind with low affinity to protein surfaces, but they are invaluable since they reveal critical interaction hotspots which otherwise would be screened in a traditional high-throughput fashion. Fragment hits, when found, are used as starting points for systematic optimization, for example, fragment growing, fragment merging, or fragment linking. This method enables researchers to build a greater number of high-affinity and selective drug candidates with great efficiency. FBDD is well suited to structural methods such as NMR spectroscopy and X-ray crystallography, which can give ample information on fragment positioning and binding. The technique has also enhanced the identification of protein kinases, epigenetic regulators, and other difficult-to-target cancer-related protein inhibitors, a fact that supports the relevance of the technique in contemporary medicinal chemistry (Bon et al., 2022).

18.3 Omics integration

18.3.1 Multiomics integration in systems oncology

Cancer is now recognized as a complex systems-level disorder caused by coordinated disruptions in genes, proteins, metabolites, cells, and their microenvironment. The examination of individual pathways is insufficient to understand important tumor phenomena such as heterogeneity, drug resistance, clonal evolution, and metastasis. Systems biology focuses on filling this gap by using genomic, transcriptomic, proteomic, metabolic, and clinical data within single, integrated computational models.

Network medicine also projects such changes onto cellular interaction networks to identify key regulatory nodes that promote malignancy. Collectively, these methods can offer a holistic way of forecasting tumor behavior and directing more specific and personalized treatment plans (Cancer cell plasticity: from cellular, molecular, and genetic mechanisms to tumor heterogeneity and drug resistance - PMC [Internet], no date).

18.3.1.1 Rationale for omics integration

Cancer is a complex disease that manifests as perturbations across multiple biological layers, including genetic mutations, epigenetic modifications, transcriptional alterations, protein dysregulation, metabolic rewiring, and changes in extracellular signaling, as shown in Fig. 18.2. Each omics layer—genomics, transcriptomics, proteomics, metabolomics, and epigenomics—represents a unique area in tumor biology. Nevertheless, no single dataset provides a complete picture of the disease. Combining several layers of omics, researchers can be able to recapitulate the interactions between these regulatory layers and understand how the convergence of molecular perturbations causes tumor initiation, progression, and resistance to therapy (Knox, 2010).

Multiomics techniques allow a number of important studies to be made: it is possible to examine how DNA mutations change RNA and protein expression, find pathways that are simultaneously disrupted at the genomic, proteomic, and metabolic levels, observe how transcriptional networks respond to therapy or relapse, and identify molecular signatures that predict prognosis or response to therapy. It is on the basis of these integrative analyses that systems-based cancer diagnostics, risk stratification, and personalized therapeutic strategies are developed (Liu et al., 2025).

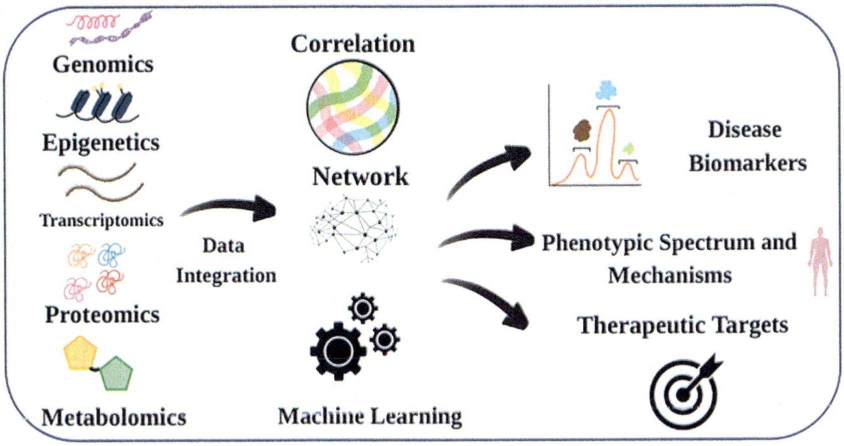

Figure 18.2 Method for multiomics drug Integration. *No Permission Required.*

18.3.1.2 Genomics

Genomic analysis is the main component of determining the molecular environment of a tumor. Such methods include whole-genome and whole-exome sequencing, which identify somatic mutations, copy number gains or losses, structural rearrangements, and chromosomal instabilities involved in malignant transformation. The most frequent genomic changes include important tumor suppressors, including TP53 and RB1; oncogenes, including previously mentioned KRAS, BRAF, and MYC; and chromatin-modifying complexes and their members, such as the SWI-SNF family. Whereas first molecular disruptions are described by genomic profiling, not all changes that are found are biologically meaningful. A significant portion of mutations do not cause a functionally significant effect, and their impact on cellular processes can be observed only when they are used in combination with other omics data. Combining the results of genomic research with transcriptomic, proteomic, and metabolomic data can assist researchers in differentiating between actual driver mutations and mere coincidental cases of passenger events and give the researcher a more detailed insight into the way in which these mutations reconstitute physiology of cells (Emerging patterns of somatic mutations in cancer - PMC [Internet], no date).

The postgenomic era has seen the molecular discovery, from clinical application, is cyclic and iterative in nature, as seen in Fig. 18.3. The initial step of genomic profiling is to reveal candidate mutations or pathways of interest. These observations are then confirmed by functional experiments, for example, gene editing, signaling, or protein-interaction experiments, to prove their biological significance. The lessons learned during these experiments are used in the creation of specific molecules, biomarkers, or diagnostic instruments. Prior to entering early-phase clinical trials in humans, candidate therapies or assays are tested in cell lines, organoids, and animal models. In turn, clinical outcomes produce new information about therapeutic response, resistance mechanisms,

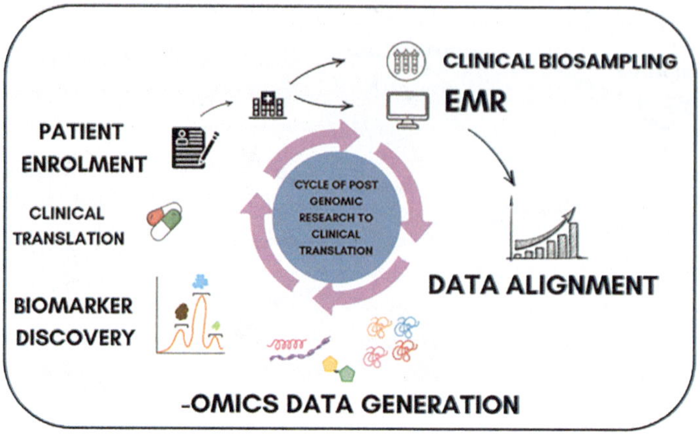

Figure 18.3 Postcycle genomic research-clinical translation. *No Permission Required.*

and patient heterogeneity. The results of these observations are integrated into genomic and multiomics studies, narrowing down the target and improving the effectiveness of other interventions. This perpetual cycle, which includes discovery, validation, translation, and feedback, is a sure way to make sure that genomic knowledge is properly translated into meaningful clinical interventions (Target discovery in the postgenomic era - PMC [Internet], no date).

18.3.1.3 Transcriptomics

Transcriptomic profiling studies the downstream effects of genomic changes by studying patterns of RNA expression. High-throughput RNA sequencing can be used to characterize the differences in gene expression, alternative splicing, and fusion transcripts, as well as dysregulation of noncoding RNAs, such as microRNAs and long noncoding RNAs.

When transcriptomics and genomics are used together, scientists can identify mutations that cause functional changes in gene expression. This integration defines two types of mutations in drivers: those that actively control transcriptional programs and those that lack a major functional consequence (passengers). Transcriptomics data also help in the understanding of pathway activation and inhibition, and inform the selection of possible therapeutic targets and resistance mechanisms (Wang et al., 2009).

18.3.1.4 Proteomics

Proteins are the functional actors in the cell, and as such, proteomics is vital in comprehending tumor behavior. Proteomics: mass spectrometry measures protein levels, posttranslational modifications (like phosphorylation and acetylation), and PPIs. In particular, phosphoproteomics sheds light on activated signaling pathways, such as PI3K-AKT, MAPK, and JAK-STAT, that play a role in tumor growth and drug resistance.

Combining proteomic data with genomic and transcriptomic profiles can also help researchers to connect mutations and transcriptional changes to the real protein-level output. The method determines aberrant signaling hubs and exposes adaptive responses to the treatment, and also enables the choice of protein targets to be targeted by a drug (Protein posttranslational modifications in health and diseases: Functions, regulatory mechanisms, and therapeutic implications - PMC [Internet], no date).

18.3.1.5 Metabolomics

Metabolic reprogramming is a characteristic of cancer, which offers energy and biosynthetic precursors needed to undergo rapid proliferation. Metabolomics measures changes in central processes, including glycolysis, glutaminolysis, one-carbon metabolism, and lipid biosynthesis. Combining metabolomic data with transcriptomic and proteomic data allows researchers to map metabolic flux changes that promote tumor survival, immune evasion, or contribute to treatment resistance. These findings are used to guide the adoption of metabolically specific strategies to address metabolic vulnerabilities (Protein posttranslational modifications in health and diseases: Functions, regulatory mechanisms, and therapeutic implications - PMC [Internet], no date).

18.3.1.5.1 Epigenomics

Epigenetic changes, such as DNA methylation, histone remodeling, and chromatin remodeling, are important in the regulation of genes in cancer. Silenced tumor-suppressor networks or activated oncogenic pathways are identified through epigenomic profiling and would not be found using genomic sequencing. Integrating epigenomic data with transcriptomic and proteomic data would allow one to gain a detailed insight into how malignancy is regulated. It further helps in the identification of epigenetic biomarkers that, in turn, make it possible to detect, prognosticate, and monitor therapeutic responses (Epigenetic regulation in cancer - PMC [Internet], no date).

18.3.1.5.2 Multiomics data fusion approaches

The integration of multiomics can be achieved using a range of complementary methods. Vertical integration is a type of integration used to analyze interactions and dependencies at multiple molecular scales within a tumor sample, with interactions and dependencies observed across several layers of the tumor. On the other hand, horizontal integration compares the same omics layer across several patients, which helps reveal shared disease signatures and common patterns. Hybrid models combine vertical and horizontal methods and allow a detailed analysis of the opportunity to study the complex interactions of molecules across separate tumors and to identify common tendencies within a patient collective. Complex datasets can be interpreted through advanced computational methods such as Bayesian modeling, deep learning networks, nonnegative matrix factorization, and similarity network fusion, which show synergy between the two. Such approaches reveal patterns and relationships that cannot be detected by single-omics analyses, which inform the discovery of novel biomarkers, treatment targets, and patient-specific treatment plans (Subramanian et al., 2020).

18.3.2 Network reconstruction and pathway modeling in systems oncology

To define cancer as a system-level disease, it is necessary to map the intricate network of molecular interactions that govern cell behavior. The reconstruction of biological networks and pathway modeling is illustrated in Table 18.2 and offers a framework by which such interactions are decoded and in which high-throughput data are integrated to reveal important nodes and pathways leading to tumorigenesis. Networks facilitate visualization of gene, protein, and metabolic interactions, and therefore help in the identification of gaps that can be exploited to initiate therapeutic interventions (Laubenbacher et al., 2009).

18.3.2.1 Principles of biological network reconstruction

Formal representations of molecular interactions are biological networks, in which nodes represent biological entities (e.g., genes, proteins, or metabolites) and edges represent relationships between them (e.g., binding, regulation, activation, or inhibition). Network reconstruction is the systematic assembly of such interactions based on experimental data, curated publicly available databases, and computational inference. There are four common categories of biological networks used in systems oncology: gene regulatory networks (GRNs), PPIs networks, metabolic networks, and signal transduction networks. All types of networks offer distinct information regarding cellular control and enable the discovery of essential nodes, which, when disturbed, can have a substantial effect on tumor development and survival (Panditrao et al., 2022).

18.3.2.2 Gene regulatory networks

GRNs explain how transcription factors and other regulatory proteins regulate gene expression. GRNs reveal the dynamics of oncogene upregulation, silencing of tumor suppressors, and reprogramming of transcriptional outputs in malignant transformation based on transcriptomic patterns. GRN reconstruction methods have been used, such as mutual information-based inference, regression modeling, Bayesian network analysis,

Table 18.2 Analytical tools for network reconstruction, modeling, and simulation in systems biology.

Tool/software	Application area	Key features	Platform/access
Cytoscape	Network reconstruction/ visualization	Integrated interaction mapping.	Open-source; https:// cytoscape.org
STRING	Protein–Protein interaction networks	Protein association insights.	Web-based; https://string-db.org
COPASI	Computational modeling/ simulation	Kinetic system modeling.	Open-source; https:// copasi.org
CellDesigner	Pathway modeling	Graphical network design.	Free academic use; https://celldesigner.org
NetPath/Pathway Commons	Signaling pathway databases	Curated pathway libraries.	https://www. pathwaycommons.org
BioModels Database	Dynamic simulation models	Validated systems models.	https://www.ebi.ac.uk/ biomodels
OmicsNet/ MetaboAnalyst	Multiomics integration	Multiomics integration.	Web-based; https://www. omicsnet.ca/https:// www.metaboanalyst.ca

No Permission Required.

and machine learning-based network inference. Such networks have been found especially useful in the process of establishing master regulators, that is, genes that have large-scale control over downstream targets and that become dysregulated to facilitate tumorigenesis. With the specificity of core oncogenic programs, precision oncology can identify such master regulators to disrupt them (Kim et al., 2023).

18.3.2.3 Protein–protein interaction networks

PPIs map the physical interactions among proteins in a cell. These networks in cancer can also show many interconnected hubs, which are proteins needed to keep cells alive or to multiply. Protein interface mutations can restructure PPIs, forming erroneous signaling networks that play a role in uncontrolled growth, apoptosis evasion, and drug resistance. Cancer-specific PPIs can be mapped to identify essential interactions that may be targeted by small molecules or biologics, enabling interventions targeted at particular risk factors (Mapping the protein–protein and genetic interactions of cancer to guide precision medicine - PMC [Internet], no date).

18.3.2.4 Metabolic network models

Enzyme activities, metabolite concentrations, and pathway connectivity combine to simulate tumor metabolism at a systems level using metabolic networks. Metabolic pathway rewiring is a common behavior among cancer cells, enabling them to grow rapidly, survive in hypoxic conditions, and survive stress. Genome-scale metabolic models can simulate such adaptations and predict which pathways are required to maintain a tumor. Flux balance analysis and metabolic flux modeling are among the techniques used to recognize metabolic bottlenecks or so-called chokepoints that can be targets of therapeutic intervention. As an example, enzymes that play essential roles in glycolysis or glutaminolysis can be used as weaknesses to selectively inhibit cancer cell growth (Metabolic reprogramming in cancer: Mechanisms and therapeutics - PMC [Internet], no date).

18.3.2.5 Signaling network reconstruction

Flow signaling networks represent the relationship between extracellular signals and intracellular responses, including kinase cascades, receptor activation, transcription factor control, and feedback. Real-time reconstruction of dynamic signaling behaviors is achieved through the integration of proteomic and phosphoproteomic data. These networks play a critical role in the comprehension of growth factor receptor activation, drug resistance mechanisms, and cross-talk between important pathways, including PI3K-AKT-mTOR and MAPK signaling. The modeling of these interactions allows researchers to forecast the effect of perturbation, discover compensatory pathways resulting in resistance, and generate combination therapies that destabilize multiple nodes at the same time (Profiling Cell Signaling Networks at Single-cell Resolution - PMC [Internet], no date).

18.3.2.6 Implications for therapeutic discovery

Reconstructed biological networks offer a basis for rational cancer therapy design. Network analysis, by identifying central hubs, bottlenecks, feedback loops, and so on, can reveal the most vulnerable points likely to affect tumor progression. Together with multiomics integration, network reconstruction enables the construction of patient-specific models of cancer behavior, which can provide predictive information on how to respond to targeted therapies, immune interventions, and metabolic modulators. This systems-level outlook moves oncology beyond single-gene solutions, enabling more accurate and effective treatments that account for the complexity and heterogeneity of tumors (Barabási et al., 2011).

18.3.3 Identification of key signaling nodes and feedback loops

Cancer is beginning to be viewed as a system-level disease in which network topology and dynamic feedback loops define cellular behavior, response to therapy, and disease progression. It is vital to identify the nodes that are most influential in signaling networks and how feedback networks work in order to design targeted interventions rationally. This part brings into the limelight network topology, critical nodes, feedback, and implications of network topology for therapeutic strategies (Kreeger and Lauffenburger, 2010).

18.3.3.1 Network topology and critical nodes

All nodes in a biological network are important. In cancer networks, nodes act as hubs, bottlenecks, or articulation points and exert disproportionate influence on information flow and pathway connectivity. Such nodes frequently represent oncogenic signaling and tumor maintenance proteins that are centrally positioned. The major features of the influential nodes are a high degree, which is the number of direct interactions; high betweenness centrality, which is the control of the passage of information by the node; strategic positioning at the point of the network intersection; and participation in critical pathways that govern proliferation, apoptosis, and metabolic adjustments (The Importance of Bottlenecks in Protein Networks: Correlation with Gene Essentiality and Expression Dynamics - PMC [Internet], no date).

Examples of network nodes essential in cancer include MYC, TP53, AKT1, KRAS, and EGFR. MYC is a transcriptional regulator that controls the cell cycle, metabolism, and ribosome biogenesis. TP53 is a crucial regulator of genome integrity, and its loss disrupts many downstream pathways. AKT1 integrates information about growth factors and regulates survival and metabolic processes. KRAS and EGFR are upstream signaling centers that affect various downstream signaling pathways. Compensation of the network is a risk of the approach to such central nodes, so a precautionary systems-level analysis is required, although this approach has therapeutic leverage (Rewired Metabolism Caused by the Oncogenic Deregulation of MYC as an Attractive Therapeutic Target in Cancers - PMC [Internet], no date).

18.3.3.2 Feedback loops in cancer networks

Biological networks are characterized by positive feedback, signal amplification, and negative feedback, which undergo homeostasis, and in cancer, their dysregulation facilitates ongoing oncogenic signaling and resistance to therapy (Negative feedback and adaptive resistance to the targeted therapy of cancer - PMC [Internet], no date).

18.3.3.2.1 Negative feedback in cancer networks

The negative feedback mechanisms serve to limit signaling and maintain cellular homeostasis. Disruption of these loops can extend oncogenic activity in cancer. As an illustration, the negative feedback on the receptor and RAS is expected to be regulated by the ERK feedback in the MAPK pathway. The failure to regulate this in tumors leads to persistence of MAPK signaling, leading to uncontrolled proliferation. Likewise, mTORC1 has a negative feedback mechanism through inhibition of insulin receptor substrates that inhibit excessive stimulation of downstream pathways. The interference with these negative feedback mechanisms frequently plays a role in adaptive resistance to directed therapies, with the lack of restraint leading to ongoing compensatory signaling (Negative feedback and adaptive resistance to the targeted therapy of cancer - PMC [Internet], no date).

18.3.3.2.2 Positive feedback in cancer networks

Signals in positive feedback loops are amplified, and oncogenic pathways are reinforced. Activation of AKT in PI3K-AKT signaling stimulates PI3K activity, which stimulates cell survival, growth, and metabolic adaptation. Similarly, mTORC2 also promotes AKT activation via a positive-feedback mechanism, which also promotes tumor progression. The loops are capable of supporting the aberrant signaling response in the presence of inhibition of the upstream nodes, underscoring the importance of combination therapeutic approaches that consider the compensatory response of amplification (Suppression of feedback loops mediated by PI3K/mTOR induces multiple over-activation of compensatory pathways: an unintended consequence leading to drug resistance - PMC [Internet], no date).

18.3.3.3 Network vulnerabilities and combination therapy

Synthetic lethal interactions, network topology, and feedback loops. Analysis of network topology and feedback loops reveals synthetic lethal interactions, with simultaneous disruption of two nonessential genes leading to cell death, but loss of one being tolerated. Leveraging these weaknesses allows rational combination therapy, with BRCA1/2-deficient tumors being an example of using PARP inhibitors. Compensatory feedback pathways are also targeted using network-guided strategies, which improve efficacy and reduce resistance, for examle, MEK-PI3K/AKT EGFR-MET, or HER2 combinations. Combining multiomics data, dynamic signaling models, and iterative validation ensures that data is clinically relevant. Detecting important nodes, feedback loops, and network weaknesses will give a systems-level understanding of precision oncology to predict tumor behavior, resistance to therapeutics, and construct useful combination regimens (Srivas et al., 2016).

18.4 Computational modeling and dynamic simulations

It is impossible to conduct work in systems oncology without the use of computational modeling and dynamic simulations, which allow researchers to investigate how complex cancer networks behave over time. These models combine experimental data and mathematical models to forecast cellular reactions, signaling, and treatment response.

Computational methods use simulations of molecular interactions and network perturbations to provide insights that are hard to determine solely by experimental studies, to inform hypothesis generation as well as rational drug design (Scibilia et al., 2025).

18.4.1 Deterministic and stochastic models

Deterministic models are time-dependent models of network changes, often represented by ordinary differential equations, in scenarios where fluctuations in the numbers of molecules are negligible, and the numbers are large. The models are specifically applicable in modeling the dynamics of pathways like kinase cascades, metabolic fluxes, and transcriptional regulation of high-copy systems. By contrast, stochastic models, frequently represented Gillespie algorithms, include the effects of random variation in molecular interactions that are observed when the number of molecules is small. Stochasticity, like this, has wide biological implications in cancer, affecting cell-fate choices, mutation spread, and signaling buffering. The deterministic and stochastic frameworks are combined to permit thorough simulation of all types of molecules (Frontiers | A Comparison of Deterministic and Stochastic Modeling Approaches for Biochemical Reaction Systems: On Fixed Points, Means, and Modes [Internet], no date).

18.4.2 Agent-based models

In agent-based modeling, individual cancer cells are regarded as autonomous agents that are part of their microenvironment. Agents have specific proliferation, migration, signaling, and death laws, which enable one to observe the emergent behaviors of tissues or tumors. Such models find particular application in the study of clonal selection, tumor-immune interactions, spatial heterogeneity, and gradients in drug penetration. Agent-based methods have the ability to produce complex tumor behavior by modeling the dynamics of individual cellular responses in the presence of intrinsic cellular behaviors as well as external environmental signals, thus bridging the gap between molecular network modeling and tissue-level behaviors (Agent-based modeling in cancer biomedicine: applications and tools for calibration and validation - PMC [Internet], no date).

18.4.3 Boolean and logical models

Signaling networks are simplified into binary states of ON/OFF by using Boolean and logical models to express activation or inhibition of nodes. It is this abstraction that allows the effective simulation of large networks, in which the detailed kinetic parameters might not be known. Even though simple, Boolean models are able to reflect critical transitions, including apoptosis activation, cell-cycle switching, or differentiation. Logical modeling is specifically handy for discovering network motifs, regulatory feedback loops, and key nodes, which determine system-wide behavior (Logic-Based Models for the Analysis of Cell Signaling Networks | Biochemistry [Internet], no date).

18.4.4 Machine learning and artificial intelligence-driven models

Machine learning and artificial intelligence (AI) complement computational modeling, multiomics data integration, prediction of pathway activity, biomarker identification, and drug-response simulation. Nonlinear interactions of complex biological systems can be learned by neural networks, support vector machines, and graph neural networks. Reinforcement learning methods enable time-dependent optimization of treatment regimes and simulate the effect of therapeutic interventions on network behavior. Therefore AI-directed models can be used to reinforce conventional computational models, offering predictive capability in precision-oncology and systems-level insights into tumor dynamics.

This form of integrative computation, incorporating deterministic, stochastic, agent-based, Boolean, and AI-driven models, facilitates strong simulation of cancer networks, provides a mechanistic understanding, and allows rational therapeutic design (Artificial intelligence [AI] and machine learning [ML] in precision oncology: a review on enhancing discoverability through multiomics integration - PMC [Internet], no date).

18.5 Convergence of structural and systems biology

Cancer is a result of complicated, networked structures, and not solitary, molecular alterations. The atomic-level insights into a mutation on protein architecture are offered by structural biology, and systems biology offers the effects and consequences on cells on a network-wide level. The combination of these areas allows multiscale mechanistic models of the connections between the changes in the molecules and the signaling networks and cellular functions. This is at the core of structural systems pharmacology, predictive biomarker development, and network-based modeling of drug response. Kinase inhibitor networks and PARP DNA repair interactions are examples that illustrate the application of structural and systems-level approaches to inform rational treatment choices (Protein structural context of cancer mutations reveals molecular mechanisms and candidate driver genes - PMC [Internet], no date).

18.5.1 Integrative multiscale modeling

Integrative multiscale modeling is a methodology that bridges the gap between molecular, protein, and complex structural changes and cellular pathways, tissue behavior, and tumor dynamics, giving a holistic paradigm to forecast cancer progression and therapeutic (Multiscale Cancer Modeling - PMC [Internet], no date).

18.5.1.1 Rationale for multiscale integration

Biological systems work on hierarchical levels, from atoms to tissues, where each hierarchical level has an effect on higher-order functions. Individual atoms at the atomic level compose amino acid residues, which are folded into a 3D protein structure. The interaction between these proteins produces dynamic complexes that are involved in both signaling pathways and regulatory networks. Any scale of disruption, including point mutations, structural destabilization, or aberrant complex formation, may spread through the system, alter downstream pathways, cellular phenotypes, and, ultimately, tumor progression. Multiscale modeling can be used to connect molecular-scale perturbations, such as a mutation at a kinase domain or a conformational change at a receptor, to emergent cellular behavior, such as uncontrolled proliferation, metabolic rewiring, and therapy resistance. In the absence of such integration, structural data will be isolated from functional results, and systems-level models will not have a mechanistic basis. Multiscale modeling fills this gap, relating molecular mechanisms to system behavior (Molecular Structure and Function - Opportunities in Biology - NCBI Bookshelf [Internet], no date).

18.5.1.2 Atomistic and molecular-level modeling

Structural models derived by X-ray crystallography, cryo-EM and NMR spectroscopy are at the bottom of the list of multiscale integration. Atomistic simulations are based on these high-resolution structures. Molecular dynamics [MD] studies are able to capture the dynamics of atomic fluctuations, domain movement, binding kinetics of ligands, and instabilities of mutations, and show us how proteins can behave dynamically. The detailed modeling of catalytic residues, transition states, and the changes in electron distribution caused by mutations or drug binding can be calculated using quantum mechanical calculations. As an example, oncogenic mutants of RAS, in turn, adopt a GTP-bound active conformation with limited flexibility in switch regions to maintain downstream signaling. In the same way, BRAF, EGFR, and PI3K structural perturbations cause changes in the conformational landscapes, affecting drug interactions and enzymatic activity. Atomistic modeling, therefore links genetic mutations to the behavior of functional proteins (High-resolution protein modeling through Cryo-EM and AI: current trends and future perspectives – a review - PMC [Internet], no date).

18.5.1.3 Protein complexes and mesoscale modeling

Mesoscale models are based on structural knowledge of interface geometry, binding sites, and allosteric interactions and are used to predict the formation and stability of complexes. These include HER2–HER3 receptor interactions, interaction of kinase activation loops, PARP-XRCC1 interactions of DNA repair, and transcription-related assemblies of chromatin remodeling. Mesoscale modeling determines the effects of ligand availability, therapeutic agents, or changes caused by mutations on complex behavior. This method enables the prediction of cooperative interactions, allosteric regulation, and the repercussions of perturbed interfaces and provides a mechanistic understanding of oncogenic signaling and points of therapeutic intervention.

18.5.1.4 Pathway-level and network-level modeling

At the pathway and network scale, systems biology unites structural data into computational models of protein, metabolite, and regulatory interactions. Structural parameters optimize simulations of pathways by giving correct binding affinities, catalytic rate constants, allosteric transition dynamics, and receptor-ligand activation thresholds. An example is EGFR dimerization interfaces, which are based on structural studies and guide models that predict ligand-dependent receptor activation, the effectiveness of inhibitors, and drug resistance mechanisms. By incorporating atomic-level information into networks, scientists have the capability to rewire pathways, compensate for signaling, and predict drug response (Insights into Protein–Ligand Interactions: Mechanisms, Models, and Methods [Internet], no date).

18.5.1.5 Whole-cell and tissue-level models

On the largest scale, cellular and tissue-level models are used to model growth, apoptosis, cell-cycle dynamics, clonal evolution, immune dynamics, metabolic flux, nutrient competition, and spatial gradients of oxygen or drug diffusion. Perturbations in structure spread via pathways to affect cellular outcomes. As an illustration, stabilization of the active form of a kinase mutation may induce changes in signaling flux, and/or cell-cycle progression, and/or tumor aggressiveness at the tissue level. Whole-cell and tissue-level models combine changes in the sphere of molecules with the effects on phenotypes and allow prediction of the behavior of tumors and the design of rational therapies (A comprehensive review of computational cell cycle models in guiding cancer treatment strategies - PMC [Internet], no date).

18.6 Structural systems pharmacology

18.6.1 Conceptual overview

Structural systems pharmacology is a combination of atomic-scale structural data of drug-target interactions with systems-scale pharmacodynamics and pharmacokinetics modeling. In contrast to the previous conventional methods of testing drugs using specific targets, this system evaluates the event of binding and its impact on protein folding, network functionality, and eventual therapeutic effects. This combined methodology offers important information about drug responses in cancer, where compensatory signaling and multitarget effects often compromise monotherapy efficacy. Through a dual approach to molecular and network-level impacts, structural systems pharmacology allows a predictive approach to the development of more effective, sustainable therapeutic approaches (Towards Structural Systems Pharmacology to Study Complex Diseases and Personalized Medicine - PMC [Internet], no date).

18.6.2 Linking structural variants to drug response

Cancer mutations have the potential to radically change drug responses in several ways. Direct steric hindrance can result from gatekeeper mutations in kinases that prevent inhibitor binding, whereas conformational changes may leave inhibitor-binding states inaccessible. Allosteric regulation has the ability to reposition key domains, change activity, and affect stability protein turnover and network feedback loops. Structural systems pharmacology forecasts drug sensitivity, optimal dosing, dose-response curves, emergent mechanisms of resistance, and even combination therapies by modeling structural variations into network models. As an example, in EGFR-mutant lung cancer, the T790M mutation disrupts the binding of inhibitors, and systems-level modeling identifies compensatory MET amplification, which drives resistance to therapy (Drug resistance in cancer: molecular mechanisms and emerging treatment strategies - PMC [Internet], no date).

18.6.3 Structural constraints in systems-level drug screening

Structural guidance is advantageous in high-throughput drug screening, where interfering molecules are targeted to interact well with the target sites. The use of binding-pocket geometry and structural motifs enables the prioritization of compounds with favorable interaction profiles and reduces off-target effects. The combination with network models also determines important nodes that might be inhibited twice to defeat compensatory paths. The approach is lean in drug discovery by integrating molecular accuracy with systems understanding, minimizing time and cost, and improving therapeutic relevance. Structural constraints are useful in ensuring that that are selected are not only high-affinity targets but also those with a positive effect on the overall signaling network (Comprehensive detection and characterization of human druggable pockets through binding site descriptors - PMC [Internet], no date).

18.6.4 Allosteric modulators and system-wide impact

Allosteric modulators bind outside of the active sites and rearrange the dynamic ensemble of protein conformations. The effect of these conformational changes on network flux, pathway activation, and feedback loops is modeled by structural systems pharmacology. An example of this is allosteric MEK inhibitors that regulate the Raf-MEK-ERK pathway, affecting positive and negative feedback loops and, in some cases, giving surprising adaptive outcomes. Structural systems pharmacology helps to design allosteric drugs with desirable synergy or reduced negative network effects by predicting system-wide interactions. The introduced approach focuses on the need to take into account both local structural modulations and global network interactions in contemporary cancer therapeutics in order to be able to design rational, targeted, and multitarget interventions (Abdel-Magid, 2015).

18.6.5 Predictive biomarkers and drug-response modeling

18.6.5.1 Biomarkers derived from structural features

Cancer proteins often undergo structural changes, which create biomarkers that are measurable and reflect functional change and can be used clinically. Protein conformation, generation of new binding pockets, or changes in stability, which may be diagnostic or prognostic markers, are frequently altered by mutations. Postprotein modifications may reveal new surfaces, and binding affinity alterations may be a sign of cancer vulnerability to targeted therapy. An example of how this has been achieved is that the KRAS G12C mutation introduces a druggable pocket, which can be measured through molecular assays, and BRAF V600E stabilizes the activation loop, modulating downstream phosphorylation signatures. Equally, deletions in EGFR exon 19 yield predictable patterns of inhibitor sensitivity and serve to demonstrate how structural characteristics are directly translated into quantifiable biomarkers of accurate oncology. Structural biomarkers offer atomic-scale insight into changes in functions, which allows the rational choice of therapy and the prediction of resistance mechanisms at an early stage (Passaro et al., 2024).

18.6.5.2 Network biomarkers

In addition to changes in individual proteins, systems biology also permits the determination of network biomarkers that are indicative of the overall behavior of molecular pathways. These biomarkers use multilayered data, such as coactivation of signaling hubs, feedback-loop interactions, and cross-talk between metabolic and proliferative pathways, and dynamic pathway rewiring during drug stimulation. Network biomarkers are able to better depict tumor heterogeneity as compared to their single-gene counterparts and support cellular dependencies as well as vulnerabilities in terms of cellular networks. With the combination of structural data and network-level dynamics, scholars can discover nodes of special importance, the perturbation of which is associated with therapeutic responsiveness or resistance. As an illustration, PI3K-AKT, and MAPK concomitant activation can be used as a predictive network biomarker of combination therapy efficacy. Structural and network biomarkers can also be used in combination, which improves the accuracy of prediction and allows distinguishing between patients according to the profile of molecular networks (Networks as Biomarkers: Uses and Purposes [Internet], no date).

18.6.6 Drug-response modeling using structural constraints

Drug-response models reproduce simulations of cellular outcomes resulting from interactions with different concentrations of therapeutic agents. The addition of structural data enhances these models by including correct binding kinetics, mutation-induced alterations in catalytic rates, mutation-induced allosteric modulation, and transition probabilities between active and inactive states of proteins. Structural constraints enable models to show how specific mutations affect the affinity or activity of an inhibitor or a pathway, providing information on mechanisms of resistance. Indicatively, modeling of EGFR T790M mutants can predict resistance to kinase inhibitors, enabling the discovery of combination regimens to overcome compensatory mechanisms. These types of models are useful in predicting the efficacy of therapeutic intervention, the choice of dosage, and the prioritization of patients to undergo specific therapy or a combination of drugs by connecting perturbations in structure with network outputs (Study Models of Drug–Drug Interactions Involving P-Glycoprotein: The Potential Benefit of P-Glycoprotein Modulation at the Kidney and Intestinal Levels - PMC [Internet], no date).

18.6.7 Machine learning in structural systems biology

Such machine-learning schemes are gradually incorporating structural descriptions alongside systems-level data to develop predictive cancer therapeutic models. Transcriptomic signatures, proteomic data, and pathway activation scores are combined with structural features, including surface topology, binding-pocket geometry, and predicted protein stability, to predict drug response. Examples of applications of these models include predicting drug efficacy in a specific patient, detecting resistance mutations before clinical manifestation, predicting synergistic drug combinations, and stratifying patients into molecular subgroups to provide precision in treatment. Machine learning enables the high-dimensional synthesis of structural and system-level features and allows predicting tumor behavior during therapy. In the end, this technique enables personalized medicine to be used, whereby atomic-level information on proteins is related to network-level responses in order to maximize treatment strategies (Optimized models and deep learning methods for drug response prediction in cancer treatments: a review - PMC [Internet], no date).

Structural systems pharmacology offers a comprehensive model for predicting cancer therapy outcomes by incorporating structural biomarkers, network biomarkers, and computational drug-response modeling. The multifaceted nature of this approach allows the accurate stratification of patients, the rational design of combination therapies, and the early detection of resistance mechanisms, which represents a revolutionary change in precision oncology (Computational frameworks transform antagonism to synergy in optimizing combination therapies - PMC [Internet], no date).

18.7 Illustrative examples

18.7.1 Kinase inhibitor networks

18.7.1.1 Role of kinases in signal integration

Kinases serve as key nodes in cellular signaling pathways and govern proliferation, survival, metabolism, and apoptosis. Structural studies show that mutations in kinase domains can significantly alter protein behavior, including ATP-binding sites, activation loops, and dimerization interfaces. These structural alterations affect the transmission of signals downstream, which tends to increase oncogenic signaling and pose a challenge to drug resistance (Protein Kinase C Family: Structures, Biological Functions, Diseases, and Pharmaceutical Interventions - PMC [Internet], no date).

18.7.1.2 Structural impacts of oncogenic mutations

Mutations such as BRAF V600E stabilize the kinase in an active state and facilitate constitutive activation of the MEK/ERK pathway, resulting in uncontrolled cell proliferation. On the same note, EGFR mutations increase receptor dimerization dynamics, thereby amplifying PI3K-AKT signaling. Not only do these changes cause an increase in kinase activity, but they also disrupt normal feedback and permit aberrant signaling to be sustained despite targeted inhibitors. The effective action of single-agent therapies may be restricted by adaptive feedback loops involving ERK or AKT, and it is necessary to pay attention to network-wide effects (Cosci et al., 2025).

18.7.1.3 Systems-level modeling for combination therapy

By combining structural knowledge with systems biology methods, scientists can understand the dynamics of perturbations within signaling networks and identify weak points that can be targeted therapeutically. This strategy has motivated combination therapy, for example, cotargeting EGFR and MET in lung cancer or combining BRAF and MEK inhibitors in melanoma, which successfully overcomes negative feedback responses. Systems-level modeling is used to forecast network behaviors, maximize the use of inhibitors, and enhance the durability of treatments, which gives a mechanistic explanation of multitarget therapies (Precision combination therapies based on recurrent oncogenic co-alterations - PMC [Internet], no date).

18.7.2 Poly (ADP-ribose) polymerase–DNA repair interactions

18.7.2.1 Structural basis of poly (ADP-ribose) polymerase function

PARP enzymes are of significance in the restoration of single-strand breakes in DNA. Structural studies reveal the damage recognition mechanism by which PARP interacts with damaged DNA, binds to sites of repair, and triggers ADP-ribose addition with NAD + as a substrate. PARP inhibitors operate by either pushing out NAD + or trapping the enzyme on the DNA, causing arrested repair forks. The changes in conformations upon inhibitors bind destabilize the DNA repair complex, which increases the cytotoxic stress in cancer cells (Multifaceted Role of PARP1 in Maintaining Genome Stability Through Its Binding to Alternative DNA Structures - PMC [Internet], no date).

18.7.2.2 Mechanisms of synthetic lethality

Modeling at the systems level incorporates structural knowledge to understand the rationale behind the potency of PARP inhibitors in BRCA1- or BRCA2-deficient cells. Loss of homologous recombination in such tumors renders single-strand DNA breaks lethal, resulting in synthetic lethality. Network models also indicate crosstalk with the replication stress pathway and the alternative DNA repair system, which can offset PARP inhibition in resistant cells. The knowledge of these interactions is important in the prediction of patient response and the design of effective therapies (Helleday, 2011).

18.7.2.3 Predicting resistance and optimizing therapy

PARP inhibitors often develop resistance through rewiring networks or by regaining BRCA activity. Structural systems methods enable scientists to predict the existence of compensatory pathways by modeling the effects of PARP conformational changes on downstream signaling and DNA repair pathways. This combination aids in the rational design of combination therapies, including PARP inhibitors combined with replication stress or repair complement pathway agents. Integrating system and structural-level insights, clinicians can determine the most effective drug combinations, forecast resistance, and categorize individuals most likely to respond to treatment (Mechanisms of PARP-Inhibitor-Resistance in BRCA-Mutated Breast Cancer and New Therapeutic Approaches - PMC [Internet], no date).

18.8 Applications in cancer therapy

18.8.1 Targeted therapy and rational drug design

Fig. 18.4 illustrates the advent of targeted therapy, which has revolutionized cancer treatment by enabling selective inhibition of oncogenic pathways rather than nonspecific cytotoxicity. The resurgent role of rational drug design, driven by advances in structural biology, crystallography, and computational modeling, is critical for identifying and optimizing molecules that interact with well-defined molecular targets. The atomic resolution of protein structures can be used to design small molecules with high precision to take advantage of distinct binding pockets, allosteric sites, conformational switches, and posttranslational modifications, which contribute to the maintenance of the malignant phenotype.

A core value recognized in targeted therapy is the knowledge of the molecular basis of tumorigenesis. Aberrant kinases, transcription factors, growth factor receptors, or epigenetic regulators dysregulate several types of cancer, impacting proliferation, survival, and metastasis. Structural bioinformatics assists in mapping these abnormalities onto useful functional domains to enable rational drug designers to recognize the regions that can be inhibited. Indicatively, an example of a successful application targeting in the ABL kinase domain of chronic myeloid leukemia is the structural elucidation of an inactive conformation pocket used by imatinib. Table 18.3 which

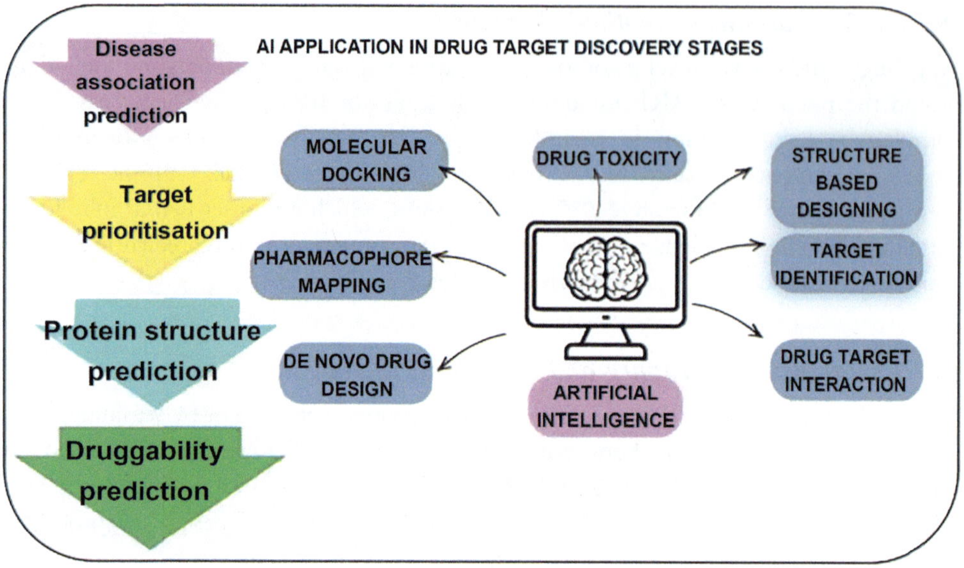

Figure 18.4 Application in drug-target discovery stages. *No Permission Required.*

Table 18.3 Selected FDA-approved structure-guided targeted cancer drugs.

Drug	Target	Structural insight used	Cancer type
Imatinib	BCR-ABL kinase	Inactive conformation binding	CML
Osimertinib	EGFR T790M mutant	Mutation-specific pocket	NSCLC
Vemurafenib	BRAF V600E	ATP-binding site conformation	Melanoma
Sotorasib	KRAS G12C	Covalent cysteine pocket	NSCLC
Palbociclib	CDK4/6	Cyclin-dependent kinase structure	Breast cancer

No Permission Required.

presents structure-guided targeted cancer drugs, demonstrates that high selectivity and sustained response may be delivered by targeting particular conformational states (Molecular targeted therapy for anticancer treatment - PMC [Internet], no date).

Modern drug design uses multiple approaches as:

18.8.1.1 Structure-based drug design

SBDD is used to position small molecules at active sites using X-ray crystallographic or cryo-EM structures. These interactions are optimized using computational docking and MD to compute binding energies, hydrogen-bond networks, and hydrophobic complementarity. In designing first-, second-, and third-generation inhibitors against kinases, structural information about gatekeeper residues, activation loops, and catalytic motifs can be used to design more potent and resistance–resistant inhibitors (Lionta et al., 2014).

18.8.1.2 Fragment-based drug discovery

FBDD uses low-molecular-weight fragments that bind the target weakly. Fragment-protein complexes are structurally analyzed to be elaborated chemically into high-affinity inhibitors. A number of cancer drugs, such as vemurafenib, were produced in such fragment-based workflows with crystallographic confirmation of fragment interactions (Li, 2020).

18.8.1.3 Allosteric modulation

Not every protein of interest in cancer has a druggable ATP-binding or catalytic site. Proteins such as MEK, the BCR-ABL myristoyl pocket, or mutant IDH can be targeted at alternative sites, referred to as allosteric sites. Mapping protein dynamics, finding conformational hotspots, and stabilizing inactive structural states are important in the rational design of allosteric inhibitors. Such inhibitors are normally more specific and less toxic off-target (Bcr-Abl Allosteric Inhibitors: Where We Are and Where We Are Going to - PMC [Internet], no date).

18.8.1.4 Overcoming drug resistance using structure-guided tools

Major challenges in targeted therapy are resistance mutations. Knowledge of the structure of resistance-conferring variants (e.g., EGFR T790M, ALK L1196M, BCR-ABL T315I) facilitates the rational design of next-generation inhibitors that can fit mutated pockets. Indicatively, osimertinib was designed based on structural knowledge to interact selectively with the T790M mutation and not the wild-type EGFR to minimize dose-limiting toxicities (Lovly et al., 2017).

18.8.1.5 Covalent and reversible covalent inhibition

Structure-guided design has been used to create covalent inhibitors that irreversibly bind nucleophilic residues, such as cysteines, in mutant proteins. KRAS G12C inhibitors [sotorasib, adagrasib] are a good example of rational covalent design: their structural modeling has found a previously inaccessible pocket only in the GDP-bound form, which then selectively binds the mutant cysteine (Evolution of structure-guided drug design strategies targeting mutations in codon 12 of KRAS - PMC [Internet], no date).

18.8.1.6 Artificial intelligence-enhanced rational design

Structural databases, docking simulations, and MD trajectories are now incorporated into AI and deep learning to predict ideal interactions, create synthetically accessible molecules, and shorten iterative wet-lab cycles. Structural coverage of undruggable proteins was greatly increased by tools such as AlphaFold, and has hastened the process of rational design of cancer targets (Aghajani et al., 2022).

18.8.2 Immunotherapy and structural epitope prediction

Immunotherapy is one of the most radical developments in oncology, using the immune system to selectively destroy malignant cells. Structural epitope prediction [recognition of antigenic sites by T-cells or antibodies] is at the core of the development of next-generation immunotherapies, such as vaccines, monoclonal antibodies, checkpoint inhibitors, and engineered immune cells (Ahuja and Zaheer, 2024).

18.8.2.1 Structural basis of antigen recognition

Major histocompatibility complex [MHC] molecules display protein antigens as either conformations or linear epitopes. The peptide-MHC-T-cell receptor (TCR) complexes exist in high-resolution structures of TCR complexes exist to identify the effect of small changes in the side chains, conformation, or anchor residues on immunogenicity. Peptide binding affinity predictions to multiple HLA alleles are calculated by means of computational modeling, which makes it possible to identify tumor-specific neoantigens produced by somatic mutations (Szeto et al., 2021).

18.8.2.2 Neoantigen discovery using structural tools

Neoantigens are specific to cancer cells and are relevant in personalized immunotherapy. Structural epitope predictor tools combine:

- HLA-binding prediction based on deep learning.
- Molecular docking to simulate peptide binding to HLA grooves.
- MD to determine stability.
- Structural flexibility analysis in order to optimize immunogenicity.

The models aid in prioritizing neoantigens that can induce a robust cytotoxic T-cell response. Such computational pipelines are used in neoantigen-based vaccines, which are currently being trialed and are based on individually generated personalized formulations (Identification of neoantigens for individualised cancer immunotherapy - PMC [Internet], no date).

18.8.2.3 Monoclonal antibody and checkpoint inhibitor optimization

Structural prediction assists in mapping the paratopes of antibodies and antigen epitopes in order to engineer affinity, specificity, and effector functionality rationally. In the case of immune checkpoints, for example, PD-1/PD-L1 and CTLA-4, structural studies inform the design of inhibitors that prevent inhibitory interactions and re-activate T-cells. Epitope maps indicate the binding modes of therapeutic antibodies (agents), as seen in Table 18.4, such as pembrolizumab or nivolumab, and can be used to engineer variants with greater blockade efficacy. Structural knowledge is also helpful in developing bispecific antibodies that target both immune and tumor antigens simultaneously (Chiu et al., 2019).

18.8.2.4 Chimeric antigen receptor T-cell engineering

Chimeric antigen receptor (CAR) T-cell therapy is based on antigen specificity on a single-chain variable fragment (scFv). Prediction of structural epitopes is used to select the optimal antigen concentration, availability, and membrane position. The scFv-antigen interaction modeling allows the design of CARs with better avidity and reduced cross-reactivity off-tumor (Chiu et al., 2019).

18.8.2.5 Artificial intelligence-driven epitope prediction

Deep learning models (e.g., transformer-based models) combine structural and sequence data to make B-cell and T-cell epitope predictions at levels never before achieved. These technologies speed up vaccine development, optimization of checkpoint inhibitors, and identification of immune-dominant regions within tumor proteomes. Therefore prediction of structural epitopes is a crucial part of contemporary immunotherapy, as it allows absolute targeting and enhances the therapeutic response

Table 18.4 Structure-guided immunotherapy agents.

Agent	Mechanism	Structural basis	Notes
Pembrolizumab	PD-1 inhibitor	Epitope mapping of PD-1	High-affinity engineering
Nivolumab	PD-1 inhibitor	PD-1/PD-L1 interface disruption	Structural optimization
Ipilimumab	CTLA-4 inhibitor	CTLA-4–B7 structural interface	Fc-enhanced variants
CAR-T (CD19)	Antigen recognition	scFv–CD19 structural model	Most widely used CAR-T

No Permission Required.

(Transformer-based deep learning enables improved B-cell epitope prediction in parasitic pathogens: A proof-of-concept study on Fasciola hepatica - PMC [Internet], no date).

18.8.3 Multitarget drug discovery using network biology

Cancer is a complex disease that is fueled by interrelated signaling pathways, metabolic re-engineering, transcriptional dyscontrol, and microenvironmental surveillance. Single-target therapies have an initial response but result in adaptive resistance by virtue of pathway redundancy and network rewiring. Multitarget drug discovery is a method to overcome these shortcomings by developing therapeutics that are able to modulate multiple nodes in cancer networks.

Network biology using omics data [genomics, transcriptomics, proteomics, metabolomics] involves the construction of interaction maps of signaling cascades, PPIs, and feedback loops, as well as cross-talk pathways. Structural systems pharmacology superimposes structural data in 3D on network nodes in order to establish conserved motifs, druggable pockets, and structural resemblances across proteins (Understanding and targeting resistance mechanisms in cancer - PMC [Internet], no date).

18.8.3.1 Network-based target identification

Essential nodes, bottlenecks, and hubs are detected using algorithms, and their inhibition prevents tumor survival. The design of polypharmacological inhibitors that bind multiple targets is guided by the structural clustering of homologous pockets within network nodes. As an example, dual PI3K/mTOR inhibitors take advantage of a shared catalytic site geometry to inhibit both pathways in a synergistic manner (Systems approaches to polypharmacology and drug discovery - PMC [Internet], no date).

18.8.3.2 Rational design of multitarget molecules

Multitarget agents may be designed using various strategic design paradigms, including linking pharmacophores, which interact with distinct proteins, hybridizing them to form combined structures, and synthesizing scaffold-hopping derivatives, which facilitate flexible and adaptive binding. The other viable approach is by stabilizing important protein–protein interfaces to be able to regulate several pathways at a time. This is aided by computational tools that simulate the interaction of the various targets and optimize the candidates based on selectivity, minimizing off-target effects, and their fit within complex biological network topologies (Prospects of multitarget drug designing strategies by linking molecular docking and molecular dynamics to explore the protein-ligand recognition process - PubMed [Internet], no date).

18.8.3.3 *Synergistic combinations and structure integration*

Structural modeling is used to design combination therapies through prediction of complementary mechanisms. Network models simulate pathway suppression when there are two inhibitors acting on different nodes. Even BRAF and MEK coinhibition in melanoma can be used as an example to overcome side effects of feedback reactivation observed with single agents (Combination therapy with BRAF and MEK inhibitors for melanoma: latest evidence and place in therapy - PMC [Internet], no date).

18.8.3.4 *Artificial intelligence and network pharmacology*

Machine learning finds latent characteristics between structural similarity, pathway position, and phenotype effects. Deep generative models are molecule inference models that predict molecules that are expected to modulate multiple targets at once. The addition of AlphaFold structures opens the space of potential network nodes. Multitarget drug discovery by networks is, therefore a powerful approach to conquer resistance, increase durability, and tackle complicated oncogenic circuitry (Combination therapy with BRAF and MEK inhibitors for melanoma: latest evidence and place in therapy - PMC [Internet], no date).

18.8.4 Case examples: checkpoint inhibitors and chimeric antigen receptor-T structural models

18.8.4.1 *Checkpoint inhibitors*

The example of immune checkpoint inhibitors is structure-guided immuno-oncology. PD-1/PD L1 interaction, as elucidated by crystallography, identified key residues mediating inhibitory signaling. Antibodies like nivolumab and pembrolizumab were created to have a high-affinity and to be specific to PD-1 and block its engagement with ligands, thus reactivating T-cells. Small-molecule inhibitor development, as well as peptide antagonists, was also made possible by the structural understanding of PD-L1. The combination of CTLA-4 and CTLA-7 interplay structures led to the design of ipilimumab in a way that maximized Fc activity to promote the elimination of regulatory T-cells in tumors (Yang et al., 2023).

18.9 Future direction and potential role of artificial intelligence

The potential of AI-based technologies to become a significant force in cancer research and treatment in the future is great, with the capabilities of this field only increasing in quantity and complexity as biomedical data continues to be collected and processed in large volumes. A key trend is the application of AI to combine and analyze multiomics data consisting of genomic, transcriptomic, proteomic, and metabolomic data. AI will assist in the discovery of new therapeutic targets, the optimization of biomarker panels,

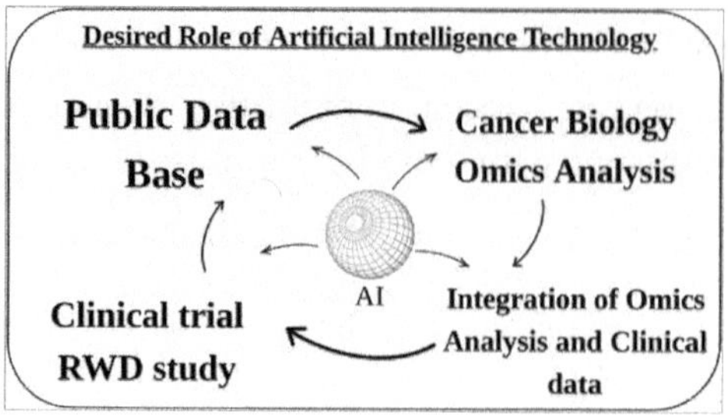

Figure 18.5 Future direction and potential role of artificial intelligence (*AI*) for the development of new treatment strategies. *No Permission Required.*

and the prediction of disease course more accurate by revealing subtle patterns and relationships, which might otherwise remain hidden using more traditional methods of analysis, as seen in Fig. 18.5A. A second potentially successful application is in the field of structural biology, where increasingly sophisticated machine-learning models can predict protein structures, structural changes, and interaction sites with extremely high accuracy. Such information can be used to accelerate rational designs of drugs, eliminate the need to use labor-intensive experimental approaches, and develop highly selective therapeutic molecules (Current AI technologies in cancer diagnostics and treatment - PMC [Internet], no date).

Personalized treatment decisions are also supposed to be dictated by AI. Clinicians can be assisted by decision-support systems trained on clinical histories, imaging, molecular data, and treatment responses, to customize interventions to the specifics of each patient and enhance efficacy and safety. Moreover, the use of AI-powered digital pathology and radiology tools is improving early diagnosis by identifying molecular and morphological patterns associated with high-risk lesions or early-stage disease. Beyond diagnostics and therapeutics, AI can also be used to understand resistance mechanisms by analyzing longitudinal patient data and simulating tumor evolution. Simultaneously, building understandable and transparent AI models will be necessary to ensure clinical trust and regulatory acceptance. There will be a need for close interaction among computational scientists, clinicians, and experimental researchers as these tools develop. In the end, AI can integrate diverse strands of biological data, optimize drug discovery, and enhance personalized cancer care at an individual-patient level, making it an indispensable tool for the future of oncology, at a truly personalized level, which is why it is a must-have in oncology in the future (Chitoran et al., 2025).

18.10 Conclusion

Structural and systems biology integration has become a new paradigm in contemporary cancer treatment, offering unprecedented understanding of the molecular and network responses that contribute to tumor formation, progression, and survival after therapy. Atomic-level information about oncogenic proteins, tumor suppressors, receptor-ligand complexes, and important signaling molecules has been elucidated through structural biology and is utilized to design highly selective inhibitors, allosteric modulators, and subsequent-generation targeted therapeutics. In parallel, systems biology builds on this molecular knowledge by reconstructing the highly complex genetic, epigenetic, transcriptomic, metabolic, and proteomic networks that contribute to shaping cancer cell behavior in both tumor-intrinsic and microenvironmental contexts. These intersections enable the field to adopt more network-aware therapeutic approaches to overcome pathway cross-talk, redundancy, and compensatory signaling, which are key obstacles to traditional monotherapies. Integrative approaches combining multiomics, computational modeling, and machine learning make systems biology even more predictive and enable the identification of key regulatory centers, biomarkers of response to therapy, and individual vulnerabilities, which form the foundation of precision oncology. Furthermore, cryo-EM, high-throughput sequencing, MD simulations, and AI have shortened the discovery cycle, speeding up the transition from molecular understanding to translation. Although recent discoveries, such as tumor heterogeneity, tumor dynamics, and data interpretation, remain challenging, the synergistic application of structural and systems biology has continued to foster innovation in drug design, the development of immunotherapies, and the planning of personalized treatment. With the development of these fields, they are providing an effective and coherent model to address the complexity of cancer, enhance early detection, overcome resistance mechanisms, and ultimately achieve complete eradication.

References

A comprehensive review of computational cell cycle models in guiding cancer treatment strategies - PMC [Internet], no date. [cited 2025 Dec 19]. Available at https://pmc.ncbi.nlm.nih.gov/articles/PMC11226463/.

A Detailed Catalogue of Multi-Omics Methodologies for Identification of Putative Biomarkers and Causal Molecular Networks in Translational Cancer Research [Internet], no date. [cited 2025 Dec 19]. Available at https://www.mdpi.com/1422-0067/22/6/2822.

Abdel-Magid, A.F., 2015. Allosteric modulators: an emerging concept in drug discovery. ACS Medicinal Chemistry Letters 6 (2), 104–107. Available at https://doi.org/10.1021/ml5005365.

Agent based modeling in cancer biomedicine: applications and tools for calibration and validation - PMC [Internet], no date. [cited 2025 Dec 19]. Available at https://pmc.ncbi.nlm.nih.gov/articles/PMC11057625/.

Aghajani, J., Farnia, P., Farnia, P., Ghanavi, J., Velayati, A.A., 2022. Molecular dynamic simulations and molecular docking as a potential way for designed new inhibitor drug without resistance. Tanaffos 21 (1), 1–14. Available at http://www.tanaffosjournal.ir/.

Ahmad, A., Imran, M., Ahsan, H., 2023. Biomarkers as biomedical bioindicators: approaches and techniques for the detection, analysis, and validation of novel biomarkers of diseases. Pharmaceutics 15 (6), 1630. Available at https://doi.org/10.3390/pharmaceutics15061630.

Ahuja, S., Zaheer, S., 2024. The evolution of cancer immunotherapy: a comprehensive review of its history and current perspectives. Korean Journal of Clinical Oncology 20 (2), 51–73. Available at https://doi.org/10.14216/kjco.24009.

Artificial intelligence [AI] and machine learning [ML] in precision oncology: a review on enhancing discoverability through multiomics integration - PMC [Internet], no date. [cited 2025 Dec 19]. Available at https://pmc.ncbi.nlm.nih.gov/articles/PMC10546458/.

Barabási, A.L., Gulbahce, N., Loscalzo, J., 2011. Network medicine: a network-based approach to human disease. Nature Reviews Genetics 12 (1), 56–68. Available at https://doi.org/10.1038/nrg2918.

Bcr-Abl Allosteric Inhibitors: Where We Are and Where We Are Going to - PMC [Internet], no date. [cited 2025 Dec 19]. Available at https://pmc.ncbi.nlm.nih.gov/articles/PMC7570842/.

Behar, M., Barken, D., Werner, S.L., Hoffmann, A., 2013. XThe dynamics of signaling as a pharmacological target. Cell 155 (2), 448. Available at https://doi.org/10.1016/j.cell.2013.09.018, https://www.sciencedirect.com/journal/cell.

Biophysical and structural analysis of KRAS switch-II pocket inhibitors reveals allele-specific binding constraints[image] - PMC [Internet], no date. [cited 2025 Dec 19]. Available at https://pmc.ncbi.nlm.nih.gov/articles/PMC12270674/.

Bon, M., Bilsland, A., Bower, J., McAulay, K., 2022. Fragment-based drug discovery—the importance of high-quality molecule libraries. Molecular Oncology 16 (21), 3761–3777. Available at https://doi.org/10.1002/1878-0261.13277, http://onlinelibrary.wiley.com/journal/10.1002/(ISSN)1878-0261.

Cancer cell plasticity: from cellular, molecular, and genetic mechanisms to tumor heterogeneity and drug resistance - PMC [Internet], no date. [cited 2025 Dec 19]. Available at https://www.pmc.ncbi.nlm.nih.gov/articles/PMC11016008/ (no date).

Chakraborty, S., Rahman, T., 2012. The difficulties in cancer treatment. Ecancermedicalscience 6, ed16 Nov 14.

Chen, C., Wang, J., Pan, D., Wang, X., Xu, Y., Yan, J., et al., 2023. Applications of multi-omics analysis in human diseases. MedComm 4 (4). Available at https://doi.org/10.1002/mco2.315.

Chitoran, E., Rotaru, V., Gelal, A., Ionescu, S.O., Gullo, G., Stefan, D.C., et al., 2025. Using artificial intelligence to develop clinical decision support systems—the evolving road of personalized oncologic therapy. Diagnostics 15 (18). Available at https://doi.org/10.3390/diagnostics15182391, http://www.mdpi.com/journal/diagnostics/.

Chiu, M.L., Goulet, D.R., Teplyakov, A., Gilliland, G.L., 2019. Antibody structure and function: The basis for engineering therapeutics. Antibodies 8 (4). Available at https://doi.org/10.3390/antib8040055, https://www.mdpi.com/2073-4468/8/4/55/pdf.

Combination therapy with BRAF and MEK inhibitors for melanoma: latest evidence and place in therapy - PMC [Internet], no date. [cited 2025 Dec 19]. Available at https://pmc.ncbi.nlm.nih.gov/articles/PMC4699264/.

Comprehensive detection and characterization of human druggable pockets through binding site descriptors - PMC [Internet], no date. [cited 2025 Dec 19]. Available at https://pmc.ncbi.nlm.nih.gov/articles/PMC11387482/.

Computational frameworks transform antagonism to synergy in optimizing combination therapies - PMC [Internet], no date. [cited 2025 Dec 19]. Available at https://pmc.ncbi.nlm.nih.gov/articles/PMC11743742/.

Conformational changes studied by cryo-electron microscopy - PubMed [Internet], no date. [cited 2025 Dec 19]. Available at https://pubmed.ncbi.nlm.nih.gov/10966635/.

Conformational Dynamics and Structural Transitions of Arginine Kinase: Implications for Catalysis and Allergen Control [Internet], no date. [cited 2025 Dec 19]. Available at https://www.mdpi.com/2075-1729/15/8/1248.

Cosci, I., Salizzato, V., Del Fiore, P., Pigozzo, J., Guarneri, V., Mocellin, S., et al., 2025. Molecular basis of BRAF inhibitor resistance in melanoma: a systematic review. Pharmaceuticals 18 (8), 1235. Available at https://doi.org/10.3390/ph18081235.

Cryo-EM of NHEJ supercomplexes provides insights into DNA repair - PMC [Internet], no date. [cited 2025 Dec 19]. Available at https://pmc.ncbi.nlm.nih.gov/articles/PMC9006396/.

Current AI technologies in cancer diagnostics and treatment - PMC [Internet], no date. [cited 2025 Dec 19]. Available at https://pmc.ncbi.nlm.nih.gov/articles/PMC12128506/.

Depending on the binding with the ligands, the extracellular part of EGFR takes different folds that allow receptor dimerization, a condition that enables receptor activation. In the intracellular kinase domain, oncogenic mutations, e.g. L858R or exon 19 deletions, support the active conformation, increasing the affinity of ATP and allowing better signaling. Comparisons between structures indicate that first-generation inhibitors select these mutant forms. - Google Search [Internet], no date. [cited 2025 Dec 19]. Available at https://www.google.com/search?q=Depending+on+the+binding+with+the+ligands%2C+the+extracellular+part+of+EGFR+takes+different+folds+that+allow+receptor+dimerization%2C+a+condition+that+enables+receptor+activation.+In+the+intracellular+kinase+domain%2C+oncogenic+mutations%2C+e.g.+L858R+or+exon+19+deletions%2C+support+the+active+conformation%2C+increasing+the+affinity+of+ATP+and+allowing+better+signaling.+Comparisons+between+structures+indicate+that+first-generation+inhibitors+select+these+mutant+forms.&rlz=1C1RXQR_enIN1183IN1183&oq=Depending+on+the+binding+with+the+ligands%2C+the+extracellular+part+of+EGFR+takes+different+folds+that+allow+receptor+dimerization%2C+a+condition+that+enables+receptor+activation.+In+the+intracellular+kinase+domain%2C+oncogenic+mutations%2C+e.g.+L858R+or+exon+19+deletions%2C+support+the+active+conformation%2C+increasing+the+affinity+of+ATP+and+allowing+better+signaling.+Comparisons+between+structures+indicate+that+first-generation+inhibitors+select+these+mutant+forms.&gs_lcrp=EgZjaHJvbWUyBggAEEUYOdIBBzI1OWowajSoAgCwAgE&sourceid=chrome&ie=UTF-8.

Developments, applications, and prospects of cryo-electron microscopy - PMC [Internet], no date. [cited 2025 Dec 19]. Available at https://pmc.ncbi.nlm.nih.gov/articles/PMC7096719/.

Dong, M., Wang, L., Hu, N., Rao, Y., Wang, Z., Zhang, Y., 2025. Integration of multi-omics approaches in exploring intra-tumoral heterogeneity. Cancer Cell International 25 (1). Available at https://doi.org/10.1186/s12935-025-03944-2.

Dong, Y., Zhou, X., Ding, Y., Luo, Y., Zhao, H., 2024. Advances in tumor microenvironment: applications and challenges of 3D bioprinting. Biochemical and Biophysical Research Communications 730, 150339. Available at https://doi.org/10.1016/j.bbrc.2024.150339.

Drug resistance in cancer: molecular mechanisms and emerging treatment strategies - PMC [Internet], no date. [cited 2025 Dec 19]. Available at https://pmc.ncbi.nlm.nih.gov/articles/PMC12623568/.

EGFR in Cancer: Signaling Mechanisms, Drugs, and Acquired Resistance - PMC [Internet], no date. [cited 2025 Dec 19]. Available at https://pmc.ncbi.nlm.nih.gov/articles/PMC8197917/.

Emerging patterns of somatic mutations in cancer - PMC [Internet], no date. [cited 2025 Dec 19]. Available at https://pmc.ncbi.nlm.nih.gov/articles/PMC4014352/.

Epigenetic regulation in cancer - PMC [Internet], no date. [cited 2025 Dec 19]. Available at https://pmc.ncbi.nlm.nih.gov/articles/PMC10876210/.

Evolution of structure-guided drug design strategies targeting mutations in codon 12 of KRAS - PMC [Internet], no date. [cited 2025 Dec 19]. Available at https://pmc.ncbi.nlm.nih.gov/articles/PMC12135073/.

Fragment-Based Drug Discovery Using NMR Spectroscopy - PMC [Internet], no date. [cited 2025 Dec 19]. Available at https://pmc.ncbi.nlm.nih.gov/articles/PMC3699969/.

Frontiers | A Comparison of Deterministic and Stochastic Modeling Approaches for Biochemical Reaction Systems: On Fixed Points, Means, and Modes [Internet], no date. [cited 2025 Dec 19]. Available at https://www.frontiersin.org/journals/genetics/articles/10.3389/fgene.2016.00157/full.

Ghoreyshi, N., Heidari, R., Farhadi, A., Chamanara, M., Farahani, N., Vahidi, M., et al., 2025. Next-generation sequencing in cancer diagnosis and treatment: clinical applications and future directions. Discover Oncology 16 (1). Available at https://doi.org/10.1007/s12672-025-01816-9.

Hao, G.F., Yang, G.F., Zhan, C.G., 2012. Structure-based methods for predicting target mutation-induced drug resistance and rational drug design to overcome the problem. Drug Discovery Today 17 (19-20), 1121–1126. Available at https://doi.org/10.1016/j.drudis.2012.06.018.

Helleday, T., 2011. The underlying mechanism for the PARP and BRCA synthetic lethality: Clearing up the misunderstandings. Molecular oncology 5 (4), 387–393. Available at https://doi.org/10.1016/j.molonc.2011.07.001.

HER2 Dimerization Inhibitor Pertuzumab – Mode of Action and Clinical Data in Breast Cancer - PMC [Internet], no date. [cited 2025 Dec 19]. Available at https://pmc.ncbi.nlm.nih.gov/articles/PMC3971793/.

Heterogeneity in precision oncology - PMC [Internet], no date. [cited 2025 Dec 19]. Available at https://pmc.ncbi.nlm.nih.gov/articles/PMC10953762/.

He, J., Zeng, X., Wang, C., Wang, E., Li, Y., 2024. Antibody–drug conjugates in cancer therapy: mechanisms and clinical studies. MedComm 5 (8). Available at https://doi.org/10.1002/mco2.671.

High-resolution protein modeling through Cryo-EM and AI: current trends and future perspectives – a review - PMC [Internet], no date. [cited 2025 Dec 19]. Available at https://pmc.ncbi.nlm.nih.gov/articles/PMC12590954/.

How Neuromembrane Lipids Modulate Membrane Proteins: Insights from G-Protein-Coupled Receptors [GPCRs] and Receptor Tyrosine Kinases [RTKs] - PMC [Internet], no date. [cited 2025 Dec 19]. Available at https://pmc.ncbi.nlm.nih.gov/articles/PMC10547395/.

Identification of neoantigens for individualised cancer immunotherapy - PMC [Internet], no date. [cited 2025 Dec 19]. Available at https://pmc.ncbi.nlm.nih.gov/articles/PMC7612664/.

Immune Surveillance and Immune Escape in Cancer: Mechanisms and Immunotherapy - PMC [Internet], no date. [cited 2025 Dec 19]. Available at https://pmc.ncbi.nlm.nih.gov/articles/PMC12514560/.

Insights into Protein–Ligand Interactions: Mechanisms, Models, and Methods [Internet], no date. [cited 2025 Dec 19]. Available at https://www.mdpi.com/1422-0067/17/2/144.

Jin, S., Long, H., Huang, A., Wang, J., Yu, X., Xu, Z., et al., 2025. Accurate prediction of synergistic drug combination using a multi-source information fusion framework. BMC Biology 23 (1). Available at https://doi.org/10.1186/s12915-025-02302-y.

Kanavos, P., 2006. The rising burden of cancer in the developing world. Annals of Oncology: Official Journal of the European Society for Medical Oncology 17 (Suppl 8), viii15–viii23.

Kim, D., Tran, A., Kim, H.J., Lin, Y., Yang, J.Y.H., Yang, P., 2023. Gene regulatory network reconstruction: harnessing the power of single-cell multi-omic data. NPJ Systems Biology and Applications 9 (1). Available at https://doi.org/10.1038/s41540-023-00312-6, https://www.nature.com/npjsba/.

Kleckner, I.R., Foster, M.P., 2011. An introduction to NMR-based approaches for measuring protein dynamics. Biochimica et Biophysica Acta - Proteins and Proteomics 1814 (8), 942–968. Available at https://doi.org/10.1016/j.bbapap.2010.10.012.

Knox, S.S., 2010. From "omics" to complex disease: a systems biology approach to gene-environment interactions in cancer. Cancer Cell International 10, 11.

Koutsogiannouli, E., Papavassiliou, A.G., Papanikolaou, N.A., 2013. Complexity in cancer biology: is systems biology the answer? Cancer Medicine 2 (2), 164–177. Available at https://doi.org/10.1002/cam4.62.

KRAS Mutation Subtypes and Their Association with Other Driver Mutations in Oncogenic Pathways - PMC [Internet], no date. [cited 2025 Dec 19]. Available at https://pmc.ncbi.nlm.nih.gov/articles/PMC11274496/.

Kreeger, P.K., Lauffenburger, D.A., 2010. Cancer systems biology: a network modeling perspective. Carcinogenesis 31 (1), 2–8. Available at https://doi.org/10.1093/carcin/bgp261.

Laubenbacher, R., Hower, V., Jarrah, A., Torti, S.V., Shulaev, V., Mendes, P., et al., 2009. A systems biology view of cancer. Biochimica et Biophysica Acta - Reviews on Cancer 1796 (2), 129–139. Available at https://doi.org/10.1016/j.bbcan.2009.06.001.

Laviana, A.A., Luckenbaugh, A.N., Resnick, M.J., 2020. Trends in the cost of cancer care: Beyond drugs. Journal of Clinical Oncology 38 (4), 316–322. Available at https://doi.org/10.1200/JCO.19.01963, https://ascopubs.org/doi/pdfdirect/10.1200/JCO.19.01963.

Leveraging artificial intelligence and machine learning in kinase inhibitor development: advances, challenges, and future prospects - PMC [Internet], no date. [cited 2025 Dec 19]. Available at https://pmc.ncbi.nlm.nih.gov/articles/PMC12412618/.

Li, Q., 2020. Application of fragment-based drug discovery to versatile targets. Frontiers in Molecular Biosciences 7. Available at https://doi.org/10.3389/fmolb.2020.00180, journal.frontiersin.org/journal/molecular-biosciences.

Lionta, E., Spyrou, G., Vassilatis, D.K., Cournia, Z., 2014. Structure-based virtual screening for drug discovery: principles, applications and recent advances. Current Topics in Medicinal Chemistry 14 (16), 1923–1938. Available at https://doi.org/10.2174/1568026614666140929124445, http://www.benthamdirect.org/pages/all_b_bypublication.php.

Liu, J., Yang, L., Liu, D., Wu, Q., Yu, Y., Huang, X., et al., 2025. The role of multi-omics in biomarker discovery, diagnosis, prognosis, and therapeutic monitoring of tissue repair and regeneration processes. Journal of Orthopaedic Translation 54, 131–151. Available at https://doi.org/10.1016/j.jot.2025.07.006.

Logic-Based Models for the Analysis of Cell Signaling Networks | Biochemistry [Internet], no date. [cited 2025 Dec 19]. Available at https://pubs.acs.org/doi/10.1021/bi902202q.

Lovly, C.M., Iyengar, P., Gainor, J.F., 2017. Managing Resistance to EFGR- and ALK-targeted therapies. American Society of Clinical Oncology Educational Book 37, 607–618. Available at https://doi.org/10.14694/EDBK_176251.

Mapping the protein–protein and genetic interactions of cancer to guide precision medicine - PMC [Internet], no date. [cited 2025 Dec 19]. Available at https://pmc.ncbi.nlm.nih.gov/articles/PMC6746583/.

Mechanisms of PARP-Inhibitor-Resistance in BRCA-Mutated Breast Cancer and New Therapeutic Approaches - PMC [Internet], no date. [cited 2025 Dec 19]. Available at https://pmc.ncbi.nlm.nih.gov/articles/PMC10378018/.

Metabolic reprogramming in cancer: Mechanisms and therapeutics - PMC [Internet], no date. [cited 2025 Dec 19]. Available at https://pmc.ncbi.nlm.nih.gov/articles/PMC10041388/.

Molecular Structure and Function - Opportunities in Biology - NCBI Bookshelf [Internet], no date. [cited 2025 Dec 19]. Available at https://www.ncbi.nlm.nih.gov/books/NBK217812/.

Molecular targeted therapy for anticancer treatment - PMC [Internet], no date. [cited 2025 Dec 19]. Available at https://pmc.ncbi.nlm.nih.gov/articles/PMC9636149/.

Multifaceted Role of PARP1 in Maintaining Genome Stability Through Its Binding to Alternative DNA Structures - PMC [Internet], no date. [cited 2025 Dec 19]. Available at https://pmc.ncbi.nlm.nih.gov/articles/PMC11552663/.

Multiscale Cancer Modeling - PMC [Internet], no date. [cited 2025 Dec 19]. Available at https://pmc.ncbi.nlm.nih.gov/articles/PMC3883359/.

Mutations in the Serine/Threonine Kinase BRAF: Oncogenic Drivers in Solid Tumors - PMC [Internet], no date. [cited 2025 Dec 19]. Available at https://pmc.ncbi.nlm.nih.gov/articles/PMC10968748/.

Negative feedback and adaptive resistance to the targeted therapy of cancer - PMC [Internet], no date. [cited 2025 Dec 19]. Available at https://pmc.ncbi.nlm.nih.gov/articles/PMC3351275/.

Neoantigen-driven personalized tumor therapy: An update from discovery to clinical application - PMC [Internet], no date. [cited 2025 Dec 19]. Available at https://pmc.ncbi.nlm.nih.gov/articles/PMC12407177/.

Networks as Biomarkers: Uses and Purposes [Internet], no date. [cited 2025 Dec 19]. Available at https://www.mdpi.com/2073-4425/14/2/429.

New Horizons in Structural Biology of Membrane Proteins: Experimental Evaluation of the Role of Conformational Dynamics and Intrinsic Flexibility [Internet], no date. [cited 2025 Dec 19]. Available at https://www.mdpi.com/2077-0375/12/2/227.

Oncogenes and tumor suppressor genes: functions and roles in cancers - PMC [Internet]. [cited 2025 Dec 19]. Available at https://pmc.ncbi.nlm.nih.gov/articles/PMC11141506/ (no date).

Optimized models and deep learning methods for drug response prediction in cancer treatments: a review - PMC [Internet], no date. [cited 2025 Dec 19]. Available at https://pmc.ncbi.nlm.nih.gov/articles/PMC11042005/.

Overcoming EGFR T790M and C797S resistance with mutant-selective allosteric inhibitors - PMC [Internet], no date. [cited 2025 Dec 19]. Available at https://pmc.ncbi.nlm.nih.gov/articles/PMC4929832/.

Panditrao, G., Bhowmick, R., Meena, C., Sarkar, R.R., 2022. Emerging landscape of molecular interaction networks: opportunities, challenges and prospects. Journal of Biosciences 47 (2). Available at https://doi.org/10.1007/s12038-022-00253-y, http://www.ias.ac.in/jbiosci/.

Pandurangan, A.P., Blundell, T.L., 2020. Prediction of impacts of mutations on protein structure and interactions: SDM, a statistical approach, and mCSM, using machine learning. Protein Science 29 (1), 247–257. Available at https://doi.org/10.1002/pro.3774, http://onlinelibrary.wiley.com/journal/10.1002/(ISSN)1469-896X.

Passaro, A., Al Bakir, M., Hamilton, E.G., Diehn, M., André, F., Roy-Chowdhuri, S., et al., 2024. Cancer biomarkers: Emerging trends and clinical implications for personalized treatment. Cell 187 (7), 1617–1635. Available at https://doi.org/10.1016/j.cell.2024.02.041, https://www.sciencedirect.com/science/journal/00928674.

Pathological Unfoldomics of Uncontrolled Chaos: Intrinsically Disordered Proteins and Human Diseases - PMC [Internet], no date. [cited 2025 Dec 19]. Available at https://pmc.ncbi.nlm.nih.gov/articles/PMC4100540/.

Precision combination therapies based on recurrent oncogenic co-alterations - PMC [Internet], no date. [cited 2025 Dec 19]. Available at https://pmc.ncbi.nlm.nih.gov/articles/PMC9524464/.

Prischi, F., Filippakopoulos, P., 2021. Editorial: structural studies of protein complexes in signaling pathways. Frontiers in Molecular Biosciences 8. Available at https://doi.org/10.3389/fmolb.2021.641932, journal.frontiersin.org/journal/molecular-biosciences.

Profiling Cell Signaling Networks at Single-cell Resolution - PMC [Internet], no date. [cited 2025 Dec 19]. Available at https://pmc.ncbi.nlm.nih.gov/articles/PMC7196580/.

Prospects of multitarget drug designing strategies by linking molecular docking and molecular dynamics to explore the protein-ligand recognition process - PubMed [Internet], no date. [cited 2025 Dec 19]. Available at https://pubmed.ncbi.nlm.nih.gov/32329098/.

Protein Kinase C Family: Structures, Biological Functions, Diseases, and Pharmaceutical Interventions - PMC [Internet], no date. [cited 2025 Dec 19]. Available at https://pmc.ncbi.nlm.nih.gov/articles/PMC12617277/.

Protein Misfolding and Aggregation in Proteinopathies: Causes, Mechanism and Cellular Response - PMC [Internet], no date. [cited 2025 Dec 19]. Available at https://pmc.ncbi.nlm.nih.gov/articles/PMC9944956/.

Protein posttranslational modifications in health and diseases: Functions, regulatory mechanisms, and therapeutic implications - PMC [Internet], no date. [cited 2025 Dec 19]. Available at https://pmc.ncbi.nlm.nih.gov/articles/PMC10152985/.

Protein structural context of cancer mutations reveals molecular mechanisms and candidate driver genes - PMC [Internet], no date. [cited 2025 Dec 19]. Available at https://pmc.ncbi.nlm.nih.gov/articles/PMC7617530/.

Rewired Metabolism Caused by the Oncogenic Deregulation of MYC as an Attractive Therapeutic Target in Cancers - PMC [Internet], no date. [cited 2025 Dec 19]. Available at https://pmc.ncbi.nlm.nih.gov/articles/PMC10341379/.

Scibilia, K.R., Gallagher, K., Masud, M.A., Robertson-Tessi, M., Gatenbee, C.D., West, J., et al., 2025. Mathematical oncology: how modeling is transforming clinical decision-making. Cancer Research 85 (24), 4866–4879. Available at https://doi.org/10.1158/0008-5472.CAN-25-0750.

Shieh, Y., Eklund, M., Sawaya, G.F., Black, W.C., Kramer, B.S., Esserman, L.J., 2016. Population-based screening for cancer: hope and hype. Nature Reviews Clinical Oncology 13 (9), 550–565. Available at https://doi.org/10.1038/nrclinonc.2016.50, http://www.nature.com/nrclinonc/archive/index.html.

Simanshu, D.K., Nissley, D.V., McCormick, F., 2017. RAS proteins and their regulators in human disease. Cell 170 (1), 17–33. Available at https://doi.org/10.1016/j.cell.2017.06.009, https://www.sciencedirect.com/journal/cell.

Smyth, M.S., Martin, J.H.J., 2000. x Ray crystallography. Journal of Clinical Pathology - Molecular Pathology 53 (1), 8–14. Available at https://doi.org/10.1136/mp.53.1.8.

Srivas, R., Shen, J.P., Yang, C.C., Sun, S.M., Li, J., Gross, A.M., et al., 2016. A network of conserved synthetic lethal interactions for exploration of precision cancer therapy. Molecular Cell 63 (3), 514–525. Available at https://doi.org/10.1016/j.molcel.2016.06.022, www.molecule.org.

Strategies for the drug development of cancer therapeutics - PMC [Internet], no date. [cited 2025 Dec 19]. Available at https://pmc.ncbi.nlm.nih.gov/articles/PMC12518319/.

Structural Variations in Cancer and the 3D Genome - PMC [Internet], no date. [cited 2025 Dec 19]. Available at https://pmc.ncbi.nlm.nih.gov/articles/PMC10423586/.

Structure and dynamics of the EGFR/HER2 heterodimer - PMC [Internet], no date. [cited 2025 Dec 19]. Available at https://pmc.ncbi.nlm.nih.gov/articles/PMC9925823/.

Structure-based drug design: aiming for a perfect fit - PMC [Internet], no date. [cited 2025 Dec 19]. Available at https://pmc.ncbi.nlm.nih.gov/articles/PMC5869280/.

Structure-Based Virtual Screening for Drug Discovery: Principles, Applications and Recent Advances - PMC [Internet], no date. [cited 2025 Dec 19]. Available at https://pmc.ncbi.nlm.nih.gov/articles/PMC4443793/.

Study Models of Drug–Drug Interactions Involving P-Glycoprotein: The Potential Benefit of P-Glycoprotein Modulation at the Kidney and Intestinal Levels - PMC [Internet], no date. [cited 2025 Dec 19]. Available at https://pmc.ncbi.nlm.nih.gov/articles/PMC10673607/.

Subramanian, I., Verma, S., Kumar, S., Jere, A., Anamika, K., 2020. Multi-omics data integration, interpretation, and its application. Bioinformatics and Biology Insights 14. Available at https://doi.org/10.1177/1177932219899051 117793221989905.

Suppression of feedback loops mediated by PI3K/mTOR induces multiple over-activation of compensatory pathways: an unintended consequence leading to drug resistance - PMC [Internet], no date. [cited 2025 Dec 19]. Available at https://pmc.ncbi.nlm.nih.gov/articles/PMC4222988/.

Systems approaches to polypharmacology and drug discovery - PMC [Internet], no date. [cited 2025 Dec 19]. Available at https://pmc.ncbi.nlm.nih.gov/articles/PMC3068535/.

Systems Biology as a Comparative Approach to Understand Complex Gene Expression in Neurological Diseases - PMC [Internet], no date. [cited 2025 Dec 19]. Available at https://pmc.ncbi.nlm.nih.gov/articles/PMC4217627/.

Szeto, C., Lobos, C.A., Nguyen, A.T., Gras, S., 2021. TCR recognition of peptide–MHC-I: rule makers and breakers. International Journal of Molecular Sciences 22 (1), 68. Available at https://doi.org/10.3390/ijms22010068.

Tamura, T., Yoshioka, Y., Sakamoto, S., Ichikawa, T., Ochiya, T., 2021. Extracellular vesicles as a promising biomarker resource in liquid biopsy for cancer. Extracellular Vesicles and Circulating Nucleic Acids 2 (2), 148–174. Available at https://doi.org/10.20517/evcna.2021.06, https://oaepublishstorage.blob.core.windows.net/f8af7577-7ad4-4756-bebc-777528328a93/4038.pdf.

Target discovery in the postgenomic era - PMC [Internet], no date. [cited 2025 Dec 19]. Available at https://pmc.ncbi.nlm.nih.gov/articles/PMC3300178/.

The effect of mutation on an aggregation-prone protein: An in vivo, in vitro, and in silico analysis - PMC [Internet], no date. [cited 2025 Dec 19]. Available at https://pmc.ncbi.nlm.nih.gov/articles/PMC9295795/.

The Importance of Bottlenecks in Protein Networks: Correlation with Gene Essentiality and Expression Dynamics - PMC [Internet], no date. [cited 2025 Dec 19]. Available at https://pmc.ncbi.nlm.nih.gov/articles/PMC1853125/.

The mechanism of activation of monomeric B-Raf V600E - PMC [Internet], no date. [cited 2025 Dec 19]. Available at https://pmc.ncbi.nlm.nih.gov/articles/PMC8215184/.

The Tumor Suppressor p53: From Structures to Drug Discovery - PMC [Internet], no date. [cited 2025 Dec 19]. Available at https://pmc.ncbi.nlm.nih.gov/articles/PMC2869527/.

Towards Structural Systems Pharmacology to Study Complex Diseases and Personalized Medicine - PMC [Internet], no date. [cited 2025 Dec 19]. Available at https://pmc.ncbi.nlm.nih.gov/articles/PMC4022462/.

Transformer-based deep learning enables improved B-cell epitope prediction in parasitic pathogens: A proof-of-concept study on Fasciola hepatica - PMC [Internet], no date. [cited 2025 Dec 19]. Available at https://pmc.ncbi.nlm.nih.gov/articles/PMC12064019/.

Understanding and targeting resistance mechanisms in cancer - PMC [Internet], no date. [cited 2025 Dec 19]. Available at https://pmc.ncbi.nlm.nih.gov/articles/PMC10203373/.

Unraveling the Guardian: p53's Multifaceted Role in the DNA Damage Response and Tumor Treatment Strategies - PMC [Internet], no date. [cited 2025 Dec 19]. Available at https://pmc.ncbi.nlm.nih.gov/articles/PMC11641486/.

Wang, Z., Gerstein, M., Snyder, M., 2009. RNA-Seq: a revolutionary tool for transcriptomics. Nature Reviews Genetics 10 (1), 57–63. Available at https://doi.org/10.1038/nrg2484.

Wlodawer, A., Minor, W., Dauter, Z., Jaskolski, M., 2008. Protein crystallography for non-crystallographers, or how to get the best (but not more) from published macromolecular structures. FEBS Journal 275 (1), 1–21. Available at https://doi.org/10.1111/j.1742-4658.2007.06178.x.

Yang, Z., Zeng, X., Zhao, Y., Chen, R., 2023. AlphaFold2 and its applications in the fields of biology and medicine. Signal Transduction and Targeted Therapy 8 (1). Available at https://doi.org/10.1038/s41392-023-01381-z.

Zhou, Z., Zhang, R., Zhou, A., Lv, J., Chen, S., Zou, H., et al., 2024. Proteomics appending a complementary dimension to precision oncotherapy. Computational and Structural Biotechnology Journal 23, 1725–1739. Available at https://doi.org/10.1016/j.csbj.2024.04.044.

CHAPTER 19

Protein structural biology in precision medicine and clinical translation

Maryam Khan[1], Minal Richi[2], and Sara Othman[3]
[1]Department of Biological Sciences, Kuwait University, Shadadiya, Al Asimah, Kuwait,
[2]Department of Biotechnology, American International University, Jahra, Kuwait,
[3]Department of Molecular Biology and Genetics, Faculty of Engineering and Natural Sciences, Istinye University, Istanbul, Turkey

List of abbreviations

bnAbs	Broadly neutralizing antibodies
cryo-EM	Cryo-electron microscopy
DL	Deep learning
EGFR	Epidermal growth factor receptor
FAK	Focal adhesion kinase
FDA	Food and Drug Administration
GPCR	G-protein coupled receptor
KRAS	Kristen rat sarcoma
mAbs	Monoclonal antibodies
MAPK	Mitogen-activated protein kinase
ML	Machine learning
NMR	Nuclear magnetic resonance
PDB	Protein Data Bank
PPI	Protein–protein interactions

19.1 Introduction

19.1.1 Overview of protein and structural biology

Almost all biological activities in a living system depend on proteins, which are among the most important macromolecules. Proteins are essential for preserving cellular and organismal homeostasis because they catalyze all metabolic reactions, control gene expression, offer structural support, and mediate cellular communication (Morris et al., 2022). Moreover, they operate as transporters, signaling molecules, structural elements, enzymes, and cellular function regulators in all biological processes (Shang et al., 2022). Proteins' remarkable functional variety stems from their capacity to adopt precise three-dimensional structures that are quite tailored to their biological functions and governed

Advances in Structural Biology
DOI: https://doi.org/10.1016/B978-0-443-44988-8.00005-8

by their amino acid sequences and the physicochemical environment in which they reside (Listov et al., 2024).

Nevertheless, understanding protein structure and function is essential for defining the molecular basis of human health and disease. Protein structural biology is the study of protein structures and how they relate to shape, dynamics, and biological function (Wang et al., 2016). Protein structure is arranged from the fundamental primary sequence to secondary, tertiary, and quaternary conformations, providing an infrastructure for understanding how proteins acquire differentiation and functional specialization (Caetano-Anollés et al., 2009).

19.1.2 Fundamentals of protein structure and function

The basic structure of a protein encompasses the linear sequence of amino acids joined by peptide bonds. This primary sequence, encoded by genetic information, regulates all higher levels of protein folding, which ultimately affect the protein function. Even a single change in the amino acid sequence can have greater consequences on the protein behavior, hence affecting the protein structure–function relationship (Schaefer and Burkhard, 2012).

Secondary structure is formed by local folding of the amino acid sequences, supported by hydrogen bonding. Mostly, the proteins in the secondary structure exist in the form of α-helices and β-sheets. These infoldings act as important motifs to assist in intricate folding and stabilize the overall protein integrity and flexibility (Debenedictis and Keten, 2019). Moreover, the tertiary structure of a protein represents the three-dimensional shape, often stabilized by various kinds of interactions like hydrophobic forces, ionic bonds, and disulfide bridges. This stage is crucial for the proteins to generate functional domains responsible for molecular recognition, enzymatic activity, and ligand binding (Babine and Bender, 1997).

In addition, there are certain proteins that exist as multisubunit complexes. These proteins combine two or more polypeptide chains to generate a quaternary structure. For example, hemoglobin is a specialized protein that has the ability to enhance the oxygen-carrying capacity of the blood, and this effectively depends on its unique quaternary structure (Marengo-Rowe, 2017). For a protein or enzyme to be biologically active and stable, these complexes must be assembled correctly. The conformational flexibility essentially required for these proteins to be functionally active is supported by various interactions like hydrogen bonding, electrostatic forces, hydrophobic bonds, and van der Waals forces (Zhou and Pang, 2018). It has been well documented in the literature that even a minute change in the structure or protein folds can lead to significant changes in the protein activity, stability, and interactions with other biomolecules (Jeffery, 2016).

Both conformational dynamics and protein folding are tightly controlled processes found inside the cell. Molecular chaperones assist some proteins in preventing misfolding and aggregation, while the majority of proteins reach their native states without any assistance (Chaudhuri and Paul, 2006). To maintain homeostasis, damaged or misfolded proteins are eliminated via cellular quality-control systems such as autophagy and proteasome degradation. Anomalies in protein structure and folding can thereby lead to a variety of clinical conditions, including cancer, metabolic problems, and neurological disorders. Thus maintaining protein structural integrity is crucial for cellular health. To understand disease mechanisms at the molecular level, additional studies in protein structural biology have lately been conducted (Huang and Horowitz, 2025).

19.1.3 Importance of protein structure in human health

The protein structure and its biological function go hand in hand, connecting genes to phenotypic traits. Any kind of structural disruption, such as mutation, misfolding, or environmental stress, can compromise cellular homeostasis, thereby leading to disease condition. Therefore in order to advance biomedical research and improve human health, it is essential to understand proteins' structural–functional relationships. Several examples quote the inseparable nature of protein structure and function. Active sites are naturally and precisely designed for enzymes to bind specific substrates and catalyze reactions (Siddikey et al., 2025). Similarly, cell signaling through receptors requires certain structural conformations to identify a particular ligand and send signals across the cellular membranes. Moreover, collagen and keratin are core structural proteins that provide mechanical strength and flexibility, demonstrating the role of protein architecture in maintaining biological integrity (Long et al., 2025).

Modern medicine and drug delivery heavily rely on binding to specific proteins, and their efficacy is dependent on precise structural configurations. Structure-based drug design has become a foundation of pharmacological science, enabling the development of tailored medicines with greater specificity and fewer side effects. Additionally, synthetic proteins and biologics, such as monoclonal antibodies (mAbs) and recombinant enzymes, have transformed the treatment of many diseases (Tang et al., 2023).

19.1.4 Techniques for protein structure and determination

Modern biomedical research relies heavily on an understanding of protein structure, especially when it comes to precision medicine and clinical translation. The capacity to resolve proteins at atomic or near-atomic levels has profoundly altered how diseases are diagnosed, classified, and managed (Gobena et al., 2024). Structural biology establishes a clear relationship between genotype and phenotype by exposing how mutations affect protein folding, stability, interactions, and function. X-ray crystallography has long been considered the gold standard for determining high-resolution protein structures (Colonna,

2025). This technique uses X-ray diffraction through a well-ordered protein crystal to generate electron density maps from which atomic models are generated (Huxford, 2013). It has given the vast majority of entries to the Protein Data Bank (PDB), providing extensive insights into enzyme processes, ligand binding, and drug-target interactions, which are critical in structure-based drug design and pharmacological optimization (Kent, 2026) (Fig. 19.1A). Nuclear magnetic resonance (NMR) spectroscopy gives structural and dynamic information on proteins in solution, which closely resembles physiological circumstances (Letertre et al., 2021). Unlike crystallography, NMR captures conformational flexibility and protein dynamics, resulting in structural ensembles rather than a single static model. This property is utilized for exploring intrinsically disordered proteins, folding mechanisms, or transitory interactions.

However, its applicability is often confined to smaller proteins due to the spectral complexity and sensitivity limits (Frueh et al., 2013). More recently, for massive macromolecular complexes and membrane proteins that are challenging to crystallize, cryo-electron microscopy (cryo-EM) has become a groundbreaking method. The steps involved in cryo-EM are illustrated in Fig. 19.1B. Cryo-EM visualizes biomolecules in their near-native states by fast-freezing them in vitreous ice and photographing them with an electron beam. Advances in direct electron detectors and image processing techniques have increased resolution to near-atomic levels, making cryo-EM a valuable tool in structural virology, neuroscience, and drug discovery (Jeyaraj et al., 2025).

Another important analytical tool in proteomics that is frequently used to separate, identify, and characterize different proteins and their changes with high sensitivity is Liquid chromatography–mass spectrometry (Alanazi, 2025). A protein sample that has

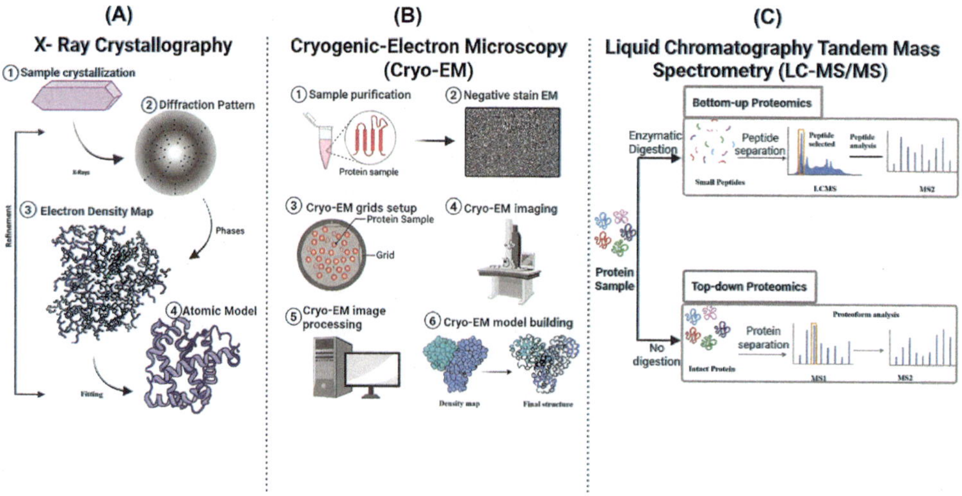

Figure 19.1 Schematic illustration of the key steps involved in protein structure determination techniques.

been extracted from cells, tissues, or biofluids is first purified and processed in a normal workflow, which frequently entails denaturation, reduction, and enzymatic digestion. In bottom-up proteomics, proteins are digested (usually using trypsin) into smaller peptides, which are then separated by liquid chromatography and subjected to mass spectrometry analysis.

Protein identity and posttranslational modifications are analyzed using peptide mass fingerprints and fragmentation spectra. The top-down approach in proteomics, on the other hand, analyzes complete proteins without breakdown. This provides direct identification of full-length proteoforms, including the sequence variants, and the use of more sophisticated instrumentation (Fig. 19.1C). When liquid chromatography-mass spectrometry-based top-down and bottom-up approaches are combined, it offers a comprehensive insight. The bottom-up approach provides precise structural and functional information at the proteoform level, while the top-down method offers high throughput and coverage.

Apart from traditional methodologies, the scope of structural biology has been redefined by innovative and integrative approaches. The low-resolution structural envelopes of proteins in solution can be analyzed by the small-angle X-ray scattering technique (Byer et al., 2023). The protein–protein interactions (PPI) and spatial restrictions within the complexes can be deeply studied by cross-linking mass spectroscopy (Piersimoni et al., 2022). Moreover, computer-based methodologies like homology modeling, molecular dynamics simulations, and AI-powered tools like AlphaFold, in recent years, have significantly impacted the validation and accuracy of protein structure prediction (Nussinov et al., 2022).

19.1.5 Fundamentals of the proteome and its role in personalized medicine

The proteome is defined as the total number of proteins expressed within a cell, tissue, or organism at any given point in time. Unlike the genome, which tends to be generally stable, the proteome is highly dynamic. The proteome can be altered in response to physiological conditions, environmental stimuli, and diseased conditions (Carbonara et al., 2021). Proteins serve as the cell's functional machinery, many times acting as enzymes, cofactors, coenzymes, structural components, and signaling molecules, thereby directly affecting biological processes. High-throughput proteomics technologies, particularly mass spectroscopy, allow the rapid analysis of thousands of proteins at a given time, therefore greatly improving our ability to study complex biological systems (Cui et al., 2024).

In addition, proteomics is an excellent tool in precision medicine for searching new biomarkers, stratifying diseases, and identifying therapeutic targets. Proteomic insights capturing real-time molecular changes help bridge the gap between genotype and phenotype. This allows for more enhanced, precise diagnoses and tailored treatment methods. Precision medicine, often called personalized medicine, is an evolving healthcare approach that includes an individual's genetics, molecular profiling, and environment/lifestyle to forecast what treatment would work best for them (Evans et al., 2024). Precision medicine is moving away from a "one size fits all" model to a successful clinical outcome revolving around "five rights": the right treatment to the right patient at the right time, in the right dose, and through the right route of administration. Proteomics plays an important role in making precision medicine a reality (Nice, 2016). It provides information related to genomic and transcriptome data, providing real-time data on protein function, posttranslational modifications, and interaction networks. This allows highly tailored treatments, accurate prognosis and diagnosis, and effective therapeutic responses (Vitorino, 2024). Fig. 19.2 illustrates the integrated role of protein structural biology in shaping precision medicine.

This chapter provides a complete framework for understanding how molecular structure can influence biological function. It gives an overview of the fundamental principles of protein structural biology and recent in silico techniques encompassing mechanistic insights into protein dysfunction and its role in clinical translation. Despite rapid developments in high-throughput sequencing, there is still a significant gap between detected genetic variants and their structural interpretation. Emphasis is placed on the relevance of protein structure in biological systems and the convergence of experimental and computational techniques, including growing artificial intelligence-driven structural prediction, to enable tailored therapy options and their implications for

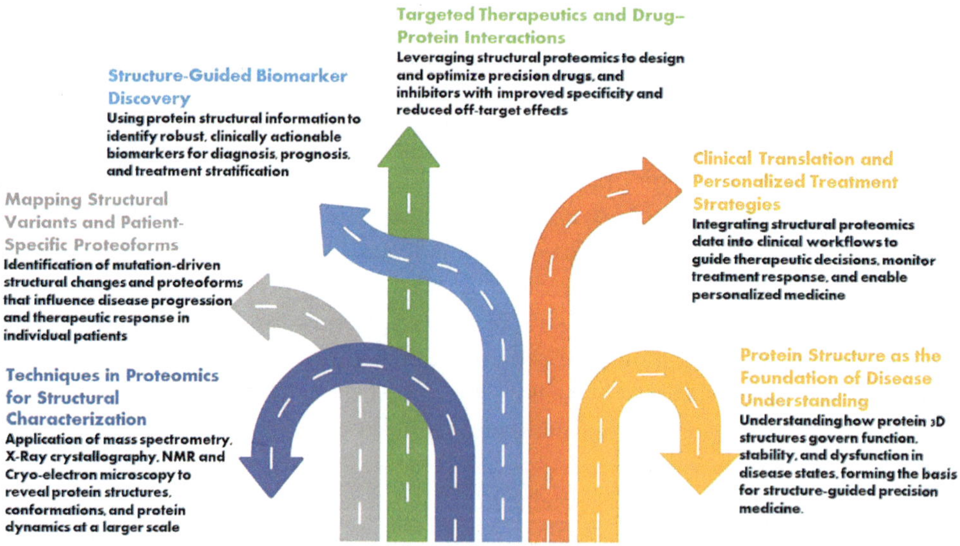

Figure 19.2 Roadmap showing the role of structural proteomics in precision medicine.

drug discovery, biomolecular engineering, and therapeutic applications. The following sections are organized to first introduce the foundational concepts of protein structure and analytical techniques, followed by an exploration of the structural basis of disease. Subsequent sections discuss structure-guided drug discovery, translational and clinical applications, and recent advances in protein modeling and engineering before concluding with future directions, challenges, and ethical considerations for the implementation of structure-based precision medicine.

19.2 Structural basis of protein dysfunction in disease

Understanding the structural basis of protein dysfunction is essential for estimating disease mechanisms and developing targeted therapies (Wang et al., 2016). Protein dysfunction is the main cause of many human diseases, as alterations in structure directly impair biological activity and cellular homeostasis. Structural biology provides a mechanistic understanding of how disruptions in folding, stability, or interactions translate into disease phenotypes (Marinko et al., 2019).

Recent structural genomic studies have shown that over 30%–40% of disease–associated missense variants lead to measurable destabilization of protein structure rather than complete loss of function. For example, large scale analyses combining protein structures with clinical variant databases revealed that pathogenic mutations often reduce thermodynamic stability, alter flexibility, or disrupt interaction interfaces, leading to partial loss of function that manifests as disease rather than lethality (Stein et al., 2019). Additionally, these findings suggest that protein dysfunction is a spectrum rather than a binary state of functional versus nonfunctional proteins. Structural biology has moved beyond basic knockout models toward more refined interpretations, where even minor destabilization, altered dynamics, or impaired molecular interactions can initiate and sustain viability while progressively weakening cellular resilience (Carugo and Djinović-Carugo, 2023).

Thus the structural basis of protein dysfunction represents not only a mechanical explanation of a disease but also provides a strategic entry point for intervention. The following sections present a more detailed analysis of variant effects, misfolding phenomenon, and disease-specific structural pathologies.

19.2.1 Impact of genetic variants on protein stability and function

Protein structure allows the prediction of the phenotypic effects of genetic variants. Genetic variants such as missense mutations, insertions, and deletions can interfere with protein conformation, destabilize protein folding and structural domains, or alter catalytic and binding properties (Akdel et al., 2022). Disease-associated variants often occur in conserved regions essential for stability or activity. Mapping these variants onto protein structures helps predict their pathogenic potential and guides therapeutic approaches, especially in monogenic disorders and cancer.

High-throughput multinational scanning studies have provided experimental evidence linking structural destabilization to disease severity. Furthermore, deep mutational scanning of enzymes and signaling proteins showed that even slight reductions in folding can decrease cellular abundance and activity due to increased proteasomal degradation (Matreyek et al., 2018). In cancer biology, structure-guided analyses of oncogenic mutations indicate that activating variants often function by stabilizing active conformations rather than simply enhancing catalytic efficiency. For example, structural analysis of Kristen rat sarcoma (KRAS) and BRAF mutants demonstrated that these oncogenic variants shift conformational equilibria toward signaling-competent states, thereby driving sustained pathway activation (Mozzarelli et al., 2024).

Overall, structural analysis indicates that protein stability and conformational dynamics should be viewed as key factors in determining variant pathogenicity. Instead of solely disrupting active sites, many disease-associated mutations act indirectly by altering folding energetics, domain coupling, or degradation susceptibility (Chaudhuri and Paul, 2006). Structural frameworks allow these indirect effects to be quantified and compared across variants, offering predictive power that complements population genetics and evolutionary conservation analyses. This is particularly relevant in clinical genomics, where a large number of variants of uncertain significance are often identified, but functional annotation is lacking (Ittisoponpisan and David, 2018). Overall, structure-based variant interpretation provides a mechanistic bridge between genotype and phenotype, reinforcing its value as a core component of precision medicine and motivating deeper exploration of misfolding- and aggregation-driven disease mechanisms.

19.2.2 Protein misfolding, aggregation, and conformational disorders

Misfolded proteins can evade cellular quality control, aggregate, and form toxic species. They expose hydrophobic surfaces that drive aggregation into oligomers or amyloid fibrils, a hallmark of conformational diseases such as Alzheimer's, Parkinson's, and Huntington's disease (Shamsi et al., 2017). Cryo-EM has revealed near-atomic structures of disease-associated protein aggregates, uncovered distinct strains, and deepened our understanding of their neurotoxic effects and propagation (Todd et al., 2024).

Recent cryo-EM studies have shown that amyloid fibrils derived from the same protein can adopt multiple structurally distinct conformations, or strains, which correlate with disease heterogeneity. For example, structural characterization of tau fibrils from Alzheimer's versus Pick's disease brains revealed fundamentally different folding topologies, providing a molecular explanation for disease-specific pathology and progression (Falcon et al., 2018; Fitzpatrick et al., 2017). Likewise, α-synuclein fibrils extracted from Parkinson's disease patients exhibit structural polymorphism that influences seeding efficiency and neurotoxicity (Schweighauser et al., 2020). It is worth noting that these findings show that protein aggregation is not a uniform process but a structurally encoded determinant of disease phenotype, highlighting the limitations of studying aggregation in simplified in vitro systems and emphasizing the need for patient-derived structural data. By resolving aggregates at near-atomic resolution, cyro-EM has enabled direct correlation between molecular architecture and disease manifestation, establishing aggregation as an encoded structural pathology rather than an incidental consequence of protein instability.

19.2.3 Structural pathology in cancer, metabolic, and neurodegenerative diseases

The principles of structural pathology manifest differently across disease classes. In cancer pathophysiology, mutations in tumor suppressors (e.g., p53) and kinases (e.g., BRAF) alter conformational states, affecting the signaling pathways and mechanisms (Tornesello, 2024). In contrast, metabolic disorders are often characterized by enzyme deficiencies that arise from structural destabilization, resulting in impaired catalytic activity and metabolic imbalance (Chaturvedi et al., 2016). Neurodegenerative disorders are characterized by conformational changes leading to toxic aggregates or impaired proteins (Höhn et al., 2020).

Different studies of P53 over the past decade have shown that many cancer-associated missense mutations do not abolish DNA binding directly but instead destabilize the core domain, increasing susceptibility to misfolding and aggregation. Further, crystallographic and NMR analyses revealed that "structural" p53 mutants lose proper conformational folding at physiological temperature, explaining the dominant-negative and gain-of-function effects in tumors (Joerger and Fersht, 2016). In metabolic diseases, cryo-EM and X-ray studies of enzymes such as phenylalanine hydroxylase and glucocerebrosidase showed that pathogenic variants reduce conformational flexibility required for catalysis, supporting the development of pharmacological chaperones as stabilizing therapies (Parenti et al., 2014). These findings proved how structural pathology underlies diverse disease classes through shared mechanisms of instability and dysregulation.

19.2.4 Host–pathogen protein interactions

PPIs between the microbiome and the host are important for host health. Pathogens frequently use structural mimicry to interact with host receptors and interfere with immune signaling pathways. Studies in structural biology have characterized these interactions at the atomic level, offering important insights for antiviral drug and vaccine development (Post and Brito, 2022). Understanding these molecular interfaces helps identify vulnerable points within infection cycles, creating targeted opportunities for therapeutic intervention.

Over the past decade, structural analysis of viral and bacterial effector proteins has shown that pathogens often evolve into host-like conformations that enable them to hijack cellular signaling pathways. For instance, cryo-EM and crystallographic studies of SARS-CoV-2 spike protein interactions with ACE2 demonstrated how subtle structural adaptations can increase receptor-binding affinity and enhance viral transmissibility, directly informing vaccine and antibody design (Walls et al., 2020). Similarly, bacterial type-III secretion effectors structurally mimic host enzymes, allowing them to rewire immune signaling pathways (de Groot et al., 2020). These studies emphasized the power of structural biology in uncovering molecular strategies of the immune system.

19.3 Protein structure-guided drug discovery

Structure-based drug discovery uses atomic-level information to design molecules that adjust protein function, effectively bridging fundamental biological research with translational pharmacology.

Structural insights have become indispensable for modern drug discovery, enabling the rational design of highly specific therapeutic techniques such as X-ray crystallography, cryo-EM, and AI-driven modeling, which allow three-dimensional structures of proteins, visualization of druggable pockets, and conformational landscapes. By finding the structure of proteins, nucleic acids, and viruses, these methods provide essential information that underpins both drug development and structural bioinformatics (Bijak et al., 2023).

Recent progress in integrating cryo-EM with AI-based structure prediction has expanded the druggable proteome. Work incorporating predicted models from AlphaFold showed that high-confidence structures can effectively guide hit identification and optimization, especially for proteins previously considered difficult to study (Jumper et al., 2021). These approaches have already sped up early-stage drug discovery workflows in areas such as oncology and infectious diseases.

19.3.1 Target identification and validation

Structural characterization plays a leading role in assessing whether proteins possess accessible binding pockets, allosteric sites, or critical interfaces suitable for drug targeting. Structure-based drug design enables faster, more cost-effective discovery, as genomic, proteomic, and structural studies continue to reveal many new therapeutic targets (Batool et al., 2019). This shift in understanding has significantly expanded the target space in cancer and neurodegenerative disease research.

By determining whether proteins exhibit ligand-like conformations or contain regulatory sites amenable to modulation, structural characterization helps distinguish practical therapeutic targets from those unlikely to yield clinical benefit. Importantly, structure-based validation goes beyond static assessment, incorporating protein dynamics, conformational plasticity, and interaction context (Kuzmanic et al., 2020). This dynamic perspective reduces false positives and improves confidence that targeting a specific protein will produce meaningful biological outcomes. As structural datasets continue to grow, they are playing an increasingly significant role in guiding target prioritization across oncology, neurodegeneration, and infectious disease research (Esmaeilpour et al., 2026).

19.3.2 Drug design and molecular docking approaches

Computer-aided drug design, including molecular docking and structure-based approaches, supports the development of therapeutically important small molecules by enabling the prediction of ligand binding and affinity. This information guides the optimization of interaction strength, selectivity, and pharmacokinetic properties while reducing the chemical space before the drug development process (Eberhardt et al., 2021).

Studies over the past decade have further demonstrated that integrating docking methods with experimental structural data improves hit discovery rates. For example, prospective docking campaigns targeting kinases and G-protein coupled receptors achieved success rates exceeding traditional methods. High-throughput screening, when combined with structural refinement and free-energy calculations (Chen et al., 2019). These results confirm docking as a vital component of modern drug discovery when it is based on accurate structural models.

Overall, this approach reduces experimental workload while increasing the probability of identifying

high-quality lead compounds. Importantly, docking performance is highly dependent on the reliability of structural inputs, highlighting the importance of experimentally derived or well-validated predicted protein models. As computational power and algorithms improve, docking has shifted from a supportive tool to a primary driver of lead discovery in various drug development programs (Nivatya et al., 2025).

19.3.3 Fragment-based and virtual screening strategies

Fragment-based drug discovery involves identifying small chemical fragments that bind weakly to targets but can be developed into more potent compounds. In parallel, virtual screening uses computational methods to assess large libraries against structural models. Both approaches are dependent on accurate protein structures and have led to the development of multiple approved therapeutics (Erlanson et al., 2004).

Recent fragment-based campaigns guided by X-ray crystallography and cryo-EM candidates against challenging targets such as KRAS G12C. Research has demonstrated that small molecular fragments use shallow pockets, which were later optimized into high-affinity inhibitors now applied in cancer therapy (Ostrem et al., 2013; Tanaka et al., 2021). Overall, fragment-based approaches illustrate how structural precision can offset the limitations of weak initial binding by guiding systematic optimization. Instead of depending on high-affinity interactions from the beginning, these approaches utilize structural insight to iteratively build potency and specificity. Their success in addressing challenging targets highlights the critical role of structural resolution and experimental validation. Together, these advances show how structural biology enables the exploration of previously inaccessible regions of chemical and target space, reinforcing its translational impact (Stachowski and Fischer, 2026).

19.3.4 Allosteric modulators and conformational targeting

Allosteric modulators target known active site regions to regulate protein function, enabling high specificity by exploiting conformational and functional transitions within proteins. Allosteric ligand conformational equilibria have been shown in many studies to offer an opportunity to target proteins that are otherwise undruggable (Nussinov and Tsai, 2013). Analysis of structural features shows that allosteric inhibitors typically act by stabilizing inactive conformations rather than directly blocking enzyme activity. Cryo-EM studies of kinases and ion channels have demonstrated that allosteric ligands can reshape global conformation landscapes, resulting in sustained functional inhibition with lower resistance (Jin et al., 2024; Vishweshwaraiah et al., 2021). This has positioned allostery as a key pillar of next-generation drug design. By targeting conformational transitions rather than catalytic residues, allosteric drugs can overcome resistance mechanisms and modulate complex signaling behaviors. This strategy exemplifies the power of structural insight to redefine what constitutes a druggable mechanism.

19.3.5 Structural basis of protein–protein and protein–ligand interactions

Protein-protein interaction governs many essential biological processes, yet its complexity has historically limited therapeutic exploitation. Structural biology has reshaped this field by enabling the identification of interaction hotspots, transient interfaces, and inducible binding modes that can be selectively targeted. The development of molecular glues and targeted degraders highlights how indirect modulation of interactions can produce highly specific therapeutic effects (Morris et al., 2022; Su et al., 2024). Structural approaches using X-ray, cryo-EM, and NMR reveal the energetic and geometric principles of protein–protein and protein–ligand interactions, offering a framework for designing inhibitors, peptides, biologics, and molecular glues to modulate key interaction networks (Shoemaker and Ando, 2015).

Understanding and targeting PPIs are important for modulating pathological or beneficial complexes. Techniques such as hotspot mapping and interface modeling guide the development of inhibitors, ligand mimetics, and other modulators, providing significant therapeutic potential in oncology, immunology, and infectious diseases, particularly where dysregulated signaling underlies pathology (Monti et al., 2023).

Recent work has further demonstrated the potential of targeting PPIs through molecular glues that stabilize or reprogram protein interactions. Structural characterization of cereblon-based degraders has shown how small molecules induce neomorphic PPIs, triggering selective protein degradation (Schreiber, 2021). Collectively, these advances represent a major shift in drug discovery philosophy, positioning structural insight as a foundation for innovative intervention strategies that extend well beyond traditional inhibition-based models.

19.4 Translational and clinical applications

Advancements in understanding structural biology have guided translational research into better resolution of protein molecules (Burley et al., 2024). Protein structural data plays a pivotal role in bridging the gap between disease mechanisms and clinical interventions. Structural discoveries help understand the complex interplay between genetic variation, protein dysfunction, and molecular interactions (Stein et al., 2019). Such data guides the foundation of the development of targeted therapies by recognizing the target sites, guiding lead optimization, and improving therapeutic specificity and safety.

In clinical research, protein-based therapies include enzymes, mAbs, and therapeutic peptides. Structure-based drug and vaccine design further accelerate translational pipelines by predicting binding-site affinity, epitope accessibility, and resistance-associated mutations (Stein et al., 2019). Most importantly, the clinical relevance of structural biology extends to the interpretation of patient-derived genetic variants. Mapping dysfunctional variants of proteins is critical for assessing pathogenicity and predicting therapeutic response (Burley et al., 2024). Furthermore, structural characteristics of proteins and protein complexes have emerged as valuable biomarkers for evaluating patient diagnosis and prognosis of disease. Together, these developments illustrate how protein structural biology is becoming an indispensable component of precision medicine, directly influencing therapeutic development, clinical decision-making, and patient outcomes.

19.4.1 Structural discovery and therapeutic development

Structural discovery plays a pivotal role in therapeutic development by linking molecular architecture to functional outcomes and clinical phenotypes. Structural discovery enables the identification and characterization of target sites and guides the design of small molecules, biologics, and protein-based therapeutics with improved efficacy and safety profiles. Integration of structural data into translational pipelines has effectively reduced development timeframes and improved the success rate of clinical evaluation.

Currently, biopharmaceutical research employs macromolecular crystallography and 3D electron microscopy for structure-guided optimization of the affinity of small-molecule screening for hit identification and lead compounds (Blundell, 2017). Moreover, NMR is used to screen fragments to identify and characterize micro-molecules (Harner et al., 2013). Structural data generated through these techniques are systematically archived in the PDB (Vallat et al., 2026). As of 2024, PDB archives 6000 atomic-level 3D structures of molecular targets, and 88% among them are referred to as new medical entities (Burley et al., 2024). In recent years, these findings have provided a foundation for the development of new antineoplastic, low-molecular drugs. According to the U.S. Food and Drug Administration (FDA), between 2019 and '23, they have approved 38 low-molecular-weight antineoplastic drugs that target human proteins based on atomic-level 3D structure. Furthermore, the U.S. FDA Novel Drug Therapy Approval Reports show that currently 50 novel drugs were approved in 2024 alone, following 46 more in 2025. This plethora of drugs encompasses several low-molecular-weight oncology agents (FDA and Cder., 2024).

Some of the approved antineoplastic molecules include low-molecular-weight oncologic drugs that target the therapeutic landscape across solid tumors and hematologic malignancies (Agrawal, 2025). The development of highly selective small-molecule inhibitors targets molecular drivers of tumorigenesis such as anaplastic lymphoma kinase (ALK), epidermal growth factor receptor (EGFR) exon 20 insertions, RET (rearranged during transfection) fusions, and menin-KMT2A interactions. Therefore, agents such as ensartinib, sunvozertinib, taletrectinib, and ziftomenib demonstrate structure-guided drug design and molecularly characterized treatment approaches (Burley et al., 2024).

For example, ensartinib is a recently approved potent second-generation tyrosine kinase inhibitor that targets the ALK domain (Agrawal, 2025). The ALK domain is known to drive activation of tumor-cell proliferation and survival via signaling pathways. Ensartinib competitively binds to the ATP-binding site of the ALK domain and inhibits its function, which subsequently inhibits downstream processing (Agrawal, 2025).

Similarly, suvozertinib is also a lower-molecular-weight drug that targets the mutant form of EGFR, particularly exon 20 insertions (Dhillon, 2023). The first- and second-generation EGFR tyrosine kinase inhibitors are designed to target exon 20 insertions.

However, steric hindrance and alterations in the ATP-binding site result in limited efficacy of these drugs (Tian et al., 2024). Unlike its predecessors, sunvozertinib is structurally designed to accommodate these conformational changes, allowing the formation of a covalent bond to the cysteine 797 residue within the ATP-binding sites of altered EGFR (Sangwan & Agarwal, 2025). A similar mode of inhibition is seen with taletrectinib, which selectively targets RET receptors by competitively binding to the RET kinase domain, blocking oncogenic signaling (Hoy, 2026; Papadopoulos et al., 2020).

Furthermore, many therapeutic approaches involve combination of treatment strategies such as targeting dual mitogen-activated protein kinase (MAPK) and KRASKR focal adhesion kinase (FAK) signaling pathways (Banerjee et al., 2025). One example is the combination of avutometinib and defactinib (Blair, 2025). Avutometinib is a dual inhibitor of M and FAK signaling pathways via inhibition of rapidly accelerated fibrosarcoma (RAF) and PD-1 (a specific kinase in MAPK pathway). Addition of defactinib selectively inhibits FAK, a nonreceptor tyrosine kinase that contributes to tumor progression, metastasis, and resistance (Banerjee et al., 2025). Hence, the combination of avuTometinib and defactinib inhibits RAF/MEK while simultaneously blocking FAK signaling, thereby inhibiting potential development of drug resistance.

The development of such strategies illustrates the evolving shift towards translational research and clinical drug design that aims at overcoming adaptive resistance and improving durability response for long-term treatment strategies.

19.4.2 Protein-based therapeutics: enzymes, antibodies, and peptides

Enzyme-based therapies are designed to replace, supplement, or inhibit specific enzyme functions, addressing deficiencies or pathological activity (Del Toro Runzer et al., 2026; Jayan et al., 2026). The mAbs act by recognizing unique epitopes on target proteins, guiding selective inhibition, immune-mediated cytotoxicity, or receptor modulation (Del Toro Runzer et al., 2026). Whereas peptide-based therapeutics are often derived from signaling motifs allowing high target specificity (Sen et al., 2026).

The design and optimization of these protein-based approaches rely heavily on structural biology, aiding the rational engineering of binding affinity, stability, and pharmacokinetic properties (Martins et al., 2025). Similarly, mAbs, representing a major class of proteins that target immune checkpoints and cell-surface antigens, have transformed cancer therapy (Martins et al., 2025). For example, recombinant enzymes such as cerliponase alfa replace deficient lysosomal activity in late infantile neuronal ceroid lipofuscinosis type 2 (Oliveira et al., 2026).

In 2024, the U.S. FDA approved 50 novel drugs, of which 16 were biologics that included mAbs, recombinant proteins, enzyme replacement agents, and vaccines (Table 19.1). Tislelizumab, also known as a humanized IgG4 kappa mAb, has been approved for metastatic esophageal squamous cell carcinoma (Lee & Keam, 2020). It targets the programmed death receptor-1 (PD-1) found on T-cells, which bind to PD-L1 and PD-L2 present on tumor cells, resulting in reduced T-cell proliferation and

Table 19.1 Protein-based treatment approved by the U.S. FDA in 2025. The lists highlight each drug's mechanism of action, primary molecular targets, therapeutic indication, and approval date.

Agent	Approval date	Mode of action	References
Monoclonal antibodies			
Narsoplimab	March 2025	Monoclonal antibody targeting MASP-2. It inhibits lectin.	Crescioli et al. (2025)
Ivonescimab	April 2025	Monoclonal antibody targeting PD-1 and VEGF. It enhances antitumor immunity and inhibits angiogenesis.	Crescioli et al. (2025)
Sibeprenlimab	May 2025	Monoclinal antibody targeting a proliferation-inducing ligand (APRIL). It reduces IgA nephropathy.	Cheung et al. (2025)
Depemokimab	June 2025	Antiinterleukin-5 monoclonal antibody. It reduces eosinophilic inflammation (asthma).	Nolasco & Crimi (2024)
Antibody-drug conjugates			
Telisotuzumab vedotin	August 2025	Antibody-drug conjugate (ADC) targeting c-Met receptor. It delivers a cytotoxic payload to tumor cells.	Somme et al. (2024)
Datopotamab deruxtecan	September 2025	ADC targeting TROP2. It delivers chemotherapy directly to cancer cells.	Schipilliti et al. (2024)
Enzyme			
Tividenofusp alfa	October 2025	Enzyme replacement therapy. It restores deficient metabolic enzyme activity.	Muenzer et al. (2026)
Peptide-based			
IcoSema	Under the regulatory stages of the FDA.	Peptide-based combination that includes insulin icodec and semaglutide. Improves glycemic control in diabetes.	Westergaard et al. (2024)

cytokine production. Tislelizumab binds to PD-1, preventing the interaction between PD-L1 and PD-L2; however, it also avoids binding to Fc gamma receptor 1 to avoid cross-linking and improving T-cell function (Lee & Keam, 2020; Martins et al., 2025; Zhang et al., 2018).

Recently, in 2024, the U.S. FDA had approved nogapendekin alfa inbakicept (N803) for unresponsive non-muscle-invasive bladder cancer (Keam, 2024). N803 is a recombinant IL-15 receptor superagonist that consists of a modified IL-15 molecule bound to an IL-15Rα/IgG1 Fc fusion protein, forming a soluble immune-stimulating complex (Lui et al., 2023). At the molecular level, this enables N803 to enhance the activation, proliferation, and cytotoxic function of natural killer cells and $CD8^+$ T-cells, promoting a strong antitumor immune response (Lui et al., 2023).

Recent research highlights peptide-based scaffolds as a promising tool against drug resistance seen in chemotherapy and immunotherapy (Zhang et al., 2026). Current research led by Sen et al. (2026) has featured two promising approaches for peptide-based therapy that include peptide–drug conjugate (PDC), cell-penetrating peptide, and pro-apoptotic and oncolytic peptides. These peptides can interfere with key resistance mechanisms such as DNA repair, cell-cycle regulation, apoptosis pathways, gene expression, and modulation of the tumor microenvironment (Sen et al., 2026). They have also highlighted the potential of these peptides in reversing chemotherapy resistance and restoring antitumor immunity. Additionally, a study led by Zhang et al. (2026) engineered enzyme-responsive PDC-proteolysis-targeting chimeras by integrating cell-penetrating and tumor-targeting peptides with a cathepsin B-cleavable linker to enhance tumor-specific delivery and intracellular release. These peptide-based complexes demonstrated increased membrane permeability, target protein degradation, antiproliferative activity, and in vivo therapeutic efficacy, hence creating approaches with advanced tumor-specific degradation platforms (Zhang et al., 2026).

19.4.3 Structure-based drug and vaccine design

Although current vaccines have achieved remarkable success, infectious diseases continue to significantly impact global health, requiring the need to expand and innovate vaccine development strategies (Huang et al., 2020; Kulp & Schief, 2013). Among various new approaches, structural vaccinology has emerged as a promising tool in the development of new and effective vaccines (Panda et al., 2020). Structural vaccinology refers to the detailed study of protein structures to design improved vaccine components or immunogens (Kulp & Schief, 2013). It combines structural and genetic sequence information from broadly neutralizing antibodies (bnAbs). Researchers have been able to design germline-targeting immunogens (Burton, 2002). These are specifically engineered to bind and activate early-stages (germline) B cells, initiating the production of potent neutralizing antibodies (Burton, 2002; Kulp & Schief, 2013). The interaction between an antigen and a neutralizing antibody defines a structural epitope on stable scaffold proteins. This is essentially done to preserve their shape or design smaller, simplified antigens that retain important protective regions while removing unnecessary parts (Burton & Hangartner, 2016). For example, HIV invades humoral immunity by masking conserved envelope trimer epitopes with variable regions and dense glycans, restricting bnAbs access (Burton & Hangartner, 2016). Despite this obstacle, a small percentage of individuals are able to develop potent bnAbs, which are currently being studied to identify recombinant envelope glycoprotein trimers. This is one example of a rational immunogen design to elicit effective bnAb responses through vaccination strategies.

19.4.4 Mapping patients' variants onto protein structures

Understanding how genetic variation influences protein structure and function is central to modern biomedical research. Mapping patient-based variants onto 3D protein structure enables insights into disease mechanism, functional disruption, and therapeutic targeting, bridging genomics with structural biology to support precision medicine and improved clinical interpretation of pathogenic mutations (Burley et al., 2024). Structural characterization of the protein interactome is critical for interpreting genetic variations (Gao et al., 2015; Nussinov et al., 2017). High-throughput sequencing has resulted in extensive human variation, including numerous missense mutations, yet experimental structures in the PDB cover only a fraction of all proteins (Berman, 2000; Burley et al., 2019). As PPI far outnumber individual proteins and are harder to resolve experimentally, computational modeling plays a central role in defining interactome structure (Stein et al., 2011; Vakser, 2014). Advances in structure prediction, particularly deep learning (DL) approaches such as AlphaFold, have transformed the field (Jumper et al., 2021). However, faster template-based methods remain valuable for large-scale analysis where approximate models are sufficient (David et al., 2022).

Multiple database systems for PPI include IntAct, BioGRID, and STRING, while UniPort aggregates genetic variation to enhance functional interpretation (Bateman et al., 2021; Orchard et al., 2014; Oughtred et al., 2021; Szklarczyk et al., 2021). Some resources, such as PDBe-KB and ADDRESS, focus on experimental structures, whereas others, including PhyreRisk, DeepSAV, and MSV3d, integrate modeled structures (Luu et al., 2012; Ofoegbu et al., 2019; Varadi et al., 2022). Databases like Interactome3D and dSysMap extended coverage by incorporating predicted quaternary complexes (Mosca et al., 2013, 2015).

19.4.5 Predicting drug response and resistance mechanisms

Predicting drug response and resistance mechanisms is crucial for the advancement of precision medicine (Gu et al., 2025). Variability in therapeutic efficacy stems from genetic mutations, altered gene expression, and variation in cell signaling pathways that influence drug targets and metabolism (Yu et al., 2025). Integrative approaches combining genomics, transcriptomics, and proteomics with machine learning (ML) enable the identification of predictive biomarkers and drug sensitivity profiles. Resistance may occur through target modification, activation of compensatory pathways, drug efflux, or epigenetic alterations. Understanding these mechanisms facilitates the development of more effective therapeutic strategies, including combination treatments, and supports the design of next-generation drugs to overcome or delay resistance in complex diseases such as cancer (Gu et al., 2025).

Further, the importance of predicting drug response and resistance mechanisms, such as drug repurposing, that particularly targets epigenetic regulators, is emerging as a complementary strategy to enhance therapeutic efficacy and overcome resistance in precision medicine (Salvati et al., 2025). Epigenetic drug repurposing strategies, also termed "drug repositing," have recently emerged as a potentially effective strategy to overcome resistance. Drug repositioning involves identifying new therapeutic uses for existing FDA-approved drugs or compounds in advanced preclinical stages, beyond their original indications, using in vitro, in vivo, or in silico approaches (Cavalcante et al., 2024; Salvati et al., 2025). Repurposed drugs can now bypass early-stage trials, as they have preestablished pharmacokinetics, pharmacodynamics, and toxicity profiles. Additionally, translational bioinformatics models enable large-scale analysis of drugs and biomarkers, while off-target studies can identify and link unexpected drug effects in cancer (Dong et al., 2025). ML and DL enhance drug repurposing by integrating multiomics datasets. Although it is largely investigational, these models provide valuable early translational research and hypothesis generation (Chaudhary et al., 2018; Ching et al., 2018). Databases such as PubChem and DrugBank support ML-based drug-target predictions (Carli et al., 2025). However, data quality issues persist, but the improvement in interpretability tools like SHAP plays an essential role in providing reliable and significant predictions (Li & Guan, 2022; Lundberg & Lee, 2017).

Similarly, DL frameworks such as DeepDTAF provide a comprehensive strategy for predicting anticancer drug response by integrating genomic mutations, gene expression, and drug structural features through a ernel-based similarity model (K. Wang et al., 2021). These features allow deep neural networks to classify cancer cell line–drug interactions based on sensitivity and resistance. Moreover, models such as "The Cancer Genome Atlas" data, along with DeepDRK, identified breast cancer patients' response to vorinostat (Y. Wang et al., 2021).

Models like DeepChrom, another convolutional neural network, can predict gene expression from five histone markers across 56 different cell types (Kundaje et al., 2015; Salvati et al., 2025; Singh et al., 2016). In silico medicine developed the multimodal AI platform PandaOmics that integrates transcriptomic, genomics, and text-based data to prioritize disease drivers and therapeutic compounds (Butler et al., 2022; Kundaje et al.,

2015). This system has successfully identified 28 repurposable targets for ALS and supported the development of the TINK inhibitor ISM001–055 for idiopathic pulmonary fibrosis, advancing it into Phase II trials (Xu et al., 2025). In parallel, graph-based approache such as the BenevolentAI Knowledge Graph successfully predicted the efficacy of baricitinib in COVID-19 patients by associating it with the viral endocytosis pathways, which was later validated in clinical trials (Stebbing et al., 2020).

19.5 Conclusion and future direction

Structural biology has evolved as a crucial component of precision medicine, giving accurate information regarding protein architecture and its direct involvement in disease outcomes. High-resolution structural images allow a better understanding of the molecular characteristics and mechanisms underlying a disease, thereby influencing disease onset and progression. This molecular-level perspective often led to accurate diagnosis and the rational creation of targeted therapies, thus providing individualized and effective treatment plans.

In addition, the latest technical developments and breakthroughs have revolutionized the field of structural biology and its therapeutic application. Recent technological advancements like cryo-electron spectroscopy, integrative structural methodologies, and artificial intelligence-driven protein modeling have considerably enhanced our ability to study complicated biomolecular systems. These advancements have demonstrated significant potential in drug development, enhanced biomarker identification and, most importantly, have enabled the integration of multiomics data to provide a detailed and comprehensive understanding of patient-specific biology. As these tools become more accessible, their integration into clinical translation is likely to increase the outcome of precision medicine.

"Omics" is an umbrella term encompassing diverse fields of science, including genomics, proteomics, and metabolomics. These collective technologies are used to explore the roles, relationships, and actions of various metabolites, thereby improving clinical decision-making and providing detailed insights into patient-specific molecular profiling. Multiomics opens greater avenues for disease prevention, diagnosis, and treatment. However, there are challenges in implementing translational omics in healthcare settings. These include, but are not limited to, managing large amounts of complex data and the lack of technical knowledge to collect that data. More recently, ML and DL have been integrated into omics, that has helped in efficient rapid collection, processing, and integration of large datasets.

Despite these potential advancements, the use of structure-based techniques, pharmacogenomics, and multiomics in clinical settings poses ethical and regulatory concerns. Ensuring data privacy, maintaining standardization of testing methodologies, and encouraging fair access to new diagnostics and therapies are all essential areas that must be addressed. Navigating these challenges is crucial for widespread adoption. Addressing these challenges not only requires the adoption of technological advancements in ongoing research but also an extensive collaboration between researchers, clinicians, and policymakers. Robust regulatory frameworks are essentially required to ensure safe and successful implementation of these advanced technologies in healthcare settings.

Contributions

MK, MR and SO contributed in writing different sections of the book chapter. MK prepared the figures and MR worked on the table. MK proofread the entire MS.

Conflict of Interest

All authors declare that there was no conflict of interest.

References

Agrawal, T., 2025. Efficacy and safety of ensartinib in the treatment of non-small cell lung cancer: A systematic review of clinical trials. Journal of Oncology Pharmacy Practice. Available at https://doi.org/10.1177/10781552251387058.

Akdel, M., Pires, D.E.V., Pardo, E.P., Jänes, J., Zalevsky, A.O., Mészáros, B., et al., 2022. A structural biology community assessment of AlphaFold2 applications. Nature Structural and Molecular Biology 29 (11), 1056–1067. Available at https://doi.org/10.1038/s41594-022-00849-w, https://www.nature.com/nsmb/.

Alanazi, S., 2025. Recent advances in liquid chromatography–mass spectrometry (LC–MS) applications in biological and applied sciences. Analytical Science Advances 6 (1). Available at https://doi.org/10.1002/ansa.70024.

Babine, R.E., Bender, S.L., 1997. Molecular recognition of protein-ligand complexes: applications to drug design. Chemical Reviews 97 (5), 1359–1472. Available at https://doi.org/10.1021/cr960370z, http://pubs.acs.org/journal/chreay.

Banerjee, S., Krebs, M.G., Greystoke, A., Garces, A.I., Perez, V.S., Terbuch, A., et al., 2025. Defactinib with avutometinib in patients with solid tumors: the phase 1 FRAME trial. Nature Medicine 31 (9), 3074–3080. Available at https://doi.org/10.1038/s41591-025-03763-y.

Bateman, A., Martin, M.-J., Orchard, S., Magrane, M., Agivetova, R., Ahmad, S., et al., 2021. UniProt: the universal protein knowledgebase in 2021. Nucleic Acids Research 49 (D1), D480–D489. Available at https://doi.org/10.1093/nar/gkaa1100.

Batool, M., Ahmad, B., Choi, S., 2019. A structure-based drug discovery paradigm. IJMS 20 (11), 2783. Available at https://doi.org/10.3390/ijms20112783.

Berman, H.M., 2000. The Protein Data Bank. Nucleic Acids Research 28 (1), 235–242. Available at https://doi.org/10.1093/nar/28.1.235.

Bijak, V., Szczygiel, M., Lenkiewicz, J., Gucwa, M., Cooper, D.R., Murzyn, K., et al., 2023. The current role and evolution of X-ray crystallography in drug discovery and development. Expert Opinion on Drug Discovery 18 (11), 1221–1230. Available at https://doi.org/10.1080/17460441.2023.2246881, https://www.tandfonline.com/loi/iedc20.

Blair, H.A., 2025. Avutometinib and Defactinib: First Approval. Drugs 85 (10), 1319–1327. Available at https://doi.org/10.1007/s40265-025-02215-8.

Blundell, T.L., 2017. Protein crystallography and drug discovery: recollections of knowledge exchange between academia and industry. IUCrJ 4 (4), 308–321. Available at https://doi.org/10.1107/S2052252517009241.

Burley, S.K., Wu-Wu, A., Dutta, S., Ganesan, S., Zheng, S.X.F., 2024. Impact of structural biology and the protein data bank on us fda new drug approvals of low molecular weight antineoplastic agents 2019–2023. Oncogene 43 (29), 2229–2243. Available at https://doi.org/10.1038/s41388-024-03077-2.

Burley, S. K., Berman, H. M., Bhikadiya, C., Bi, C., Chen, L., Di Costanzo, L., et al. 2019. RCSB protein data bank: biological macromolecular structures enabling research and education in fundamental biology, biomedicine, biotechnology and energy. Nucleic Acids Research, 47(D1), D464–D474. Available at https://doi.org/10.1093/nar/gky1004.

Burton, D.R., 2002. Antibodies, viruses and vaccines. Nature Reviews. Immunology 2 (9), 706–713. Available at https://doi.org/10.1038/nri891.

Burton, D.R., Hangartner, L., 2016. Broadly Neutralizing Antibodies to HIV and Their Role in Vaccine Design. Annual Review of Immunology 34 (1), 635–659. Available at https://doi.org/10.1146/annurev-immunol-041015-055515.

Butler, R., Bradford, D., Rodgers, K.E., 2022. Analysis of shared underlying mechanism in neurodegenerative disease. Frontiers in Aging Neuroscience 14. Available at https://doi.org/10.3389/fnagi.2022.1006089.

Byer, A.S., Pei, X., Patterson, M.G., Ando, N., 2023. Small-angle X-ray scattering studies of enzymes. Current Opinion in Chemical Biology 72. Available at https://doi.org/10.1016/j.cbpa.2022.102232, http://www.elsevier.com/locate/cbi.

Caetano-Anollés, G., Wang, M., Caetano-Anollés, D., Mittenthal, J.E., 2009. The origin, evolution and structure of the protein world. Biochemical Journal 417 (3), 621–637. Available at https://doi.org/10.1042/BJ20082063, http://www.biochemj.org/bj/417/0621/4170621.pdf.

Carbonara, K., Andonovski, M., Coorssen, J.R., 2021. Proteomes are of proteoforms: embracing the complexity. Proteomes 9 (3). Available at https://doi.org/10.3390/proteomes9030038.

Carli, F., Di Chiaro, P., Morelli, M., Arora, C., Bisceglia, L., De Oliveira Rosa, N., et al., 2025. Learning and actioning general principles of cancer cell drug sensitivity. Nature Communications 16 (1), 1654. Available at https://doi.org/10.1038/s41467-025-56827-5.

Carugo, O., Djinović-Carugo, K., 2023. Structural biology: a golden era. PLoS Biology 21 (6), e3002187. Available at https://doi.org/10.1371/journal.pbio.3002187.

Cavalcante, B.R.R., Freitas, R.D., Siquara da Rocha, L.de O., Santos, R.de S.B.D., Souza, B.S.de F., Ramos, P.I.P., et al., 2024. In silico approaches for drug repurposing in oncology: a scoping review. Frontiers in Pharmacology 15. Available at https://doi.org/10.3389/fphar.2024.1400029.

Chaudhuri, T.K., Paul, S., 2006. Protein-misfolding diseases and chaperone-based therapeutic approaches. FEBS Journal 273 (7), 1331–1349. Available at https://doi.org/10.1111/j.1742-4658.2006.05181.x.

Chaudhary, K., Poirion, O.B., Lu, L., Garmire, L.X., 2018. Deep Learning–Based Multi-Omics Integration Robustly Predicts Survival in Liver Cancer. Clinical Cancer Research 24 (6), 1248–1259. Available at https://doi.org/10.1158/1078-0432.CCR-17-0853.

Chaturvedi, S., Singh, A.K., Keshari, A.K., Maity, S., Sarkar, S., Saha, S., 2016. Human metabolic enzymes deficiency: a genetic mutation based approach. Scientifica 2016. Available at https://doi.org/10.1155/2016/9828672, www.hindawi.com/journals/scientifica/.

Chen, H., Fu, W., Wang, Z., Wang, X., Lei, T., Zhu, F., et al., 2019. Reliability of docking-based virtual screening for GPCR ligands with homology modeled structures: a case study of the angiotensin II type I receptor. ACS Chemical Neuroscience 10 (1), 677–689. Available at https://doi.org/10.1021/acschemneuro.8b00489.

Cheung, C.K., Alexander, S., Reich, H.N., Selvaskandan, H., Zhang, H., Barratt, J., 2025. The pathogenesis of IgA nephropathy and implications for treatment. Nature Reviews Nephrology 21 (1), 9–23. Available at https://doi.org/10.1038/s41581-024-00885-3.

Ching, T., Himmelstein, D.S., Beaulieu-Jones, B.K., Kalinin, A.A., Do, B.T., Way, G.P., et al., 2018. Opportunities and obstacles for deep learning in biology and medicine. Journal of The Royal Society Interface 15 (141). Available at https://doi.org/10.1098/rsif.2017.0387.

Crescioli, S., Kaplon, H., Wang, L., Visweswaraiah, J., Kapoor, V., Reichert, J.M., 2025. Antibodies to watch in 2025. MAbs 17 (1), 2443538. Available at https://doi.org/10.1080/19420862.2024.2443538.

Colonna, G., 2025. Protein biophysics: current limitations and prospects. AIMS Biophysics 12 (3), 333–374. Available at https://doi.org/10.3934/biophy.2025018.

Cui, M., Deng, F., Disis, M.L., Cheng, C., Zhang, L., 2024. Advances in the clinical application of high-throughput proteomics. Exploratory Research and Hypothesis in Medicine. Available at https://doi.org/10.14218/erhm.2024.00006, **000** (000) 000–000.

David, A., Islam, S., Tankhilevich, E., Sternberg, M.J.E., 2022. The AlphaFold Database of Protein Structures: A Biologist's Guide. Journal of Molecular Biology 434 (2), 167336. Available at https://doi.org/10.1016/j.jmb.2021.167336.

Debenedictis, E.P., Keten, S., 2019. Mechanical unfolding of alpha- and beta-helical protein motifs. Soft Matter 15 (6), 1243–1252. Available at https://doi.org/10.1039/c8sm02046a, http://pubs.rsc.org/en/journals/journal/sm.

Dhillon, S., 2023. Sunvozertinib: First Approval. Drugs 83 (17), 1629–1634. Available at https://doi.org/10.1007/s40265-023-01959-5.

Dong, X., Liu, H., Tong, T., Wu, L., Wang, J., You, T., et al., 2025. Personalized prediction of anticancer potential of non-oncology drugs through learning from genome derived molecular pathways. Npj Precision Oncology 9 (1), 36. Available at https://doi.org/10.1038/s41698-025-00813-z.

Eberhardt, J., Santos-Martins, D., Tillack, A.F., Forli, S., 2021. AutoDock Vina 1.2.0: new docking methods, expanded force field, and python bindings. Journal of Chemical Information and Modeling 61 (8), 3891–3898. Available at https://doi.org/10.1021/acs.jcim.1c00203, http://pubs.acs.org/journal/jcisd8.

Erlanson, D.A., McDowell, R.S., O'Brien, T., 2004. Fragment-based drug discovery. Journal of Medicinal Chemistry 47 (14), 3463–3482. Available at https://doi.org/10.1021/jm040031v.

Esmaeilpour, D., Hamblin, M.R., Cheng, J., Khosravi, A., Liu, J., Zarepour, A., et al., 2026. Artificial intelligence driven protein design and sustainable nanomedicine for advanced theranostics. Bioactive Materials 60, 425–455. Available at https://doi.org/10.1016/j.bioactmat.2026.01.036, https://www.sciencedirect.com/science/journal/2452199X.

Evans, W., Meslin, E.M., Kai, J., Qureshi, N., 2024. Precision medicine—are we there yet? A narrative review of precision medicine's applicability in primary care. Journal of Personalized Medicine 14 (4). Available at https://doi.org/10.3390/jpm14040418, http://www.mdpi.com/journal/jpm.

Falcon, B., Zhang, W., Murzin, A.G., Murshudov, G., Garringer, H.J., Vidal, R., et al., 2018. Structures of filaments from Pick's disease reveal a novel tau protein fold. Nature 561 (7721), 137–140. Available at https://doi.org/10.1038/s41586-018-0454-y, http://www.nature.com/nature/index.html.

FDA, & Cder, 2024. New Drug Therapy Approvals 2024. January, 1–30. www.fda.gov.

Fitzpatrick, A.W.P., Falcon, B., He, S., Murzin, A.G., Murshudov, G., Garringer, H.J., et al., 2017. Cryo-EM structures of tau filaments from Alzheimer's disease. Nature 547 (7662), 185–190. Available at https://doi.org/10.1038/nature23002, http://www.nature.com/nature/index.html.

Frueh, D.P., Goodrich, A.C., Mishra, S.H., Nichols, S.R., 2013. NMR methods for structural studies of large monomeric and multimeric proteins. Current Opinion in Structural Biology 23 (5), 734–739. Available at https://doi.org/10.1016/j.sbi.2013.06.016.

Gao, M., Zhou, H., Skolnick, J., 2015. Insights into Disease-Associated Mutations in the Human Proteome through Protein Structural Analysis. Structure 23 (7), 1362–1369. Available at https://doi.org/10.1016/j.str.2015.03.028.

Gobena, S., Admassu, B., Kinde, M.Z., Gessese, A.T., 2024. Proteomics and its current application in biomedical area: concise review. Scientific World Journal 2024. Available at https://doi.org/10.1155/2024/4454744, http://www.hindawi.com/journals/tswj/.

de Groot, N.S., Burgas, M.T., Elofsson, A., 2020. Bacteria use structural imperfect mimicry to hijack the host interactome. PLoS Computational Biology 16 (12), e1008395. Available at https://doi.org/10.1371/journal.pcbi.1008395.

Gu, Y., Zhou, Y., Xie, C., He, G., Zhang, M., 2025. Identification of key ferroptosis-related genes and therapeutic target in nasopharyngeal carcinoma. Frontiers in Genetics 16. Available at https://doi.org/10.3389/fgene.2025.1595456.

Harner, M.J., Frank, A.O., Fesik, S.W., 2013. Fragment-based drug discovery using NMR spectroscopy. Journal of Biomolecular NMR 56 (2), 65–75. Available at https://doi.org/10.1007/s10858-013-9740-z.

Hoy, S.M., 2026. Taletrectinib: First Approval. Drugs 86 (2), 249–257. Available at https://doi.org/10.1007/s40265-025-02256-z.

Huang, C., Wang, Y., Li, X., Ren, L., Zhao, J., Hu, Y., et al., 2020. Clinical features of patients infected with 2019 novel coronavirus in Wuhan, China. The Lancet 395 (10223), 497–506. Available at https://doi.org/10.1016/S0140-6736(20)30183-5.

Huang, Z., Horowitz, S., 2025. Catalyzing protein folding by chaperones. Biology 14 (10), 1450. Available at https://doi.org/10.3390/biology14101450.

Huxford, T., 2013. X-ray crystallography. Brenner's Encyclopedia of Genetics:, Second Edition. Elsevier Inc.,, United States, pp. 366–368. Available at http://doi.org/10.1016/B978-0-12-374984-0.01657-0, http://www.sciencedirect.com/science/book/9780080961569.

Höhn, A., Tramutola, A., Cascella, R., 2020. Proteostasis failure in neurodegenerative diseases: focus on oxidative stress. Oxidative Medicine and Cellular Longevity 2020. Available at https://doi.org/10.1155/2020/5497046.

Ittisoponpisan, S., David, A., 2018. Structural biology helps interpret variants of uncertain significance in genes causing endocrine and metabolic disorders. Journal of the Endocrine Society 2 (8), 842–854. Available at https://doi.org/10.1210/js.2018-00077.

Jayan, M., Vani Raju, M., Kaniyur Chandrasekaran, M., Velliyur Kanniappan, G., Panneerselvam, N., 2026. Recombinant Enzyme Therapies: A Pioneering Approach to Combat Oxidative Stress-Related Diseases. Recent Advances in Oxidative Stress Associated Chronic Diseases Volume 1. Springer Nature Singapore, pp. 323–356. Available at https://doi.org/10.1007/978-981-96-9670-3_13.

Jeffery, C.J., 2016. Protein species and moonlighting proteins: very small changes in a protein's covalent structure can change its biochemical function. Journal of Proteomics 134, 19–24. Available at https://doi.org/10.1016/j.jprot.2015.10.003, http://www.elsevier.com.

Jeyaraj, G., Rajendran, A.K., Sathishkumar, K., Almutairi, B.O., Vadivelu, A., Chokkakula, S., et al., 2025. High-resolution protein modeling through Cryo-EM and AI: current trends and future perspectives – a review. Frontiers in Molecular Biosciences 12. Available at https://doi.org/10.3389/fmolb.2025.1688455, journal.frontiersin.org/journal/molecular-biosciences.

Jin, M., Seed, R.I., Cai, G., Shing, T., Wang, L., Ito, S., et al., 2024. Dynamic allostery drives autocrine and paracrine TGF-β signaling. Cell 187 (22), 6200. Available at https://doi.org/10.1016/j.cell.2024.08.036, https://www.sciencedirect.com/science/journal/00928674.

Joerger, A.C., Fersht, A.R., 2016. The p53 pathway: origins, inactivation in cancer, and emerging therapeutic approaches. Annual Review of Biochemistry 85, 375–404. Available at https://doi.org/10.1146/annurev-biochem-060815-014710, http://arjournals.annualreviews.org/loi/biochem.

Jumper, J., Evans, R., Pritzel, A., Green, T., Figurnov, M., Ronneberger, O., et al., 2021. Highly accurate protein structure prediction with AlphaFold. Nature 596 (7873), 583–589. Available at https://doi.org/10.1038/s41586-021-03819-2, http://www.nature.com/nature/index.html.

Keam, S.J., 2024. Nogapendekin alfa Inbakicept: First Approval. Drugs 84 (7), 867–874. Available at https://doi.org/10.1007/s40265-024-02060-1.

Kent, S.B.H., 2026. Protein structure determination by racemic and quasi-racemic x-ray crystallography. Chembiochem: A European Journal of Chemical Biology 27 (4). Available at https://doi.org/10.1002/cbic.202500950, http://onlinelibrary.wiley.com/journal/10.1002/(ISSN)1439-7633.

Kulp, D.W., Schief, W.R., 2013. Advances in structure-based vaccine design. Current Opinion in Virology 3 (3), 322–331. Available at https://doi.org/10.1016/j.coviro.2013.05.010.

Kundaje, A., Meuleman, W., Ernst, J., Bilenky, M., Yen, A., Heravi-Moussavi, A., et al., 2015. Integrative analysis of 111 reference human epigenomes. Nature 518 (7539), 317–330. Available at https://doi.org/10.1038/nature14248.

Kuzmanic, A., Bowman, G.R., Juarez-Jimenez, J., Michel, J., Gervasio, F.L., 2020. Investigating cryptic binding sites by molecular dynamics simulations. Accounts of Chemical Research 53 (3), 654–661. Available at https://doi.org/10.1021/acs.accounts.9b00613, http://pubs.acs.org/journal/achre4.

Lee, A., Keam, S.J., 2020. Tislelizumab: First Approval. Drugs 80 (6), 617–624. Available at https://doi.org/10.1007/s40265-020-01286-z.

Letertre, M.P.M., Giraudeau, P., de Tullio, P., 2021. Nuclear magnetic resonance spectroscopy in clinical metabolomics and personalized medicine: current challenges and perspectives. Frontiers in Molecular Biosciences 8 (September), 1–25. Available at https://doi.org/10.3389/fmolb.2021.698337.

Li, H., Guan, Y., 2022. Asymmetric predictive relationships across histone modifications. Nature Machine Intelligence 4 (3), 288–299. Available at https://doi.org/10.1038/s42256-022-00455-x.

Listov, D., Goverde, C.A., Correia, B.E., Fleishman, S.J., 2024. Opportunities and challenges in design and optimization of protein function. Nature Reviews. Molecular Cell Biology 25 (8), 639–653. Available at https://doi.org/10.1038/s41580-024-00718-y, https://www.nature.com/molcellbio.

Long, Y., Wang, P., Lei, J., Su, B., Wei, Q., Liu, X., 2025. Mechanical signaling: molecular mechanisms, biological functions, diseases, and therapeutic targets. MedComm 6 (12). Available at https://doi.org/10.1002/mco2.70523.

Lui, G., Minnar, C.M., Soon-Shiong, P., Schlom, J., Gameiro, S.R., 2023. Exploiting an Interleukin-15 Heterodimeric Agonist (N803) for Effective Immunotherapy of Solid Malignancies. Cells 12 (12), 1611. Available at https://doi.org/10.3390/cells12121611.

Lundberg, S., & Lee, S.-I. (2017). *A Unified Approach to Interpreting Model Predictions*.

Luu, T.-D., Rusu, A.-M., Walter, V., Ripp, R., Moulinier, L., Muller, J., et al., 2012. MSV3d: database of human MisSense variants mapped to 3D protein structure. Database 2012 (0), bas018. Available at https://doi.org/10.1093/database/bas018.

Martins, A.C., Oshiro, M.Y., Schiavon, B.N., de Jesus, G.A., de la Torre, B.G., Albericio, F., 2025. Monoclonal Antibodies (mAbs) and Proteins: The Biologic Drugs Approved by the Food and Drug Administration (FDA) in 2024. Biomedicines 13 (8), 1962. Available at https://doi.org/10.3390/biomedicines13081962.

Marengo-Rowe, A.J., 2017. Structure-function relations of human hemoglobins. Baylor University Medical Center Proceedings 19 (3), 239–245. Available at https://doi.org/10.1080/08998280.2006.11928171.

Marinko, J.T., Huang, H., Penn, W.D., Capra, J.A., Schlebach, J.P., Sanders, C.R., 2019. Folding and misfolding of human membrane proteins in health and disease: from single molecules to cellular proteostasis. Chemical Reviews 119 (9), 5537–5606. Available at https://doi.org/10.1021/acs.chemrev.8b00532, http://pubs.acs.org/journal/chreay.

Matreyek, K.A., Starita, L.M., Stephany, J.J., Martin, B., Chiasson, M.A., Gray, V.E., et al., 2018. Multiplex assessment of protein variant abundance by massively parallel sequencing. Nature Genetics 50 (6), 874–882. Available at https://doi.org/10.1038/s41588-018-0122-z, http://www.nature.com/ng/index.html.

Monti, A., Vitagliano, L., Caporale, A., Ruvo, M., Doti, N., 2023. Targeting protein–protein interfaces with peptides: the contribution of chemical combinatorial peptide library approaches. International Journal of Molecular Sciences 24 (9), 7842. Available at https://doi.org/10.3390/ijms24097842.

Morris, R., Black, K.A., Stollar, E.J., 2022. Uncovering protein function: from classification to complexes. Essays in Biochemistry 66 (2), 255–285. Available at https://doi.org/10.1042/EBC20200108.

Mosca, R., Céol, A., Aloy, P., 2013. Interactome3D: adding structural details to protein networks. Nature Methods 10 (1), 47–53. Available at https://doi.org/10.1038/nmeth.2289.

Mosca, R., Tenorio-Laranga, J., Olivella, R., Alcalde, V., Céol, A., Soler-López, M., et al., 2015. dSysMap: exploring the edgetic role of disease mutations. Nature Methods 12 (3), 167–168. Available at https://doi.org/10.1038/nmeth.3289.

Mozzarelli, A.M., Simanshu, D.K., Castel, P., 2024. Functional and structural insights into RAS effector proteins. Molecular Cell 84 (15), 2807–2821. Available at https://doi.org/10.1016/j.molcel.2024.06.027.

Muenzer, J., Burton, B.K., Harmatz, P., Rajan, D.S., Jones, S.A., van den Hout, J.M.P., et al., 2026. An Intravenous Brain-Penetrant Enzyme Therapy for Mucopolysaccharidosis II. New England Journal of Medicine 394 (1), 39–50. Available at https://doi.org/10.1056/NEJMoa2508681.

Nice, E.C., 2016. From proteomics to personalized medicine: the road ahead. Expert Review of Proteomics 13 (4), 341–343. Available at https://doi.org/10.1586/14789450.2016.1158107.

Nivatya, H.K., Singh, A., Kumar, N., Sonam, Sharma, L., Singh, V., et al., 2025. Assessing molecular docking tools: understanding drug discovery and design. Future Journal of Pharmaceutical Sciences 11 (1). Available at https://doi.org/10.1186/s43094-025-00862-y.

Nolasco, S., Crimi, C., 2024. Depemokimab, the first ultra-long-acting anti-IL-5 monoclonal antibody for severe eosinophilic asthma. Med 5 (12), 1452–1455. Available at https://doi.org/10.1016/j.medj.2024.10.022.

Nussinov, R., Tsai, C.J., 2013. Allostery in disease and in drug discovery. Cell 153 (2), 293–305. Available at https://doi.org/10.1016/j.cell.2013.03.034, https://www.sciencedirect.com/journal/cell.

Nussinov, R., Zhang, M., Liu, Y., Jang, H., 2022. AlphaFold, artificial intelligence (AI), and allostery. Journal of Physical Chemistry B 126 (34), 6372–6383. Available at https://doi.org/10.1021/acs.jpcb.2c04346, http://pubs.acs.org/journal/jpcbfk.

Nussinov, R., Papin, J.A., Vakser, I., 2017. Computing the Dynamic Supramolecular Structural Proteome. PLoS Computational Biology 13 (1), e1005290. Available at https://doi.org/10.1371/journal.pcbi.1005290.

Ofoegbu, T.C., David, A., Kelley, L.A., Mezulis, S., Islam, S.A., Mersmann, S.F., et al., 2019. PhyreRisk: A Dynamic Web Application to Bridge Genomics, Proteomics and 3D Structural Data to Guide Interpretation of Human Genetic Variants. Journal of Molecular Biology 431 (13), 2460–2466. Available at https://doi.org/10.1016/j.jmb.2019.04.043.

Oliveira, J.A., Saddique, M.N., Qadri, M., Shahid, F., de Jesus, A.C.F.S., Sarmento, F.P., 2026. Treatment of Neuronal Ceroid Lipofuscinosis Type 2 with Cerliponase Alfa: A Systematic Review and Single-Arm Meta-Analysis of Two Studies. Journal of Child Neurology 41 (3), 430–436. Available at https://doi.org/10.1177/08830738251389180.

Orchard, S., Ammari, M., Aranda, B., Breuza, L., Briganti, L., Broackes-Carter, F., et al., 2014. The MIntAct project—IntAct as a common curation platform for 11 molecular interaction databases. Nucleic Acids Research 42 (D1), D358–D363. Available at https://doi.org/10.1093/nar/gkt1115.

Ostrem, J.M., Peters, U., Sos, M.L., Wells, J.A., Shokat, K.M., 2013. K-Ras(G12C) inhibitors allosterically control GTP affinity and effector interactions. Nature 503 (7477), 548–551. Available at https://doi.org/10.1038/nature12796.

Oughtred, R., Rust, J., Chang, C., Breitkreutz, B., Stark, C., Willems, A., et al., 2021. The BioGRID database: A comprehensive biomedical resource of curated protein, genetic, and chemical interactions. Protein Science 30 (1), 187–200. Available at https://doi.org/10.1002/pro.3978.

Panda, P.K., Arul, M.N., Patel, P., Verma, S.K., Luo, W., Rubahn, H.-G., et al., 2020. Structure-based drug designing and immunoinformatics approach for SARS-CoV-2. Science Advances 6 (28). Available at https://doi.org/10.1126/sciadv.abb8097.

Papadopoulos, K.P., Borazanci, E., Shaw, A.T., Katayama, R., Shimizu, Y., Zhu, V.W., et al., 2020. U.S. Phase I First-in-human Study of Taletrectinib (DS-6051b/AB-106), a ROS1/TRK Inhibitor, in Patients with Advanced Solid Tumors. Clinical Cancer Research 26 (18), 4785–4794. Available at https://doi.org/10.1158/1078-0432.CCR-20-1630.

Parenti, G., Moracci, M., Fecarotta, S., Andria, G., 2014. Pharmacological chaperone therapy for lysosomal storage diseases. Future Medicinal Chemistry 6 (9), 1031–1045. Available at https://doi.org/10.4155/fmc.14.40.

Piersimoni, L., Kastritis, P.L., Arlt, C., Sinz, A., 2022. Cross-linking mass spectrometry for investigating protein conformations and protein-protein interactions-a method for all seasons. Chemical Reviews 122 (8), 7500–7531. Available at https://doi.org/10.1021/acs.chemrev.1c00786, http://pubs.acs.org/journal/chreay.

Post, S.E., Brito, I.L., 2022. Structural insight into protein–protein interactions between intestinal microbiome and host. Current Opinion in Structural Biology 74. Available at https://doi.org/10.1016/j.sbi.2022.102354, http://www.elsevier.com/locate/sbi.

Salvati, A., Melone, V., Giordano, A., Lamberti, J., Palumbo, D., Palo, L., et al., 2025. Multi-omics based and AI-driven drug repositioning for epigenetic therapy in female malignancies. Journal of Translational Medicine 23 (1), 837. Available at https://doi.org/10.1186/s12967-025-06856-x.

Sangwan, K., Agarwal, G., 2025. Sunvozertinib: a Promising Oral EGFR Inhibitor Approved for NSCLC with EGFR Ex20ins Mutations. Investigational New Drugs 43 (5), 1177–1189. Available at https://doi.org/10.1007/s10637-025-01588-7.

Schaefer, C., Burkhard, R., 2012. Predict impact of single amino acid change upon protein structure. BMC Genomics 13, 54.

Schipilliti, F.M., Drittone, D., Mazzuca, F., La Forgia, D., Guven, D.C., Rizzo, A., 2024. Datopotamab deruxtecan: A novel antibody drug conjugate for triple-negative breast cancer. Heliyon 10 (7), e28385. Available at https://doi.org/10.1016/j.heliyon.2024.e28385.

Schreiber, S.L., 2021. The rise of molecular glues. Cell 184 (1), 3–9. Available at https://doi.org/10.1016/j.cell.2020.12.020, https://www.sciencedirect.com/science/journal/00928674.

Schweighauser, M., Shi, Y., Tarutani, A., Kametani, F., Murzin, A.G., Ghetti, B., et al., 2020. Structures of α-synuclein filaments from multiple system atrophy. Nature 585 (7825), 464–469. Available at https://doi.org/10.1038/s41586-020-2317-6, https://www.nature.com/nature/.

Sen, S., Mitra, S., Karati, D., 2026. Targeting drug resistance in colorectal cancer using peptide-based scaffolds: new frontiers in colorectal cancer therapy. Nanomedicine: Nanotechnology, Biology, and Medicine 1–13. Available at https://doi.org/10.1080/17435889.2026.2628241.

Shamsi, T.N., Athar, T., Parveen, R., Fatima, S., 2017. A review on protein misfolding, aggregation and strategies to prevent related ailments. International Journal of Biological Macromolecules 105, 993–1000. Available at https://doi.org/10.1016/j.ijbiomac.2017.07.116, www.elsevier.com/locate/ijbiomac.

Shang, S., Liu, J., Hua, F., 2022. Protein acylation: mechanisms, biological functions and therapeutic targets. Signal Transduction and Targeted Therapy 7 (1). Available at https://doi.org/10.1038/s41392-022-01245-y.

Shoemaker, S.C., Ando, N., 2015. X-rays in the cryo-EM era: structural biology's dynamic future. Biochemistry 91 (2), 165–171. Available at https://doi.org/10.1021/acs.biochem.7b01031.X-rays.

Siddikey, F., Jahan, M.I., Hormoni, Hasan, M.T., Nishi, N.J., Hasan, S.M.K., et al., 2025. Enzyme technology in the food industry: molecular mechanisms, applications, and sustainable innovations. Food Science and Nutrition 13 (9). Available at https://doi.org/10.1002/fsn3.70927, onlinelibrary.wiley.com/journal/10.1002/%28ISSN%292048-7177.

Singh, R., Lanchantin, J., Robins, G., Qi, Y., 2016. DeepChrome: deep-learning for predicting gene expression from histone modifications. Bioinformatics 32 (17), i639–i648. Available at https://doi.org/10.1093/bioinformatics/btw427.

Somme, L.B., Chouaid, C., Moinard-Butot, F., Barbe-Richaud, J.-B., Greillier, L., Schott, R., 2024. Antibody-Drug Conjugates as Novel Therapeutic Agents for Non-Small Cell Lung Carcinoma with or without Alterations in Oncogenic Drivers. BioDrugs: Clinical Immunotherapeutics, Biopharmaceuticals and Gene Therapy 38 (4), 487–497. Available at https://doi.org/10.1007/s40259-024-00660-7.

Stebbing, J., Krishnan, V., de Bono, S., Ottaviani, S., Casalini, G., Richardson, P.J., et al., 2020. Mechanism of baricitinib supports artificial intelligence-predicted testing in COVID-19 patients. EMBO Molecular Medicine 12 (8). Available at https://doi.org/10.15252/emmm.202012697.

Stein, A., Mosca, R., Aloy, P., 2011. Three-dimensional modeling of protein interactions and complexes is going 'omics. Current Opinion in Structural Biology 21 (2), 200–208. Available at https://doi.org/10.1016/j.sbi.2011.01.005.

Su, J., Yang, L., Sun, Z., Zhan, X., 2024. Personalized Drug Therapy: Innovative Concept Guided With Proteoformics. Molecular and Cellular Proteomics 23 (3), 100737. Available at https://doi.org/10.1016/j.mcpro.2024.100737.

Szklarczyk, D., Gable, A.L., Nastou, K.C., Lyon, D., Kirsch, R., Pyysalo, S., et al., 2021. The STRING database in 2021: customizable protein–protein networks, and functional characterization of user-uploaded gene/measurement sets. Nucleic Acids Research 49 (D1), D605–D612. Available at https://doi.org/10.1093/nar/gkaa1074.

Stachowski, T.R., Fischer, M., 2026. Unfreezing structural biology for drug discovery. Nature Chemical Biology 22 (2), 169–179. Available at https://doi.org/10.1038/s41589-025-02047-3, https://www.nature.com/nchembio.

Stein, A., Fowler, D.M., Hartmann-Petersen, R., Lindorff-Larsen, K., 2019. Biophysical and mechanistic models for disease-causing protein variants. Trends in Biochemical Sciences 44 (7), 575–588. Available at https://doi.org/10.1016/j.tibs.2019.01.003, www.elsevier.com/locate/tibs.

Tanaka, N., Lin, J.J., Li, C., Ryan, M.B., Zhang, J., Kiedrowski, L.A., et al., 2021. Clinical acquired resistance to krasg12c inhibition through a novel kras switch-ii pocket mutation and polyclonal alterations converging on ras–mapk reactivation. Cancer Discovery 11 (8), 1913–1922. Available at https://doi.org/10.1158/2159-8290.CD-21-0365, https://cancerdiscovery.aacrjournals.org/content/11/8/1913.full.pdf.

Tang, C., Wang, L., Zang, L., Wang, Q., Qi, D., Dai, Z., 2023. On-demand biomanufacturing through synthetic biology approach. Materials Today Bio 18, 100518. Available at https://doi.org/10.1016/j.mtbio.2022.100518.

Tian, M., Wang, N., Dou, Z., Song, X., Zhang, X., 2024. [EGFR Exon 20 Insertion Mutation: Research Status and New Treatment Strategies]. Zhongguo Fei Ai Za Zhi Chinese Journal of Lung Cancer 27 (8), 579–592. Available at https://doi.org/10.3779/j.issn.1009-3419.2024.106.19.

Todd, T.W., Islam, N.N., Cook, C.N., Caulfield, T.R., Petrucelli, L., 2024. Cryo-EM structures of pathogenic fibrils and their impact on neurodegenerative disease research. Neuron 112 (14), 2269–2288. Available at https://doi.org/10.1016/j.neuron.2024.05.012, https://www.sciencedirect.com/science/journal/08966273.

Del Toro Runzer, C., Balmayor, E.R., van Griensven, M., 2026. Biologics for bone regeneration: advances in cell, protein, gene, and mRNA therapies. Bone Research 14 (1), 5. Available at https://doi.org/10.1038/s41413-025-00487-0.

Tornesello, M., 2024. TP53 mutations in cancer: molecular features and therapeutic opportunities (Review). International Journal of Molecular Medicine 55 (1). Available at https://doi.org/10.3892/ijmm.2024.5448.

Vakser, I.A., 2014. Protein-Protein Docking: From Interaction to Interactome. Biophysical Journal 107 (8), 1785–1793. Available at https://doi.org/10.1016/j.bpj.2014.08.033.

Vallat, B., Rose, Y., Piehl, D. W., Duarte, J. M., Bittrich, S., Bi, C., Segura, J., Zalevsky, A., Sekharan, M. R., Webb, B. M., Bhikadiya, C., Voigt, M., Henry, J., Chao, H., Pingale, A., Raymond, D., Sagendorf, J., Peisach, E., Feng, Z., … Burley, S. K. (2026). RCSB Protein Data Bank: Delivering integrative structures alongside experimental structures and computed structure models. Nucleic Acids Research, 54(D1), D489–D498. Available at https://doi.org/10.1093/nar/gkaf1187.

Varadi, M., Anyango, S., Armstrong, D., Berrisford, J., Choudhary, P., Deshpande, M., et al., 2022. PDBe-KB: collaboratively defining the biological context of structural data. Nucleic Acids Research 50 (D1), D534–D542. Available at https://doi.org/10.1093/nar/gkab988.

Vishweshwaraiah, Y.L., Chen, J., Dokholyan, N.V., 2021. Engineering an allosteric control of protein function. The Journal of Physical Chemistry. B 125 (7), 1806–1814. Available at https://doi.org/10.1021/acs.jpcb.0c11640.

Vitorino, R., 2024. Transforming clinical research: the power of high-throughput omics integration. Proteomes 12 (3), 25. Available at https://doi.org/10.3390/proteomes12030025.

Walls, A.C., Park, Y.J., Tortorici, M.A., Wall, A., McGuire, A.T., Veesler, D., 2020. Structure, function, and antigenicity of the SARS-CoV-2 spike glycoprotein. Cell 181 (2), 281. Available at https://doi.org/10.1016/j.cell.2020.02.058, https://www.sciencedirect.com/journal/cell.

Wang, J., Luttrell, J., Zhang, N., Khan, S., Shi, N., Wang, M.X., et al., 2016. Exploring human diseases and biological mechanisms by protein structure prediction and modeling. In: In: Shen, In. B., Tang, H., Jiang, X. (Eds.), Translational Biomedical Informatics 939. Springer Singapore, pp. 39–61 https://doi.org/10.1007/978-981-10-1503-8_3.

Wang, K., Zhou, R., Li, Y., Li, M., 2021. DeepDTAF: a deep learning method to predict protein–ligand binding affinity. Briefings in Bioinformatics 22 (5). Available at https://doi.org/10.1093/bib/bbab072.

Westergaard, L., Alifrangis, L., Buckley, S.T., Coester, H.V., Klitgaard, T., Kristensen, N.R., et al., 2024. Pharmacokinetic Properties of a Once-Weekly Fixed-Ratio Combination of Insulin Icodec and Semaglutide Compared with Separate Administration of Each Component in Individuals with Type 2 Diabetes Mellitus. Clinical Drug Investigation 44 (11), 849–861. Available at https://doi.org/10.1007/s40261-024-01405-8.

Xu, Z., Ren, F., Wang, P., Cao, J., Tan, C., Ma, D., et al., 2025. A generative AI-discovered TNIK inhibitor for idiopathic pulmonary fibrosis: a randomized phase 2a trial. Nature Medicine 31 (8), 2602–2610. Available at https://doi.org/10.1038/s41591-025-03743-2.

Yu, M., Cui, W., Hu, Y., Song, W., Wang, Z., Chen, W., et al., 2025. Ferroptosis research based on bibliometric and visual analysis: mechanism exploration and clinical application prospects in gastric cancer, prostate cancer, leukemia, and brain tumors. Frontiers in Medicine 12. Available at https://doi.org/10.3389/fmed.2025.1640497.

Zhang, T., Song, X., Xu, L., Ma, J., Zhang, Y., Gong, W., et al., 2018. The binding of an anti-PD-1 antibody to FcγRI has a profound impact on its biological functions. Cancer Immunology, Immunotherapy 67 (7), 1079–1090. Available at https://doi.org/10.1007/s00262-018-2160-x.

Zhang, Q., Liu, Y., Zhang, J., Dong, H., Pan, X., Zhang, J., 2026. Development and mechanistic evaluation of cathepsin B-activatable peptide-drug conjugate PROTACs for targeted protein degradation in malignant tumors: advancing precision protein degradation therapeutics in oncology. Small (Weinheim an der Bergstrasse, Germany) 22 (4). Available at https://doi.org/10.1002/smll.202506562.

Zhou, H.X., Pang, X., 2018. Electrostatic interactions in protein structure, folding, binding, and condensation. Chemical Reviews 118 (4), 1691–1741. Available at https://doi.org/10.1021/acs.chemrev.7b00305, http://pubs.acs.org/journal/chreay.

CHAPTER 20

Future trends in structural biology

Saleha Anwar[1], Sumaiya Khan[2], Sahar Afzal Khan[3], Khuzin Dinislam[4], Suhel Parvez[1], and Anas Shamsi[5]
[1]Department of Toxicology, School of Chemical and Life Sciences, Jamia Hamdard, New Delhi, Delhi, India,
[2]Centre for Interdisciplinary Research in Basic Sciences, Jamia Millia Islamia, New Delhi, India,
[3]Department of Biotechnology, Jamia Hamdard, New Delhi, Delhi, India,
[4]Department of General Chemistry, Bashkir State Medical University, Ufa, Republic of Bashkortostan, Russia,
[5]Centre of Medical and Bio-allied Health Sciences Research, Ajman University, Ajman, United Arab Emirates

20.1 Introduction

Structural biology emerged and evolved in the 1950s, shortly after the discovery of the double helical structure of DNA by James Watson and Francis Crick, which revolutionized the field of structural biology and genetics. In the early 1950s, Pauling and Corey established the fundamentals of polypeptide chain organization within protein molecules in the form of α-helices and β-sheets. Half a decade later, the structure of the first protein was documented, that is, myoglobin showed multiple helices in its structure, as indicated earlier by Pauling and Corey. With advancements in biophysics and the development of high-resolution structures of hemoglobin (Hb), the oxygen transportation and role of Hb in human physiology became much more understandable (Finch et al., 1973). In fact, this work also produced the first understanding of the molecular causes of a disease associated with an abnormal protein conformation, sickle cell disease (Finch et al., 1973). The field of structural biology and biophysics is now crucial, given the rising incidence of neurodegenerative diseases, protein misfolding, etc. (Arar et al., 2025; Nedelsky and Taylor, 2019; Nuka, 2025). After lysozyme's structure was determined in the middle of the 1960s, the first atomic-resolution image of an enzyme was obtained, and studies of its complexes with substrate analogs quickly followed, culminating in a suggestion for its catalytic mechanism. The enormous promise of the developing science of structural biology was recognized because of these early demonstrations that an understanding of molecular structure may so clearly expose the nature of biological activity. The development of Bragg's rule in 1913 allowed for the use of X-ray diffraction to determine structures, initially those of simple inorganic salts, which paved the way for all these revolutionary developments (Bragg and Bragg, 1913). The structure of lysozyme was discovered in 1965, demonstrating the evolution of X-ray crystallography from a physical technique to a potent biological instrument.

The revolutionary power of X-ray diffraction was shown by Lawrence Bragg, who directly evaluated electron density maps to design the first enzyme structure decades

Advances in Structural Biology
DOI: https://doi.org/10.1016/B978-0-443-44988-8.00017-4

earlier. This significant event marked the start of the broad application of X-ray crystallography to proteins and nucleic acids, which deepened our understanding of basic biological processes and allowed for the determination of thousands of more complicated biological structures (Bragg, 1913). X-ray crystallography, although it was developed as a technique in physics, evolved progressively as a vital tool in biology, enabling researchers to discover the intricate structures of biological molecules. This structural understanding has provided profound insight into the operation of basic biological systems. Over time, the technique has uncovered the structures of ever-more complex systems, most notably the ribosome. These high-resolution structures have revolutionized our knowledge of protein synthesis and the workings of antibiotics, directly facilitating the development of logical drugs to combat pathogens that are resistant to antibiotics (Dobson, 2019; Lin et al., 2018).

The extensive use of physics in biology gave rise to the field of biophysics. X-ray crystallography emerged as the first major technique employed in biology, followed by some other major techniques such as nuclear magnetic resonance (NMR) spectroscopy, mass spectrometry (MS), and optical methods like circular dichroism and fluorescence spectroscopy. Over time, these techniques have undergone dramatic technological and methodological improvements, and each advance has revolutionized our ability to study biological structure (Elmlund and Elmlund, 2015; Mandala et al., 2018; Tycko, 2011). Protein structure analysis with X-ray diffraction helped in recognizing that proteins that tend to have a high level of similarity in their sequences eventually have similarities in their structures. The prediction of a model of α-lactalbumin from that of lysozyme showed that the sequence shows high similarity; however, the physiological functions of both proteins are entirely different, which was an evident example of molecular homology modeling (Browne et al., 1969). This approach led to expansions in the Protein Data Bank and UniProt's protein sequences, assisting in a better understanding of the associations in structure and protein sequences.

Furthermore, the application of computational techniques such as structure prediction and molecular dynamics simulations (MDS) provides a detailed insight into the properties and interactions of the macromolecules that are difficult to predict by experimental models. The advancements in biophysics can be attributed to the development and enhancements of computational software and techniques, providing high precision (Dror et al., 2012). Recent advancements in computational biology have aided in the prediction of structural and functional properties of proteins and protein complexes. Advancements in in silico technologies and algorithms, including machine learning (ML), have enabled protein structure prediction by using multiomics databases (Lin et al., 2023; Tunyasuvunakool et al., 2021). The field of biophysics has contributed significantly to structure-based drug design (SBDD). SBDD utilizes the three-dimensional structure of macromolecules, including nucleic acids, proteins, and other targets related to disorders and diseases. SBDD utilizes the noncovalent interactions between the biological target and the ligand designed or predicted with computational biology techniques. Biophysical techniques such as fluorescence spectroscopy, isothermal titration calorimetry, and circular dichroism are utilized to further validate the results (Ahmadi et al., 2025; Shahwan et al., 2023; Shamsi et al., 2024). In this book chapter, we aim to integrate the past and future of tools in structural biology, advancements in structural biology, and therapeutic outcomes.

20.2 Advancements in hybrid and integrative structural biology

Modern cryo-electron microscopy (cryo-EM) is powered by a combination of sophisticated image processing and high-resolution cryo-electron microscopes with direct electron detectors. Because they convert 2D microscope images into intricate 3D models, image processing and 3D reconstruction are crucial. The 2D projection images created by a transmission electron microscope capture every interior detail of the three-dimensional object from various perspectives. By integrating these various 2D images using back-projection, a complete 3D representation of the item may be recreated and is applicable to both cryo-electron tomography (cryo-ET) and single-particle cryo-EM (Orlov et al., 2017). Some of the recent examples of structures revealed by cryo-EM includes; structures of yeast and human CMG replisomes stalled at a DNA G-quadruplex (G4), showing how the DNA replication engine is blocked and revealing a new "helical inchworm" translocation mechanism, and cryo-EM structures of the Smc5/6 complex (Batra et al., 2025; Li et al., 2024).

20.2.1 Cryo-EM

Cryogenic electron microscopy, also known as cryo-EM, is one of the most disruptive technologies in current structural biology, and a new way in which we are able to visualize macromolecular structures, a new way in which we are able to study the intricate dance of biological complexes on a scale that has been previously unavailable to us (Fukuda et al., 2023). Compared to conventional structural methods, such as X-ray crystallography, which requires highly ordered crystals, or NMR spectroscopy, which is limited to the size of the molecule, cryo-EM allows the determination of high-resolution structures even of heterogeneous, flexible, or large biological complexes under conditions that approach their native environment (Nogales and Mahamid, 2024). Fig. 20.1 illustrates some of the tools in structural biology. Cryo-EM has undergone a boom period in technology in the past ten years, which we commonly call the resolution revolution (Glaeser, 2021). This has been accelerated by new developments in instrumentation, the introduction of new advanced electron detectors, the development of improved image-processing complexities, and improved protocols in sample preparation. The essence of cryo-EM is the quick freezing of wet biological

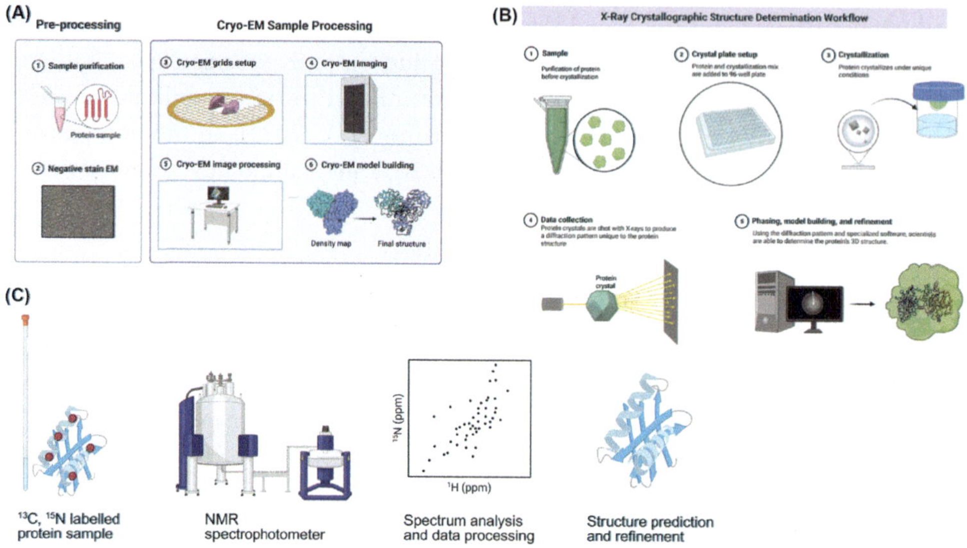

Figure 20.1 Illustration of major approaches in structural biology for the determination of protein structure. (A) Cryo-electron microscopy workflow; (B) X-ray crystallography workflow for protein purification; (C) Nuclear magnetic resonance spectroscopy process.

samples, which are then maintained in a coating of noncrystalline amorphous ice, thus retaining the sample in a near native state without the artifacts created by freeze fracture or fixatives that could alter the sample (Weissenberger et al., 2021). The vitrification is commonly attained by plunge freezing, in which an EM grid holding the sample is promptly immersed in a bath of liquid ethane cooled swiftly by liquid nitrogen, and it is then directly transferred to vitrified ice (Weissenberger et al., 2021). This very rapid cooling regime can essentially eliminate the presence of harmful ice crystals, which freeze biomolecules in their close to native structures, essentially providing us with snapshots of their structural landscapes without the necessity of chemical fixation or staining (Glaeser, 2021).

After being subjected to vitrification, samples are treated under a transmission electron microscope, heating to cryogenic temperatures (usually about—180°C) (Glaeser, 2021). The sample is illuminated by an electron beam, and the scattered electrons are registered using a detector that produces two-dimensional projection images (Weissenberger et al., 2021). Due to the necessity to keep the dose of electrons extremely low in order to reduce radiation effects, such images are inherently noisy (Weissenberger et al., 2021). Noise notwithstanding, computational algorithms can match, sort, and mean thousands to millions of particle images to recreate a three-dimensional electron density map of the biomolecule (Nogales and Mahamid, 2024). Such a strategy has made cryo-EM a revolutionary instrument, shifting the field out of stasis, as a structural snapshot of a structure into a mechanistic view of macromolecular systems. Cryo-EM has become an important part of all biomedical sciences by enabling the structural analysis of complexes that are key to processes such as transcription, translation, viral infection, immunological recognition, and intracellular transport (Nogales and Mahamid, 2024). The greatest contribution of it is the capability to visualize molecules in various functional states, including enzymes in motion, assembly intermediates, and conformational changes that underlie biological regulation (Weissenberger et al., 2021).

20.2.2 Cryo-electron tomography

Cryo-electron tomography, also known as cryo-ET, has become one of the most popular methodologies in structural biology, especially the imaging of macromolecular structures in their natural cellular environment (Young and Villa, 2023). Unlike single-particle cryo-EM, which requires purified, isolated specimens, cryo-ET provides a true three-dimensional representation of cellular structures, organelles, complexes, and other higher-order supramolecular structures without using crystallization, staining, or fixation methods (Nogales and Mahamid, 2024). The application of cryogenic preservation with electron tomography reduces the separation in resolution between high-resolution structural biology and conventional cellular imaging, which makes cryo-ET an effective approach to provide nanoscale information about the living system in its natural state (Premaraj et al., 2025). Cryo-ET relies on a tilt series of images, in which a series of two-dimensional images is captured as the specimen is gradually rotated in the electron microscope (Premaraj et al., 2025). The specimen is tilted between -60 and +60 degrees typically, but in many systems, a wider range of tilt is allowed (Martone et al., 2000). Every tilted picture provides an image of the specimen as projected in a particular orientation (Martone et al., 2000). These projections are then computationally aligned and reassembled to produce a three-dimensional tomogram, and thus allow visualization of cellular structure in a near-native vitrified state (Martone et al., 2000). Since cryo-ET does not use the averaging methods used by single-particle examination, it is ideally suitable for pleomorphic and heterogeneous samples or when they are embedded within complex surroundings (Khanna, 2024). This method maintains both the molecular context and spatial distribution of the native structure, making it an essential method of investigating structures that cannot be purified or isolated without altering their activity (Khanna, 2024). Cryo-ET opens the perspectives of structural biology to the cellular and tissue levels, whereby three-dimensional visualization of macromolecular assemblies in their natural environment are being offered (Khanna, 2024). The unique combination of preservation as found in the native state with volumetric fixation, with compatibility with subtomogram averaging as well, makes it a necessity in the elucidation of biological functioning in situ. With improved detector technology, automation, specimen preparation, and computational analysis technologies, cryo-ET can become a pillar of future integrative and cellular structural biology (Khanna, 2024).

20.2.3 Time-resolved imaging

Time-resolved imaging is a game-changing technique in structural biology, capable of visualizing dynamic molecular events at biologically relevant time scales (Schmidt, 2023). Unlike conventional structural methods (including X-ray crystallography, NMR spectroscopy, or classical cryo-EM) that provide essentially static information about cellular macromolecular complexes, time-resolved methods use transient intermediates and conformational pathways to elucidate cellular processes at the underlying mechanisms (Schmidt, 2023). This capability is essential to a mechanistic explanation of enzymatic catalysis, dynamics of motor proteins, ligand recognition, protein folding dynamics, and the dynamics of assembling macromolecules (Poddar et al., 2022). These approaches create a true molecular movie of structural change by coordinating rapid reaction initiation with either ultrafast vitrification or nanosecond-resolution X-ray or electron irradiation, thereby increasing our ability to interrogate dynamic biology at an atomic or near-atomic scale (Poddar et al., 2022).

The outline of time-resolved structural methods lies on two elements that form its basis:

- Accurate competence of a biological process, which is tightly regulated by time.
- Freezing and fixing of the molecular state at time intervals.

Fast immobilization of intermediates is necessary once a reaction has been initiated.

Time-resolved cryo-EM uses ultrafast plunge freezing to freeze molecular motion within milliseconds, and therefore maintains transient states prior to decay (Glaeser, 2021). XFEL-based methods use femtosecond X-ray pulses to freeze the positions of molecules using the diffraction before destruction principle (Tschentscher et al., 2017). Combined, they can offer unprecedented insight into the structural foundation of molecular dynamics, a paradigm shift in structural biology: now scientists can study molecular processes as dynamic processes instead of giving them snapshots. With advances in cryo-EM, cryo-ET, XFEL crystallography, and single-molecule imaging, scientists can now measure reaction intermediates, conformational transitions, and cellular events with much greater temporal and spatial resolution than ever before (Banari et al., 2025). With the development of instrumentation, the techniques of computation, and integrative workflows, time-resolved imaging will take center stage in understanding the dynamic structure of life (Van Den Bedem and Fraser, 2015). X-ray crystallography has been a classic component of structural biology, providing atomic-scale structures of macromolecular assemblies, nucleic acids, and proteins

(Muruganantham and Veerabathiran, 2024). Nevertheless, classical crystallography is constrained by the size of large crystals and requires large, well-ordered crystals, radiation damage, crystal-packing effects, and failure to measure molecular dynamics (Carvalho et al., 2009). The past ten years have witnessed the introduction of next-generation crystallographic methods, such as micro-crystallography, serial crystallography, and XFEL-based methods, that have dramatically broadened the application of diffraction imaging (Sui et al., 2016). These developments have allowed structural determination of microcrystals and nanocrystals, visualization of transient intermediates, minimization of radiation-induced damage, and structural dynamics at femtosecond scales (Martin-Garcia et al., 2016).

20.2.4 Limitations and principles of classical crystallography

Traditional crystallography is based on exposing the crystals of macromolecules to synchrotron X-rays (Martin-Garcia et al., 2016). The radiation damage caused by repeated irradiation is always present, thus limiting the amount of radiation and, thus the achievable resolution. Further, there is a plethora of biological systems, such as membrane proteins, large complexes, and natural assemblies that are already flexible, which are impossible to crystallization even in high volumes to facilitate diffraction (Muruganantham and Veerabathiran, 2024). The thick crystal lattice often enforces some packing rules, limiting the number of molecular conformations, and hiding the dynamics or other physiologically significant states.

20.2.5 Next-generation crystallography and X-ray free-electron laser-based approaches

Next-generation crystallography represents a paradigm change to the traditional practice of X-ray crystallography, with innovative beamlines, detectors, specimen preparation methods, and data-collection schemes that all overcome many of the long-standing limitations of the traditional methods (Muruganantham and Veerabathiran, 2024). Historically, structural biology has relied on the growth of large, well-ordered crystals and has tolerated the radiation damage that is inevitably received when long exposures are made. Since the introduction of micro-focused beamlines, serial data-collection platforms, X-ray free-electron laser (XFEL), and advanced computational pipelines, the field has way more capabilities in terms of versatility and responsiveness to time (Muruganantham and Veerabathiran, 2024). These state-of-the-art techniques allow the structure to be elucidated using micro- and nano-crystalline samples, make measurements at ambient temperature without changing the native physiological conformations, and perform time-resolved studies to understand the mechanistic mechanism of biomolecular activity (Carvalho et al., 2009).

20.2.6 Disadvantages of classical crystallography

- In classical measurements, a macromolecular crystal is placed under the synchrotron radiation whilst rotating, and patterns of diffraction are later combined to produce a complete set of data. Despite the impressive achievements that this strategy has brought, it is being crippled by several inherent drawbacks.
- Most proteins only form microcrystals or disordered samples that cannot be utilized using rotation-based methods.
- Prolonged exposure to X-rays will cause cumulative structural perturbations and limit the dose that can be administered, and in some cases, high-resolution imaging may not be possible.
- Due to the rigid lattice packing available with a crystal, accessible conformational states can be limited, inhibiting the study of dynamic biological processes.
- Single-crystal data acquisition is time-consuming when it is required, especially when it is difficult to grow crystals or when these crystals are low-yielding.
- To overcome these stumbling blocks, next-generation crystallography uses novel instrumentation, scalable sample-delivery plans, and novel data-processing algorithms.

20.2.7 Serial synchrotron crystallography

Serial crystallography is a paradigm shift in the representation of the X-ray data. Instead of having to rotate a single crystal, a large number of microcrystals are placed individually and exposed to brief exposure events (Gati et al., 2014). Throughout the series of snapshots, a partial dataset is provided, and when these snapshots are combined, they provide complete structural data (Gati et al., 2014).

20.2.7.1 Key features of serial synchrotron crystallography

- Less radiation damage: Each crystal is penetrated by the beam once or a few times.
- Room temperature data collection: In this modality, scientists are allowed to observe biologically relevant conformational states and dynamic fluctuations.
- High throughput: Suited appropriately to large-scale screening programs, ligand-soaking experiments, and fragment-based drug discovery.

20.2.7.2 Scientific impact of serial synchrotron crystallography

Serial synchrotron crystallography (SSX) has preempted the groundbreaking developments, such as:

- Structural determination of G-protein coupled receptors and other difficult-to-determine membrane proteins.

- Conformational ensembles are observed, which are suppressed at cryogenic temperatures when crystals are held at cryogenic temperatures.
- Exploration of radiation-based metalloenzymes and redox-based proteins.
- SSX is also capable of providing a path to truly time-resolved crystallography at synchrotron sources, and this is accomplished in pump-probe activation schemes and through rapid mixing technologies.

20.2.8 Time-resolved crystallography and dynamic studies

Next-generation crystallography is no longer limited to the field of static imaging but has moved into the area of dynamic structural biology (Gruner and Lattman, 2015). Time-resolved crystallography allows the observer to measure changes in structure in real-time in biochemically relevant reactions, over a range of microseconds to milliseconds (Brändén and Neutze, 2021). This can be used to measure transient reaction intermediates by recording diffraction data with controlled time delays among them, and build up molecular movies that define catalysis, protein folding or conformational changes (Brändén and Neutze, 2021). Though time-resolved synchrotron crystallography usually provides temporal resolution in the microsecond-to-millisecond range, the mechanistic information it provides cannot be recreated from snapshots alone (Brändén and Neutze, 2021). Next-generation crystallography has been enabled by advances in detector technology and beamline instrumentation (Singhapong et al., 2024). Modern synchrotrons combine high-quality optics (including Kirkpatrick-Baez mirrors and multilayer devices) to achieve nanometer spatial resolution in the beam, thereby enhancing diffraction signal intensity even when analyzing nanoscale crystal specimens (Singhapong et al., 2024). Robot arms, automated loop-mounting systems, and software-controlled positioning mechanisms improve reproducibility and reduce beamtime waste (Galchenkova, 2024). All these innovations significantly enhance throughput, resolving, and the quality of data. Massive-scale serial experiments produce terabytes of data, which require more complex computational pipelines. Software packages such as DIALS, CrystFEL, XDS, and the cctbx library have become invaluable for analyzing data from both micro-crystallography and serial crystallography projects (Galchenkova, 2024). The next generation of crystallography has radically transformed the structural biology landscape by breaking the dependence on large, perfect crystals and through its ability to probe larger and more complex, as well as dynamic, systems (Zheng et al., 2015). It is a transitional step between traditional crystallographic activity and ultrafast XFEL-based imaging by combining serial data collection, microfocus beamlines, improved detectors, and advanced computational pipelines (Van Thor, 2019). With constant upgrades in synchrotron facilities, automation, and the development of the hybrid design of an experiment, next-generation crystallography will continue to be the focus of high-resolution structural determination and mechanistic discovery (Van Thor, 2019).

20.2.9 X-ray free-electron laser-based approaches

The XFEL has transformed the field of present-day structural biology by enabling ultrafast, high-brilliance X-ray experiments that are no longer bound by the limits of traditional crystallography (Adams et al., 2019). XFEL radiation is very short, coherent jolts of femtosecond length, which allows gathering of diffraction information before the radiation damage appears, the phenomenon known as diffraction before destruction (Adams et al., 2019). The given attribute has allowed obtaining atomic-resolution structures of micro and nanocrystals, operative measurements at ambient temperature, as well as, most importantly, the direct observation of ultrafast molecular dynamics (Pfau and Eisebitt, 2023). Thus XFEL-based methodologies have extended the scope of structural biology to include dynamic and nonequilibrium processes previously unreachable (Pfau and Eisebitt, 2023).

In XFELs, X-ray pulses are produced by using accelerated electrons generated by high-energy electron beams made to oscillate in undulators, which are arrays of magnetic elements that force electrons into oscillatory paths (Geloni et al., 2017). The emission of the coherent X-rays is induced by this motion and enhanced by self-amplified spontaneous emission. XFELs are highly applicable in the study of radiation-sensitive samples, rapid reactions, as well as small crystals (Johansson et al., 2017). The largest XFEL facilities include the Linac Coherent Light Source, based in the United States, the SACLA, located in Japan, the European XFEL in Germany, PAL-XFEL in South Korea, and SwissFEL in Switzerland (Johansson et al., 2017). XFEL strategies are a paradigm shift in structural biology (Gati et al., 2014). XFELs make a difference by combining the methods of serial delivery of ultrabright femtosecond X-ray pulses and pump-probe methods, and atomic precision in the study of biological dynamics becomes achievable (Gati et al., 2014). These technologies make it possible to visualize reaction intermediates, enzymatic pathways, and molecular movements which were previously experimentally inaccessible (Liu and Spence, 2016). With the development of beamline facilities, automation, and computational strategies, XFEL methods will take a central place in the new generation of integrative and dynamic structural biology (Liu and Spence, 2016).

20.2.10 In-cell and solid-state nuclear magnetic resonance innovations

NMR spectroscopy is an essentially inseparable component of the methodological foundation for molecular structure, dynamics, and interactions at atomic scales (Lambert et al., 2019). Particularly, solution-state NMR has played a central role in questioning the proteins in physiological conditions; however, it is inherently limited by spectral multiplicity and size effects as well as the inability to place the proteins in their native cellular microenvironment (Kovermann et al., 2016). During the last ten years, some interesting methodological advances have been made, such as in-cell NMR and solid-state NMR (ssNMR), which have significantly expanded the scope of NMR beyond these traditional restrictions (Emwas et al., 2020). These breakthroughs make possible the study of biomolecules in crowded cell environments, systems embedded in membranes, large complexes, and material systems that cannot be studied by crystallography or conventional solution NMR (Emwas et al., 2020). Together, they indicate a revolutionary change in high-resolution structural analysis in physiologically relevant conditions in systems considered intractable before.

In-cell NMR provides the direct imaging of biomolecules in live cells, thus providing structural and dynamic data in a truly natural environment (Emwas et al., 2020). This method overcomes a significant challenge in structural biology, the difficulty of studying molecular behavior in the very crowded and heterogeneous internal environment of the cell, where interactions, posttranslational modifications (PTMs), and conformational equilibria can be significantly divergent (Zhao et al., 2022). The basis of in-cell NMR rests on the use of isotopically labeled proteins (typically, 15 N and/or 13 C) in intact cells (through either potential expression systems or direct delivery) and the NMR spectrophotometry of intact cells (Zhao et al., 2022). Since it is the labeled protein that provides the resulting signals, in-cell NMR provides residue-level information even in the complicated cellular environment (Luchinat and Banci, 2023).

Solid-state NMR (ssNMR) also allows structural studies of proteins in noncrystalline, immobilized, or membrane-bound conditions (Ubbink et al., 2018). Within ten years, ssNMR has developed into a high-resolution analysis platform that is able to provide atomic-level structural data of systems between amyloid fibrils and membrane channels, as well as intact viral assemblies (Hong, 2025). Solution NMR, as compared to ssNMR, requires samples that are not tumbling rapidly (Hong, 2025). Anisotropic interactions are averaged to produce high-resolution spectra using magic-angle spinning, a rotation of the sample by some angle of about 54.7 degrees of the magnetic field. These breakthroughs in fast MAS (~60 kHz) and ultrafast MAS (~100 kHz) have provided very high resolutions and sensitivities (Nishiyama et al., 2023). DNP-enhanced ssNMR at the cellular level provides unparalleled sensitivity, enabling atomic-level insights in the natural biological environment (Nishiyama et al., 2023). The intersection of in-cell and solid-state NMR has significantly expanded the domain of NMR spectroscopy, making it easy to study proteins in native, crowded, or immobilized conditions as well as giving access to atomic-level information that was previously unavailable to experimentation (Narasimhan et al., 2020). Together, these methods comprise a vast set of tools to study biomolecular organization, dynamics, interactions, and activity under physiologically relevant conditions, in addition, to cryo-

EM, XFEL crystallography, and solution NMR, and are a core part of integrative structural biology today (Table 20.1).

20.3 Mass spectrometry

MS has evolved as a method of protein identification and quantification to one of the most powerful methods of investigating macromolecular structure, dynamics, interactions, and assembly (Konermann et al., 2014). Current MS-based structural biology uses more powerful ionization techniques, native MS, cross-linking schemes, hydrogen-deuterium exchange, ion mobility, and hybrid MS-computational techniques to reveal the structural details complementary and extending those of cryo-EM, crystallography, and NMR (Konermann et al., 2014). Its flexibility, sensitivity, and ability to study heterogeneous, transient, or dynamic systems make MS an absolute necessity for the structural analysis of biomolecular assemblies. Standard MS is used to determine the mass-to-charge ($m\,z^{-1}$) ratio of the ionized molecules and gives the mass information with a high level of accuracy (Konermann et al., 2014). Structural MS uses other methods that indicate protein folding, subunit assembly as complexes, and interactions between the ligand and the cofactor targets (Chait et al., 2016). The following methods are particularly helpful in the study. Structural MS has no restriction to crystallization, molecular size, or solubility, which means that it is an essential tool of integrative structural biology (Chait et al., 2016).

20.3.1 Native mass spectrometry

Native MS maintains noncovalent interactions upon ionization, and it allows the analysis of intact protein assemblies, membrane assemblies, and ribonucleoproteins in the gas phase (Heck, 2008). With mild conditions (e.g., ammonium acetate buffers, controlled voltages), proteins maintain their native like conformation and provide information about composition, stoichiometry, and topology. Native MS plays a key role in the characterization of ribosomes, proteasomes, viral capsids, and receptor complexes (Heck, 2008). IM-MS ion mobility-mass spectrometry technique separates ions based on size, shape, and charge by determining their drift with a gas in an electric field (Christofi and Barran, 2023). In combination with MS, IM-MS gives collision cross sections that are indicative of the three-dimensional structure of biomolecules (Christofi and Barran, 2023). Hydrogen-Deuterium Exchange Mass Spectrometry (HDX-MS) quantifies the dynamics of proteins, solvent accessibility, and changes in conformation by observing the exchange of backbone amide hydrogen and deuterium in D_2O. Areas of high flexibility or solvent availability have fast uptake of deuterium, whereas buried or hydrogen-bonded areas are more slowly exchanged. MS is now a ubiquitous part of structural biology that presents an unrivaled level of flexibility in the

Table 20.1 Comparison of classical and next-generation structural biology techniques.

Technique	Sample requirements	Resolution	Advantages	Limitations	Applications	References
X-ray crystallography (classical)	Large, well-ordered crystals	Atomic resolution (static structures)	High resolution, well-established	Radiation damage, crystal-packing effects, poor dynamics	Enzymes, stable protein complexes	Carvalho et al. (2009)
Cryo-EM (single particle)	Purified macromolecules, vitrified	Near-atomic to atomic resolution	No crystallization, captures multiple conformations	Limited for very small proteins, beam damage	Large complexes, ribosomes, viruses	Weissenberger et al. (2021)
Cryo-Electron Tomography (Cryo-ET)	Intact cells or cellular sections	Nanometer-scale, 3D cellular context	In situ visualization, native environment	Lower resolution than single-particle EM	Cellular assemblies, organelles	Young and Villa (2023)
Time-resolved Cryo-EM	Rapidly triggered reactions	Millisecond structural intermediates	Captures transient states	Technical complexity	Enzyme dynamics, conformational changes	Brändén and Neutze (2021)
Serial synchrotron crystallography (SSX)	Micro-/nano-crystals	High resolution, room temperature	Reduced radiation damage, high throughput	Large data volumes	G-protein coupled receptors, ligand screening	Gati et al. (2014)

(*Continued*)

Table 20.1 (Continued)

Technique	Sample requirements	Resolution	Advantages	Limitations	Applications	References
XFEL-based crystallography	Micro-/nano-crystals	Atomic resolution, femtosecond timescale	Diffraction before destruction, ultrafast dynamics	Limited facility access, cost	Reaction intermediates, ultrafast processes	Johansson et al. (2017)
In-cell NMR	Isotopically labeled proteins in cells	Residue-level dynamics	True cellular environment	Signal crowding, sensitivity limits	Protein folding, PTMs in vivo	Narasimhan et al. (2020)
Solid-state NMR (ssNMR)	Immobilized or membrane proteins	Atomic-level structure and dynamics	No crystallization required	Specialized instrumentation	Amyloids, membrane proteins	Hong (2025)
Mass spectrometry (structural MS)	Minimal sample prep	Assembly, dynamics, interactions	Works on heterogeneous systems	Indirect structural information	Complex stoichiometry, conformational changes	Konermann et al. (2014)

analysis of assembly, dynamics, conformational heterogeneity, and macromolecular interactions. As native MS, HDX-MS, ion mobility, cross-linking, and integrative modeling technologies have advanced, MS fills the gap between high-resolution structural technology on one hand and dynamic and heterogeneous biological systems on the other (Tamara et al., 2022). With the ongoing innovations, MS will still be at the heart of the changing face of hybrid and integrative structural biology.

20.4 Structural biology as a foundation for translational research

Structural biology has come up as a fundamental framework through which one can understand the molecular pathophysiology of human disease as well as convert the information into a therapeutic approach. Structural biology brings together basic mechanistic knowledge with clinical roles by revealing the architectural form of structures of proteins, nucleic acids, and macromolecular complexes over three dimensions (Heppner et al., 2024). X-ray crystallography, cryo-EM, NMR spectroscopy, and integrative hybrid techniques are used to provide high-resolution structures that allow researchers to see dynamical structures, ligand binding modes, and disease-related structural changes (Carvalho et al., 2009). The understanding of these has been pivotal in the choice of targets, elucidation of mechanisms, and rational drug design, and they have been a cornerstone of contemporary translational research (Carvalho et al., 2009). Knowledge of the atomic structure of biomolecules enables researchers to relate pathophysiology and molecular functionality. Structural data may highlight the role of mutations, misfolding, or aberrant interactions in the pathogenesis of disease and form the basis of therapeutic hypotheses.

20.4.1 Major contributions in therapeutic discovery

Mapping functional domains: Structural analysis determines functional binding pockets, catalytic sites, regulatory regions, and interfaces between proteins that are important in their functioning.

Pathogenic mechanisms: disease-related mutations frequently cause alterations in stability, conformational dynamics, or binding specificity, information that can be used to develop activators or inhibitors.

Explaining ligand-protein: Cocrystal structures demonstrate the way small molecules, peptides, or antibodies interact with their targets, which are used as templates to optimize.

Informed choice of the therapeutic modality: Structures can be used to guide the choice of the best therapeutic modality between small molecules, biologics, nucleic acid therapeutics, or proteolysis-targeting chimeras (PROTACs). Therefore the molecular structures can be taken as a blueprint for therapeutic innovation, which connects the simple discoveries to the therapeutic ones.

20.4.2 Target identification and validation

The first steps in drug discovery are target identification and validation, so it must be established that therapeutic intervention at a selected molecular site will be associated with clinical effects. This process is reinforced by structural biology, which offers mechanistic clarity and predictive ability (Yuan et al., 2025). The repertoire of annotated protein structures has been increased through high-throughput structural genomics initiatives, making it fast to be able to compare homologous targets as well as find conserved functional motifs. Structures demonstrate the relations between the mutation or dysregulated conformations and the disease phenotypes (Yuan et al., 2025). Structural studies can be used to ensure that biological effects are on-target and not off-target. Cryo-EM, NMR chemical shift perturbation, and ligand-bound crystallography are the techniques that enable visualization of target-drug interactions in near-physiologic conditions and increase confidence in target validity (Carvalho et al., 2009). Predictive modeling employs structural data to model biomolecular behavior and predicts how molecular events affect disease progression. The combination of computational tools and structural biology has made a mechanistic interpretation of complicated biological systems possible (Singh et al., 2026).

20.4.3 Translational biomarkers and diagnostics

Translational biomarkers and diagnostics are very important tools that can be used to bridge basic biological findings and clinical decisions. Biomarkers help in early disease diagnostics, patient stratification, therapeutic choice, and treatment follow-up on physiological or pathological processes. The highly specific biomarkers emerging in the era of precision medicine have been made possible by structural biology and the latest technologies in analytical methods, which have allowed the identification of highly specific biomarkers that are based on molecular mechanisms. Such insights can be used to create diagnostics that are sensitive, precise, and predictive of clinical outcomes, ensuring that therapies are specific to individual patients. The biomarkers are especially useful in multifactorial diseases, which are often complex and neurodegeneration, cancer, and metabolic disorders, where earlier diagnosis and mechanistic knowledge are important to enhance treatment outcomes. Translational biomarkers are multidimensional biological biomarkers which can be used more reliably to inform clinical pathways and optimally to improve patient care, based on information emitted by molecular signatures and imaging-based readouts to clinical measures. Their development starts with a mechanistic understanding of preclinical studies development and goes on to clinical validation. Structural biology has a groundbreaking role in the identification of biomarkers, as it helps in providing the molecular impacts of the changes associated with disease. Detailed structural information sheds light on how mutations, misfolded proteins, or aberrant interactions generate signals that are detectable and can be used as biomarkers. Recent developments in cryo-EM, MS, and structural proteomics have expedited the identification of conformation-specific biomarkers in a variety of diseases. The contemporary diagnostics is based on the combination of biomarker science and high-level detection platforms. The specificity and design of these technologies are made more specific through structural knowledge. Biomarkers improve the predictability of preclinical models as well as regulatory approval because they bridge the relationship between molecular action and clinical outcome.

20.4.4 Structural biology of intrinsically disordered proteins

The intrinsically disordered proteins (IDPs) and intrinsically disordered regions (IDRs) are the new paradigm of structural biology (Fung et al., 2018). IDPs are dynamic ensembles of interconverting conformations, unlike classical globular proteins, which only fold to a single thermodynamically stable structure (Fung et al., 2018). They lack structural heterogeneity as their adaptive characteristic, which offers them functional plasticity, rapid regulation, and versatility of interaction. Over the last 10 years, IDPs have ceased to be regarded as exceptions and have been recognized as key players in cell signaling, gene regulation, and stress response, as well as phase separation (Utami et al., 2025). The disorder offers special mechanistic properties that structured proteins can hardly accomplish. The lack of a defined tertiary structure reduces the energy required to make conformational transitions, which allows IDPs to be highly specific and low-affinity binders of many partners—a critical feature of regulatory networks. IDPs usually engage in momentary associations, molecular recognition, and structure within dynamic complexes or biomolecular condensates (Utami et al., 2025). The characteristics of IDPs are low hydrophobicity, high net charge, and increased polar, aromatic, and disorder-promoting residues in the form of Pro, Lys, Arg, and Gln (Utami et al., 2025). These characteristics block hydrophobic collapses and go into an extension of large and loosened conformations. IDRs will often have low-complexity domains (LCDs), repetitive motifs, and PTM are, frequently embedded within IDRs, enabling them to be rapidly modulated in their structural ensembles (Utami et al., 2025). Since IDPs do not fold into a single structure, their structural code is not on stable secondary/tertiary motifs, but on sequence patterns, charge distribution, and transiently sampled conformations (Fung et al., 2018). In their disordered condition, IDPs do have residual structural propensities (Fung et al., 2018). On picosecond-millisecond scales, short α-helices, β-strands, or polyproline II may come together and dissociate. Such structural microstates often stabilize on binding ligands in what is called the mechanic coupled folding and binding mechanism. Other possibilities are a folding upon binding, a binding-upon-folding, and a fuzzy annex, in which there is still disorder in the bound form (Utami et al., 2025). The provided adaptability allows IDPs to identify various partners, cellular signal integrators, and to promote multistep procedures. The analysis of IDPs involves methods that can capture ensembles, but not snapshots. NMR spectroscopy that gives information about the residues on the chemical shift level, the retained dipolar couplings, paramagnetic relaxation enhancements, and nanosecond-millisecond scales (Emwas et al., 2020). Small-angle X-ray scattering, which provides low-resolution global shapes, radius of gyration distributions (Galchenkova, 2024). The change in the distance and conformational heterogeneity is determined by single-molecule FRET. Flexible fitting cryo-EM, which makes it possible to see disorder in larger complexes, can be used together with computational models (McGreevy et al., 2016). MS/cross-linking, which limits the sampling of ensembles to transient proximities (Tamara et al., 2022). These methods are frequently integrated within the structural biology paradigms, which involve integrating in silico computing to settle the favorable experimental observation. IDPs are often points of protein-protein interfaces, where many signaling inputs can be brought together to enable diversification to occur rapidly, as IDRs can collect mutations without destabilizing a stable fold, and allow fine-tuned evolutionary changes across the species (Fung et al., 2018).

20.4.4.1 Intrinsically disordered proteins in health and disease

IDPs are central regulators of cellular physiology, enabling flexible, quick, and reversible responses to both environmental and signaling changes (Fung et al., 2018). Neurodegeneration (e.g., tau, α-synuclein), cancer (e.g., p53, c-Myc), and viral pathogenesis are also involved in the implicated aberrant behavior of IDPs (Xiao and Xia, 2025). Unregulated phase separation has the potential to transform liquid-like condensates into gel-like or solid aggregates, highlighting the fine line between functional disorder and pathology (Abyzov et al., 2022). Their active conformational readings, multivalent interplay, and sensitivity to PTMs render them irreplaceable in gene regulation, signal transduction, stress adaptation, and cellular division (Abyzov et al., 2022). Nevertheless, the same properties that make IDPs feature functional versatility make them vulnerable to misregulation, pathological aggregation, and aberrant phase transitions (). Consequently, IDPs are the center of many human pathologies, specifically neurodegeneration, cancer, heart diseases, viral infection, and metabolic maladaptation (Abyzov et al., 2022).

20.4.4.2 Intrinsically disordered proteins in neurodegenerative diseases

The most evident type of disease connected with the dysfunction of IDPs is neurodegenerative disorders (Khan et al., 2023). There are a number of hallmark proteins that are intrinsically disordered or that have disordered regions that become aggregation-prone under pathological conditions (Table 20.2). Tau is a protein that is highly disordered, a microtubule-associated protein, and is a dynamic stabilizer of the microtubules as well as a regulator of axonal transportation (Singh et al., 2025). In AD and other tauopathies, tau hyperphosphorylates and assumes a β-sheet-rich structure, forming neurofibrillary tangles (Khan et al., 2025b). A dysregulated phase separation of tau, transitioning of liquid droplets into solid fibrils, has now been identified as a major pathological process. α-synuclein is an IDP that regulates synaptic vesicles (Basurto-Islas et al., 2025; Khan et al., 2025a). It can change under oxidative stress (mutations A53T, A30P), environmental stimuli, and reorganize into fibrillar aggregates, resulting in the formation of Lewy bodies (Khan et al., 2025a). Early condensate-like assemblies have been suggested as intermediates that template pathogenic fibrils. LCDs are parts of TDP-43 and FUS, which facilitate phase segregation in RNA processing and in the development of stress granules (Portz et al., 2021). The mutations in these areas interfere with the properties of condensate materials, leading to the irreversible solidification and the formation of inclusions in amyotrophic lateral sclerosis and frontotemporal dementia (Portz et al., 2021). Oncoproteins, tumor suppressors, and transcriptional regulators are highly represented by IDPs, and it is indicative of their networks of interactions (Xiao and Xia, 2025). Extensive IDRs mediate DNA binding, recruitment of coactivators, and integration of PTMs, and are present in many transcription factors, including c-Myc, p53, CREB, NF-kB, and the STATs (Elrakaybi et al., 2022). Kinases, phosphatases, scaffold proteins, and E3 ligases are reconfigured rapidly by IDRs. Alterations in mutated or modified PTM patterns have the capacity to rewire pathways and result in uncontrolled proliferation (Abyzov et al., 2022). In other instances, intrinsic disorder in itself is required to play a tumor-suppressive role (Abyzov et al., 2022). The regulation of the IDRs can be impaired by stabilization or forced folding, which contributes to oncogenesis (Xiao and Xia, 2025).

20.5 Structure-guided therapeutics and drug discovery

Medicinal chemists often harness a "structural basis" to clarify the action of small molecules and inform structure-based optimization in drug discovery (Malhotra et al., 2016). The application of protein crystallography in SBDD supports the detection of possible inhibitor-docking sites and the optimization of contacts between target-specific hits and lead compounds (Gupta et al., 2023). Recently, a combination of Fragment-based drug discovery guided by structural information, which reduced screening library size through decreased chemical complexity, thereby improving ligand efficiency in drug design and enabling the progression of three oncology treatments from clinical trials to FDA approval. In silico methods enable rapid identification and design of drug candidates and analysis of target interactions (Montinari et al., 2019).

Artificial intelligence (AI) has been widely used in several phases of drug discovery in recent years, arising from ongoing advancements in ML algorithms (Patel et al., 2025;

Table 20.2 Intrinsically disordered proteins (*IDPs*): structural features, biological roles, and disease associations.

IDP/IDR	Structural characteristics	Function	Disease association	Pathological mechanism	Refrences
Tau	Highly disordered, PTM-rich	Microtubule stabilization, axonal transport	Alzheimer's disease, tauopathies	Hyperphosphorylation → β-sheet aggregation → neurofibrillary tangles	Khan et al. (2023)
α-Synuclein	Disordered monomer, aggregation-prone	Synaptic vesicle regulation	Parkinson's disease	Misfolding and fibril formation → lewy bodies	Khan et al. (2025a)
TDP-43	Low-complexity domains	RNA processing, stress granules	ALS, FTD	Aberrant phase separation → insoluble inclusions	Portz et al. (2021)
FUS	Intrinsically disordered regions	Transcription, RNA metabolism	ALS, FTD	Altered condensate dynamics → aggregation	Portz et al. (2021)
p53	Disordered N- and C-termini	Tumor suppression, DNA damage response	Cancer	Mutation-induced loss of regulatory flexibility	Xiao and Xia (2025)
c-Myc	Highly disordered transcription factor	Gene regulation, cell growth	Cancer	Dysregulated interactions → uncontrolled proliferation	Xiao and Xia (2025)

Vemula et al., 2023). The big data era in drug discovery has just recently begun. ML has gradually evolved into a deep learning (DL) methodology with strong extrapolation capabilities and other efficient large-scale data processing. This subsequently encourages the incorporation of AI with computer-aided drug design, thereby speeding up the design and discovery of the newest drugs (Zhang et al., 2025). Drug discovery can be done in a variety of ways. Serendipity is the oldest. Serendipity refers to accidental discovery through trial and error, such as the discovery of penicillin (Butkiewicz et al., 2017; Klebe, 2025). Chemically altering well-known medications or natural ingredients is another method for finding new pharmaceuticals. Chemical alteration led to the discovery of aspirin (Klebe, 2025; Vane and Botting, 2003). Another method for finding new medications is chemically altering existing medications or natural items. Another method for finding new medications is to screen databases, either virtually or using high-throughput assays. Random in vitro screening led to the discovery of Prontosil, the first sulfonamide medication (Lawson et al., 2024).

AI techniques are applied across the drug discovery pipeline, including *de novo* molecular design and preclinical and clinical development, to improve efficiency and outcomes (Jorgensen, 2009). An integrated computational workflow combining these approaches is presented in Fig. 20.2. Recent developments in cryo-EM have ushered in a new age of high-resolution structural characterization of biological macromolecules and supramolecular assemblies, making cryo-EM a promising tool for researching ligand-protein interactions. Structure-based drug discovery is a logical strategy to drug creation that considers the biomolecular target's 3D structure and seeks to forecast the binding properties of potential ligands. Structures having a high enough resolution (usually less than 2.5 Å) make it easier to find the right ligand positions and bound water molecules, both of which are important for drug creation (Abel et al., 2008). The number of new ligands in cryo EM derived structures is growing rapidly, so it's critical to look into the best ways to validate them to guarantee the best modeled ligand quality using a variety of metrics like model-to-map fit, ligand geometry scores, and local interactions with ions, waters, protein, or nucleic acid components (Abel et al., 2008).

The significance of hydration in lead optimization is further highlighted by recent computational predictions that show remarkable consistency with empirically determined structures (Ichihara et al., 2014; Jorgensen, 2009). Various softwares are developed to dock chemical compounds into the biological target's structure (Jilla, 2025; Li, 2026; Pagadala et al., 2017). These programs estimate the affinity of compound binding and the geometry it takes on when bound. It can also assist in identifying the structural process *via* which a ligand influences the target's function. Additionally, docking also facilitates more effective ligand fine-tuning, a pivotal step in the drug discovery process that starts with a ligand and looks for related molecules with better qualities, such as stronger binding, increased efficacy, less toxicity, and fewer side effects (Kulkarni et al., 2025; Muhammed and Aki-Yalcin, 2024).

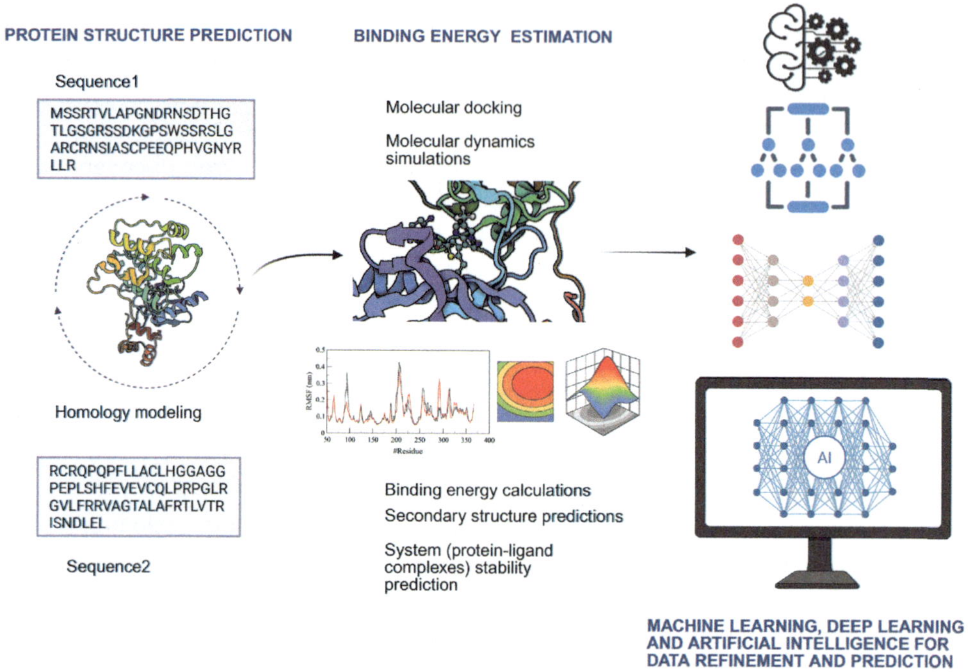

Figure 20.2 Integrated computational method for estimating binding energy and predicting protein structure. Molecular docking and molecular dynamics simulations are used to assess protein and ligand interactions after homology modeling is used to model protein sequences into 3D structures. Stability analysis, secondary structure prediction, and binding energy calculations are carried out, and data refinement, pattern identification, and increased forecast accuracy are made possible by machine learning and artificial intelligence tools.

Finding entirely new ligands can help avoid several problems with current ligands, including side effects brought on by off-target action. Furthermore, many identified ligands have distinct impacts on the target's function, increasing the range of medications and chemical probes that are available (Chaachouay and Zidane, 2024; Sadybekov and Katritch, 2023). Like how one can utilize an experimentally discovered ligand-bound structure to guide optimization of a ligand's binding affinity and adjustment of other ligand properties, docking can be used to predict binding poses. Medicinal chemists frequently employ predicted binding postures to guide their intuitive, visual analysis. Chemical modifications, deletions, or substitutions that could improve the ligand's compatibility with the binding site can be evaluated by chemists using anticipated poses (Lenz et al., 2000; Surade and Blundell, 2012). More generally, given the context of the binding pocket, scientists can determine which ligand components might be altered to accomplish a wide range of additional design objectives (Shamsi et al., 2022; Khan et al., 2025a). Docking can also help with several different ligand design objectives in addition, to enhancing binding affinity (Pantsar and Poso, 2018). Predicted poses, for instance, are used to guide the linking of compounds to form bivalent ligands. These include compounds that target sites on two distinct molecules, such as proteolysis-targeting chimeras (Li and Crews, 2022; Aboelez et al., 2022), and ligands that selectively bind to one member of a group of related targets (subtype selectivity) (Chen et al., 2023).

Docking can be used to predict the location and orientation of a ligand interacting with a protein, also referred to as its binding pose or mode. Lead optimization is the process of using this data to create more potent and specific analogs (Verma and Pathak, 2021). Fischer's lock-and-key theory serves as the foundation for the original docking method (Gupta et al., 2025; Spassov, 2024). According to this assumption, the affinity of the ligand and the receptor is proportionate to their geometric shapes, and both are treated as rigid substances. The ligand is sampled in a six-dimensional rotational and translational space to align with the binding site in many simple rigid-body systems. Subsequently, Koshland put out the induced-fit idea, which suggests that ligand interactions would constantly alter a receptor's active site. The docking process is essentially regarded as dynamic and adaptable (Wang, 2024).

MolAICal uses a DL model with a classical method to create three-dimensional structural ligands in protein pockets. Additionally, scoring functions have been successfully constructed using ML and DL techniques. ML and DL built on scoring functions can learn the protein-ligand interaction information from bigger training datasets than classical score functions (Wang et al., 2023). Furthermore, both ML and DL are more effective at handling the nonlinear relationship between features and labels, which helps explain why their use for SFs is becoming increasingly common. It has been demonstrated that the majority of widely cited ML- and DL- based SFs, including RF-Score, NNScore, AGL, OnionNet, OnionNet-2, OnionNet-SFCT, HPC/HWPC-GBT, and OPRC-GBT, among others, have great effectiveness in predicting protein-ligand binding affinities (Wang, 2024). Experimental techniques (such as isothermal titration calorimetry, surface plasmon resonance, and then MTT assay) can be used to measure the strength of protein-ligand interactions using a variety of indicators, including inhibition constant (Ki), dissociation constant (K_d), Michaelis constant (K_m), half-maximal inhibitory concentration (IC_{50}), or median effective concentration (EC_{50}). The protein-ligand binding affinities increase with decreasing values. Measuring the binding affinities of several protein-ligand pairings is difficult, though, because these experimental techniques are complicated, costly, and time-consuming (Si et al., 2019).

20.6 New paradigms in fragment-based and allosteric drug discovery

In contemporary drug development, fragment-based drug design (FBDD) has proven to be a viable strategy that speeds up and enhances lead optimization while significantly reducing the high attrition rates across the whole drug development process (Erlanson et al., 2019). The FDA has approved several medications, including vemurafenib, an oncogenic B-Raf kinase inhibitor generated from a fragment-based method (Cheng et al., 2007). This method has gained widespread recognition in drug development as more drugs developed from FBDD go into various phases of clinical trials. The druggability of targets is assessed using a variety of techniques, which is crucial for finding new drugs (Basarab et al., 2013). Typically, a druggable target has a hydrophobic pocket that helps it bind to hydrophobic substances. Undruggable targets, on the other hand, either lack a pocket or have a shallow, very hydrophilic pocket. Unstructured proteins that are crucial in the control of disease are among these targets. Applying FBDD to these druggable targets can yield a high hit rate, providing more possibilities for future research. For instance, a few research teams have created drugs with various scaffolds by using FBDD against the bacterium Gyrase-B (Christopoulos, 2002; Geromichalos et al., 2016).

Allosteric medications provide several distinct pharmacological advantages to conventional orthosteric medications, such as improved physicochemical characteristics, less off-target toxicity, and increased selectivity. Orthosteric medications are drugs that bind directly to the primary, active site (orthosteric site) of a receptor or enzyme, competing directly with natural, endogenous ligands (Barresi et al., 2021). Following the FDA's approval of several allosteric medications, including trametinib and MK2206, which are MEK and AKT inhibitors, correspondingly, the pharmaceutical industry has made significant investments in creating therapeutic molecules for a variety of targets that utilize allosteric regulation. The FDA has authorized 19 medications for a variety of illnesses that use allosteric processes to produce their therapeutic benefits (Lu et al., 2018). Andrographolide derivatives were found to bind to the secret allosteric site of the challenging-to-target pharmacologically oncogenic K-Ras4B protein, inhibiting nucleotide exchange and oncogenic K-Ras4B-dependent signaling (Li et al., 2020). However, it is challenging to find hidden allosteric sites since they are not evident from a single apo crystal structure and are frequently exclusively discovered by chance when a particular tiny molecule attaches to and reinforces a particular geometry that the molecule prefers (Fournier et al., 2025).

Allosteric modulation, which has historically been studied in relation to small molecules, is becoming more popular as a primary mechanism in the field of antibodies, which provide improved selectivity and reduced toxicity (Changeux, 2013). Modulation of the target protein that causes an antibody to attach to an allosteric site or epitope and affect its functional dynamics is known as antibody-mediated allosteric effect (Fournier et al., 2025). Alteration of the conformation in an active or inactive state, activation or inhibition of activity, enhancement or blocking binding to a partner, encompassing dimerization, receptor internalization, and target complex disruption, are all examples of these effects (Guarnera and Berezovsky, 2020). Muromonab-CD3 (OKT3), the first authorized antibody, was released in 1986. It targets the CD3 epsilon unit, which is a component of the T cell receptor on T cell surfaces (Kjer-Nielsen et al., 2004). Although a lot of research has been conducted on allostery still the identification of allosteric drugs and explanation of the allosteric mechanisms are substantial challenges. In the future, allostery can be an emerging field of research in structural biology-based drug discovery with a focus on current challenges such as identification of allosteric sites, mutations, and mechanisms in biological systems (Lu et al., 2019).

20.7 Structural biology in vaccine and antiviral design

The structures of whole viruses, viral proteins, and antigen-antibody complexes can now be determined using structural biology techniques (Handa et al., 2025; Meloen et al., 2000). Important information about the tertiary conformation and the location of antigenic sites on viral proteins is provided by 3D structures (Area et al., 2004; McGreig et al., 2022). Linear epitopes on a protein can be identified using experimental methods such as phage display libraries or peptide microarrays (Buus et al., 2012; Wang et al., 2021). In peptide microarrays, overlapping peptides spanning the protein sequence are synthesized and screened with an antibody to determine binding sites (Meng et al., 2018). Bacteriophages displaying peptide sequences on their surface that may be tested for antibody binding are used in phage display libraries. Peptide mimics, or mimotopes, can replace the natural epitopes of various chemical species, such as proteins, polysaccharides, and DNA. These mimotopes can later be applied in vaccines and diagnostics (Palma, 2023). The development of SARS-CoV-2 mutations is causing the COVID-19 vaccine-induced protection to decline. The accumulation of mutations in its S proteins, particularly those affecting immune evasion, while maintaining its ability to bind the human ACE2 receptor (SARS-CoV, 2021). Only after a comprehensive structural analysis of the altered complex formed between S proteins and neutralizing antibody epitopes should the vaccines be modified as outlined in Fig. 20.3.

20.8 Structural systems biology and network modeling

A huge amount of data has been contributed by the Human Genome Project, now it has become a challenge to link different cords of this data and determine their sequence, to understand how this information works together to achieve a particular task. Systems biology takes this challenge and decodes the relationship between different data cords (Ayton and Voth, 2009). Some molecular and signaling pathway databases such as Kyoto encyclopedia of genes and genomes, signal transduction knowledge environment, and BioCarta have been created in the past decade (Ayton et al., 2007). These databases serve as a platform representing the relationship between molecules in the context of their reactivity, inhibition, or activation inside the cell. Intracellular signaling pathways contribute to the decoding of the specificity of cellular responses through propagating spatial and temporal signals. Stimuli, cellular responses, and cell fate share a complex relationship, which could be analyzed through computational models and network analysis tools (Ayton et al., 2007). The information about molecular interactions is best represented by 3D structures, because they provide details such as binding specificity and molecular recognition, but

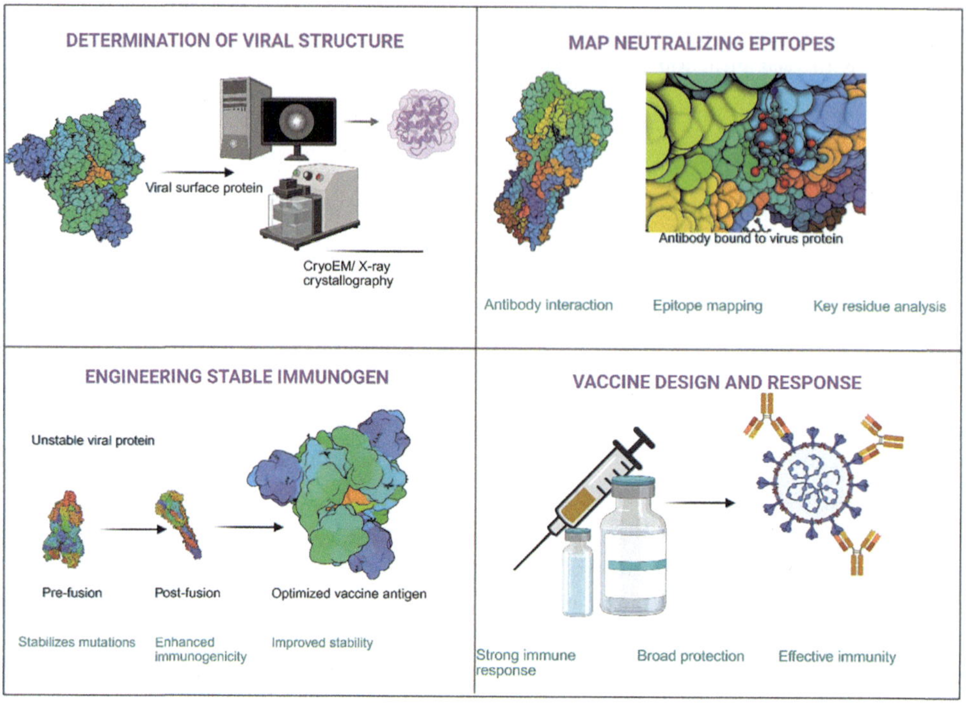

Figure 20.3 Workflow for designing vaccines guided by structure. High-resolution methods like X-ray crystallography and cryo-electron microscopy are used to characterize viral surface proteins. Neutralizing epitopes are found using crucial residue analysis and antibody-antigen interaction mapping. Mapping of these neutralizing regions guides the redesign of unstable viral antigens into conformationally stable, immunogenic forms. The optimized immunogen is then formulated into a vaccine that triggers robust, broad-spectrum immune protection against the target virus.

determination of macromolecular (protein-protein, protein-nucleic acid) interfaces and dealing with large protein complexes pose limitations for structural biology. Here, multiscale computational modeling comes into play, which helps in achieving the study of molecular complexes and their propagation at the cellular signaling scale. Multiscale modeling helps in the determination of the effect caused by perturbations of molecular structure on downstream signaling events. This is achievable because multiscale modeling bridges the scales of structural and systems biology (Chodera et al., 2007).

20.8.1 Multiscale modeling from proteins to pathways

An extensive range of spatial and temporal scales makes protein modeling a challenge. Modeling biological processes, like membrane pore assembly, requires descriptions at both the molecular scale (e.g., membrane proteins) and the macroscopic scale (e.g., lipids) (Chodera et al., 2007). As microseconds to milliseconds is the timescale of occurrence of many biological processes, in comparison to atomistic MDS, it is much longer. So, it becomes accessible to stimulate complex protein systems at atomic resolution while capturing biologically relevant time scales. Simplifying details in the length of the time scale addresses the above challenge. The common and effective approach to connect the time scale gap is to simplify the protein representations in the length scale (Alhumaydhi et al., 2021; Bashir et al., 2020). Simulations can integrate multiple length-scale resolutions in a stepwise or parallel manner. Model construction from short atomistic simulations is the other solution for bridging the time scale to forecast dynamics over long timescales by effectively simplifying the temporal resolution. To investigate protein folding kinetics on microsecond or even millisecond timescales by combining numerous shorter (nanosecond-scale) atomistic simulations, recently a method has been developed, known as the discrete time and discrete state Markov state model (Ayton et al., 2007).

20.8.2 Coarse graining in the length domain

The governing entity underlying the protein dynamics is a rugged free energy landscape. As the system size increases, the landscape contains an exponentially increasing number of local minima. That's why computing power is the limiting factor for a comprehensive exploration of the protein free energy surface using atomistic MDS. The sampling problem is efficiently alleviated by coarse-grained modeling, in which the grouping of atoms into a single site takes place. The reduction in the degree of freedom of the system and smoothness of the potential energy surface is provided by coarse-grained models (Sherwood et al., 2008). Thermodynamic experiments, structural databases, and information from atomic simulations can be exploited for coarse grained simulations (Shamsi et al., 2022; Waseem et al., 2021). Different resolutions are employed in series rather than in parallel in these simulations. The development of coarse-grained potentials for proteins was started in the 1970s. In 1975, Levitt pioneered a work, which included the representation of protein residues with three points: one for the side chain and two for the main chain. Many protein coarse-grained models have been developed (Izvekov and Voth, 2005).

20.8.3 Predictive simulations for disease progression and drug response

A mathematical model used to describe how a disease changes and develops over time is called a disease progression model (Cook and Bies, 2016). The characterization of treatment and placebo effects can be done through the disease Progression Model, and these effects can be integrated with other models to determine dose selection and optimization (Ito and Romero, 2020; Venuto et al., 2016). Critical aspects, such as when to drop out from the trial, the determination of the power of study, duration of study, or other trial design enhancements, could be achieved through disease progression models (Kale et al., 2024). The integration of information, including clinical trial, translational, and real-world data, combined with multidisciplinary domain expertise and clinical knowledge, is done by a disease progression model to guide decision-making across the medical product development lifecycle (Venuto et al., 2016; Wang et al., 2014). For example, the early identification of biomarkers for disease modifiers in drug development and quantification of exposure-response to enable safe extrapolation or support dose selection across population (Inojosa et al., 2021; Samtani et al., 2013; Slate and Turnbull, 2000). The full potential of the disease progression model's applications is yet to be defined and appreciated.

20.9 Conclusions

The integration of structural biology with computational and AI–driven methodologies has fundamentally reshaped modern biomedical research. High-resolution structural determination has enabled precise mapping of active sites, conformational landscapes, and macromolecular assemblies, providing mechanistic insights into protein function across diverse biological systems. Classical scoring functions laid the groundwork for SBDD; however, their limitations in capturing nonlinear relationships and complex interaction networks prompted the adoption of ML, and deep learning models. Recent advances in AI-driven structure prediction, such as AlphaFold and RoseTTAFold, demonstrate that computational approaches can now approximate experimental structural accuracy. These developments reduce dependency on time-intensive experimental techniques while complementing them. Furthermore, ML-based scoring functions and graph neural networks have significantly improved binding affinity prediction and virtual screening performance. Such tools enhance lead optimization and reduce attrition rates in pharmaceutical pipelines.

Looking forward, integration of structural datasets with systems biology, multiomics platforms, and high-throughput screening will enable predictive modeling of biological responses at a systems level. Cryo-ET and time-resolved structural methods will provide insights into dynamic biomolecular processes within native cellular environments. Additionally, explainable AI frameworks are expected to improve the interpretability and reliability of ML-driven drug design models. Cloud-based structural repositories and collaborative computational infrastructures will further democratize access to structural data, accelerating innovation across academia and industry. As structural biology increasingly converges with data science, quantum chemistry, and advanced simulation techniques, future therapeutic development will rely on an integrated structural–computational ecosystem. This multidisciplinary synergy will enhance precision medicine strategies, enable rational targeting of complex diseases, and establish a robust foundation for next-generation biomedical innovation.

Acknowledgement

A.S. is grateful to Ajman University, UAE for supporting the publication through 2025-IRG-DRG-10. S.A. acknowledges the financial support from the Department of Biotechnology (DBT), Ministry of Science and Technology, Government of India, through the DBT Research Associateship (DBT-RA) Programme (Award No. DBT-RA/2022/July/N/2482).

References

Abel, R., Young, T., Farid, R., Berne, B.J., Friesner, R.A., 2008. Role of the active-site solvent in the thermodynamics of factor Xa ligand binding. Journal of the American Chemical Society 130 (9), 2817–2831. Available at https://doi.org/10.1021/ja0771033.

Abyzov, A., Blackledge, M., Zweckstetter, M., 2022. Conformational dynamics of intrinsically disordered proteins regulate biomolecular condensate chemistry. Chemical Reviews 122 (6), 6719–6748. Available at https://doi.org/10.1021/acs.chemrev.1c00774.

Adams, B., Aeppli, G., Allison, T., Baron, A.Q.R., Bucksbaum, P., Chumakov, A.I., et al., 2019. Scientific opportunities with an X-ray free-electron laser oscillator. arXiv, United States arXiv. Available at: https://arxiv.org. https://doi.org/10.48550/arxiv.1903.09317.

Ahmadi, A., Gupta, S., Menon, V., Baudry, J., 2025. Linking machine learning and biophysical structural features in drug discovery. Frontiers in Molecular Biosciences 11. Available at https://doi.org/10.3389/fmolb.2024.1305272.

Alhumaydhi, F.A., Aljasir, M.A., Aljohani, A.S.M., Alsagaby, S.A., Alwashmi, A.S.S., Shahwan, M., et al., 2021. Probing the interaction of memantine, an important Alzheimer's drug, with human serum albumin: in silico and in vitro approach. Journal of Molecular Liquids 340. Available at https://doi.org/10.1016/j.molliq.2021.116888, https://www.journals.elsevier.com/journal-of-molecular-liquids.

Arar, S., Haque, M.A., Kayed, R., 2025. Protein aggregation and neurodegenerative disease: structural outlook for the novel therapeutics. Proteins: Structure, Function and Bioinformatics 93 (8), 1314–1329. Available at https://doi.org/10.1002/prot.26561, http://onlinelibrary.wiley.com/journal/10.1002/(ISSN)1097-0134.

Area, E., Martín-Benito, J., Gastaminza, P., Torreira, E., Valpuesta, J.M., Carrascosa, J.L., et al., 2004. 3D structure of the influenza virus polymerase complex: localization of subunit domains. Proceedings of the National Academy of Sciences 101 (1), 308–313. Available at https://doi.org/10.1073/pnas.0307127101.

Ayton, G.S., Voth, G.A., 2009. Systematic multiscale simulation of membrane protein systems. Current Opinion in Structural Biology 19 (2), 138–144. Available at https://doi.org/10.1016/j.sbi.2009.03.001.

Ayton, G.S., Noid, W.G., Voth, G.A., 2007. Multiscale modeling of biomolecular systems: in serial and in parallel. Current Opinion in Structural Biology 17 (2), 192–198. Available at https://doi.org/10.1016/j.sbi.2007.03.004.

Banari, A., Samanta, A.K., Munke, A., Laugks, T., Bajt, S., Grünewald, K., et al., 2025. Advancing time-resolved structural biology: latest strategies in cryo-EM and X-ray crystallography. Nature Methods 22 (7), 1420–1435. Available at https://doi.org/10.1038/s41592-025-02659-6, https://www.nature.com/nmeth/.

Barresi, E., Martini, C., Da Settimo, F., Greco, G., Taliani, S., Giacomelli, C., et al., 2021. Allosterism vs. orthosterism: recent findings and future perspectives on A2B AR physio-pathological implications. Frontiers in Pharmacology 12. Available at https://doi.org/10.3389/fphar.2021.652121, http://www.frontiersin.org/Pharmacology.

Basarab, G.S., Manchester, J.I., Bist, S., Boriack-Sjodin, P.A., Dangel, B., Illingworth, R., et al., 2013. Fragment-to-hit-to-lead discovery of a novel pyridylurea scaffold of ATP competitive dual targeting type II topoisomerase inhibiting antibacterial agents. Journal of Medicinal Chemistry 56 (21), 8712–8735. Available at https://doi.org/10.1021/jm401208b.

Bashir, S., Shamsi, A., Ahmad, F., Hassan, M.I., Kamal, M.A., Islam, A., 2020. Biophysical elucidation of fibrillation inhibition by sugar osmolytes in α-lactalbumin: multispectroscopic and molecular docking approaches. ACS Omega 5 (41), 26871–26882. Available at https://doi.org/10.1021/acsomega.0c04062, pubs.acs.org/journal/acsodf.

Basurto-Islas, G., Diaz, M.C., Ocampo, L.M.Z., Martínez-Herrera, M., López-Camacho, P.Y., 2025. Natural products against tau hyperphosphorylation-induced aggregates: potential therapies for Alzheimer's disease. Archiv der Pharmazie 358 (1). Available at https://doi.org/10.1002/ardp.202400721, http://onlinelibrary.wiley.com/journal/10.1002/(ISSN)1521-4184.

Batra, S., Allwein, B., Kumar, C., Devbhandari, S., Brüning, J.G., Bahng, S., et al., 2025. G-quadruplex–stalled eukaryotic replisome structure reveals helical inchworm DNA translocation. Science (New York, N.Y.) 387 (6738). Available at https://doi.org/10.1126/science.adt1978, https://www.science.org/toc/science/387/6738.

Bragg, W.H., Bragg, W.L., 1913. The reflection of X-rays by crystals. Proceedings of the Royal Society of London. Series A, Containing Papers of a Mathematical and Physical Character. 88 (605), 428–438. Available at https://doi.org/10.1098/rspa.1913.0040.

Bragg, W.L., 1913. The structure of some crystals as indicated by their diffraction of X-rays. Proceedings of the Royal Society of London. Series A, Containing Papers of a Mathematical and Physical Character. 89 (610), 248–277. Available at https://doi.org/10.1098/rspa.1913.0083.

Brändén, G., Neutze, R., 2021. Advances and challenges in time-resolved macromolecular crystallography. Science (New York, N.Y.) 373 (6558). Available at https://doi.org/10.1126/science.aba0954.

Browne, W.J., North, A.C.T., Phillips, D.C., Brew, K., Vanaman, T.C., Hill, R.L., 1969. A possible three-dimensional structure of bovine α-lactalbumin based on that of hen's egg-white lysozyme. Journal of Molecular Biology 42 (1), 65. Available at https://doi.org/10.1016/0022-2836(69)90487-2.

Butkiewicz, Wang, Bryant, Lowe, Weaver, Meiler, 2017. High-throughput screening assay datasets from the pubchem database. Chemical informatics (Wilmington, Del.) 3 (1).

Buus, S., Rockberg, J., Forsström, B., Nilsson, P., Uhlen, M., Schafer-Nielsen, C., 2012. High-resolution mapping of linear antibody epitopes using ultrahigh-density peptide microarrays. Molecular & Cellular Proteomics 11 (12), 1790–1800. Available at https://doi.org/10.1074/mcp.m112.020800.

Carvalho, A.L., Trincão, J., Romão, M.J., 2009. X-ray crystallography in drug discovery. Ligand-macromolecular interactions in drug discovery: methods and protocols, pp. 31–56.

Chaachouay, N., Zidane, L., 2024. Plant-derived natural products: a source for drug discovery and development. Drugs and Drug Candidates 3 (1), 184–207. Available at https://doi.org/10.3390/ddc3010011.

Chait, B.T., Cadene, M., Olinares, P.D., Rout, M.P., Shi, Y., 2016. Revealing Higher order protein structure using mass spectrometry. Journal of the American Society for Mass Spectrometry 27 (6), 952–965. Available at https://doi.org/10.1007/s13361-016-1385-1, https://pubs.acs.org/loi/jamsef.

Changeux, J.P., 2013. The concept of allosteric modulation: an overview. Drug Discovery Today: Technologies 10 (2), e223. Available at https://doi.org/10.1016/j.ddtec.2012.07.007.

Chen, T., Shu, X., Zhou, H., Beckford, F.A., Misir, M., 2023. Algorithm selection for protein–ligand docking: strategies and analysis on ACE. Scientific Reports 13 (1). Available at https://doi.org/10.1038/s41598-023-35132-5, https://www.nature.com/srep/.

Cheng, A.C., Coleman, R.G., Smyth, K.T., Cao, Q., Soulard, P., Caffrey, D.R., et al., 2007. Structure-based maximal affinity model predicts small-molecule druggability. Nature Biotechnology 25 (1), 71–75. Available at https://doi.org/10.1038/nbt1273.

Chodera, J.D., Singhal, N., Pande, V.S., Dill, K.A., Swope, W.C., 2007. Automatic discovery of metastable states for the construction of Markov models of macromolecular conformational dynamics. Journal of Chemical Physics 126 (15). Available at https://doi.org/10.1063/1.2714538.

Christofi, E., Barran, P., 2023. Ion mobility mass spectrometry (IM-MS) for structural biology: insights gained by measuring mass, charge, and collision cross section. Chemical Reviews 123 (6), 2902–2949. Available at https://doi.org/10.1021/acs.chemrev.2c00600.

Christopoulos, A., 2002. Allosteric binding sites on cell-surface receptors: novel targets for drug discovery. Nature Reviews. Drug Discovery 1 (3), 198–210. Available at https://doi.org/10.1038/nrd746.

Cook, S.F., Bies, R.R., 2016. Disease progression modeling: key concepts and recent developments. Current Pharmacology Reports 2 (5), 221–230. Available at https://doi.org/10.1007/s40495-016-0066-x, http://www.springer.com/biomed/pharmacology+%26+toxicology/journal/40495.

Dobson, C.M., 2019. Biophysical techniques in structural biology. Annual Review of Biochemistry 88, 25–33. Available at https://doi.org/10.1146/annurev-biochem-013118-111947, http://www.annualreviews.org/journal/biochem.

Dror, R.O., Dirks, R.M., Grossman, J.P., Xu, H., Shaw, D.E., 2012. Biomolecular simulation: a computational microscope for molecular biology. Annual Review of Biophysics 41 (1), 429–452. Available at https://doi.org/10.1146/annurev-biophys-042910-155245, http://www.annualreviews.org/journal/biophys.

Elmlund, D., Elmlund, H., 2015. Cryogenic electron microscopy and single-particle analysis. Annual Review of Biochemistry 84 (1), 499–517. Available at https://doi.org/10.1146/annurev-biochem-060614-034226.

Elrakaybi, A., Ruess, D.A., Lübbert, M., Quante, M., Becker, H., 2022. Epigenetics in pancreatic ductal adenocarcinoma: impact on biology and utilization in diagnostics and treatment. Cancers 14 (23). Available at https://doi.org/10.3390/cancers14235926, http://www.mdpi.com/journal/cancers/.

Emwas, A.H., Szczepski, K., Poulson, B.G., Chandra, K., McKay, R.T., Dhahri, M., et al., 2020. NMR as a "gold standard" method in drug design and discovery. Molecules (Basel, Switzerland) 25 (20). Available at https://doi.org/10.3390/molecules25204597, https://www.mdpi.com/1420-3049/25/20/4597.

Erlanson, D.A., Davis, B.J., Jahnke, W., 2019. Fragment-based drug discovery: advancing fragments in the absence of crystal structures. Cell Chemical Biology 26 (1), 9–15. Available at https://doi.org/10.1016/j.chembiol.2018.10.001.

Finch, J.T., Perutz, M.F., Bertles, J.F., Döbler, J., 1973. Structure of sickled erythrocytes and of sickle-cell hemoglobin fibers. Proceedings of the National Academy of Sciences 70 (3), 718–722. Available at https://doi.org/10.1073/pnas.70.3.718.

Fournier, L., Guarnera, E., Kolmar, H., Becker, S., 2025. Allosteric antibodies: a novel paradigm in drug discovery. Trends in Pharmacological Sciences 46 (4), 311–323. Available at https://doi.org/10.1016/j.tips.2024.10.007.

Fukuda, Y., Stapleton, K., Kato, T., 2023. Progress in spatial resolution of structural analysis by cryo-EM. Microscopy: The Journal of the Quekett Microscopical Club 72 (2), 135–143. Available at https://doi.org/10.1093/jmicro/dfac053.

Fung, H.Y.J., Birol, M., Rhoades, E., 2018. IDPs in macromolecular complexes: the roles of multivalent interactions in diverse assemblies. Current Opinion in Structural Biology 49, 36–43. Available at https://doi.org/10.1016/j.sbi.2017.12.007, http://www.elsevier.com/locate/sbi.

Galchenkova, M., 2024. Advancements in Diffraction Analysis Methods and Data Reduction Techniques for Serial Crystallography. Universitaet Hamburg (Germany).

Gati, C., Bourenkov, G., Klinge, M., Rehders, D., Stellato, F., Oberthür, D., et al., 2014. Serial crystallography on in vivo grown microcrystals using synchrotron radiation. IUCrJ 1 (2), 87–94. Available at https://doi.org/10.1107/S2052252513033939.

Geloni, G., Huang, Z., Pellegrini, C., 2017. The Physics and Status of X-ray free-electron lasers.

Geromichalos, G.D., Alifieris, C.E., Geromichalou, E.G., Trafalis, D.T., 2016. Overview on the current status of virtual high-throughput screening and combinatorial chemistry approaches in multi-target anticancer drug discovery; Part I. Zerbinis Publications. Greece Journal of B.U.ON. 21 (4), 764–779. Available at https://www.jbuon.com/web/index.php/arc?id=43.

Glaeser, R.M., 2021. Preparing better samples for cryo-electron microscopy: biochemical challenges do not end with isolation and purification. Annual Review of Biochemistry 90, 451–474. Available at https://doi.org/10.1146/annurev-biochem-072020-020231, http://www.annualreviews.org/journal/biochem.

Gruner, S.M., Lattman, E.E., 2015. Biostructural science inspired by next-generation x-ray sources. Annual Review of Biophysics 44, 33–51. Available at https://doi.org/10.1146/annurev-biophys-060414-033813, http://www.annualreviews.org/journal/biophys.

Guarnera, E., Berezovsky, I.N., 2020. Allosteric drugs and mutations: chances, challenges, and necessity. Current Opinion in Structural Biology 62, 149–157. Available at https://doi.org/10.1016/j.sbi.2020.01.010, http://www.elsevier.com/locate/sbi.

Gupta, Y., Savytskyi, O.V., Coban, M., Venugopal, A., Pleqi, V., Weber, C.A., et al., 2023. Protein structure-based in-silico approaches to drug discovery: guide to COVID-19 therapeutics. Molecular Aspects of Medicine 91. Available at https://doi.org/10.1016/j.mam.2022.101151, www.elsevier.com/locate/mam.

Gupta, M.N., Tripathi, T., Uversky, V.N., 2025. Binding of a Substrate ("Lock and Key") and Conformational Adaption ("Induced Fit") are Different Stages of Enzyme Action. Elsevier BV, pp. 31–43. Available at https://doi.org/10.1016/b978-0-443-21809-5.00017-x.

Handa, T., Saha, A., Narayanan, A., Ronzier, E., Kumar, P., Singla, J., et al., 2025. Structural virology: the key determinants in development of antiviral therapeutics. Viruses 17 (3), 417. Available at https://doi.org/10.3390/v17030417.

Heck, A.J.R., 2008. Native mass spectrometry: a bridge between interactomics and structural biology. Nature Methods 5 (11), 927–933. Available at https://doi.org/10.1038/nmeth.1265, https://www.nature.com/nmeth/.

Heppner, D.E., Bigi-Botterill, S.V., Riley, A.P., Chibale, K., Lindsley, C.W., 2024. Structural biology in drug discovery and development: call for papers. Journal of Medicinal Chemistry 67 (15), 12461–12462. Available at https://doi.org/10.1021/acs.jmedchem.4c01492, http://pubs.acs.org/jmc.

Hong, M., 2025. Solid-state NMR of virus membrane proteins. Accounts of Chemical Research 58 (6), 847–860. Available at https://doi.org/10.1021/acs.accounts.4c00800.

Ichihara, O., Shimada, Y., Yoshidome, D., 2014. The importance of hydration thermodynamics in fragment-to-lead optimization. ChemMedChem 9 (12), 2708–2717. Available at https://doi.org/10.1002/cmdc.201402207.

Inojosa, H., Proschmann, U., Akgün, K., Ziemssen, T., 2021. Should we use clinical tools to identify disease progression? Frontiers in Neurology 11. Available at https://doi.org/10.3389/fneur.2020.628542.

Ito, K., Romero, K., 2020. Placebo effect in subjects with cognitive impairment. International Review of Neurobiology 153, 213–230. Available at https://doi.org/10.1016/bs.irn.2020.03.032, http://www.elsevier.com/wps/find/bookdescription.cws_home/704657/description#description.

Izvekov, S., Voth, G.A., 2005. A multiscale coarse-graining method for biomolecular systems. Journal of Physical Chemistry B 109 (7), 2469–2473. Available at https://doi.org/10.1021/jp044629q.

Jilla, T., 2025. The role of molecular docking in drug discovery: current approaches and innovations. International Journal of Zoology, Environment and Life Sciences 16–20. Available at https://doi.org/10.70604/ijzels.v2i2.55.

Johansson, L.C., Stauch, B., Ishchenko, A., Cherezov, V., 2017. A bright future for serial femtosecond crystallography with XFELs. Trends in Biochemical Sciences 42 (9), 749–762. Available at https://doi.org/10.1016/j.tibs.2017.06.007, www.elsevier.com/locate/tibs.

Jorgensen, W.L., 2009. Efficient drug lead discovery and optimization. Accounts of Chemical Research 42 (6), 724–733. Available at https://doi.org/10.1021/ar800236t, http://pubs.acs.org/doi/pdfplus/10.1021/ar800236t.

Kale, M., Wankhede, N., Pawar, R., Ballal, S., Kumawat, R., Goswami, M., et al., 2024. AI-driven innovations in Alzheimer's disease: integrating early diagnosis, personalized treatment, and prognostic modelling. Ageing Research Reviews 101, 102497. Available at https://doi.org/10.1016/j.arr.2024.102497.

Khan, M.S., Shamsi, A., Shahwan, M., Dinislam, K., Yadav, D.K., 2025a. Targeting programmed death ligand 1 for anticancer therapy using computational drug repurposing and molecular simulations. Scientific Reports 15 (1). Available at https://doi.org/10.1038/s41598-025-14503-0, https://www.nature.com/srep/.

Khan, S., Choudhury, A., Mohammad, T., Shamsi, A., Hassan, M.I., Islam, A., 2025b. Structure-guided screening identified bioactive phytoconstituents Hernandonine and Anolobine as potential inhibitors of dual specificity protein kinase CLK1. Scientific Reports 15 (1). Available at https://doi.org/10.1038/s41598-025-97753-2, https://www.nature.com/srep/.

Khan, S., Hassan, M.I., Shahid, M., Islam, A., 2023. Nature's toolbox against tau aggregation: an updated review of current research. Ageing Research Reviews 87, 101924. Available at https://doi.org/10.1016/j.arr.2023.101924.

Khanna, K., 2024. Emerging technologies in cryo-electron tomography. Cryo-Electron Microscopy in Structural Biology 395–406.

Kjer-Nielsen, L., Dunstone, M.A., Kostenko, L., Ely, L.K., Beddoe, T., Mifsud, N.A., et al., 2004. Crystal structure of the human T cell receptor CD3εγ heterodimer complexed to the therapeutic mAb OKT3. Proceedings of the National Academy of Sciences 101 (20), 7675–7680. Available at https://doi.org/10.1073/pnas.0402295101.

Klebe, G., 2025. In the beginning, there was serendipity. Drug Design: From Structure and Mode-of-Action to Rational Design Concepts. Springer, pp. 15–22.

Konermann, L., Vahidi, S., Sowole, M.A., 2014. Mass spectrometry methods for studying structure and dynamics of biological macromolecules. Analytical Chemistry 86 (1), 213–232. Available at https://doi.org/10.1021/ac4039306.

Kovermann, M., Rogne, P., Wolf-Watz, M., 2016. Protein dynamics and function from solution state NMR spectroscopy. Quarterly Reviews of Biophysics 49, e6. Available at https://doi.org/10.1017/S0033583516000019.

Kulkarni, N., Momin, S., Jain, L., More, R., Jadhav, S., Lungase, M., 2025. A comprehensive review on molecular docking in drug discovery. International Journal of Scientific Research and Technology.

Lambert, J.B., Mazzola, E.P., Ridge, C.D., 2019. Nuclear Magnetic Resonance Spectroscopy: an Introduction to Principles, Applications, and Experimental Methods. John Wiley & Sons.

Lawson, C.L., Kryshtafovych, A., Pintilie, G.D., Burley, S.K., Černý, J., Chen, V.B., et al., 2024. Outcomes of the EMDataResource cryo-EM ligand modeling challenge. Nature Methods 21 (7), 1340–1348. Available at https://doi.org/10.1038/s41592-024-02321-7, https://www.nature.com/nmeth/.

Lenz, G.R., Nash, H.M., Jindal, S., 2000. Chemical ligands, genomics and drug discovery. Drug Discovery Today 5 (4), 145–156. Available at https://doi.org/10.1016/S1359-6446(00)01468-9.

Li, G., 2026. Integration of artificial intelligence in molecular docking and virtual screening methods. Artificial Intelligence for Drug Design. Springer, pp. 731–743.

Li, K., Crews, C.M., 2022. PROTACs: past, present and future. Chemical Society Reviews 51 (12), 5214–5236. Available at https://doi.org/10.1039/d2cs00193d, http://pubs.rsc.org/en/journals/journal/cs.

Li, X., Dai, J., Ni, D., He, X., Zhang, H., Zhang, J., et al., 2020. Insight into the mechanism of allosteric activation of PI3Kα by oncoprotein K-Ras4B. International Journal of Biological Macromolecules 144, 643–655. Available at https://doi.org/10.1016/j.ijbiomac.2019.12.020.

Li, Q., Zhang, J., Haluska, C., Zhang, X., Wang, L., Liu, G., et al., 2024. Cryo-EM structures of Smc5/6 in multiple states reveal its assembly and functional mechanisms. Nature Structural & Molecular Biology 31 (10), 1532–1542. Available at https://doi.org/10.1038/s41594-024-01319-1.

Lin, J., Zhou, D., Steitz, T.A., Polikanov, Y.S., Gagnon, M.G., 2018. Ribosome-targeting antibiotics: modes of action, mechanisms of resistance, and implications for drug design. Annual Review of Biochemistry 87, 451–478. Available at https://doi.org/10.1146/annurev-biochem-062917-011942, http://www.annualreviews.org/journal/biochem.

Lin, Z., Akin, H., Rao, R., Hie, B., Zhu, Z., Lu, W., et al., 2023. Evolutionary-scale prediction of atomic-level protein structure with a language model. Science (New York, N.Y.) 379 (6637), 1123–1130. Available at https://doi.org/10.1126/science.ade2574.

Liu, H., Spence, J.C.H., 2016. XFEL data analysis for structural biology. Quantitative Biology 4 (3), 159–176. Available at https://doi.org/10.1007/s40484-016-0076-z.

Lu, S., Ji, M., Ni, D., Zhang, J., 2018. Discovery of hidden allosteric sites as novel targets for allosteric drug design. Drug Discovery Today 23 (2), 359–365. Available at https://doi.org/10.1016/j.drudis.2017.10.001.

Lu, S., Shen, Q., Zhang, J., 2019. Allosteric methods and their applications: facilitating the discovery of allosteric drugs and the investigation of allosteric mechanisms. Accounts of Chemical Research 52 (2), 492–500. Available at https://doi.org/10.1021/acs.accounts.8b00570.

Luchinat, E., Banci, L., 2023. In-cell NMR: recent progresses and future challenges. Rendiconti Lincei. Scienze Fisiche e Naturali. 34 (3), 653–661. Available at https://doi.org/10.1007/s12210-023-01168-y.

Malhotra, S., E. Thomas, S., Ochoa Montano, B., L. Blundell, T., 2016. Structure-guided, target-based drug discovery – exploiting genome information from HIV to mycobacterial infections. Postępy Biochemii 62 (3), 262–272. Available at https://doi.org/10.18388/pb.2016_25.

Mandala, V.S., Williams, J.K., Hong, M., 2018. Structure and dynamics of membrane proteins from solid-state NMR. Annual Review of Biophysics 47, 201–222. Available at https://doi.org/10.1146/annurev-biophys-070816-033712, http://www.annualreviews.org/journal/biophys.

Martin-Garcia, J.M., Conrad, C.E., Coe, J., Roy-Chowdhury, S., Fromme, P., 2016. Serial femtosecond crystallography: a revolution in structural biology. Archives of Biochemistry and Biophysics 602, 32–47. Available at https://doi.org/10.1016/j.abb.2016.03.036, http://www.elsevier.com/inca/publications/store/6/2/2/7/8/7/index.htt.

Martone, M.E., Deerinck, T.J., Yamada, N., Bushong, E., Ellisman, M.H., 2000. Correlated 3D light and electron microscopy: use of high voltage electron microscopy and electron tomography for imaging large biological structures. Journal of Histotechnology 23 (3), 261–270. Available at https://doi.org/10.1179/his.2000.23.3.261, http://www.tandfonline.com/loi/yhis20#.VwHgl01f1Qs.

McGreevy, R., Teo, I., Singharoy, A., Schulten, K., 2016. Advances in the molecular dynamics flexible fitting method for cryo-EM modeling. Methods (San Diego, Calif.) 100, 50–60. Available at https://doi.org/10.1016/j.ymeth.2016.01.009.

McGreig, J.E., Uri, H., Antczak, M., Sternberg, M.J.E., Michaelis, M., Wass, M.N., 2022. 3DligandSite: structure-based prediction of protein-ligand binding sites. Nucleic Acids Research 50 (1), W13. Available at https://doi.org/10.1093/nar/gkac250, https://academic.oup.com/nar/issue.

Meloen, R.H., Puijk, W.C., Slootstra, J.W., 2000. Mimotopes: realization of an unlikely concept. Journal of Molecular Recognition 13 (6), 352–359. Available at https://doi.org/10.1002/1099-1352(200011/12)13:6<352::aid-jmr509>3.0.co;2-c.

Meng, X., Wei, J., Wang, Y., Zhang, H., Wang, Z., 2018. The role of peptide microarrays in biomedical research. Analytical Methods 10 (38), 4614–4624. Available at https://doi.org/10.1039/c8ay01442f.

Montinari, M.R., Minelli, S., De Caterina, R., 2019. The first 3500 years of aspirin history from its roots – A concise summary. Vascular Pharmacology 113, 1–8. Available at https://doi.org/10.1016/j.vph.2018.10.008, www.elsevier.com/locate/genpharm.

Muhammed, M.T., Aki-Yalcin, E., 2024. Molecular docking: principles, advances, and its applications in drug discovery. Letters in Drug Design and Discovery 21 (3), 480–495. Available at https://doi.org/10.2174/1570180819666220922103109, https://www.eurekaselect.com/article/126512.

Muruganantham, J.K., Veerabathiran, R., 2024. The use of X-ray crystallography in the analysis of a variety of protein-lipid complexes. Biochemical Techniques for Analyzing Protein-Lipid Interactions. Springer Nature, India, pp. 55–69. Available at https://doi.org/10.1007/978-981-97-5437-3_4, https://link.springer.com/book/10.1007/978-981-97-5437-3.

Narasimhan, S., Folkers, G.E., Baldus, M., 2020. When small becomes too big: expanding the use of in-cell solid-state NMR spectroscopy. ChemPlusChem 85 (4), 760–768. Available at https://doi.org/10.1002/cplu.202000167, http://onlinelibrary.wiley.com/journal/10.1002/(ISSN)2192-6506.

Nishiyama, Y., Hou, G., Agarwal, V., Su, Y., Ramamoorthy, A., 2023. Ultrafast magic angle spinning solid-state nmr spectroscopy: advances in methodology and applications. Chemical Reviews 123 (3), 918–988. Available at https://doi.org/10.1021/acs.chemrev.2c00197.

Nedelsky, N.B., Taylor, J.P., 2019. Bridging biophysics and neurology: aberrant phase transitions in neurodegenerative disease. Nature Reviews Neurology 15 (5), 272–286. Available at https://doi.org/10.1038/s41582-019-0157-5, http://www.nature.com/nrneurol/archive/index.html.

Nuka, S.T., 2025. AI-driven drug discovery: transforming neurological and neurodegenerative disease treatment through bioinformatics and genomic research. American Journal of Psychiatric Rehabilitation 124–135. Available at https://doi.org/10.69980/ajpr.v28i1.71.

Nogales, E., Mahamid, J., 2024. Bridging structural and cell biology with cryo-electron microscopy. Nature 628 (8006), 47–56. Available at https://doi.org/10.1038/s41586-024-07198-2.

Aboelez, M.O., Belal, A., Xiang, G., Ma, X., 2022. Design, synthesis, and molecular docking studies of novel pomalidomide-based PROTACs as potential anti-cancer agents targeting EGFR WT and EGFR T790M. Journal of Enzyme Inhibition and Medicinal Chemistry 37 (1), 1196–1211. Available at https://doi.org/10.1080/14756366.2022.2062338.

Orlov, I., Myasnikov, A.G., Andronov, L., Natchiar, S.K., Khatter, H., Beinsteiner, B., et al., 2017. The integrative role of cryo electron microscopy in molecular and cellular structural biology. Biology of the Cell. 109 (2), 81–93. Available at https://doi.org/10.1111/boc.201600042, http://onlinelibrary.wiley.com/journal/10.1111/(ISSN)1768-322X.

Pagadala, N.S., Syed, K., Tuszynski, J., 2017. Software for molecular docking: a review. Biophysical Reviews 9 (2), 91 102. Available at https://doi.org/10.1007/s12551-016-0247-1, http://www.springer.com/life+sciences/biochemistry+%26+biophysics/journal/12551.

Palma, M., 2023. Epitopes and mimotopes identification using phage display for vaccine development against infectious pathogens. Vaccines 11 (7), 1176. Available at https://doi.org/10.3390/vaccines11071176.

Pantsar, T., Poso, A., 2018. Binding affinity via docking: fact and fiction. Molecules (Basel, Switzerland) 23 (8), 1899. Available at https://doi.org/10.3390/molecules23081899.

Patel, D.D., Pathak, R.S., Patel, K.S., Bhatt, H.G., Patel, P.K., 2025. The future of medicine: AI and ML driven drug discovery advancements. Current Topics in Medicinal Chemistry 25 (16), 1957–1978. Available at https://doi.org/10.2174/0115680266346722250401191232, http://www.eurekaselect.com/615/journal/current-topics-medicinal-chemistry.

Pfau, B., Eisebitt, S., 2023. X-ray Resonant Scattering and Holography with Application to Magnetization Dynamics, 25. Royal Society of Chemistry (RSC), pp. 254–300. Available at https://doi.org/10.1039/bk9781837671564-00254.

Poddar, H., Heyes, D.J., Schirò, G., Weik, M., Leys, D., Scrutton, N.S., 2022. A guide to time-resolved structural analysis of light-activated proteins. FEBS Journal 289 (3), 576–595. Available at https://doi.org/10.1111/febs.15880, http://onlinelibrary.wiley.com/journal/10.1111/(ISSN)1742-4658.

Portz, B., Lee, B.L., Shorter, J., 2021. FUS and TDP-43 phases in health and disease. Trends in Biochemical Sciences 46 (7), 550–563. Available at https://doi.org/10.1016/j.tibs.2020.12.005, www.elsevier.com/locate/tibs.

Premaraj, N., Heeren, R.M.A., Ravelli, R.B.G., Knoops, K., 2025. Advancing cryo-EM and cryo ET through innovation in sample carriers: a perspective. Analytical Chemistry 97 (23), 11959–11967. Available at https://doi.org/10.1021/acs.analchem.5c01534.

Sadybekov, A.V., Katritch, V., 2023. Computational approaches streamlining drug discovery. Nature 616 (7958), 673–685. Available at https://doi.org/10.1038/s41586-023-05905-z, https://www.nature.com/nature/.

Samtani, M.N., Raghavan, N., Shi, Y., Novak, G., Farnum, M., Lobanov, V., et al., 2013. Disease progression model in subjects with mild cognitive impairment from the Alzheimer's disease neuroimaging initiative: CSF biomarkers predict population subtypes. British Journal of Clinical Pharmacology 75 (1), 146–161. Available at https://doi.org/10.1111/j.1365-2125.2012.04308.x.

SARS-CoV, P., 2021. Variants of Concern and Variants Under Investigation in England–Technical Briefing 15. Public Health England (PHE), London.

Schmidt, M., 2023. Biological function investigated by time-resolved structure determination. Structural Dynamics 10 (1). Available at https://doi.org/10.1063/4.0000177.

Shahwan, M., Alhumaydhi, F.A., Sharaf, S.E., Alghamdi, B.S., Baeesa, S., Tayeb, H.O., et al., 2023. Computational insight into the binding of bryostatin 1 with ferritin: implication of natural compounds in Alzheimer's disease therapeutics. Journal of Biomolecular Structure and Dynamics 41 (12), 5635–5645. Available at https://doi.org/10.1080/07391102.2022.2092552, http://www.tandfonline.com/loi/tbsd20.

Shamsi, A., Shahwan, M., Khan, M.S., Alhumaydhi, F.A., Alsagaby, S.A., Abdulmonem, W.A., et al., 2022. Mechanistic insight into binding of huperzine a with human serum albumin: computational and spectroscopic approaches. Molecules (Basel, Switzerland) 27 (3). Available at https://doi.org/10.3390/molecules27030797, https://www.mdpi.com/1420-3049/27/3/797/pdf.

Shamsi, A., Khan, M.S., Yadav, D.K., Shahwan, M., Furkan, M., Khan, R.H., 2024. Structure-based drug-development study against fibroblast growth factor receptor 2: molecular docking and Molecular dynamics simulation approaches. Scientific Reports 14 (1). Available at https://doi.org/10.1038/s41598-024-69850-1, https://www.nature.com/srep/.

Sherwood, P., Brooks, B.R., Sansom, M.S., 2008. Multiscale methods for macromolecular simulations. Current Opinion in Structural Biology 18 (5), 630–640. Available at https://doi.org/10.1016/j.sbi.2008.07.003.

Si, Y., Xu, D., Bum-Erdene, K., Ghozayel, M.K., Yang, B., Clemons, P.A., et al., 2019. Chemical space overlap with critical protein–protein interface residues in commercial and specialized small-molecule libraries. ChemMedChem 14 (1), 119–131. Available at https://doi.org/10.1002/cmdc.201800537, http://onlinelibrary.wiley.com/journal/10.1002/(ISSN)1860-7187.

Singh, S., Khan, S., Shahid, M., Sardar, M., Hassan, M.I., Islam, A., 2025. Targeting tau in Alzheimer's and beyond: insights into pathology and therapeutic strategies. Ageing Research Reviews 104, 102639. Available at https://doi.org/10.1016/j.arr.2024.102639.

Singh, V., Sinha, S., Verma, J., 2026. Bioinformatics models in drug delivery: predicting biomaterial-biological interactions for targeted therapies. Next Nanotechnology 9, 100335. Available at https://doi.org/10.1016/j.nxnano.2025.100335.

Singhapong, W., Bowen, C., Wang, H., Sawhney, K., Lunt, A.J.G., 2024. Multilayer optics for synchrotron applications. Advanced Materials Technologies 9 (18). Available at https://doi.org/10.1002/admt.202302187.

Slate, E.H., Turnbull, B.W., 2000. Statistical models for longitudinal biomarkers of disease onset. Statistics in Medicine 19, 617–637. Available at https://doi.org/10.1002/(SICI)1097-0258(20000229)19:4<617::AID-SIM360>3.0.CO;2-R 4.

Spassov, D.S., 2024. Binding affinity determination in drug design: insights from lock and key, induced fit, conformational selection, and inhibitor trapping models. International Journal of Molecular Sciences 25 (13), 7124. Available at https://doi.org/10.3390/ijms25137124.

Sui, S., Wang, Y., Kolewe, K.W., Srajer, V., Henning, R., Schiffman, J.D., et al., 2016. Graphene-based microfluidics for serial crystallography. Lab on a Chip 16 (16), 3082–3096. Available at https://doi.org/10.1039/c6lc00451b, http://pubs.rsc.org/en/journals/journal/lc.

Surade, S., Blundell, T.L., 2012. Structural biology and drug discovery of difficult targets: the limits of ligandability. Chemistry & Biology 19 (1), 42–50. Available at https://doi.org/10.1016/j.chembiol.2011.12.013.

Tamara, S., Den Boer, M.A., Heck, A.J.R., 2022. High-resolution native mass spectrometry. Chemical Reviews 122 (8), 7269–7326. Available at https://doi.org/10.1021/acs.chemrev.1c00212, http://pubs.acs.org/journal/chreay.

Tschentscher, T., Bressler, C., Grünert, J., Madsen, A., Mancuso, A., Meyer, M., et al., 2017. Photon beam transport and scientific instruments at the European XFEL. Applied Sciences 7 (6), 592. Available at https://doi.org/10.3390/app7060592.

Tunyasuvunakool, K., Adler, J., Wu, Z., Green, T., Zielinski, M., Žídek, A., et al., 2021. Highly accurate protein structure prediction for the human proteome. Nature 596 (7873), 590–596. Available at https://doi.org/10.1038/s41586-021-03828-1.

Tycko, R., 2011. Solid-state NMR studies of amyloid fibril structure. Annual Review of Physical Chemistry 62 (1), 279–299. Available at https://doi.org/10.1146/annurev-physchem-032210-103539.

Ubbink, M., Perrakis, A., Patrick, C.A., 2018. van der Wel, New applications of solid-state NMR in structural biology. Emerging Topics in Life Sciences 2 (1), 57–67. Available at https://doi.org/10.1042/etls20170088.

Utami, K.H., Morimoto, S., Mitsukura, Y., Okano, H., 2025. The roles of intrinsically disordered proteins in neurodegeneration. Biochimica et Biophysica Acta (BBA) - General Subjects 1869 (4), 130772. Available at https://doi.org/10.1016/j.bbagen.2025.130772.

Van Den Bedem, H., Fraser, J.S., 2015. Integrative, dynamic structural biology at atomic resolution - It's about time. Nature Methods 12 (4), 307–318. Available at https://doi.org/10.1038/nmeth.3324, http://www.nature.com/nmeth/.

Van Thor, J.J., 2019. Advances and opportunities in ultrafast X-ray crystallography and ultrafast structural optical crystallography of nuclear and electronic protein dynamics. Structural Dynamics 6 (5). Available at https://doi.org/10.1063/1.5110685, http://scitation.aip.org/content/aca/journal/sdy.

Vane, J.R., Botting, R.M., 2003. The mechanism of action of aspirin. Thrombosis Research 110 (5-6), 255–258. Available at https://doi.org/10.1016/s0049-3848(03)00379-7.

Vemula, D., Jayasurya, P., Sushmitha, V., Kumar, Y.N., Bhandari, V., 2023. CADD, AI and ML in drug discovery: a comprehensive review. European Journal of Pharmaceutical Sciences 181. Available at https://doi.org/10.1016/j.ejps.2022.106324, www.elsevier.com/locate/ejps.

Venuto, C.S., Potter, N.B., Ray Dorsey, E., Kieburtz, K., 2016. A review of disease progression models of Parkinson's disease and applications in clinical trials. Movement Disorders 31 (7), 947–956. Available at https://doi.org/10.1002/mds.26644, www.interscience.wiley.com/jpages/0885-3185.

Verma, S., Pathak, R.K., 2021. Discovery and optimization of lead molecules in drug designing. Bioinformatics: Methods and Applications. Elsevier, India, pp. 253–267. Available at https://doi.org/10.1016/B978-0-323-89775-4.00004-3, https://www.sciencedirect.com/book/9780323897754.

Wang, H., 2024. Prediction of protein–ligand binding affinity via deep learning models. Briefings in Bioinformatics 25 (2). Available at https://doi.org/10.1093/bib/bbae081.

Wang, X., Sontag, D., Wang, F., 2014. Association for Computing Machinery United States Unsupervised learning of disease progression models. Proceedings of the ACM SIGKDD International Conference on Knowledge Discovery and Data Mining 9781450329569, pp. 85–94. 10.1145/2623330.2623754.

Wang, Q., Sun, Z., Li, J., Qin, T., Ma, H., Chen, S., et al., 2021. Identification of a universal antigen epitope of influenza A virus using peptide microarray. BMC Veterinary Research 17 (1). Available at https://doi.org/10.1186/s12917-020-02725-5.

Wang, Z., Zheng, L., Wang, S., Lin, M., Wang, Z., Kong, A.W.K., et al., 2023. A fully differentiable ligand pose optimization framework guided by deep learning and a traditional scoring function. Briefings in Bioinformatics 24 (1). Available at https://doi.org/10.1093/bib/bbac520, http://bib.oxfordjournals.org.

Waseem, R., Anwar, S., Khan, S., Shamsi, A., Hassan, M.I., Anjum, F., et al., 2021. Map/microtubule affinity regulating kinase 4 inhibitory potential of irisin: a new therapeutic strategy to combat cancer and alzheimer's disease. International Journal of Molecular Sciences 22 (20). Available at https://doi.org/10.3390/ijms222010986, https://www.mdpi.com/1422-0067/22/20/10986/pdf.

Weissenberger, G., Henderikx, R.J.M., Peters, P.J., 2021. Understanding the invisible hands of sample preparation for cryo-EM. Nature Methods 18 (5), 463–471. Available at https://doi.org/10.1038/s41592-021-01130-6, http://www.nature.com/nmeth/.

Xiao, L., Xia, K., 2025. Functions of intrinsically disordered regions. Biology 14 (7), 810. Available at https://doi.org/10.3390/biology14070810.

Young, L.N., Villa, E., 2023. Bringing structure to cell biology with cryo-electron tomography. Annual Review of Biophysics 52, 573–595. Available at https://doi.org/10.1146/annurev-biophys-111622-091327, http://www.annualreviews.org/journal/biophys.

Yuan, R., Zhang, J., Zhou, J., Cong, Q., 2025. Recent progress and future challenges in structure-based protein-protein interaction prediction. Molecular Therapy 33 (5), 2252–2268. Available at https://doi.org/10.1016/j.ymthe.2025.04.003.

Zhang, K., Yang, X., Wang, Y., Yu, Y., Huang, N., Li, G., et al., 2025. Artificial intelligence in drug development. Nature Medicine 31 (1), 45–59. Available at https://doi.org/10.1038/s41591-024-03434-4.

Zhao, W., Deligey, F., Chandra Shekar, S., Mentink-Vigier, F., Wang, T., 2022. Current limitations of solid-state NMR in carbohydrate and cell wall research. Journal of Magnetic Resonance 341, 107263. Available at https://doi.org/10.1016/j.jmr.2022.107263.

Zheng, H., Handing, K.B., Zimmerman, M.D., Shabalin, I.G., Almo, S.C., Minor, W., 2015. X-ray crystallography over the past decade for novel drug discovery -where are we heading next? Expert Opinion on Drug Discovery 10 (9), 975–989. Available at https://doi.org/10.1517/17460441.2015.1061991.

Index

Note: Page numbers followed by "*f*" and "*t*" refer to figures and tables, respectively.

B

D

H

J

K

L

M

O

P

Z

9780443449888